Materials Forming, Machining and Tribology

Series Editor

J. Paulo Davim, Department of Mechanical Engineering, University of Aveiro, Aveiro, Portugal

This series fosters information exchange and discussion on all aspects of materials forming, machining and tribology. This series focuses on materials forming and machining processes, namely, metal casting, rolling, forging, extrusion, drawing, sheet metal forming, microforming, hydroforming, thermoforming, incremental forming, joining, powder metallurgy and ceramics processing, shaping processes for plastics/composites, traditional machining (turning, drilling, miling, broaching, etc.), non-traditional machining (EDM, ECM, USM, LAM, etc.), grinding and others abrasive processes, hard part machining, high speed machining, high efficiency machining, micro and nanomachining, among others. The formability and machinability of all materials will be considered, including metals, polymers, ceramics, composites, biomaterials, nanomaterials, special materials, etc. The series covers the full range of tribological aspects such as surface integrity, friction and wear, lubrication and multiscale tribology including biomedical systems and manufacturing processes. It also covers modelling and optimization techniques applied in materials forming, machining and tribology. Contributions to this book series are welcome on all subjects of "green" materials forming, machining and tribology. To submit a proposal or request further information, please contact Dr. Mayra Castro, Publishing Editor Applied Sciences, via mayra.castro@springer.com or Professor J. Paulo Davim, Book Series Editor, via pdavim@ua.pt.

More information about this series at http://www.springer.com/series/11181

Jitendra Kumar Katiyar · P. Ramkumar ·
T. V. V. L. N. Rao · J. Paulo Davim
Editors

Tribology in Materials and Applications

Springer

Editors
Jitendra Kumar Katiyar
Department of Mechanical Engineering
SRM Institute of Science and Technology
Kattankulathur Campus
Chennai, Tamil Nadu, India

T. V. V. L. N. Rao
SRM Institute of Science and Technology
Chennai, Tamil Nadu, India

P. Ramkumar
Indian Institute of Technology Madras
Chennai, Tamil Nadu, India

J. Paulo Davim
Department of Mechanical Engineering
University of Aveiro
Aveiro, Portugal

ISSN 2195-0911 ISSN 2195-092X (electronic)
Materials Forming, Machining and Tribology
ISBN 978-3-030-47450-8 ISBN 978-3-030-47451-5 (eBook)
https://doi.org/10.1007/978-3-030-47451-5

This Springer imprint is published by the registered company Springer Nature Switzerland AG
The registered company address is: Gewerbestrasse 11, 6330 Cham, Switzerland

Preface

Tribology is the study of science and technology of two surfaces which are in relative motion. The relative motion causes friction at the interface, and friction eventually causes wear of one or both the surfaces. Friction and wear between two bodies can be overcome by the application of lubricants. A number of examples available where tribology is very important. Those are different kinds of bearing, gears, engines, orthopaedic joints, and micro-machines.

The primary objective of this book is to broaden the knowledge of tribology. This book is evolved out of current research trends on tribological performance of systems related to nano tribology, rheology, engines, polymer brushes, composite materials, erosive wear and lubrication. The book deals with enhancing the ideas on tribological properties, the different types of wear phenomenon and lubrication enhancement. Further, the tribological performance of systems, whether nano, micro or macro-scale, depends upon a large number of external parameters and important among them are temperature, contact pressure and relative speed. Thus, the book focuses on the theoretical aspects to industrial applications of tribology.

Students, academicians, researchers, practising engineers, who are working in the field of tribology, design, materials and manufacturing will be interested to enrich their ideas about ongoing tribological research themes.

Chennai, India — Jitendra Kumar Katiyar
Chennai, India — P. Ramkumar
Chennai, India — T. V. V. L. N. Rao
Aveiro, Portugal — J. Paulo Davim

Contents

Editors and Contributors

About the Editors

Dr. Jitendra Kumar Katiyar presently working as a research assistant professor, in Department of Mechanical Engineering, SRM Institute of Science and Technology, Kattankulathur, Chennai, India. His research interests include tribology of carbon materials, polymer composites, self-lubricating polymers, lubrication tribology and coatings for advanced technologies. He obtained his Ph.D., from the Indian Institute of Technology Kanpur in 2017 and masters from the same institution in 2010. He obtained his bachelor degree from UPTU, Lucknow, with honours in 2007. He has life professional memberships such as Tribology Society of India, Malaysian Society of Tribology and The Indian Society for Technical Education (ISTE). He has published more than two dozen papers in reputed journals and international conferences. His two books are in press, one is *Engineering Thermodynamics* for UG level in Khanna Publication and another is *Automotive Tribology* in Springer. He actively participated in so many activities related to tribology.

Dr. P. Ramkumar presently working as an associate professor in the Department of Mechanical Engineering, IIT Madras. Prior to coming to IIT Madras, he was an associate professor at SSN College of Engineering, Chennai. He received his B.E. in Mechanical Engineering from College of Engineering, Guindy, Anna University, and his M.Tech. (Industrial Tribology) from IIT Madras. Later, he pursued Ph.D. from University of Southampton, UK, and did his postdoctoral fellowship in University of Leicester, UK. His research interests are in the field of tribology, including developing new materials and coatings for engine tribology, nano-lubrication, gearbox design, electrostatic condition monitoring and developing new materials for brake application.

Dr. T. V. V. L. N. Rao is currently a research associate professor in School of Mechanical Engineering at SRM Institute of Science and Technology, since December 2017. He received PhD in tribology of fluid film bearings from the Indian Institute of Technology Delhi in 2000 and M.Tech. in Mechanical Manufacturing Technology from the National Institute of Technology Calicut in 1994. Prior to joining SRM, he served as a faculty member at LNMIIT, Universiti Teknologi PETRONAS, and BITS at Pilani and Dubai campuses. He has authored over 100 publications. He is a member of Society of Tribologists and Lubrication Engineers, Malaysian Tribology Society and Tribology Society of India. His research interests are in tribology, lubrication and bearings.

Prof. J. Paulo Davim received his Ph.D. degree in Mechanical Engineering in 1997, M.Sc. degree in Mechanical Engineering (materials and manufacturing processes) in 1991, Mechanical Engineering degree (5 years) in 1986, from the University of Porto (FEUP), the Aggregate title (Full Habilitation) from the University of Coimbra in 2005 and the D.Sc. (Higher Doctorate) from London Metropolitan University in 2013. He is a senior chartered engineer by the Portuguese Institution of Engineers with an MBA and Specialist titles in Engineering and Industrial Management as well as in Metrology. He is also Eur Ing by FEANI-Brussels and Fellow (FIET) of

IET-London. Currently, he is a professor at the Department of Mechanical Engineering of the University of Aveiro, Portugal. He is also distinguished as a honorary professor in several universities/colleges. He has more than 30 years of teaching and research experience in Manufacturing, Materials, Mechanical and Industrial Engineering, with special emphasis on machining and tribology. He has also interested in management, engineering education and higher education for sustainability. He has guided large numbers of postdoc, Ph.D. and master's students as well as has coordinated and participated in several financed research projects. He has received several scientific awards and honours. He has worked as an evaluator of projects for ERC-European Research Council and other international research agencies as well as examiner of Ph.D. thesis for many universities in different countries. He is the editor in chief of several international journals, guest editor of journals, books editor, book series editor and scientific advisory for many international journals and conferences. Presently, he is an editorial board member of 30 international journals and acts as a reviewer for more than 100 prestigious Web of Science journals. In addition, he has also published as editor (and co-editor) more than 125 books and as an author (and co-author) more than ten books, 80 chapters and 400 articles in journals and conferences (more than 250 articles in journals indexed in Web of Science core collection/h-index 55+/9500+ citations, SCOPUS/h-index 59+/11500+ citations, Google Scholar/h-index 76+/19000+).

Contributors

Aydar Akchurin ASML, Veldhoven, The Netherlands

Mokhtar Awang Department of Mechanical Engineering, Universiti Teknologi PETRONAS, Bandar Seri Iskandar, Perak Darul Ridzuan, Malaysia

Deepak Bhardwaj Department of Mechanical and Automation Engineering, Dr. Akhilesh Das Gupta Institute of Technology and Management, New Delhi, India

Ajay Chouhan CSIR-Indian Institute of Petroleum, Dehradun, India

Febin Cyriac Institute of Chemical and Engineering Sciences, A*STAR, Singapore, Singapore

Amit Rai Dixit Department of Mechanical Engineering, Indian Institute of Technology (ISM), Dhanbad, India

Fakhruldin Mohd Hashim Department of Mechanical Engineering, Universiti Teknologi PETRONAS, Bandar Seri Iskandar, Perak Darul Ridzuan, Malaysia

T. Jagadeesha Department of Mechanical Engineering, National Institute of Technology, Calicut, Kerala, India

K. Jegadeesan Department of Mechanical Engineering, SRM Institute of Science and Technology, Kattankulathur, India

Rakesh Jinaga Department of Mechanical Engineering, National Institute of Technology, NIT Calicut, Calicut, Kerala, India

Vikram G. Kamble Leibniz Institute for Polymer Research, Dresden, Germany

Saravanan Karuppanan Department of Mechanical Engineering, Universiti Teknologi PETRONAS, Bandar Seri Iskandar, Perak, Malaysia

Jitendra Kumar Katiyar Department of Mechanical Engineering, SRM Institute of Science and Technology, Kattankulathur Campus, Chennai, Tamil Nadu, India

Om P. Khatri CSIR-Indian Institute of Petroleum, Dehradun, India

Shreedhar Kolekar Department of Mechanical Engineering, SCOEM Satara, Satara, Maharashtra, India;
Department of Mechanical Engineering, National Institute of Technology, NIT Calicut, Calicut, Kerala, India

Sangita Kumari CSIR-Indian Institute of Petroleum, Dehradun, India

Norani Muti Mohamed Department of Fundamental and Applied Sciences, Universiti Teknologi PETRONAS, Bandar Seri Iskandar, Perak Darul Ridzuan, Malaysia

Manidipto Mukherjee CAMM, CSIR-Central Mechanical Engineering Research Institute, Durgapur, West Bengal, India

H. S. Panda Ballistic Center, Proof & Experimental Establishment (PXE) Defence R & D Organization, Balasore, Odisha, India

Vinay Kumar Patel Department of Mechanical Engineering, Govind Ballabh Pant Institute of Engineering and Technology Ghurdauri, Pauri Garhwal, Uttarakhand, India

Santosh S. Patil Department of Mechanical Engineering, Manipal University Jaipur, Jaipur, India

Ramkumar Penchaliah Machine Design Section, Department of Mechanical Engineering, Indian Institute of Technology Madras, Chennai, India

Deepak K. Prajapati Department of Mechanical Engineering, Indian Institute of Technology, Patna, India

Ahmad Majdi Abdul Rani Department of Mechanical Engineering, Universiti Teknologi PETRONAS, Bandar Seri Iskandar, Perak Darul Ridzuan, Malaysia

T. V. V. L. N. Rao Department of Mechanical Engineering, SRM Institute of Science and Technology, Kattankulathur, India

Anand Singh Rathaur Department of Mechanical Engineering, Govind Ballabh Pant Institute of Engineering and Technology Ghurdauri, Pauri Garhwal, Uttarakhand, India

Anuj Kumar Sharma Mechatronics, Centre for Advanced Studies, Dr. APJ Abdul Kalam Technical University, Lucknow, India

C. Shravankumar Department of Mechanical Engineering, SRM Institute of Science and Technology, Kattankulathur, India

Jayant Singh Department of Mechanical and Automation Engineering, Dr. Akhilesh Das Gupta Institute of Technology and Management, New Delhi, India

Manjesh K. Singh Indian Institute of Technology, Kanpur, India

Rabesh Kumar Singh Mechatronics, Centre for Advanced Studies, Dr. APJ Abdul Kalam Technical University, Lucknow, India

Arun Kumar Tiwari Department of Mechanical Engineering, Institute of Engineering and Technology, Dr. APJ Abdul Kalam Technical University, Lucknow, India

Mayank Tiwari Department of Mechanical Engineering, Indian Institute of Technology, Patna, India

Hamdan Haji Ya Department of Mechanical Engineering, Universiti Teknologi PETRONAS, Bandar Seri Iskandar, Perak Darul Ridzuan, Malaysia

Chapter 1
Introduction and Applications of Tribology

Anand Singh Rathaur, Jitendra Kumar Katiyar, and Vinay Kumar Patel

Abstract Present chapter describes about the basic terms involve in tribology and their application in different fields. When two bodies are in a relative motion, at the interface rubbing occurred, due to rubbing, friction produced and friction eventually causes the wear on one or both the materials. Therefore, the role of lubricant comes into the picture. Friction and wear between two bodies can be overcome by the application of lubricants. In the environment, air and water may act as a lubricant to reduce friction and improve wear resistant property. Furthermore, friction and wear performance of any structure either nano, micro or macro-scale is depends upon a large member of external parameters and important among them are relative speed, temperature, and contact pressure. This chapter also describes the important application of tribology in brief.

1.1 Introduction

The word "Tribology" has been derived from the Greek word "Tribos" which means "rubbing" or to rub and suffix "Ology" means "the study of". Hence, tribology is "the study of rubbing of two materials" [1]. Tribology involves the study and the applications of the principles of friction, wear, adhesion, lubrication and surface modification which is diagrammatically represented in Fig. 1.1.

The relative motion of two bodies cause friction at the interface and this eventually origins wear on either one or both the bodies. Friction and wear between two bodies can be overcome by the application of lubricants. In the environment, air and

A. S. Rathaur · V. K. Patel
Department of Mechanical Engineering, Govind Ballabh Pant Institute of Engineering and Technology Ghurdauri, Pauri Garhwal, Uttarakhand 246194, India

J. K. Katiyar (✉)
Department of Mechanical Engineering, SRM Institute of Science and Technology, Kattankulathur Campus, Chennai, Tamil Nadu 603203, India
e-mail: jitendrv@srmist.edu.in

J. K. Katiyar et al. (eds.), *Tribology in Materials and Applications*,
Materials Forming, Machining and Tribology,
https://doi.org/10.1007/978-3-030-47451-5_1

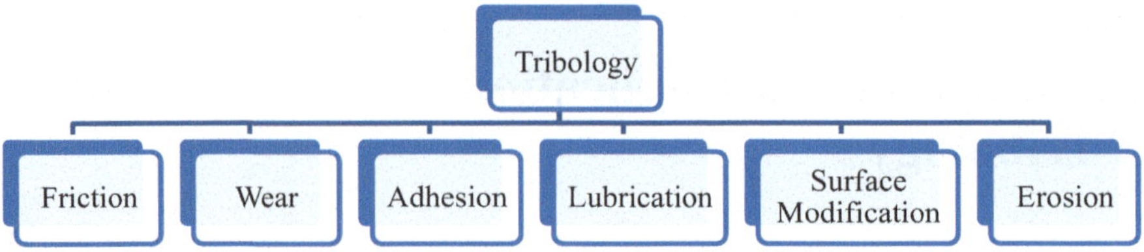

Fig. 1.1 Structure of tribology

water may act as a lubricant to decrease friction and improve wear resistant property. Comprehensive explanation of friction, wear and lubrication is specified in the following sections.

1.1.1 Friction

The friction is known as a resistance force experienced by the different surfaces like solid surface, fluid layers etc. in relative motion which is shown in Fig. 1.2. Friction is divided into two categories, dry and fluid film friction. Dry friction is defined when there are two solid surfaces sliding against each other and energy is completely intemperate between the surfaces. Further, it is again subdivided into two parts; static and kinetic friction.

When the energy dissipation taken place within a thin liquid layer between the two surfaces then this type of friction is known as fluid film friction. Essentially, friction is a result of the energy dissipation at the interface which can take place in the form of material deformation, asperity-asperity interactions, fracture, viscous flow of fluid film, interatomic interactions etc.

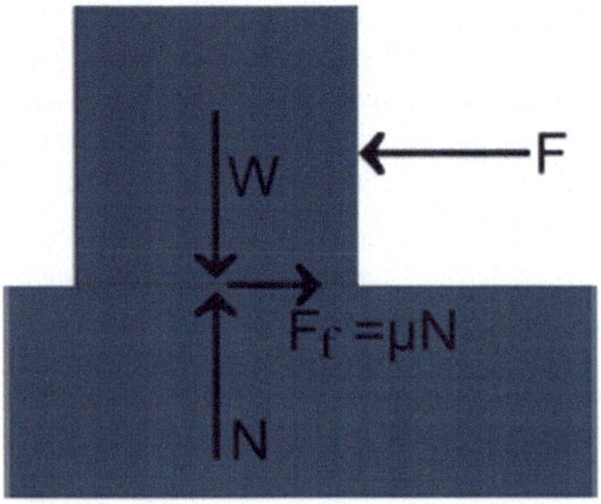

Fig. 1.2 Schematic of friction

1.1.2 Wear

Because of rubbing action in relative motion of solid bodies, material is continuously lost at the interface. This phenomenon of loss of material is known as Wear. One example of wear of the surface is shown in Fig. 1.3. Form the schematic; it is clearly observed that bottom surface have initially V-notches and top surface rubbing over it, resulting in the abrasion of V-notches. Wear processes are classified into various types such as adhesive wear, abrasive wear, surface fatigue and erosion which are further classified into sub groups as shown in Fig. 1.4 [2]. Detailed description of wear mechanisms are given in the following sections.

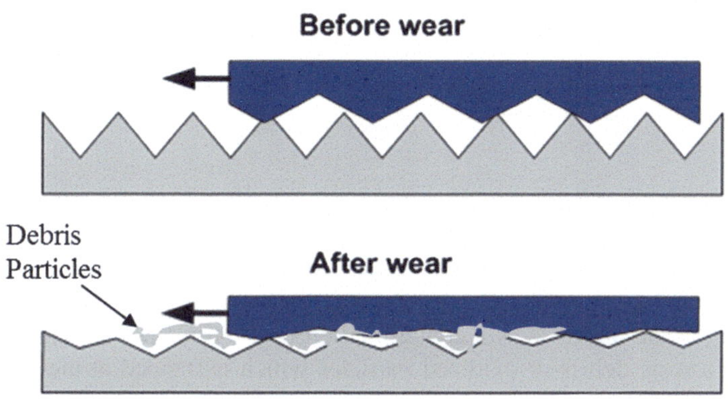

Fig. 1.3 Schematic of wear

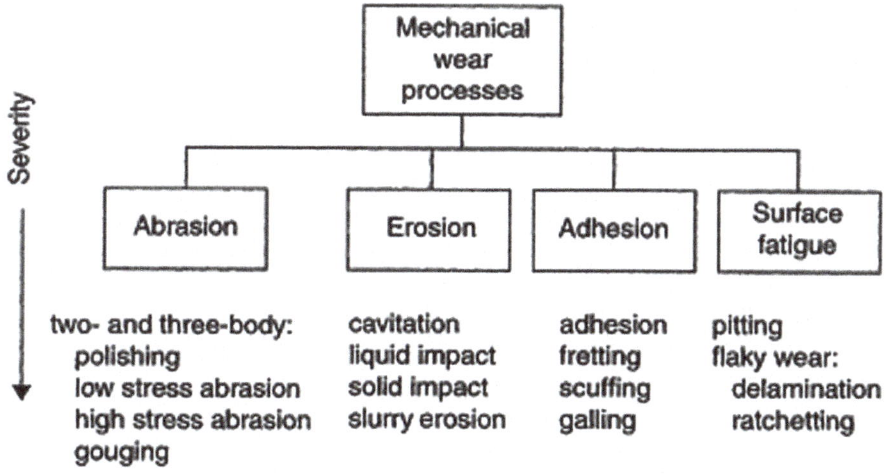

Fig. 1.4 Classification of wear processes. Reproduced with permission of Ref. [2]

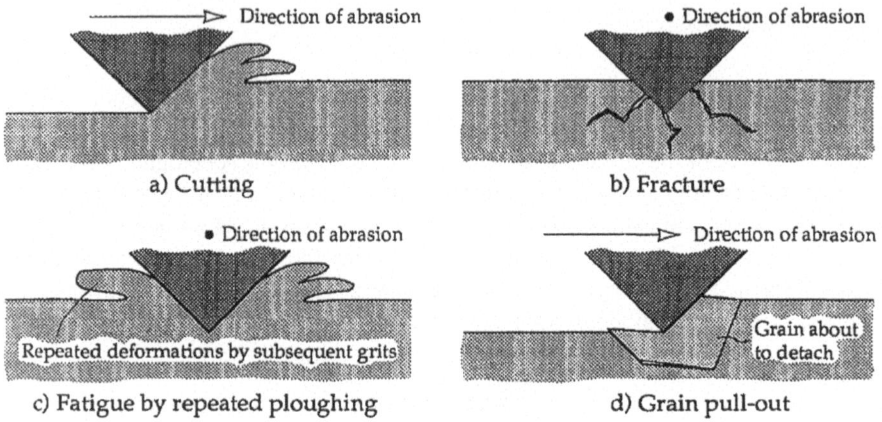

Fig. 1.5 Types of abrasive wear: micro-cutting, fracture, fatigue and grain pull-out. Reproduced with permission of Ref. [3]

1.1.2.1 Abrasive Wear

When two bodies come into a direct contact in which one is soft and another one is hard then abrasive wear occurred. Further, it might also confirm that hard particles are existing at the interface as a contamination or as "third body". The third body defined as a wear debris or oxidized particles which is trapped at the interface. A few examples of abrasive wear are shown in Fig. 1.5 which schematically show the generation of wear particles by the processes such as micro-cutting, micro-fracture, pull-out of individual grains or accelerated fatigue by repeated deformations. The actual micrograph of abrasive wear is shown in Fig. 1.6.

1.1.2.2 Adhesive Wear

When large forces applied on interface of two bodies then materials at the interface experience plastic deformation and removal in the form of highly deformed flakes. This type of wear is known as adhesive wear. The high amount of adhesive forces can also lead to cold welding at confined contact points. The cold-welded part can also tear-off local material. Adhesive wear can happen in bearing surfaces made of metals where lubrication is not present or not effective. The Mechanism of adhesive wear is shown in Fig. 1.7 and the actual micrograph of adhesive wear is shown in Fig. 1.8.

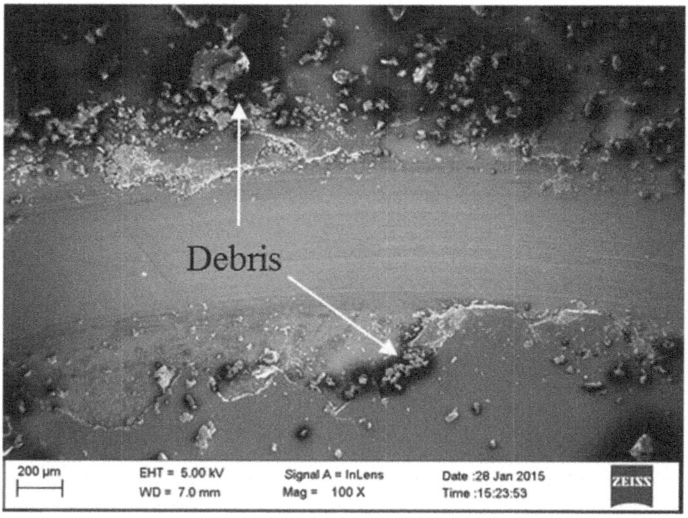

Fig. 1.6 SU-8/graphite composite after tribotest at 600 RPM and 2 N applied load

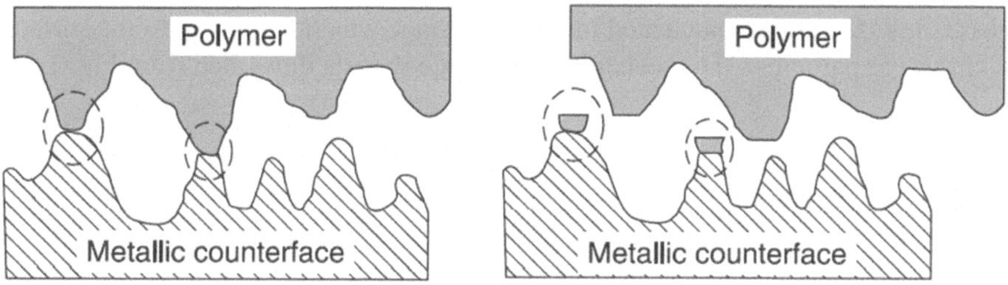

Fig. 1.7 Mechanism of adhesive wear. Reproduced with permission of Ref. [4]

1.1.2.3 Erosive Wear

When the deformation on the solid surface might yield because of the impact of a very small solid or liquid particle and if such impressions are continual over and over again then there will be elimination of local material which is known as erosive wear. It depends upon material of particle (hardness), the impingement angle, the impact velocity, and the size of particle. The mechanism of erosive wear is illustrated in Fig. 1.9.

1.1.2.4 Fatigue Wear

Fatigue wear occurs primarily in rolling contact and in sliding with lubrication. It is a result of non-conformal contact where the contact stress is large but the shear

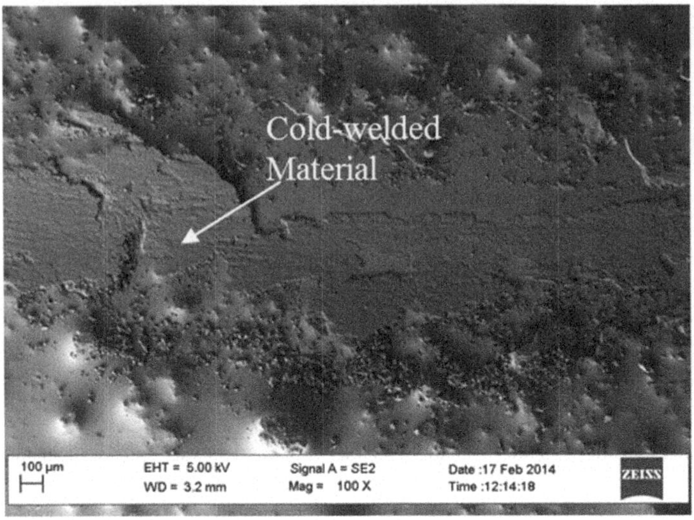

Fig. 1.8 SU-8/graphene composite after tribotest at 200 RPM and 1.5 N applied load

stress is low. Cracks are nucleated in the subsurface which than grow to the surface making wear particles. The mechanism of fatigue wear is illustrated from Fig. 1.10.

1.1.3 Lubrication

Lubrication is very important agent which reduce the friction and wear between two surfaces in relative motion. The correct lubrication is chosen by the application and condition where it is applying. The main functions of lubricants are to control the friction and wear, interfacial temperature, and corrosion. Also, it can be used for removing of unwanted particles between interface and form a fluid film to diminish the friction and wear. There are various parameters which are affected the act of lubricant. These are its physical, chemical, and rheological properties. Apart from that, it also depends on the some external parameters levied such as surface temperature, pressure between contacts, and relative speed. Because of the development of fluid film at the counterface, There are various lubrication regimes are defined which are

1. Boundary lubrication
2. Mixed Lubrication
3. Elasto-hydrodynamic lubrication
4. Hydrodynamic Lubrication.

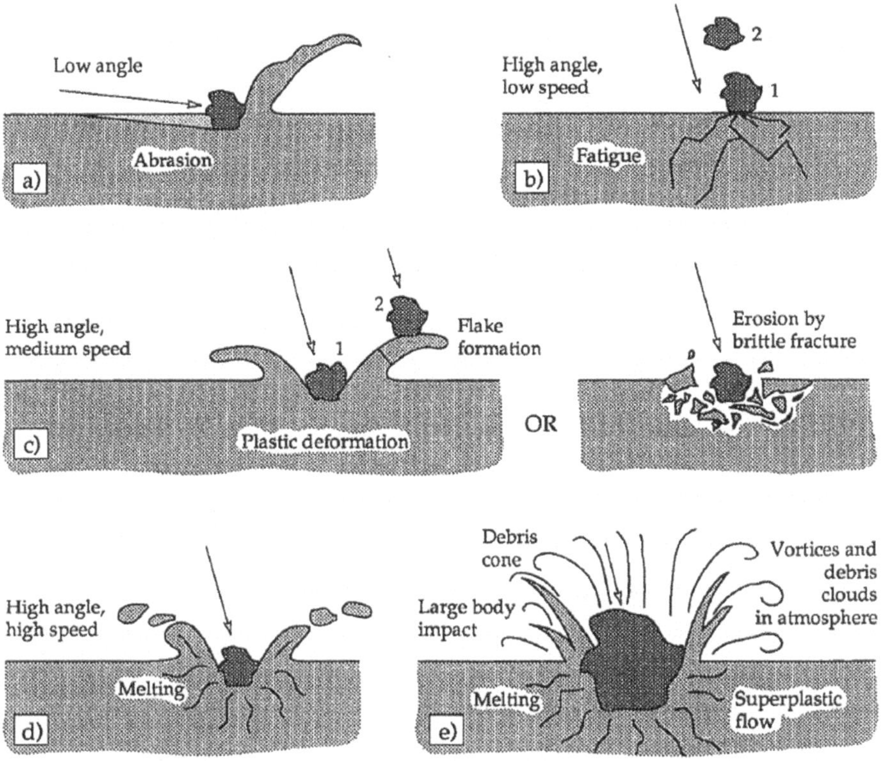

Fig. 1.9 Possible mechanisms of erosion; **a** abrasion at low impact angles, **b** surface fatigue during low speed, high impingement angle impact, **c** brittle fracture or multiple plastic deformation during medium speed, large impingement angle impact, **d** surface melting at high impact speeds, **e** macroscopic erosion with secondary effects. Reproduced with permission of Ref. [3]

Fig. 1.10 Mechanism of fatigue wear. Reproduced with permission of Ref. [4]

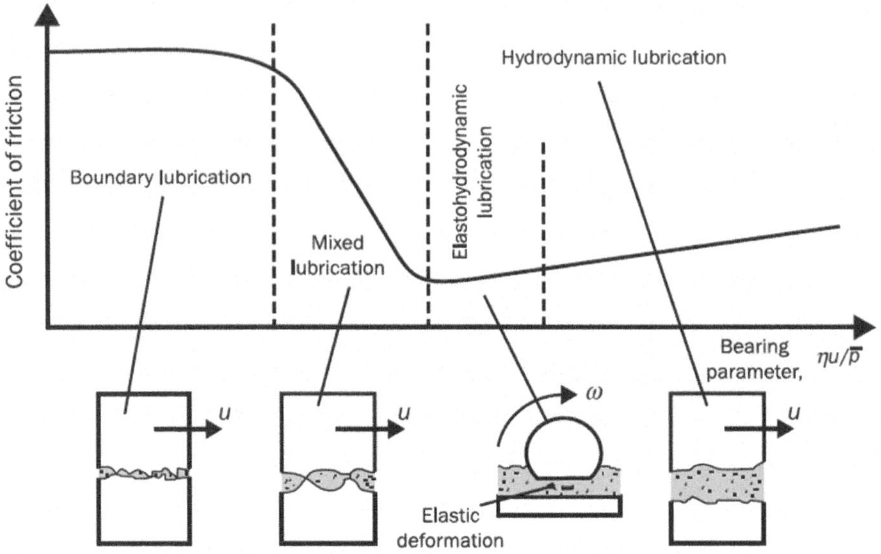

Fig. 1.11 Stribeck curve and mechanism of lubrication. Reproduced with permission of Ref. [6]

The mechanisms of lubrication are illustrated in Fig. 1.11. The lubrication is provided by either liquid (e.g. natural oil, organic oil and inorganic oil) or solid (e.g. Grease, graphite, graphene etc.).

The liquid lubrication can be studied by a plot known as the "Stribeck curve" or "Stribeck-Hersey curve". The "Stribeck curve" or "Stribeck–Hersey curve" (named after Richard Stribeck and Mayo D. Hersey) (shown in Fig. 1.11) was developed in the first half of the twentieth century to categorize the friction properties between two liquid lubricated surfaces [5]. The curve is plotted with friction coefficient as a function of a parameter given by hV/P; where h is the viscosity of the lubricant, V is the relative velocity and P is the contact pressure. The Stribeck curve contains three main regimes known as Boundary lubrication, Mixed lubrication and Hydrodynamic lubrication (HL) with Elasto-hydrodynamic lubrication (EHL) regimes in between mixed and HL regimes.

The boundary lubrication regime is characterized by solid-solid interactions even though there is presence of liquid lubricant. This condition exits when the contact pressure is high and/or the relative speed is low. The hydrodynamic lubrication regime is characterized by whole separation of the two mating surfaces by a thin layer of lubricant. This thin lubricant film is maintained by the fluid as it enters the minimum gap between the two mating parts. The smallest film thickness must be greater that the surfaces roughness. The mixed lubrication regime is categorised by partial solid-solid interaction at asperity level with partial fluid film separation. The Elasto-hydrodynamic lubrication (EHL) is a distinct case of HL where there is elastic deformation of one or both solid surfaces in contact and thus facilitating the formation of a fluid film in-between. The coefficient of friction is high when there is

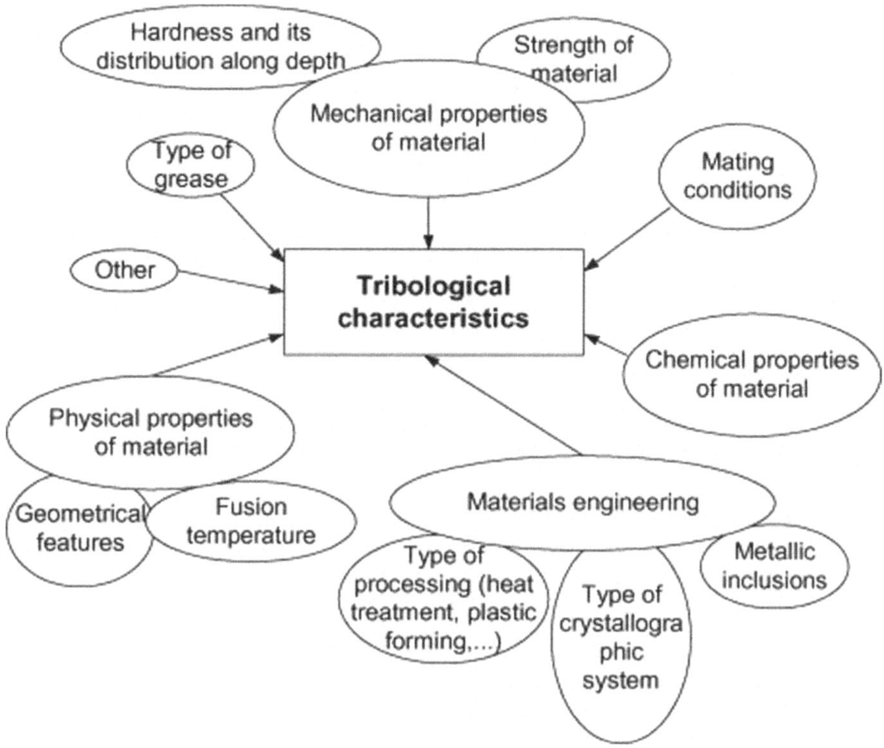

Fig. 1.12 Factors determining the wear and friction behavior. Reproduced with permission of Ref. [7]

solid-solid interaction at asperity level as in the boundary lubrication regime. It can be extremely low (in the range of 0.001) in the hydrodynamic film regime.

1.1.4 Factors Affecting Tribology Performance

Figure 1.12 illustrates the factors which disturb the performance of any tribological system.

1.1.5 Tribology in Materials

Materials play a very important parts in technological development of any country. The driving force behind any advancement in any materials are various social,

Table 1.1 Tribological characteristic in relation to material types

Mass forces	$F_{polymer} < F_{ceramics} < F_{metal}$
Hertzian pressure	$P_{polymer} < P_{ceramics} < P_{metal}$
Friction induced temperature increases	$T_{metal} < T_{polymer} < T_{ceramics}$
Adhesion energy (surface tension)	$Ad_{ceramics} < Ad_{metal} < Ad_{polymer}$
Abrasion	$Ab_{ceramics} < Ab_{metal} < Ab_{polymer}$
Tribo-chemical reactivity	$R_{polymer} < R_{ceramics} < R_{metal}$

Reproduced with permission of Ref. [8]

environmental and technological requirements which provides reliability of any engineering system, durability of any products, higher efficiency, light weight and high strength structure, higher productivity and miniaturization of components [8]. Furthermore, developed new materials are showing very promising use in various applications because friction and wear is directly related to the properties of materials which is shown in Table 1.1.

1.1.6 Applications of Tribology

The tribological performance of any system either nano, micro or macro-scale is depends upon a large member of external parameters such as temperature, contact pressure and relative speed etc. There are various examples, where tribology is very imperative. These are various kinds of bearings, gears, engines, orthopaedic joints and micro-machines. The brief description is given in following sections

1.1.6.1 Bio-tribology

The word bio-tribology, first time introduced by Dowson in 1970 [9]. It deals with the tribological aspects related to the biological systems. This is a promptly rising area of tribology and spreads well outside the predictable boundaries. These systems include an extensive range of synthetic materials and natural tissues, including cartilage, blood vessels, heart, tendons, ligaments, and skin. The materials used in biological systems work in a complex cooperative biological environments. The researchers incorporate concepts of friction, wear, and lubrication of these biological surfaces in numerous applications, such as the design of joints and prosthetic devices, the wear of plates and screws in bone fracture repair, wear of denture and restorative materials, wear of replacement heart valves, and even the tribology of contact lenses.

1.1.6.2 Micro/Nano Tribology

Micro Electro-Mechanical System (MEMS)/Nano Electro-Mechanical System (NEMS) devices continue to find new applications in technology. Some of the successful MEMS are micro-reservoir, micro-pumps, cantilever, micro-pillars for holding the mirror devices in projectors, rotors, channels, valves, sensors etc. In majority of the cases, polycrystalline silicon is used as the structural material for MEMS fabrication because of the micro fabrication process knowledge acquired from the semi-conductor industry. Recently, polymer materials have also been used for MEMS fabrication. Some of the polymers used as structural material are acrylic (PMMA), PDMS and the epoxy-based SU-8. SU-8 is a negative photoresist which is UV curable and has excellent mechanical properties over other polymers. However, when compared to silicon, SU-8 is mechanically inferior. SU-8 has excellent thermal stability. Despite many processing advantages, the bulk mechanical and tribological properties of SU-8 are the main limitations in making it is a versatile MEMS material.

The development of sophisticated scanning probe technologies and computational methods has given upswing to the field of nanotribology for examinations of processes at the atomic, molecular, and microscopic scale. Nano tribological studies are facilitating to develop important understanding of surface counterfaces in micro/nanostructures used in a variety of current applications.

Some of these applications comprise chemical and biodetectors, advanced drug delivery systems, information recording layers, molecular sieves, systems on a chip, nanoparticle reinforced materials, and a new generation of lasers.

1.1.6.3 Wind Turbines

Wind turbines are widely used for power generation using non-renewable energy such as wind. It is a feasible substitute energy resource. While they have made expansions in trustworthiness in the past decade, they are subject to friction and wear problems. Due to these problems, it is difficult and costly to repair. Further, it can drastically reduce their predictable lifecycles. Two key areas of concern are reliability of gearboxes and turbine lubrication. To report these problems, extensive studies of lubricants have been carried out which are used in wind-turbine gearboxes and hydraulic systems such as the blade pitch control, drive train brake subsystems, and bearings to expand the turbine reliability in thrilling environments. The possible failure modes in wind turbines are micro pitting (that is also called frosting) because of high tangential stresses, scuffing because of sever plastic deformation, electric discharge damage because of faulty insulation and microstructural alteration (i.e. also known as white etching area cracks) because of the insufficient design consideration of bearing [10].

1.1.6.4 Nano-lubricants

One of the recent areas in the arena of tribology is nano-lubricants, also known as nano fluids. The nano lubricants got its name from its composition i.e. nano particles dispersed in lubricants. It has been observed that the tribological properties of a lubricant enhanced significantly with the addition of nano particles. These nano particles added lubricants protects the parent materials from contact with each other in either of the following mechanisms: formation of a protective layer over the parent material, filling the crevices, polishing the mating pairs or acting as spacers in the form of rollers between the mating parts.

1.1.6.5 Automotive Tribology

Now a days the automotive technologies are developing very fast. The main concern in these technologies are customer satisfaction and environmental protection. Automobile consists various mechanisms which shake the efficiency of engine mostly such as bearing, piston ring, valve train and crank shaft etc. to increase the efficiency of automobile, various researchers had tried numerous methods which are very effective such as surface texturing, surface coating, modification in lubricant etc. for improving the efficiency of engine. Also, they concluded that the efficiency of any engine is mostly affected by friction.

1.1.6.6 Bearings

In bearing industries, steel is most commonly used material. The industries are fabricating various kinds of bearings such as ball bearing, roller bearing and thrust bearing etc. These bearings have been used in diverse applications of automobile; power driven machine etc. [11]. The coefficient of friction in steel on steel interface has revealed a very high (~0.7–1.0) [12]. Due to very high friction coefficient, a substantial damage on the surfaces and subsurface of bearing balls occurs. These are in the form of crack propagation, cavity formation etc. which affects the life of bearing destructively. Therefore, the reduction in friction coefficient is extremely desirable. To attain low coefficient of friction, the enterprises are exhausting excessive quantity of lubricant which distresses the environment and produces the health risk. Hence, the scientist and researchers have moved their devotion towards some alternate solutions which can reduce the consumption of lubricant. Because of that reason, the few researchers invented a self-lubricating polymer composite coating using solid friction modifier or additives which obsessed lower coefficient of friction [13, 14]. Further, researchers are trying to develop polymer composite bearing balls which is the one step ahead in reduction of lubrication. These polymer composite bearing balls have shown lower friction coefficient as well as higher wear resistance at low load [15].

1.1.6.7 Green Tribology

The concept of "green tribology" was also introduced by Jost, who defined it as, "The science and technology of the tribological aspects of ecological balance and of environmental and biological impacts." There are a numeral of complications which can be addressed using green tribology. The unambiguous field of green or environment-friendly tribology highlights the features of mating surfaces in relative motion. This is having importance for energy or environmental sustainability or which have influence upon today's environment.

In this field, Nosonovsky and Bhushan suggested the 12 principles [16] as the minimization of (1) friction and (2) wear, (3) the reduction or complete elimination of lubrication, including self-lubrication, (4) natural and (5) biodegradable lubrication, (6) using sustainable chemistry and engineering principles, (7) biomimetic approaches, (8) surface texturing, (9) environmental implications of coatings, (10) real-time monitoring, (11) design for degradation, and (12) sustainable energy applications.

1.2 Summary

Tribology is the study of science and technology of two rubbing systems. Due to rubbing, friction occurs which causes wear at the interface. To reduce these effect from interface, a protective layer of lubricant is applied. Therefore, the development of materials are very important in any tribological systems because friction and wear is widely affected by the properties of bulk materials. According to the use of materials, there are various applications of tribology. Therefore, tribology is also known as interdisciplinary branch of the science.

References

1. D. Duncan, *History of Tribology*, 2nd edn. (Professional Engineering Publishing, 1997). ISBN 1-86058-070-X
2. J.A. Williams, Wear and wear particles—some fundamentals. Tribol. Int. **38**(10), 863–870 (2005)
3. G.W. Stachowiak, A.W. Batchelor, Engineering tribology: abrasive, erosive and cavitation wear. Tribol. Ser. **24**, 557–612 (1993)
4. A. Abdelbary, *Wear of Polymers and Composites: Polymer Tribology* (Elsevier, 2013), pp. 1–36
5. R. Stribeck, Die wesentlichen Eigenschaften der Gleit- und Rollenlager (Characteristics of plain and roller bearings), Zeit. des VDI 46 (1902)
6. L. Burstein, *Lubrication and Roughness, Tribology for Engineers, A Practical Guide* (Elsevier, 2011)
7. T. Leśniewski, S. Krawiec, The effect of ball hardness on four-ball wear test results. Wear **264**(7–8), 662–670 (2008)
8. H. Czichos, D. Klaffke, E. Santner, M. Woydt, Advances in tribology: the materials point of view. Wear **190**, 155–161 (1995)

9. D. Dowson, V. Wright, Bio-tribology, in *Proceeding of the Conference on the Rheology of Lubrication* (The Institute of Petroleum, The Institution of Mechanical Engineers, and the British Society of Rheology, London, 1973), pp. 81–88
10. A. Greco, S. Sheng, J. Keller, A. Erdemira, Material wear and fatigue in wind turbine ss. Wear **302**(1–2), 1583–1591 (2013)
11. H. Bangert, C. Eisenmenger-Sittner, A. Bergauer, Deposition and structural properties of two-component metal coatings for tribological applications. Surf. Coat. Technol. **80**(1–2), 162–170 (1996)
12. B. Bilyeu, W. Brostow, K.P. Menard, Epoxy thermosets and their applications I: chemical structures and applications'. J. Mater. Educ. **21**(5–6), 281–286 (1999)
13. A.S. Rathaur, J.K. Katiyar, V.K. Patel, S. Bhaumik, A.K. Sharma, A comparative study of tribological and mechanical properties of composite polymer coatings on bearing steel. Int. J. Surf. Sci. Eng. **12**(5/6), 379–401 (2018)
14. J.K. Katiyar, S.K. Sinha, A. Kumar, In situ lubrication of SU-8/Talc composite with base oil (SN150) and perfluoropolyether as fillers. Tribol. Lett. **64**(1), 5 (2016)
15. A.S. Rathaur, J.K. Katiyar, V.K. Patel, Tribo-mechanical properties of graphite/talc modified polymer composite bearing balls. Mater. Res. Express **7**(1), 15305 (2019)
16. M. Nosonovsky, B. Bhushan, Green tribology: principles, research areas and challenges. Philos. Trans. R. Soc. A **368**, 4677–4694 (2010)

Chapter 2
Polymer Brush Based Tribology

Manjesh K. Singh

Abstract Polymer chains with one of their ends grafted on a surface, stretch out in a good solvent to take a brush-like formation when the grafting density, $\rho_g > 1/(\pi R_g^2)$, where R_g is the radius of gyration of a chain in a good solvent. The equilibrium height of such a polymer brush is larger than the unperturbed size (R_g) of the corresponding polymer chain in a bulk solution. Polymer brushes find applications in the fields of tribology, rheology, biology and colloid-stabilization. Polymer brush based tribology is a recent attempt to mimic glycoproteins based lubrication found in nature. Higher coefficients of friction are observed due to as perity contact when hard surfaces are brought in contact and sheared against each-other. In contrast, when polymer-brush bearing surfaces are brought in contact with each-other and sheared in the presence of a good-solvent, much lower coefficients of friction are observed. Due to entropic reasons opposing polymer brushes avoid inter-digitation even under high compression enabling development of a thin fluid film between the brushes. Such a formation helps polymer brushes to support relatively high applied normal load while the thin fluid film in-between helps in reducing the friction. Tribological behavior of polymer brushes can be tuned by changing the grafting-density (ρ_g), chain-length (L_c), chain-stiffness (K_b), solvent-quality and cross-linking of chains. Possibility of designing lubricant with specific tribological properties make polymer-brushes an interesting topic of research. In this chapter we will go through the effects of different polymer-brush architectures on the tribological behavior of polymer brushes.

2.1 Introduction

Nature has a very complex method of lubricating sliding surfaces in an aqueous medium with the help of bio-polymers such as glycoproteins. In recent times humankind has attempted to understand and imitate this using polymer brushes.

M. K. Singh (✉)
Indian Institute of Technology, Kanpur, India
e-mail: manjesh@iitk.ac.in

J. K. Katiyar et al. (eds.), *Tribology in Materials and Applications*,
Materials Forming, Machining and Tribology,
https://doi.org/10.1007/978-3-030-47451-5_2

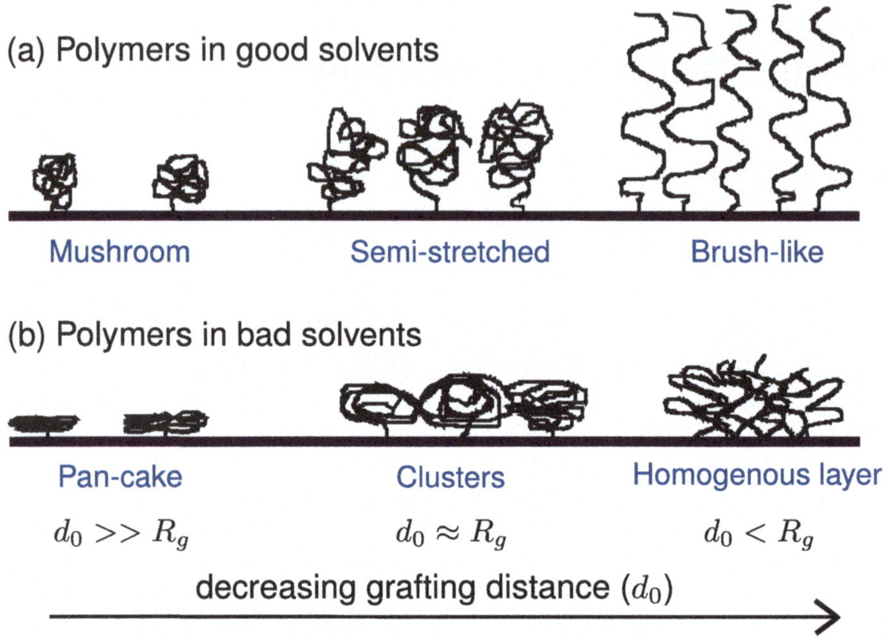

Fig. 2.1 Schematic of polymer brushes

When polymer chains are grafted to any surface at a sufficiently high grafting density (ρ_g), in presence of a good solvent, the chains stretch out to form a brush-like structure known as polymer brush.

The structure of grafted polymer chains depends on the grafting density (ρ_g) and solvent-quality. Figure 2.1 shows the different forms grafted chains take at different grafting densities (ρ) in presence of good and bad solvents. When the distance (d_0) between grafting sites of adjacent polymer chains is greater than radius of gyration (R_g) of the polymer chains in the bulk solution, they take mushroom-like form in good solvent and pan-cake like form in bad solvent. When $d_0 \approx R_g$, polymer chains take semi-stretched form in a good solvent whereas in a bad solvent, chains form clusters. When $d_0 < R_g$, the grafted polymer chains stretch out to form a polymer brush in a good solvent, whereas the chains form a homogenous layer in a bad solvent. Hence we see that the critical grafting density (ρ_g^*) to form a brush in a good solvent is $\rho_g \sim d_0^{-2} < R_g^{-2}$ [1–3].

2.1.1 *Preparation of Polymer Brushes*

Polymer brushes are synthesised using "garfting-from" and "grafting-to" approaches. Both the methods have advantages and limitations (Fig. 2.2).

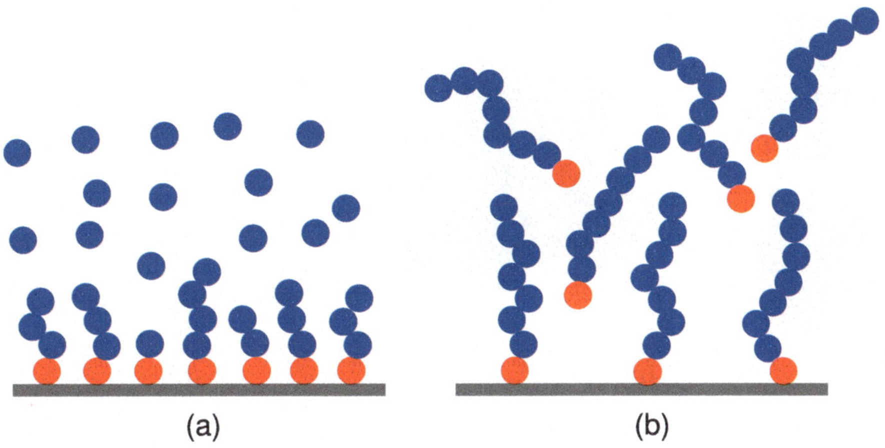

Fig. 2.2 Method to synthesis polymer brushes: **a** "grafting-from", and **b** "grafting-to"

In "grafting-from" approach, initiators are attached to the grafting surface and polymer chains are grown out of the initiators to form polymer brushes. In the "grafting-to" approach, solutions containing polymer chains is poured on the grafting surface and chains get attached by one end to the grafting surface to form polymer brushes. Higher grafting densities can be achieved using "grafting-from" approach in comparison to the "grafting-to" method. The latter has limitations in achieving higher grafting density because already adsorbed chains block other chains in the solution from reaching the grafting surface. The advantage of grafting-to approach is that chain detached (worn-off) from the grafting surface during sliding are replaced by other chains present in the solution, so this method has a "self-healing" characteristics. In recent times "grafting-from" approach has become more popular because it gives flexibility by modulating the feed monomers to synthesize block co-polymers based brushes and switching on or off the crosslinking during growth to synthesize layered structures of polymer brushes and gels (crosslinked polymer brushes).

2.2 Polymer Brush Mediated Lubrication

Tribological behavior of polymer brushes have been studied extensively using experiments and simulations to understand the origin of frictional forces at different loads and shear rates. Boundary and hydrodynamic lubrication have been established and studied [4]. Figure 2.3a shows the schematic of brush-against-brush and brush-against-wall systems. The density profile of brush-against-brush system (Fig. 2.3b) shows that as the density of polymers deplete moving away from the grafting surfaces, solvent density goes up and at the point where monomer-density is minimum, solvent density is maximum [5]. Hence, an effective fluid layer is created between the two opposite brushes (or between brush and wall in brush-against-wall system).

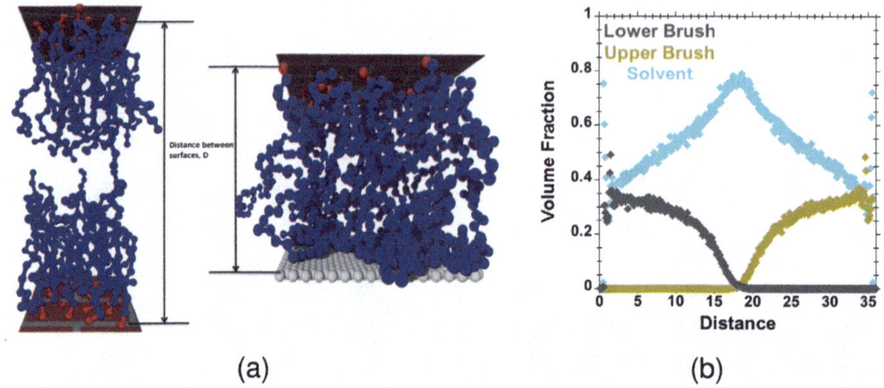

Fig. 2.3 Polymer brushes and density profile. **a** Snapshots of polymer-brush systems: symmetric brush-brush configuration and asymmetric brush-against-wall configuration. Red beads are tethered, blue beads are non-tethered, grey beads are explicit wall particles, and impenetrable grafting-surfaces (walls) are flat and indicated as "brick walls". There are $M = 20$ chains tethered on each grafting surface; each chain has $N = 50$ beads. Reprinted from Ref. [5] after permission. **b** Density profile for symmetric brush-brush configuration having explicit solvent beads. Reprinted from Ref. [6] after permission

When hard surfaces are brought in contact, due to asperity-asperity contact high friction force is observed. On contrary when polymer brush bearing surfaces are brought in contact in presence of a good solvent, due to entropic reason and excluded volume effect, polymers from the two surfaces avoid interpenetration even under high load. The resulting thin fluid film created in between as explained above, enhances the lubrication.

The use of polymer brushes for lubrication has been gaining in acceptance over recent years, and a wealth of useful observations have been made for simple brush systems. Recently, interesting tribological properties have been achieved with novel brush architectures, involving relatively stiff polymer chains (e.g. dextran), various degrees of crosslinking, different layers within the brush with different properties, and interactions between the polymer chains (e.g. hydrogen bonding or electrostatic). Modeling studies of simple brush systems have been successfully carried out by a number of groups, but a comprehensive study of the correlation between brush architectures/polymer-chain properties and tribological behavior (both friction and the fracture phenomena that lead to wear) is yet to be performed.

First experiments on polymer brushes under shear were carried out by Klein et al. [7–9] reporting that the resulting friction coefficients may be orders of magnitude smaller than those found in dry friction. Such observations indicated the importance of polymer brushes in lubrication applications and the realization of such potential applications triggered further studies in the same direction [10–14]. For further improvement on applications, a better understanding at the molecular level is essential. Due to limitations in the experiments such molecular insight is still lacking.

However, considerable grounds have been covered using theoretical and computational approaches. Many theoretical [15–17] and computer simulation [18–21] studies have been made for the interpretation of the observed phenomena.

Figure 2.4 shows the experimental and molecular dynamics simulation studies of tribological behavior of dextran brushes in HEPES solvent for a brush-against-wall system. The experimental study establishes that at low shear speed, coefficient of friction (CoF) does not change significantly with speed but at higher speed CoF increases linearly with speed in log-log scale. Subsequently MD simulation [5] has helped to understand the experimental observation and it is found that attractive interaction between brush and counter-wall surface is the reason for plateau in the CoF values at lower speed.

Friction and normal forces arise due to the interaction between polymers and solvent (polymer, solvent and wall-surface for brush-against wall system). There is an attractive van der Waals force present between the brush and the wall, which reduces the overall repulsion, but it does not lead to net attractive interaction. The above simulation results for the coefficient of friction are not affected by adhesion between the wall and the polymer brush. Van der Waals interactions between polymer brushes and surfaces are considered as "bridging forces" and can be specific or non-specific. Israelachvili [22] has explained various attractive forces, termed as "Intersegment", "Bridging" and "Depletion" forces, in detail between polymers and counter surfaces. Under suitable conditions, "Bridging forces" can lead to an overall attractive force.

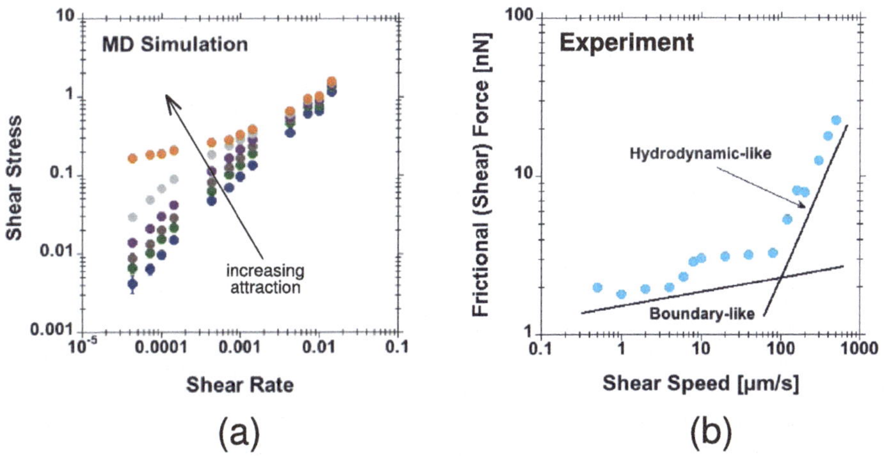

Fig. 2.4 Forces and stresses against shear speed divided by gap distance, i.e. shear rate in (**a**) and shear speed in (**b**), for brush-against wall systems. **a** MD simulation results for modified brush-wall interaction, making it more attractive from blue circles to orange circle in increasing order, **b** LFM experiment results for dextran polymer brushes in HEPES. The simulation results are in reduced Lennard-Jones units. Reprinted from Ref. [5] after permission

Polymer-brush based aqueous lubrication is a good candidate to replace oil based lubrication for its positive impact on environment. Water has some excellent properties such as high thermal capacity and environment-friendly. It makes water an attractive candidate for lubrication but due to its poor viscosity at high pressure, water can not support load and thus it is not a good lubricant on its own. In polymerbrush-based aqueous lubrication, polymers rather than hydrodynamic forces, support the normal load and thin liquid of water inbetween reduces the friction and thus the combination together works as an excellent lubricant. Problems such as eco-toxicity, bioaccumulation and renewability of lubricants can be in principle solved using such polymer based aqueous lubricants.

Tribological behavior of polymer brushes depend hugely on grafting density, length of polymer-chains, stiffness of chains, solvent quality and crosslinking of polymers. Thus polymer brush is a designers material and provides excellent opportunities to tune tribological behavior by playing with different brush-architectures. We would see in the following sections how different architectures affect frictional behavior of polymer brushes based on previous research works on this topic. But before getting into that it is important to introduce some of the important definitions which are already used and will again be used in this chapter.

(a) Degree of polymerisation (N): Polymers consist of repeated units called monomers. The number of monomers in a polymer chain is referred to as degree of polymerisation, N. N can be estimated from molecular weight of polymers which can be measured using different experimental techniques such as gel permeation chromatography (GPC).

(b) Dry thickness (t_d), wet thickness (t_w) and swelling-ratio ($S_{W)}$: Dry thickness (t_d) and wet thickness (t_w) of a surface-grafted polymer refer to the thickness in absence and presence of solvent(s) respectively. Swelling-ratio ($S_{W)}$ is defined as:

$$S_W = \frac{t_w}{t_d} - 1 \quad (2.1)$$

t_d and t_w are experimentally measurable quantities. Ellipsometry is one of the most commonly used techniques for these measurements. Wet-thickness can also be estimated from atomic force microscope (AFM) and surface force apparatus (SFA) experiments [23, 24].

(c) Grafting density (ρ_g): Grafting-density (ρ_g) of polymer brushes is defined as: $\rho_g = M/A$, where M is the number of polymer chains grafted and A is the area of grafting surface. It is estimated from the measured value of thickness using the relation:

$$M_n \approx \frac{6.023\Omega t_d}{\rho_g} \times 100 \quad (2.2)$$

where M_n is the molecular weight (in g/mol), Ω the polymer mass density (in g/cm^3), t_d the polymer brush dry thickness (nm), and ρ_g the grafting density (in chains/nm^2) [25].

(d) Persistence length (L_p): For modeling chain-stiffness the following discrete version of a bending potential is commonly used

$$\frac{U_b}{k_B T} = k_b \sum_{i=1}^{N-1} (1 - u_i \cdot u_{i+1}), \tag{2.3}$$

where k_b is the dimensionless bending stiffness and u_i denotes the unit segment vector connecting the ith with the $(i + 1)$th bead along the polymer chain. The bending stiffness k_b and persistence length L_p are interrelated via $L_p = -\frac{b_0}{\ln \mathcal{L}(k_b)}$ with the Langevin function $|\mathcal{L}(k_b) = \coth(k_b) - \frac{1}{k_b}$ and a bond length $b_0 \approx 1$ [6].

(e) Degree of crosslinking (p): By degree of crosslinking we mean,

$$p = \frac{2 \times N_{cross}}{M \times N} \times 100\% \tag{2.4}$$

where N_{cross}, M and N are number of crosslinkers, number of polymer chains and degree of polymerization respectively in surface grafted polymer gels. Degree of crosslinking in crosslinked polymer brushes is estimated by the change in thickness with amount of crosslinkers added during the crosslinking stage [25].

2.3 Effect of Grafting-Density (ρ_g)

Conformations of surface-grafted polymer chains depend strongly on grafting density (ρ_g). The equilibrium height of a polymer-brush (its thickness), $h \propto \rho_g^{1/3}$: this is confirmed in Fig. 2.5 which shows the effect of grafting density on the equilibrium brush height (h) as calculated from molecular dynamics simulations of the brush bearing surfaces. Grafting density also plays an important role in the tribological behavior of polymer brushes. Rosenberg et al. [26] experimentally studied the effect of grafting density on tribological behavior of PLL-g-dextran polymer brushes and concluded that there exist two distinct regimes: (i) at lower loads, the friction coefficient decreases with increasing grafting density, and (ii) at higher load the friction coefficient increases with increasing grafting density. So there exist a "transition-load" in-between for polymer brushes beyond which their frictional behavior switches with ρ_g.

The effect of grafting density on tribological behavior of polymer brushes have also been studied using MD simulations [3, 6]. Figure 2.5b shows coefficient of

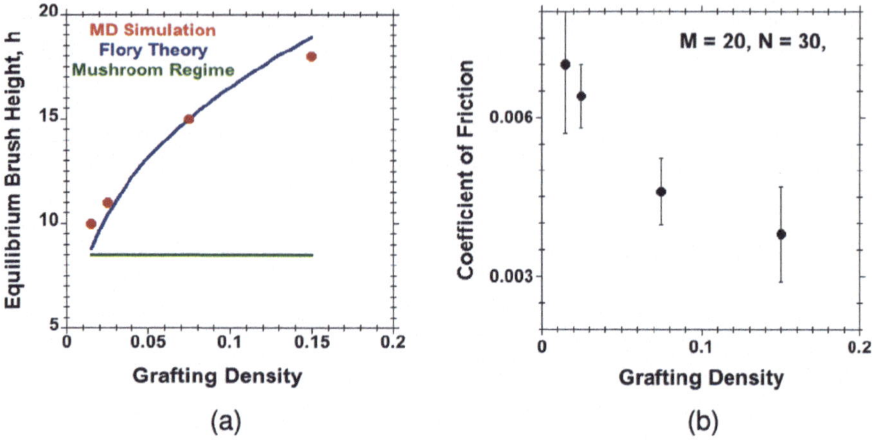

Fig. 2.5 Effect of grafting density. **a** Equilibrium-brush height (h) and **b** coefficient of friction versus grafting density (ρ_g) for M = 20 flexible chains with N = 30 beads per chain under good-solvent conditions. The ρ_g-independent reference result for the mushroom regime is marked by the horizontal line. The simulation results are in reduced Lennard-Jones units. Reprinted from Ref. [6] after permission

friction as a function of grafting density of polymer brushes obtained from the simulations. The coefficient of friction is observed to be decreasing with increase in the grafting density. Singh et al. [6] explained the effect of grafting density on tribological behavior of polymer brushes as follows. Polymer brushes with higher grafting densities experience higher repulsive forces acting between opposite chains. At moderate load the system with the higher grafting density has a more pronounced solvent layer between the oppositely facing brushes which results in less friction. For a given normal load, the polymer brush within the system with lower grafting density cannot support the load as well as brushes with higher ρ_g, resulting in interpenetration between chains and a corresponding higher friction coefficient. As the magnitude of the normal stress increases further, the system with higher grafting density starts to show signatures of interpenetration between opposite chains as well. Because of the higher grafting density, there is more interaction between opposing polymer chains, resulting in even higher friction at higher load compared to the system with lower grafting density. The simulation work has limitations in the sense that it cannot see a clear “transition load”. Using dissipitive particle dynamics (DPD) approach, Mayoral et al. [27] also studied the effect of grafting density on the rheological and tribological behavior of polymer brushes. While their study showed a monotonus increase in viscosity with increase in ρ_g, the decrease in coefficient of friction is found to be nonmonotous. Up to a certain value of grafting density $\rho_{g,min}$, coefficient of friction decreased with increasing grafting-density but after that it did not change significantly, rather, in some cases the trend reversed.

2.4 Solvent Effect

The size and conformations of polymer chains in a solvent depend on the net interaction of the polymers with the surrounding solvent molecules. In simple words, if polymer molecules like itself more than solvent, polymer chains collapse to form globule like structures and there is a phase separation between polymer and solvent molecules. Such a solvent is referred to as a bad solvent for the polymer. On the other hand if the net interaction between the polymer chains and solvent is attractive i.e. the polymers prefer solvent molecules to itself, the chains open up to take a coil like form. In that case the solvent is referred to as a good solvent for the said polymer. The interaction between solvent and polymer is described by the mean field theory based Flory-Huggins parameter χ. χ measures the interaction energy between polymer segments and solvent molecules [28, 29]. The χ parameter is obtained from mean field calculations where the configuration of polymer chain surrounded by the solvent in a lattice space is described through its configurational entropy and enthalpy of mixing. Flory [28] and de Gennes [30] describe the excluded volume as $\omega = (1 - 2\chi)a^3$, a being the size of a monomer. Depending on the value of the parameter χ, the interaction between polymer segments and solvent molecules can be classified in following cases [31]:

(a) A thermal solvents: $\chi = 0$, the interaction between polymer and solvent is very strong leading to large chain-expansion or swelling. The monomer makes no energetic difference between other monomers and solvent molecules.
(b) Good solvents: $\chi < 0.5$, the limit of a thermal solvent where the excluded volume value is less than the a thermal case. The size of single polymer chain in a good solvent is given by $\langle R \rangle \propto N^{0.588}$, N being degree of polymerization.
(c) Theta solvent: $\chi = 0.5$, implying excluded volume parameter becomes zero. The polymer chains behave ideally. The size of a chain in a θ-solvent is given by $\langle R \rangle \propto N^{0.5}$.
(d) Bad solvent: $\chi > 0.5$, implying excluded volume parameter becomes negative. The polymer chains collapse to form globular like structure and the size of a chain in a bad solvent is given by $\langle R \rangle \propto N^{1/3}$.

2.4.1 Mixing Good and Bad Solvents

Understanding the effect of solvent quality is important to tune tribological behavior of polymer brushes. As discussed in the Sect. 2.1 grafted polymer chains stretch out in a good solvent whereas in a bad solvent they collapse to form a homgeneous layer on the grafting surface. Researchers have mixed a good solvent and a bad solvent and studied the swelling and frictional behavior of polymer brushes over a range of composition of good and bad solvents in the mixture [4, 14, 32]. Nomura et al. [14] studied polystyrene (PS) polymer brushes in a mixture of toluene (good solvent for PS) and 2-propanol (bad solvent for PS). They controlled the swelling of highly concentrated PS brushes by changing the fraction of 2-propanol in the mixture and thus tuned the friction behavior. As the fraction of bad solvent (2-propanol) increases, swelling of polymer brushes decrease and frictional coefficient increases. Different lubrication regimes boundary and hydrodynamic were observed at different compositions of good and bad solvents in the mixture. In 100% 2-propanol, only boundary regime (friction depends strongly on the physical or chemical properties of the outermost surface and weakly on applied shear speed) high frictional coefficient was observed. As the fraction of toluene was increased in the mixture, coefficient of friction starts decreasing and hydrodynamic regime (strongly shear speed dependent frictional behavior) start to appear. In the intermediate composition only hydrodynamic lubrication regime was observed. In 100% toluene boundary lubrication along with extremely low friction coefficient was observed at lower speed and at higher shear speeds hydrodynamic lubrication regime was observed. Nalam et al. investigated frictional properties of PEG and dextran brushes in mixed solvents using colloid-probe lateral force microscopy [4]. Water miscible glycerol and ethylene glycol are bad solvents for hydrophilic (waterloving) poly (L-lysine)-graft-poly(ethylene glycol) (PLL-g-PEG) and poly(Llysine)-graft-dextran (PLL-g-dextran). Figure 2.6 shows coefficient of friction against shear-speed multiplied viscosity for bare silica,

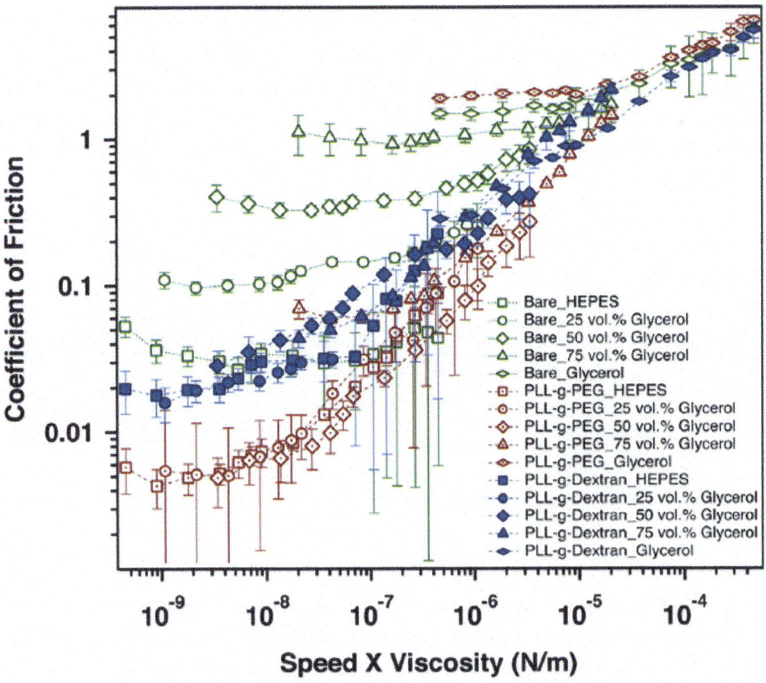

Fig. 2.6 Dependence of coefficient of friction (μ) on speed multiplied viscosity for bare silica (open green), PEG (open dot red) and dextran (filled blue)-coated surfaces at a normal load of 100 nN with varying viscosity of aqueous glycerol mixtures. Reprinted from Ref. [4] after permission

PLL-gPEG and PLL-g-dextran against silica micro-sphere for different composition of aqueous glycerol solvent mixture. Aqueous glycerol mixtures are viscous. The frictional forces as a function of shear speeds is studied at the nano scale. The presence of polymers at the interface prevents contact between the surfaces and allows for formation of a hydration layer in between. It results in reduction of friction. The coefficient of friction values for polymer-coated-surfaces are found to be order of magnitude less than the bare silica surface at lower speed for most of the compositions mixed solvents. The effect of polymer conformations with varying solvent quality was also studied by measuring pull-off force. The collapsed brushes exhibited an increase in adhesion.

The study of Nalam et al. [4] at the nano-scale has concluded that shorter and stiffer dextran brushes lead to higher frictional forces than PEG brushes. The study further showed that for polymer brushes, the lowest friction is achieved in the boundary regime at the nano-scale. It is in contrast to the situation in the macroscopic Stribeck curve where lowest friction is observed in the mixed regime. This is explained as follow: friction in the boundary regime at the macro scale is dominated by surface roughness, whereas on the nano-scale it is dominated by adhesion. Adhesion is negligible for brushes in good solvents, so lower friction in the boundary regime at the nanoscale. If sufficiently long brushes are used, a low-friction boundary regime may also be achieved at the macro-scale [23].

2.4.2 Co-solvency and Co-non-solvency

Polymers exhibit co-solvency and co-non-solvency in mixture of solvents. Co-solvency occurs if a mixture of two bad solvents causes swelling of polymer leading to globule-coil transition in a certain range of composition of these two solvents. A typical example of co-solvency is Poly(methyl methacrylate) (PMMA) in aqueous alcohol mixture. Both water and alcohol are bad solvents for PMMA but together they cause swelling of PMMA. Co-non-solvency is mirror effect of co-solvency. In co-non-solvency, mixture of two good solvents cause collapse of polymer chain to take globular conformation. An example of co-non-solvency is poly(Nisopropylacrylamide) (PNiPAAm) in aqueous alcohol mixture. Co-non-solvency leads to segregation of polymer solutions into a polymer-rich phase in a certain range of compositions of two good solvents [33–36].

Several studies have been performed to tune structure, mechanical and tribological behavior of polymer brushes [37–40]. de Beer et al. [40] have shown that tunable friction is provided by co-nonsolvency of a PNIPAM brush in ethanol-water mixtures. Both water and ethanol are good solvents for PNIPAM. The PNIPAM brush is swollen in pure water and acts as a lubricant layer. As volume fraction of ethanol increases in the ethanol-water mixture, the PNIPAM brush starts collapsing. The PNIPAM brush is partially collapsed at 10% ethanol in the ethanol-water mixture and a large friction force is observed. It is due to stretching of the partly collapsed brush chains when the colloid probe is moved away from the initial point of contact. The strongest collapse of the PNIPAM brush occurs in 30% volume fraction of ethanol. In this state, the friction is higher than in pure water. The friction force value at 30% ethanol-content is, however, smaller than the friction force observed at the 10% ethanol-content. This observation is attributed to the fact that the grafted chains do not stretch (at 30% ethanol-content) during the sliding motion. At higher volume fractions of ethanol, the brush re-swells. At these solvent compositions, a lubricant layer is again formed. The friction force gradually decreases after reaching a peak value at 10% ethanol-content, with further increasing ethanol-content and reaches a minimum in pure ethanol. At this minimum, typical values of the friction force are two orders of magnitude smaller than the forces measured at 10% ethanol in the ethanol-water mixture. The study shows that friction can be tuned in a straightforward manner by varying the relative amount of ethanol. Controlling friction properties by employing co-non-solvency of PNIPAM brush in ethanol-water mixtures has potential application in walking robots, and switchable tweezers.

2.5 Effect of Crosslinking

Crosslinking of polymer chains in a polymer brush is used to tune its swelling, mechanical and tribological behavior. Crosslinking among chains in polymer brush can be achieved using two different methods. First the in situ method in which crosslinking is performed while growing the chains out of grafting surface. In the second ex situ method, first polymer chains are grown to prepare polymer brush and then crosslinking is performed in subsequent step. Crosslinked polymer brushes are called as polymer brush-gels or simply gels. Polymer gels can swell either in water (hydrogels) or oil (lipogels) making them highly suitable candidates for applications in the fields of drug delivery, pharmaceuticals, tissue engineering and other bio-medical applications. Tribological studies of polymer gels have been focussed on the effect of degree of crosslinking and length of crosslinkers [13, 24, 25, 41, 42]. All the studies have confirmed: crosslinking leads to increase in friction between polymer brushes and counter-surfaces, (ii) decrease in swelling ratio and (iii) increase in modulus. Studies have also confirmed that crosslinking leads to improvement in wear behavior of polymer brushes [24, 43, 44]. In Ref. [25] effect of crosslinking was studied using complementary experimental and MD simulations approaches. The experiments were performed on (poly)glycdyl-methacrylate (PGMA) brushes and gels in the presence of DMF solvent. The tribological and mechanical behavior experiments were performed using silica microspheres under the colloid-probe-based lateral force microscopy (LFM) technique. The PGMA brushes showed a remarkable decrease in friction forces when compared to bare silicon surfaces. The study showed that crosslinking in general leads to an increase in friction between polymer brushes and a counter-surface. It is primarily because crosslinked polymer brushes are more resistant to shear compared to their non-crosslinked counterparts. Figure 2.7 shows that the coefficient of friction increases with increasing degree of crosslinking and decreases with increasing length of the crosslinker chains.

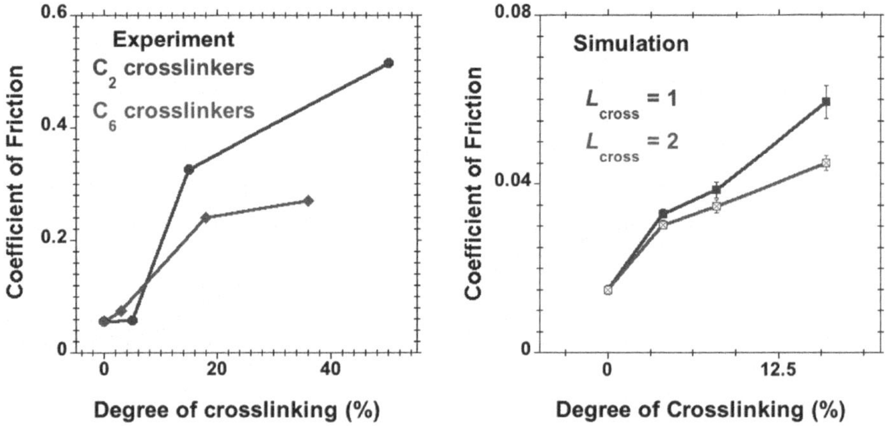

Fig. 2.7 Coefficient of friction against degree of crosslinking for **a** Experimental results for systems with shorter crosslinkers C_2 (magenta lines) and longer crosslinkers C_6 (gray lines) at a shear speed of 1 μm/s and **b** Simulation results for systems with $M = 50$ chains of length N "50 for different lengths of crosslinkers, $L_{cross} = 1$ (magenta lines) and $L_{cross} = 2$ (gray lines) at a shear speed, $v = 0.001$ for a brush-against-wall system. Reprinted from Ref. [25] after permission

AFM-based indentation of PGMA brushes and gels in DMF solvent showed a decrease in their swelling ratio with increasing degree of crosslinking and can very well explain the tribological response of gels at different degrees of crosslinking for different lengths of crosslinkers. Molecular dynamics study on the tribological behavior of crosslinked polymer brushes gave results consistent with the experimental findings: increase in coefficient of friction with increasing crosslinking degree and decrease in coefficient of friction with increasing crosslinker length. A careful look at the density profiles obtained from MD simulations of polymer brushes and gels explains the effect of degree of crosslinking and length of crossliners. As the degree of crosslinking increases keeping the crosslinker-length constant, the polymer concentration in the outer layer that can participate in brush-assisted lubrication is reduced. The reverse happens when crosslinker-length was increased keeping degree of crosslinking constant. The study concluded that that the brush-forming polymer chains in the outer layer play a significant role in reducing friction at the interface. As the concentration of brush forming chains in the outer layer decreases, the coefficient of friction at the interface goes up and vice versa.

2.6 Effect of Chain Stiffness

Increase in chain stiffness leads to increase in persistence length of polymers. Molecular dynamics study mentioned in details in Ref. [6] concluded that increase in persistence length of polymers causes decrease in coefficient of friction in polymer-brush-based lubrication. Study of chain-stiffness on tribological behavior of polymer brushes are particularly relevant in the context of DNA-coated nanoparticle crystallization.

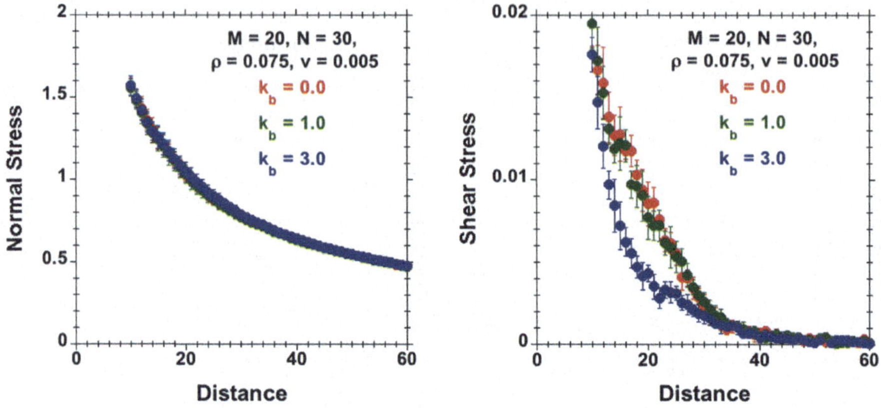

Fig. 2.8 Steady-state **a** normal stress and **b** shear stress against separation for brush-against-brush systems ($M = 20$, $N = 30$, good solvent conditions, $\rho_g = 0.075$) subjected to shear velocity $S = 0.005$ for chains with varying stiffness: $K_b = 0$ (red), $K_b = 1$ (green) and $K_b = 3$ (blue). For intermediate distances there is a remarkable drop of shear stress for the stiffer chains, while the normal stress remains basically unaffected by K_b. The simulation results are in reduced Lennard-Jones units. Reprinted from Ref. [6] after permission

Figure 2.8 shows the effect of chain stiffness on normal stress and shear stress at different separation between brush bearing surfaces at a constant shear speed. There is no change observed in normal stresses with change in semi-flexibility of chains but lower shear stress was observed for chains with higher stiffness. The observed results were explained by the fact that even though at the same separation the polymer-brushes with higher stiffness showed higher interpenetration between opposite chains due to increase equilibrium brush height (h), brushes having stiffer chains align themselves along the shear direction easily, so the friction decreased rather than increasing.

2.7 Effect of Chain-Length

Degree of polymerization (N) or chain-length is one of most important parameters, affecting the physical, thermal and mechanical properties of polymers. For example glass- transition temperature (T_g), melting temperature (T_m), viscosoity (ν) and mechanical strength of polymers increase with increase in chain length or degree of polymerization. There have also been several experimental studies on the effect of chain length or brush thickness on the friction behavior of polymer brushes [45–47]. Zhang et al. [47] observed for poly (2-methacryloyloxy) ethylphosphorylcholine (PMPC) brushes with a thickness greater than 200 nm, that the coefficient of friction decreases with increasing brush thickness. For shorter PMPC brushes, little variation was observed in the coefficient of friction upon varying brush thickness. The findings are consistent with the study of chain length effect in 46. The observation were attributed to an increase in the amount of bound solvent with the increase in brush thickness, which leads to an increase in osmotic pressure, causing less deformation at any given applied load. McNamee et al. [45] studied tribological behavior of poly (ethylene glycol) (PEG) polymer brush for different chain length. The study found higher friction for poly (ethylene glycol) (PEG) polymer brush systems with longer chains and they attributed this to higher degree of entanglements in longer brushes. The grafting density (ρ) was not kept fixed in this study. Still there are open questions on the effect of chain length on tribological behavior of polymer brushes.

2.8 Summary

Tribological behavior of polymer brushes can be tuned by controlling different parameters. Based on the studies cited in this chapter, it can be concluded that thickness of polymer brushes is the most important factor in deciding their frictional behavior. As the brush thickness increases, the bound solvent increases and coefficient of friction decreases. Polymer brush can be designed to achieve specific tribological properties for application in biotribology. Polymers have been grafted successfully on steel surfaces to study potential applications in metal working and manufacturing. Polymer brush based lubrication has been shown to be working in the range of hundreds of MPa load under which bearing works [48]. Aqueous lubrication using hydrophilic polymer brush has potential to replace oil based lubricants for environment-friendly green lubrication. Ionic-liquid type polymer brush (ILPBs) based lubrications exhibited low friction and wear behavior at high pressure of 540 MPa [49]. Synthesizing stratified layers of "brush-gel" or "gel-brush" structure by controlling crosslinking can lead to fabrication of coating of specific mechanical and tribological behavior [50]. The layer of gel at the bottom would help to improve modulus and wear-resistant properties of coating whereas brush layer at the top will help to reduce friction. Inspired by nature, polymer brush based lubrication exhibit ultra low friction at low contact loads and has potential application in many fields. There is still a lot of research and study required to use polymer brushes for lubrication in different industries. Role of entanglements among grafted polymer chains is still not well-understood and needs further studies.

Acknowledgements The author thanks the American Chemical Society publications, USA and Multidisciplinary Digital Publishing Institute, Switzerland for granting permission to reprint the graphs and images. The author also thanks Dr. Tapan Chandra Adhyapak (IISER Tirupati) and Dr. Arghya Dutta (MPIP Mainz) for critically reading the manuscript.

References

1. N.D. Spencer, *Tailoring Surfaces: Modifying Surface Composition and Structure for Applications in Tribology, Biology and Catalysis*, vol. 5 (World Scientific, 2011)
2. M.K. Singh, *Simulation and Experimental Studies of Polymer-Brushes under Shear*. Ph.D. thesis (ETH Zurich, 2016)
3. A. Galuschko, L. Spirin, T. Kreer, A. Johner, C. Pastorino, J. Wittmer, J. Baschnagel, Frictional forces between strongly compressed, nonentangled polymer brushes: molecular dynamics simulations and scaling theory. Langmuir **26**(9), 6418–6429 (2010)
4. P.C. Nalam, S.N. Ramakrishna, R.M. Espinosa-Marzal, N.D. Spencer, Exploring lubrication regimes at the nanoscale: nanotribological characterization of silica and polymer brushes in viscous solvents. Langmuir **29**(32), 10149–10158 (2013)
5. M.K. Singh, P. Ilg, R.M. Espinosa-Marzal, M. Kröger, N.D. Spencer, Polymer brushes under shear: molecular dynamics simulations compared to experiments. Langmuir **31**(16), 4798–4805 (2015)
6. M.K. Singh, P. Ilg, R.M. Espinosa-Marzal, N.D. Spencer, M. Kröger, Influence of chain stiffness, grafting density and normal load on the tribological and structural behavior of polymer brushes: a nonequilibrium-molecular-dynamics study. Polymers **8**(7), 254 (2016)
7. J. Klein, D. Perahia, S. Warburg, Forces between polymer-bearing surfaces undergoing shear. Nature **352**(6331), 143 (1991)
8. J. Klein, E. Kumacheva, D. Mahalu, D. Perahia, L.J. Fetters, Reduction of frictional forces between solid surfaces bearing polymer brushes. Nature **370**(6491), 634 (1994)
9. J. Klein, Shear, friction, and lubrication forces between polymer-bearing surfaces. Annu. Rev. Mater. Sci. **26**(1), 581–612 (1996)
10. S. Lee, N.D. Spencer, Sweet, hairy, soft, and slippery. Science **319**(5863), 575 (2008)
11. S. Lee, M. Müller, M. Ratoi-Salagean, J. Vörös, S. Pasche, S.M. De Paul, H.A. Spikes, M. Textor, N.D. Spencer, Boundary lubrication of oxide surfaces by poly (l-lysine)-g-poly (ethylene glycol)(pll-g-peg) in aqueous media. Tribol. Lett. **15**(3), 231–239 (2003)
12. M. Müller, S. Lee, H.A. Spikes, N.D. Spencer, The influence of molecular architecture on the macroscopic lubrication properties of the brush-like co-polyelectrolyte poly (l-lysine)-g-poly (ethylene glycol)(pll-g-peg) adsorbed on oxide surfaces. Tribol. Lett. **15**(4), 395–405 (2003)
13. V.P. Bavaresco, C.A.C. Zavaglia, M.C. Reis, J.R. Gomes, Study on the tribological properties of phema hydrogels for use in artificial articular cartilage. Wear **265**(3–4), 269–277 (2008)
14. A. Nomura, K. Okayasu, K. Ohno, T. Fukuda, Y. Tsujii, Lubrication mechanism of concentrated polymer brushes in solvents: effect of solvent quality and thereby swelling state. Macromolecules **44**(12), 5013–5019 (2011)
15. P.-G. de Gennes, Conformations of polymers attached to an interface. Macromolecules **13**(5), 1069–1075 (1980)
16. S.T. Milner, T.A. Witten, M.E. Cates, Theory of the grafted polymer brush. Macromolecules **21**(8), 2610–2619 (1988)
17. Y.B. Zhulina, V.A. Pryamitsyn, O.V. Borisov, Structure and conformational transitions in grafted polymer chain layers. A new theory. Polym. Sci. USSR **31**(1), 205–216 (1989)
18. G.S. Grest, Computer simulations of shear and friction between polymer brushes. Curr. Opin. Colloid Interface Sci. **2**(3), 271–277 (1997)
19. R.S. Hoy, G.S. Grest, Entanglements of an endgrafted polymer brush in a polymeric matrix. Macromolecules **40**(23), 8389–8395 (2007)
20. G.S. Grest, Interfacial sliding of polymer brushes: a molecular dynamics simulation. Phys. Rev. Lett. **76**(26), 4979 (1996)
21. M. Murat, G.S. Grest, Molecular dynamics simulations of the force between a polymer brush and an afm tip. Macromolecules **29**(25), 8282–8284 (1996)
22. J.N. Israelachvili, *Intermolecular and Surface Forces* (Academic press, Cambridge, 2011)
23. P.C. Nalam, *Polymer Brushes in Aqueous Solvent Mixtures: Impact of Polymer Conformation on Tribological Properties*. Ph.D. thesis (ETH Zurich, 2012)

24. A. Li, E.M. Benetti, D. Tranchida, J.N. Clasohm, H. Schönherr, N.D. Spencer, Surface-grafted, covalently cross-linked hydrogel brushes with tunable interfacial and bulk properties. Macromolecules **44**(13), 5344–5351 (2011)
25. M.K. Singh, C. Kang, P. Ilg, R. Crockett, M. Kröger, N.D. Spencer, Combined experimental and simulation studies of cross-linked polymer brushes under shear. Macromolecules **51**(24), 10174–10183 (2018)
26. K.J. Rosenberg, T. Goren, R. Crockett, N.D. Spencer, Load-induced transitions in the lubricity of adsorbed poly (l-lysine)-g-dextran as a function of polysaccharide chain density. ACS Appl. Mater. Interfaces **3**(8), 3020–3025 (2011)
27. E. Mayoral, J. Klapp, A. Gama Goicochea, Scaling features of the tribology of polymer brushes of increasing grafting density around the mushroom-to-brush transition. Phys. Rev. E **95**(1), 012505 (2017)
28. P.J. Flory, *Principles of Polymer Chemistry* (Cornell University Press, Ithaca, 1953)
29. M.L. Huggins, Some properties of solutions of long-chain compounds. J. Phys. Chem. **46**(1), 151–158 (1942)
30. P.-G. de Gennes, *Scaling Concepts in Polymer Physics* (Cornell University Press, Ithaca, 1979)
31. M. Rubinstein, R.H. Colby, *Polymer Physics*, vol. 23 (Oxford University Press, New York, 2003)
32. R.M. Espinosa-Marzal, P.C. Nalam, S. Bolisetty, N.D. Spencer, Impact of solvation on equilibrium conformation of polymer brushes in solvent mixtures. Soft Matter **9**(15), 4045–4057 (2013)
33. D. Mukherji, K. Kremer, Coil–globule–coil transition of pnipam in aqueous methanol: Coupling all-atom simulations to semi-grand canonical coarse-grained reservoir. Macromolecules **46**(22), 9158–9163 (2013)
34. D. Mukherji, C.M. Marques, T. Stuehn, K. Kremer, Depleted depletion drives polymer swelling in poor solvent mixtures. Nat. Commun. **8**(1), 1374 (2017)
35. D. Mukherji, C.M. Marques, K. Kremer, Collapse in two good solvents, swelling in two poor solvents: defying the laws of polymer solubility? J. Phys. Condens. Matter **30**(2), 024002 (2017)
36. D. Mukherji, C.M. Marques, K. Kremer, Polymer collapse in miscible good solvents is a generic phenomenon driven by preferential adsorption. Nat. Commun. **5**, 4882 (2014)
37. H. Yong, S. Rauch, K.-J. Eichhorn, P. Uhlmann, A. Fery, J.-U. Sommer, Cononsolvency transition of polymer brushes: a combined experimental and theoretical study. Materials **11**(6), 991 (2018)
38. J.-U. Sommer, Adsorption-attraction model for co-nonsolvency in polymer brushes. Macromolecules **50**(5), 2219–2228 (2017)
39. Y. Yu, B.D. Kieviet, E. Kutnyanszky, G.J. Vancso, S. de Beer, Cosolvency-induced switching of the adhesion between poly (methyl methacrylate) brushes. ACS Macro Lett. **4**(1), 75–79 (2014)
40. Y. Yu, M. Cirelli, B.D. Kieviet, E.S. Kooij, G.J. Vancso, S. de Beer, Tunable friction by employment of co-non-solvency of pnipam brushes. Polymer **102**, 372–378 (2016)
41. J. Gong, M. Higa, Y. Iwasaki, Y. Katsuyama, Y. Osada, Friction of gels. J. Phys. Chem. B **101**(28), 5487–5489 (1997)
42. M.K. Singh, P. Ilg, R.M. Espinosa-Marzal, M. Kröger, N.D. Spencer, Effect of crosslinking on the microtribological behavior of model polymer brushes. Tribol. Lett. **63**(2), 17 (2016)
43. Motoyasu Kobayashi, Masami Terada, Atsushi Takahara, Polyelectrolyte brushes: a novel stable lubrication system in aqueous conditions. Faraday Discuss. **156**(1), 403–412 (2012)
44. Q. Zhang, L.A. Archer, Interfacial friction and adhesion of cross-linked polymer thin films swollen with linear chains. Langmuir **23**(14), 7562–7570 (2007)
45. C.E. McNamee, S. Yamamoto, K. Higashitani, Preparation and characterization of pure and mixed monolayers of poly (ethylene glycol) brushes chemically adsorbed to silica surfaces. Langmuir **23**(8), 4389–4399 (2007)
46. K. Kitano, Y. Inoue, R. Matsuno, M. Takai, K. Ishihara, Nanoscale evaluation of lubricity on welldefined polymer brush surfaces using qcm-d and afm. Colloids Surf., B **74**(1), 350–357 (2009)

47. Z. Zhang, A.J. Morse, S.P. Armes, A.L. Lewis, M. Geoghegan, G.J. Leggett, Effect of brush thickness and solvent composition on the friction force response of poly (2-(methacryloyloxy) ethylphosphorylcholine) brushes. Langmuir **27**(6), 2514–2521 (2011)
48. S. Watson, S. Dennington, L. Wang, M. Nie, S. Hinder, K. Stokes, Polymer brush lubrication of the silicon nitride–steel contact: a colloidal force microscopy study. Rsc Adv. **7**(68), 42667–42676 (2017)
49. M. Belin, H. Arafune, T. Kamijo, J. PerretLiaudet, T. Morinaga, S. Honma, T. Sato, Low friction, lubricity, and durability of polymer brush coatings, characterized using the relaxation tribometer technique. Lubricants **6**(2), 52 (2018)
50. A. Li, S.N. Ramakrishna, P.C. Nalam, E.M. Benetti, N.D. Spencer, Stratified polymer grafts: synthesis and characterization of layered 'brush' and 'gel' structures. Adv. Mater. Interfaces **1**(1), 1300007 (2014)

Chapter 3
Thin Film Lubrication, Lubricants and Additives

Febin Cyriac and Aydar Akchurin

Abstract Friction will be generated when two solid bodies are pressed over or slide against each other, and it acts opposite to the direction of relative motion. Lubricants are frequently used to reduce friction which otherwise may result in high machine wear and energy losses. Depending upon the phenomenon, lubrication can be classified into four different regimes: boundary, mixed, elastohydrodynamic and hydrodynamic. In boundary regime, the frictional response is mainly governed by the properties of the surfaces and it generally involves adsorption of lubricant molecules onto the mating surfaces. Therefore, in this regime, properties other than bulk properties of the lubricants play a significant role in determining the frictional response. Mixed or thin film lubrication (TFL) is a bridge that mark the transition from boundary to Elasto-Hydrodynamic (EHL) [or hydrodynamic (HL)] regimes. In TFL the load is partly supported by direct contact of the surface asperities and partly by the fluid. EHL regime is a type of HL regime which is characterized by the formation of sufficiently thick fluid film which fully separates the surfaces from direct contact thus reducing friction. Elastic deflections of the surfaces in contact in EHL regime influence the shape and thickness of the lubricant film significantly. HL differs from EHL due to negligible elastic deformation of the surfaces at the contact interface. In EHL/HL, load is fully supported by the lubricant where the bulk property of the lubricant and entrainment velocity of the tribo pairs determines the film thickness and friction. Transition between different lubrication regimes is well described by Stribeck curve. In this chapter, the mechanism of transition between different regimes and factors influencing the frictional response, different types of lubricants and additives types and their key features will be covered.

F. Cyriac (✉)
Institute of Chemical and Engineering Sciences, A*STAR, Singapore, Singapore
e-mail: febin_cyriac@ices.a-star.edu.sg

A. Akchurin
ASML, Veldhoven, The Netherlands

J. K. Katiyar et al. (eds.), *Tribology in Materials and Applications*,
Materials Forming, Machining and Tribology,
https://doi.org/10.1007/978-3-030-47451-5_3

3.1 Thin Film Lubrication

3.1.1 (Elasto-) Hydrodynamic Lubrication

From the everyday life experience, it has been long ago recognized that dry contacts often produce much higher friction than lubricated contacts. Already in 2400 BC, Egyptians knew the concept of lubrication. During the transport of heavy objects, such as statues, stones, etc. Egyptians used liquids (oil/water) to reduce friction. The purpose of lubrication is to reduce the friction and wear of the contacting surfaces, where a thin lubricant film of low shear stress carries the load. However, the theoretical ground for this phenomenon was put in 1886 by Osborne Reynolds in his seminal work 'On the theory of lubrication and its application to Mr. Beauchamp Tower's experiments, including experimental determination of the viscosity of olive oil' [1]. In this work, Reynolds derived a simple equation that links the lubricant properties (viscosity), experiment conditions and the geometry of the surfaces. He also gives first analytical solutions of the equation. This work is considered to be the birth of the lubrication theory and the equation of lubrication carries his name. This equation is the basis of the modern lubrication theory. Later on systematically this equation was generalized to consider more sophisticated lubrication problems, such as gas lubrication, turbulent flows, Non-Newtonian lubricants, etc.

Important steps in the understanding of the lubrication theory were achieved by solving the Reynolds equation. Under the assumption of rigid surfaces and isoviscous lubricants, Martin [2] obtained a closed form solution for a minimum film thickness and pressure for a cylinder and plane geometry in 1916. This solution showed a direct proportionality of the film thickness to the viscosity of the oil, sliding speed and radius of a cylinder and inverse proportionality with normal load. But as it was shown experimentally later, this solution did not match with film thickness measurements in case of high loads.

Divergence of experimental and theoretical results obtained by Martin for high loads leaded researchers to the conclusion that elastic distortion and pressure-viscosity dependence play a significant role in lubrication. In 1949, by combining elastic deformation and lubricant hydrodynamic oil flow, Grubin [3] obtained a solution for so called elasto-hydrodynamic lubrication (EHL) line contact problem with certain simplifications. His analysis was recognized as extremely useful as it showed for the first time the importance of elastic deformation of the bodies in contact. His solution also included variance of viscosity with pressure, but did not satisfy both elastic and hydrodynamic equations of EHL. This solution is considered the birth of Elasto-Hydrodynamic theory.

Petrusevich [4] was actually first to obtain the pressure spike predicted by Grubin in his numerical analysis of the EHL line contact problem. For this reason, the spike sometimes is referred to as "Petrushevich" spike. Petrushevich solution that was obtained in 1951 was the first to account elastic distortion, fluid flow and pressure-viscosity dependency equations. In the thesis, it was noticed that pressure spike does not occur, if only constant pressure is considered. It should be emphasized, that development of a pressure spike in a tribological contact to be closely related to the variance of viscosity with pressure. The elastic properties of materials in contact and relative speed between the tribopairs was also found to exert an influence.

Remarkable analytical solution of a point contact problem was derived by a famous Russian scientist P. L. Kapitsa [5] in assumptions of rigid substrates, isoviscous fluid and half-Sommerfeld boundary conditions. Under these conditions, he obtained a closed form relation of minimum film thickness on load and other parameters of the problem. By that moment, newly introduced computer technologies became available. Further development of the solution approaches was closely related to the growth of computer capabilities and evolution of numerical methods.

In 1959 Dowson and Higginson [6] obtained a regression formula for a minimum film thickness by using computed series of numerical solutions for EHL line contact problems. They also introduced three non-dimensional groups of parameters, namely "load", "speed" and "materials" parameters. Dowson and Higginson equation that is still being widely used is given by:

$$\frac{h_0}{R} = 2.65\overline{U}^{0.3}\overline{G}^{0.54}\overline{W}_L^{-0.13} \tag{1}$$

where $\overline{W}_L = F_L/E'R$ is the load parameter, $\overline{G} = \alpha E'$ is the materials parameter and $\overline{U} = U\mu_0/E'R$ is the speed parameter.

Later, several approaches were introduced to solve the system of EHL equations. Direct methods are used to obtain the pressure distribution for a known fluid film thickness and inverse methods vice versa: for a fixed pressure distribution, fluid film thickness is calculated [7]. Multigrid approach introduced by Lubrecht [8] and further improved by Venner and Lubrecht [9] significantly increased efficiency of numerical methods and allowed researchers analyze more complex problems. In early 2000, a fully implicit approach along with Newton's method was reexamined in a light of development of differential deflection approach [10]. Advanced computational methods combined with sophisticated hardware nowadays allow tribologists to explore transient problem [11], incorporate plastic effects [12], surface roughness[13], study thermal problems [14], mixed EHL [15] and wear processes [16]. Recently, CFD [17] and molecular dynamic simulations [18] were employed in EHL problems.

3.1.2 Boundary Lubrication

In many practical applications, hydrodynamic or elasto-hydrodynamic regimes of lubrication cannot be established. Other physical and chemical processes occur at the interface and control friction and wear. This regime is called boundary lubrication regime. It typically occurs in highly loaded contacts and at low sliding speeds. Viscosity of a lubricant has a small impact on friction and additives are commonly used to provide the desirable properties and interaction with the interface. Thin layers of the thickness raging from monolayer to several monolayers typically cover surfaces.

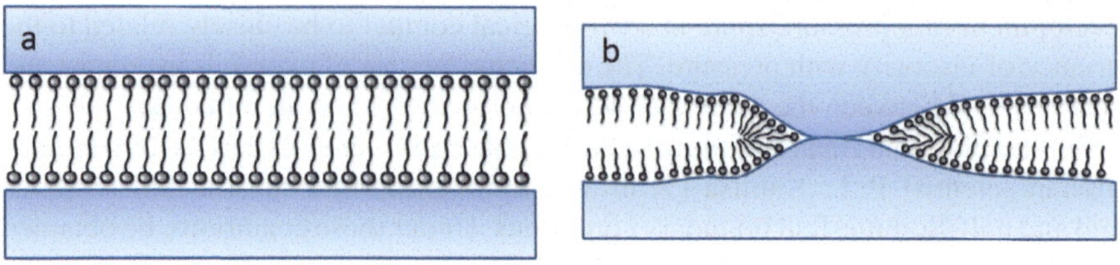

Fig. 3.1 **a** Boundary film on the surfaces; **b** broken boundary film. Reprinted from [20]

These layers prevent the surfaces from direct contact in most of the situations; however, if the loads are too high, direct contact of asperity may occur. If the boundary film is broken in many locations, severe wear will occur (typically catastrophic [19]). Schematically, the boundary films are shown in Fig. 3.1.

In boundary lubrication conditions, surfaces come into direct contact at the microscopic level (asperity level) and these collisions lead to deformation of the surfaces (elastic and plastic), rise of temperature, fracture and wear. Besides mechanical processes, chemical reactions between the surface atoms and lubricant molecules occur. These chemical reactions are driven by the additives in the oil and typically result in formation of protective films, which are referred to as boundary films or tribofilms. The details of the physical and chemical processes in the tribology contacts are not yet well understood.

The boundary films are formed by physisorption, chemisorption, and chemical reaction. The physisorbed film can be of either monomolecular (typically <3 nm) or polymolecular thickness. Physical adsorption or physisorption is the classical adsorption form. In physisorption van der Walls or dispersion forces are responsible for the bonding between substrate and adsorbates. This type of films can provide low friction, however, is relatively weak, since the bonding process is reversible without any changes to the surfaces. The chemisorbed films are monomolecular and a more stable films. Chemisorption or 'chemical adsorption' is an irreversible or partially irreversible form of adsorption and it involves a certain degree of chemical bonding between adsorbate and substrate. These films are irreversibly connected to the surface, which makes them different from physically adsorbed films. Chemical reaction films are thick films, reaching hundreds of nanometers in thickness. For example, ZDDP tribofilms are up to 200 nm thick [21]. The tribofilms formed by chemical reaction are the most stable and wear resistant, much more durable than the adsorbed films. They form due to chemical reactions occurring between the ingredients of the lubricant, such as extreme pressure additives, containing sulfur, phosphorus, chlorine, etc., with the metal surfaces. During sliding, the reaction films are constantly worn, but also formed. Schematic representation of different films is shown in Fig. 3.2.

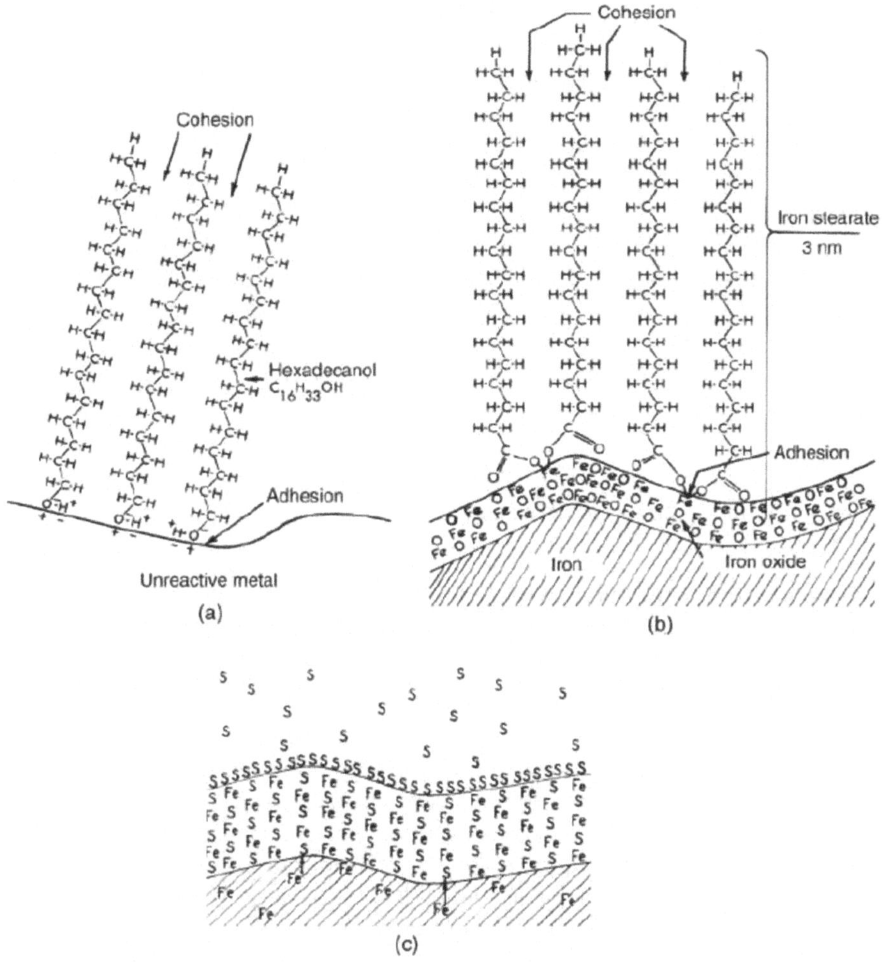

Fig. 3.2 Schematic diagram representing different mechanism of boundary film formation on metal surface **a** physisorption of hexadecanol; **b** chemisorption of stearic acid leading to the formation of iron stearate on the surface; **c** chemical reaction of sulfur with iron to form iron sulfide. Reprinted from [22]

3.1.3 Mixed Lubrication

According to hydrodynamic lubrication theory, the lubricant film thickness decreases with the increase of applied load. Correspondingly, in certain locations direct asperity contacts occur and lubrication fail. In classical lubrication theory, transition regime between BL and (E)-HL is noted as Mixed Lubrication regime. In this regime, part of the load is carried by the direct contact of the asperities and part by the lubricant [23]. The different lubrication regimes can be recognized by their associated coefficient

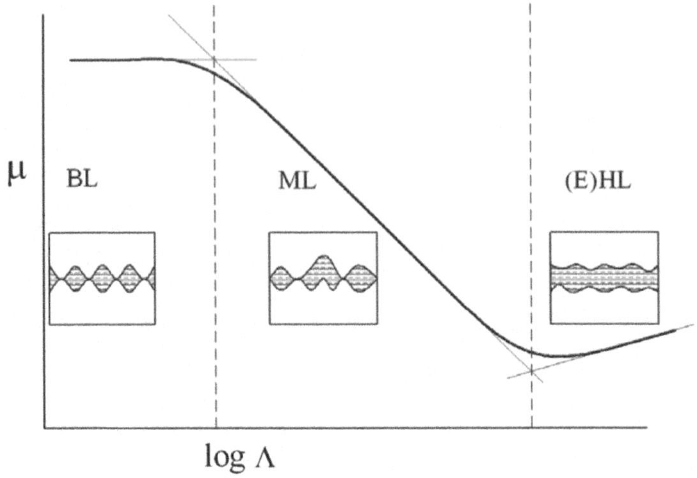

Fig. 3.3 Lubrication regimes. Reprinted from [19]

of friction in a Stribeck curve as shown in Fig. 3.3. The friction coefficient f_C is related to the ratio $\Lambda = h/R_q$, where h is the hydrodynamic film thickness and R_q is the standard deviation of the surface heights [23]. Typically, three regimes can be identified based on Λ:

1. $\Lambda > 3$, full film lubrication.
2. $\Lambda < 0.1$, boundary lubrication; friction is fully determined by the shear strength of the surface boundary layers and is typically characterized by a boundary friction coefficient (relatively constant, characteristic value of a given system).
3. $0.1 < \Lambda < 3$, mixed lubrication; friction is influenced by both the lubricant properties and those of the "solid" contacts. The total load is partly carried by the lubricant and partly by the solid contacts.

It should be noted that the above analysis was performed under assumptions of the classical EHL theory. With further development of the mechanical components and their surface finishing, the thin-film lubrication was proposed as a new lubrication state among EHL (HL), boundary lubrication and mixed lubrication.

3.1.4 Thin Film Lubrication

The Stribeck curve gives the variation of the lubrication regions as a function of operating conditions. (E-)HL and BL film lubrication regimes have been extensively studied and are relatively well understood. However, transition between these lubrication regimes, which is usually referred to as mixed lubrication, is not yet well studied. In well-lubricated modern gears and bearings, the calculated EHL thickness is composed only of several layers of lubricant molecules. This raises a question

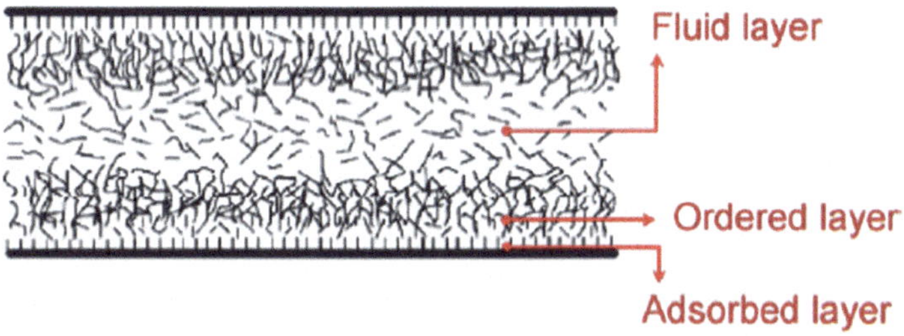

Fig. 3.4 Thin film lubrication model. Reprinted from [25]

regarding the limits of the applicability of EHL theory, since it relies on continuum assumptions. When the film thickness is only few nanometres, these assumptions are no longer valid. Besides, when the thickness of the lubricant layers is only few manometers, the behaviour of the lubricant is determined by various phenomena, the effect of which is negligible for thicker films [24]. Thin film lubrication (TFL), a lubrication regime that falls between BL and EHL regimes, was proposed 20 years ago [25]. In this regime, the lubricant thickness is in the range from several to tens nanometres and consisting of three layers, namely, adsorbed layer (or a monolayer), fluid layer and ordered layer, see Fig. 3.4. The adsorbed layer consists of molecules adsorbed and packed on the solid surfaces. The fluid layer is the layer with freely moving molecules. The ordered layer is located in between the other two layers. In thin film lubrication, the behaviour of a film thickness deviates from the one predicted by EHL due to relatively high impact of the adsorbed and ordered layers.

3.1.5 Factors Affecting Thin Film Lubrication

There are several factors influencing the behaviour of a thin film lubricant. In case of EHL, one of the most critical parameters is the rolling/sliding speed. Clearly, a significant dependence of the film thickness on speed is preserved also for the thin film regime. However, this behaviour is distinct from the classical EHL. The difference comes due to molecular interactions, and is especially obvious for lubricant molecules with polar ends. Molecular polarity is another important factor influencing the lubricant film behaviour in TFL. Polarity of the molecules determine the ordering of the molecules near the surfaces. Besides, the surface energy and surface tension of the surfaces has a clear impact on the TFL. Both play a role in adsorption processes and they were shown to have a significant impact on the film thickness in TFL. The surfaces with lower surface energy show smaller films. It was also shown, that the surfaces with similar surface energies show similar film thicknesses, indicating that surface energy is a key factor in TFL, as shown in Fig. 3.5.

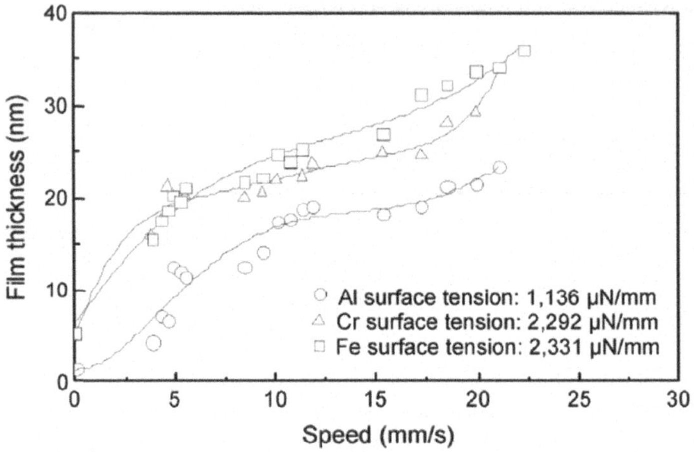

Fig. 3.5 Film thickness as a function of sliding speed and surface tension. Reprinted from [25]

3.2 Lubricants

3.2.1 Introduction

It is estimated that around 20% of the global energy is used for overcoming frictional losses in machine components [26]. Therefore, there is a continuous demand for the increased reliability and performance of machine components. Achieving lower friction can result in fuel savings and reduction in wear can lead to increased durability of machine components. One of the most cost effective and feasible method to improve the frictional response and to reduce wear related material wastage is by using lubricants. Lubricants reduce friction and wear by either preventing/reducing the direct contact between various rolling and siding elements. In addition, supplementary functions served by the lubricants includes reduction in noise and vibration, corrosion prevention, heat transfer, removal of contaminants and debris, increase in efficiency and lifetime of mechanical components. Lubricants are classified in several ways; based on the physical state as liquid, semi-solid and solid lubricants. Another classification is based on the type of base oil used in the formulation such as mineral oil, synthetic/semi-synthetic and environmentally friendly oils. The most common type of lubricant is the motor oil that is used for the protection of internal combustion engines. Common industrial applications of lubricants includes bearings, gears, hydraulics, turbines, wind turbines, air compressors etc. However, there are myriads of other applications that requires specifically tailored lubricants which is illustrated by over 10,000 different types of lubricants that satisfy more than 90% of all lubricant applications used around the world [27]. Therefore, irrespective of their chemical architecture, lubricants can also be conveniently classified based on their intended applications.

Lubricants today are classified into automotive oils, industrial oils, process oils, metal working fluids and greases [28]. In terms of volume, automotive oils account for the largest share at 57% of total market volume, followed by industrial oils at 26%, with rest comprising of metal working fluids, process oils and greases. The global lubricants consumption was estimated at over 40 million tons in 2016 and is expected to register a compound annual growth rate of 2.8% from 2019 to 2025. This translates to an increase in market value from USD 118.89 billion in 2016 to approximately 166 billion USD by 2025. Asia-Pacific, Africa and Middle East are the regions expected to drive the lubricant market and may collectively account for nearly two third of the global market by 2025. An important factor that is driving the lubricant growth is rapid industrialization and rising manufacturing output. Even though, the interval for oil change in automobiles has significantly improved due to the introduction of long lasting high-performance lubricants, increasing automobile production in Asia-Pacific and Europe is expected to support the growing demands. During the forecast period, additional gains are expected due to an increase in demand for nonconventional lubricants to provide enhanced vehicle performance, better fuel efficiency and lower emissions. Additionally, adoption of biobased lubricants due to regulatory sanctions on conventional lubricants has also lead to an increase in market growth. In addition, a surge in demand for lubricants are envisaged as new industries are emerging and new technologies and applications are being explored.

3.3 Solid Lubricants

Solid or dry lubricants are materials used either in the form of dispersed particles or surface films for reducing friction and wear of surfaces in relative motion even under high loads. Solid lubricants are specially meant for applications such as space technology and ultra-high vacuum where liquid lubricants malfunction or where contamination by liquid lubricants must be avoided [29]. Solid lubricants are used to combat friction and wear in applications involving extreme temperature conditions, or when the tribological contacts need to be efficiently separated under higher loads and lower hydrodynamic speeds. They are also used in combination with liquid lubricants as they are expected to have beneficial synergistic effect on friction and wear characteristics [30]. For example, they are conveniently used in tribological applications as extreme pressure or anti-wear additives where fluids are frequently used as carriers or as a lubricant [31]. Even though, they possess many advantages, poor thermal conductivity, the low stability of the lubrication film, oxidation and aging related degradation, higher friction and wear compared to hydrodynamic regimes are some of their potential shortcomings.

3.3.1 Classification

Solid lubricants can be classified into (a) Lamellar solids, (b) Polymers, (c) Soft metals and (d) oxides. The most prominent lamellar solid lubricants are transition-metal dichalcogenides (WS_2 and MoS_2), hexagonal boron nitride (HBN) and graphite. The low friction characteristics of these inorganic materials is mainly attributed to their lamellar or layered crystal structure consisting of hexagonal rings forming thin parallel planes. Within the planes, the atoms are bonded to each other by strong covalent bonds whereas weak van der Waals bonds the planes, which are relatively far apart, to each other (Fig. 3.6). Within the tribological contact, these planes can align in the direction of relative motion and allows sliding movement of the parallel planes due to low shear strength, thus facilitating low friction. However, a favorable crystal structure does not guarantee intrinsic lubricity. The shear strength of the solid lubricants at the siding interfaces is very sensitive to test environment, nature and type of counter face materials, temperature, contact pressures, microstructure and chemistry of the tribofilm. In addition, the deposition or lubrication application methods and presence or absence of certain chemical adsorbates can also play an important role in determining the friction [32, 33]. For example, graphite will not function without humidity in the surrounding air and others like MoS_2 will not lubricate under humid conditions [34, 35].

Some soft metals such as Indium (In), lead (Pb), gold (Au), silver (Ag) etc. in the pure form or in the form of coatings on relatively hard substrates offers low friction due to their low shear strength and high plasticity. The friction coefficient of soft metals ranges from 0.1 to 0.4, depending on the test conditions and counteracting

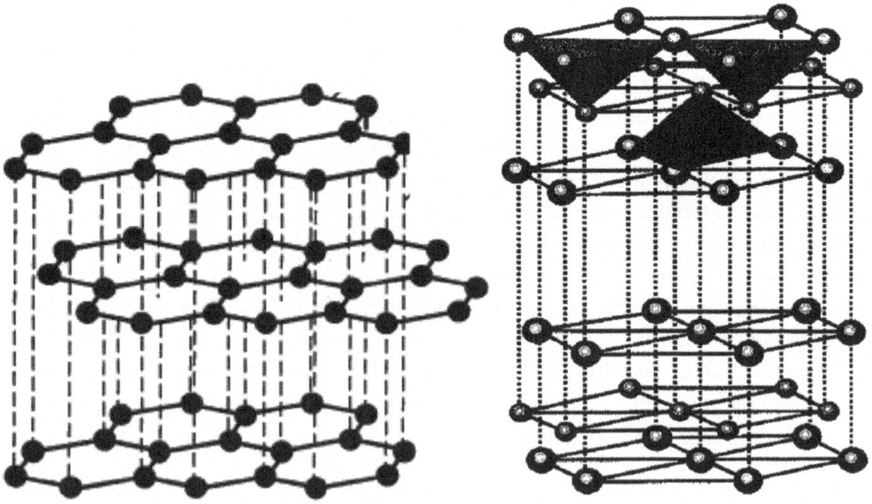

Fig. 3.6 Lamellar structure of graphite, reprinted from [43] and hexagonal tungsten disulfide. Reprinted with permission from [44]

surfaces. However, they exhibit significantly higher friction than lamellar solids. This is because for soft metals the lubrication occur by shear within the film whereas in solids transfer of thin layer of material from the solid to the counterface (transfer film) is the predominant mechanism [36]. The thickness of the film is also an important parameter in determining friction and wear within the contact. A thicker film can result in larger real area of contact and thus higher friction. However, too thin films may tend to wear out quickly exposing the bare counteracting surfaces resulting in high friction and wear [36]. Therefore, an optimum value of friction and wear are obtained for 0.5–1 μm thick films produced by the methods of electroplating, vapor deposition, thermal spraying etc. [37]. Soft metals are mainly used for high temperature rolling applications as most of them shows a decrease in friction coefficient with temperature because of softening and rapid recovery from strain hardening [30]. However, they should not be used in the presence of sulfur and chlorine as they may undergo rapid corrosive wear.

Recently, there in a growing interest of using plastics in tribology, mostly as solid lubricants in pure form or in composites or laminated structures and as coatings [38]. Polytetrafluoroethylene (PTFE) is the most commonly used polymer based tribological material that is known for its anti-friction property [39]. Other important polymers that are used in tribological applications include ultra-high molecular weight polyethylene (UHMWPE), nylon, polychlorofluoroethylene, polyimide etc. The good lubrication property of the polymer materials is closely associated to its structural characteristics. The molecular structure of these materials consist of long chain molecules parallel to each other with weak bonding strength. Therefore, the chains may slide against each other at low shear stress resulting in low friction coefficient [40]. The frictional characteristic of PTFE is dependent on the initial adhesion, where the large lumps and slabs are formed in the high static friction regime due to bulk fracture. The low friction during further interaction is due to shifting of active shear from metal-polymer interface to polymer-polymer interface due to the development of coherent and continuous transfer layer on the counter surface. As the adhesive bonding of the transfer layer with the counter surface is of Van der Waals type, the film can only anchor to the surface under mild conditions [41]. Therefore, fresh films will be created with each pass leading to high wear rate when PTFE is rubbed or slid against a hard surface. As it is impossible to obtain the desired combination of friction and wear using PTFE alone, fillers are added to the PTFE matrix for increased wear resistance [42]. Even though, PTFE has a low friction both in vacuum and atmosphere and shows a decrease in friction with increasing contact stress, its low thermal conductivity has limited its use to low sliding speed applications, which otherwise may result in premature failure due to melting [36].

Certain Oxides such as MoO_2, ZnO, TiO_2, B_2O_3 when applied as coatings can provide long wear life and reduced friction for high temperature applications. The reduced frictional response of these materials at higher temperature is due to the formation of soft and highly shareable lubricious surfaces [45]. However, they are very susceptible to fracture due to their inherent brittleness.

Solid lubricant technology has grown considerably in the last decade. Carbon-based materials like DLCs (diamond like carbon compounds) and graphene are being extensively studied as thin films coatings on tribological components. Fundamental tribological analysis of these materials has revealed existence of superlubricity—the most ideal tribological condition where the frictional force almost vanishes between the solid surfaces. Superlubricity is currently a hot topic in tribology as achieving ultra-low friction can contribute to reduced maintenance, increased reliability and reduced energy consumption. One aspect of research is based on experimental studies with the aim of finding suitable superlubric materials whereas theoretical studies were focused on investigating the mechanism of superlubricity.

DLCs are diamond-like metastable amorphous materials with a wide range of compositions, properties, and performance. They exhibit unusual combination of tribological and mechanical properties such as high elastic modulus and hardness with low friction and wear [46]. DLCs can also be doped with lightweight elements (nitrogen, silicon, and silicon oxide) and transition metals (Ti, W and Cr) for improving their mechanical strength, wear resistance and hardness [36]. Even though, they are shown to exhibit super-low friction and low wear rates, the reduced state of friction was found to depend on the test conditions (contact stress, sliding velocity, temperature, and counterface material), environmental conditions and nature of the coatings as determined by the deposition process [47–49].

Graphene is a one-atom-thick layer of carbon atoms arranged in a hexagonal lattice which is 200 times stronger than steel. In the last few years, graphene has attracted a significant attention in the field of tribology due to its exotic mechanical, electrical, chemical and thermal properties. Research based on computational methods have predicted the mechanisms of superlubricity of graphene at different scales. The atomically thin nature of graphene and its ability to conformally coat nano to micro scale tribological contacts have made it a convenient material for reducing friction and wear of NEMS/MEMS systems [50]. Recently, a friction coefficient of 0.004 was achieved by using graphene in combination with crystalline diamond nanoparticles and diamond-like carbon [51], which is at least an order of magnitude lower than what, can be achieved using conventional lubricants. However, macroscale superlubricity was only realized under specialized operating conditions, which impeded its industrial applications. The fascinating properties of graphene and its application in multidisciplinary area acted as an impetus for the development of a new class of materials known as 2D materials (Fig. 3.7). 2D materials, which are mostly separated from their bulk material, can be modified to enhance their properties for a wide range of engineering application [52]. Theoretical studies of these materials have revealed that they might exhibit physical properties similar to graphene [52]. Therefore, some of these materials has the potential of replacing conventional lubricants for future tribology applications (as surface coatings or as additives in oil/dispersions) by offering effective lubrication with minimum quantities of material.

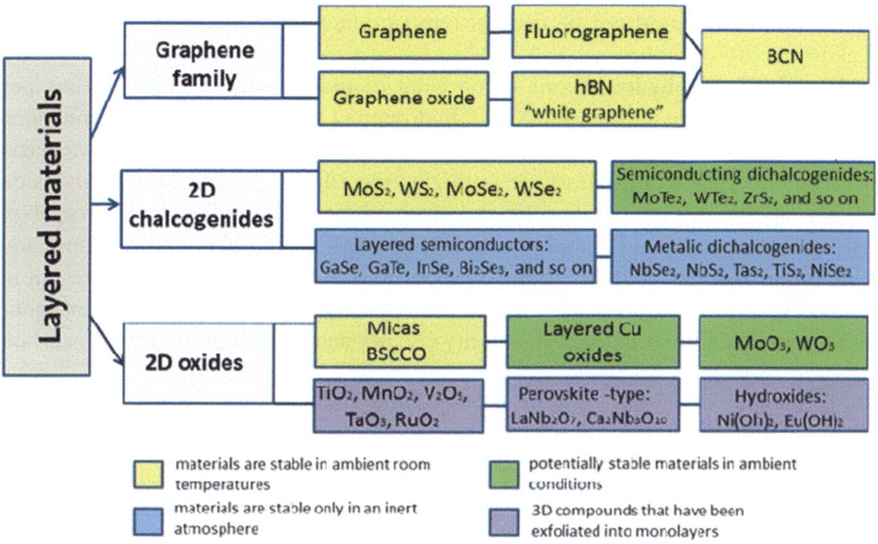

Fig. 3.7 2D materials family. Reprinted from [53]

3.4 Liquid Lubricants

Liquid lubricants are generally referred to as oils and has been around for a long time. Compared to solid lubricants they possess a number of advantages, such as long-term endurance, low mechanical noise and low friction under EHL regime. They are classified based on the source from which they are extracted as vegetable oil (canola, castor, rapeseed, soybean etc.) animal oil (whale oil, shark oil), mineral oil and synthetic oil. Mineral oils are refined from naturally occurring crude oil and is the most commonly used oil because of its low cost. In addition, the feasibility of manufacturing them at varying viscosities makes it a convenient choice for a wide range of applications. Synthetic lubricants are manufactured and is therefore more expensive than mineral oil. They are used in applications were mineral oils are inadequate. The performance characteristics of base oils are determined by four physical properties [54]: (a) pour point (the lowest temperature below which the oil cease to flow), (b) viscosity (resistance to flow), (c) Viscosity Index (variation of viscosity with temperature) and (d) purity.

3.4.1 Natural Oils

Vegetable oils and animal oils are mainly used in applications where risk of contamination need to be reduced to minimum. However, they are highly susceptible to oxidation and should not be used at temperatures exceeding 120 °C due to oil breakdown.

Mineral oils are hydrocarbons comprising of more than 20 carbon atoms per molecule and include carbon (84–87%), hydrogen (11–14), sulfur (0–3%), nitrogen (0–0.6), oxygen (0–0.5%), metals (0–0.2%) and other elements. The three basic forms of mineral oil are paraffinic (straight linear/linear hydrocarbons with side chains), naphthenic (saturated cyclic carbon chains consisting of cyclic hydrocarbon rings) and aromatic (conjugated or alternating single or double bond carbons) and are distinguished based on the relative proportions of above components present in the oil (Fig. 3.8). The properties of the mineral oils such as viscosity index, pour point, thermal stability, oxidation stability etc. are determined by their hydrocarbon

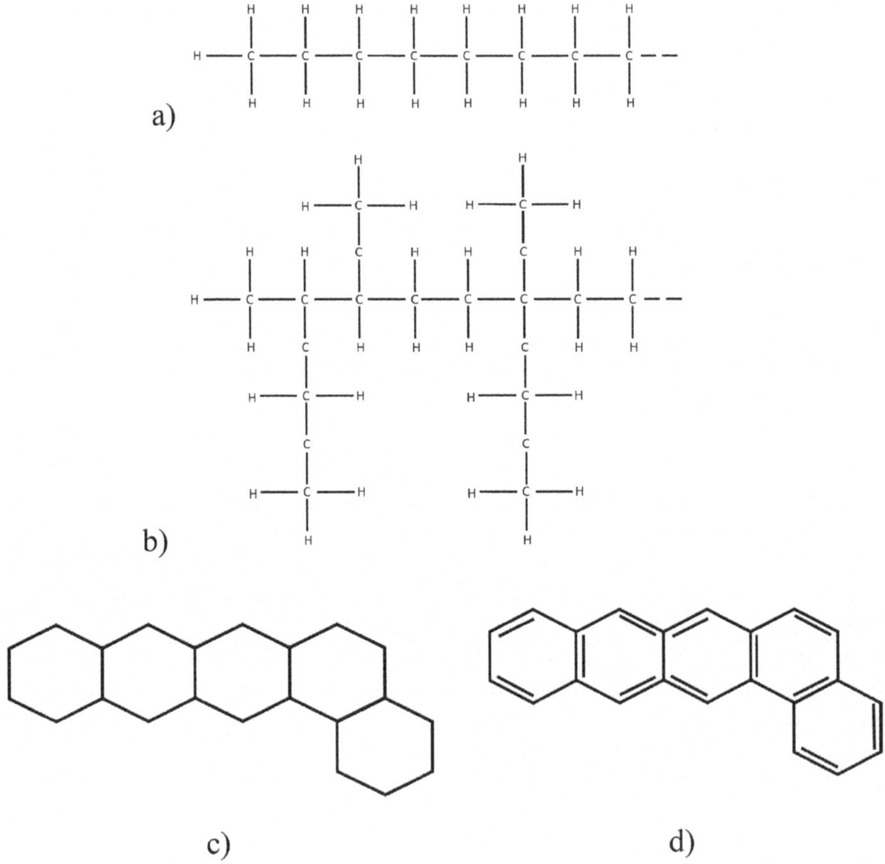

Fig. 3.8 Main types of mineral oil **a** linear paraffins, **b** branched paraffins, **c** naphthenes and **d** aromatics

Table 3.1 Some properties of three basic forms of mineral oils

Property	Paraffinic	Naphthenic	Aromatic
Pour point	High	Low	Low
Viscosity index	High	Low	Low
Thermal stability	Low	Low/medium	High
Oxidation stability	High	High	Low
Flash point	High	Low	Low/medium
Volatility	Low	Medium	High

chemistry and molecular weight distribution (Table 3.1). In general, properties that are more predictable can be obtained using a narrow molecular distribution. The maximum operating temperature of mineral oils is 130 °C, but super-refined oils can be used up to 200 °C.

3.4.2 Synthetic Oils

The demand for high performance lubricants has led to the development of synthetic lubricants and have been around for 70 years. They are formulated by chemical reactions through the precise application of pressure and temperature where low molecular weight materials are combined to produce high molecular weight fractions with uniform property and targeted performance properties. Synthetic lubricants can withstand environment with extreme temperature, extreme pressure, and high humidity and are less of fire hazard. However, they are inherently more expensive than mineral oils. Synthetics are generally superior to refined oils in terms of better oxidation stability, better viscosity index, much lower pour point, and low coefficient of friction. Therefore, they are meant for specialized applications such as gas turbine engines, jet engines, vacuum pumps etc. where they overcome the challenges better than mineral-based oils. Commonly used synthetic oils in lubricant formulations are PAOs, esters (diesters, polyol esters) and PAGs. Other synthetic fluids used as lubricants are phosphate esters, poly-isobutylene (PIB), silicones, silanes, Polyphenyl ethers etc.

3.4.3 Polyalphaolefins

Polyalphaolefins are synthetic hydrocarbons with relatively uniform structure (Fig. 3.9). They are manufactured by the catalytic oligomerization of linear alphaolefins and is composed of hydrogenated olefin oligomers having the general formula of C_nH_{2n+2} [55]. The well-defined molecular structure of PAO implies more consistent lubrication properties than their mineral oil counterparts.

Fig. 3.9 Structure of PAO

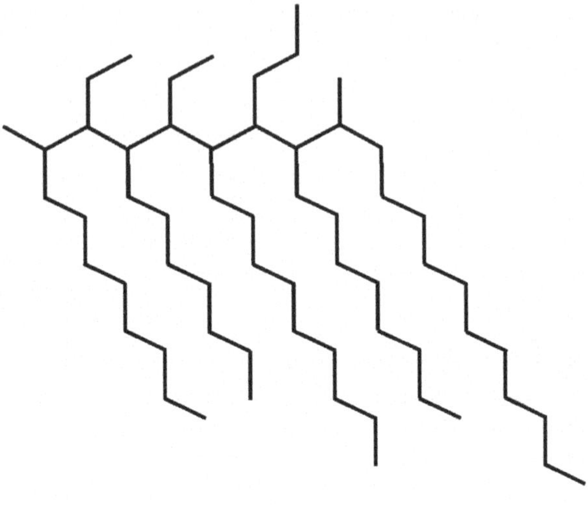

Polyalphaolefins are used extensively in automotive fluids and industrial applications including transmission fluids, hydraulic fluids and motor oils. They offer superior lubricating performance versus mineral oil in terms of higher viscosity index, greater thermal stability, better oxidative stability, low temperature fluidity, lower volatility, and lower traction force. Their superior performance is due to the absence of ring structures, double bonds, nitrogen, sulphur or waxy hydrocarbons. However, they have difficulties in dissolving common oil additives and have poor fire resistance and biodegradability.

3.4.4 Esters

Esters are produced by reacting alcohol with organic or inorganic acids where R1 and R2 are hydrocarbons with different number of carbon atoms, Fig. 3.10. The high bond energy of ester linkages make them more stable than those of typical hydrocarbons with their C–C bonds. Esters exhibit good oxidation stability and excellent viscosity-temperature and volatility characteristics. They provide high performance lubrication as engine oils, compressor oils, metalworking fluids, transmission fluids and many more. They are also used in environmentally sensitive areas due to their biodegradability and low toxicity. They can be tailored for different applications and environments. An enormous variety of esters such as fluoro esters, phosphate

$$R_1 - OH + R_2 - COOH \leftrightarrow R_1 - COO - R_2 + H_2O$$

Fig. 3.10 The esterification reaction

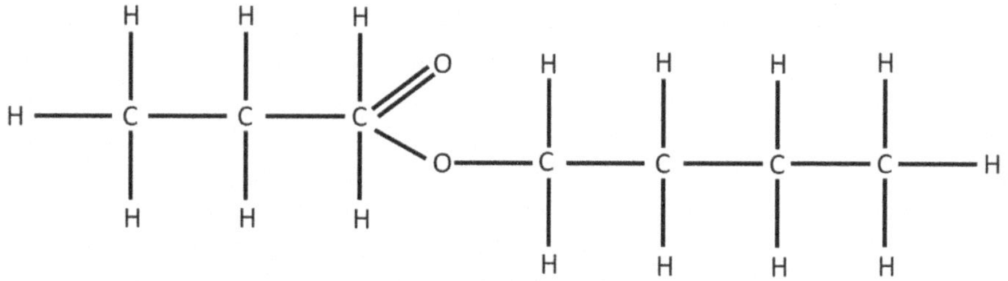

Fig. 3.11 Structure of ester

esters, silicate esters, dibasic acid esters, neopentyl polyol esters and many more are possible by different combination of acids and alcohols.

Diesters are produced by the reaction of dibasic acid and mono-functional alcohol. However, in the presence of heat and water, diesters can decompose back to acid and alcohol. Diesters have excellent volatility and good low temperature properties. However, the major drawback of using them as lubricants is their low viscosities and high seal swell. Polyol esters are produced by reacting a highly branched di-functional alcohol with a monobasic acid. The highly branched ester has properties similar to diesters, but offer a wide viscosity range. They also have a higher operating temperature limit (250 °C) compared to diesters (200 °C). However, they are nearly 50% more expensive than diesters are (Fig. 3.11).

3.4.5 Polyalkylene Glycols (PAGs)

PAGs/polyglycol (Fig. 3.12) are produced by reacting an alcohol with one or more alkylene oxides. They are intrinsically clean lubricants, leaving no residue unlike petroleum products that would build tars and sludges. The structure of PAGs can be tailored for varying their solubility from water-soluble to water insoluble and in a wide range of viscosity grades. However, water-soluble polyglycol are not compatible with mineral oils. PAGs offer high lubricity, high viscosity index (180–280), low traction properties, low flammability and low pour point. PAGs are widely used as metal working fluids, quenchants, gear oils, air compressor fluids, natural gas

Fig. 3.12 Structure of oil soluble PAG. Reprinted from [56]

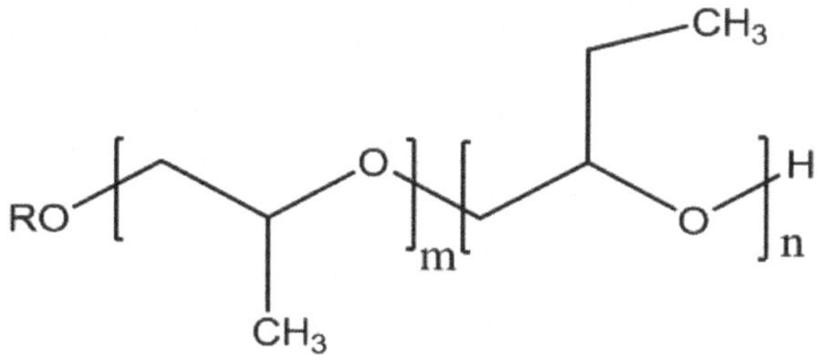

compressor fluids etc. As they exhibit high tolerance for natural rubbers and other elastomers, they are also used for the lubrication of the rubber parts such as O-rings, joints etc. They are also tolerant to small quantities of water and is therefore widely used in applications susceptible to water contamination. Even though, they are expensive, their use is justified by their performance advantage.

The merits and demerits that an oil possess determines its suitability for a given application. For example, in the field of lubrication, a significant quantity of lubricants are lost to the environment, contaminating air, water, and land [58]. So in the unlikelihood of spillage or leakage as in case of railway lubrication, industry lubrication, automotive and water craft lubrication these materials present a serious threat to aquatic, marine, animals and human health alike. Therefore, recent trend is towards use of oils that exhibit high degree of biodegradability than mineral-based oils for environmentally sensitive applications. Even though, vegetable oils and esters exhibit good lubricity in comparison to mineral oils, they possess many disadvantages that have limited their industrial application, Table 3.2. For example, vegetable oils are susceptible to oxidation and therefore has service restrictions at high temperatures. Therefore, fundamental structural changes or chemical modification of these oils is needed for enhancing their performance as a lubricant.

Table 3.2 Comparison of properties of different base stocks

Properties	Mineral	Vegetable	PAOs	Diesters	Polyol esters	PAGs
Viscosity	Fair	Good	Good	Fair	Very good	Very good
Low temperature properties	Poor	Poor	Very good	Good	Good	Good
Oxidation stability	Fair	Poor	Very good	Good	Good	Good
Low volatility	Fair	Good	Excellent	Excellent	Excellent	Good
Lubricating properties	Good	Very good	Good	Very good	Very good	Good
Hydrolytic stability	Excellent	Poor	Excellent	Fair	Fair	Very good
Thermal stability	Fair	Fair	Fair	Good	Good	Good
Additive solubility	Excellent	Excellent	Fair	Very good	Very good	Fair
Biodegradability	Poor	Excellent	Poor	Good	Good	Poor

Source [57]

3.4.6 API Classification

The American Petroleum Institute (API) has categorized base oils into five categories based on the sulfur content, degree of oil saturation and performance characteristics (Table 3.3). Group I to III are refined from petroleum crude oil. Group IV is synthetic oils and Group V comprises of all other base oils that are not included in Group I through V.

Group I oils are prepared by solvent extraction and dewaxing process and has a viscosity index of 80–120. They contain less than 90% saturates and sulfur greater than 0.03%. They are the cheapest base oils on the market as they are obtained using a simpler solvent refining process. Group I oils are mostly meant for less demanding applications as the mix of different hydrocarbon chains with little uniformity from the solvent refining can lead to high friction. Even though, they remain the dominant base oil for lubricant blending (Fig. 3.13), they are falling out of favor to Group II and Group III because of growing environmental concern and demand for better fuel economy.

Group II oils are often manufactured by hydrocracking process at very high pressures where larger hydrocarbon molecules are broken down into smaller ones. They have more than 90% saturates, less than 0.03% sulfur and a viscosity index of 80–120. As the hydrocarbon molecules of Group II oils are saturated, they offer better antioxidation properties and have fair to good performance in the areas of volatility,

Table 3.3 API classification of base oils

Base oil category	Sulfur (%)		Saturates (%)	Viscosity index
Group I	>0.03	And/or	<90	80–120
Group II	<0.03	And	>90	80–120
Group III	<0.03	And	>90	>120
Group IV	PAO synthetic lubricants			
Group V	Base oils not included in group I through group IV			

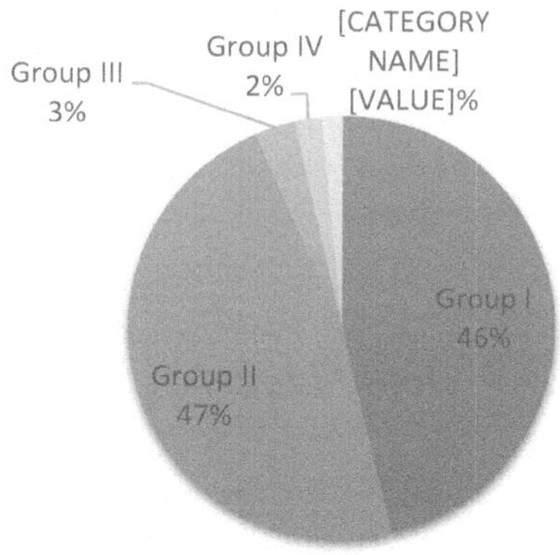

Fig. 3.13 Base oil production percentage

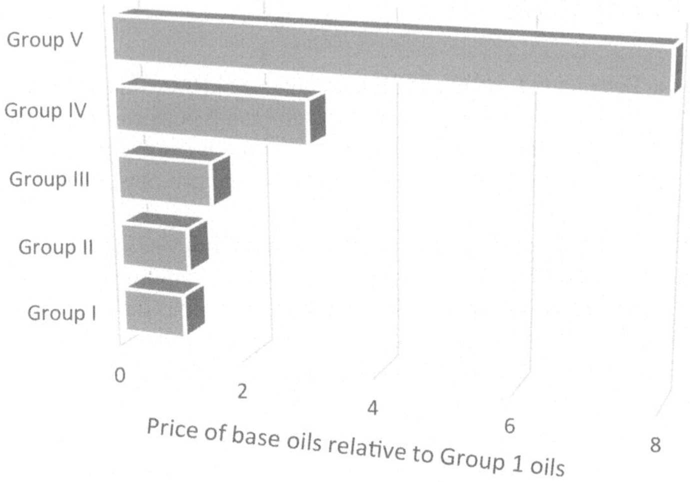

Fig. 3.14 Base oil production and price of base fluids relative to group 1 oils

wear prevention and flash/fire point. As they are priced close to Group I, they are becoming more common in the market. Group III oils undergo a longer, more severe hydrocracking process than Group II oils. More pressure and heat applied during the process result in a purer oil with good molecular uniformity and stability. They have more than 90% saturates, less than 0.03% sulfur and have a viscosity index above 120. These synthesized hydrocarbons are used in the production of synthetic and semi-synthetic lubricants. Group III oils are also becoming common and demand for Groups II and III combined is forecast to exceed Group I demand by 2030.

Group IV base oils are polyalphaolefins made from small uniform molecules through synthesizing process. The biggest advantage of Group IV oil is that they can be tailored to have structure with predictable properties. They offer excellent stability, molecular uniformity and is therefore meant for extreme cold conditions and high heat applications. But, the use of these oils is limited to exotic applications due to the cost and availability of base stock [54] (Fig. 3.14). Group V comprises of all other base oils that are not included in Group I through V, ranging from very low-quality naphthenics to very exotic synthetic oils. Typical examples of Group V among others include silicone, polyalkylene glycol (PAG), polyolester and phosphate ester. Group V oils are commonly used as an additive to enhance the properties of the base stock.

3.4.7 *Ionic Liquids*

Ionic Liquids (ILs) are ionic salts, which are liquid below 100 °C or even at room temperature (room temperature ionic liquids, RTIL's). Typical ILs are composed of

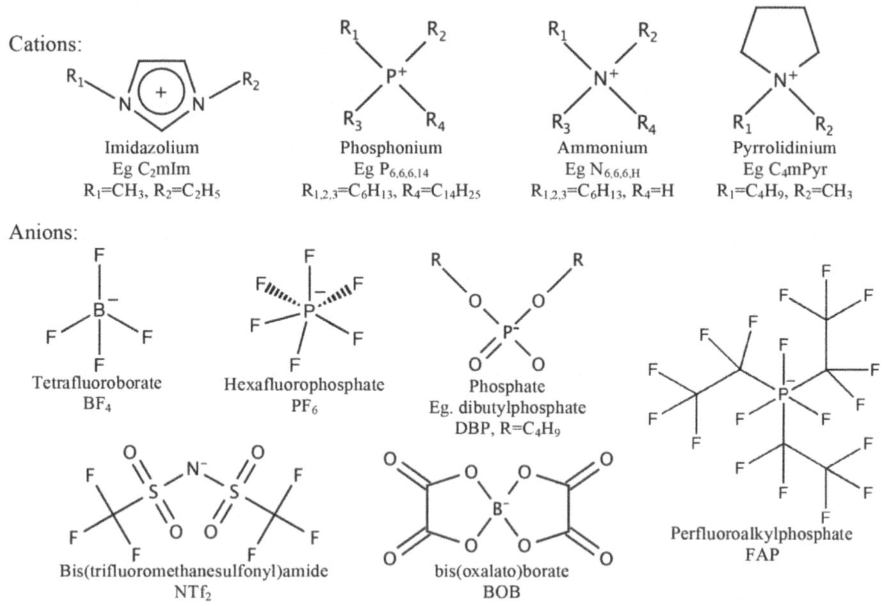

Fig. 3.15 Typical molecular structure of cations and anions. Reprinted with permission from [61]

large asymmetric organic cations (usually a lengthy alkyl chain) and organic or inorganic anions that determines their fundamental properties [59]. Strong electrostatic forces hold these ions together where at least one ion has a delocalized charge that prevents the formation of a stable crystal lattice. As a result, ILs are liquid below 100 °C or even at room temperature. The molecular structure of typical cations and are shown in Fig. 3.15. The melting point and viscosity of the ILs depends on their molecular structure, cations, the types and length of alkyl chains and anions [59, 60] making them an ideal candidate for tribological application oriented molecular design.

Ionic liquids have been extensively investigated since 2001 as base oils [62–65], additives [66–70] and thin films [71–74]. They have attracted much attention in the field of tribology due to its unique physical and chemical properties such as, high thermal and electrochemical stability, very low vapor pressure, high electric conductivity, non-volatility and non-flammability [75–80]. In addition, they also offer flexibility of creating tailor made lubricants and additives by having a choice of cations and anions in the liquid [59]. They are neat lubricants and does not present environmental problems unlike lubricants derived from petroleum. Another characteristic that distinguishes ILs from conventional lubricants is high thermal stability and chemical inertness. They can also be used in a wide temperature range due to its low pour point and much higher decomposition temperature than conventional lubricating oils. The low volatility of IL lubrication oils also makes it a convenient choice for vacuum, especially spacecraft applications [81–85]. Unlike conventional lubricants that are electrically insulating, the high electrical conductivity of ILs also makes them suitable for various electrical applications. However, the most notable feature of ILs is its high polarity that contributes to the formation of very strong adsorption film with superior anti-wear capability making them attractive for various material sliding pairs such as metal-metal, metal-ceramic and ceramic-ceramic [72]. They are also shown to reduce friction and wear in the hydrodynamic and mixed hydrodynamic regimes [67]. Even though, interest in the tribological applications of ILs has increased significantly over the past decade, their wide use is limited by their high cost, thermo-oxidation and corrosive properties [86, 87]. In addition, they often have to be optimized by additive technology to meet the requirements of task specific practical applications.

3.5 Semi-solid Lubricants

Semi-solid lubricants are gel like substances for e.g. waxes, vaseline and greases that are used for reducing friction and wear between sliding and rolling components. Grease is the most common semi-solid lubricant where the primary lubricant is oil.

The ASTM definition of lubricating grease is:

> A solid or semi-solid lubricant consisting of a thickener agent in a liquid lubricant. Other ingredients imparting special properties may be included [88].

A commercial lubricating grease usually consists of a thickener (5–35 wt%), base oil (65–95 wt%) and various additives (0–10 wt%) [89]. The length of the thickener fibers is typically about 1–100 μm and have a length to diameter ratio of 10–100. The thickener forms a solid like network and acts as an oil reservoir (Fig. 3.16). The base oil is kept inside the thickener structure by a combination of Van der Waals and capillary forces [90]. The consistency of the grease is mostly governed by the physical interaction of the thickener with the base oil.

The National Lubricating Grease Institute (NLGI) has standardized a numerical scale for classifying the consistency of lubricating greases. Consistency, which is

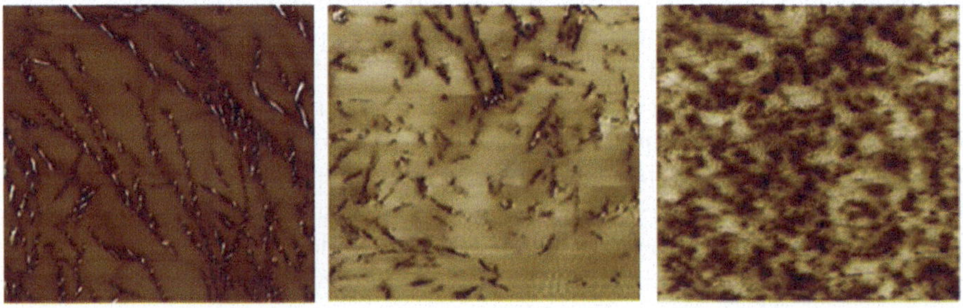

Fig. 3.16 AFM images of lithium, lithium complex and calcium sulfonate grease. Reprinted from [91]

Table 3.4 NLGI consistency grades and worked penetration range

NLGI grade	Worked penetration range	Description
000	445–475	Fluid
00	400–430	Semi-fluid
0	355–385	Very fluid
1	310–340	Soft
2	265–295	Medium
3	220–250	Medium-hard
4	175–205	Hard
5	130–160	Very hard
6	85–115	Block

the measure of relative hardness or softness, is measured by ASTM D27 worked penetration ranging from 000 for semifluid to 6 for block greases (Table 3.4). The length to diameter ratio for a given concentration of thickener has also been correlated to the consistency of greases [92].

Industrial grease are generally classified based on the base oil composition as mineral oil-based, synthetic/semi-synthetic oil-based and environmentally friendly greases. Lubricating greases can also be classified based on the thickener that constitutes the grease matrix. The most widely used greases are based on lithium as the thickening agent, Fig. 3.17. Lithium complex grease, which is made by combining conventional lithium soap and a low-molecular-weight organic acid, is also widely used. Others used as thickeners in greases include aluminium, polyurea, clay, calcium soaps etc. either alone or in combination. There is also 'solid grease', which essentially contains grease and ultra-high polymer polyethylene. The operational temperature window and grease selection criterion is usually determined by

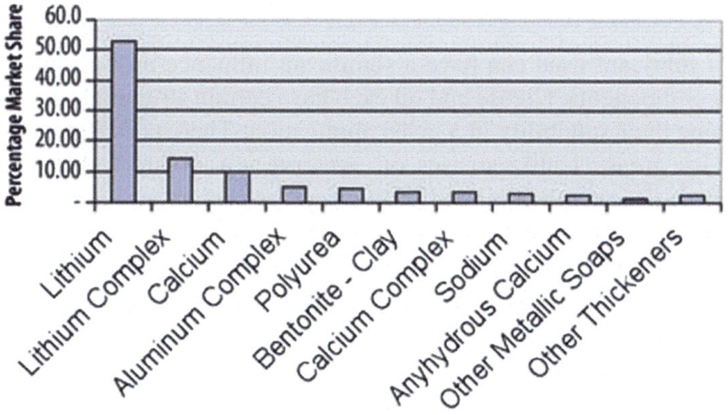

Fig. 3.17 Market distribution of greases by thickener type. Reprinted from [95]

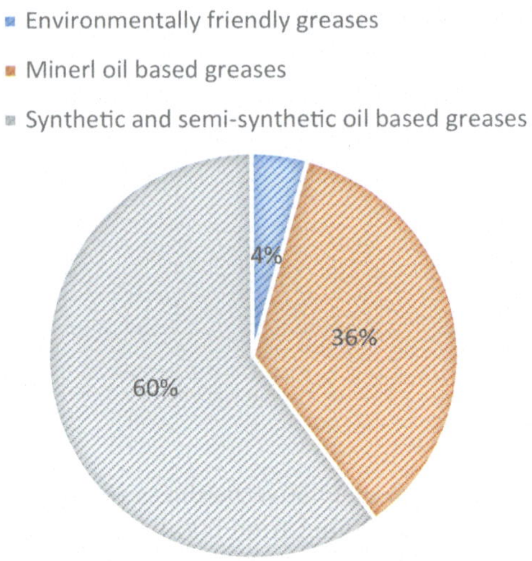

Fig. 3.18 Global grease market share

the properties of the thickeners and base oils [93]. However, the lubricity is mainly governed by the properties of the oil and the additives [94].

In global grease market, the end users are quite price sensitive. Therefore, majority of industrial greases manufactured today are either based on non-biodegradable mineral or synthetic/semi-synthetic oils, Fig. 3.18. In addition, for a given industrial application, they guarantee improved performance than their environmentally friendly counterparts. Therefore, reduced price and improved performance are the key factors for increasing the market share of biodegradable greases.

3.5.1 Grease Versus Oil Lubrication

The type of lubricant used can have a significant influence on the performance of mechanical components. Grease and oil each have certain strengths and weaknesses that determine their suitability in a given application. There are clearly some disadvantages to use grease. Unlike greases, oil can serve as a coolant. Lubricating greases are more prone to aging than oils. In addition, in the case of oil-lubricated bearings, contaminants may be less detrimental to the operational life as they can be filtered out easily. Grease lubrication also has some clear benefits. A lubricating grease can act as an effective sealant that prevents the entry of contaminants. The requirements on sealing are less stringent since grease does not easily leak out and therefore can be used in application in vertically mounter positions [93]. This makes it possible to use low friction seals or even non-contacting shields. Finally, the friction torque and therefore energy losses in grease-lubricated bearings is lower than for oil-lubricated bearings. As a result, around 80–90% of the rolling bearings are grease lubricated.

3.6 Lubricant Rheology

Viscosity, which is measurement of fluid internal resistance to flow, is the most important characteristic of a lubricant in fluid film lubrication. The viscosity selected for a given application should be sufficient to separate the interacting surfaces by generating a thin (EHL/HL) film but without inducing excessive viscous drag. For Newtonian fluids such as lubricating oils, viscosity is the coefficient relating the shear stress and shear strain (Fig. 3.19). However, a lubricating oil may show Non-Newtonian behavior at high shear rates and pressure [93]. The viscosity of the oil is usually measured using viscometers and reported in a unit called the centistoke (cSt). Lubricating oils are classified according to their viscosity at 40 °C by authorized bodies such as SAE, ISO etc. Comparative classification of oils based on viscosity is shown in Fig. 3.20.

Lubricant viscosity is a strong function of temperature and pressure. Viscosity Index (VI) is an arbitrary scalar value that is used to characterize viscosity-temperature behavior of lubricating oils. VI is calculated from kinetic viscosity at 40 and 100 °C (Eq. 2), where U is the kinematic viscosity of oil at 40 °C, L and H are viscosities of oils at 40 °C that has the same viscosity of the test oil at 100 °C. L and H can be found in ASTM D2270. A high VI-value indicates smaller viscosity changes with temperature.

$$\mathrm{VI} = 100\frac{L - U}{L - H} \tag{2}$$

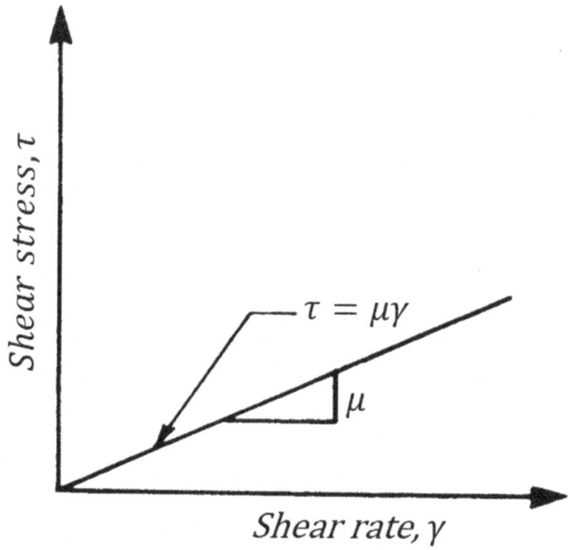

Fig. 3.19 Shear stress versus shear rate for a Newtonian fluid

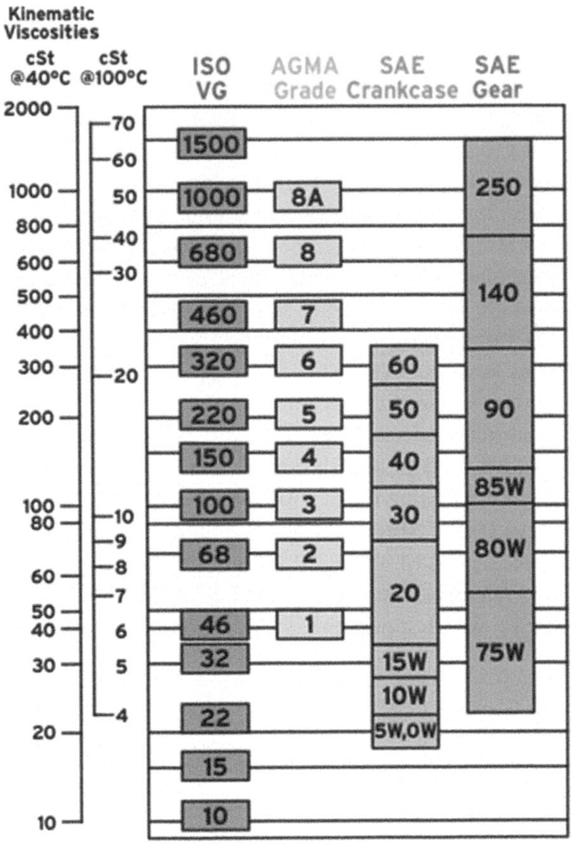

Fig. 3.20 Comparative classification of oils based on viscosity

An increase in temperature will weakens the intermolecular bonds in a fluid, leading to a decrease in viscosity (Fig. 3.21). The viscosity of the base oil at a given temperature can be obtained using Walther's equation [96]. The equation, which is the basis for ASTM, DIN and ISO charts, is defined as:

$$\mu = 10^{10(A - B*\log_{10}(T))} - 0.7 \tag{3}$$

where μ is kinematic viscosity in mm^2/s, T is the temperature in Kelvin, A and B are unknown constants calculated from the viscosity at two temperatures $\mu_1(T_1)$ and $\mu_2(T_2)$.

$$A = \log_{10}(\log_{10}(\mu_1 + 0.7)) + B \log_{10} T_1 \tag{4}$$

$$B = \frac{\log_{10}(\log_{10}(\mu_1 + 0.7)) - \log_{10}(\log_{10}(\log_{10}(\mu_2 + 0.7)))}{\log_{10} T_2 - \log_{10} T_1} \tag{5}$$

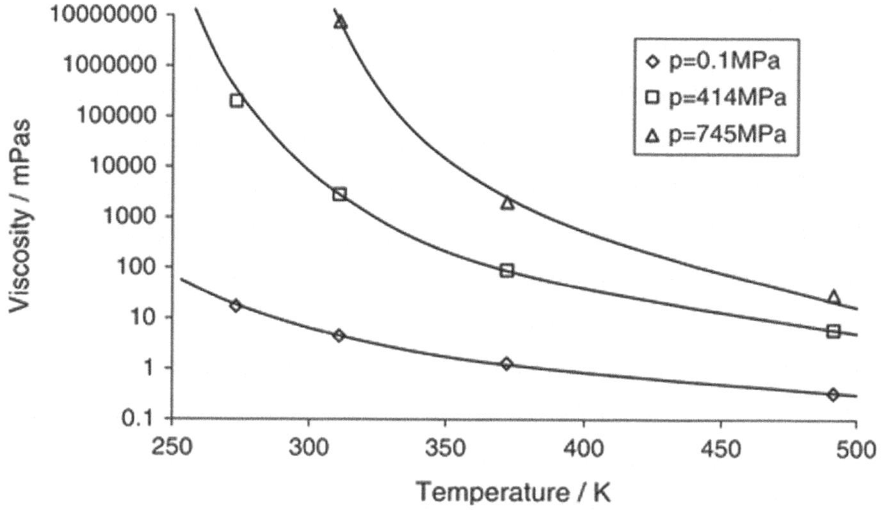

Fig. 3.21 Variation of viscosity versus temperature for a mineral oil at three different pressures. Reprinted from [97]

The viscosity of lubricating oil is strongly dependent on the pressure (Fig. 3.21). Therefore, calculation of viscosity at the inlet and Hertzian contacts is needed for the prediction of lubricant film thickness and friction. A commonly used equation for pressure viscosity relation is Barus equation [98] and is given by.

$$\eta(p) = \eta_0 \exp(\alpha p) \tag{6}$$

where η_0 is the viscosity at ambient pressure and α is the pressure viscosity coefficient.

Even though, Barus equation is widely used because of its simplicity, it significantly over predicts the viscosity at high pressures. There are several equations for calculating viscosity as function of temperature (Vogel equation, Reynolds equation etc.) and pressure (McEwen [99], Roelands equation [100], Yasutomi [101] etc.). However, their choice should be based on the balance between required accuracy and complexity.

In reality, either most of the fluids are Non-Newtonian where the fluid have a yield stress, or exhibit a non-linear relation between shear stress or shear rate, or the viscosity will depend on time or deformation history. All the above is applicable for lubricating greases. Grease exhibits visco-elastic, time dependent (Thixotropic) and shear thinning (non-Newtonian) behavior. The grease may behave as a solid and/or as a liquid depending upon the applied shear and shear rate. At very small deformations, grease behaves as an elastic solid. When the deformation exceeds a critical value, it becomes a viscous liquid. With increasing shear rate the viscosity decreases monotonically over a few decades of shear rates before finally reaching a plateau viscosity, which is for lubricating grease close to the base oil viscosity (Fig. 3.22). In addition, the timescale plays a role. The grease exhibits 'high stiffness' in the case of short and quick deformations while it behaves as a soft solid for long and slow deformations. In other words, a lubricating grease may exhibit visco-plastic and visco-elastic flow behavior depending upon the applied stress and strain. The transition from visco-elastic to visco-plastic behavior is characterized by a yield stress. When the stress exceeds the yield stress, the material will lose its solid-like properties. At low shear, lubricating greases can be modeled as visco-elastic liquids. One-dimensional mechanical models (Maxwell, Kelvin-Voigt) comprising of springs (elasticity) and dashpots (viscosity) can be used for describing the visco-elastic behavior of a lubricating grease.

For lubricating greases, Cross model [102] is a widely used equation for fitting the experimentally obtained viscosity (using a rheometer) from extremely low to high shear rates (Fig. 3.22). The Cross model reads:

$$\eta = \frac{\eta_0 - \eta_\infty}{\left(1 + \frac{\dot{\gamma}}{\dot{\gamma}_c}\right)^m} + \eta_\infty \tag{7}$$

where η_0 is the zero shear rate viscosity, η_∞ is the viscosity at very high shear rates (approximated by the base oil viscosity), $\dot{\gamma}_c$ is the critical shear rate corresponding to transition from Newtonian to shear thinning behavior and m is the shear thinning index.

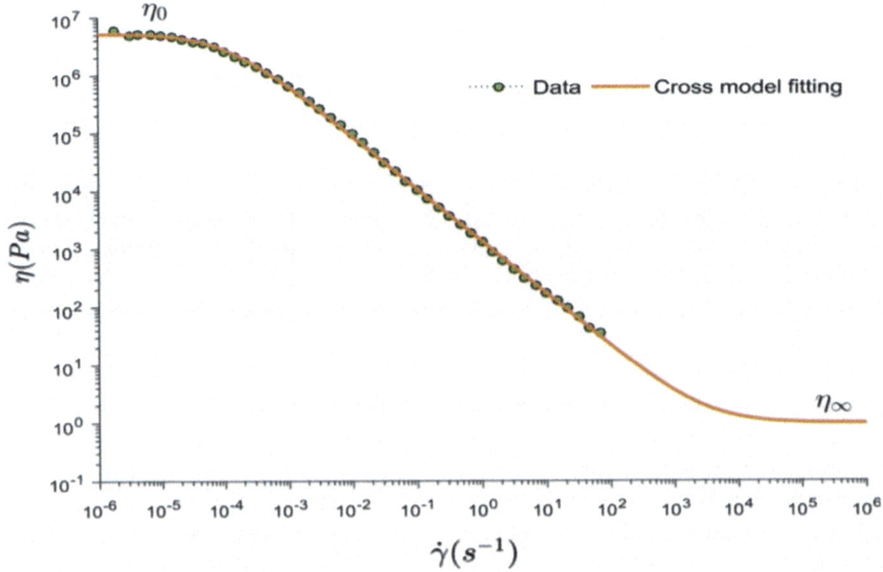

Fig. 3.22 Cross model fitting of viscosity data obtained using a commercial rheometer for a lithium-based grease. Reprinted from [103]

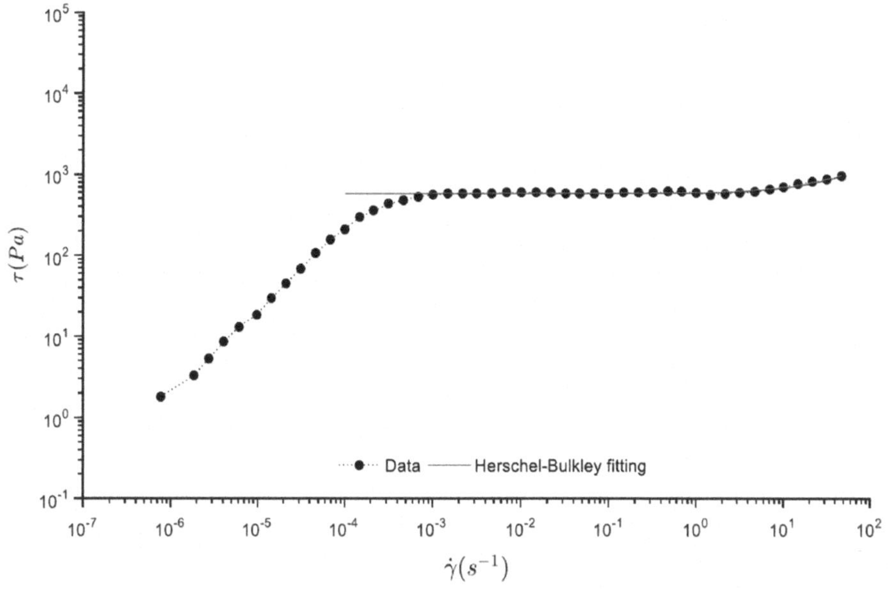

Fig. 3.23 Herschel-Bulkley model fitting of experimental data for estimating the yield stress. Reprinted from [107]

The high zero-shear-rate-viscosity is often considered as apparent elastic behavior. In that case, the yield stress, marking the transition from predominant elastic to viscous behavior [104, 105] is then given by the stress corresponding to the shear rate at the end of the first viscosity plateau in Fig. 3.22. This is shown in Fig. 3.23. Since this particular definition does not represent a true transition from elastic to plastic behavior, but more the stress which marks the onset of flow, this definition of the yield stress is regarded as being an engineering concept [106]. The most widely used model to describe the shear stress–shear rate relation is the Herschel-Bulkley model and it reads.

$$\tau = \tau_y + K\dot{\gamma}^n \tag{8}$$

In the above equation, τ_y is the yield stress, K is the consistency index (Pa s^{-n}), $\dot{\gamma}$ is the shear rate (s^{-1}), n—the shear-thinning index and τ is the shear stress.

3.7 Additives

3.7.1 *Introduction*

Commercial lubricating oils on an average consist of about 75–85% base oil and remainder is a package of additives to meet the performance demands of various applications [108]. Viscosity modifiers, corrosion inhibitors, antioxidants, friction modifiers etc. are a few of the additives used for ensuring the performance demands of modern machineries and automotive engines. Therefore, the final performance of a formulation depends on the ability of the base oil and the additives to work in tandem. These additives enhance the performance of the lubricating oil either by imparting new properties or by suppressing or eliminating the undesirable characteristics.

The additives can be classified into two types: surface additives and bulk oil additives (Fig. 3.24). Surface additives are those active at the contact surface. Examples are anti-wear additives, friction modifiers and extreme pressure additives. These additives are collectively known as tribo-improvers. Bulk oil additives are those that affect the rheological, interfacial and chemical properties of the lubricant by acting within the lubricant phase.

3.7.2 *Viscosity Modifiers*

Viscosity modifiers (VM), formerly called as viscosity index improvers are polymers used as additives in lubricants for minimizing variation in viscosity as a function of temperatures [109]. The solubility of high molecular weight polymers with chain like molecules depends on its chain length, structure and chemical composition. At low temperatures, the polymer chains contracts forming coils of low hydrodynamic volume without influencing the lubricant viscosity. At high temperatures, the molecules expands and an increase in viscosity occurs (Fig. 3.25). As they increase the temperature range of the lubricants by retaining the low temperature characteristics, they

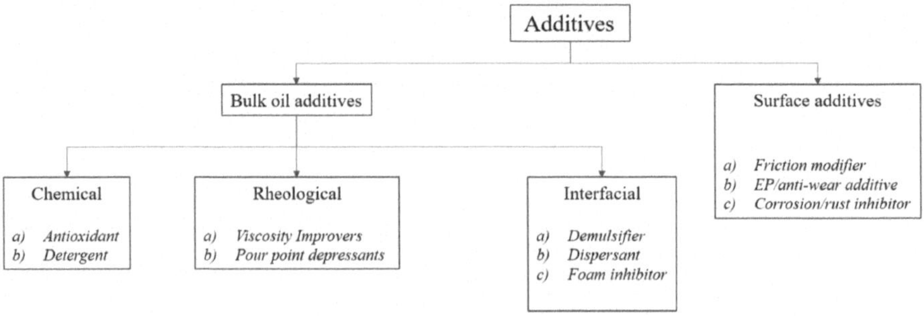

Fig. 3.24 Classification of additives based on the working site

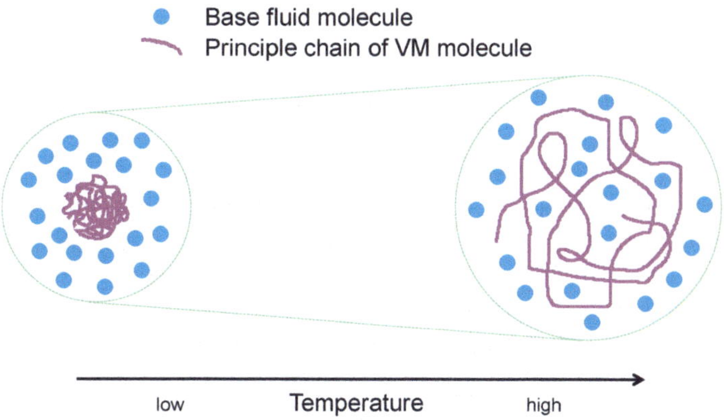

Fig. 3.25 Mechanism of operation of VM. Reprinted from [110]

are primarily used in multigrade engine oils, gear oils, hydraulic fluids, automatic transmission fluids etc. The increase in viscosity depends on the type of polymer, its molecular weight and concentration in the lubricant. The most commonly used VM are olefin copolymer (OCP) and Poly meth acrylates (PMA). Others include hydrogenated styrene–butadiene (SBR), polyisobutylene (PIB) and hydrogenated styrene–isoprene copolymers (SIP).

Olefin copolymer VM are oil soluble copolymers comprising ethylene and propylene and may contain a noncojugated diene as a third monomer. They are commonly used in engine oil and hydraulic oil formulations due to their high thickening efficiency and relatively low cost. In addition, their antioxidant activity, dispersancy and low-temperature viscosity can be enhanced by grafting chemical functional groups to the OCP backbone [111]. A methacrylate is a linear polymer constructed from hydrocarbons with three distinct chain lengths: short (influence the viscosity index of the polymer in oil), intermediate (determine the solubility of the polymer in lubricating oil) and long chain (impart pour point depressing properties), the specific ratio of which determine its functional properties in the oil solution [111]. PMA are commonly used in applications where high viscosity index and enhanced low temperature properties are required.

One of the major disadvantage of VM is they are prone to either temporary or permanent shear degradation that may result in loss of thickening. Higher molecular weight polymers have higher thickening efficiency but is less resistant to mechanical shear. Low molecular weight polymers exhibit more shear resistance, but has to be used in higher concentration to maintain improved viscosity at higher temperatures. Depending on the application and polymer, the mass fraction of VM in an oil usually varies from 3 to 30%.

3.7.3 Friction Modifiers

Friction modifiers (FMs) have been around for over a century and is widely being introduced into engine crankcase lubricants for improving the fuel efficiency [112]. The main types of FMs that are typically used in lubricant formulations are nanoparticles, organomolybdenum compounds and organic friction modifiers (OFMs). Nanoparticles based on carbon compounds, metals, metal oxides, metal borate, metal carbonate etc. in the size range of 2–120 nm are shown to exhibit reduction in friction and wear depending on their crystallinity, size, shape and concentration [113]. The friction reduction of nanoparticles is either due to polishing effect, mending effect, rolling effect or protective film (Fig. 3.26). Organomolybdenum compounds is mostly based on sulphur and phosphorus containing compounds (with the exception of molybdate ester) (e.g. molybdenum dialkyldithiophosphates (MoDTP), molybdenum dithiocarbamates (MoDTC) and MoS_2). As the market focus is also in ensuring low SAPS (sulphated ash, phosphorus and sulphur), OFMs containing carbon, hydrogen, oxygen and nitrogen are starting to replace the former.

OFMs are of the following categories: ester, alcohol, amine, amide, imide, carboxylic acids, phosphate, borate, ionic liquids and their derivatives [112]. OFMs are usually long slender molecules having a straight hydrocarbon chain of at least ten carbon atoms and a polar group at the end [115]. The polar end group either gets physisorbed or chemisorbed on to the surface while hydrocarbon chain remain solubilized into the lubricant, perpendicular to the metal surfaces (Fig. 3.27) [116, 117]. Later, clusters will be formed by dipole-dipole interactions between the adjacent polar head of OFM molecules and Van der Waals forces results in parallel alignment of hydrocarbon chains leading to the formation of multi-molecular cluster. Finally, a multi-layer matrix of OFM will be formed by molecular stacking on the adsorbed monolayer as shown in Fig. 3.27. The adsorbed layers thus formed are difficult to compress, but the layers can easily be sheard off at the hydrocarbon tail interface leading to reduced friction and wear. Even though, many researches have ascribed

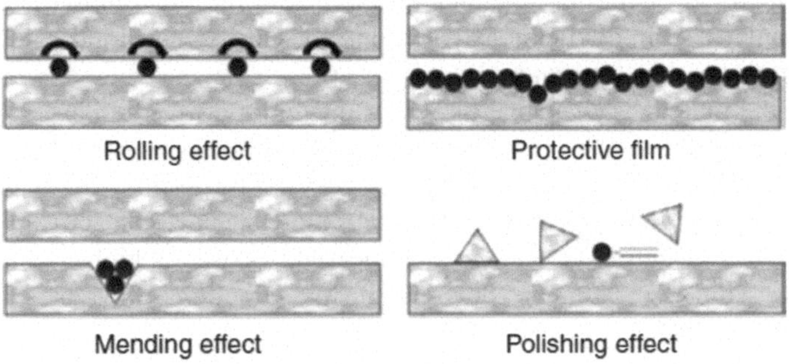

Fig. 3.26 Possible lubrication mechanism of nanoparticles as FMs. Reprinted with permission from [114]

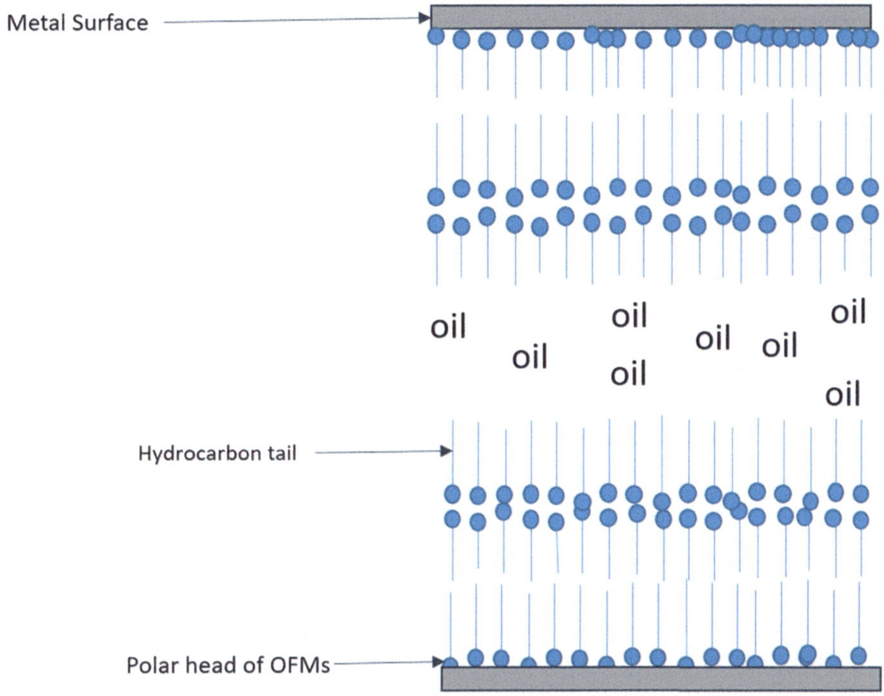

Fig. 3.27 Mechanism of operation of organic friction modifiers

reduced friction to the formation of monolayers some studies have confirmed formation of viscous layer deposits of tens to hundred nanometers thick on the metal surface [118].

A number of factors influence the effectiveness and thickness of the OFM films formed on the surface. Usually, slim molecules with longer chain lengths are preferred as they have direct implication on the packing density and thickness of the adsorbed film [119]. Increased polarity of the end groups with higher intermolecular hydrogen bonding capability is also critical for enhanced surface interactions and reduced friction [120]. In addition, base oil characteristics (polarity), competing additives, contaminants, metallurgy, temperature and concentration also has significant impact [115, 121]. Stronger films are usually obtained when the chain length of OFMs are similar to that of the base oil. Other polar additives (AW/EP, anticorrosion additives etc.) in the oil and contaminants formed by oxidative degradation of lubricants with more affinity to metal surface can have an antagonistic effect on the friction modifying characteristics of OFMs. Lower temperatures are favorable for efficient adsorption of OFM molecules on to the surface, whereas higher temperature may lead to decreased tenacity and desorption of OFM tribo-layers. An increase in concentration of OFMs in the oil does not necessarily mean improved frictional performance. OFM concentration of about 0.5 up to 1% are shown to be cost effective in reducing friction and only a marginal improvement in friction is usually observed above a concentration of 1%.

3.8 Anti-wear and Extreme Pressure Additives

Currently, market drive is towards using lubricants of lower viscosity as a strategic means of improving fuel efficiency by reducing viscous drag. However, this has resulted in the shifting of film thickness from full film regime to boundary and mixed regime where the friction and wear are particularly higher. Wear is the process of physical loss of the material because of local mechanical failure of highly stressed interfacial zones [122]. In a sliding contact wear arise from adhesion, abrasion, corrosion and contact fatigue. Anti-wear additives (AW) are load-carrying additives that reduce wear and prolong the life of tribological contact by modifying the metal surfaces by the process of surface film formation. AW are usually active in the mixed lubrication regime and requires higher temperature, load and shear to be active compared to FMs. Under moderately loaded sliding conditions the AW react with the asperities to form a thin film and aid the oil film by reducing the metallic contacts [123].

Extreme Pressure (EP) additives are another type of load carrying additives that modify the metal surface in the boundary surface regime and prevent failures such as scuffing, galling and seizure [124, 125]. There is no clear distinction between AW/EP as many are classed as either AW or EP depending upon the application. However, compared to the former, EP are expected to operate by reacting rapidly with the surface under severe surface distress such as under high load, high speed or high temperature operation [123]. EP usually forms a tougher metal compound (iron sulfide, metal phosphate etc.) by reacting with the metal surface where the sulfur/phosphor content of the additive determine its effectiveness. There are mild and strong EP additives. Mild EP are effective under low speed and high loads whilst severe EP are effective for high speed and high temperature applications. EP additives should only be used for demanding applications were contacts are subjected to severe distress as they have an adverse effect on oxidative stability of oils. In addition, they can also lead to corrosion of nonferrous materials.

There are variety of classes of compounds that are used as AW/EP additives such as sulfur, phosphorus, sulfur-phosphorus, sulfur-nitrogen, phosphorus-nitrogen, ZDDP etc. They are extensively used in engine oils, hydraulic oils and transmission fluids at 1–3%. However, due to environmental constraints, the future trend is towards additives that is free of metal and phosphorous such as graphite, cyclic amides, boron derivatives, dimer acids etc.

3.9 Miscellaneous Additives

3.9.1 *Pour Point Depressants*

Crude oils and their products contain substantial amount of oil waxes called paraffins. The paraffin crystals form a crystalline network by trapping molecules of liquid hydrocarbon with decrease in temperature leading to cessation of flow [126]. The lowest temperature at which the oil become semi-solid and loses its flow characteristics is called the pour point. Pour point is an important property in lubrication of any system as it provides the lower bound for the operating temperature.

Pour point depressants (PPDs) are the same compounds as viscosity improvers that maintain the fluidity of the lubricant on the lower temperature side by preventing the formation of wax crystals by dislocating them (Fig. 3.28). Therefore, they increase the operating temperature range of the base stock. The major difference between PPDs and VI is their application concentration, selection of monomer compounds, chain length, chain number and distribution of side chains [27] (Fig. 3.29). They are usually added at 0.1–2% and above or below the optical concentration level, they may become less effective.

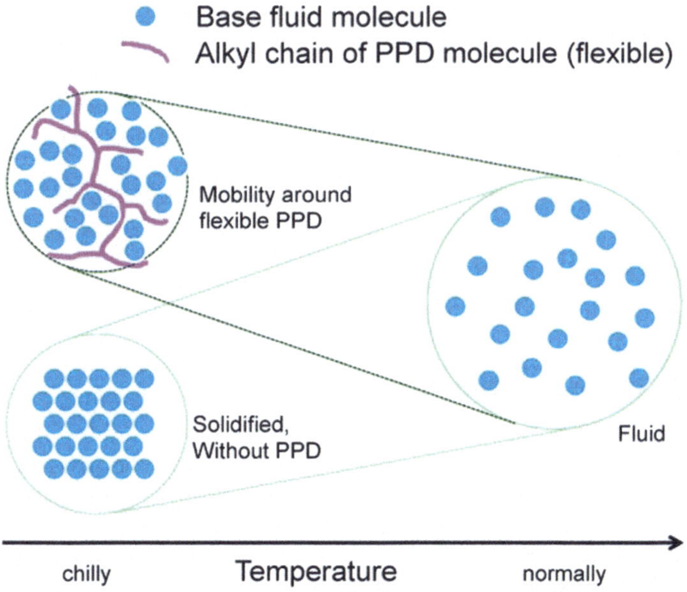

Fig. 3.28 Mechanism of operation of PPD. Reprinted from [110]

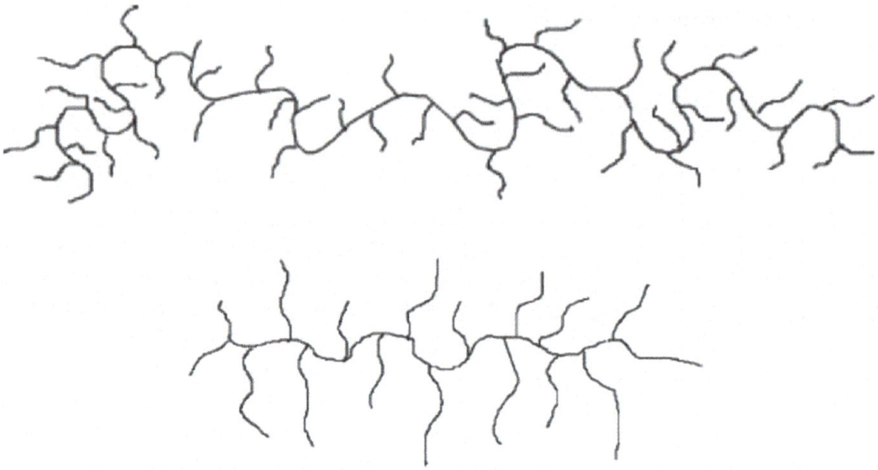

Fig. 3.29 Structural difference between VI (long chains with short branches) and PPDs (short chains with long branches). Reprinted from [127]

3.9.2 Corrosion Inhibitors

The main goal of corrosion inhibitors is to prevent corrosion and rusting of metal parts in corrosive environments. A corrosion inhibitor is a chemical substance that when added to a corrosive environment decreases the corrosion rate of the exposed metallic materials. These chemicals adsorb on metal surfaces to provide a protective film and/or neutralize corrosive acids (Fig. 3.30). This layer prevents oxygen and water from reaching the surface. A variety of organic compounds act as corrosion inhibitors with typical compounds being zinc dithiophosphates, metal phenolates, basic metal sulfonates, fatty acids and amines.

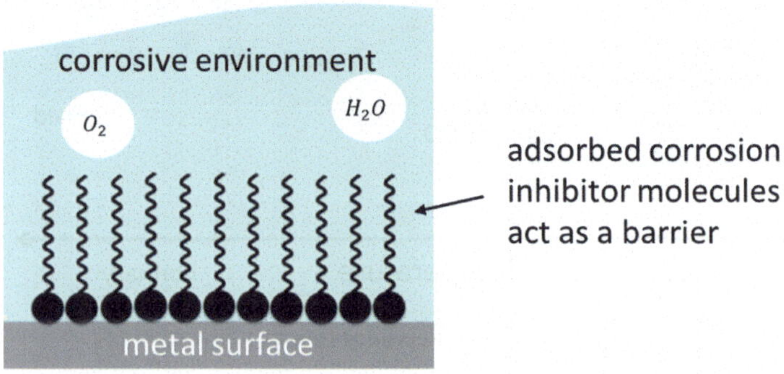

Fig. 3.30 Schematic representation of corrosion inhibition mechanism. Absorbed corrosion inhibitor molecules protect the surface from reacting with oxygen and water molecules

A group of additives generally termed as lubricant life additives contains the following types [128].

3.9.3 *Antioxidants*

One of the oil degradation paths is related to oxidation. Through oxidation the oil is subjected to a complex chain of reactions that eventually lead to the formation of insoluble sludge. This sludge may interfere with the lubrication and cooling functions of the oil. The process is accelerated by high temperatures, presence of catalysts (e.g., wear particles, metals), contamination by water, etc. Oxidation occurs though a three-stage process: initiation, propagation and termination. At first, a radical is formed due to an external force or event: this process involves a bond breakage between an organic species and hydrogen. The radical in propagation step reacts with oxygen to form peroxide radical, as well as other radicals that lead to further lubricant degradation. Finally, in the termination stage, radicals form a stable compound, thus terminating the process.

Antioxidants are specific agents that react with radicals to disrupt the described process. Anti-oxidants may act in two different ways, namely, by peroxide inhibition and radical scavenging [129].

The common chemicals used as anti-oxidants includes organo-copper compounds, aromatic amines, sulfur compounds, sulfur-nitrogen compounds, phosphorus compounds and sulfur-phosphorous compounds [130].

These additives are consumed during the operation of the machine. Therefore, it is very important to monitor the oxidation of oils (ASTM D 2272) to determine the remaining useful life of lubricating oil.

3.9.4 *Detergents and Dispersants*

Dispersants are special agents that keep insoluble contaminants (e.g., diesel engine soot) dispersed in the lubricant. These additives prevent contamination particles from agglomeration which otherwise could cause mechanical damage. Dispersant also keep the surfaces clean. Additive molecules are typically organic, contain organic groups and envelope nonpolar molecules (Fig. 3.31) so that they stay suspended in the lubricant [131, 132].

There are two main functions for detergent agents, namely, neutralizing acids formed in the oil and suspending the contamination particles (together with dispersants) [133, 134]. The metal salts of following acids are being used as detergents:

- arylsulfonic acids
- alkylphenols

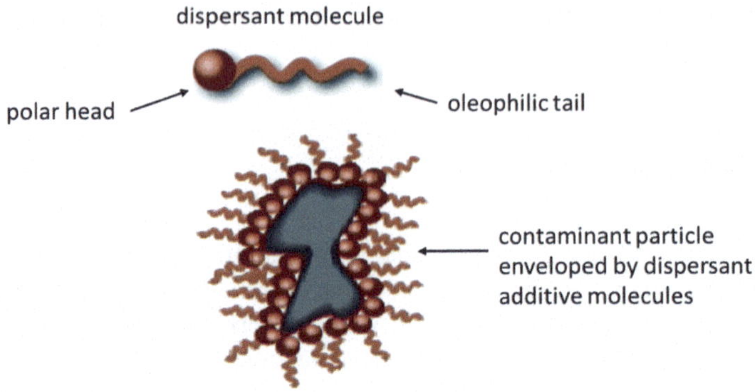

Fig. 3.31 Dispersant molecules envelope a contamination particle

- carboxylic acids
- petroleum oxidates

Both detergents and dispersants are consumed and therefore eventually need to be replenished.

3.9.5 Antifoam Agents

These additives prevent the lubricant from forming a persistent foam. These chemicals, typically silicone polymers or polyacrylates, accelerate coalescence of air bubbles by reducing surface tension of the bubble. This process is schematically shown in Fig. 3.32.

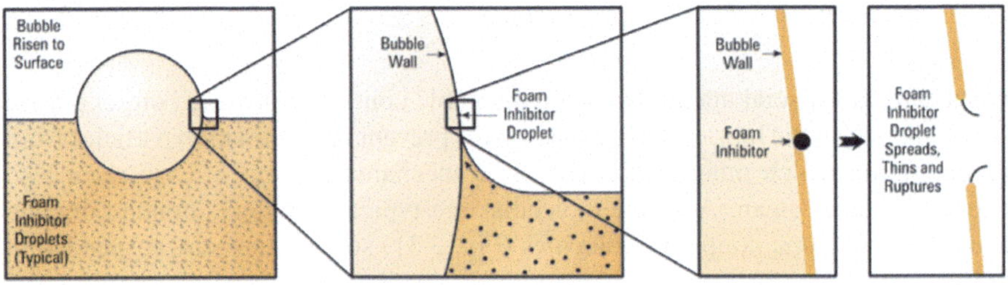

Fig. 3.32 Mechanism of anti-foam agent action. Reprinted from [135]

References

1. O. Reynolds, IV. On the theory of lubrication and its application to Mr. Beauchamp tower's experiments, including an experimental determination of the viscosity of olive oil. Philos. Trans. R. Soc. Lond. **177**, 157–234 (1886)
2. H.M. Martin, Lubrication of gear teeth. Engineering (London) **102**, 119–121 (1916)
3. A.N. Grubin, *Investigation of the Contact of Machine Components* (Central Scientific Research Institute for Technology and Mechanical Engineering, Moscow, 1949)
4. A.f.I. Petrusevich, *Principal Conclusions from Contact-Hydrodynamic Theory of Lubrication* (Associated Technical Services, East Orange, NJ, 1950)
5. A. Cameron, C. Mc Ettles, *Basic Lubrication Theory* (E. Horwood, Chichester, 1976)
6. D. Dowson, G. Higginson, A numerical solution to the elasto-hydrodynamic problem. J. Mech. Eng. Sci. **1**(1), 6–15 (1959)
7. R. Gohar, *Elastohydrodynamics* (World Scientific, Singapore, 2001)
8. A.A. Lubrecht, *The numerical solution of elastohydrodynamic lubricated line and point contact problems using multigrid techniques*, Ph.D. Thesis, University of Twente, The Netherlands (1987)
9. C.H. Venner, A.A. Lubrecht, *Multi-level Methods in Lubrication*, vol. 37 (Elsevier, Burlington, 2000)
10. B. Briscoe, D. Evans, D. Tabor, The influence of contact pressure and saponification on the sliding behavior of stearic acid monolayers. J. Colloid Interface Sci. **61**(1), 9–13 (1977)
11. Y.H. Wijnant, *Contact Dynamics in the Field of Elastohydrodynamic Lubrication* (Department of Mechanical Engineering, University of Twente, 1998)
12. N. Ren et al., Plasto-elastohydrodynamic lubrication (PEHL) in point contacts. J. Tribol. **132**(3), 031501 (2010)
13. D. Zhu, Y.-Z. Hu, Effects of rough surface topography and orientation on the characteristics of EHD and mixed lubrication in both circular and elliptical contacts. Tribol. Trans. **44**(3), 391–398 (2001)
14. H. Khan, P. Sinha, A. Saxena, A simple algorithm for thermo-elasto-hydrodynamic lubrication problems. Int. J. Res. Rev. Appl. Sci. **1**(3), 265–279 (2009)
15. Y.-Z. Hu et al., A computer model of mixed lubrication in point contacts. Tribol. Int. **34**(1), 65–73 (2001)
16. D. Zhu et al., Simulation of sliding wear in mixed lubrication. J. Tribol. **129**(3), 544–552 (2007)
17. M. Hartinger et al., CFD modelling of elastohydrodynamic lubrication, in *World Tribology Congress III* (American Society of Mechanical Engineers, 2005)
18. H. Spikes, Sixty years of EHL. Lubr. Sci. **18**(4), 265–291 (2006)
19. R. Bosman, Mild wear modeling in the boundary lubrication regime (2011)
20. J. Zhang, Y. Meng, Boundary lubrication by adsorption film. Friction **3**(2), 115–147 (2015)
21. A. Akchurin, R. Bosman, A deterministic stress-activated model for tribo-film growth and wear simulation. Tribol. Lett. **65**(2), 59 (2017)
22. P. Ku, Interdisciplinary approach to the lubrication of concentrated contacts. NASA SP-237. NASA Special Publication, 237, 1970
23. H. Spikes, A. Olver, Basics of mixed lubrication. Lubr. Sci. **16**(1), 1–28 (2003)
24. W. Shizhu, P. Huang, *Principle of Tribology* (Wiley, United States, 2012)
25. L. Ma, J. Luo, Thin film lubrication in the past 20 years. Friction **4**(4), 280–302 (2016)
26. K. Holmberg, A. Erdemir, Global impact of friction on energy consumption, economy and environment. FME Trans. **43**(3), 181–185 (2015)
27. P.L. Menezes, C.J. Reeves, M.R. Lovell, Fundamentals of lubrication, in *Tribology for scientists and engineers* (Springer, Cham, 2013), pp. 295–340
28. T. Mang, A. Gosalia, Lubricants and their market. *Lubricants and Lubrication* (Wiley, Weinheim, 2017), pp. 1–10
29. C.E. Campañá, M.H. Müser, Theoretical studies of superlubricity, *Superlubricity* (Elsevier, New York, 2007), pp. 39–56

30. B. Bhushan, Solid lubricants and self-lubricating films, in *Modern Tribology Handbook, Two Volume Set* (CRC Press, Boca Raton, 2000), pp. 817–856
31. A. Erdemir, *Lubrication from Mixture of Boric Acid with Oils and Greases* (Argonne National Laboratory (ANL), Argonne, IL, 1995)
32. M. Kanakia, M. Peterson, *Literature Review of Solid Lubrication Mechanisms* (Southwest Research Inst San Antonio TX Belvoir Fuels and Lubricants Research, 1987)
33. R.H. Savage, Graphite lubrication. J. Appl. Phys. **19**(1), 1–10 (1948)
34. J. Lancaster, A review of the influence of environmental humidity and water on friction, lubrication and wear. Tribol. Int. **23**(6), 371–389 (1990)
35. C. Pritchard, J. Midgley, The effect of humidity on the friction and life of unbonded molybdenum disulphide films. Wear **13**(1), 39–50 (1969)
36. T. Scharf, S. Prasad, Solid lubricants: a review. J. Mater. Sci. **48**(2), 511–531 (2013)
37. M. El-Sherbiny, F. Salem, Tribological properties of PVD silver films. ASLE Trans. **29**(2), 223–228 (1986)
38. N. Myshkin, M. Petrokovets, A. Kovalev, Tribology of polymers: adhesion, friction, wear, and mass-transfer. Tribol. Int. **38**(11–12), 910–921 (2005)
39. F.P. Bowden, D. Tabor, *The Friction and Lubrication of Solids*, vol. 1 (Oxford University Press, Oxford, 2001)
40. C.M. Pooley, D. Tabor, Friction and molecular structure: the behaviour of some thermoplastics. Proc. R. Soc. Lond. A **329**(1578), 251–274 (1972)
41. K.R. Makinson, D. Tabor, The friction and transfer of polytetrafluoroethylene. Proc. R. Soc. Lond. A **281**(1384), 49–61 (1964)
42. S. Bahadur, D. Tabor, The wear of filled polytetrafluoroethylene. Wear **98**, 1–13 (1984)
43. M. Brehob et al., The potential of carbon-based memory systems, in *Records of the 1999 IEEE International Workshop on Memory Technology, Design and Testing, 1999* (IEEE, 1999)
44. S. Prasad, J. Zabinski, Lubricants: super slippery solids. Nature **387**(6635), 761 (1997)
45. P. John, J. Zabinski, Sulfate based coatings for use as high temperature lubricants. Tribol. Lett. **7**(1), 31–37 (1999)
46. C. Donnet, A. Erdemir, *Tribology of diamond-like carbon films: fundamentals and applications* (Springer, Cham, 2007)
47. I. Sugimoto, S. Miyake, Oriented hydrocarbons transferred from a high performance lubricative amorphous C: H: Si film during sliding in a vacuum. Appl. Phys. Lett. **56**(19), 1868–1870 (1990)
48. C. Donnet et al., Diamond-like carbon-based functionally gradient coatings for space tribology. Surf. Coat. Technol. **120**, 548–554 (1999)
49. J. Andersson, R. Erck, A. Erdemir, Friction of diamond-like carbon films in different atmospheres. Wear **254**(11), 1070–1075 (2003)
50. D. Berman, A. Erdemir, A.V. Sumant, Graphene: a new emerging lubricant. Mater. Today **17**(1), 31–42 (2014)
51. D. Berman et al., Macroscale superlubricity enabled by graphene nanoscroll formation. Science **348**(6239), 1118–1122 (2015)
52. A. Gupta, T. Sakthivel, S. Seal, Recent development in 2D materials beyond graphene. Prog. Mater. Sci. **73**, 44–126 (2015)
53. K. Shavanova et al., Application of 2D non-graphene materials and 2D oxide nanostructures for biosensing technology. Sensors **16**(2), 223 (2016)
54. S.F. Brown, Base oil groups: manufacture, properties and performance. Tribol. Lubr. Technol. **71**(4), 32 (2015)
55. T. Zolper et al., Lubrication properties of polyalphaolefin and polysiloxane lubricants: molecular structure-tribology relationships. Tribol. Lett. **48**(3), 355–365 (2012)
56. M. Greaves, Oil soluble synthetic polyalkylene glycols a new type of group V base oil. Lube Mag. **104**, 21–24 (2011)
57. D.M. Pirro, M. Webster, E. Daschner, *Lubrication Fundamentals, Revised and Expanded* (CRC Press, Boca Raton, 2016)

58. P. Nagendramma, S. Kaul, Development of ecofriendly/biodegradable lubricants: an overview. Renew. Sustain. Energy Rev. **16**(1), 764–774 (2012)
59. A. Pensado, M. Comunas, J. Fernández, The pressure–viscosity coefficient of several ionic liquids. Tribol. Lett. **31**(2), 107–118 (2008)
60. J.G. Huddleston et al., Characterization and comparison of hydrophilic and hydrophobic room temperature ionic liquids incorporating the imidazolium cation. Green Chem. **3**(4), 156–164 (2001)
61. A.E. Somers et al., A review of ionic liquid lubricants. Lubricants **1**(1), 3–21 (2013)
62. A.H. Battez et al., Phosphonium cation-based ionic liquids as neat lubricants: physicochemical and tribological performance. Tribol. Int. **95**, 118–131 (2016)
63. J. Qu et al., Ionic liquids as novel lubricants and additives for diesel engine applications. Tribol. Lett. **35**(3), 181–189 (2009)
64. M.-D. Bermúdez et al., Ionic liquids as advanced lubricant fluids. Molecules **14**(8), 2888–2908 (2009)
65. A. Jimenez et al., Room temperature ionic liquids as lubricant additives in steel–aluminium contacts: influence of sliding velocity, normal load and temperature. Wear **261**(3–4), 347–359 (2006)
66. A.H. Battez et al., Tribological behaviour of two imidazolium ionic liquids as lubricant additives for steel/steel contacts. Wear **266**(11–12), 1224–1228 (2009)
67. B. Phillips, J. Zabinski, Ionic liquid lubrication effects on ceramics in a water environment. Tribol. Lett. **17**(3), 533–541 (2004)
68. J. Qu et al., Ionic liquids with ammonium cations as lubricants or additives. Tribol. Lett. **22**(3), 207–214 (2006)
69. A.H. Battez et al., Two phosphonium cation-based ionic liquids used as lubricant additive: Part I: Film thickness and friction characteristics. Tribol. Int. **107**, 233–239 (2017)
70. G. Goindi et al., Investigation of ionic liquids as additives to canola oil in minimum quantity lubrication milling of plain medium carbon steel. Int. J. Adv. Manuf. Technol. **94**(1–4), 881–896 (2018)
71. B. Bhushan, M. Palacio, B. Kinzig, AFM-based nanotribological and electrical characterization of ultrathin wear-resistant ionic liquid films. J. Colloid Interface Sci. **317**(1), 275–287 (2008)
72. F. Zhou, Y. Liang, W. Liu, Ionic liquid lubricants: designed chemistry for engineering applications. Chem. Soc. Rev. **38**(9), 2590–2599 (2009)
73. R. Gusain et al., Self-assembled thin film of imidazolium ionic liquid on a silicon surface: Low friction and remarkable wear-resistivity. Appl. Surf. Sci. **364**, 878–885 (2016)
74. M. Palacio, B. Bhushan, Ultrathin wear-resistant ionic liquid films for novel MEMS/NEMS applications. Adv. Mater. **20**(6), 1194–1198 (2008)
75. S. Zhang et al., Physical properties of ionic liquids: database and evaluation. J. Phys. Chem. Ref. Data **35**(4), 1475–1517 (2006)
76. H. Tokuda et al., Physicochemical properties and structures of room temperature ionic liquids. 1. Variation of anionic species. J. Phys. Chem. B **108**(42), 16593–16600 (2004)
77. H. Tokuda et al., Physicochemical properties and structures of room temperature ionic liquids. 2. Variation of alkyl chain length in imidazolium cation. J. Phys. Chem. B **109**(13), 6103–6110 (2005)
78. H.L. Ngo et al., Thermal properties of imidazolium ionic liquids. Thermochim. Acta **357**, 97–102 (2000)
79. T. Welton, Room-temperature ionic liquids. Solvents for synthesis and catalysis. Chem. Rev. **99**(8), 2071–2084 (1999)
80. R. Hagiwara, Y. Ito, Room temperature ionic liquids of alkylimidazolium cations and fluoroanions. J. Fluor. Chem. **105**(2), 221–227 (2000)
81. W. Liu et al., Tribological performance of room-temperature ionic liquids as lubricant. Tribol. Lett. **13**(2), 81–85 (2002)
82. S. Zhang et al., Vacuum tribological performance of phosphonium-based ionic liquids as lubricants and lubricant additives of multialkylated cyclopentanes. Tribol. Int. **66**, 289–295 (2013)

83. A. Suzuki, Y. Shinka, M. Masuko, Tribological characteristics of imidazolium-based room temperature ionic liquids under high vacuum. Tribol. Lett. **27**(3), 307–313 (2007)
84. K.W. Street Jr. et al., Evaluation of vapor pressure and ultra-high vacuum tribological properties of ionic liquids. Tribol. Trans. **54**(6), 911–919 (2011)
85. W. Morales et al., Tribological testing and thermal analysis of an alkyl sulfate series of ionic liquids for use as aerospace lubricants. Tribol. Trans. **55**(6), 815–821 (2012)
86. H. Xiao, Ionic liquid lubricants: basics and applications. Tribol. Trans. **60**(1), 20–30 (2017)
87. I. Perissi et al., High temperature corrosion properties of ionic liquids. Corros. Sci. **48**(9), 2349–2362 (2006)
88. ASTM D288-61 Definitions of terms relating to petroleum, 1978
89. K. Bauer, D. Garbe, H. Surburg, Ullmann's encyclopedia of industrial chemistry. *Ullmann's Encyclopedia of Industrial Chemistry*, vol. 11 (Wiley, Weinheim, 1988)
90. W.H. Bauer, A.P. Finkelstein, S.E. Wiberley, Flow properties of lithium stearate-oil model greases as functions of soap concentration and temperature. ASLE Trans. **3**(2), 215–224 (1960)
91. F. Cyriac et al., Effect of thickener particle geometry and concentration on the grease EHL film thickness at medium speeds. Tribol. Lett. **61**(2), 18 (2016)
92. N. Scarlett, Paper 21: Use of grease in rolling bearings, in *Proceedings of the Institution of Mechanical Engineers, Conference Proceedings* (SAGE Publications Sage, London, 1967)
93. P.M. Lugt, *Grease Lubrication in Rolling Bearings* (Wiley, Chichester, 2012)
94. P.M. Lugt, Modern advancements in lubricating grease technology. Tribol. Int. **97**, 467–477 (2016)
95. C. Mike Johnson, C. Contributing, Understanding grease construction and function. Tribol. Lubr. Technol. **3**, 3 (2008)
96. C. Walther, The evaluation of viscosity data. Erdol Teer **7**, 382–384 (1931)
97. S. Bair, Temperature and pressure dependence of viscosity, in *Encyclopedia of Tribology*, ed. by Q.J. Wang, Y.-W. Chung (Springer, New York, 2013), pp. 3533–3538
98. C. Barus, ART. X.—Isothermals, isopiestics and isometrics relative to viscosity. Am. J. Sci. (1880–1910) **45**(266), 87 (1893)
99. E. McEwen, The effect of variation of viscosity with pressure on the load-carrying capacity of the oil film between gear-teeth. J. Inst. Pet. **38**(344–345), 646–672 (1952)
100. C. Roelands, J. Vlugter, H. Waterman, The viscosity-temperature-pressure relationship of lubricating oils and its correlation with chemical constitution. J. Basic Eng. **85**(4), 601–607 (1963)
101. S. Yasutomi, S. Bair, W. Winer, An application of a free volume model to lubricant rheology I—dependence of viscosity on temperature and pressure. J. Tribol. **106**(2), 291–302 (1984)
102. M.M. Cross, Rheology of non-Newtonian fluids: a new flow equation for pseudoplastic systems. J. Colloid Interface Sci. **20**, 417–437 (1965)
103. F. Cyriac, P.M. Lugt, R. Bosman, Impact of water on the rheology of lubricating greases. Tribol. Trans. **59**(4), 679–689 (2016)
104. R. Houwink, H.K. De Decker, H.K. DeDecker, *Elasticity, Plasticity and Structure of Matter* (Cambridge University Press, Cambridge, 1971)
105. H. Green, *Industrial Rheology and Rheological Structures* (Wiley, New York, 1949)
106. H. Barnes, K. Walters, The yield stress myth? Rheol. Acta **24**(4), 323–326 (1985)
107. F. Cyriac, On rheology, film-build-up and water in grease lubricated bearings, 2016
108. J.C. Bart, E. Gucciardi, S. Cavallaro, *Biolubricants: Science and Technology* (Elsevier, Amsterdam, 2012)
109. P. Ghosh et al., Shear stability of polymers used as viscosity modifiers in lubricating oils, 1998
110. I. Minami, Molecular science of lubricant additives. Appl. Sci. **7**(5), 445 (2017)
111. L.R. Rudnick, *Lubricant Additives: Chemistry and Applications* (CRC Press, Boca Raton, 2017)
112. Z. Tang, S. Li, A review of recent developments of friction modifiers for liquid lubricants (2007–present). Curr. Opin. Solid State Mater. Sci. **18**(3), 119–139 (2014)

113. M. Akbulut, Nanoparticle-based lubrication systems. J. Powder Metall. Min. **1**, e101 (2012)
114. K. Lee et al., Understanding the role of nanoparticles in nano-oil lubrication. Tribol. Lett. **35**(2), 127–131 (2009)
115. D. Kenbeek, T. Buenemann, H. Rieffe, Review of organic friction modifiers-contribution to fuel efficiency? SAE Technical Paper, 2000
116. M. Beltzer, S. Jahanmir, Effect of additive molecular structure on friction. Lubr. Sci. **1**(1), 3–26 (1988)
117. S. Jahanmir, M. Beltzer, Effect of additive molecular structure on friction coefficient and adsorption. J. Tribol. **108**(1), 109–116 (1986)
118. C. Allen, E. Drauglis, Boundary layer lubrication: monolayer or multilayer. Wear **14**(5), 363–384 (1969)
119. A.S. Akhmatov, *Molecular Physics of Boundary Friction*, vol. 2108 (Israel Program for Scientific Translations, 1966)
120. J. Davidson et al., Molecular dynamics simulations to aid the rational design of organic friction modifiers. J. Mol. Graph. Model. **25**(4), 495–506 (2006)
121. R. Castle, C. Bovington, The behaviour of friction modifiers under boundary and mixed EHD conditions. Lubr. Sci. **15**(3), 253–263 (2003)
122. C. Bovington, Friction, wear and the role of additives in controlling them, in *Chemistry and Technology of Lubricants* (Springer, New York, 2010), pp. 77–105
123. E.S. Forbes, Antiwear and extreme pressure additives for lubricants. Tribology **3**(3), 145–152 (1970)
124. W. Piekoszewski, M. Szczerek, W. Tuszynski, The action of lubricants under extreme pressure conditions in a modified four-ball tester. Wear **249**(3–4), 188–193 (2001)
125. M. Kawamura, K. Fujita, Organic sulphur and phosphorus compounds as extreme pressure additives. Wear **72**(1), 45–53 (1981)
126. R.A. Soldi et al., Polymethacrylates: pour point depressants in diesel oil. Eur. Polym. J. **43**(8), 3671–3678 (2007)
127. R.M. Nasser, *The Behavior of Some Acrylate Copolymers as Lubricating Oil Additives* (LAP LAMBERT Academic Publishing, 2015)
128. A.V. Beek, *Advanced Engineering Design. Lifetime Performance and Reliability* (TU Delft, Delft, 2012)
129. A.M. Barnes, K. Bartle, V.R.A. Thibon, A review of zinc dialkyldithiophosphates (ZDDPS): characterisation and role in the lubricating oil. Tribol. Int. **34**(6), 389–395 (2001)
130. S. Shahnazar, S. Bagheri, S.B. Abd Hamid, Enhancing lubricant properties by nanoparticle additives. Int. J. Hydrogen Energy **41**(4), 3153–3170 (2015)
131. Á. Beck, G. Pölczmann, Z. Eller, J. Hancsók, Investigation of the effect of detergent-dispersant additives on the oxidation stability of biodiesel, diesel fuel and their blends. Biomass Bioenergy **66**, 328–336 (2014)
132. P. Sassiat, G. Machtalere, F. Hui, H. Kolodziejczyk, R. Rosset, Liquid chromatographic determination of base oil composition and content in lubricating oils containing dispersants of the polybutenylsuccinimide type. Anal. Chim. Acta **306**(1), 73–79 (1995)
133. L.K. Hudson, J. Eastoe, P.J. Dowding, Nanotechnology in action: overbased nanodetergents as lubricant oil additives. Adv. Colloid Interface Sci. **123**, 123–126 (2006)
134. M. Reyes, A. Neville, The effect of anti-wear additives, detergents and friction modifiers in boundary lubrication of traditional Fe-base materials. Tribol. Ser. **41**, 57–65 (2003)
135. Z.E. Dadach, Applied research: foaming in sea water cooling tower, 2015

Chapter 4
Rheological Behaviour of Hybrid Nanofluids: A Review

Anuj Kumar Sharma, Rabesh Kumar Singh, Arun Kumar Tiwari, Amit Rai Dixit, and Jitendra Kumar Katiyar

Abstract A colloidal mixture of two different nanoparticles into conventional fluid (water, oil and metal working fluids etc.) called as hybrid nanofluids. Hybrid nanofluids are considered as the most promising and emerging as heat transfer fluid in cooling applications as compared to the conventional fluid and a well as mono type nanofluids. Mixture of solid particles and fluid is also called as two phase fluids. Mixing of nano meter-sized particles into conventional heat transfer fluid enhance the performance of the newly developed hybrid nanofluid. In last few years, it has taken the researchers attention to work on the mixing of two or more nano-sized particles in conventional heat transfer fluids. Few of the studies shows that hybrid nanofluid perform better as compared to single nanofluids and has the ability to replace the single nanoparticles mixed fluid. However, in the published literature several studies has shown that there are number of parameters such as concentration of nanoparticles, shape, size, temperature, intensity of ultra-sonication, pH, and stability affecting the performance of the nanoparticles mixed cutting fluids. In the present paper, a comprehensive study has been carried out to show the recent development related to the hybrid nanofluids.

A. K. Sharma · R. K. Singh (✉)
Mechatronics, Centre for Advanced Studies, Dr. APJ Abdul Kalam Technical University, Lucknow 226031, India
e-mail: rasm.singh@gmail.com

A. K. Tiwari
Department of Mechanical Engineering, Institute of Engineering and Technology, Dr. APJ Abdul Kalam Technical University, Lucknow 226021, India

A. R. Dixit
Department of Mechanical Engineering, Indian Institute of Technology (ISM), Dhanbad 826004, India

J. K. Katiyar
Department of Mechanical Engineering, SRM Institute of Science and Technology, Kattankulathur, Chennai, Tamil Nadu 603203, India

J. K. Katiyar et al. (eds.), *Tribology in Materials and Applications*,
Materials Forming, Machining and Tribology,
https://doi.org/10.1007/978-3-030-47451-5_4

4.1 Introduction

In the age of industrial globalization and competitive world the heat transfer fluids has a significant contribution in the manufacturing, electronic cooling, and energy sector etc. Furthermore, conventional cutting fluids have the limited cooling and heat transfer capabilities due to their lower thermo-physical properties such as ethylene glycol, oil and water [1, 2]. The poor properties limits their use in heating and coolants applications. Presently, there are various ways are available to improve the heat carrying capacity of commonly used heat transfer fluids. Therefore, thermal properties of metal working fluid becomes essential part of the research. At present one such technique may be the mixing of nano-meter sized in the commercially available heat transfer fluids that can enhance the heat carrying capacity. About two decades ago, a new type of heat transfer fluid has been developed and firstly the term as nanofluid [3], which has the higher heat transfer capabilities. Nanofluids are the uniform dispersion of the small particles (less than 100 nm) in a conventional heat transfer fluids. The nanoparticles can be of carbonic, carbide, ceramics, oxides, metallic and non-metallic or the composite of two and more (hybrid) nanoparticles or sometimes nano-sized fluid droplets. For the conventional heat transfer fluid any high viscous fluids like mineral oil, ethylene glycol, or a combination of two types of fluids (water + EG, propylene glycol + water etc.) or a low viscous fluid such as, refrigerant and water may be considered. In the published literature, there are number of researchers have noticed a significant increase in the heat transfer capabilities with nanofluid over base fluid. These developed nanofluid have various advantages like better thermo-physical properties, stability and less pressure drop over conventional fluid. Moreover, there are various types of nanofluids has been used as cutting fluids in the different conventional machining process like turning [4, 5], drilling [6], grinding [7, 8] and milling [9, 10]. Application of different nanoparticles mixed cutting fluids in various metal removal processes improves the performance of the processes. Furthermore, number of scientists have experimentally studies the performance of nanoparticles mixed heat transfer fluid in conventional fluids and significant improvement in their thermal conductivity was noticed over base fluid [11–14]. Saidur et al. [15] observed that increase in the volumetric fraction of nanoparticles improve thermal properties of nanofluids over conventional fluid. In another study mixing of nanoparticles in base fluid enhance the thermo-physical properties of developed nanofluid as compared to base fluids [16, 17]. However, few experimental studies have reported that viscosity of nanofluid increases with increase in the thermal conductivity, which increases the pumping power [18, 19]. In furthermore investigation, Tiwari et al. [20, 21] noticed an enhancement in the density and viscosity of nanofluid at higher nanoparticles concentration, which in turn to increase the pumping power. Figure 4.1 exhibits that shear viscosity of nanofluids and nanoparticle concentration has a directly proportional relationship. In another investigation it has been found that significant enhancement in the density and viscosity of developed nanofluid at higher volumetric fraction of nanoparticles as compared to conventional fluid [22]. An enhancement of approximately 13.9 and 91% in the density and viscosity was recorded at 6 vol% alumina

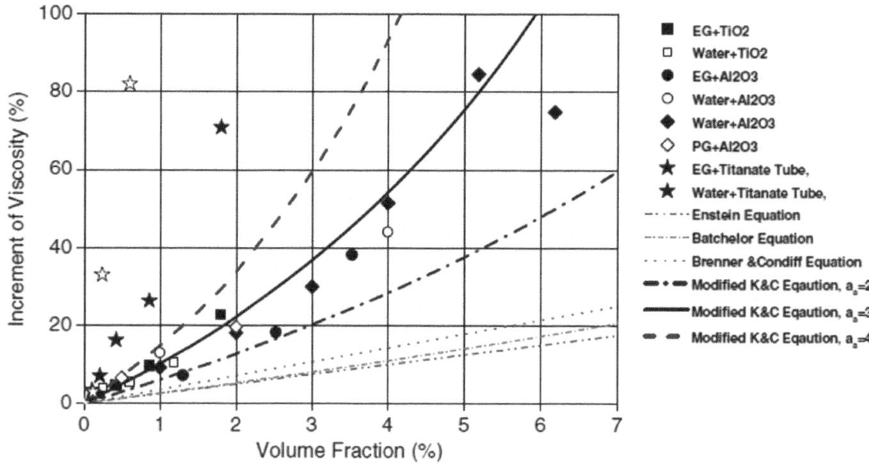

Fig. 4.1 Effect of nanoparticles concentration on shear viscosity [reproduce with the permission of Ref. 23]

oxide mixed in ethylene glycol and water based nanofluid. A comparative analysis was also conducted to assess the effectiveness of different oxide based nanofluids (CuO, Al_2O_3 and SiO_2) and less pumping power was required at 2 vol% for all nanofluids over base fluid at a given temperature. Furthermore, enhancement in the viscosity of nanofluid sample may sustain a penalty in the form of pressure drop that will rise in the pumping power. Therefore, viscosity of the nanofluids may be a deciding factor for the selection of nanofluids for different applications.

Addition of mono type or single nanoparticle in conventional fluid enhance the viscosity and thermal conductivity. However, better results may be obtained with the hybrid nanofluids or composite nanofluids. Hybrid nanofluid is a mixture of two or more nanoparticles in the same conventional fluid. Furthermore, addition of two nanoparticles enhance the heat transfer capability of nano-composite fluid due to its synergistic effect of solid particles as compared to single or mono type of nanofluids. From the few published literatures, it has been noticed that combination of two nanoparticles can enhance the thermo-physical properties as compared to single nanoparticle mixed fluid and base fluid [24, 25]. In another study Jana et al. [26] conducted a comparative study of thermo-physical properties of single and combination of two or more nano-composites. However, in few studies interesting results have been reported, where mixing of CNT-AuNP and CNT-CuNP hybrid nanofluid lowers the thermal conductivity over mono type or single nanofluid (CNT, Cu and Au) and base fluid. After few years, another study was conducted to investigate the performance of similar type of hybrid nanofluid, which was a combination of silver and multiwall carbon nanotube and result shows a substantial increase in the thermal conductivity over mono-type nanofluids [27].

References	Base fluids	Nano-particle 1	Nano-particle 2	Method of preparation	Outcomes/findings
Jana and Zhong [26]	Deionize water	CNTs, CUNPs, AuNPs	CUNP, AuNP		From the experimental results it has been noticed that hybrid nanofluid exhibits lower value of thermal conductivity over nanofluids containing the single nanoparticles due to non-compatibility of the individual CNTs, CuNPs, and AuNPs nanoparticles
Chopkar et al. [28]	Deionize water and ethylene glycol	Al_2Cu	Ag_2Al	Two step method	Thin Sheet shaped nanoparticles structure exhibit better performance as compared to cylindrical and spherical shaped nanoparticles. Furthermore, Ag_2Al hybrid nanoparticles shows better performance than Al_2Cu
Chen et al. [29]	Deionize water	Fe_2O_3	MWCNT	Two step method	Significant enhancement in the thermal conductivity with both the hybrid nanofluid has been recorded with increase in the concentration of both the MWCNT and Fe_2O_3 in the hybrid mixture. However, Fe_2O_3 outperform thermal performance beyond 0.02% concentration over MWCNT nanofluid alone

(continued)

(continued)

References	Base fluids	Nano-particle 1	Nano-particle 2	Method of preparation	Outcomes/findings
Munkhbayar et al. [30]	Deionize water	Ag	MWCNT	Single step method	The composite of Ag/MWCNT noticed a significant improvement in thermal conductivity. However, better result of thermal conductivity has been observed with purified MWCNT as compared to basic MWCNT
Baby and Ramaprabhu [31]	Deionize water and ethylene glycol (EG)	Ag	HEG	Two step method	An increase in the thermal conductivity was noticed at higher temperature and nanoparticles concentrations. Furthermore, deionized water base hybrid nanofluid performs better over EG based hybrid nanofluid due to higher viscosity
Baby and Ramaprubhu [32]	Ethylene glycol and Deionize water	CuO	HEG	Two step method	Enhancement of 28% in thermal conductivity was noticed at 0.05% concentration of the hybrid nanofluid at 25 °C while 90% increment was obtain at 50 °C
Aravind and Ramaprabhu [33], Singh et al. [34]	Deionize water and ethylene glycol	Graphene	MWCNT	*Two step method	MWCNT mixed nanofluid shows better enhancement due to their high aspect ratio and this enhancement leads to the better thermal conduction at higher temperature and concentration of the nanoparticles

(continued)

(continued)

References	Base fluids	Nano-particle 1	Nano-particle 2	Method of preparation	Outcomes/findings
Suresh et al. [35]	Deionize water	Al_2O_3	Cu	*	It has been noticed that considerable increment in the viscosity. Furthermore, the thermal conductivity of hybrid nanofluid increases linearly with increase in the nanoparticles concentration
Batmunkh et al. [36]	Deionize water	Ag	TiO_2	*	Considerable enhancement has been noticed with Ag/TiO_2 hybrid nanofluid over TiO_2 nanofluid alone. The crushing of Ag nanoparticles improves the aspect ratio of the composite nanoparticles in the suspension
Han et al. [37]	poly-alpha-olefin	Sphere	CNT	*	At 0.2% an enhancement of 21% in the thermal conductivity was recorded at room temperature. The composite mixture of sphere and CNT nanoparticles lowers the thermal conduction among the nanoparticles
Botha et al. [38]	Oil	Ag	Silica	*	Significant enhancement in the thermal conductivity was noticed at higher concentrations.

(continued)

(continued)

References	Base fluids	Nano-particle 1	Nano-particle 2	Method of preparation	Outcomes/findings
Paul et al. [39]	Ethylene glycol	Al	Zn	*	From the experimental result it has been observed that as nanoparticles grain size reduces the thermal conductivity increases. Furthermore, nearly 100% enhancement has been observed at 70 °C
Abbasi et al. [40]	Distilled water	γ-Al_2O_3	MWCNT	*	From the experimental result it has been noticed that higher concentration of surfactant lowers the stability of nanofluids. Therefore, further treatment of MWCNT reduces thermal conductivity of nanofluid due to excessive deterioration in the aspect ratio
Nine et al. [41]	Distilled water	Al_2O_3	MWCNTs	*	The cylindrical shape nanoparticles have shown better results over spherical due to higher aspect ratio. Therefore, hybridization on nanoparticles improve the thermal conductivity over Al_2O_3 nanofluids alone
Afrand [42]	Ethylene glycol	MgO	f-MWCNT	Two step method	From the study a correlation has been developed between the thermal conductivity of prepared hybrid nanofluid and temperature (25–50 °C) and nanoparticles concentration (0–6%)

(continued)

(continued)

References	Base fluids	Nano-particle 1	Nano-particle 2	Method of preparation	Outcomes/findings
Mohammad and Omid [43]	Ethylene glycol	MgO	MWCNT	Two step method	A new correlation has been developed between the viscosity and hybrid nanofluid at different fraction of nanoparticles (0.1–1%) and at varying temperature (30–60 °C)
Sunder et al. [44]	Water	ND	Fe_3O_4	Single step method	Form the experimental studied a new the correlation has been developed for viscosity and conductivity with hybrid nanofluid at varying nanoparticles concentration (0.05–0.2%) and temperature (20–60 °C)
Saeed et al. [45]	Ethylene glycol	f-MWCNT	Fe_3O_4	Two step method	To studies the influence of temperature (25–50 °C) and nanoparticles fraction (0.1–2.3%) on the thermal conductivity, a new correlation has been developed with the experimental studies for developed hybrid nanofluid
Eshgarf and Afrand [46]	Water, Ethylene glycol	MWCNT	SiO_2	*	Experimental studies have been perform to test the thermo-physical properties of hybrid nanofluid for different heat transfer applications at different temperature (25–50 °C) and varying fraction of nanoparticle (0.0625–2%)

(continued)

(continued)

References	Base fluids	Nano-particle 1	Nano-particle 2	Method of preparation	Outcomes/findings
Afrand et al. [47]	Ethylene glycol	Fe_3O_4	Ag	Two step method	From the experimental study it has been observed that temperature and nanoparticles concentration has significant effect on the rheological behaviour of prepared hybrid nanofluid
Hemmat Esfe et al. [48]	Water	Ag	MgO	*	A new correlation has been developed between the thermal conductivity and dynamic viscosity and prepared hybrid nanofluid for varying nanoparticles concentration (0–2%)
Hemmat Esfe et al. [49]	Water, Ethylene glycol	DWCNT	ZnO	*	A significant enhancement in the thermal conductivity was recorded with newly developed hybrid nanofluid at varying concentration (0.25–1%) and temperature (25–50 °C)
Hemmat Esfe et al. [50]	Water, Ethylene glycol	Cu	TiO_2	Two step method	A new modelling and correlation has been developed between the thermo-physical properties and prepared hybrid heat transfer fluid using Artificial Neural Network (ANN)

(continued)

(continued)

References	Base fluids	Nano-particle 1	Nano-particle 2	Method of preparation	Outcomes/findings
Sunadr et al. [51]	Water, Ethylene glycol	CNT	Fe_3O_4	In situ method	At lower volume fraction an enhancement in the heat transfer was recorded for the newly prepared hybrid nanofluids under turbulent flow in a tube with twisted tape inserts
Sunadr et al. [52]	Water, Ethylene glycol	ND	NI	In situ method	From the experimental results, an enhancement in the thermo-physical for newly developed nano-composite heat transfer fluid recorded as compared to base fluid
Baghbanzadeh et al. [53]	Water	Silica	MWCNT	Wet chemical method.	Experimentally investigated the effect of hybrid nanofluid on the rheological properties such as density and viscosity
Takabi and Salehi [54]	Water	Al_2O_3	Cu	*	Studied the heating and cooling performance of the prepared hybrid nanofluids for a sinusoidal corrugated enclosure
Xuan et al. [55]	Water	TiO_2	Ag	Photochemical impregnation method.	Application of hybrid nanofluids significantly improves the absorption rate in the solar collector
Madhesh et al. [56]	Water	Cu	TiO_2	*	Addition of nano-composite (Cu/TiO_2) in base fluid improves the convective heat transfer and rheological behaviour in tube heat exchanger

(continued)

(continued)

References	Base fluids	Nano-particle 1	Nano-particle 2	Method of preparation	Outcomes/findings
Singh et al. [57] Sharma et al. [58]	Water-based emulsion	Al_2O_3	Graphene	Two step method	The experimental result shows an increase in the nanoparticles volumetric fraction, enhances both, the thermal conductivity and viscosity over base fluid. However, of the hybrid nanofluid has lower thermal conductivity over the Al_2O_3 and the viscosity lies in between them
Sharma et al. [59, 60]	Water-based emulsion	Al_2O_3	MoS_2	Two step method	An enhancement in the thermo-physical properties were noticed at varying concentration (0.25–1.25%) at different temperatures
Sharma et al. [61, 62]	Water-based emulsion	Al_2O_3	MWCNT	Two step method	Experimentally investigated the thermo-physical properties of hybrid heat transfer fluid at varying concentration (0.25–1.25%) at different temperatures

*Not mentioned

4.2 Effect of Various Parameters of Hybrid Nanofluid

Composite of two or more nanoparticles in heat transfer fluid termed as Hybrid nanofluid, which possesses the excellent thermo-physical properties. However, proper dispersion and stability are the most influencing parameters which effects the performance of hybrid heat transfer fluids.

4.2.1 Stability of the Hybrid Nanofluids

From the previous studies it has been noticed that stability of the nanoparticles mixture is one of the critical parameter which influences the performance of hybrid nanofluid [63]. Suresh et al. [64] explored the stability of the hybrid nanofluid (Al_2O_3-Cu) at varying nanoparticle concentration. Lack of stability of can degrade the thermal properties of the nano-composite fluids, which results in the poor performance in the cooling applications. Therefore, several researchers have developed different methods (such as, light scattering methods, Electron microscopy, Spectral analysis method, Zeta potential analysis, Sedimentation method and centrifugation method) analyse and maintain the uniformity in the hybrid suspension. Zeta potential the most commonly used method to assess the stability of the nanoparticles mixed suspension. It is measuring the potential difference between the nanoparticles and base fluid particles, which shows the degree of repulsion between the two charged particles in the colloidal suspensions. Zhu et al. [65] and Kim et al. [66] investigated the stability of the hybrid nanofluid (Au-Al_2O_3) using centrifugation method of zeta potential. To assess the stability of the nanofluid another method UV–V is spectrophotometer was used, which analyses the dynamics of the nanoparticles and base fluid particles. It is work on the basis of absorption capacity of the suspended and mixed nanoparticles [63]. Furthermore, there are two other general techniques such as light scattering techniques and particle size distribution by microscopy are available to measure the distribution of the nanoparticles in conventional fluid. However, to get the high resolution micrographs of the dispersion transmission electron microscopy and scanning electron microscopy were used.

4.2.2 Effect of pH on Nanofluids

The pH of the prepared fluid mixture have significant effect on the stability of nanofluid. Therefore, pH control can increase the stability of the nanofluid mixture due to strong repulsive forces. Wen et al. [67] noticed that even simple acid washing of the carbon nanotubes can increase the stability of the nanofluid samples. In another study Xie et al. [68] and Fovet et al. [68] studies different pH values for alumina, copper and graphite and noticed that modification in the pH values reduces the agglomeration in the nanofluid samples.

4.2.3 Role of Ultrasonic Intensity

It is very difficult to get the uniform dispersion of the nanofluid samples. Therefore, ultra-sonication plays a significant role by changing the morphological and dimensional properties of the nano materials. Chattopadhyay and Gupta [69] investigated the effect of the intensity of ultrasound on different sample of nanofluid. It has been found that increase in the ultrasound intensity increases the remarkable cavitation with in the particles, which increases the collapsed cavity within the nanofluid mixture and generates the shockwave that reduces the size of the nanoparticles and improve the stability of the nanofluids.

4.3 Enhancement in the Thermo-Physical Properties of the Hybrid Nanofluid

4.3.1 Enhancement in the Thermal Conductivity and Viscosity

In the published literature more than a few researchers have investigated the enhancement in the thermo-physical properties of the single and hybrid nanofluid. In the recent study Sundar et al. [70] studies the thermal conductivity and viscosity of the single graphene (GO) and hybrid (GO/Co_3O_4) hybrid nanofluid. In the experimental study it has been noticed that hybrid nanofluid has the higher thermal conductivity over single nanofluid due to synergistic effect of the composite nanoparticles. However, no enhancement was recorded in the viscosity of hybrid nanofluid over single (GO) of nanofluids. In another study Chopkar et al. [28] examined the thermal conductivities of the synthesized Al_2Cu and Ag_2Al hybrid nanofluids. An enhancement of 70 and 50% in the thermal conductivity of hybrid nanofluid were recorded at 1 vol% for water and ethylene glycol as a base fluid respectively as compared to base fluid. Furthermore, addition of thin sheet and cylindrical shaped nanoparticles in base fluid shows better enhancement in terms of thermal conductivity than spherical shaped mixed nanoparticles. An enhancement of 28 and 90% in the thermal conductivity was achieved with the synthesized nano-composite of copper oxide and graphene via reduction method at 0.05 vol% at 25 °C and 50 °C respectively [32]. Aravind and Ramaprabhu [33] investigated the performance of the deionize water based graphene and graphene-MWCNT hybrid nanofluids and it has been noticed an enhancement of 9.2 and 73% in thermal conductivity at 0.04 vol% at of 25 °C and

50 °C respectively. Furthermore, Chen et al. [32] synthesized the Iron oxide/multi-walled carbon nanotube (Fe_2O_3/MWCNTs) hybrid nanofluid and an enhancement of 27.7% in thermal conductivity was obtained at 0.02 wt% Fe_2O_3 and 0.05 wt% MWCNT. In another study aluminium oxide/copper (Al_2O_3/Cu) hybrid nanofluid shows directly proposal relationship between the thermal conductivity and nanoparticles concentration. Addition of Al_2O_3/Cu nano-composite in water improve the thermal conductivity as compared to alumina alone nanofluid.

4.3.2 Influence of the Nanoparticles Concentration and Temperature on Thermal Conductivity and Viscosity

In the published literature, there are number of researches have investigated the influence of temperature and volumetric fraction of nanoparticles on thermo-physical properties of hybrid nanofluids. In one of the study baby and Sundra [31] experimentally investigated the influence of temperature and volumetric fraction of nanoparticles of Ag/HEG mixed nano-composite. An enhancements of 25 and 86% in thermal conductivity was observed at 0.05% concentration and temperature of 25 °C and 70 °C respectively. Furthermore, Masoud et al. [46, 71] noticed an enhancement in the viscosity of ethylene glycol and water based Fe-CuO hybrid nanofluid with increase in the nanoparticles concentration, while reduce with increase in the temperature. However, the Newtonian behaviour has been observed at lower concentration of nanoparticles and non-Newtonian behaviour has been observed at higher concentrations. In another study, Sayed et al. [72] noticed increase in the temperature and volumetric fraction of nanoparticles increases the thermal conductivity of CuO/SWCNT ethylene glycol and water hybrid nanofluids.

4.4 Conclusions

The present work is mainly focused on the method of synthesis of the hybrid nanofluids including the effect of stability, ultrasonic intensity, pH value on the performance of different nanofluids. Moreover, the effect of different base fluids, temperature and nanoparticles concentration and their shape on thermo-physical properties was also discussed. Based on the available literature, few concluding points have been drawn:

- There are various techniques used to assess the stability of the nanoparticles suspension.
- Proper ultra-sonication and control of pH in the nanoparticles mixture improves the stability of the hybrid nanofluids.

- An enhancement of 9.2 and 73% in thermal properties of deionize water based graphene-MWCNT hybrid nanofluids noticed at 0.04 vol% and at of 25 °C and 50 °C respectively.
- Addition of sheet like and cylindrical like structure nanoparticles in base fluid exhibit superior performance over spherical like nanoparticles.
- The Newtonian behaviour has been observed at lower concentration of nanoparticles while non-Newtonian behaviour has been observed at higher concentrations.

References

1. N. Azwadi, C. Sidik, M. Noor, W. Mohd, R. Mamat, A review on the application of nano fluids in vehicle engine cooling system. Int Commun. Heat Mass Transf. **68**, 85–90 (2015)
2. S. Zainal, C. Tan, C.J. Sian, ANSYS simulation for Ag/HEG hybrid nanofluid in turbulent circular pipe. J. Adv. Res. Appl. Mech. **23**, 20–35 (2016)
3. S. Lee, S.U.S. Choi, Measuring thermal conductivity of fluids containing oxide nanoparticles. J. Heat Transf. Trans. ASME **121**, 280–289 (1999)
4. S. Khandekar, M.R. Sankar, V. Agnihotri, J. Ramkumar, Nano-cutting fluid for enhancement of metal cutting performance. Mater. Manuf. Process. **27**(9), 963–967 (2012)
5. M. Sayuti, O.M. Erh, A.A.D. Sarhan, M. Hamdi, Investigation on the morphology of the machined surface in end milling of aerospace AL6061-T6 for novel uses of SiO_2 nanolubrication system. J. Clean. Prod. **66**, 655–663 (2014)
6. J.S. Nam, P.-H. Lee, S.W. Lee, Experimental characterization of micro-drilling process using nanofluid minimum quantity lubrication. Int. J. Mach. Tools Manuf **51**(7–8), 649–652 (2011)
7. R.K. Singh, A.R. Dixit, A. Mandal, A.K. Sharma, Emerging application of nanoparticle-enriched cutting fluid in metal removal processes: a review. J. Brazilian Soc. Mech. Sci. Eng. **39**(11), 4677–4717 (2017)
8. S. Prabhu, B.K. Vinayagam, AFM investigation in grinding process with nanofluids using taguchi analysis. Int. J. Adv. Manuf. Technol. **60**(1), 149–160 (2012)
9. B. Rahmati, A.A.D. Sarhan, M. Sayuti, Investigating the optimum molybdenum disulfide (MoS_2) nanolubrication parameters in CNC milling of AL6061-T6 alloy. Int. J. Adv. Manuf. Technol. **70**(5–8), 1143–1155 (2014)
10. K.-H. Park, B. Ewald, P.Y. Kwon, Effect of Nano-enhanced lubricant in minimum quantity lubrication balling milling. J. Tribol. **133**(3), 031803 (2011)
11. E.O.L. Ettefaghi, A. Rashidi, H. Ahmadi, S.S. Mohtasebi, M. Pourkhalil, Thermal and rheological properties of oil-based nanofluids from different carbon nanostructures. Int. Commun. Heat Mass Transf. **48**, 178–182 (2013)
12. A. Nasiri, M. Shariaty-Niasar, A.M. Rashidi, R. Khodafarin, Effect of CNT structures on thermal conductivity and stability of nanofluid. Int. J. Heat Mass Transf. **55**(5–6), 1529–1535 (2012)
13. A.K. Tiwari, P. Ghosh, J. Sarkar, Performance comparison of the plate heat exchanger using different nanofluids. Exp. Therm. Fluid Sci. **49**, 141–151 (2013)
14. S.M.S. Murshed, K.C. Leong, C. Yang, Enhanced thermal conductivity of TiO_2—water based nanofluids. Int. J. Therm. Sci. **44**(4), 367–373 (2005)
15. R. Saidur, K.Y. Leong, H.A. Mohammad, A review on applications and challenges of nanofluids. Renew. Sustain. Energy Rev. **15**(3), 1646–1668 (2011)
16. X.Q. Wang, A.S. Mujumdar, Heat transfer characteristics of nanofluids: a review. Int. J. Therm. Sci. **46**(1), 1–19 (2007)
17. R.K. Singh, A.K. Sharma, A.R. Dixit, A. Mandal, A.K. Tiwari, Experimental investigation of thermal conductivity and specific heat of nanoparticles mixed cutting fluids. Mater. Today Proc. **4**(8), 8587–8596 (2017)

18. T. Chang, S. Syu, Y. Yang, International Journal of heat and mass transfer effects of particle volume fraction on spray heat transfer performance of Al_2O_3—water nanofluid. Int. J. Heat Mass Transf. **55**(4), 1014–1021 (2012)
19. J. Sarkar, A critical review on convective heat transfer correlations of nanofluids. Renew. Sustain. Energy Rev. **15**(6), 3271–3277 (2011)
20. A.K. Tiwari, P. Ghosh, J. Sarkar, Heat transfer and pressure drop characteristics of CeO_2/water nanofluid in plate heat exchanger. Appl. Therm. Eng. **57**(1–2), 24–32 (2013)
21. A.K. Tiwari, G. Pradyumna, S. Jahar, Investigation of thermal conductivity and viscosity of nanofluids. J. Environ. Res. Dev. **7**(2), 768–777 (2012)
22. R.S. Vajjha, D.K. Das, A review and analysis on influence of temperature and concentration of nanofluids on thermophysical properties, heat transfer and pumping power. Int. J. Heat Mass Transf. **55**(15–16), 4063–4078 (2012)
23. H. Chen, Y. Ding, Heat transfer and rheological behaviour of nanofluids—a review, in *Advances in Transport Phenomena: 2009*, ed. by L. Wang (Springer, Heidelberg, 2009), pp. 135–177
24. J. Sarkar, P. Ghosh, A. Adil, A review on hybrid nanofluids: recent research, development and applications. Renew. Sustain. Energy Rev. **43**, 164–177 (2015)
25. C. Sinz, H. Woei, M. Khalis, S.A. Abbas, Akademia baru numerical study on turbulent force convective heat transfer of hybrid nanofluid, Ag/HEG in a circular channel with constant heat flux. J. Adv. Res. Fluid Mech. Therm. Sci. **24**(1), 1–11 (2016)
26. S. Jana, A. Salehi-Khojin, W.H. Zhong, Enhancement of fluid thermal conductivity by the addition of single and hybrid nano-additives. Thermochim. Acta **462**(1–2), 45–55 (2007)
27. N. Jha, S. Ramaprabhu, Thermal conductivity studies of metal dispersed multiwalled carbon nanotubes in water and ethylene glycol based nanofluids Thermal conductivity studies of metal dispersed multiwalled carbon nanotubes in water and ethylene glycol based nanofluids. J. Appl. Phys. **106**, 084317 (2009)
28. M. Chopkar, S. Kumar, D.R. Bhandari, P.K. Das, I. Manna, Development and characterization of Al_2Cu and Ag_2Al nanoparticle dispersed water and ethylene glycol based nanofluid. Mater. Sci. Eng.: B **139**, 141–148 (2007)
29. L.F. Chen, M. Cheng, D.J. Yang, L. Yang, Enhanced thermal conductivity of nanofluid by synergistic effect of multi-walled carbon nanotubes and Fe_2O_3 nanoparticles. Appl. Mech. Mater. **548–549**, 118–123 (2014)
30. B. Munkhbayar, M.R. Tanshen, J. Jeoun, H. Chung, H. Jeong, Surfactant-free dispersion of silver nanoparticles into MWCNT-aqueous nanofluids prepared by one-step technique and their thermal characteristics. Ceram. Int. **39**(6), 6415–6425 (2013)
31. T.T. Baby, S. Ramaprabhu, Synthesis and nanofluid application of silver nanoparticles decorated graphene. J. Mater. Chem. **21**, 9702–9709 (2011)
32. T.T. Baby, R. Sundara, Synthesis and transport properties of metal oxide decorated graphene dispersed nanofluids. J. Phys. Chem. **115**(17), 8527–8533 (2011)
33. S.S.J. Aravind, S. Ramaprabhu, Graphene-multiwalled carbon nanotube-based nanofluids for improved heat dissipation. RSC Adv. **3**(13), 4199–4206 (2013)
34. R.K. Singh, A.R. Dixit, A.K. Sharma, A.K. Tiwari, V. Mandal, A. Pramanik, Influence of graphene and multi-walled carbon nanotube additives on tribological behaviour of lubricants. Int. J. Surf. Sci. Eng. **12**(3), 207–227 (2018)
35. S. Suresh, K.P. Venkitaraj, P. Selvakumar, M. Chandrasekar, Synthesis of Al_2O_3-Cu/water hybrid nanofluids using two step method and its thermo physical properties. Colloids Surf. A Physicochem. Eng. Asp. **388**(1–3), 41–48 (2011)
36. M. Batmunkh, M. Myekhlai, H. Choi, H. Chung, H. Jeong, Thermal conductivity of TiO_2 nanoparticles based aqueous nano fluids with an addition of a modified silver particle. Ind. Eng. Chem. Res. **53**, 8445–8451 (2014)
37. Z.H. Han, B. Yang, S.H. Kim, M.R. Zachariah, Application of hybrid sphere/carbon nanotube particles in nanofluids. Nanotechnology **18**(10), 105701 (2007)
38. S.S. Botha, P. Ndungu, B.J. Bladergroen, Physicochemical properties of oil-based nanofluids containing hybrid structures of silver nanoparticles supported on silica. Ind. Eng. Chem. Res. **50**(6), 3071–3077 (2011)

39. G. Paul, J. Philip, B. Raj, P.K. Das, I. Manna, Synthesis, characterization, and thermal property measurement of nano-$Al_{95}Zn_{05}$ dispersed nanofluid prepared by a two-step process. Int. J. Heat Mass Transf. **54**(15–16), 3783–3788 (2011)
40. S.M. Abbasi, A. Rashidi, A. Nemati, K. Arzani, The effect of functionalisation method on the stability and the thermal conductivity of nanofluid hybrids of carbon nanotubes/gamma alumina. Ceram. Int. **39**(4), 3885–3891 (2013)
41. M.J. Nine, M. Batmunkh, J.-H. Kim, H.-S. Chung, H.-M. Jeong, Investigation of Al_2O_3-MWCNTs hybrid dispersion in water and their thermal characterization. J. Nanosci. Nanotechnol. **12**(6), 4553–4559 (2012)
42. M. Afrand, Experimental study on thermal conductivity of ethylene glycol containing hybrid nano-additives and development of a new correlation. Appl. Therm. Eng. **110**, 1111–1119 (2017)
43. O. Soltani, M. Akbari, Effects of temperature and particles concentration on the dynamic viscosity of MgO-MWCNT/ethylene glycol hybrid nano fluid: Experimental study. Phys. E Low-dimensional Syst. Nanostruct. **84**, 564–570 (2016)
44. L.S. Sundar, E.V. Ramana, M.P.F. Graça, M.K. Singh, A.C.M. Sousa, Nanodiamond-Fe_3O nanofluids: preparation and measurement of viscosity, electrical and thermal conductivities. Int. Commun. Heat Mass Transf. **73**, 62–74 (2016)
45. H. Saeed Sarbolookzadeh, A. Karimipour, M. Afrand, M. Akbari, A.D. Orazio, An experimental study on thermal conductivity of F-MWCNTs—Fe_3O_4/EG hybrid nano fluid : Effects of temperature and concentration. Int. Commun. Heat Mass Transf. **76**, 171–177 (2016)
46. H. Eshgarf, M. Afrand, An experimental study on rheological behavior of non-newtonian hybrid nano-coolant for application in cooling and heating systems. Exp. Therm. Fluid Sci. **76**, 221–227 (2016)
47. M. Afrand, D. Toghraie, B. Ruhani, Effects of temperature and nanoparticles concentration on rheological behavior of Fe_3O_4–Ag/EG hybrid nanofluid: an experimental study. Exp. Therm. Fluid Sci. **77**, 38–44 (2016)
48. M. Hemmat Esfe, A. Akbar, A. Arani, M. Rezaie, W. Yan, A. Karimipour, Experimental determination of thermal conductivity and dynamic viscosity of Ag–MgO/water hybrid nano fluid. Int. Commun. Heat Mass Transf. **66**, 189–195 (2015)
49. M.H. Esfe, W. Yan, M. Akbari, A. Karimipour, M. Hassani, Experimental study on thermal conductivity of DWCNT-ZnO/water-EG nanofluids. Int. Commun. Heat Mass Transf. **68**, 248–251 (2015)
50. M. Hemmat Esfe et al., Thermal conductivity of Cu/TiO_2–water/EG hybrid nanofluid: experimental data and modeling using artificial neural network and correlation. Int. Commun. Heat Mass Transf. **66**, 100–104 (2015)
51. L.S. Sundar, A.C.M. Sousa, M.K. Singh, Heat transfer enhancement of low volume concentration of hybrid nanofluids in a tube with twisted tape inserts under turbulent flow. J Therm. Sci. Eng. Appl. **7**(2), 021015 (1–12) (2015)
52. L.S. Sundar, M.K. Singh, E.V. Ramana, B. Singh, J. Grácio, A.C.M. Sousa, Enhanced thermal conductivity and viscosity of nanodiamond-nickel nanocomposite nanofluids. Sci. Rep. **4**, 4039 (1–14) (2014)
53. M. Baghbanzadeh, A. Rashidi, A.H. Soleimanisalim, D. Rashtchian, Investigating the rheological properties of nanofluids of water/hybrid nanostructure of spherical silica/MWCNT. Thermochim. Acta **578**, 53–58 (2014)
54. B. Takabi, S. Salehi, Augmentation of the heat transfer performance of a sinusoidal corrugated enclosure by employing hybrid nanofluid. Adv. Mech. Eng. **6**, Article ID 147059 (1–16) (2014)
55. Y. Xuan, H. Duan, Q. Li, Enhancement of solar energy absorption using a plasmonic nanofluid based on TiO_2/Ag composite nanoparticles. RSC Adv. **4**(31), 6206–16213 (2014)
56. D. Madhesh, R. Parameshwaran, S. Kalaiselvam, Experimental investigation on convective heat transfer and rheological characteristics of Cu-TiO_2 hybrid nanofluids. Exp. Therm. Fluid Sci. **52**, 104–115 (2014)
57. R.K. Singh, A.K. Sharma, A.R. Dixit, A.K. Tiwari, A. Pramanik, A. Mandal, Performance evaluation of alumina-graphene hybrid nano-cutting fluid in hard turning. J. Clean. Prod. **162**, 830–845 (2017)

58. A.K. Sharma, A.K. Tiwari, A.R. Dixit, R.K. Singh, M. Singh, Novel uses of alumina/graphene hybrid nanoparticle additives for improved tribological properties of lubricant in turning operation. Tribol. Int. **119**(September 2017), 99–111 (2018)
59. A.K. Sharma, R.K. Singh, A.R. Dixit, A.K. Tiwari, Novel uses of alumina-MoS_2 hybrid nanoparticle enriched cutting fluid in hard turning of AISI 304 steel. J. Manuf. Process. **30**, 467–482 (2017)
60. A.K. Sharma, R.K. Singh, A.R. Dixit, An investigation on tool flank wear using alumina/MoS_2 hybrid nanofluid in turning operation, in *Advances in Manufacturing Engineering and Materials. Lecture Notes in Mechanical Engineering* (Springer, Berlin, 2019), pp. 213–219
61. A.K. Sharma, J.K. Katiyar, S. Bhaumik, S. Roy, Influence of alumina/MWCNT hybrid nanoparticle additives on tribological properties of lubricants in turning operations. Friction **7**(2), 153–168 (2019)
62. A.K. Sharma, A.K. Tiwari, A.R. Dixit, Prediction of temperature distribution over cutting tool with alumina-MWCNT hybrid nanofluid using computational fluid dynamics (CFD) analysis. Int. J. Adv. Manuf. Technol. **97**(1–4), 427–439 (2018)
63. Y. Hwang et al., Stability and thermal conductivity characteristics of nanofluids. Thermochimica Acta **455**, 70–74 (2007)
64. S. Suresh, K.P. Venkitaraj, P. Selvakumar, Synthesis, characterisation of Al_2O_3-Cu nano composite powder and water based nanofluids. Adv. Mater. Res. **328–330**, 1560–1567 (2011)
65. H. Zhu, Y. Lin, Y. Yin, A novel one-step chemical method for preparation of copper nanofluids. J. Colloid Interface Sci. **277**, 100–103 (2004)
66. H. Jin, I. Cheol, J. Onoe, Characteristic stability of bare Au-water nanofluids fabricated by pulsed laser ablation in liquids. Opt. Lasers in Eng. **47**, 532–538 (2009)
67. D. Wen, G. Lin, S. Vafaei, K. Zhang, Review of nanofluids for heat transfer applications. Particuology **7**(2), 141–150 (2009)
68. H. Xie et al., Nanofluids containing multiwalled carbon nanotubes and their enhanced thermal conductivities nanofluids containing multiwalled carbon nanotubes and their enhanced thermal conductivities. **94**(8), 4967–4971 (2003)
69. P. Chattopadhyay, R.B. Gupta, Production of griseofulvin nanoparticles using supercritical CO_2 antisolvent with enhanced mass transfer. Int. J. Pharm. **228**, 19–31 (2001)
70. L.S. Sundar, M.K. Singh, M.C. Ferro, A.C.M. Sousa, Experimental investigation of the thermal transport properties of graphene oxide/CO_3O_4 hybrid nano fluids. Int. Commun. Heat Mass Transf. **84**, 1–10 (2017)
71. M. Bahrami, M. Akbari, A. Karimipour, M. Afrand, An experimental study on rheological behavior of hybrid nanofluids made of iron and copper oxide in a binary mixture of water and ethylene glycol: non-newtonian behavior. Exp. Therm. Fluid Sci. **79**, 231–237 (2016)
72. S.H. Rostamian, M. Biglari, S. Saedodin, M. Hemmat Esfe, An inspection of thermal conductivity of CuO-SWCNTs hybrid nanofluid versus temperature and concentration using experimental data, ANN modeling and new correlation. J. Mol. Liq. **231**, 364–369 (2017)

Chapter 5
Energy Efficient Graphene Based Nano-composite Grease

Jayant Singh, Deepak Bhardwaj, and Jitendra Kumar Katiyar

Abstract Performance of grease lubricated point contact under elasto-hydrodynamics lubrication (EHL) regime is complex and critical in many practical and engineering applications. This article deals with the detailed rheology, film formation and elastic recovery characteristics of the developed nano-composite grease. The composite grease is prepared by dispersing the reduced graphene oxide nano-sheets to bare grease. The detailed microstructural morphology of the nano-additives and different grease samples is evaluated using high resolution transmission electron microscopy (HRTEM). The rotational rheometer under oscillatory mode is used to evaluate the storage (G') and loss moduli (G"). The film formation is recorded using elasto-hydrodynamic rig for range of speed at constant load and different temperatures. The results shows that modification in microstructure due to dispersion of reduced graphene oxide (rGO) nano-sheets, improves the rheological and tribological behavior. Based on film formation and recovery or reflow characteristics, the rGO dispersed composite grease is efficient. The rGO dispersed composite grease registers 92% of elastic recovery.

5.1 Introduction

Around 85% of the grease lubricated rolling element bearings are designed to operate under fully flooded conditions [1]. As detected by microscopic techniques [2], the grease has been classified as colloidal suspension of metal soap dispersed in liquid lubricant. The entangled fibers of the thickener determines the rheological nature of the lubricating greases. The thickener is basically utilized to prevent the loss of

J. Singh (✉) · D. Bhardwaj
Department of Mechanical and Automation Engineering, Dr. Akhilesh Das Gupta Institute of Technology and Management, New Delhi 110053, India
e-mail: singhjayant11@gmail.com

J. K. Katiyar
Department of Mechanical Engineering, SRM Institute of Science and Technology, Kattankulathur Campus, Chennai, Tamil Nadu 603203, India

J. K. Katiyar et al. (eds.), *Tribology in Materials and Applications*,
Materials Forming, Machining and Tribology,
https://doi.org/10.1007/978-3-030-47451-5_5

lubricant during operation. This allows the greases to share the benefits of both solid and liquid lubricants. However, it is incapable of yielding the significant resistance to the motion of the mating surfaces. Along with author's study [3], there are wide variety of flow models describing the complex non-Newtonian behavior of the lubricating greases [2, 4]. Many models have not considered the visco-elastic nature of the greases. Which, nowadays became the area of research and open new avenues for characterizing the rheology of greases. The grease lubricating mechanism is commonly studied on ball on disc geometry. Although, there lies the difference between actual conditions, yet the ball on disc pair is efficient in simulating the operating conditions (slide roll ratios (SRR), operating contact pressures, temperatures etc.). The coated transparent disc utilizing the interferometry technique can efficiently measure the film thickness both in starved and fully flooded conditions [5–7].

The flow of lubricant inside the elastohydrodynamic contact is very perplexing and challenging. The flow exists for very short interval of $\approx$1 ms under the strain rate of $\approx 10^{-7}$ s^{-1} [8]. This results into frequent variation of operating pressures and temperatures and hence influence the lubricant properties. Thus, it is very challenging to study the chemical, physical and rheological properties of the lubricant inside the contact.

The soap in greases are responsible for avoiding the contaminants, sealing effect and water to enter the contact zone, without affecting the lubricating characteristics provided by the base oil. The previous studies [2] suggest that greases consist of fibers matrix in which soap forms the entangled fibers network with oil suspended in it. Hence the multi-phase system of greases allows them to share both the advantages of lubricating oil and solid lubricant. This makes the greases, the attractive lubricants and to explore their rheological behavior. The rheological characteristics of the greases could better be explored by analyzing their viscoelastic nature. The fluidity of greases are controlled by their viscosity and elastic nature controls their leakage from the contact zone. The previous outcomes [9–11] also supports that the thickener plays an significant role in grease lubrication.

Finding by Cann [11] and Kanazawa et al. [10] recommend that the film formation is controlled by type of the thickener content and free of base oil viscosity. Thus, while entering the contact, the side leakage [12], the shear degradation of soap content [13] etc. attributes to the change in rheological behavior and finally modifies the film formation between the contact. Also, the study by Couronne et al. [2] suggest that grease's elastic moduli governs the film formation between the contact. According to their findings all the experiments were done at 20 °C for short interval of time. While, ignoring the effect of high temperatures.

With the approach of nano-innovations, nano-additives are utilized to upgrade the lubricating characteristics of greases. The bulk properties and grease structure gets modified by the addition of nano-additives [14]. The lamellar structure of graphene with low shear strength between layers allows graphene to be most suitable for Tribological applications. Also, graphene exhibits few novel physical properties like—zero effectice mass, low shear strength, high thermal conductivity etc. [15–19].

The present work aims to establish relation between the rheological behaviors of greases with film forming ability of the greases. The different experimental conditions are planned like varying rolling speed, temperatures, reflow or recovery etc. to give concluding remarks on the lubricating ability of the developed nano-composite grease.

5.2 Experimental Details

5.2.1 *Materials and Methods*

Commercial bare grease (BG) with no additive content, comprising of lithium 12-hydroxy stearate as soap and mineral oil are utilized for the formulation of composite grease. The rGO nano-sheets with interlayer spacing of ~0.35 nm and average size of around 500 nm were received from Indian Institute of Petroleum, Dehradun, India. The methodology for preparation of rGO nano-sheets is as mentioned by Chaudhary et al. [20]. The chromium coated glass disc with Ra $\approx$ 0.015 μm and steel ball with Ra $\approx$ 0.025 μm are used for studying film thickness behavior.

5.2.2 *Development of Nano-composite Grease*

Prior to dispersion of rGO nano-sheets to BG, they are sonicated in toluene, to confirm the minimal clustering. The drop-wise addition of sonicated liquid is followed with continuous mechanical stirring of BG (maintained at 110 °C). Thereafter, the mixture is then passed through triple-roll-mill for five times. The complete study for developing the rGO doped nano-composite grease is elaborated in the author's previous articles [3, 21].

5.2.3 *Characterization*

The morphological and microstructural characterization of rGO nano-sheets, BG and rGO doped nano-composite grease was depicted using HRTEM (high resolution transmission electron microscopy) (Tecnai G^2 F 20 TWIN, Tokyo, Japan). The size of rGO nano-sheets in toluene was estimated through dynamic light scattering (DLS) (Micromatrics, Norcross, GA). The flow characteristics of BG and rGO doped composite grease was measured on HAAKE Viscotester-550 for strain rate of 10 s^{-1}. All the other physical parameters like Grease consistency through Penetrometer apparatus, drop point through drop point apparatus, as per the ASTM-D566 and refractive index through refractometer are listed in Table 5.1.

Table 5.1 The various physical characteristics of different grease samples

Grease samples	τ_y (Pa)	$k(Pa\ s)^n$	n	Correlation coefficient R	Unworked penetration (10^{-1} mm)	Refractive index at 25 °C	Drop point
BG	251	189	0.66	0.915	294	1.436	210
BG + 0.1% rGO	564.2	261	0.57	0.927	281	1.441	223
BG + 0.2% rGO	630.8	257	0.32	0.939	277	1.440	223
BG + 0.4% rGO	926.1	336	0.25	0.956	275	1.443	225
BG + 0.6% rGO	901.6	328	0.43	0.932	280	1.446	225

5.2.4 *Rheological Measurements*

Rheological characteristics of the grease samples were measured using Anton Paar rheometer (MCR 102 model) with parallel plates of 25 mm diameter and gap of 0.5 mm. The following two conditions were planned to estimate the viscoelastic behavior under oscillatory mode: (a) constant frequency of 1 s^{-1} and by varying shear stresses of 1–500 Pa, and (b) varying frequency from 10^{-1} to 10^2 s^{-1} under constant strain rate of 0.1%. Further, to detect the recovery behavior of greases the complex modulus (G*) is measured under the varying stress plan of 10–1000–10 Pa.

5.2.5 *Film Thickness Measurements*

Film thickness of the grease samples was measured on EHD rig. The rolling point contact is formed between steel ball and glass disc coated with Cr and SiO_2 spacer layer. The grease lubricated rolling point contact is shown in Fig. 5.1. Optical interferometry technique is utilized for the contact film measurement. Under the constant load of 20 N (Hertzian pressure, $p_H = 0.5$ GPa), the contact operates for pure rolling condition (SRR = 0%) with the varying speed (0.001–4 m/s) and at different temperatures (30, 70 and 130 °C), respectively. The film thickness was evaluated for (i) *initial stage* (the film thickness data was recorded from the beginning of the experimentation) and, (ii) *matured stage* (the film thickness data was recorded after 1 h of extended running). Before the start of every experiment, the thin layer of ~1 mm was uniformly distributed over the glass disc. As most of the greases is thrown out

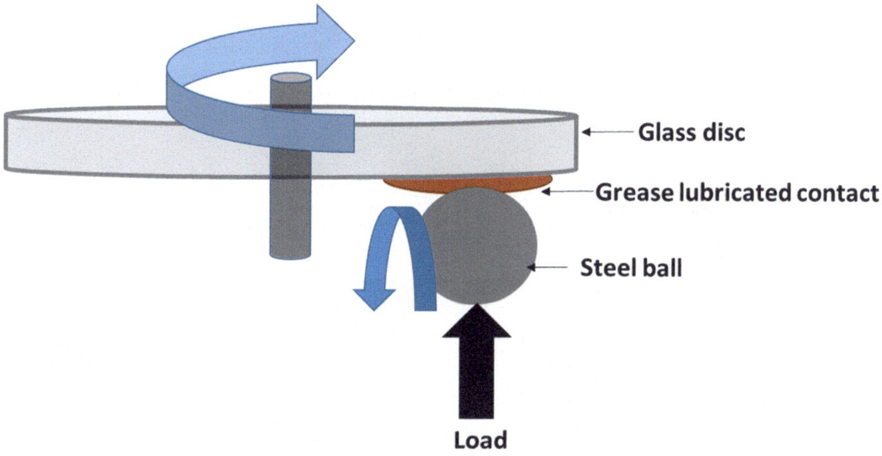

Fig. 5.1 Ball on disc configuration with grease lubricated contact

of the contact due to large centrifugal forces and no other source present for resupplying the grease to the contact, eventually makes the contact starved. Further, under starved condition and to better analyze the reflow characteristics the film thickness measurements are scheduled for 2 h. The 4 slot were planned for the test. During, the first three slots (1–3) 20 min running was involved with 5 min successive halts to estimate if any reflow occurs. At last, in the fourth slot the starved rolling contact operates for 1 h to assess the film thickness lubricated with BG and 0.4% (w/w) rGO doped nano-composite grease.

5.3 Results

5.3.1 Physical Characteristics

The evaluation of yielding stress (τ_y), shear thinning index (n) and consistency factor (k) show significant variation with respect to different greases. On comparing to BG, rGO doped greases possess higher τ_y and k, while registers lower values for shear thinning index (n). BG registers higher value of cone penetration depth as compared to rGO doped greases. The changes in the values of refractive index and drop point are marginal. The parameters are listed in Table 5.1. The possible explanation for this different values of parameters is structural modification of greases, as discussed in the previous article [3].

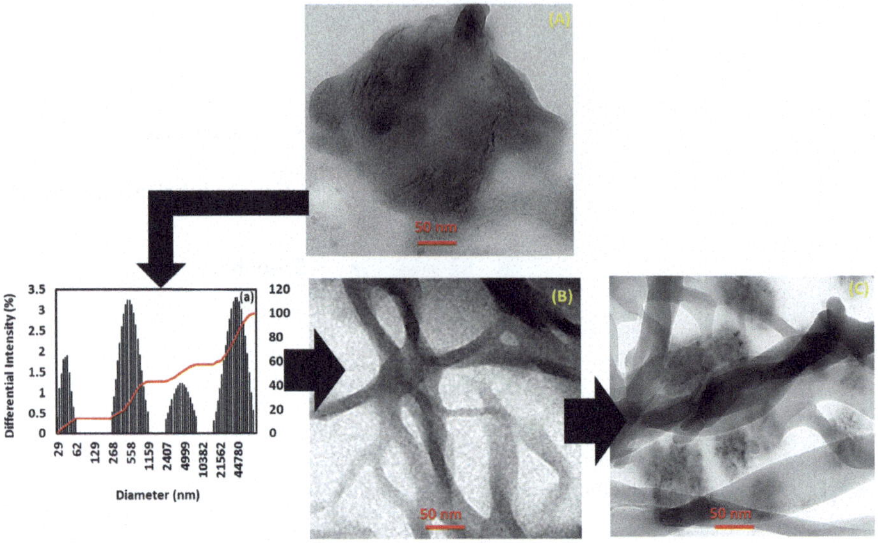

Fig. 5.2 HRTEM micrographs of **A** rGO nano-sheets with their DLS spectrum (a), **B** fibrous structure of BG and **C** rGO doped BG after removing the oil content

5.3.2 Microstructural Characteristics

The rGO nano-sheets with its microstructural features are shown in Fig. 5.2. Figure 5.2A shows the rippled, folded regions and layered arrangement of rGO sheets. After sonicating them in toluene, the solution is evaluated through DLS technique. Along with nano-sized sheets of rGO, some micro sized are also present in the solution (Fig. 5.2A). The mixture is then dispersed in BG (Fig. 5.2B). After dispersion it seems that rGO nano-sheets get adhered by wrapping themselves to the fibers of BG and also modifies the fibers by thickening them (Fig. 5.2C).

5.3.3 Rheological Characteristics

The variation of frequency with the strain rate (0.1%) under linear viscoelastic (LVE) region are represented in Fig. 5.3. At lower frequencies, G' and G" are almost parallel to each other. While, at higher frequencies they start converging. Also, 0.4% (w/w) is the optimal concentration of rGO nano-sheets in BG. Figure 5.4 represents shear stress sweeps for G' and G" under oscillatory mode in LVE region. Initially, at lower shear stress, G" is lower than G'. There exists the cross-over point at which G" overlaps G'. Loss modulus (G") is responsible for flow behavior. The critical shear stress (τ_{co}) corresponds to the cross-over point. The cross-over point for rGO nano-sheets (0.4% w/w) doped grease lies above BG.

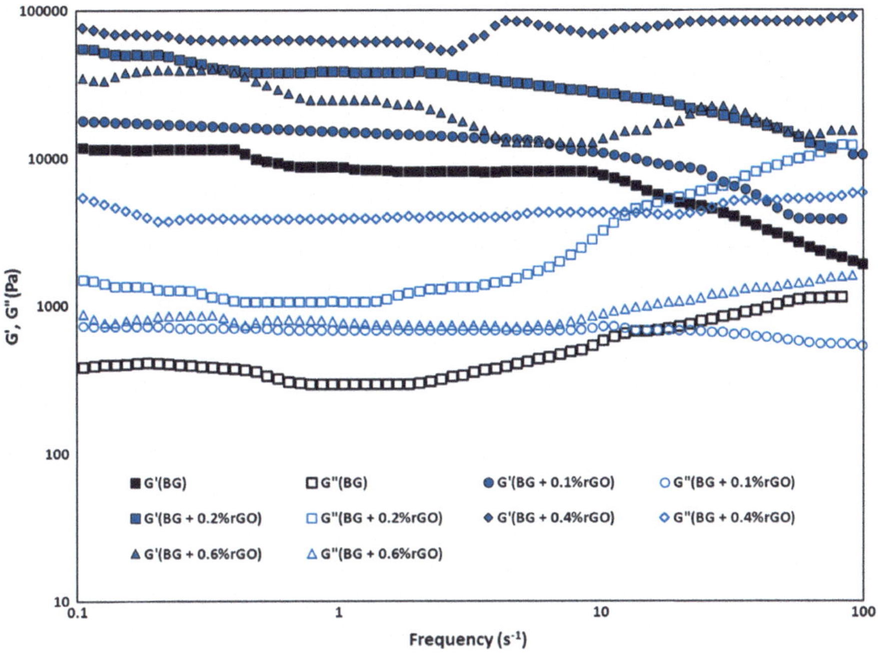

Fig. 5.3 Storage (G') and loss moduli (G") variation with frequency for BG and rGO doped BG

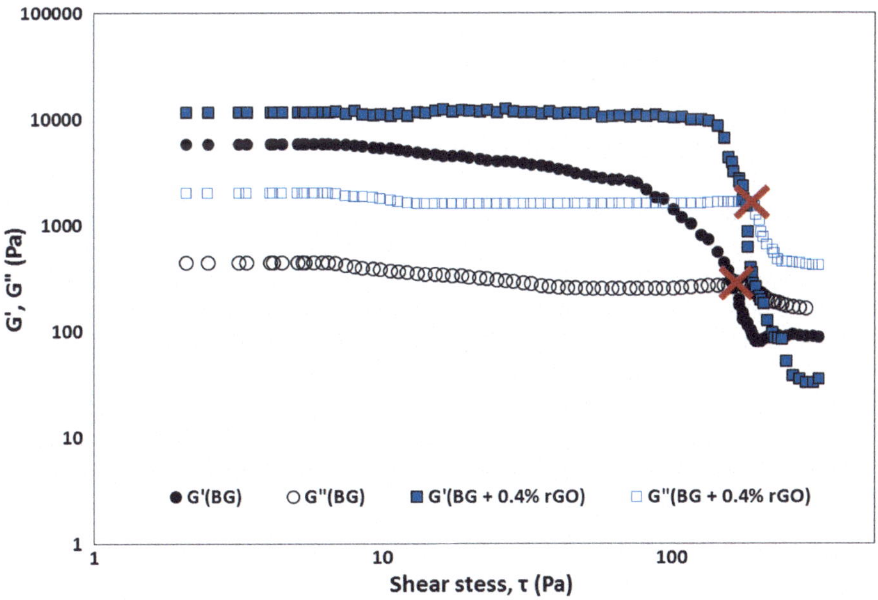

Fig. 5.4 Shear stress sweeps of storage and loss moduli under LVE region of BG and 0.4% (wt./wt.) rGO nano-sheets doped BG. ✕ represents the cross-over point

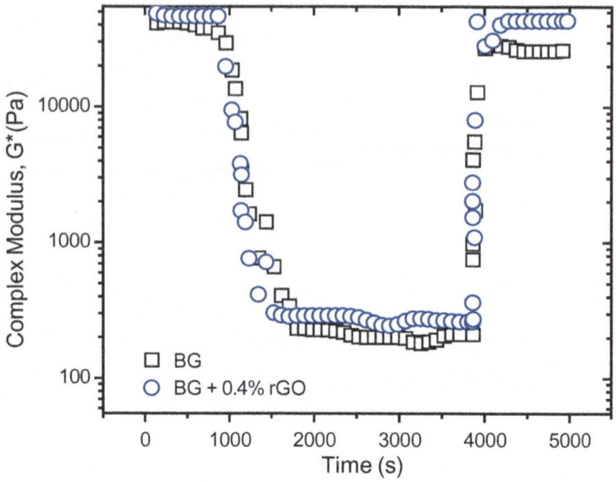

Fig. 5.5 Complex modulus (G*) decay of BG and 0.4% (wt./wt.) rGO nano-sheets doped BG under varying shear stress plan (10–1000–10 Pa) indicating the recovery of greases

Figure 5.5 shows the variation of complex modulus (G*) with respect to time. The greases undergo varying stress plans: 10 Pa under viscoelastic region (LVE) to 1000 Pa under non-viscoelastic region (NVE) and again former stress of 10 Pa is applied under viscoelastic region (LVE). The level of recovery is analyzed for both the greases. The rGO nano-sheets (0.4% w/w) doped grease registers the recovery of ≈92%, while the BG recovers ≈67%.

5.3.4 *Film Formation*

5.3.4.1 Effect of Speed and Temperature

The film forming ability at the rolling contact of different grease samples is recorded for varying speed (0.001–4 m/s) and temperatures (30, 70 and 130 °C), at both the stages—primary stage and matured stage. From Fig. 5.6a, c, e, it may be visualized that during primary stage, the film thickness increases with rise in temperatures (30, 70 and 130 °C), respectively. The rGO (0.4% wt./wt.) doped grease registers the higher film thickness, compared to BG. While, from Fig. 5.6b, d, f (representing the matured stages) it can be depicted that film thickness falls steeply with increase in temperatures and rolling speed. The rGO dispersed nano-grease shows the higher and better film forming ability in all cases. Further, the shift of the lubrication regime to the lower speed range at 130 °C can be visualized.

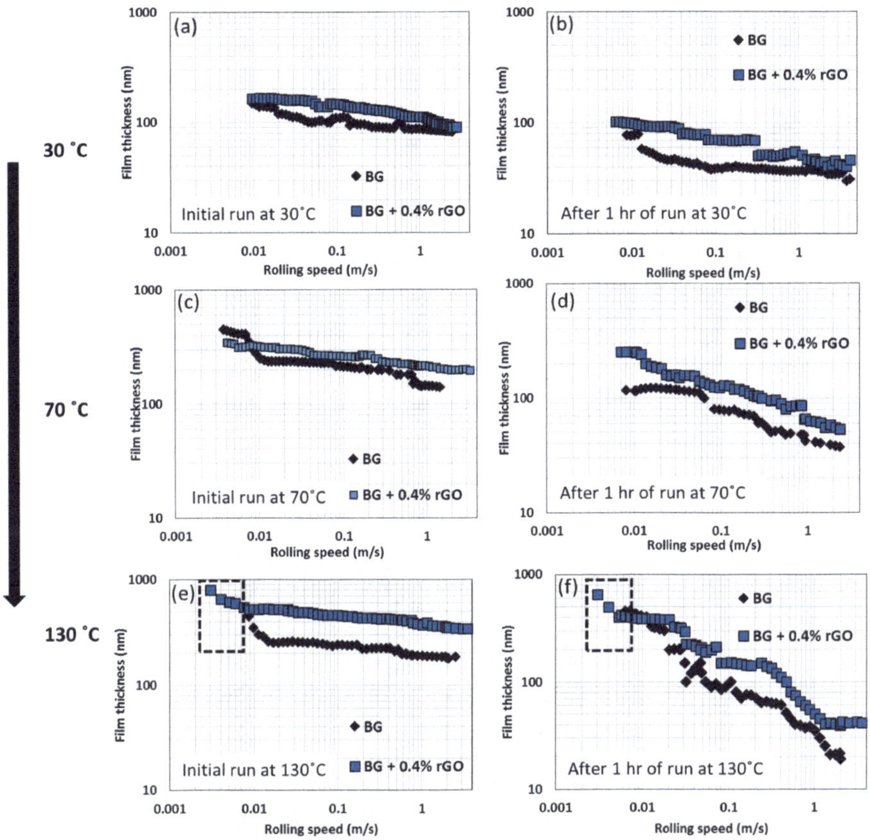

Fig. 5.6 Variation of film thickness of BG and 0.4% (wt./wt.) of rGO (optimized concentration) in BG with varying rolling speed (0.001–4 m/s) at different temperatures 30, 70 and 130 °C, respectively. The shifting of lubrication regime is shown by black dotted rectangle to the lower speed range

5.3.4.2 Film Thickness Recovery Under Starved Conditions

The reduction of film thickness over time attributes to the starvation of grease lubricated contact [22]. It is also well known that the grease lubricated contact often replenishes with its ability to release oil [23]. The reflow characteristic of BG and the developed nano-composite grease was analyzed by evaluating the film thickness into 4 different slots. Slots 1–3 involve 20 min running with subsequent 5 min halts, to evaluate the reflow behavior. While, slot 4 involve 60 min continuous running to evaluate the starvation effects. From Fig. 5.7 it is depicted that rGO nano-sheets doped grease registers the maximum film thickness and recovery after 5 min halts. Also, the greases show the continuous decay of film thickness in slot 4 with respect to time, with rGO doped grease registering the higher film thickness.

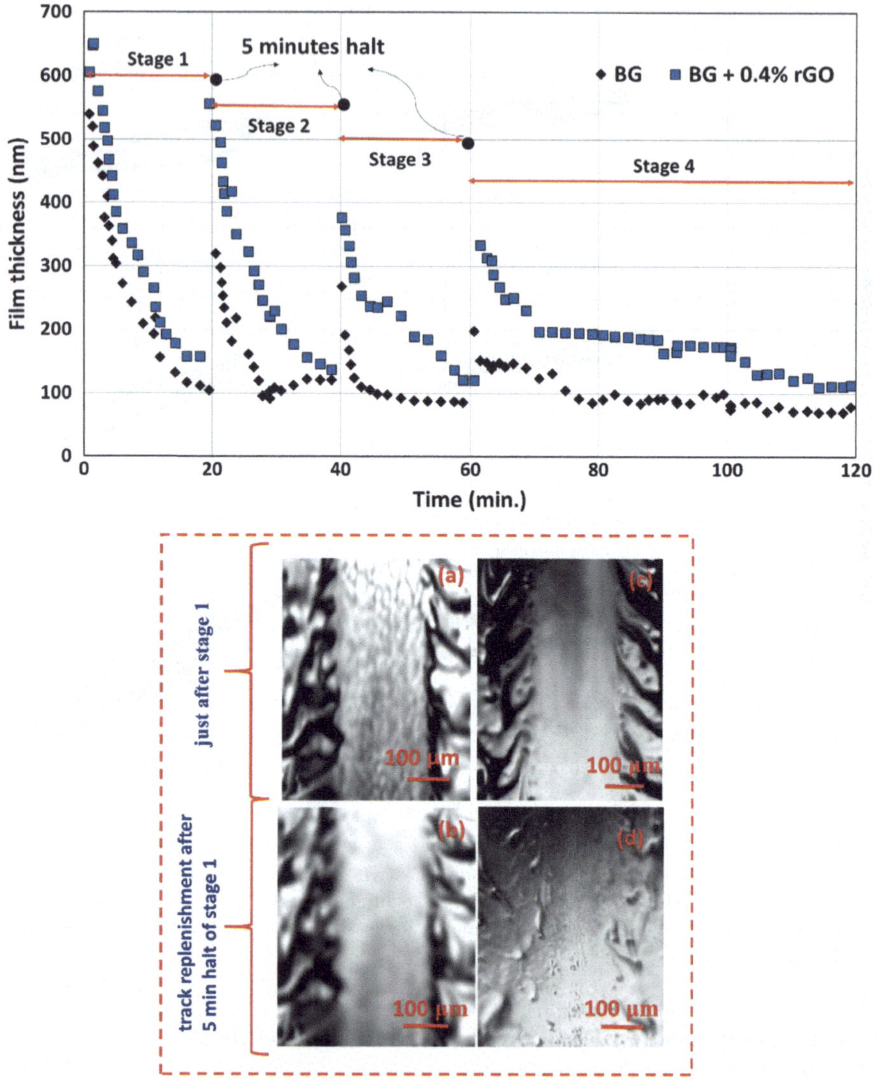

Fig. 5.7 Variation in film thickness for 2 h under rolling speed of 1 m/s at 30 °C. Slots 1–3 shows 1 h running with 20 min for each slot. The 5 min successive halt is provided after the completion of each slot (shown by black circular dot), after which the film thickness measurement is again started. Slot 4 represent the continuous running for 1 h. The red dotted rectangle represents the rolled track for (**a**, **b**) BG and (**c**, **d**) 0.4% (wt./wt.) rGO nano-sheets doped grease after completion of slot 1 and grease recovery on track after 5 min halt of slot 1

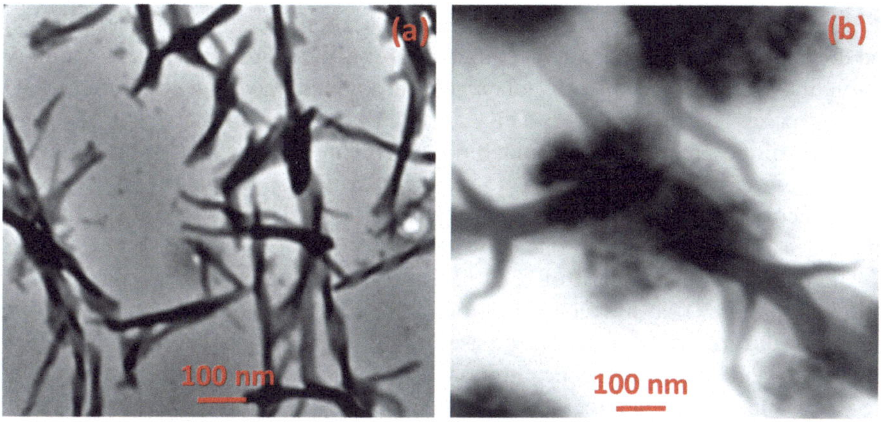

Fig. 5.8 HRTEM microstructures of used samples (a) BG and 0.4% rGO doped grease after 1 h of matured run at 130 °C

5.3.5 Microstructural Characteristics of Used Grease Samples

The extent of grease fibers degradation of used greases can be estimated through their appearance and morphology of soap fibers after the experimentation. The grease samples are collected after 1 h of extended running at 130 °C (representing the *matured stage at 130* °C). Figure 5.8 depicts the HRTEM micrographs of the utilized grease samples. On comparing with Fig. 5.1b it is depicted that there is minor damage to the bare grease structure (thin, broken and discontinuous fibers). Further, comparing Fig. 5.8b with Fig. 5.2C, it is clear that there is not much damage to the rGO doped grease. Also, rGO doped grease retains the thickened fibers after use, with rGO nano-sheets wrapped around the fibers.

5.4 Discussion

The film formation and rheology are the two determining factors for the successful operation of grease lubricated EHL point contact. Both the factors are dependent on grease's microstructure. The additives play intriguing role in modifying the soap structure [24]. The film thickness increases with the rise in temperature (30–130 °C) during the primary stage (Fig. 5.6a, c, e) can be related to enhanced movement of the fibers. Similar effects were also studied by Vengudusamy et al. [25]. The ability of rGO nano-sheets (0.4% wt./wt.) to form thicker films (Fig. 5.6) lies in their microstructural improvement of the thicker fibers with improved rheological response (G' and G") compared to BG (Figs. 5.3 and 5.4).

The determining factors for grease's recovery after operation (Fig. 5.7) are oil and thickener interaction and oil bleeding rates [26]. With the help of micro capillary action, the local replenishment is taken care through the modified structure of rGO doped grease. The film recovery (Fig. 5.7) can be related to the elastic recovery (Fig. 5.5) of the grease samples undergoing different stress plans. The rGO doped composite grease registers the elastic recovery of $\approx 92\%$, while BG registers the elastic recovery of $\approx 67\%$. This suggests that rGO doped grease is more stable and reports least damage to the fibrous network (Fig. 5.8b).

The optimal concentration of rGO nano-sheets can be understood through the following points: (i) at low concentration, there is scarcity of rGO nano-sheets in modifying the grease structure. The grease structure lacks proper interaction between oil and thickener due to low concentration of nano-sheets. (ii) at high concentration, the rGO nano-sheets agglomerates and find difficulty in entering the contact zone. The optimality in concentration of nano-additives is also explained by many authors [3, 27]. Thus, there exists the optimal concentration of rGO nano-sheets for bare grease micro-structure.

5.5 Conclusion

The dispersion of micro and nano additives modifies the grease structure. This modification leads to change in rheological characteristics. In the present work doping of rGO nano-sheets to the bare grease improves the thickener quality. The rGO nano-sheets get wrapped around the thickener fibers and thickens them. Thus, rGO nano-sheets doped grease registers the higher film thickness which could sustain severe operating conditions like high rolling speed and temperatures. The highest elastic recovery of $\approx 92\%$ is also registered by rGO doped grease, along with the local replenishment of the contact. It is also clear from the present work that film thickness of the greases falls with the increase in rolling speeds.

Acknowledgements The author is thankful to Indian Institute of Technology Delhi for providing the experimental facilities.

References

1. P.M. Lugt, A review on grease lubrication in rolling bearings. Tribol. Trans. **52**, 470–480 (2009)
2. I. Couronne, P. Vergne, D. Mazuyer, N. Truong-Dinh, D. Girodin, Effects of grease composition and structure on film thickness in rolling contact. Tribol. Trans. **46**, 31–36 (2003)
3. J. Singh, D. Kumar, N. Tandon, Development of nanocomposite grease: microstructure, flow, and tribological studies. J. Tribol. **139**, 1–9 (2017)
4. M. Yonggang, Z. Jie, Rheological model for lithium lubricating grease. Tribol. Int. **31**, 619–625 (1998)

5. T. Cousseau, M. Bjorling, B. Graca, A. Campos, J. Seabra, R. Larsson, Film thickness in a ball-on-disc contact lubricated with greases, bleed oils and base oils. Tribol. Int. **53**, 53–60 (2012)
6. P.M. Lugt, Modern advancements in lubricating grease technology. Tribol. Int. **97**, 467–477 (2016)
7. P.M. Cann, H.A. Spikes, J. Hutchinson, The development of a spacer layer imaging method (SLIM) for mapping elastohydrodynamic contacts. Tribol. Trans. **39**, 915–921 (1996)
8. D. Gonçalves, R. Marques, B. Graça, A.V. Campos, J.H.O. Seabra, J. Leckner, Westbroek R (2015) Formulation, rheology and thermal aging of polymer greases—Part II: influence of the co-thickener content. Tribol. Int. **87**, 171–177 (2015)
9. M. Kaneta, T. Ogata, Y. Takubo, M. Naka, Effects of a thickener structure on grease elastohydrodynamic lubrication films. Proc. Inst. Mech. Eng. Part J J. Eng. Tribol. **214**, 327–336 (2000)
10. Y. Kanazawa, R.S. Sayles, A. Kadiric, Film formation and friction in grease lubricated rolling-sliding non-conformal contacts. Tribol. Int. **109**, 505–518 (2017)
11. P.M. Cann, Grease lubrication of rolling element bearings—role of the grease thickener. Lubr. Sci. **19**, 183–196 (2007)
12. V. Wikström, B. Jacobson, Loss of lubricant from oil-lubricated near-starved spherical roller bearings. Proc. Inst. Mech. Eng. Part J J. Eng. Tribol. **211**, 51–66 (1997)
13. H. Cen, P.M. Lugt, G. Morales-Espejel, Film thickness of mechanically worked lubricating grease at very low speeds. Tribol. Trans. **57**, 1066–1071 (2014)
14. H.B. Silver, I.R. Stanley, The effect of the thickener on the efficiency of load-carrying additives in greases. Tribol. Int. **7**, 113–118 (1974)
15. H.P. Mungse, O.P. Khatri, Chemically functionalized reduced graphene oxide as a novel material for reduction of friction and wear. J. Phys. Chem. C **118**, 14394–14402 (2014)
16. C. Lee, Q. Li, W. Kalb, X.-Z. Liu, H. Berger, R.W. Carpick, J. Hone, Frictional characteristics of atomically thin sheets. Science **328**, 76–80 (2010)
17. S. Choudhary, H.P. Mungse, O.P. Khatri, Dispersion of alkylated graphene in organic solvents and its potential for lubrication applications. J. Mater. Chem. **22**, 21032–21039 (2012)
18. D. Berman, S.A. Deshmukh, S.K.R.S. Sankaranarayanan, A. Erdemir, A.V. Sumant, Extraordinary macroscale wear resistance of one atom thick graphene layer. Adv. Func. Mater. **24**, 6640–6646 (2014)
19. M. Nasrollahzadeh, F. Babaei, P. Fakhri, B. Jaleh, Synthesis, characterization, structural, optical properties and catalytic activity of reduced graphene oxide/copper nanocomposites. Rsc Adv. **5**, 10782–10789 (2015)
20. S. Choudhary, H.P. Mungse, O.P. Khatri, Hydrothermal deoxygenation of graphene oxide: chemical and structural evolution. Chem. Asian J. **8**, 2070–2078 (2013)
21. J. Singh, D. Kumar, N. Tandon, Tribological and vibration studies on newly developed nano-composite greases under boundary lubrication regime. J. Tribol. **140**, 32001–32010 (2017)
22. P.M.E. Cann, B. Damiens, A.A. Lubrecht, The transition between fully flooded and starved regimes in EHL. Tribol. Int. **37**, 859–864 (2004)
23. J.S. Mérieux, S. Hurley, A.A. Lubrecht, P.M. Cann, Shear-degradation of grease and base oil availability in starved EHL lubrication. Tribol. Ser. **38**, 581–588 (2000)
24. I. Couronné, P. Vergne, L. Ponsonnet, N. Truong-Dinh, D. Girodin, Influence of grease composition on its structure and its rheological behaviour. Tribol. Ser. **38**, 425–432 (2000)
25. B. Vengudusamy, M. Kuhn, M. Rankl, R. Spallek, Film forming behavior of greases under starved and fully flooded EHL conditions. Tribol. Trans. **59**, 62–71 (2016)
26. A.E. Baker, Grease bleeding—a factor in ball bearing performance. NLGI Spokesman **22**, 271–277 (1958)
27. M. Kumar, J. Bijwe, S.S.V. Ramakumar, Tribology international nano-PTFE: New entrant as a very promising EP additive. Tribiol. Int. **87**, 121–131 (2015)

Chapter 6
Synthesis of Magneto Rheological Fluids Using Nickel Particles and Study on Their Rheological Behaviour

Vikram G. Kamble, H. S. Panda, Shreedhar Kolekar, and T. Jagadeesha

Abstract Magnetorheological fluids are the best options for the many of the problems related to engineering to medical applications. This is the reason why many researchers are concentrated to make their study on these fluids. Fundamental study on these fluids had been considered as the most significant aspect of this field. Till now many researchers made their study on magneto rheological fluid synthesized by using iron (Fe) particles as their magnetic particles. Now in our study we have made an attempt to replace the iron particles by Nickel (2–5 μ) one, silicone oil is used as a carrier oil and oleic acid as surfactant. Prepared MR sample was tested under rheometer for the rheological properties thixotropic and regression analysis is made for different samples.

Acronyms

MR	Magnetorheological
St	Straight

V. G. Kamble (✉)
Leibniz Institute for Polymer Research, 01069 Dresden, Germany
e-mail: iknowvikramkamble@gmail.com

H. S. Panda
Ballistic Center, Proof & Experimental Establishment (PXE) Defence R & D Organization, Balasore, Odisha 756025, India

S. Kolekar
Department of Mechanical Engineering, SCOEM Satara, Satara, Maharashtra 415015, India

T. Jagadeesha
Department of Mechanical Engineering, National Institute of Technology, Calicut, Kerala 673601, India

J. K. Katiyar et al. (eds.), *Tribology in Materials and Applications*, Materials Forming, Machining and Tribology,
https://doi.org/10.1007/978-3-030-47451-5_6

Nomenclature

τ	Fluid stress
τ_y	Yield stress
H	Magnetic field
γ	Shear rate
η	Viscosity
G	Complex modulus
U	Velocity
N_1	σ_x–σ_y (primary stress difference)
N_2	σ_z–σ_y (secondary stress difference)
Ψ_1 and Ψ_2	Stress difference coefficients

6.1 Introduction

Magnetorheological [MR] fluids are the main solutions to most of the engineering problems. MR fluids are highly successful in many disciplines of science, ranging from mechanical, automobile, civil structures to many major medical applications. This is the reason why many researchers are motivated to MR fluids. Magnetorheological fluids (MRFs) are the suspension of magnetic particles in matrix fluids. They exhibit the outstanding properties that their operating behaviour can be reversibly controlled by applied external magnetic field. Controlling of fluid by external magnetic field takes milliseconds, which makes these fluids superior to other comparative smart materials. This quick responsive behaviour is the core of many MR fluids devices. For example, damping systems and detailed explanation on MR fluid preparation, characterization and application are well explained in many articles [1–6].

When MR fluids are subjected to magnetic field, they undergo variations in their viscosity and can change themselves like solid state. The time taken to change their state (liquid → solid and solid → liquid) is few milliseconds. The interactions of induced dipoles results the particle to form a chain like structure, parallel to the acting magnetic field. Required energy to yield the increase in microstructure as the field increases resulting in the field dependent yield stress. So the act of controlled fluids is expressed as Bingham plastic model having variable yield strength [7], in these models the flow is ruled by the Bingham's formula:

$$\tau = \tau_y(\mathrm{H}) + \eta\gamma, \quad \tau < \tau_y \tag{6.1}$$

Yield stress of material behaves Visco-elastically:

$$\tau = \mathrm{G}\,\gamma, \quad \tau < \tau_y \tag{6.2}$$

As we know the complex material (G) modulus is also a field dependent parameter. Real MR fluid expresses some mentionable departure from the Bingham plastic model. So this departure includes the non-Newtonian characters of MR fluids in the absence of magnetic field. Still, if we used exactly, Eq. 6.1 gives us a useful basic for the designing MR fluid-based applications. Aggregates of Nickel particles are constructed by the attractive dipolar forces between particles. The aggregate sixe of nickel particles is attained by these forces and so it is seen as inversely proportional to root square of shear rate. However, such character is good within certain values of shear stress and magnetic fields. At this stage slender body theory [8] can be used for developing safe MR fluid applications. The viscosity is the measure of the "resistance to the flow" of a liquid. This is explained by taking an experiment in which the gap between the two plates of the rheometer, of area (A) and separation gap (L), is completely filled with our fluid. Among the two plates of rheometer one is fixed and another is capable of rotating in its axis. Rotating plate moves with the velocity (U) relative to the fixed plate. The force required is directly proportional to the area of the plate; it is to define the stress (τ). This total concept is given by the equation:

$$\tau = \eta(A/L) \tag{6.3}$$

The coefficient η is defined as the viscosity of the fluid. The stress has the units of Pascal (Pa) or Newton/m^2 (N/m^2). The strain rate has unit of s^{-1} and velocity has the unit of meter/second (m/s). So the viscosity of the fluid is given by Pascal Second (Pa-s) [9]. In Table 6.1 we have shown the values of some important properties of MR fluids. The method preparation of MR Fluid is explained in our recent articles [10–14]. Most of the researchers are only concentrated on the iron (Fe) particles as magnetic particles. But in our present study we made use of Nickel particles to study the Rheological behaviour of MR fluids. The specific concept of this chapter is to study the characteristics of the MR fluids at desired operating parameters. Mainly in this study authors concentrate to recognise the characters of these smart MR fluids which is under high shear rate, shear stress and varying viscosities. Concluding

Table 6.1 Properties of MR fluids

Properties	Values
Viscosity (Pa-s)	Variable (0.03–0.05)
(B) density (gm/cm^3)	3–4
Size of nickel particles (μm)	2–5
(τ) max. yield stress (kPa)	10–150
(H) max. Field strength(kA/m)	250
Operating temperature (°C)	−40 to +200
Stability	Wont react to most of the impurities

results of our study will surely serve in designing MR fluid applied devices in various applications. Moreover, our defined characters will lead to future developments in MR fluid discipline.

6.2 Working Principle

If there is no applied field, then the fluid shows the Newtonian behaviour. If MR fluid is under some magnetic field, the magnetic particles get magnetized and they start behaving like magnets with their own dipoles. Then all the particles aligned along the direction of magnetic field lines. The interaction between the dipoles will result in the formation of wire like structures in the carrier fluid. These wire-like structures are also known as chain like structures which is shown in Fig. 6.1.

The wire like structures offers high resistance to the carrier fluid. So, the viscosity term comes into the scene. High resistance offered by wire like structure is directly proportional to the higher viscosity. With increasing the applied magnetic field (H) the non-linearity of fluids increases, which results dependent yield stress.

6.2.1 Composition of MR Fluids

Normally MR fluids consist of three main types: suspension particles, carrier oil and surfactants or additives. A magnetic particle mainly includes iron (Fe), cobalt (Co), Nickel (Ni), Fe_3O_4, Fe_3 and sometimes rhodium also. Among these, Fe particles are widely used as suspension particles, but the iron cobalt alloy will have the highest magnetic intensity and its intensity can reach up to 4 T. Whereas the magnetic intensity of iron and carbonyl iron particles can reach up to 2 T. Magnetic suspension particle size may vary between 3 and 5 μm but in some cases it will be in nano scale.

The carrier oil is generally synthetic oil. It is important component of MR fluid. Carrier oil acts as the working environment for the suspension particles. Synthetic oil, mineral oil and water based fluids may used as carrier oil. Carrier oil must exhibit some mandatory characters:

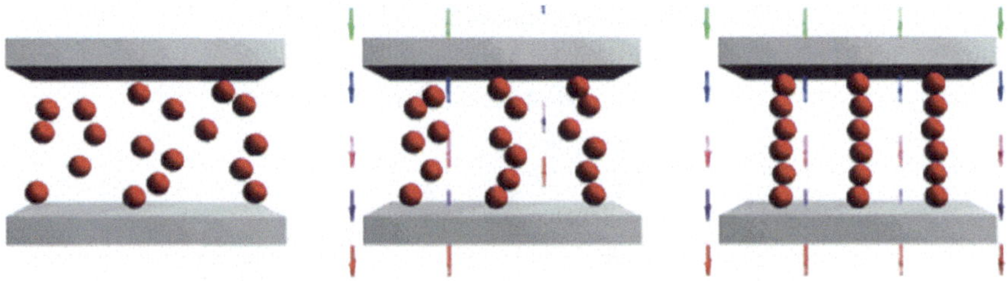

Fig. 6.1 Wire like structures in MR fluid (www.lord.com)

Table 6.2 Composition of MR fluid experimental sample

Magnetic particle (wt%)	Carrier oil (wt%)	Oleic acid (wt%)
40	60	15
35	65	15
30	70	15

- Low freezing point.
- High boiling point.
- Required viscosity.

Lower freezing point and higher boiling point ensures the MR fluids have very wide range of operational capabilities. Viscosity of MR fluid must be as per requirement because zero magnetic field of MR fluid should have lower viscosity. It is to note that alluviation of suspension particle is the major drawback when viscosity is low.

Surfactants are mainly used for preventing the alluviation of magnetic particles and improving stability, dispersion, and shear yield strength of the MR fluids. Surfactant material may be of fine lithium grease, oleic acid, citric acid, naphthenate, sulfonate, stearic, alcohols, silica, glycerol monoleate etc. alluviation preventing agents incudes dense polymers, organometallic silicon copolymer and hydrophilic silicon oligomer etc. [15].

In our experimental samples we made use of Engine oil as carrier oil, Nickel particles as suspension particles and oleic acid as surfactant. And Table 6.2 will give an explanation for our experimental samples. All the measurements are made by weight percentage.

6.3 Experimental Background

Equation which depicts the linear relationship between x and y curve is shown in Fig. 6.2 and it is represented as:

$$y = mx + c \tag{6.4}$$

where m denotes the slope of the line and c denotes the y-intercept. So, 2 additional curves are introduced to elucidate the rheological properties of the MR fluids. There are many laws to explain the linear relationships, power law is one among them and shows the following mathematical differences:

(i) the y-intercept isn't any c, and
(ii) x will be power.

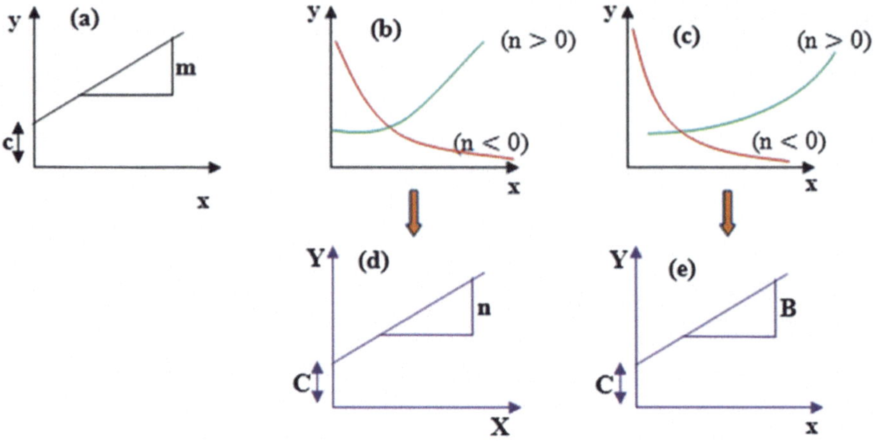

Fig. 6.2 Power law model: representation of *x* and *y* relationships

Therefore, according to the curve in Fig. 6.2b, the equation is given as

$$y = mx^n \tag{6.5}$$

By taking log from above equation,

$$\log y = \log(mx^n) \tag{6.6}$$

$\log y = n \log x + \log m$, so, if log y is equal to Y, $\log x$ is equal to X and, $\log m$ is equal to C. By which we obtain as follows and it is depicted in curve shown in Fig. 6.2d.

$$Y = nX + C \tag{6.7}$$

Another curve is graphical record, as shown in Fig. 6.2c, that which look like the power law curve, however the curve approaches zero at short route of x axis. Thus, its equation is given as,

$$y = Ae^{Bx} \tag{6.8}$$

By considering log,

$$\log y = \log\left(Ae^{Bx}\right) \tag{6.9}$$

$\log y = Bx + \log A$, with log y equal to Y, and $\log A$ equal to C, it can be modified as follows; its representative curve is shown in Fig. 6.2e.

$$Y = Bx + C \tag{6.10}$$

Facts prove that the power law of fluid is similar to Newtonian fluid as Eq. 6.11 shown below:

$$\tau = k\sigma^{n} \tag{6.11}$$

where σ is the shear rate, K is the flow consistency index and n is the flow behaviour index. An effective or apparent viscosity (μ_{eff}) as a function of σ is shown as

$$\mu_{eff} = k\sigma^{n-1} \tag{6.12}$$

This relationship is useful, because of its simplicity, however they conjointly describes non-Newtonian working behaviour. For example, if the n is less than 1, the forecasts of power law explains that the total viscosity roughly would drop back with increasing in shear rate indefinitely which requires n fluid with infinite range viscosity at rest and zero viscosity. A real working fluid has both maximum and minimum effective viscosity that depends on the chemistry of the molecules at molecular level. The power law holds best of fluid behaviour at various ranges of the shear rates to that the coefficients were fitted accordingly. Broadly explaining, the power law holds 2 conditions: (1) flow index, a measurement of non-newtonian-ness, (2) consistency (K), that is not any over the viscosity at shear rate of one s^{-1}.

When we talk about shear thinning fluid, the flow index varies from 0 to 1. For newtonian fluid, the power law is 1 but its higher than 1 for shear thickening fluid.

6.4 Results

6.4.1 Shear Stress as a Function of Shear Rate (Visco-Elastic Behavior)

For an elastic body model, the stress in a sheared state is directly proportional to strain. The relative Hooke's law is used and the constant proportionality is the normal complex modulus (G). It is written by the Eq. (6.2). Here the shearing action gives rise to the stress. The resulting stresses are also proportional to shear rate. That is given by

$$\Psi_1 = N_1/\gamma^2 \tag{6.13}$$

$$\Psi_2 = N_2/\gamma^2 \tag{6.14}$$

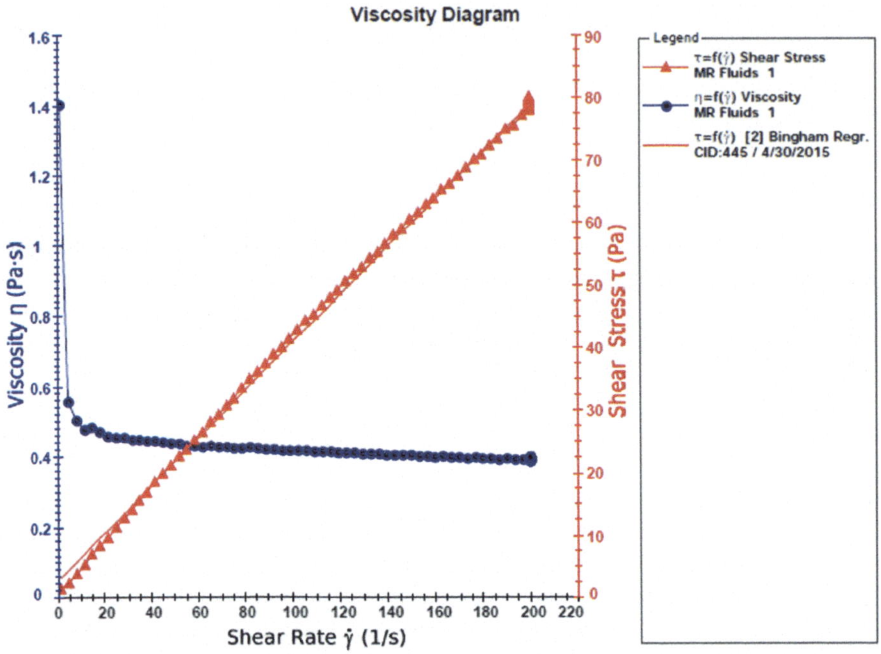

Fig. 6.3 Viscosity diagram of sample 1

The actual rates of changes of N_1 and N_2 with rate vary from one to other system and we can see some important observations here. Normally, the decrease of Ψ_1 and Ψ_2 with γ is greater than the viscosity. At lower rates, N_1 is expected to vary as γ^2. That is Ψ_1 will get a constant value in this range. The same is shown in Figs. 6.3, 6.4 and 6.5.

By our results we can observe that the resulted stress seem to be independent of shear rates. This is due to generated stress caused by the particles are nearly proportional to the yielded shear rates. We can see that shear stress increases monotonically with the shear rate. Concerning the shear stress, we observe that it follows a Bingham rheological law, with the yield stress.

6.4.2 Viscosity as a Function of Shear Rate

The viscosity of the MR fluid samples was measured by cone and plate type rheometer. In Figs. 6.3, 6.4 and 6.5 we can see the viscosity of MR fluid as function of shear rate. The viscosity observed decreasing with increase in the applied shear rates, which leading towards the term called "shear thinning" characteristics although the terms loss of viscosity and pseudo plasticity have seen. Even we observed that the viscosity will be constant at very low shear stresses.

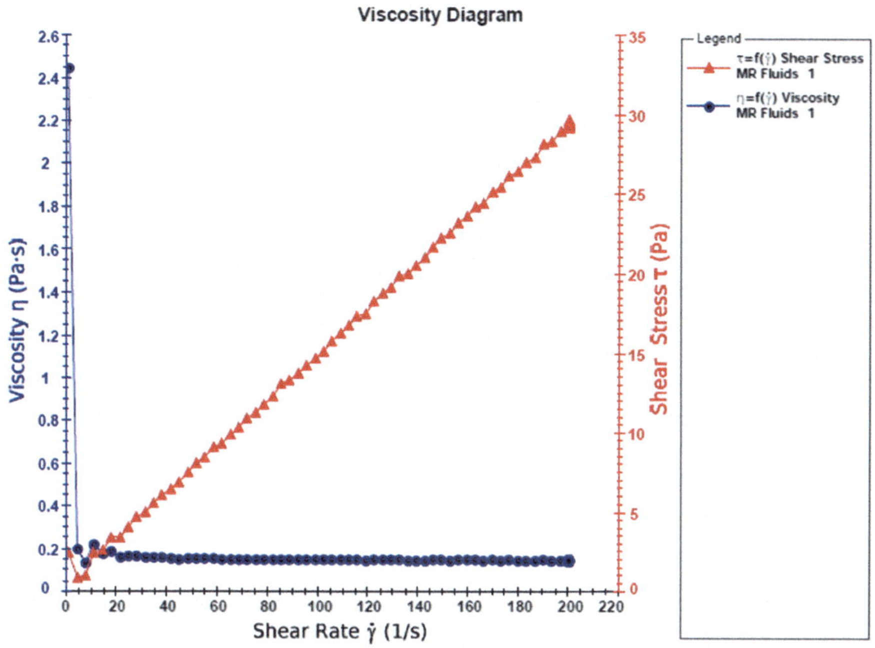

Fig. 6.4 Viscosity diagram of sample 2

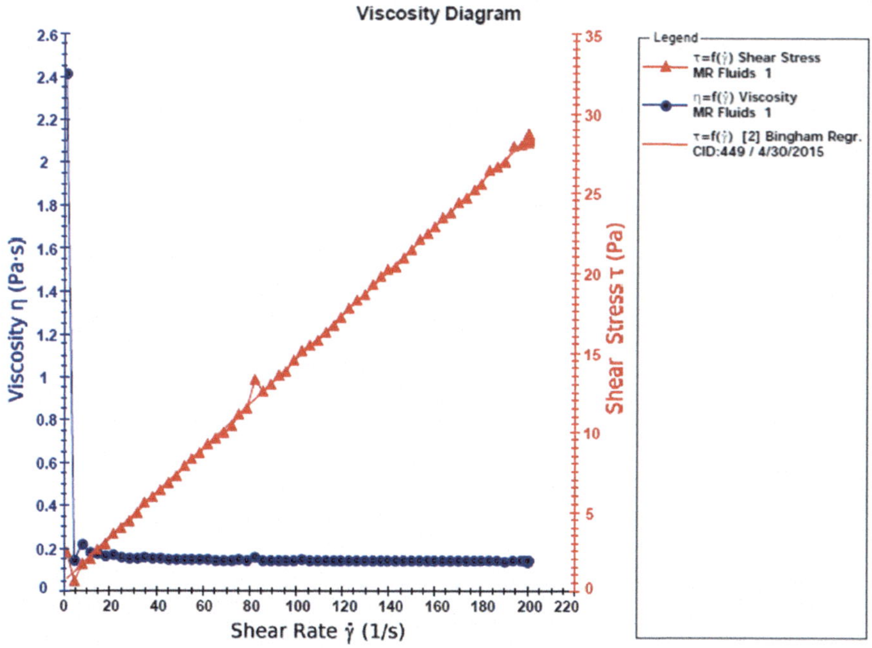

Fig. 6.5 Viscosity diagram of sample 3

6.4.3 Thixotropic Analysis

The material said to be thixotropy, when it is sheared at a constant rate. Its viscosity can be determined by equation,

$$\eta = \sigma / \gamma \tag{6.15}$$

Viscosity decreases with the duration of continuous shearing. High viscosity will be offered by chain like structures of Nickel particles. In some conditions, the failure of structures may be reversible that is when we remove the external shear stress the fluid may come back to its initial viscosity.

Thixotropy describes behaviour of material to change from high viscous fluid to a low viscous fluid as the result of extreme high shear during test period. The thixotropy analysis is shown in Figs. 6.6, 6.7 and 6.8. The term Thixotropy comes

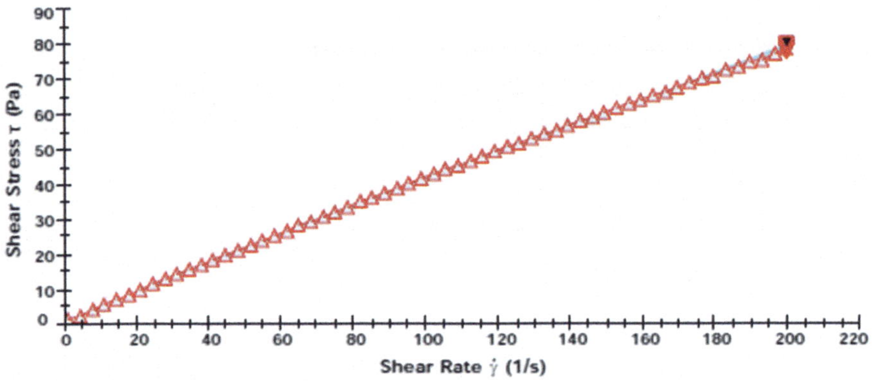

Fig. 6.6 Thixotropic analysis of sample 1

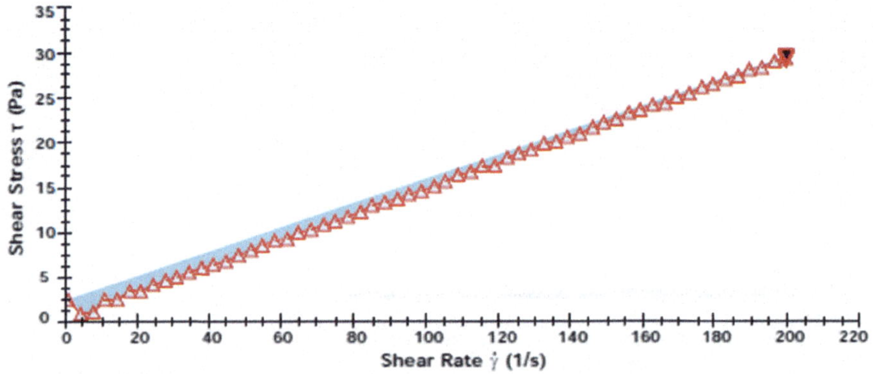

Fig. 6.7 Thixotropic analysis of sample 2

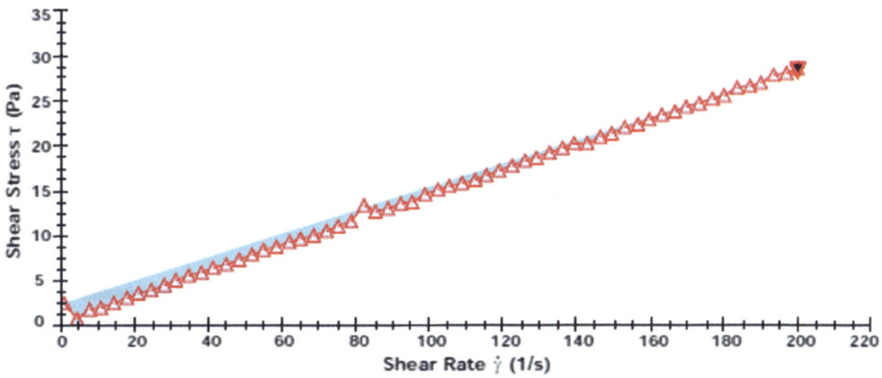

Fig. 6.8 Thixotropic analysis of sample 3

into the picture when the high viscous fluid loses its viscosity on continuous act of shear on it. Even this phenomenon is main reason for the formation of shear thinning process.

6.5 Regression Analysis

Regression analysis is a process of establishing a relationship between a dependent variable and independent variable. A relationship is to get by plotting a graph between values of the two variables. The relationship may be linear, parabolic, cubic, exponential or logarithmic. But in this case studies; we initially assume it is linear and calculate the constants involved in it and calculate the correlation index and test whether linear relationship holds good. In Figs. 6.9, 6.10 and 6.11 we can see the

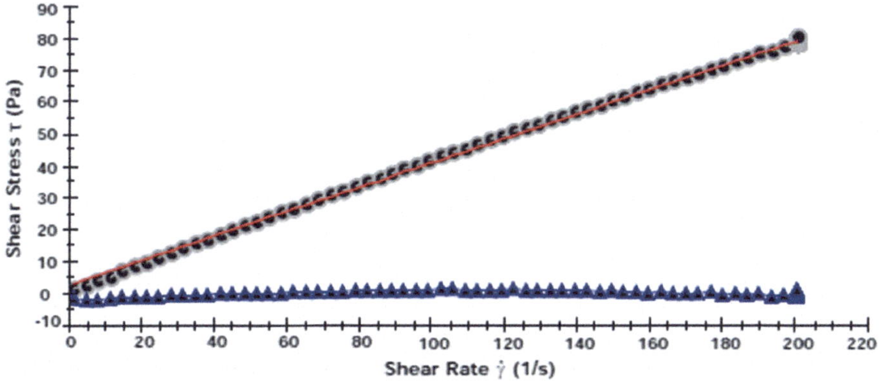

Fig. 6.9 Regression analysis of sample 1

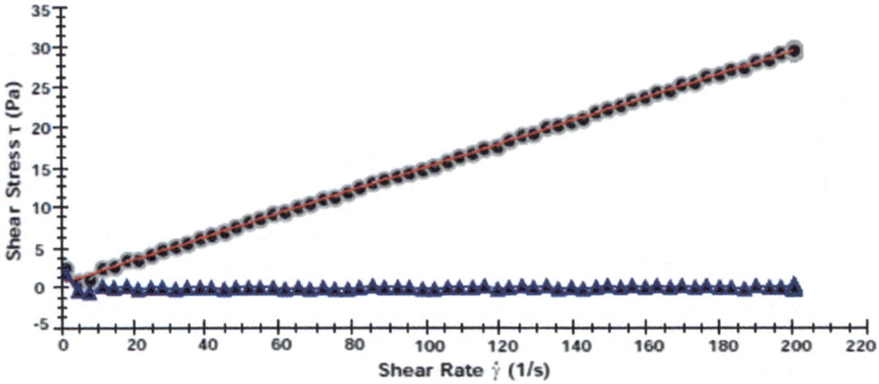

Fig. 6.10 Regression analysis of sample 2

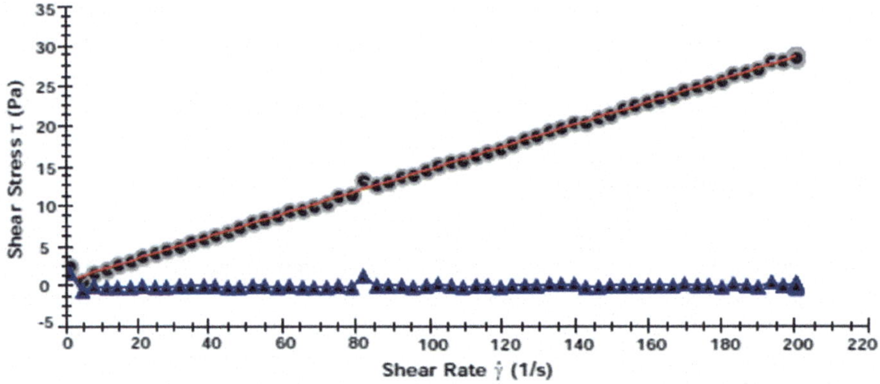

Fig. 6.11 Regression analysis of sample 3

regression fit curves. Correlation indexes (C) of all three samples are nearly 0.9994 which is close to 1 hence the relationship for samples can be confirmed as linear.

The relationship between Shear Stress (σ) and Shear Rate (γ) is given by

$$\sigma = A\gamma + \mathrm{B} \tag{6.16}$$

6.6 Conclusion

As most of the researchers are concentrated on Fe particle based MR fluids but we made an attempt to prepare Ni particle based MR fluids and made fundamental study on them. We have presented the experimental rheological behaviours of MR fluids, in which Nickel particles are used as magnetic particles. In future one can compare our Ni particle results with the Fe particle results and can review the results for further developments in the field.

Regression Analysis Calculation

Where A and B are constants whose values are to be calculated:
For sample 1 by calculation we get

A = 0.3809 and B = 2.7471
Hence we get
$\sigma = 0.3809\gamma + 2.7471$

For sample 2 by calculation we get

A = 0.1438 and B = 0.5862
Hence we get
$\sigma = 0.1438\gamma + 0.5862$

For sample 3 by calculation we get

A = 0.1395 and B = 0.6509
Hence we get
$\sigma = 0.1385\gamma + 0.6509$

References

1. O. Ashour, C.A. Rogers, W. Kordonsky, Magnetorheological fluids: materials, characterization and devices. J. Intell. Mater. Syst. Struct. **7**, 123–130 (1996)
2. G. Bossis, O. Volkova, S. Lacis, A. Meunier, in *"Ferrofluids," Magnetorheology: Fluids, Structures and Rheology*, ed. by S. Odenbach (Springer, Berlin, 2002)
3. J.M. Ginder, Behavior of magnetorheological fluids. MRS Bull. **23**, 26–29 (1998)
4. R.G. Larson, *The Structure and RHEOLOGY of Complex Fluids* (Oxford University Press, New York, 1999)
5. P. Phule, Synthesis of novel magnetorheological fluids. MRS Bull. **23**, 23–25 (1998)
6. P.J. Rankin, J.M. Ginder, D.J. Klingenberg, Electro- and magneto-rheology. Curr. Opin. Colloid Interface Sci. **3**, 373–381 (1998)
7. R.W. Phillips, Engineering applications of fluids with a variable yield stress. Ph.D. thesis, University of California, Berkley, 1969
8. G.K. Batchelor, Slender-body theory for particles of arbitrary cross-section in Stokes flow. J. Fluid Mech. **44**, 419–440 (1970)
9. A.P. Deshpande et al., *Rheology of Complex Fluids* (Springer, Berlin, 2010). ISBN 978-1-4419-6493-9
10. V.G. Kamble et al., Int. J. ChemTech Res. **7**(2), 639–646 (2014–2015)
11. G. Vikram, Kamble and Shreedhar Kolekar. Am. J. Nanotechnol. **5**(2), 12–16 (2014)

12. S. Kolekar, Preparation of MR fluid and study on its rheological properties. Int. J. Nanosci. **13**(2), 1450009 (6pp) (2014). https://doi.org/10.1142/S0219581X14500094
13. V.G. Kamble, V. Revadi, Am. J. Nanotechnol. **6**(1), 1–6 (2015). https://doi.org/10.3844/ajntsp.2015.1.6
14. V.G. Kamble et al., Am. J. Nanotechnol. **6**(1), 7–15 (2015). https://doi.org/10.3844/ajntsp.2015.7.15
15. F. Weibang, Review on MR fluid technology. Sh. Electr. Technol. **32**(6), 15–18 (2012)

Chapter 7
Tribology of Intelligent Magnetorheological Materials

Rakesh Jinaga, Shreedhar Kolekar, and T. Jagadeesha

Abstract Magneto rheological (MR) fluid are categorized as one of smart materials, where the viscosity of the fluid enhances significantly under the influence of applied magnetic field. The fluids are set up by scattering micron scale magnetic particles into a fluid media called as carrier fluid with added substances for improving the rheological characteristics of fluid. The fundamental element of these fluid is the capacity to undergo change from fluidized state to semisolid state under controllable yield stress within couple of milliseconds in the wake of externally activated magnetic field. Lower magneto rheological impact and sedimentation of particles in MR fluids are the most challenging topics against the broad applications of MR fluid revolution in current ventures. Different techniques have been proposed and utilized by analysts to enhance the magneto rheological impact and stability of these liquids against the sedimentation. The primary focal point of this brief is to show a thorough survey on various strategies for synthesis and reduction in sedimentation rate of MR fluids. Besides, rheological models and use of MR liquids are talked about along this compilation.

7.1 Introduction

Smart materials the need of the shifting technology where energy feasting and effectiveness being critical aspects of any mechanical systems, any increase in efficiency and reduction in power consumption is appreciable. The mechanical systems is the combination of number of links and mechanism, the efficiency reduces with increase in number of connecting elements. Induction of smart material (fluid) in a system and actuation by wire is an attempt to reduce the number of moving element in the system and increase the dependability of the system. System is combination of electrical and mechanical apparatuses makes it lighter, effective, active and reliable system. The

R. Jinaga · S. Kolekar · T. Jagadeesha (✉)
Department of Mechanical Engineering, National Institute of Technology, NIT Calicut, Calicut, Kerala 673601, India
e-mail: jagdishsg@nitc.ac.in

J. K. Katiyar et al. (eds.), *Tribology in Materials and Applications*,
Materials Forming, Machining and Tribology,
https://doi.org/10.1007/978-3-030-47451-5_7

electromechanical system consist of a mechanism which is triggered by special class of fluid called Magnetorheological (MR) fluid the smart fluids.

Magneto-rheology is a stream of rheology which studies the behavior of material under distortion and flow under influence of a magnetic field. In the year 1949 discovery and coining of the term Magneto-rheological fluid is credited to Jacob Rabiinow. The Magneto-rheological fluids are the suspension of a non-colloidal approximating from 0.05 to 10 μm, magnetically soft particles and multi-domain in an aqueous or organic liquids. Numerous dissimilar to ceramic materials and compounds has been characterized and can be employed for the preparation of MR fluid as long as this exhibits low level of magnetic coercivity and the particles are magnetically multi-domain. Particle shape, size, density, saturation magnetization, coercive field and particle size distribution are some of the significant characteristics of magnetically active dispersion phase. In addition to magnetic particles, surfactants, base fluids, and anti-corrosion additive are also dominating aspects which influence the stability, re-dispersibility and rheological properties of MR fluid.

In terms of consistency under off-state condition the MR fluid behaves as same as that of carrier fluid and demonstrates the same level of apparent viscosity i.e. ranging from 0.1 to 1 Pa s at lower shear rate [1]. Whereas the apparent viscosity of the fluid significantly changes 105–106 folds, within a fraction of a seconds and usually measured in terms of milliseconds as soon as magnetic field is activated. Viscosity variation in the fluid can be entirely reversed on deactivation of the applied field. As magnetic field is activated leading to the development of dipole within the magnetic particles resulting in movement of particles along applied field.

The magnetic interactions among the particles results in inter particle-forces resulting into a material of elevated apparent viscosity. The structures resembling chain along magnetic field direction is the resultant of dipole interaction among the magnetic particles in the non-colloidal suspension.

It can be concluded that the magnetic force between the particles, leading to chain like structures also this force determines the strength under apparent viscosity. The particles which are bonded together under the magnetic force and also to some extent the chain like structures also contribute to shear stress without undergoing yield which makes it to behave similar to a solid. The structure formed by particles under magnetic field is distorted and fluid gains its fluidity at a point where the shear stress exceeds the critical value. The effect of MR fluid is characterized by using a Bingham Plastic model, where the value of shear stress leads to distortion of the structure is termed as "apparent yield stress" of the fluid. A yield stress for iron based fluid 40% by volume under the flux density of 1 T is reported to be approximately 100 kPa [2] also, yield stress of 90–100 kPa for a MR fluid of uncertain concentration under magnetic field of 3 T [3]. The table for converting the magnetic parameters are listed in Table 7.1 (Fig. 7.1).

Table 7.1 CGS to SI unit conversion of magnetic units

For converting the CGS value in	CGS value is multiplied with	Resulting SI units in
Oersteds	79.58	A/m
Maxwells	10^{-8}	Wb
Gausses	10^{-4}	Wb/m^2
Gausses	0.7958	A/m
Gilberts/cm	79.58	A/m
Gausses	10^{-4}	T

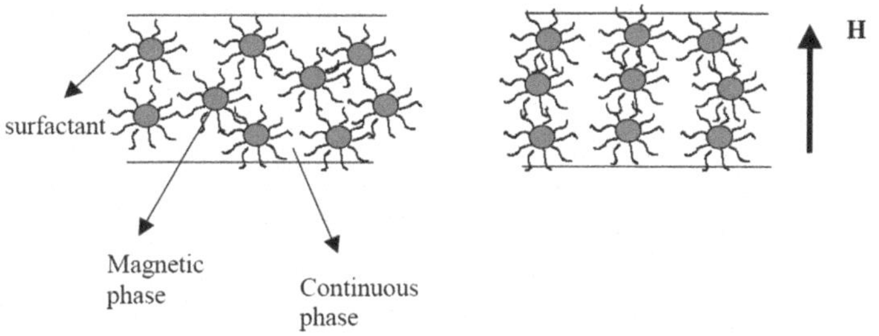

Fig. 7.1 Representation of magnetic particles in "OFF state" and "ON state" of the magnetic field

7.1.1 Electro-rheological Fluid

ER fluid is a non-colloidal suspension of electrically chargeable fragments suspended in the electrically insulating fluid [4]. ER fluids are normally made out of 0.5–100 μm sized particles of semiconductors or barium titanate, silica, cornstarch [5]. The particles for example, silica, polyelectrolytes should be added in order to enhance water adsorption over the particulate matter to improve the ER impact, therefore enhancing the electrostatic force among the particles. The conductive layer along the particle surface by the water resulting in floating of ions in water under the influence of electric field [6]. The impact resulting from interfacial polarization of the ER fluid is the cause of extrinsic polarizable material present in the fluid. The impact of ER diminishes as the water adsorbed on the surface reduces. In this manner at 50 °C, the ER movement of ions diminishes altogether hence, along these lines the higher temperature shakiness restrains the range of applications of the ER fluid.

Materials such as, metals, semiconductor polymers, ferroelectrics, inorganics, fluid gems and covered transmitters has additionally been accounted for delivering water free ER suspensions [6] also these materials are potentially termed as intrinsically polarizable material and also they function by implementing interfacial or mass polarization. The lower thermal coefficient at yield stress which aids in enhancing the range of temperature for the functioning of ER fluids.

Similar to MR fluids the excitation is aided by electric field where in the particles develop charge and lead to distortion of electric field. Ability of the elements to get charged up is enhanced by relocation of mobilized ions from the region of lower field concentration towards the higher concentration. Which results in higher dipole moments which attracts each other and results elements to build chain like structure along the direction of applied field. Also ER fluids are characterized using Bingham Plastic Model, the change in the viscosity influenced by the elemental chain structures under shear leads to the yield stress. For an electric field of 4.1 kV/mm Weiss recorded yield stress value of 3.6 kPa using Lord Corporation's (VersaFlo ER 200) ER Fluid. ER fluids are mostly used in clutches, valves, dampers, mounts and brakes. However, further development is essential for commercialization.

7.1.2 Ferrofluid

Ferrofluid commonly called as magnetic fluids which is the colloidal dispersion of ultra-fine magnetic particles usually ranging from 4 to 10 nm, such as Mn-Zn ferrites, iron oxides are the few of magnetic particles [7], Co and Fe in either fluid or semi-solid form. Under normal magnetic field strength, thermal agitation leads to Brownien motion which is sufficient to overcome the dipole alignment as the size of particles in the magnetic phase is very small. Thus, we have the relation MR fluids are based on ferrofluids and ferrofluids are based on superparamagnetic materials. Instead, magnetic field gradient is proportional to the ferrofluids experiencing force on the magnetic material in the fluid. Though ferro-fluid exhibit no yield stress $\tau_y = 0$ they exhibit field dependent viscosity. Ferro-fluids are used in bearings, magnetic motor, dampers, and rotary seals [8, 9].

Another class of the ferro-fluids is called as inverse ferro-fluids, commonly known as magnetic holes. Where non-magnetic materials commonly one or more order of larger magnitude suspended in magnetic particles in ferro-fluid. Hence, medium that is continuously magnetic is experienced by the non-magnetic particles.

7.1.3 Comparison of Field Responsive Fluids

In advancing stages MR fluid has gained significantly additional consideration than their equivalent ferro-fluids and electrorheological fluids due to its ability to produce significantly higher yield stress with minimal power requirement.

The higher yield stress of MR fluid as compared to ER fluid is add on advantage of MR fluids. For MR fluid high magneto-static energy is the cause of attaining higher yield stress compared to electrostatic energy density of ER fluids. High temperature stability ranging from −40 to +150 °C also the low voltage power requirement for MR fluid make it more reliable materials compared to ER fluid. Ferro-fluids exhibit negligible yield stress, however show elevated viscosity under magnetic field. Under applied magnetic field the viscosity is elevated to almost twice as compared to the viscosity under absence of magnetic field. Compared to MR fluid synthesized

Table 7.2 Property assessment MR, ferro-fluid and ER fluid

	MR fluids	Ferro-fluid	ER fluid
Particulate material	Iron, ferrites, etc.	Ceramics, ferrites, cobalt, iron, etc.	Polymers, zeolites, SiO_2, $BaTiO_3$
Required field	~3 kOe	~1 kOe	3 kV/mm
Density (g/cc)	3–5	1–2	1–2
Suspending fluid	Nonpolar oils, polar liquids, water and other	Oils, water	Oils
Particle size	0.1–10 μm	2–10 nm	0.1–10 μm
viscosity (mPa s)	100–1000	2–500	50–1000
Device excitation	Permanent magnets or electromagnets	Permanent magnet	High voltage
Field induced changes	τ_y $(B) \sim 100$ kPa	$[\Delta\eta(B))/(\eta(0)] \cong 2$	τ_y $(E) \sim 10$ kPa

by non-colloidal magnetic particles, the ferro-fluid exhibits more stability as these are developed from colloidal magnetic particles. The properties assessment of MR, ferro-fluid, and ER fluid is tabulated in Table 7.2.

Comparing the three smart fluids namely magneto-rheological, electro-rheological and ferro-fluid we see that the MR fluids are suitable for most of the applications with promising results.

7.1.4 Constituents of Magnetorheological Fluid

MR suspensions are synthesized using an extremely basic technique which is combining every one of the constituents. The significant issue is identification and optimizing the components. A MR fluid as a rule has three fundamental constituents: base fluid, stabilizer added substances, and magnetic particles. The base fluid carries on like a transporter which contains metallic particles. Basically, this is where the metallic particles are suspended in it. Magnetic particles, with the key part in magneto rheological impact, are scattered in the base fluid. Stabilizer added substances are used to effectively reduce sedimentation issue of substantial magnetic particles. A basic and proficient magneto rheological fluid is obtained by scattering of iron particles in silicone oil and utilizing some surfactants, for example, stearic acid to dodge irreversible coagulation [9]. Forward is reason, one needs to scatter magnetic particles into the base fluid in a productive ways and a homogenized suspension might be obtained. A standout amongst the best courses for acquiring a stable MRF is to utilize stabilizer added substances before adding magnetic particles to the base fluid. This procedure may prompt proficient MRF planning. In the accompanying areas, the MRF fixings have been examined in subtle elements.

7.2 Carrier Fluid (Base Oil)

Base fluid has to serve the purpose of the carrier, naturally greasing combined with additive and damping features. For the maximum magnetorheological effect the viscosity of MR fluid should be maintained as low as possible in OFF state and also the dependency of viscosity with respect to temperature should be maintained as low as possible. Along these lines the MRF effect will be the prevailing impact contrasting the variation of natural physical viscosity along temperature and shear stress. Principally in the off-state i.e. in the absence of magnetic field MR fluid carries on like carrier fluid conferring to the chemical composition. There are some unique kinds of liquids which can be utilized as a base fluid such as silicon oil, mineral oil or hydrocarbon oil. The magnetic particle dispersed in the carrier fluid results in the overall elevation in the viscosity of the solution, so the liquid will have with powder a characterized offset viscosity value in off-state. Commonly the dynamic viscosity of the base fluid at atmospheric temperature is near 100 mPa.

The essential purpose of a carrier fluid in a MR fluid is to provide a nonmagnetic and low permeability surrounding the magnetic particles, where in the magnetically activated particles can remain dispersed. Also, the fluid needs to have low permeability in order to aid the polarization of particles with the utmost effectiveness. Thus improving the MR effect [2]. Mineral oils, silicone oils, paraffin oils etc. are the examples for suitable carrier fluid. Silicone oil is the most regularly used carrier fluid following hydrocarbon oils. With a specific goal to keep the off-state viscosity low, 100 cSt viscosity silicone oil corresponding to 0.95 Pa s of dynamic viscosity, is used for development of MR Fluid samples. The magnetic particles are dispersed into nonmagnetic fluid called the base fluid. The base or carrier fluid should possess the ability of natural lubrication and damping features. For better execution of MR fluid viscosity should be as low as possible and also the viscosity should not be grately influenced by the temperature variation. This is essential so that the amplitude of externally applied magnetic field is the primary function for change in viscosity of the fluid. Due to the presence of magnetic particles dispersed in the carrier fluid, it becomes thicker. The carrier fluids commonly used are silicon oils, mineral oils and hydrocarbon oils.

For the maximal MRF effect the viscosity of the liquid ought to be small and relatively free of temperature, 50–80% constituent of MR fluid is carrier fluid. The most promising carrier fluids are minerals and synthetic oil. In the case of mineral oil, the rate of change in viscosity as a function of temperature is relatively high, which leads to a restricted use of mineral oil as a base oil in the development of the MR fluid and in the use of low-temperature applications. Important properties, such as no thickening at high temperatures, high flash point, lower friction, high shear strength and high viscosity index, the critical properties of the synthetic oil are best on the market for MR fluid synthesis Additional properties good oxidation resistance heat transfer characteristics, high flash points, and very low vapour pressure, on other hand sealing silicon oil is a challenging task [3]. Over wide range of temperature change there is minute change in the physical properties and a relative constant viscosity

Table 7.3 Properties of carrier oils

Properties	Synthetic oil	Silicone oil	Mineral oil
Density at 25 °C (kg/m^3)	873–894	760	825
Specific gravity	0.817	0.9124	0.818–0.95
Fire point (°C)	350	500	260–330
Flash point (°C)	230	>300	171–185
Cloud point (°C)	−20	−20	−15
Pour point (°C)	−30 to −50	50	−25 to −50
Viscosity @ 40 °C (Pa s)	0.1068	0.1100	0.028

temperature slope and the oil finds its application in temperature ranging −40 to 294 °C. The various characteristics of different base fluid are assessed and tabulated in Table 7.3.

Carrier liquid being the largest constituent in MR fluid ranging 50–80% by volume of the total fluid, which largely impacts the rheological properties of MR fluid. Its fundamental function is to give a medium to magnetic active particles to remain dispersed during the absence of a magnetic field and to facilitate rearrangement once magnetic field is activated. But depending on the application of the MR fluid type of carrier fluid employed in the synthesis may differ. Till date various carrier fluids have been used and considered in the synthesis of MR fluid and few are described here. These are naphtha thickened kerosene and polyvinyl butyl [5], mineral oils of light grade [8], silicon oil [6, 7], a blend of organic liquids, synthetic oil and water [9] etc. MR fluid are also accessible commercially which are silicon based MR fluid, hydro-carbon-based MR fluid, etc. in any case, these choices are exorbitant.

Before selecting the base fluid various parameters/characteristics of the fluid such as lower viscosity, lower density, lower freezing point, ease of availability, cheap in cost, higher flash point and fire point and good thermal stability.

7.2.1 *Dispersed Phase (Magnetic Particles)*

The prerequisites set on the selection of elemental material are that the particle must possess magnetically multi domain and display low level of magnetic coercivity, additionally enhancing the inter particle forces and along this line boosting the MR impact which can be achieved by determination of the elemental material for the saturation magnetization Js (Tesla). Higher the saturation magnetization Js, the higher the inter particle forces and the higher the MR impact. The material most utilized today is highly pure carbonyl iron powder, synthesized by using compound vapor deposition i.e. CVD for iron Penta carbonyl $Fe(CO)_5$.

The various consideration to employ this process are,

- High level of chemical purity >99.9%, leading to lower domain pinning.
- Mesoscale dimensions, which have many magnetic domains.
- Spherical shape particles, minimizing the magnetically shape anisotropy.
- High magnetic saturation Js = 2.4 T.
- Particles magnetically soft hence non-abrasive.

The existing particles are compounds of cobalt and iron which has magnetic saturation of 2.4 T. Tragically, the combinations is restrictively costly for most of the low cost application. The best viable particles are basically pure iron, as this exhibits a saturation magnetization of 2.16 T. Practically all different metals, compounds and oxides has the magnetic saturation essentially lesser than iron, bringing about generously weaker MR fluid.

Metal particles subjected to magnetic field (in on State) are guided by the magnetic field lines to form a chain like structure. The resultant chain like structures obstructs the flow of the fluid thereby changing the rheological properties of the fluid. This counteracting of the carrier fluid against the structure is the MR effect. The metal particles are commonly derived from powder iron, carbonyl iron, cobalt and iron compounds to attain larger magnetic saturation. The quantity of metal particles in MR fluid by volume can be up to 50%. The particles are measured in μ meters and the size varies as function of manufacturing process employed. The particle shape and size can be selected and also can be combined differently based on the application requirements. The particle size for carbonyl iron powder is of the range 1–10 μm. Essentially higher volume fraction of magnetic particles and larger particle size in MR fluid eventually in on state condition result in higher torque value. For ideal usage of this innovation we require particles which can be magnetized effortlessly and rapidly subsequently we lean toward metal particles. The MR fluid employs Metal particles of very fine grade. The particle size usually ranges from 1 to 7 μm frequently employed metal particles are powder iron, carbonyl iron, and cobalt iron compound. The properties of the metal particles of such materials has the ability to attain extreme magnetic saturation because of which they are capable of forming strong chain like structure. The maximum concentration of the particles in the base fluid can be up to 50% by volume.

The particles suspended in the MR fluids must poses low magnetic coercivity and high magnetic saturation. Nickel Zink ferrite [10], Carbonyl iron powder [11–15], Iron Cobalt alloy [16, 17] and Iron oxide coated polymer composite particles [18] are considered best with the requirement. Magnetic saturation of Iron powder, Nickel Zink ferrite and alloys of cobalt and iron are 2.1, 0.4 and 2.4 T respectively [16, 19] the Iron Cobalt alloys possesses the highest magnetic saturation on the other hand the value of density is 8.1 g/cm^3 compared to Iron hence it is prone to settling due to gravitation. As all the above discussed magnetic materials are expensive and hence restricts the low cost synthesis of MR fluid and its application at basic level. In order to substitute the expensive magnetic particles with approximately similar properties iron powder obtained by electrolytic process with a higher level of purity at an economical expense of 10 US$ per kg of Carbonyl Iron powder.

All the literatures available till date are dealing with particle size ranging from 5 to 10 μm. As the particle size goes smaller, Brownien motion in particles leading to reduced strength MR fluid though it enhances the stability against gravitational sedimentation. Many study contemplated the impact of adding nano sized particle on rheology of MR fluids. Lemaire in the year 1995 [20] expressed the effect of magnetic particle size on rheological characteristics and concluded that the ratio thermal energy to magnetic interaction energy is lower than unity, the particle size has positive impact over the yield stress. Kormann in the year 1996 [21] discovered non changing using nano particles and polar liquids, and reported that the fluid has very low yield stress. Also Rosenfeld in the year 2002 [22] synthesized a fluid using micron and nano sized iron powder along with hybrid fluid and reported that the fluid with bigger particle size exhibits higher yield stress. On the other hand Chaudhuri in 2005 [23] and Wereley in 2006 [24] discovered that micron particle replacing nano magnetic particles in small ratio, tends to enhance the field dependent yield stress considerably, also the presence of nano particles increases the sedimentation stability. Similar study and observations were made by Song, Choi and Park in 2009 [25] whereas Burguera in 2008 [26] observed reduction in yield stress and improvement in sedimentation stability by increasing the nano particle concentration concluding that the yield stress and sedimentation stability are inversely proportional to each other and hence an optimal volume ratio of micron and nano sized particles is to be found. Fang in the year 2009 [27] studied effect on sedimentation stability and yield stress using carbon nanotubes. Lopez-Lopez in the year 2009 [28] experimented and studied the variation in rheological properties of cobalt particle size ranging from 50 nm to 1 μm and conclude that size of the particle doesn't impact the response of MR fluid for particle size more than 100 nm. Considering the discussion it can be concluded that development of MR fluid which is stable over wide range of temperature variation and sedimentation is essential.

The MR fluids ingredients should be selected in such a way that it gives balance between shear yield stress and off state viscosity. According to previous observation on MR fluid study, including nano sized particle, has detailed on enhancement of sedimentation stability and yield strength. Hence a comprehensive study and investigation on field induced off-state viscosity versus the yield stress for various composition of fluid.

7.2.2 Additives

The third main component in MR fluid preparation is Additives. The dipole polarization between the particles being the mechanism for MR fluid, the rheological characteristics of fluid remain not affected by action of surfactant. Hence it comparatively upfront and advisable to use additive for enhancement of rheological behaviour such as:

- Reduction in the rate of particle sedimentation.
- Reduction in particle coagulation.
- To retain the coating of particles to enhance re-dispersibility.
- To enhance anti-oxidation property.

The pH value is controlled by using liquid additives to the water based MR fluids. Commonly used surfactants in the synthesis of MR fluid are as follows.

- Soy lecithin
- Citric acid
- White lithium-based grease
- Tetra methyl ammonium hydroxide
- Oleic acid.

Additives being the critical constituent for the synthesis of MR Fluids to full fill various requirements of the fluid such as stoppage and reduction of particle sedimentation, stoppage and reduction of particle coagulating, uphold the coating of the particles so as to improve re-dispersibility and anti-oxidation. The reduction in particle sedimentation is the most critical aspect of MR fluid application. In case of practical applications, the rate of particle sedimentation has be to be observed as lower as possible level. One of the good additive along silicon oil as carrier fluid is lithium White grease. The lithium White grease being easily available at many automotive stores and finds application in automotive application. Lithium White grease can be thus used for synthesis of MR fluid with silicon oil as carrier fluid.

In order to improve the particle stability in fluid material with higher viscosity like thixotropic additives or grease are used for enhancing the stability of particles. Ferrous oleate or ferrous naphthenate are being used for dispersion of particles and also the metal soap like sodium stearate or lithium stearate are used as thixotropic additives. The magnetizable particles are usually layered using material coating such as polystyrene, gaur gum etc. in order to reduces particle coagulation and to lower the density CI particles in order to enhance the sedimentation stability of the MR fluid.

The avoidance of magnetic particle sedimentation being most important criteria. The sedimentation rate can affect the rheological properties on broad scale over the time. Commonly employed additives for serving this purpose are surfactants such as xanthan gum and thixotropic agents, carboxylic acid, silica gel, and stearates. Thixotropic network interrupts flow at a ultra-lower shear rates where in viscosity attains lower value, but the shear rate is increased the value of viscosity drops. Stearates leads to the network of swollen strands usually when combined with synthetic esters and mineral oil which entraps the particle and immobilizes it. Also the fine carbon fibers has be experimented to serve the requirement. The use of fibres leads to enhancement of viscosity value resulting from physical entanglement of the fibres on the other hand exhibits shear thinning property caused by shear induced alignment. The contribution of various additives in order to maintain particles dispersed in carrier fluid. By means of the above discussed methods it was observed that there is minimal change in rheological properties over period of time. Hence it can be concluded that for every application and working conditions and loading conditions an especially composed MR fluid has to be employed.

Distinct conditions and factors need to be considered and experimented for the selection of an additive considering parameters such as loading condition, atmospheric condition, operational conditions etc. the selection of additive is carried out considering chemical compatibility. Many combinations of additives have been tried for the manufacturing of MR fluid. Usually thixotropic agent is preferred to improve the sedimentation of particle. Along with sedimentation, anti-corrosion, anti-wear, antioxidant agent and friction modifier are being used. For the synthesis of MR fluid, the major additives used along with the purpose served are listed below:

- To reduce sedimentation—Fumed Silica [29], Aluminium distearate, Thiophosphorus, Lithium stearate, Guar gum, Phosphorus, Thiocarbomate [30], Poly vinyl pyrrolidone [31], Organoclay [32], Poly vinyl butyl [33].
- To reduce agglomeration—Fibrous carbon [34], Fumed silica [29], Stearic acid [32], Viscoplastic media [31], Sodium dodecyl sulphate [33].
- To enhance abrasion resistance—ZDDP i.e. Zink dialkyl dithio phosphate [28, 29].
- To bring down oxidation rate—ZDDP [30].
- To bring down friction resistance—Organo-molybdenums (Moly) [29].

7.3 Operational Modes

Magneto-rheological liquids have a wide range of technical applications. The principle of MR fluids is divided into three classes [35]. As shown in Fig. 7.2, the "flow mode" is shown where the flow occurs, along the limited surface is also normal to the applied magnetic field, as shown in Fig. 7.2a, while "shear mode" is categorized when one of the surfaces moves at speed in one direction along the flow with respect to another solid surface and the surfaces are normal to the applied magnetic field, as shown in Fig. 7.2b. The "compression mode", like the shear mode, the surface

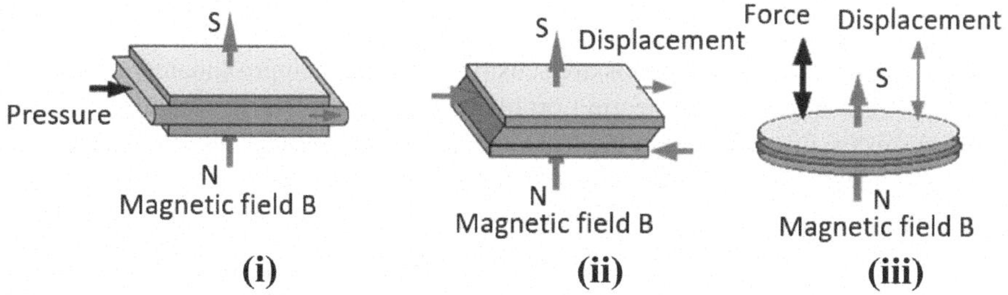

Fig. 7.2 Modes of operation; **a** flow mode; **b** shear mode; **c** squeeze mode [35]

possess a difference in speed along the normal to the other surface which is held parallel to the applied external magnetic field, as shown in Fig. 7.2c.

The magnetorheological fluid offers a solution for many technical applications. The realization of the MR fluid can be observed in many disciplines, ranging from automobile and building construction to biomedical construction applications. There are countless studies that highlight the benefits or use of MRI scanners in various fields. This global realization of MR fluids continues to convince the current and future use of MR fluids.

The significant success of MR fluid devices is generally due to advances in innovation in fluids. The existing MR fluids are apposite for an elastic stress higher than 80 kPa, which means that the wide range of dynamic operations and better control over the fluid. Liquids have also shown improved particle stability properties. In addition, the strength and shelf life of the liquid have increased with the ultimate goal that the liquid can be considered for marketing. The implementation of current MR fluids is the result of an extraordinary number of investigations which distinguish the properties and behavior of MR fluids. There are a variety of studies and experiences with work related to the production of MR fluid or the production of certain MR fluid devices. In various studies, information on the performance of the MR device precedes an exhaustive record of the behavior and function of the liquid in the device.

MR devices intended for a particular application often find its use in as a replacement application, where in the working conditions of the liquid differ considerably from the first application. An example of this "learned" innovation is the many uses of the Motion Master TM shock from Lord Corporation. This shock absorber, which was originally proposed for situational suspensions in trucks and transport vehicles, was also considered for structural design applications and prosthetic fasteners. The working conditions of the liquid change impressively in each of these applications. An even more unpleasant case would be the use of MR automatic shock absorbers, which are intended for vehicle-related suspensions in shock or impact load applications. Here too, the conditions inside the shock absorber differ significantly.

7.4 Rheological Behavior of MR Fluid

Without magnetic field, MR fluids are sensibly very much approximated as Newtonian fluids. For the most engineering problems a straightforward, Bingham plastic-model is successful at depicting the fundamental, field depended fluid characterization. A Bingham plastic is a non-Newtonian fluid where the yield stress has to exceed in order to have a flow of the fluid. From that point the shear rate versus shear stress plot is observed linear, the total yield stress is expressed as (7.1):

$$\zeta = \zeta_0(H)\eta\gamma' \tag{7.1}$$

ζ_0—yield stress resulting from the magnetic field (Pa)
H—Strength of magnetic field (A/m)
γ'—Shear rate (s^{-1})
η—Plastic viscosity (Pa s).

Characteristic of the MR fluid is it can accomplish yield strength ranging from 50 to 100 kPa when the applied magnetic field strength around 150–250 kA/m. It is discovered that surface roughness at contact along the fluid is essential characteristic for yield strengths, particularly at lower magnetic fields. For lower strain rate to the yield, the shear modulus of the MR fluid additionally demonstrates an extensive increment along increasing magnetic field. MR fluids the smart fluids in the long run achieve a saturation point where increase in applied magnetic field quality don't change in the yield strength of MR fluid. This marvel usually happens around 310 kA/m.

The MR impact is promptly revocable magnetic field is diminished, with a reaction times of 6.5 ms. MR fluids developed are prone to no major change in rheological characteristics under the temperature ranges from −50 to 150 °C. Slight changes have been recorded in the volume fraction and henceforth decrease in the yield strength at given temperatures. Likewise size and size dissemination of the suspended particles influence the variation in properties of the MR fluid when subjected to applied magnetic field [1]. Critical increment in the yield stress of a MR fluids can be acquired at a given volume fraction of magnetic particles by utilizing as particulate constituent a blend as the first constituent of moderately large particles and a second constituent of generally smaller particles with the end goal that the mean size of the larger particles is no less than 5 times the mean diameter of smaller particles. The blend of smaller and larger particles gives a considerable increment in the yield stress without an expansion in the consistency of the blend without a magnetic field [35]. The reliance of the normal stresses over shear rate and also, the strength of magnetic field in shear flow mode of MR fluid have been considered experimentally. The undeniable normal stresses have been seen when the external magnetic field is greater than an elementary esteem. Normal stresses increment impressively along escalation of applied shear rate and external magnetic field, diminishes all of a sudden and fundamentally over the beginning of shear thickening. Also, the proportion of the shear stress value to normal stress value, a simple over the contact coefficient, increases with increment over shear rate, yet the descends with increment of externally applied magnetic field [2].

Dynamic structures produced under shear are examined utilizing initial a rotational rheometer then an optical imaging framework. A microscope enable us to envision the distinctive patterns subsequently produced and contrast these outcomes with the estimations acquired with the rheometer. It additionally empowers us to dissect the different introductory static structures framed by a remotely applied field. Other than the outstanding segment ruled and bowed wall commanded static structures we could watch novel, concentric rings like developments under shear. A dynamic yield stress investigation is led for each kind of structure alongside a statistical approach to deal with additionally characteristics and differentiate different sorts of alignment of magnetic particles [3]. The impact of particle flakes on the rheological properties of MR fluid with iron powder as magnetisable particles, constituted of two diverse volume fraction of particles dispersed in a carrier fluid is considered. To look at the MR impact, carbonyl particles (spherical) iron based MR fluid is considered. In both MR fluids, linear viscoelastic property has been widely explored utilizing magnetic sweep test and smaller oscillatory investigation, in the absence and presence of applied magnetic field H. The magnetic sweep test uncover that flake based MR fluid demonstrates a higher storage modulus contrasted with spherical based MR fluid and saturates at a lower field strength [4].

7.5 Mathematical Modelling

7.5.1 Uncompressible Viscos Flow with Pressure Increase

Before we look at mathematical modeling of magnetorheological fluids, we can recognise the flow behavior of a parallel plates of Newtanian fluids based on the pressure gradient. The parallel plate flow can be closely correlated with the flow as fluid flow in a damping application when fluid flows through the perforations in the piston. The flow between parallel plates with a pressure gradient is also referred to as the Poiseuille flow.

Let us consider an incompressible two-dimensional viscous laminar flow along the two parallel plates, which are separated by a distance of 2 h (see Fig. 7.3), the plates being fixed, but the pressure fluctuating along the x-axis. It is assumed that the plates are too wide and too long, so the flow is essentially axial, i.e. ($u \leq 0$, $v = w = 0$). Using the continuity equation, he suggests an end similar to that of $u = u(y)$. Furthermore, if gravity is ignored and $v = w = 0$, the momentum equation with respective axis are derived as follows:

According to Navier Stroke equation

$$\rho\left\{\frac{\partial u}{\partial t} + u\frac{\partial u}{\partial x} + v\frac{\partial u}{\partial y} + w\frac{\partial u}{\partial z}\right\} = \rho g_x - \frac{\partial p}{\partial x} + \mu\left\{\frac{\partial^2 u}{\partial x^2} + \frac{\partial^2 u}{\partial y^2} + \frac{\partial^2 u}{\partial z^2}\right\} \quad (7.2)$$

Fig. 7.3 The parallel plates viscous incompressible flow

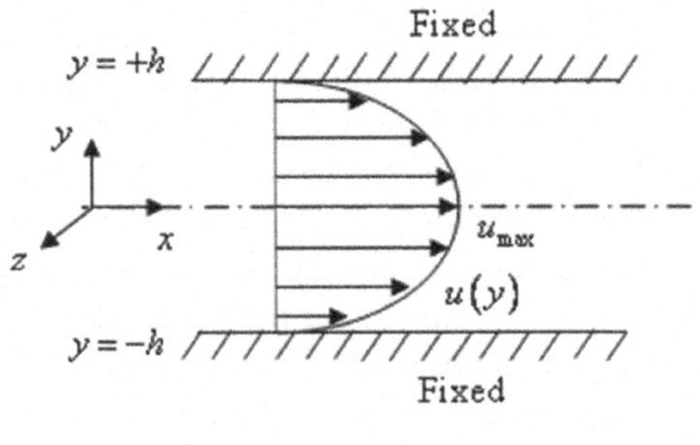

We know that,

Y directional Momentum $\frac{\partial p}{\partial y} = 0$, and z directional momentum $\frac{\partial p}{\partial z} = 0$; and $p = p(x)$.

$$X \text{ directional momentum } \mu \frac{\partial^2 u}{\partial y^2} = \frac{\partial p}{\partial x} = \frac{dp}{dx} \tag{7.3}$$

In the equation of impulse x we can see that the left side comprises of its variation along u_y, whereas the right hand side implies the dissimilarity with p_x. It essential causes the same constant value, if not they would not have been independent of its respective sides. Subsequently the flow has to overcome the shear stress of the wall and the pressure has to decrease in the direction of flow, the constant must be a negative value. This type of pressure-driven flow is called the Poiseuille flow, which is fundamental in particular with water-driven frames, brakes in automobiles, etc. The last type of equation that is obtained for a pressure gradient flow along two fixed parallel plates is given by.

$$\mu \frac{\partial^2 u}{\partial y^2} = \frac{dp}{dx} = Constant < 0 \tag{7.4}$$

By double integrating Eq. 7.4 the solution can be obtained as

$$u = \frac{1}{\mu}\left(\frac{dp}{dx}\right)\left(\frac{y^2}{2}\right) + c_1 y + c_2 \tag{7.5}$$

The constant values are obtained using no-slip condition

$$\text{At} \quad y = +h; \; u = 0 \Rightarrow c_1 = 0 \quad \text{and} \quad c_2 = -\frac{dp}{dx}\left(\frac{h^2}{2\mu}\right) \tag{7.6}$$

The general solution for Eq. 7.5 is obtained by substituting constants and solving as

$$u = \frac{1}{\mu} - \left(\frac{dp}{dx}\right)\left(\frac{h^2}{2\mu}\right)\left(1 - \frac{y^2}{h^2}\right) \tag{7.7}$$

The flow represented by the equation follows the Poiseuille parabola with a constant curvature and the maximum speed $u_{\max}$ along the median line $y = 0$

$$u_{\max} = \frac{1}{\mu} - \left(\frac{dp}{dx}\right)\left(\frac{h^2}{2\mu}\right) \tag{7.8}$$

Volume flow rate q flowing between the parallel plates per unit depth is given by

$$q = \int_{-h}^{h} \left(\frac{1}{2\mu}\right)\left(\frac{dp}{dx}\right)(h^2 - y^2)dy = \frac{2h^3}{3\mu}\left(\frac{dp}{dx}\right) \tag{7.9}$$

The pressure-drop between two point Δp at a distance l along x direction, then

$$q = \frac{2h^3}{3\mu}\left(\frac{\Delta p}{l}\right) \tag{7.10}$$

Average velocity is expressed as

$$u_{avg} = \frac{q}{2h} = \frac{h^2}{3\mu}\left(\frac{\Delta p}{l}\right) = \frac{3}{2}u_{\max} \tag{7.11}$$

The shear stress obtained using the Newtonian fluid,

$$\tau_w = \mu\left(\frac{\partial u}{\partial y} + \frac{\partial v}{\partial x}\right)_{y=\pm h} = \mu\frac{\partial}{\partial y}\left[\left(-\frac{dp}{dx}\right)\left(\frac{h^2}{2\mu}\right)\left(1 - \frac{y^2}{h^2}\right)\right]_{y=\pm h}$$

$$\tau_w = \pm\left(\frac{dp}{dx}\right)h = \pm\frac{2\mu u_{\max}}{h} \tag{7.12}$$

The equation signifies shear stress along parallel plate pressure predisposed Newtonian fluid flow.

7.5.2 Rheological Magnetic Fluid Models

The modeling of magnetorheological fluids (MR) has a very important character in the preparation of MR fluids and its applications whose working is purely based on its fluid performance, since many factors and properties of the constituents are in control for the overall behavior. Among various variables which determines the behavior of MR fluid, hence it becomes very important aspect to model and analyze the mathematical model that can predict and simulate performance of magnetorheological fluid.

The subsequent study details two models for the functioning of the MR fluids, the first model details over the visco-plastic performance employing two mathematical models, whereby the Navier-Stokes equations also use a quasi-stable state model to simulate MR fluid over a fixed parallel plates, along the evolution of this equation can be employed to validate these latest investigational results.

7.5.3 Visco Plastic Model

The visco plastic model along which the characteristic dependent on the field of the liquid MR is usually developed is the plastic model from Bingham. Bingham's model can be expressed as

$$\tau = \pm\tau_0 + \mu\gamma \tag{7.13}$$

where in τ_0—the yield stress, μ—viscosity and γ—shear rate.

The fluid begins to flow at a given point where elastic limit exceeds the shear stress value, i.e. $\tau < \tau_0 \Rightarrow \gamma = 0$. Figure 7.4 the plastic model Bingham is shown, which effectively shows the behavior of the elastic limit as the function of the externally applied magnetic field.

The second model that replaces the existing Bingham model with a little modifications is the Herschel-Bulkley's model, wherein it deals with dilution after shearing the MR fluid. The Herschel-Bulkley's model is presented as

$$\tau = \pm\tau_0 + K|\mu|^{\frac{1}{m}} \tag{7.14}$$

The thinning by shearing the fluid when m takes the value larger than the unit, while the equation when m taking the value smaller than the unit also represents the thickening by shearing for the value m which is equal to the unit becomes the equation reduced to the Bingham plastic model.

The Bingham and Herschel-Bulkley models have been used in a number of models to represent the direction of certain MR fluid applications. Many models of this type are preserved thanks to the work of Phillips. Some experimental studies have

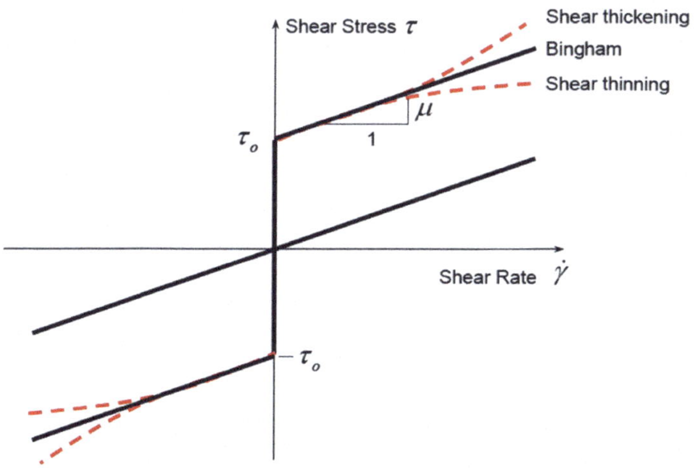

Fig. 7.4 Viscoplastic models that are often used to describe MR fluids [35]

expanded Phillip's work to include models of axially symmetrical dampers [1, 36]. As a rule, the size of the space is very small in contrast to the diameter of the annular space, the axisymmetric problem arises from the approximation of the parallel plate [35, 36]. It has been shown that most errors between the approximation of the parallel plate and the axisymmetric model are less than 5% [36]. The simplicity of the parallel plate model and the small error justify its use in the 20 most shock absorbing models. In addition, the parallel flow of MR fluid is the reason for modeling the MR fluid device that operates in shear mode or in valve mode.

7.6 Quasi-steady Parallel Flow by MR Fluid

The essential material science of fluid flow can be explained by the conditions of momentum and mass conservation. The equation of motion for an incompressible Newtonian fluid, i.e. with constant density and single phase, can be put together as Eq. 7.2.

The MR fluid comprises three essential elements, in particular the carrier fluid or the base fluid, the magnetic particles and the additive or the stabilizer, in order to improve the dispersion of the magnetic particles in the carrier fluid and to improve the performance of the fluid when it is exposed to a specific application, In all cases, MR fluids without an applied magnetic field act similarly to Newtonian fluid when the fluid is exposed to an applied magnetic field, which leads to the development of stresses and the fluid behaves like a Bingham fluid, which can be expressed by the relationship

$$\tau = \pm\tau_0 + \mu\frac{du}{dy} \tag{7.15}$$

where τ_0—the elastic limit depends on the field, *du*/*dy*—the shear rate and μ—the viscosity.

Figure 7.5 shows an almost stable flow of MR fluid through restricted parallel plates. It should be noted that this flow behavior is used regularly to describe the flow of liquid through an MR valve. The flow behavior shown in Fig. 7.5 can be divided into three specific sections. Check that the flow does not start until the limit point has been exceeded. In this way, the flow in the channel can be divided into three different sections. In sections I and II, where the shear rate is high, the liquid flows similarly to the Newtonian flow. In zone III, the liquid crosses the channel as strongly or appropriately as possible. In this zone, the elastic limit τ_0 has not been exceeded and the liquid is then not sheared.

With the specific end goal of determining a connection for the pressure drop caused by the flow characteristics (see Fig. 7.5). The technique shown begins with the Navier-Stokes equation. The velocity profile is calculated by implementing boundary conditions for the shear rate and the velocity in relation to the channel geometry. An instantaneous equilibrium of the fluid leads to a plug geometry and takes into account

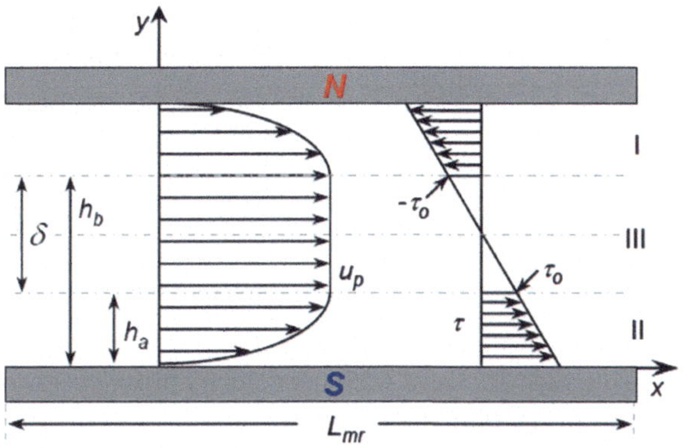

Fig. 7.5 Flow of MR fluid through fixed parallel plates [1]

the reduction in the speed profile. Finally, the average speed and the volume flow result in a cubic joint for the pressure gradient. The solution of the cubic equation is expressed in closed form.

The reduced form can be written as

$$\rho\left(\frac{\partial u}{\partial t}+u\frac{\partial u}{\partial x}\right)=\rho g_x-\frac{\partial p}{\partial x}+\mu\frac{\partial^2 u}{\partial y^2} \tag{7.16}$$

Considering horizontal and developed flow, momentum equation is written as

$$\frac{\partial^2 u}{\partial y^2}=\frac{1}{\mu}\frac{dp}{dx} \tag{7.17}$$

Integration the above equation and applying the boundary conditions

$$\frac{du}{dy}(h_a)=0 \qquad 0\le y\le h_a \tag{7.18}$$

$$\frac{du}{dy}=0 \qquad h_a\le y\le h_b \tag{7.19}$$

$$\frac{du}{dy}(h_b)=0 \qquad h_b\le y\le h \tag{7.20}$$

Hence we have equations for shear rate. Also the shear rates for region I and II are expressed using Eqs. 7.18 and 7.20 as

$$\frac{du}{dy}=\frac{1}{\mu}\frac{dp}{dx}(y-h_a) \qquad 0\le y\le h_a \tag{7.21}$$

$$\frac{du}{dy}=\frac{1}{\mu}\frac{dp}{dx}(y-h_b) \qquad h_b\le y\le h \tag{7.22}$$

Whereas the shear rate, velocity in various sections can be determined by integrating again against y and applying the following boundary conditions

$$u(0) = 0 \qquad 0 \le y \le h_a \tag{7.23}$$

$$u = u_p \qquad h_a \le y \le h_b \tag{7.24}$$

$$u(h) = 0 \qquad h_b \le y \le h \tag{7.25}$$

Using Eqs. 7.23 and 7.25, the velocity profile in the regions I and II are determined in terms of the plug geometry as

$$u = \frac{1}{2\mu}\frac{dp}{dx}y(y - 2h_a) \qquad 0 \le y \le h_a \tag{7.26}$$

$$u = \frac{1}{2\mu}\frac{dp}{dx}\left[\left(y^2 - h^2\right) - 2h_b(y - h)\right] \qquad h_b \le y \le h \tag{7.27}$$

7.7 Plug Geometry

To fully determine the velocity profile, the unknowns up, h_a and h_b must be determined. Using one of the equations, 7.26 or 7.27, the setting speed can be determined using the conditions uI $(h_a) = u_p$ or uII $(h_b) = u_p$. Assessing the velocity profile in region I at $y = h_a$, we have the expression given for the plug velocity as

$$u_p = -\frac{h_a^2}{2\mu}\frac{dp}{dx} \tag{7.28}$$

To obtain the expression for the thickness of the plug $\delta = h_b - h_a$, is calculated by applying the condition of equilibrium to the plug. Consider the fluid element as shown in Fig. 7.6. The forces acting on the fluid element are the shear forces and the pressure gradient.

Force balance over fluid element shown in Fig. 7.6 hence,

$$\Delta p dy dz = -2\tau_0 dx dz \tag{7.29}$$

The clockwise rotation of element is due to shear stress is expressed as positive [9]. Also in terms of the plug section, Eq. 7.29 can be expressed as

$$\frac{dp}{dx}\delta = -2\tau_0 \tag{7.30}$$

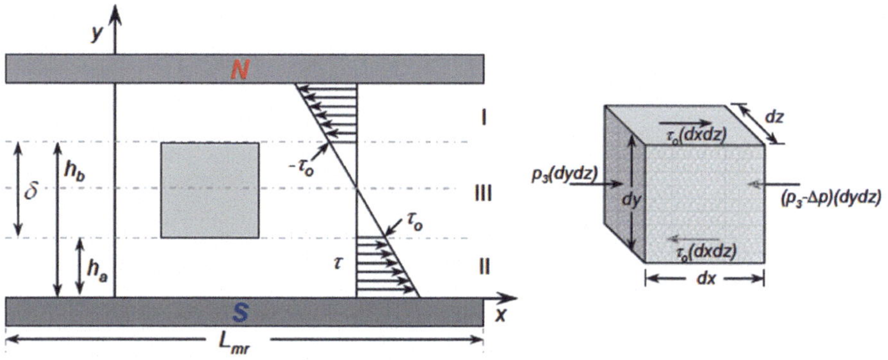

Fig. 7.6 Force balance over fluid element in plug section [1]

The plug thickness is then

$$\frac{dp}{dx}\delta = h_b - h_a = -\frac{2\tau_0}{dp/dx} \tag{7.31}$$

The velocity profile being symmetry we have $h_b = h - h_a$ and thus from Eq. 7.31 we estimate the values as

$$h_a = \frac{h}{2} + \frac{\tau_0}{dp/dx} \tag{7.32}$$

$$h_b = \frac{h}{2} - \frac{\tau_0}{dp/dx} \tag{7.33}$$

In terms of τ_0 and h the velocity profiles can be rewritten as

$$u = \frac{1}{2\mu}\frac{dp}{dx}y(y-h) - \frac{\tau_0}{\mu}y \qquad 0 \le y \le h_a \tag{7.34}$$

$$u = \frac{1}{2\mu}\frac{dp}{dx}y(y-h) + \frac{\tau_0}{\mu}(y-h) \qquad h_b \le y \le h \tag{7.35}$$

Similarly, the plug velocity can be written as

$$u_p = -\frac{1}{8\mu}\frac{dp}{dx}h^2 - \frac{\tau_0}{2\mu}\left\{h + \frac{\tau_0}{dp/dx}\right\} \tag{7.36}$$

By integrating the velocity profile along thickness of channel mean velocity can be written as

$$u_m = \frac{1}{h}\int_0^h u dy = \frac{1}{h}\left[\int_0^{h_a} u dy + \int_{h_a}^{h_b} u dy + \int_{h_b}^{h} u dy\right] \tag{7.37}$$

Substituting the expression for h_a from Eq. 7.33, we have

$$u_m = -\frac{1}{12\mu}\frac{dp}{dx}h^2 - \frac{h\tau_0}{4\mu} + \frac{1}{3\mu h}\frac{\tau_0^3}{(dp/dx)^2} \tag{7.38}$$

Also it can be seen that for $\tau_0 = 0$, Eq. 7.38 deduces the mean velocity for the Newtonian flow. Also, by knowing τ_0 and u_m, the above equation can be written in third order equation for pressure gradient.

$$\left(\frac{dp}{dx}\right)^3 + \left(\frac{12u_m\mu}{h^2} + \frac{3\tau_0}{h}\right)\left(\frac{dp}{dx}\right)^2 - \frac{4\tau_0^3}{h^3} = 0 \tag{7.39}$$

Again for $\tau_0 = 0$, Eq. 7.39 follows Newtonian flow

$$\frac{dp}{dx} = -\frac{12u_m\mu}{h^2} \tag{7.40}$$

Also take into account the opposite sides where there is no flow due to the formation of plugs of width h. If $\delta = h$, Eq. 7.40 can be expressed as a critical pressure drop

$$\frac{dp}{dx_c} = -\frac{2\tau_0}{h} \tag{7.41}$$

It is the lowest pressure gradient that can still spread the flow along parallel plates. In order for the parallel plates to circulate, the following condition must be fulfilled:

$$\frac{dp}{dx} \geq \frac{dp}{dx_c} \tag{7.42}$$

7.8 Pressure Gradient Solution

For solving Eq. 7.42 for pressure gradient, we must normalize with respect to Newtonian flow

$$\left(\frac{dp}{dx}\Big/\frac{dp}{dx_N}\right)^3 - \left(1 + \frac{3\tau_0 h}{12u_m\mu}\right)\left(\frac{dp}{dx}\Big/\frac{dp}{dx_N}\right)^2 + 4\left(\frac{\tau_0 h}{12u_m\mu}\right)^3 = 0 \tag{7.43}$$

Also the two non-dimensional parameters [9]

$$\mathcal{P} = -\frac{dp}{dx}\frac{h^2}{12u_m\mu} \tag{7.44}$$

$$\mathcal{T} = \frac{\tau_0 h}{12u_m\mu} \tag{7.45}$$

Equation 7.43 reduces to the form presented by Phillips [9].

$$\mathcal{P}^3 - (1 + 3\mathcal{T})\mathcal{P}^2 + 4\mathcal{T}^3 = 0 \tag{7.46}$$

The above equation can be written as

$$\mathcal{P}^3 + a\mathcal{P}^2 + b = 0 \tag{7.47}$$

where $a = -(1 + 3\mathcal{T})$ and $b = 4\mathcal{T}^3$.

Substituting and solving Eq. 7.47 we have

$$m = \frac{1}{3}(1 + 3\mathcal{T})^2 \tag{7.48}$$

$$\theta = \cos^{-1}\left\{\frac{54\mathcal{T}^3}{(1 + 3\mathcal{T})^{5/2}}\right\} \tag{7.49}$$

$$g = -\frac{2}{3}(1 + 3\mathcal{T}) \tag{7.50}$$

From the transformation, we find the solution for $\mathcal{P}$

$$\mathcal{P}(\mathcal{T}) = \frac{2}{3}(1 + 3\mathcal{T})\left\{\cos\left(\frac{1}{3}\cos^{-1}\left(1 - \frac{54\mathcal{T}^3}{(1 + 3\mathcal{T})^{5/2}}\right)\right) + \frac{1}{2}\right\} \tag{7.51}$$

The solution for the non-dimensional pressure gradient P is expressed by Eq. 7.51 as a non-dimensional elastic limit T. Along the solution also an equation for the pressure drop in the duct flow can be written in relation to the duct geometry.

Equation 7.52 represents the pressure drop that develops in the MR fluid flowing along two solid plates. The pressure drop is responsible for the controllability of the MR fluid flow.

The solution for the non-dimensional pressure gradient, $\mathcal{P}$ is expressed by Eq. 7.51 as a non-dimensional elastic limit, $\mathcal{T}$. Along the solution also an equation for the pressure drop in the duct flow can be written in relation to the duct geometry

$$\Delta P_\tau = \frac{12\mathcal{P}u_m\mu}{h^2}L_{mr} \tag{7.52}$$

Equation 7.52 represents the pressure drop that develops in the MR fluid flowing along two solid plates. The pressure drop is responsible for the controllability of the MR fluid flow.

7.9 Synthesis of Fluid

There enormous number of procedure being followed in synthesis of the magic fluid known as MR fluid with a basic principle of getting a complete uniform mixture of base fluid, particles additives and surfactants. The properties of each constituent contributes to the complete performance of the fluid with required yield strength, but the elementary property of the fluid which is essential to attain is nominal sedimentation ratio i.e. stable MR fluid. Here we will deliberate two of the procedure for the synthesis of fluid (Fig. 7.7).

To get MR fluid prepared, first the suitable measure of additives like anti-wear, rust, corrosion, friction modifier and anti-oxidant agent is added to the deliberate amount of carrier fluid and blended fittingly for 15 min using mechanical stirrer. The uncoated Iron particles were then straightforwardly scattered through measured volume fraction in the prepared blend. This blend was homogenized by creating disturbance using a mechanical stirrer at 1000 RPM for 24 h to make homogeneous MR fluid.

Light paraffin oil, conoco LVT, mineral, silicone, kerosene oil with viscosities ranging from 1.5 to 700 cP are some of the examples of the oils which can serve the purpose of carrier or base fluids in development of MR fluids. Colloidal silica along with glycol ethylene mono-propyl ether also called as NPC-ST and Poly-dimethyl siloxane fluid i.e. PDMS of kinematic viscosity 100 cSt which is equivalent to dynamic viscosity of 0.96 Pa s, are used in development of the MR fluids. The concentration of silica in NPC-ST is estimated to 20% by weight.

In MR fluid with PDMS as carrier fluid, the hydroxy terminated PDMS with a viscosity value ranging from 90 to 150 cSt is used as surfactant is added so is to prevent iron particles agglomeration and caking also this helps in providing steric stabilization.

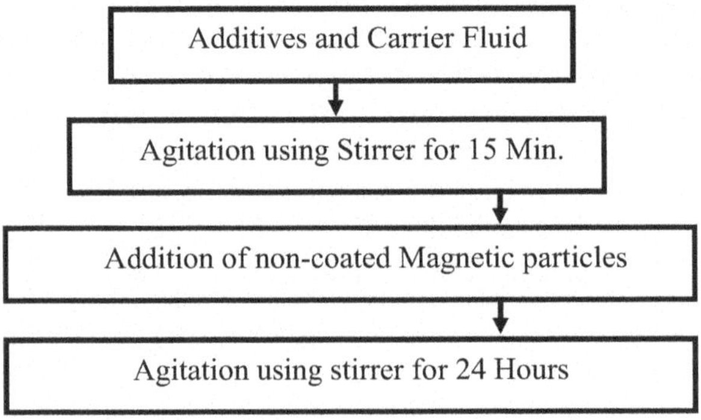

Fig. 7.7 Schematic representation for synthesis route of magneto rheological fluid

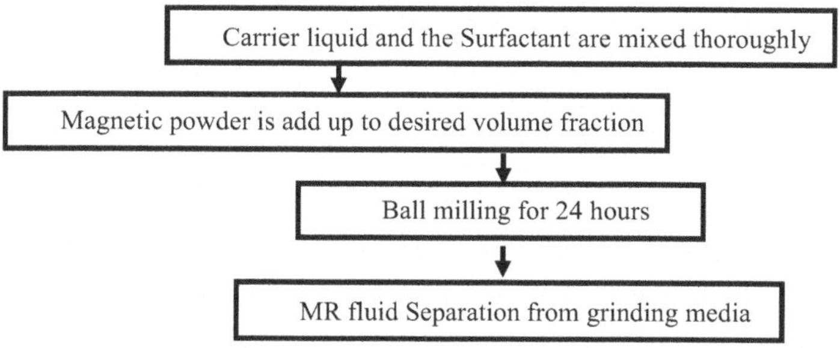

Fig. 7.8 Schematic representation of steps involved in synthesizing magneto rheological fluid

The MR fluids with base fluid as glycol ether, to form network by adsorbing on the surface of iron particles and silica poly-vinyl pyrrolidone dissolved along octanol is added to the blend. In groundwork of the fluid, a ceramic processing methods were employed. The flowchart presenting synthesis method of PDMS based MR fluids is presented Fig. 7.8, desired amounts of surfactant and base oil were agitated for 1 min using a flask. The iron particles were then poured and dispersed for 15 min at 1500 RPM. The blend is subjected to ball-milling for 24 h. The ball milling process was carried out using 5 mm stabilized zirconia balls. Also for the glycol ether based MR fluids similar process is followed.

7.10 Characterization of Fluid

Characterization of fluid is carried out in a standard rheometer with a magnetic attachment with a standard gap of 1 mm is employed for parting apart the parallel disks. Also the magnetic setup is incorporated such that magnetic flux generated are perpendicular to the parallel disk. Magnetic cell should be equipped with continuous magnetic field parallel to disks with variable magnetic strength.

All the study is carried out at 25 °C maintaining temperature constant for all the studies, the shear rate is maintained constant for few seconds until the value of shear stress reaches steady state value, this is done to ensure consistency throughout study the RPM is also maintained at a constant value.

The following tests can be carried out

1. Test rheological properties like strain, viscosity, Shear stress and shear rate using Magneto-rheometer.

    (i) Rheological properties under controlled shear rate
    Shear stress versus shear rate, Shear viscosity versus Shear rate, Shear Viscosity versus shear stress, under different magnetic field conditions like 0, 86, 171 and 342 kA/m
    (ii) Oscillatory Shear Test (Dynamic Test)

(a) Amplitude sweep results at different magnetic field

   (i) Storage Modulus versus Strain
   (ii) Loss modulus versus Strain

(b) Frequency sweep results at different magnetic field

   (i) Storage Modulus versus Strain
   (ii) Loss modulus versus Strain

2. Yield stress measurement under different magnetic field conditions
   Yield stress versus Magnetic field strength.

7.10.1 Sedimentation Test

Sedimentation is the affinity of particles in a colloidal solution to come to rest at the bottom after being dispersed in a liquid. This is the result of Brownian motion in the liquid in response to the forces acting on the particles. The cause of the forces can be a gravitational force, a magnetic field or centrifugal forces. The sedimentation rate in the MR fluid is measured by ignoring the magnetic force on the MR fluid particle. In addition, the characteristic difference in density of the carrier liquid and the iron particles leads to sedimentation of the particles, which leads to a larger clarified volume, which is referred to as the supernatant liquid above the sludge line. The mud lines form the boundary between the supernatant liquid and the resulting turbulent part in colloidal solution. The observations are recorded, followed by a visual observation of the change in position of the sedimentation in the MR fluid sample. The configuration used for observation and data acquisition is shown in Fig. 7.9. The configuration consists of a transparent measuring tube and a measuring ruler to record the protrusion. The supernatant increases and the cloudy part decreases over time.

The sedimentation levels are recorded after a visual inspection during the day to obtain a clear observation in order to record the cloudy part and the level of the supernatant of the MR fluid. The synthesized liquid is poured up to 10 cm into a test tube and allowed to stand in an upright position for a few hours. The level of excess liquid (H_s) is measured with a stopwatch and a ruler after a specified time interval.

The sedimentation rate is the ratio between the height of the supernatant liquid and the test height of the MR liquid taken for observation and is expressed in

$$S_R = \frac{H_s}{H_T} \times 100\%$$

where S_R is the sedimentation rate,
H_T is the total height of the liquid to be observed in the test tube,
H_s is the height of the excess liquid.

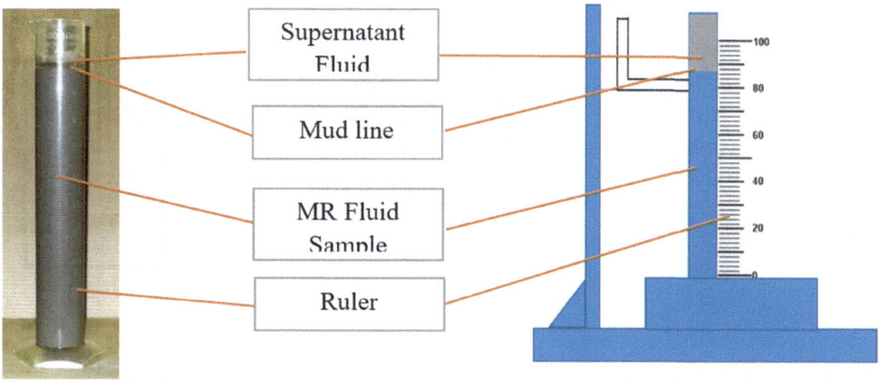

Fig. 7.9 **a** Measurement of sedimentation and **b** pictorial clarification used for determination of sedimentation-ratio

7.11 Applications by MR Fluid

The potential uses of MR fluids can be found in devices that require rapid and reversible change and persistent rheological properties [2]. Magneto-rheological devices have triggered an impressive enthusiasm in the past few decades, since a magneto-rheological fluid coordinates mechanical devices with an electronic frame and thus permanently guarantees the mechanical properties of the device. Part of these devices that use magnetorheological fluids is another era of clutches, brakes and shock absorbers. Magneto-rheological shock absorbers, especially as shock absorbers, are the most commonly used devices of this type [37]. Power steering pumps, control valves, artificial joints, alternators, engine mounts, chemical detection, sound propagation applications and others are among these precedents [38, 39]. Drug transport and tumor solution treatment strategies are advanced uses of magnetic suspensions [25].

In a generally thorough review, Bica et al. [40] investigated the possible use of magnetorheological dampers. They distinguished important variables that affect the execution of MR fluids in a particular application. In another article, Wang and Meng [41] assessed different properties and uses of magnetorheological fluids. Their study found that the three main problems against the widespread use of MR fluid innovations in many advanced devices are their stabilization ability, durability, and cost. Since not all applications are dealt with in the following, the probably most important uses of magnetorheological fluids are briefly discussed below. Kciuk and Turczyn [33] examined the basic properties of MR fluids and their widespread use in various companies. In another study, Olabi and Grunwald [42] examined the properties of magnetorheological fluids and their application. As can be seen from their overview, the promising strengths of the innovation of MR fluids, such as the fast response, the basic interface between electrical power and mechanical efficiency

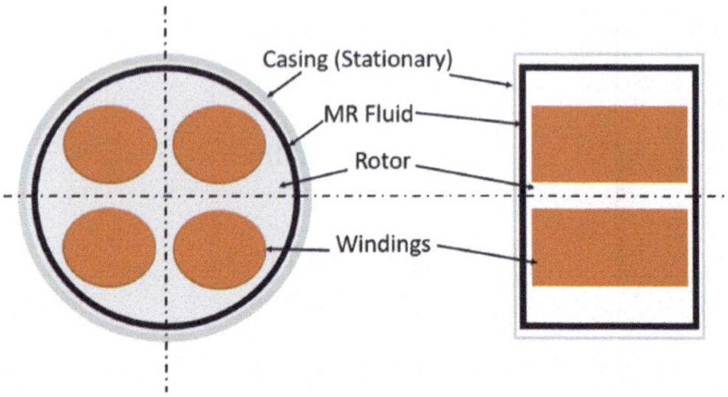

Fig. 7.10 Schematic representation of the magnetorheological braking system

and, in addition, the precise control capacity, determine the next innovation for some applications.

Due to the large variation in process and design parameters, the internally developed fluid can be synthesized in order to render it for a specific technical application. The main areas of application for the use of MR fluids are MR brakes, MR dampers, MR plain bearings, MR surface finishing.

7.11.1 Magneto-rheological Braking System

A MR break is a system comprising of consists of rotor with a rotating disk with in MR fluid which is filled up in the narrow slit between stator and rotor. The rotor rotates along the fluid imparting very low resistance to the rotor, as the magnetic field is activated increases the viscosity intern increasing yield stress due to shear resisting rotation and hence breaking is achieved. The breaking effect can be achieved with in 65 ms which make the system a quick and reliable. The investigators have already fabricated prototype MR fluid based brake. Detailed analysis of fluid flow interaction with magnetic field will be carried to optimize the brake design with a braking time of 65 ms (Fig. 7.10).

7.11.2 Magneto-rheological Journal Bearing

A journal bearing with MR fluid can be used for high speed and load application as the fluid film between journal and bearing is very low. The only separation which can be achieved to reduce the friction is by using viscosity which is achieved by step by step increasing applied magnetic field strength using electromagnetic coil. As the journal start rotating the temperature of fluid increases resulting in reduction in viscosity a controller monitoring the variation in temperature with a feedback to current source together can maintain required film thickness and hence a smart journal bearing operation can be achieved. The journal bearing will be redesigned

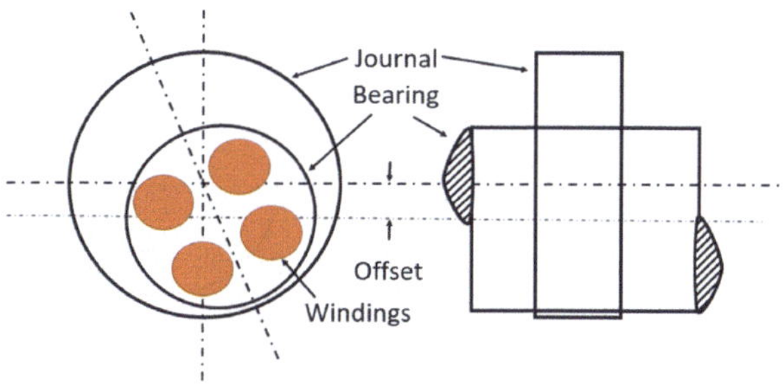

Fig. 7.11 Schematic representation of magnetorheological journal bearing

to accommodate the magnetic components with the help of Industrial collaborators. Standard MR fluid Journal bearing setup will be the outcome (Fig. 7.11).

7.11.3 Magneto-rheological Heat Conduction

Magneto-rheological fluid can be used as the heat transfer media with a variable thermal conductivity of the MR fluid which is also the function of applied current in turn induced magnetic field. The heat transfer rate largely depends upon the magnetic particles used and size of the particles used in the synthesis process (Fig. 7.12).

Under off-state condition the heat conduction is ruled by the thermal conductivity of the carrier fluid and as the intensity of applied magnetic field increases due to dipole forces between the particles they are aligned to form chain like construction i.e. semi-solid state of fluid. In this state the heat conduction is directed by thermal conductivity of the carrier fluid and also the thermal conductivity of the metal also adds up, leading to increases in overall conductivity with rapid cooling.

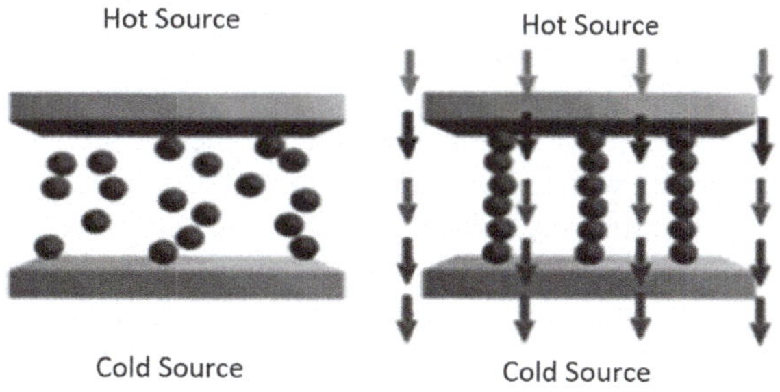

Fig. 7.12 Schematic representation of magneto-rheological heat conduction

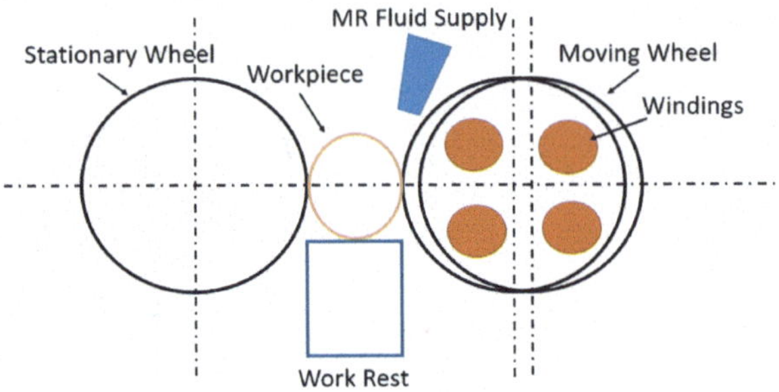

Fig. 7.13 Schematic representation of magneto-rheological grinding system

7.11.4 Magneto-rheological Surface Finishing

Magneto-rheological fluid can be used as a grinding fluid with variable magnetic field arrangement and a rolling element on the top similar to a grinding machine without any abrasive particles on the roller. The MR fluid is spread on the surface to be machined and the electric coils are energized due to dipole moment the particles align along the applied magnetic field and act as abrasive particle. This process can be used for machining of any material varying from polymers to conductors. The material harder than the iron particles needs additional appropriate abrasive particles to felicitate the machining process. In this phase in-house setup will be developed with a facility to vary magnetic field during machining (Fig. 7.13).

7.12 Conclusion

Presently use of magneto rheological fluids with satisfactory control capacity possess unlimited Importance in numerous applications. Magneto rheological fluid which are suspension of magnetic particles in a carrier fluid with additives as stabilizer, exhibit characteristics of a semi solid material under the influence of magnetic field. Acquiring incredible yield stress i.e. higher magneto rheological impact and minimizing the sedimentation of magnetic particles, under the influence of gravity, are two imperative difficulties with regards to MR fluids. These issues, which primarily rely upon shape, type, volume portion and size of the magnetic particles, impressively influences the rheological characteristics of these fluids. Selection of magnetic particles relies upon different variables like compatibility with base fluid, requested MR impact, stability requirements etc. However, carbonyl iron micro particles are the most encouraging particles for magnetic scattered phase in MR fluids. Choice of carbonyl iron particles adds to low coercivity, high saturation of magnetization, widespread availability and the relatively low pricing. Till date bunches of techniques have been acquainted with minimized sedimentation and also enhanced MR effect. The most essential of these strategies are: diminishing the density of magnetic particles by employing the process of coating, expanding the viscosity of the base fluid by utilizing high viscose fluid and utilizing nano structure-materials and altering molecule surface by including stabilizer surfactant.

Coating over the magnetic particles is one conceivable technique to decrease interactions among the particles and additionally reducing the density. By utilizing this strategy particle stability can be improved significantly, despite the fact that the MR impact will be brought down undesirably. Among every one of the materials utilized in this technique, carbonyl iron particles coated with polymeric materials have been utilized broadly. Another sedimentation technique that is a standout amongst the most encouraging ones, which likewise upgrades magneto rheological impact, is related with another structure of magneto rheological fluids derived from suspending magnetic small scale particles in a ferro fluid. Utilizing a combination of micro particles and nano particles in MR fluid makes bonds among these standard particles which thusly improve the yield stress. In addition, presence of nano particles in the base fluid expands consistency of the base fluid and consequently diminishes sedimentation. Besides, a few examinations show that the base fluid has a significant job in magneto rheology. Utilizing fluids with greater viscosity adds to limit the viscosity distinction between the base fluid and magnetic particles, as the primary powerful factor in sedimentation and in dependability of these suspensions. Unfortunately, viscosity for fluid is elevated under nonmagnetic condition called as off state which isn't alluring. In any case, silicon oil is as yet known as the main selection for the magneto rheological fluids synthesis because of its accessibility, low viscosity, minimal cost, and so on. A few research explores on utilizing legitimate additive that can enhance both MR impact and stability. Up until this point, materials that can give these prerequisites have not been distinguished. By and by, it appears that among all surfactant and additive, organo clay, fumed silica and stearic acid are appropriate materials. Unfortunately, huge numbers of the stabilizing strategies decrease yield stress and along these lines magneto rheological impact, and extra research in this setting is by all accounts essential. Regardless of various examinations directed around there, since the disclosure of magneto rheological fluids, despite everything it appears to be important to locate another composition of these substances which can yield higher stability and furthermore minimal cost. Advancement of magneto rheological fluid application, require extra research for enhancing the MR impact and additionally stability of these fluids against sedimentation. Among all examined strategies, utilizing a mixture of magneto rheological fluids and ferro fluids, utilizing nanowires and furthermore including legitimate stabilizers might be accounted as appropriate stabilizing techniques which can enhance both magneto rheological impact and stability of MR fluids. Assurance of as far as possible increase in nanoparticles and nanowires to magneto rheological fluids, and utilizing appropriate stabilizers needs additional analyzes which can be the subject of numerous ongoing researches.

References

1. W. Kordonsky, O. Ashour, C.A. Rogers, Magnetorheological fluids: materials, characterization, and devices. J. Intell. Mater. Syst. Struct. **7**, 123–130 (1996)
2. P.P. Phulé, J.M. Ginder, A.D. Jatkar, Synthesis and properties of magnetorheological fluids for active vibration control, in *Materials for Smart Systems II* (Materials Research Society, Boston, MA, 1996)
3. K.D. Weiss et al., High strength magneto- and electro-rheological fluids. SAE Trans. **102**, 425–430 (1993)
4. R.W. Phillips, Engineering applications of fluid with a variable yield stress, 1969
5. K.P. Tan, R. Stanway, W.A. Bullough, Braking responses of inertia/load by using an electro-rheological (ER) brake. J. Mechatron. **17**, 277–289 (2007)
6. E. Park, D. Stoikov, L. Falcao de Luz, A. Suleman, A performance evaluation of an automotive magnetorheological brake design with a sliding mode controller. Mechatronics **16**(16), 405 (2006)
7. E.J. Park, L. Falcao da Luz, A. Suleman, Multidisciplinary design optimization of an automotive magnetorheological brake design. Comput. Struct. **86**, 207 (2007)
8. Y. Shiao, Q.A. Nguyen, Structural analysis and validation of the multi-pole magnetorheological brake for motorcycles. Procedia Eng. **76**, 24 (2014)
9. D.J. Carlson, Magnetorheological brake with integrated flywheel. U.S. Patent 6,186,290 Bl, 2001
10. K. Karakoc, E.J. Park, A. Suleman, Design considerations for an automotive magneto-rheological brake. Mechatronics **18**(8), 434 (2008)
11. G.M. Webb, Exercise apparatus and associated method including rheological fluid brake. U.S. Patent 5,810,696, 1998
12. S. Bydon, *Construction and Operation of Magnetorheological Rotary Brake* (Archiwum Process Control, 2002), p. 20
13. B. Liu, W.H. Li, P.B. Kosasih, X.Z. Zhang, Development of an MR brake based haptic device. Smart Mater. Struct. **15**, 1960 (2006)
14. C.M. Chew, G.S. Hong, Development of a compact double-disk magnetorheological fluid brake. Robotica **25**, 493 (2006)
15. W. Zhou, C.M. Chew, G.S. Hong, Development of a compact double-disk magnetorheological fluid brake. Robotica **55**, 493 (2006)
16. J.D. Carlson, What makes a good MR fluid? J. Intell. Mater. Syst. Struct. **13**, 431 (2002)
17. P.P. Phule, A.D. Jatkar, Synthesis and processing magnetic iron cobalt alloy particles for high strength magnetorheological fluids, in *6th International Conference on ER Fluids, MR Suspensions and Their Applications* (World Scientific, Yonezawa, Japan, 1997), pp. 503–510
18. Q. Nguyen, S. Choi, Optimal design of an automotive magnetorheological brake considering geometric dimensions and zero-field friction heat. Smart Mater. Struct. **19** (2010)
19. P.P. Phule, J.M. Ginder, Synthesis and properties of novel magnetorheological fluids having improved stability and redispersibility, in *6th International Conference on ER Fluids, MR Suspensions and Their Applications* (World Scientific, Yonezawa, Japan, 1997), pp. 445–453
20. E. Lemaire, A. Meunier, G. Bossis, J. Liu, D. Felt, P. Bahtovoi, N. Matoussevitch, Influence of the particle size on the rheology of magnetorheological fluids. J. Rheol. **39**, 1011–1020
21. C. Kormann, H. Laun, H. Ritcher, MR fluids with nanosized magnetic panicles. Int. J. Mod. Phys. B **10**, 3167–3172 (1996)
22. N. Rosenfeld, N.M. Wereley, R. Radhakrishnan, T.S. Sudarshan, Behavior of magnetorheological fluids utilizing nanopowder iron. Int. J. Mod. Phys. B **16**(17–18), 2392–2398 (2002)
23. A. Chaudhuri, G. Nang, N.M. Wereley, V. Tasovksi, R. Radhakrislman, Substitution of micron by nanometer scale powders in magnetorheological fluids. Int. J. Mod. Phys. B: Condens. Matter Phys. **19**(7–9), 1374–1380 (2005)

24. N.M. Werely, A. Chaudhuri, J.H. Yoo, S. John, S. Kotha, A. Suggs, R. Radhakrishnan, B.J. Love, T.S. Sudarshan, Bidisperse magnetorheological fluids using Fe particles at nanometer and micron scale. J. Intell. Mater. Syst. Struct. **17**, 393–401 (2006)
25. B.J. Park, K.H. Song, H.J. Choi, Magnetic carbonyl lion nanoparticle based magnetorheological suspension and its characteristics. Mater. Lett. **63**(15), 1350–1352 (2009)
26. E.F. Burguera, B.J. Love, R. Sahul, G. Ngatu, N.M. Wereley, A physical basis for stability in bimodal dispersions including micrometer-sized particles and nanoparticles using both linear and non-linear models to describe yield. J. Intell. Mater. Syst. Struct. **19**(11), 1361–1367 (2008)
27. F.F. Fang, H.J. Choi, M.S. Jhon, Magnetorheology of soft magnetic carbonyl iron suspension with single-walled carbon nanotube additive and its yield stress scaling function. Colloids Surf. A **351**(1–3), 46–51 (2009)
28. M.T. Lopez-Lopez, P. Kuzhir, A. Meunier, G. Bossis, Synthesis and magneto rheology of suspensions of cobalt particles with tunable particle size. J. Phys.: Conf. Ser. **149**, 012073 (2009)
29. S.T. Lim, M.S. Cho, I.B. Jang, H.J. Choi, Magnetorheological characterization of carbonyl iron suspension stabilized by finned silica. J. Magn. Magn. Mater. **282**, 170–173 (2004)
30. C. Fang, B.Y. Zhao, L.S. Chen, Q. Wu, N. Liu, K.A. Ku, The effect of the green additive guar gum on the properties of magnetorheological fluid. Smart Mater. Struct. **14**, N1–N5 (2005)
31. P. Pilule, Magnetorheological fluid. U.S. Patent 5,985,168, 1999
32. V.R. Foista Iyanger, S.M. Yugelevic, Stabilization of magnetorheological fluid suspensions using a mixture of organoclays. U.S. Patent 6_io/7-P1
33. B. Jang, H.B. Kim, J.Y. Lee, J.L. You, Role of organic coating on carbonyl iron suspended particles in magnetorheological fluids. J. Appl. Phys. **97**, 1–3 (2005)
34. P. Phule, Synthesis of novel magnetorheological fluids. MRS Bull. **23**, 23–24 (1998)
35. C.W. Macosko, *Rheology: Principles, Measurements, and Applications* (VCH Publishers Inc., New York, 1994)
36. J. Rabinow, Magnetic fluid torque and force transmitting device. U.S. Patent 1951, USA
37. N.M. Wereley, J.U. Cho, Y.T. Choi, S.B. Choi, Magnetorheological dampers in shear mode. Smart Mater. Struct. **17**(1), 015022 (2008)
38. M.R. Jolly, J.W. Bender, R.T. Mathers, Indirect measurement of micro structural development in magnetorheological fluids, in *6th International Conference on ER Fluids, MR Suspensions and Their Applications* (World Scientific, Yonezawa, Japan, 1997), pp. 471–477
39. M. Kciuk, R. Turczyn, Properties and application of magnetorheological fluids. J. Achiev. Mater. Manuf. Eng. **18**, 127–130 (2006)
40. P.J. Rankin, A.T. Horvath, D.J. Klingenberg, Magnetorheology in viscoplastic media. Rheol. Acta **38**, 471–477 (1999)
41. V.R. Iyanger, Durable magnetorheological fluid compositions. U.S. Patent 6,818,143, 2004
42. M.A. Golden, J.C. Ulieny, K.S. Snavely, A.L. Smith, Magnetorheological fluids. U.S. Patent 6,932,917, 2005
43. J. Rabinow, The magnetic fluid clutch. AIEE Trans. **67**, 1308 (1948)
44. P. Poddar, J.L. Wilson, H. Srikanth, J.-H. Yoo, N.N. Wereley, S. Kotha, L. Barthouty, R. Radhakrishnan, Nanocomposite magneto-rheological fluids with uniformly dispersed Fe nanoparticles. J. Nanosci. Nanotechnol. **4**(1–2), 192–196 (2004)
45. J.D. Carlson, M.R. Jolly, MR fluid, foam and elastomer devices. Mechatronics **10**, 555–569 (2000)
46. J.D. Vicente, D.J. Klingenberg, R. Hidalgo-Alvarez, Magnetorheological fluids: a review. Soft Matter **7**, 3701–3710 (2011)
47. I. Bica, Y.D. Liu, H.J. Choi, Physical characteristics of magnetorheological suspensions and their applications. J. Ind. Eng. Chem. **19**, 394–406 (2013)
48. J. Wang, G. Meng, Magnetorheological fluid devices: principles, characteristics and applications in mechanical engineering (Part L). Proc. Inst. Mech. Eng. **215**, 165–174 (2001)
49. A.G. Olabi, A. Grunwald, Design and application of magneto-rheological fluid. Mater. Des. **28**, 2658–2664 (2007)

50. W.W. Chooi, S.O. Oyadiji, Design, modelling and testing of magnetorheological (MR) dampers using analytical flow solutions. Comput. Struct. **86**, 473–482 (2008)
51. C. Guerrero-Sanchez, T. Lara-Ceniceros, E. Jimenez-Regalado, M. Rasa, U.S. Schubert, Magnetorheological fluids based on ionic liquids. Adv. Mater. **19**, 1740–1747 (2007)
52. J.H. Park, B.D. Chin, O.O. Park, Rheological properties and stabilization of magnetorheological fluids in a water-in-oil emulsion. J. Colloid Interface Sci. **240**, 349–354 (2001)

Chapter 8
Nanostructured Layered Materials as Novel Lubricant Additives for Tribological Applications

Sangita Kumari, Ajay Chouhan, and Om P. Khatri

Abstract Development of novel lubricant materials to reduce the friction and wear for a diversified range of tribological applications is a subject of great interest. Over the last one decade, two-dimensional (2D) nanostructured layered materials have attracted immense interests for a wide range of applications including enhancement of tribological properties. The fascinating and favorable characteristics of 2D nanomaterials viz. excellent thermal conductivity to take away the heat from contact interfaces, exceptionally high mechanical strength to provide remarkable wear-resistivity, low shear strength to reduce the friction, and high surface area to facilitate their dispersion have made the 2D nanomaterials highly promising candidates for solid thin films and additives to liquid lubricants. The chapter discusses the tribological applications of graphene, MoS_2, WS_2, and h-BN as additives to automotive lubricants, greases, and metalworking fluids. The chemical functionalization and dispersion aspects of 2D nanostructured layered materials, tribo-performance using variable contact geometries, and analyses of worn surfaces etc. are presented to outlines the lubrication mechanism.

8.1 Friction, Wear and Lubrication

Tribology term was coined by Peter Jost and it originated from the Greek word ‘tribos’, which stands for “rub”. In the modern era, the tribology covers various aspects of wear and friction generated during the relative movement of rubbing surfaces; and the lubrication, a process/technique to minimize the friction and wear. The tribological interfaces carry incredibly complex surface interactions. Thus, the tribological phenomenon requires the understanding of mechanical engineering, surface chemistry, contact-physics, and materials science to have precise interpretation about the nature of contact interfaces [1]. Many efforts have been directed to control friction and wear since the early age of civilization. The wheel was recognized

S. Kumari · A. Chouhan · O. P. Khatri (✉)
CSIR-Indian Institute of Petroleum, Dehradun 248005, India
e-mail: opkhatri@iip.res.in

J. K. Katiyar et al. (eds.), *Tribology in Materials and Applications*,
Materials Forming, Machining and Tribology,
https://doi.org/10.1007/978-3-030-47451-5_8

as an excellent invention to minimize friction between the carrier and the place of transportation. The vegetable oils and animal fats have been well established liquid lubricants even before the arrival of the modern age. In the fifteenth century, Leonardo da Vinci conducted the first scientific work to understand the nature of friction [2]. In the nineteenth century, Reynolds proposed a scientific theory for the lubrication [3]. Since then, tribology has gained significant momentum and a lot of work has been carried out addressing fundamental aspects to industrial applications based on friction, wear and lubrication.

Friction is the force, which resist the movement of two or more surfaces in contact with each other and it leads to dissipation of energy in the form of heat. Friction is a system response yet it governs by material properties. It can be either advantageous or detrimental depending on the systems. Friction is beneficial for several daily-life activities such as walking, use of breaks to stop the car, writing on notebook and so on. However, friction is undesirable in sliding, rolling, and rotating movements in several machineries. Friction is a major source of energy loss and it heats up the contact surfaces, which facilitates the plastic deformation and eventually wearing of contact surfaces. Wear is the material loss from the tribo-surfaces of engineering systems. Wear usually occurs at the points where the surface asperities are in direct contact with each others. Working conditions like load, speed, lubricants etc. influence the wear of contact interfaces. Wear is also recognized as useful or detrimental. The productive wear is a controlled phenomenon leading to advantageous achievements in many processes like shaving, polishing, machining, and writing. Wearing of materials in cams, seals, gears, pistons, bearings, and so on are detrimental events and cause a big threat to engineering systems. The material losses and surface damages eventually lead to failure of machinery parts. Therefore, the selection and designing of machineries parts to have minimum friction and wear during their relative motions are very important to conserve the energy and extend the life of machinery parts/assemblies [1]. Lubricants are the substances (liquid, solid, or semi-solid forms) commonly used for reductions of friction and wear of engineering systems. The lubricants are applied on contact interfaces to facilitate the motion and avoid their direct contact to minimize the wear. Therefore, the lubrication is an imperative approach for energy conservation by decreasing the friction and extends the life of machinery parts by reducing the wear.

8.2 Impacts of Friction and Wear on Energy, Environment and Economy

Friction is a loss of energy in the form of heat and it compromises the efficiency of engineering systems. A wide range of industries including transportation, mining, cement, power-generation etc. lose significant fraction of energy because of high friction. In the recent years, considerable efforts have been made to improve the fuel efficiency of automotive engines by advancing the lubricant systems and monitoring the materials and design of engineering surfaces. Almost one-third of fuel energy in automotive engine goes to overcome the friction. In a four-seater passenger car, energy losses due to friction are majorly found in engine (piston assembly, valve train, seals, bearing, pumping etc., 35%), transmission (gears, bearing etc., 15%), breaks (15%), and tyres (rolling resistance, 35%) [4]. The 1% friction reduction in the engine of four-seater car using advanced lubricant technology can annually save the 1 billion liters of liquid fuel [5]. Accordingly, significant reduction in CO_2 footprint because of fuel saving. The high efficiency of automotives and machineries could achieve the target projected for global limit of CO_2 emissions. The transportation sector emission adds ~18% to the greenhouse gas in the environment [6, 7]. The paper industries consumed 15–25% of energy to overcome the friction. Worldwide, 381,000 TJ annual energy was consumed by paper industries to overcome the frictional losses. The advancement of lubricant technologies in paper industries could decrease the frictional losses by 11 and 23.6% in short (~10 years) and long-terms (20–25 years). It would be equal to savings of 36,000 and 78,000 GWh electricity, and reduction of CO_2 emission by 10.6 and 22.7 million tonnes, respectively. The proposed advancement of lubricant technologies in paper industries includes the application of highly durable coatings, surface engineering and texturing, use of low shear lubricants, and so on [8].

Human-made, natural, and biological systems generate friction during sliding, rolling and rotating motions. High friction leads to damage of engineering surfaces, which eventually cause failure of machineries parts. The elimination of root causes of friction is essential to increase the durability, performance and productivity of engineering systems [9]. Lubrication has been an efficient approach to reduce the frictional barrier. Lubricants minimize the friction by forming a boundary film of low shear strength over the tribo-interfaces and prevent their direct contact under extreme pressure conditions of boundary lubrication regime, where high friction and material losses are prime threats for engineering systems. In the engine, various components work efficiently because of lubricants. The liquid lubricants are supplied to the contacting interfaces either by a pumping or by oil delivery mechanisms. The solid lubricants have been supplied in the form of thin films by means of physical vapor deposition (PVD), chemical vapor deposition (CVD), spray coating and so on. The low friction and high durability of machinery parts in engineering systems/operations needed to conserve the energy, reduce the emission, protect the environment, and bring-down the cost of operation.

8.3 Solid and Thin Film Lubrication

Lubricants are materials/substances employed to reduce friction and wear of engineering systems, where two or more bodies are in mutual contact and moving relative to each other. The types of lubricants being used for different applications are govern by nature of engineering systems, operating parameters (load, temperature, speed, and contact geometry), designs and materials of machineries/tools etc. The extreme tribological conditions like high temperature, severe contact pressure and high vacuum require solid lubricants-based solutions for controlling the friction and wear [10]. Solid lubricants are usually applied on moving surfaces by sprinkling, rubbing or burnishing their fine powder [11, 12]. Solid lubricants blended in an aerosol carrier are directly sprayed on the interfaces of engineering surfaces. The powder of solid lubricants are strongly adhered to the contact surfaces by use of appropriate adhesives like epoxy resins and extends the life of sliding surfaces [13]. Currently, the thin films of solid lubricants are preferred over the powders or bonded forms. Mostly, the solid lubricants are deposited as thin films over contacting interfaces using a wide range of deposition techniques, including sputtering, ion-beam-assisted deposition, ion plating etc. The application of these techniques provide the thin films of uniform thickness. Which increase the bonding strength; resulting in dense microstructure and furnishes a longer life to engineering surfaces [14, 15]. The finite thickness limits the life of the thin films. A self-replenishment property of solid lubricants is required to maintain the consistency and sustainability of thin films.

Over the last one decade, nanostructured two-dimensional (2D) layered materials viz. graphene, hexagonal boron nitride (*h*-BN) and transition-metal dichalcogenides (MoS_2 and WS_2) have been gaining significant interest as lubricious materials for thin films and liquid lubricants. The remarkable mechanical, electronic, and thermal properties beside their high surface area makes them potential candidates for lubricant applications. Figure 8.1 illustrates atomic structure of graphene, MoS_2, and *h*-BN nanosheets along with their characteristics interlamellar distances. In the each lamella, atoms are closely linked with each other by a covalent interaction, while these lamellae are pilled-up on each other via weak van der Waals interactions. Under the tribo-stress, the weak van der Waals force between these lamellae eases the shearing in the sliding direction and reduces the friction. The strong interatomic covalent linkages in each lamella furnishes high mechanical strength and it contributes for antiwear properties to subsidize the materials lose of tribo-surfaces.

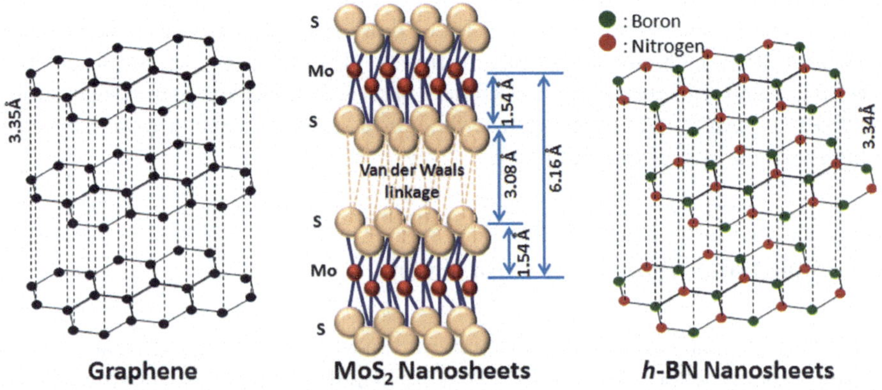

Fig. 8.1 Atomic structure of graphene, MoS_2 and *h*-BN nanosheets illustrating their lamellar view along with interlayer spacing

The transition-metal dichalcogenides exhibit optimum lubrication performance in the vacuum or dry conditions while moist and oxidizing environments deteriorate their tribological properties [16, 17]. The MoS_2 and WS_2 provide significant low coefficient of friction (0.002–0.05) while they are used as solid lubricants under dry, inert and vacuum environments. The coefficient of friction increased sharply to 0.2 in the presence of humid atmosphere. The solid lubricants interact with tribo-interfaces under the high pressure and temperature, and form the adherent tribo thin film of low shear strength, which not only reduced the friction but also decreased the wear [18]. As a result; MoS_2 has been widely explored as solid lubricants for dry lubrication applications [19].

The 2D nanostructured layered materials possess ultralow friction owing to their unique structural features. They display entirely distinct frictional properties in comparison to their bulk counterparts. The frictional properties of 2D nanostructured layered materials controlled by several parameters including the number of lamellae in the sheet, type of underlying substrate, chemical modification, nature and roughness of counter surface, the presence of defects etc. Lee et al. demonstrated the effect of the number of lamellae in graphene deposited onto SiO_2/Si substrate for their frictional characteristics [20]. The friction force microscopic measurement (Fig. 8.2) revealed that friction monotonically increased as the number of layers decreased for graphene, *h*-BN, MoS_2, and $NbSe_2$ nanosheets deposited on SiO_2 substrate [21]. The CVD-grown graphene thin film transferred onto a SiO_2/Si substrate showed significant reduction in adhesion and friction under micro- and nanoscale contact. The tribological studies revealed that graphene film promised a great potential as thinnest solid lubricant for reduction of adhesion and frictions besides extending the wonderful mechanical properties to protect the underlying substrate [22]. The 2D layered nanomaterials exhibit various types of structural defects such as grain boundaries, edges, stone-wales defects, sp^3 sites and so on. The friction of 2D layered nanomaterials such as graphene and MoS_2 increased across the edges [23]. The edges of nanomaterials are high energy sites and they showed good adhesion with counter surfaces. The strong adhesion between the two surfaces increases the friction [24]. The energy of edge surface could be effectively minimized by introducing various functional groups and absorbing alkyl chain-constituted molecules. The interlayer interaction, pulling direction and corresponding registry of every single layer influence the interlayer lubricity of graphite. The coefficient of friction is also influenced by interlayer stacking structures and loading force [25].

8.4 Liquid Lubricants and Additives

Liquids lubricants are made of lube base oil and variable additives. Lubricants function as a cooling agent to transport the heat from the tribo-interfaces, disperse the debris and oxidized products, reduce the friction, and protect the tribo-surfaces. The additives are the substances which can increase or suppress the targeted properties of lubricants. The additives also add the new desired properties to the lubricants. The

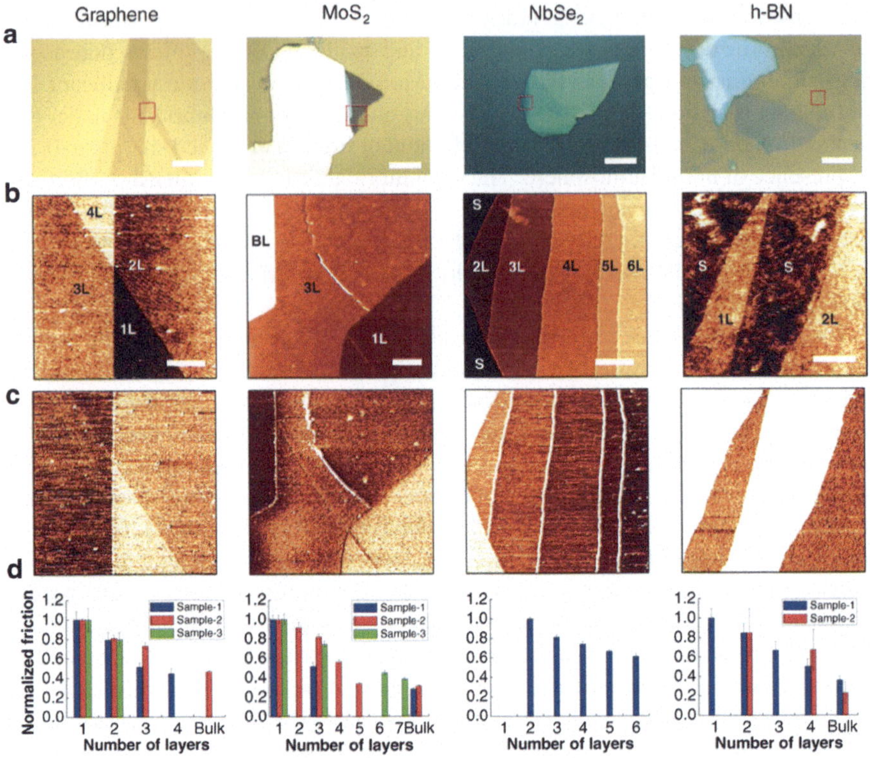

Fig. 8.2 Optical and atomic force microscopic (AFM) images of graphene, MoS_2, $NbSe_2$, and *h*-BN thin sheets deposited on silicon oxide. **a** Bright-field optical images of thin sample flakes; scale bars: 10 μm. The AFM scan area of each sample is represented by red square. **b** and **c** Topographic and corresponding area friction images of graphene, MoS_2, $NbSe_2$, and *h*-BN thin sheets; scale bars: 1 μm. 1L, 2L, 3L, etc. represent one, two, three, etc. number of lamellae in the sheets of individual sample, whereas BL and S denotes an area of bulk-like thick sheet and bare SiO_2 substrate, respectively. **d** The variation in normalized friction for different number of lamellae in individual samples. Reproduced with the permission of Ref. [21]

quantity and type of additives are determined by the applications for which they are used. The lubricants in engine oils must have excellent oxidative-resistance property, high viscosity index, and good dispersing ability; whereas the metalworking lubricants should exhibit good corrosion control and cooling ability.

Lube base oils are broadly classified based on their origin from mineral, synthetic or vegetable oils. Mineral oils are the most commonly used lube base oils in the lubricant industry. The properties of mineral lube base oil are governed by the processes used for its production which can be solvent extraction, hydrofinishing, dewaxing, and so on. The mineral lube base oils primarily contain *n*-paraffinic, iso-paraffinic, and naphthenic components. Higher components of *n*-paraffinic with longer chain furnish high viscosity and good oxidation stability, whereas iso-paraffinic components favor the low viscosity. The synthetic lube base oils are made by chemical processing of hydrocarbon-based components of refinery streams and they possess significantly improved properties of lube base oils compared to the mineral lube base oils. The polyalphaolefines (PAO) of variable viscosity grades and polyol-based esters are most commonly used synthesis lube base oils for lubricant applications. Vegetable oils-based lubricants are also gaining significant interest as renewable lube base oils. However, their poor oxidation stability has been a critical challenge for lubricant applications.

Additives are the substances/materials, which are added in lubricants to enhance the lubrication performance. Engine oils contain several additives like friction-modifier, extreme-pressure/anti-wear additive, antioxidant, viscosity index (VI) improver, pour point depressant, dispersants, corrosion-inhibitor etc. [26]. These additives extend the drain time of lube oils by inhibiting oxidation events and dispersing the soot particles along with oxidized products, reduce the friction, maintain the fluidity at low and high temperatures, and protect the contact surfaces by forming the tribo thin film. The friction-modifier, extreme-pressure/anti-wear additive, and VI improver are important for tribo-performance because they control the fuel economy by reducing the friction and controlling the wear in the engine operation. The amount of additives in lubricants varies from 5 to 30%, depending on their use for type of applications.

Hydrocarbon components of lube base oils are prone to be oxidized under the tribological conditions. High temperature, extreme pressure and metallic worn particles catalyze the oxidation events. The oxidation of lube oils leads to undesirable events like oil thickening and sludge/deposit formation. The use of antioxidants delays the oxidative events of lube base oils and extends the drain time. Hindered phenolic and aromatic amines, sulfur- and phosphorous-containing compounds are widely used as antioxidant additives to lubricants. Corrosion inhibitors protect the engineering surfaces against the corrosive molecules by their adsorption and forming a protective film. The amine succinates and alkaline earth sulfonates are commonly used corrosion inhibitors for the engine oils. Polymeric compounds are used to interfere with the crystallization process of paraffinic contents in the lube oil for improving the flow properties at low temperatures. The viscosity index modifiers are used to subsidize the oil thinning at higher temperatures. High molecular polymers increased the viscosity of the oil at high temperatures by their steric effect. Poly-alkyl-methacrylate and olefin's copolymer are widely used as VI improvers.

Detergents such as sulfonates, phenates, salicylates and phosphonates keep the tribo-surfaces clean by preventing the formation of carbon/polymeric-based deposits at high temperature. They exhibit a micellar structure with an alkaline core surrounded by surfactant chains. They are having either neutral pH or over-based to neutralize the acid compounds. The dispersants are non-metallic or ashless cleaning agents that solubilize and disperse various contaminants including oxidized products, soot particles, and worn matrix. They are mostly organic compounds with a polar head and a hydrocarbon tail (non-polar). The succinimides are commonly used dispersants for lube applications.

Friction modifier additives are used to reduce the friction between the engineering surfaces. The organic friction modifiers consist of long alkyl chain along with a polar functional group. The interaction between the polar functional group of friction modifier and the surface of engineering bodies governs the frictional properties. The various long alkyl chains fatty acids and their esters, amines, amides, phosphonic acids, organic polymers etc. interact with tribo-interfaces and forms the thin films of low shear strength, which reduce the friction. The Zinc dialkyldithiophosphate (ZDDP)-based compounds are most established multi-functional additive with primarily role as extreme-pressure/anti-wear additive besides the antioxidant and corrosion-inhibition. The ZDDP forms the tribo-chemical thin film on the steel-based engineering surfaces under the boundary lubrication regime, which protects the contact interfaces against the wear and reduces the friction. The ZDDP was invented as oxidation- and corrosion inhibitor during early of 1940s. The antiwear properties of ZDDP for engine oil were investigated in the 1950s. It bears the extreme pressure conditions and protects the engineering surfaces [27]. H. Spikes illustrated the antiwear thin film formation of ZDDP molecules on the steel surface under severe tribological contact conditions. The antiwear thin film is continuously ingested and reproduced consistently to prevent direct contact between steel bodies. The ZDDP tribofilm is mostly composed of amorphous zinc/iron polyphosphates of variable alkyl chain lengths with inclusions of ZnS and FeS. The anti-wear properties of ZDDP are associated with the polyphosphate glasses which are formed to dissolve the iron oxides. The ZDDP molecules are initially absorbed on the metallic surfaces via physisorption, which is followed chemical interaction between the phosphate/phosphothionic moieties of ZDDP with the iron surface driven by boundary lubrication conditions [28].

The molybdenum dialkyldithiocarbamate (MoDTC) has been a good friction-modifier in motor oils. In the mixed and boundary lubrication regimes, the MoDTC shows a low coefficient of friction. The MoDTC undergoes complex tribo-chemical reactions and form a thin film of molybdenum disulfide (MoS_2), which reduces the friction. The tribo-chemical reactions plausibly followed a two-steps mechanism; first step involved the formation of free radicals via transfer of electron on Mo-S chemical bond and these active radicals recombine and yielded the MoO_3 and MoS_2 as tribo-chemical products during the second step [29]. The friction reducing properties of MoS_2 are attributed to its lamellar structure. However, the MoDTC is an effective friction-modifier at high concentration and high temperature. The most of additives being used for enhancement of tribological properties carries several challenges. Furthermore, increasing environmental and emission regulations have posed numerous concerns for using the ZDDPs and MoDTC-based additives, since they carry zinc, phosphorus, and sulfur, which are hazardous to environment and living-beings. The ZDDPs-based additives also function as poisonous compounds for catalytic convertor of engine exhaust system and have poor compatibility with non-ferrous and ceramic based engineering surfaces.

8.5 Nanostructured Layered Materials as Lubricant Additives

In the recent past, several studies have been made for exploring the alternatives to the ZDDPs. In this context, nanomaterials including 2D layered materials have gained significant attention for tribological application. The 2D layered nanomaterials viz. graphene, MoS_2, WS_2 and *h*-BN have been explored as lubricious thin films and additives to lubricants for reduction of friction and wear. The high surface area of 2D layered nanomaterials furnishes greater coverage to contact surfaces and minimizes the direct contact between sliding surfaces. The controlled surface modification of 2D layered materials is very crucial to attain their stable dispersion in the liquid lubricants.

8.5.1 Dispersion of Nanostructured Layered Materials

With the advent of nanotechnological solutions, the 2D nanostructured layered materials have gained considerable attention for tribological applications. However, their poor dispersion stability has been a great concern for their application as additives to liquid lubricants. The smaller size and high surface area of 2D nanostructured layered materials leads to agglomerate because of high cohesive interaction; consequently, the larger size of agglomerates are eventually settle-down in the vessel/sump [30]. The sonication, rigorous starring, high-pressure homogenizer etc. are commonly used practices to disperse nanomaterials in various types of fluids/solvents [31]. The use of surfactants, dispersing agent, and chemical functionalization of nanomaterials facilitates their dispersion in the targeted solvents/fluids. The chemical modifiers can be polymers, surfactants, and organic molecules. The chemical functionalization having the charge-based polar functionalities furnishes stable dispersion based on electrostatic repulsive forces.

Chemical functionalization is effective than the physisorption-based approaches to increase the dispersion stability of 2D nanomaterials in liquid lubricants. Over the recent past, various studies have been made to understand the interaction between the grafting molecules and 2D nanomaterials for enhancement of their dispersion stability in various types of solvents, fluids, and lubricants. The interaction between the grafting molecules and the 2D materials are govern by several parameters viz. chemical structure of grafting molecules and 2D nanomaterials, presence of defects, surface functional groups and surface area of 2D nanomaterials, availability of π-electron rich structure, and so on. Graphene is one of the excellent examples of π-electron rich structure, which facilitates their agglomeration because of high cohesive interaction. In the recent past, several approaches have been addressed for functionalization of 2D nanostructured materials to make them dispersible.

Grafting of alkyl chains is one of the most explored approaches to disperse the 2D nanostructured layered materials in a diversified range of liquid lubricants. The use of amphiphilic or bipolar molecules having polar functional groups to interact with 2D nanomaterials, while the non-polar alkyl chains make them dispersible in the liquid lubricants. The presence of active functional sites on 2D nanostructured materials enhances their potential for chemical functionalization [32–34]. Yang et al. modified graphene nanosheets with alkylamines which showed homogeneous dispersion in organic solvents [35]. The alkylamines with variable alkyl chain length could be grafted selectively targeting the oxygen functionalities in the basal plane and along the edges of graphene-based materials. The octadecylamine was successfully grafted on carboxylate functional groups of graphene oxide (GO) and reduced graphene oxide (rGO) via covalent interaction yielding the amide linkages. The octadecyl chains grafted on GO and rGO facilitates their dispersion in the *n*-hexadecane, mineral lube base oil and fully formulated engine oil. The enhancement of dispersibility of alkylated graphene in lubricants was attributed to the cohesive interaction between the hydrocarbon components of lube base oils and alkyl chains grafted on the GO and rGO [36]. Figure 8.3 demonstrates the chemical structure of alkylated-rGO along with dispersion stability in the engine oil for one month. Fan et al. modified the GO nanosheets with imidazolium ionic liquids by a simple mixing and stirring method. The ionic liquid modified-GO exhibited improved tribo-performance in the multi alkylated cyclopentane lube base oil [37]. The covalent grafting of imidazolium ionic liquids facilitates the dispersion of graphene in the polyethylene glycol (PEG 200) synthetic lube base oil. Three variable anions bis(salicylato)borate, oleate, and

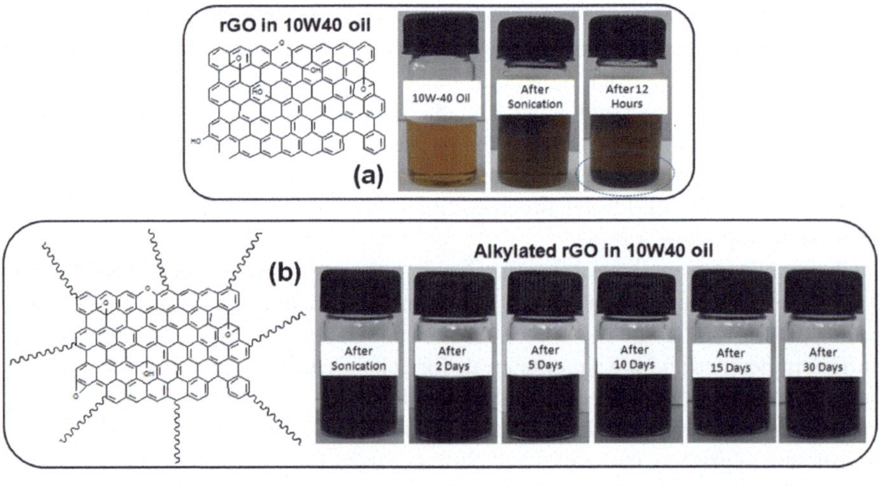

Fig. 8.3 **a** Atomic structure of rGO lamella along with digital images of 10W-40 engine oil and dispersion of rGO. The agglomerates of rGO can be seen at the bottom of the dispersion as highlighted by a circle. **b** Atomic structure of alkylated-rGO lamella along with digital images of dispersion of alkylated-rGO in the 10W-40 engine oil for one month. Time for each dispersion sample is noted on the respective vial. Concentration of graphene-based sample: 40 ppm. Reproduced with the permission of Ref. [32]

hexafluorophosphate were used for exploring the effect of anion on tribo-performance of ionic liquids-functionalized graphene in PEG 200 lube base oil [38].

Grafting of long alkyl chains on MoS_2 nanosheets facilitates their dispersion in the lubricating oils. The structural defects and vacant sulfur sites in the MoS_2 show good interaction with alkanethiol via Mo-S dative linkage and furnish long alkyl chain grafted MoS_2 nanosheets, which extend the dispersion stability of MoS_2 in the polyol ester lube base oil [39]. The lack of active functionalities in the *h*-BN makes it challenging material for chemical functionalization. Recently, a limited number of oxygen functionalities are introduced in the *h*-BN via harsh oxidation. The octadecyltriethoxysilane showed interaction with oxygen functionalities on defect and edge sites of *h*-BN nanosheets via covalent bonding and yielded the alkylated *h*-BN nanosheets. The presence of octadecyl chains on *h*-BN nanosheets enhanced the dispersion stability in polyol lube base oil [40]. The boron atoms in *h*-BN nanosheets function as Lewis acids and shows good affinity with amine compounds, which are enriched with easily accessible lone pair electrons (Lewis bases). The grafting of alkylamines on the *h*-BN nanosheets based on Lewis acid-Lewis base interaction facilities the dispersion of alkylated *h*-BN in SN-500 mineral base oil [41]. These stable dispersion shows good potential for enhancement of tribological properties.

8.5.2 *Lubricant Applications of Nanostructured Layered Materials*

The 2D nanostructured layered materials have been explored as potential candidates for different types of lubricants to enhance the tribological properties. The size, presence of chemical functionalities, and dispersibility of 2D materials besides the nature of engineering surfaces and tribo-parameters plays important roles and govern the tribological properties. The 2D nanostructured layered materials have been explored for various lubricant applications including automotive oils, cutting fluids, greases, and so on.

8.5.2.1 Automotive Lubricants

High friction between the piston rings and the cylinder liner leads to dissipation of energy. The surface properties and material of piston rings along with composition and viscosity of engine lubricants plays indispensible role to reduce the friction, consequently enhancement of engine efficiency [4, 42]. The piston rings in the automotive engines transform the heat energy into mechanical power. The increasing frictional energy component requires more fuel to drive the engine [43]. The piston ring-liner contacting interfaces show boundary and mixed lubrication regimes on the top and bottom ends, whereas the hydrodynamic lubrication is observed at mid-stroke. Considering the boundary and mixed lubrication regimes, particularly during piston operation, surface active additives are required, which can form the boundary thin film under the severe tribological conditions and reduce the friction and wear [44]. The nanomaterials as additives to engine oils not only improve the tribological properties but also enhance the thermal conductivity for dissipation of heat from the contact interfaces [45]. The nanomaterials as additives are smeared or deposited on the contact interfaces to form a thin-film of low shear strength which prevents the direct contact and minimize the wear. The nanomaterials as additive also furnishes balls-bearing model for enhancement of tribological properties.

Over the recent past, graphene has received extensive interests as lubricious material. The presence of defects, number of lamellae, interlayer spacing, and dispersion stability of graphene govern the friction and wear characteristics of lubricating oils [46]. Eswaraiah et al. demonstrated the significant enhancement of lubrication characteristics of engine oil in the presence of graphene prepared by solar exfoliation of GO. The 25 ppm dose of graphene reduced coefficient of friction, wear and extreme pressure properties by 80, 33, and 43%, respectively, compared to corresponding engine-oil (Fig. 8.4). The enhancement of tribo-performance was attributed to the nano-bearing mechanism and high mechanical strength of graphene [47]. The nanocomposite of rGO and ZrO_2 nanoparticles as additive to lubricant oil exhibited significant reduction in the friction [48]. The hybrid and nanocomposite materials revealed enhanced tribological properties as a consequence of the synergistic effect between the constituent nanomaterials [49]. A wide range of lubrication mechanisms have been demonstrated for enhancement of tribological properties using the nanomaterials; driven by hardness and mechanical strength of nanomaterials and tribo-interfaces besides the roughness of engineering surfaces and size/shape of nano-additives. These mechanisms include sliding-rolling friction [50], formation of protective tribo-film [51], mending [52] and polishing effect [53].

The chemically functionalized graphene having long alkyl chains exhibited good dispersion in the lubricating oils and furnished remarkable improvement in tribological properties of corresponding lube oils. The minute dose (60 ppm) of

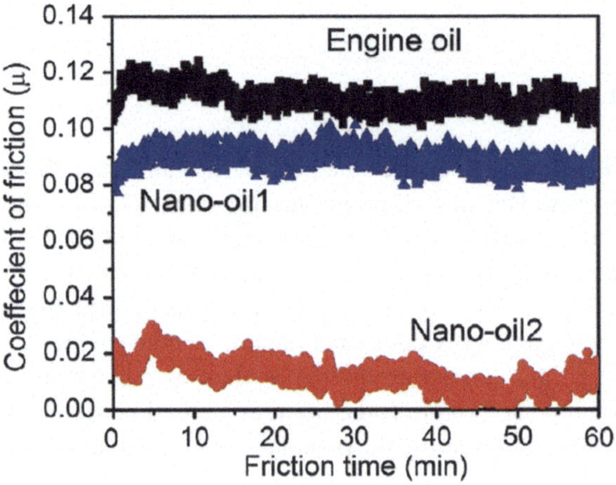

Fig. 8.4 Changes in coefficient of friction for engine oil and graphene-based engine oil nano fluids having variable doses of graphene. Nano-oil 1 and 2 contains 12.5 and 25 ppm of graphene. Reproduced with the permission of Ref. [47]

octadecylamine-grafted GO showed 26 and 9% reductions in the friction and wear of steel tribopair compared to hexadecane lube oil [36]. The alkylated rGO with significantly low degree of residual oxygen functionalities furnished stable dispersion in the 10W40 engine oil and reduced the friction and wear by 36 and 35%, respectively under the sliding contact between the steel balls (Fig. 8.5) [32]. The enhancement of tribo-performance was attributed to uninterrupted supply of alkylated rGO to contact interfaces, which avoided the direct contact between the mating surfaces and provided the easy shearing. Sahoo et al. have demonstrated that grafting of octadecylamine makes GO hydrophobic and facilitates a good compatibility with heavy paraffin oil [54]. The 1 wt% dose of highly dispersible octadecyl-functionalized GO decreased the friction by ~81% under the mean Hertzian contact pressure of 0.9 MPa. The tribo-stress induced assembly with the oil molecules and the alkyl chains of GO-ODA yielded the kind of lubricious cushion under the low contact, providing the easy shearing to reduce the friction. However, at high contact stress the cushion effect broken away providing direct contact of graphene sheets with the contact interfaces; consequently the coefficient of friction increased and became equivalent to that of graphene [54].

The MoS_2 has received interest as an additive to lubricants during early of nineteenth century [55]. The functionalized graphene composites with MoS_2 extended the dispersion stability compared to bare MoS_2. In addition to excellent dispersion in oil, the functionalized graphene with MoS_2 yielded enhanced tribological properties in term of friction reduction. The base MoS_2 showed 12.4% reduction in coefficient of friction. The functionalized graphene composites with MoS_2 reduced the friction by 15.8% using 0.8 wt% dosage in the oil. The proposed lubrication mechanism suggests the formation of a protective thin film to prevent the direct contact of rubbing surfaces. The van der Waals interaction between S-S planes in the MoS_2 structure eases the shearing and reduces the friction [56]. The MoS_2 efficiently reduces the friction and wear in the lubricating oils under boundary conditions, increases load-bearing

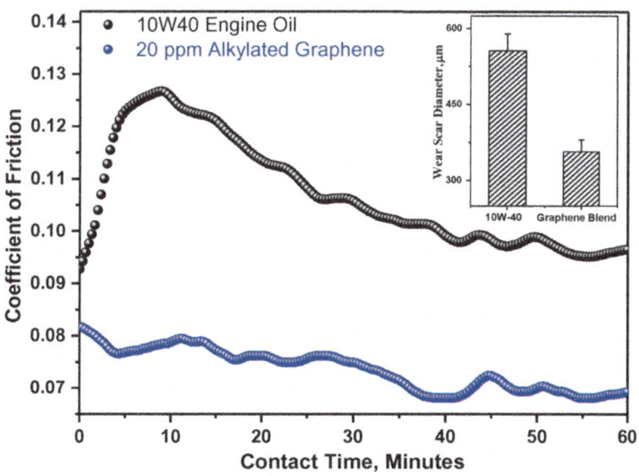

Fig. 8.5 Coefficient of friction for 10W40 engine oil and alkylated graphene-blended engine oil (dose: 20 ppm). The inset graph shows the wear scar diameter of steel balls. Tribo parameters: load: 392 N; temperature: 75 °C; rotational speed: 1200 rpm; run time: 1 h. Reproduced with the permission of Ref. [32]

capacity, and prevent catastrophic seizure of contact surfaces. The hierarchical microspheres of MoS_2 nanosheets as additive (40 ppm) to 10W40 engine oil subsidized the friction and wear by 21 and 42%, respectively. The enhanced tribological performance originated from the lamellar structure of MoS_2 and deposited thin film on the contact interfaces under the tribo-stressed conditions [57]. The octadecanethiol functionalized-MoS_2 nanosheets enhanced the antiwear properties of polyol lube base oil. The steel disc showed a wear rate of 0.0738 $mm^3\ N^{-1}\ min^{-1}$ when lubricated with polyol lube oil and it reduced to 0.0512 $mm^3\ N^{-1}\ min^{-1}$ in the presence of 50 ppm dosage of chemically functionalized-MoS_2 nanosheets in the polyol lube base oil [41]. The WS_2 nanoparticles anchored graphene (WS_2-GP) showed improved dispersion in the base oil than individual graphene and WS_2 nanoparticles. The 0.02 wt% dosage of WS_2-GP in the lubricating oil reduced the friction and wear rate by 70.2 and 65.8%, respectively. The WS_2-GP nanocomposite is believed to be deposited over the contacting interfaces and yielded a tribo thin film, which avoided the direct contact of the rubbing interfaces [58]. The *h*-BN nanosheets have also received considerable attention as an additive to lubricating oils. The lubrication characteristics of octadecyltriethoxysilane functionalized *h*-BN nanoplatelets were probed in polyol lube oil using the steel-steel contact geometry. The tribological results revealed that octadecyltriethoxysilane functionalized *h*-BN nanoplatelets, as additive to polyol ester reduced both the friction and wear of steel tribo-pair [40]. The trioctylamine functionalized *h*-BN nanoplatelets exhibited long-term dispersion stability in the mineral lube base oil. The minute dose (20 ppm) of trioctylamine functionalized *h*-BN nanoplatelets improved the tribological properties of the mineral lube base oil by reducing the friction (35%) and wear (25%). The microscopic results of contact interfaces revealed the uniform distribution of boron and nitrogen which suggests the formation of a tribo-chemical thin film based on *h*-BN nanosheets on the tribo-interfaces. The formation of tribo-chemical thin films not only reduced the friction but also protected the contact interfaces against undesirable wear events [41].

8.5.2.2 Metalworking Fluids

The coolants used for dissipation of heat from the machining process are termed as metalworking fluids. They contain mostly three phases: solid, liquid and interface. The most common cooling fluids are water, ethylene glycol, and mineral base oils. The nanoparticles contain high surface area which help to produces boundary layers and enhance the thermo-physical properties of metalworking fluids. Various types of nanoparticles viz. copper, silver, iron, graphite, copper oxide, aluminum oxide, carbon nanotubes, silicone carbides, graphene, *h*-boron nitrite, and molybdenum disulfide shows good potential as additives to metal working fluids for enhancing the coolant and lubrication properties [59]. The lateral dimension in microscale and the nanoscopic thickness along with excellent thermal conductivity (conductivity > 1000 W/mK) makes graphene a suitable candidate as a performance-enhancing additive to metalworking fluids. The performance enhancement by graphene suspension as metal working fluids was attributed to the increased wettability of the cutting fluid, which facilitates the penetration of graphene into the tool-work piece interface providing efficient lubrication and they also take away the heat from the cutting zone [60]. Heat transport is one of the major prerequisites to enhance the performance of metalworking nanofluids. The GO containing metalworking fluids based on ethylene glycol found to be stable and the thermal conductivity of resultant metalworking fluid remains almost constant for 7 days. The 5% GO nanosheets increased the thermal conductivity by 61% compared to those containing metallic oxides [61]. These results express good potential of graphene-based materials for the metalworking fluids.

8.5.2.3 Greases

Greases are semi-solid lubricants and they are prepared by synergistic blending of oil/lubricating fluid with a thickener/soap. The greases carry several disadvantages including their applications for boundary lubrication conditions, where grease-based lubricant couldn't perform well. Over the recent past various types of nanomaterials have been incorporated in greases to improve the wear and friction characteristics in boundary and mix lubrication regimes by forming a lubrication thin film on tribo-interfaces. Fan et al. demonstrated that 0.1 wt% graphene as an additive to the Bentone grease not only reduced the friction and wear but also significantly improved the load-bearing capacity. The improved tribo-performance including load-bearing capacity was attributed to the formation of graphene-based lubricating thin film [62]. The 2% graphene in the commercially available lithium grease of NLGI grade 2 showed significant decrease in both friction and wear of tribopair [63]. The graphene nanoplatelets as an additive (0.5%) to standard lubricating grease showed 8.5% reduction in the coefficient of friction, which further reduced to 16.3% with increasing dosage of graphene nanoplatelets. The friction reduction was attributed to the formation of boundary thin film and the nano-bearing effect by graphene nanoplatelets [64].

The interaction of MoS_2 with iron/steel at high temperature (>700 °C) forms the intermetallic compounds to protect the tribo-surfaces [65]. As a result, MoS_2 nanoparticles are gaining good attention as additive for the grease lubrication. The MoS_2 nanoparticles encapsulated by soap molecules in the grease matrix should be easily available on the tribo-interfaces for their efficient performance. The MoS_2 nanoparticles as additive to lithium grease improved the load bearing capacity from 100 to 220 kg load [66]. The dosage of MoS_2 nanoparticles, applied load, and temperature governs the tribo-performance of MoS_2 nanoparticles encapsulated in the grease. Under the high temperature and pressure, the encapsulated MoS_2 nanoparticles in the grease are released and established direct contact with the tribo-interfaces and gradually forms a tribo thin film as shown in Fig. 8.6. These events reduce the friction and protect the contact interfaces at higher loads [67]. The *h*-BN has been gaining large attention as a promising additive to grease because of excellent

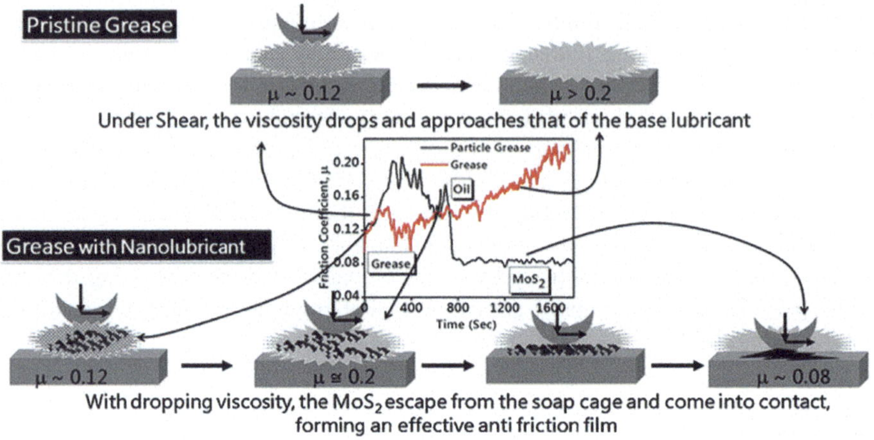

Fig. 8.6 Pictorial illustration of plausible shearing behavior of the pristine and MoS_2-blended grease based on their friction profile. Reproduced with the permission of Ref. [67]

thermal conductivity, remarkable mechanical properties, high electrical resistance, and chemically inertness. The *h*-BN as a solid lubricant is capable of replacing graphene/graphite-based additive in Al-forming operations. The tribo-performance, stability of lubrication thin film, quality of contact surfaces etc. are governed by the size and dosage of *h*-BN particles [68].

8.5.3 Lubrication Mechanisms of Nanostructured Layered Materials

Several studies have been made on lubrication aspects of 2D nanostructured layered materials as additives to lubricating oils and demonstrated the lubrication mechanism emphasizing the role of 2D nanomaterials. Liu et al. have demonstrated that 2D nanomaterials furnish lubrication properties in four different ways: (a) entering the contact zone of engineering surfaces, (b) formation of tribo thin film, (c) fillings the gaps and pits of tribo-interfaces, and (d) affecting the viscosity and fluid drag of lubricating oils [69]. Four different types of lubrication mechanism by 2D layered materials are demonstrated in Fig. 8.7.

The 2D layered nanomaterials having lamellar structure with ultra low thickness easily enters to contacting zone. Under the mild contact, the 2D nanomaterials easily sheared because of sliding stress and provide low friction [70, 71]. The tribo surfaces carry roughness at the micro-scale level with several peaks and valleys across the contact zones. The 2D layered nanomaterials fill the valleys over the sliding events and make surfaces smoother [47]. The smoother surfaces have fewer asperities and reduces the localized contact pressure leading to decreases in plastic deformation and minimizes the wear.

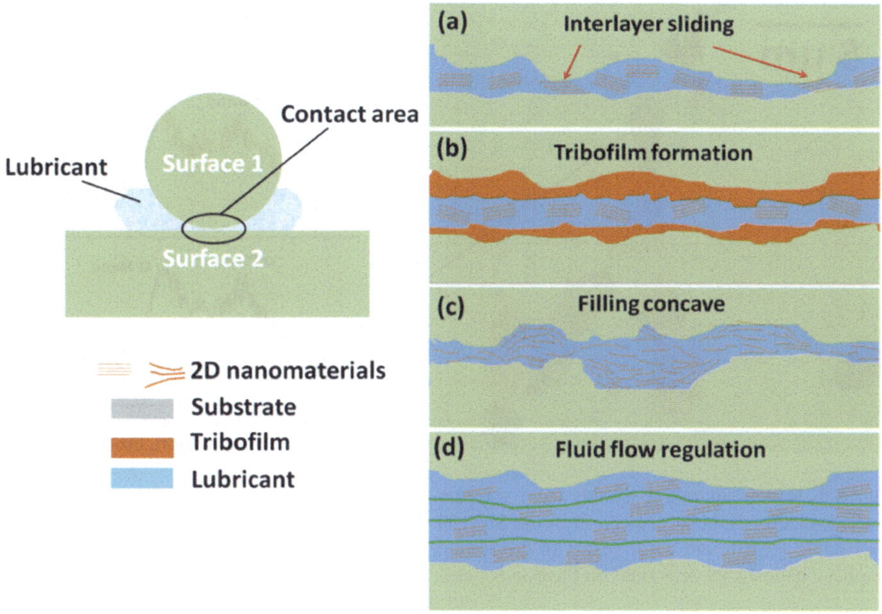

Fig. 8.7 Plausible lubrication mechanism emphasizing the roles of 2D layered materials in the contact interfaces. Reproduced with permission of Ref. [69]

The high load catalyzed the film formation over the tribo-interfaces. The tribo thin film formation involves two steps lubrication mechanism. First step comprises a physically adsorbed conformal protective film formation on the tribo interfaces due to high surface energy and high load-induced shearing of 2D nanomaterials. The protective film separates two contacting bodies and prevents their direct contact. Accordingly, the friction and wear are reduced. Furthermore, physically adsorbed protective film under the high pressure, sliding stress and temperature catalyzed the chemical reaction between the tribo-surfaces, 2D nanomaterials, and lubricants, which gradually stimulates the formation of a new tribofilm replacing the physically adsorbed film over the localized contacting surfaces as well as in the matrix of the substrate. The newly formed tribo-film helps to enhance the tribological properties [72]. The Raman results of tribo-interfaces revealed that alkylated GO dispersed in lube oil are deposited on the contact interfaces under the sliding stress. The sheared contact leads to delamination of graphene layers and then gradually transferred to the tribo-interfaces. The deposited graphene thin film easily sheared with graphene dispersed in the lube oil and reduced the friction [34]. The Raman microscopic mapping of contact interfaces suggested that the graphene-based thin film is deposited in the form of irregular patches as shown in Fig. 8.8. Plausibly, the deposition of graphene lamellae on the contact interfaces is governed by the roughness of the tribo-interfaces and load distribution, consequently formation of irregular patches of variable density [73]. Biswas et al. have demonstrated that deposition and then

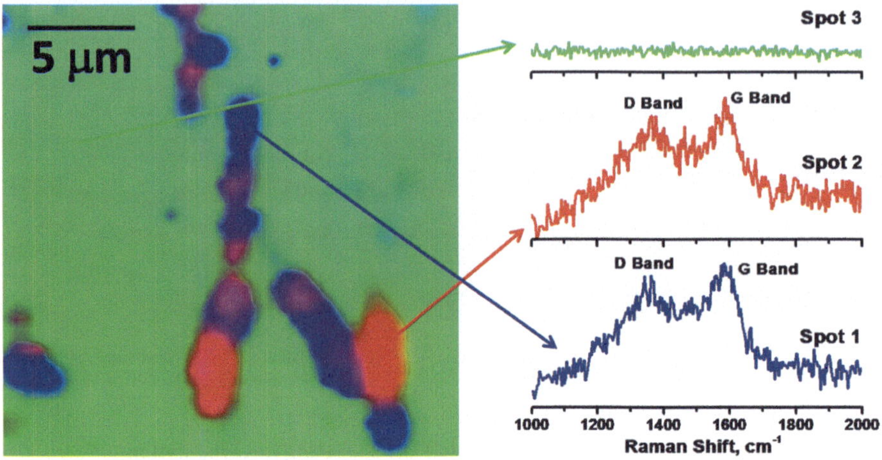

Fig. 8.8 Raman mapping of the tribo interface after the lubrication test using chemically-functionalized graphene dispersion in the SN-150 mineral lube base oil. The Raman spectra corresponding to different color pixels in the Raman map revealed the distribution variation of the graphene nanosheets deposited on the contact area. Reproduced with permission of Ref. [73]

removal of MoS_2 nanosheets from the tribo interfaces are continuous process under the sliding contact stress [74].

8.6 Summary

High friction leads to energy dissipation; consequently, more power is required for the functioning of machinery and engine. The contacts between the engineering bodies, particularly under the boundary lubrication regime experiences deformation of asperities leading to plastic flow of materials in the form of wear, which compromises the life and sustainability of engineering parts. Lubricant, an essential element between the most of engineering surfaces plays a critical role to reduce the friction, protect the contact surfaces against the wear, and dissipate the heat from contact zones. The lubricant can be a solid thin film, semi-solid grease, or liquid lubes, depending on their applications. The friction-modifier, antiwear and extreme pressure additives besides the viscosity-modifier are essential constituents of liquid lubricants, which control the tribological properties.

Two-dimensional (2D) nanostructured layered materials viz. graphene, *h*-BN, MoS_2, WS_2, etc. have been gaining significant interest for tribological applications. The remarkable characteristics such as excellent thermal conductivity, high chemical and thermal stability, and exceptionally high mechanical strength make 2D nanomaterials promising candidates as solid thin films and additives to liquid lubricants. The weak van der Waals interaction between atomic-thick lamellae of 2D layered materials provides low resistance-to-shear along the basal plane under the tribo-stress, which reduces the friction. The tribological applications of 2D nanostructured layered materials as additives to automotive lubricant, greases, and metalworking fluids are covered emphasizing their roles in performance enhancement. The dispersion aspects, tribo-performance using variable contact geometries, and worn area analyses etc. are discussed to understand the lubrication mechanism highlighting the role of 2D nanostructured layered materials for reduction of friction and wear. The 2D nanostructured layered materials could be an alternative to conventional additive like ZDDP, subjected to their prolong dispersion stability in the lubricants and it carries immense opportunity to be explored in near-term.

References

1. B. Bhushan, *Principles and Applications of Tribology* (Wiley, New York, 2013), p. 28
2. G.W. Stachowiak, A.W. Batchelor, *Engineering Tribology* (Elsevier Butterworth-Heinemann, 2005).
3. A. Oron, S.H. Davis, S.G. Bankoff, Long-scale evolution of thin liquid films. Rev. Mod. Phys. **69**, 932–976 (1997)
4. K. Holmberg, P. Andersson, A. Erdemir, Global energy consumption due to friction in passenger cars. Tribol. Int. **47**, 221–234 (2012)
5. U.D. Schwarz, Tracking antiwear film formation. Science **348**, 40–41 (2015)
6. G. Fontaras, Z. Samaras, On the way to 130 g CO_2/km—estimating the future characteristics of the average European passenger car. Energy Policy **38**, 1826–1833 (2010)
7. *Energy Technology Perspectives, Scenarios & Strategies to 2050, International Energy Agency, Paris* (2010)
8. K. Holmberg, R. Siilasto, T. Laitinen, P. Andersson, J. Ari, Global energy consumption due to friction in paper machines. Tribol. Int. **62**, 58–77 (2013)
9. A. Erdemir, J.-M. Martin, *Supralubricity* (Elsevier, 2007)
10. B. Bhushan, *Introduction to Tribology*, 2nd edn. (Wiley, New York, 2013).
11. H. Heshmat, C.A. Heshmat, The effect of slider geometry on the performance of a powder lubricated bearing. Tribol. Trans. **42**, 640–646 (1999)
12. C.F. Higgs, C.A. Heshmat, H. Heshmat, Comparative evaluation of MoS_2 and WS_2 as powder lubricants. J. Tribol. **121**, 625–630 (1999)
13. I. Booser, E. Richard, *Handbook of Lubrication Theory and Practice of Tribology Volume 1—Application and Maintenance*, vol. 1 (1983)
14. T. Spalvins, Lubrication with sputtered MoS_2 films. ASLE Trans. **14**, 267–274 (1971)
15. T. Spalvins, Tribological properties of sputtered MoS_2 films in relation to film morphology. Thin Solid Films **73**, 291–297 (1980)
16. W.O. Winer, Molybdenum disulfide as a lubricant: a review of the fundamental knowledge. Wear **10**, 422–452 (1967)
17. J.P.G. Farr, Molybdenum disulphide in lubrication. A review. Wear **35**, 1–22 (1975)
18. K.C. Mutyala, H. Singh, J.A. Fouts, R.D. Evans, G.L. Doll, Influence of MoS_2 on the rolling contact performance of bearing steels in boundary lubrication: a different approach. Tribol. Lett. **61**, 20 (2016)
19. L. Cizaire et al., Mechanisms of ultra-low friction by hollow inorganic fullerene-like MoS_2 nanoparticles. Surf. Coat. Technol. **160**, 282–287 (2002)
20. H. Lee, N. Lee, Y. Seo, J. Eom, S. Lee, Comparison of frictional forces on graphene and graphite. Nanotechnology **20**, 325701 (2009)
21. C. Lee et al., Frictional characteristics of atomically thin sheets. Science **328**, 76–80 (2010)

22. K.S. Kim, H.J. Lee, C. Lee, S.K. Lee, H. Jang, J.H. Ahn, J.H. Kim, H.J. Lee, Chemical vapor deposition-grown graphene: the thinnest solid lubricant. ACS Nano **5**(6), 5107–5114 (2011)
23. H. Hölscher, D. Ebeling, U.D. Schwarz, Friction at atomic-scale surface steps: experiment and theory. Phys. Rev. Lett. **101**, 1–4 (2008)
24. R.F. Deacon, J.F. Goodman, Lubrication by lamellar solids. Proc. R. Soc. A Math. Phys. Eng. Sci. **243**, 464–482 (1958)
25. K. Matsushita, H. Matsukawa, N. Sasaki, Atomic scale friction between clean graphite surfaces. Solid State Commun. **136**, 51–55 (2005)
26. V.W. Wong, S.C. Tung, Overview of automotive engine friction and reduction trends—effects of surface, material, and lubricant-additive technologies. Friction **4**, 1–28 (2016)
27. H. Spikes, The history and mechanisms of ZDDP. Tribol. Lett. **17**, 469–489 (2004)
28. P.A. Willermet, D.P. Dailey, R.O. Carter III, P.J. Schmitz, W. Zhu, Mechanism of formation of antiwear films from zinc dialkyldithiophosphates. Tribol. Int. **28**, 177–187 (1995)
29. C. Grossiord, K. Varlot, J. Martin, T.Le Mogne, C. Esnouf, MoS_2 single sheet lubrication by molybdenum dithiocarbamate. Tribol. Int. **31**, 737–743 (1998)
30. C. Choi, M. Jung, Extreme pressure properties of multi-component oil-based nanofluids. J. Nanosci. Nanotechnol. **12**, 3237–3241 (2012)
31. Y. Hwang et al., Production and dispersion stability of nanoparticles in nanofluids. Powder Technol. **186**, 145–153 (2008)
32. H.P. Mungse, O.P. Khatri, Chemically functionalized reduced graphene oxide as a novel material for reduction of friction and wear. J. Phys. Chem. C **118**, 14394–14402 (2014)
33. W. Zhang et al., Tribological properties of oleic acid-modified graphene as lubricant oil additives. J. Phys. D Appl. Phys. **44**, 205303 (2011)
34. H.P. Mungse, N. Kumar, O.P. Khatri, Synthesis, dispersion and lubrication potential of basal plane functionalized alkylated graphene nanosheets. RSC Adv. **5**, 25565–25571 (2015)
35. X. Yang et al., Synthesis and characterization of alkylamine-functionalized graphene for polyolefin-based nanocomposites. Appl. Surf. Sci. **305**, 725–731 (2014)
36. S. Choudhary, H.P. Mungse, O.P. Khatri, Dispersion of alkylated graphene in organic solvents and its potential for lubrication applications. J. Mater. Chem. **22**, 21032–21039 (2012)
37. X. Fan, L. Wang, High-performance lubricant additives based on modified graphene oxide by ionic liquids. J. Colloid Interface Sci. **452**, 98–108 (2015)
38. R. Gusain, H.P. Mungse, N. Kumar, T.R. Ravindran, R. Pandian, H. Sugimura, O.P. Khatri, Covalently attached graphene–ionic liquid hybrid nanomaterials: synthesis, characterization and tribological application. J. Mater. Chem. A **4**, 926–937 (2016)
39. S. Kumari et al., Octadecanethiol-grafted molybdenum disulfide nanosheets as oil-dispersible additive for reduction of friction and wear. FlatChem **3**, 16–25 (2017)
40. S. Kumari et al., Alkyl-chain-grafted hexagonal boron nitride nanoplatelets as oil-dispersible additives for friction and wear reduction. ACS Appl. Mater. Interfaces **7**, 3708–3716 (2015)
41. S. Kumari, O.P. Sharma, O.P. Khatri, Alkylamine-functionalized hexagonal boron nitride nanoplatelets as a novel material for the reduction of friction and wear. Phys. Chem. Chem. Phys. **18**, 22879–22888 (2016)
42. Y. Kligerman, I. Etsion, A. Shinkarenko, Improving tribological performance of piston rings by partial surface texturing. J. Tribol. **127**, 632 (2005)
43. N. Tiruvenkadam, P.R. Thyla, M. Senthilkumar, M. Bharathiraja, A. Murugesan, Synthesis of new aluminum nano hybrid composite liner for energy saving in diesel engines. Energy Convers. Manage. **98**, 440–448 (2015)
44. S. Ingole et al., Tribological behavior of nano TiO_2 as an additive in base oil. Wear **301**, 776–785 (2013)
45. V. Zin et al., Tribological properties of engine oil with carbon nano-horns as nano-additives. Tribol. Lett. **55**, 45–53 (2014)
46. Y. Guo, W. Guo, C. Chen, Modifying atomic-scale friction between two graphene sheets: a molecular-force-field study. Phys. Rev. B **76**(76), 155429–155434 (2007)
47. V. Eswaraiah, V. Sankaranarayanan, S. Ramaprabhu, Graphene-Based engine oil nanofluids for tribological applications. ACS Appl. Mater. Interfaces **1**, 4221–4227 (2011)

48. Q. Zhou et al., Preparation of a reduced graphene oxide/zirconia nanocomposite and its application as a novel lubricant oil additive. RSC Adv. **5**, 91802–91812 (2015)
49. J. Sarkar, P. Ghosh, A. Adil, A review on hybrid nanofluids: recent research, development and applications. Renew. Sustain. Energy Rev. **43**, 164–177 (2015)
50. F. Chiñas-Castillo, H.A. Spikes, Mechanism of action of colloidal solid dispersions. J. Tribol. **125**, 552 (2003)
51. Z.S. Hu et al., Preparation and tribological properties of nanometer magnesium borate as lubricating oil additive. Wear **252**, 370–374 (2002)
52. G. Liu et al., Investigation of the mending effect and mechanism of copper nano-particles on a tribologically stressed surface. Tribol. Lett. **17**, 961–966 (2004)
53. L. Rapoport et al., Tribological properties of WS_2 nanoparticles under mixed lubrication. Wear **255**, 785–793 (2003)
54. S. Samanta, S. Singh, R.R. Sahoo, Effect of thermal annealing on the physico-chemical and tribological performance of hydrophobic alkylated graphene sheets. New J. Chem. **43**, 2624–2639 (2019)
55. A.R. Lansdown, *Molybdenum Disulphide Lubrication* (Elsevier Science, Amsterdam, 1999)
56. M. Farsadi, S. Bagheri, N.A. Ismail, Nanocomposite of functionalized graphene and molybdenum disulfide as friction modifier additive for lubricant. J. Mol. Liq. **244**, 304–308 (2017)
57. S. Kumari, R. Gusain, N. Kumar, O.P. Khatri, PEG-mediated hydrothermal synthesis of hierarchical microspheres of MoS2 nanosheets and their potential for lubrication application. J. Ind. Eng. Chem. **42**, 87–94 (2016)
58. D. Zheng, Y.P. Wu, Z.Y. Li, Z.B. Cai, Tribological properties of WS_2/graphene nanocomposites as lubricating oil additives. RSC Adv. **7**, 14060–14068 (2017)
59. G. Biresaw, K.L. Mittal, *Surfactants in Tribology* (CRC Press, Taylor & Francis Group, 2017)
60. J. Samuel, J. Rafiee, P. Dhiman, Z.-Z. Yu, N. Koratkar, Graphene colloidal suspensions as high performance semi-synthetic metal-working fluids. J. Phys. Chem. C **115**, 3410–3415 (2011)
61. W. Yu, H. Xie, D. Bao, Enhanced thermal conductivities of nanofluids containing graphene oxide nanosheets. Nanotechnol **21**, 055705 (2010)
62. X. Fan, Y. Xia, L. Wang, W. Li, Multilayer graphene as a lubricating additive in bentone grease. Tribol. Lett. **55**, 455–464 (2014)
63. T. Missala, R. Szewczyk, W. Winiarski, M. Hamela, Study on tribological properties of lubricating grease with additive of graphene. Adv. Intell. Syst. Comput. **3**, 181–187 (2015).
64. A. Mura, F. Curà, F. Adamo, Evaluation of graphene grease compound as lubricant for spline couplings. Tribol. Int. **117**, 162-167 (2017).
65. J. Gansheimer, R. Holinski, Molybdenum disulfide in oils and greases under boundary conditions. Am. Soc. Mech. Eng. 242–246 (1973)
66. J. Gansheimer, R. Holinski, A study of solid lubricants in oils and greases under boundary conditions. Wear **19**, 439–449 (1972)
67. R.R. Sahoo, S.K. Biswas, Effect of layered MoS_2 nanoparticles on the frictional behavior and microstructure of lubricating greases. Tribol. Lett. **53**, 157–171 (2014)
68. B. Podgornik, T. Kosec, A. Kocijan, Č. Donik, Tribological behaviour and lubrication performance of hexagonal boron nitride (h-BN) as a replacement for graphite in aluminium forming. Tribiol. Int. **81**, 267-275 (2015).
69. H. Xiao, S. Liu, 2D nanomaterials as lubricant additive: a review. Mater. Des. **135**, 319–332 (2017)
70. D. Berman, A. Erdemir, A.V. Sumant, Graphene: a new emerging lubricant. Mater. Today **17**, 31–42 (2014)
71. M.R. Hilton et al., Structural and tribological studies of MoS2 solid lubricant films having tailored metal-multilayer nanostructures. Surf. Coat. Technol. **53**, 13–23 (1992)
72. M. Ratoi, V. Bogdan, J. Walker, J. Zekonyte, Mechanism of action of WS_2 lubricant nanoadditives in high-pressure contacts. Tribol. Lett. **52**, 81–91 (2013)

73. A. Chouhan, H.P. Mungse, O.P. Sharma, R.K. Singh, O.P. Khatri, Chemically functionalized graphene for lubricant applications: microscopic and spectroscopic studies of contact interfaces to probe the role of graphene for enhanced tribo-performance. J. Colloid Interface Sci. **513**, 666–676 (2018)
74. M. Praveena, C.D. Bain, V. Jayaram, S.K. Biswas, Total internal reflection (TIR) Raman tribometer: a new tool for in situ study of friction-induced material transfer. RSC Adv. **3**, 5401–5411 (2013)

Chapter 9
Evolution of Surface Topography During Wear Process

Deepak K. Prajapati and Mayank Tiwari

Abstract There is an increasing demand for high power density (power throughput/weight) machines such as wind turbines, gear boxes, electric drive trains, and turbines. This requires design of heavily loaded tribological components such as bearings, gears, CVT etc., for preventing surface failure. Tribology analyzes surface contacts between two bodies which are in relative motion. To separate contacting surfaces lubricant is supplied between contacts which form a film. In lubricated non-conformal concentrated contacts (e.g. gears, bearings, cams etc.), it is always desirable to run the components in elastohydrodynamic regime for longer life and negligible wear. However, what is achieved is mostly boundary and mixed lubrication regimes. This is because it is almost impossible to create a surface with negligible roughness. The reason is manufacturing and machining processes by which solid components are produced. During the material processing, texture form on the surface of the components which is in form of roughness, waviness and form. The cavities, voids, inclusions are also induced on the surface during heat treatment. When components are subjected to rolling/sliding motion, these defects (roughness and inclusions) act as stress raisers which is responsible for the crack initiation and crack propagation and ultimately material fails due to material degradations in form of tiny particles (e.g. micro pitting) or in form of spall (macro pitting). Recently, the research has been focused on effect of surface topography on the life of tribological components. This chapter demonstrates the determination of important surface topography parameters by using statistical and fractal methods. Later on, evolution of topography parameters during the wear process is explored in detail and it will be shown that surface topography parameters significantly vary during wear.

D. K. Prajapati · M. Tiwari (✉)
Department of Mechanical Engineering, Indian Institute of Technology, Patna 801106, India
e-mail: mayankt@iitp.ac.in

J. K. Katiyar et al. (eds.), *Tribology in Materials and Applications*,
Materials Forming, Machining and Tribology,
https://doi.org/10.1007/978-3-030-47451-5_9

9.1 Introduction

Due to increasing demand of high efficiency machines, understanding of asperity failure mechanisms which rely on surface topography is required. Mostly, fuel-consuming industries like transport, automobiles, mining, OHV's, wind power etc., in which tribological components run under extreme conditions and cause energy losses, require an in-depth understanding of properties of tribological surfaces. It is well known that particularly in mixed-EHL regime, surface topography significantly affects the performance of tribological components due to composite surface roughness. So, the correct estimation of contact characteristics is very important to get insight on asperity based mechanisms. Three dimensional (3D) characterization of rough surfaces is more relevant and provides more meaningful information than 2D profile [1, 2]. Two methods (i) statistical method and (ii) fractal method are used for rough surface characterization. In statistical method, topography parameters are determined by utilizing statistical data of roughness heights. Whereas, in fractal method, topography parameters are solely determined by utilizing power spectral density. Solid surfaces produced by machining processes are rough at microscale and instruments such as the optical profiler, AFM, stylus profiler are extensively used for measuring the roughness [1, 2]. During the relative motion, contacting surfaces evolve with operational cycles leading to change in asperity characteristics. The change in asperity characteristics can be determined by these roughness parameters. This chapters deals with determination of important topography parameters and their variation during the operation.

9.2 Characterization of Rough Surfaces

9.2.1 Statistical Characterization

It is known that various machining processes such as grinding, milling, turning, produce anisotropic and non-Gaussian texture. Due to anisotropic texture of rough surfaces, topography parameters are different when calculated at different directions [3]. Sayles and Thomas [4] showed that anisotropic rough surfaces can be considered as directional anisotropic rough surfaces. Figure 9.1 shows the isometric view of the simulated directional anisotropic rough surface. The simulated surface consists of 512 × 512 sampling points at the sampling interval of 1 μm (in both x and y directions).

Thomas and Sayles [4] developed an expression for determining the topography parameters by using the concept of equivalent moments. The final equations for determining surface moments is given in Eqs. 9.1–9.3.

$$M_{00} = M_0^1$$

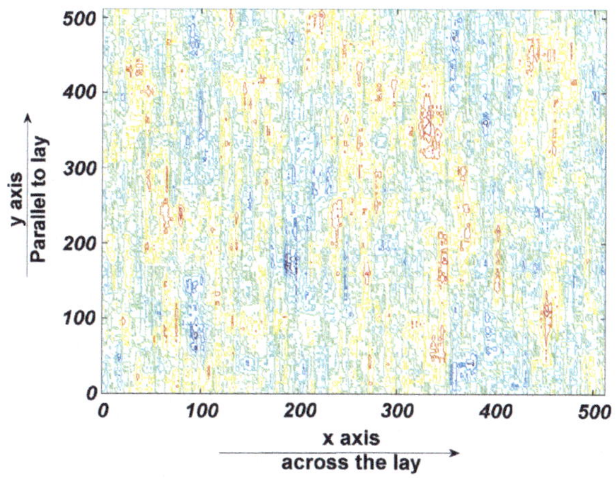

Fig. 9.1 Isotropic view of anisotropic surface (512 × 512 sampling points, $\lambda_x = 10\ \mu m$, $\lambda_y = 50\ \mu m$)

$$
\begin{aligned}
M_{20} &= M_2^1 \\
M_{02} &= M_2^4 \\
M_{40} &= M_4^1 \\
M_{04} &= M_4^4 \\
M_{11} = M_{13} = M_{31} &= 0
\end{aligned}
\tag{9.1}
$$

where, M_{20}, M_{02}, M_{40}, M_{04} is the second and fourth order surface moments across and parallel to the lay respectively and M_{00} is the 0th order surface moments, M_2^1 = second order moment, across the lay, M_2^4 = second order moment, parallel to the lay, M_4^1 = fourth order moment, across the lay M_4^4 = fourth order moment, parallel to the lay, $M_{11} = M_{13} = M_{31}$ = odd order surface moments

Thus, for directional rough surfaces, expressions for determining equivalent moments (M_{2e}, M_{4e}), summit curvature (β) and summit radius (R) are given in Eqs. 9.2–9.4 [3].

$$M_{2e} = (M_{20}M_{02})^{0.5} \tag{9.2}$$

$$M_{4e} = (M_{04}M_{40})^{0.5},\ 1/\text{mm}^2 \tag{9.3}$$

$$\beta = 1.5048\sqrt{M_{4e}},\ 1/\text{mm} \tag{9.4}$$

$$R = \frac{1}{\beta}\ \mu\text{m} \tag{9.5}$$

The variation of the summit Radius (R) with pattern ratio (γ) is presented in Fig. 9.2. The summit radius (R) is determined using Eq. 9.5. Pattern ratio ($\gamma = \lambda_x/\lambda_y$) is defined as the ratio of correlation length in x direction (λ_x) to correlation

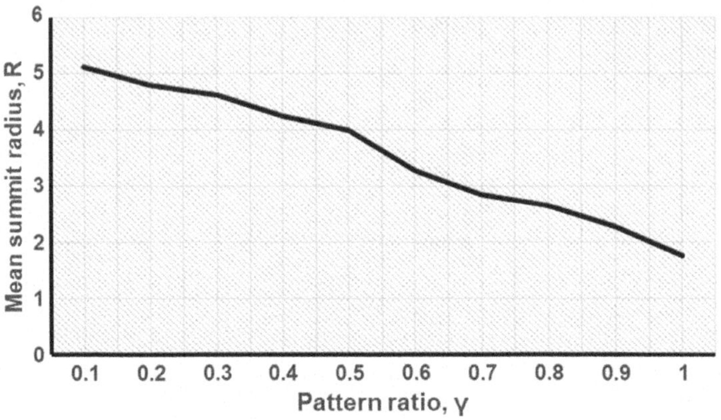

Fig. 9.2 Variation of mean summit radius (R) with pattern ratio (γ), $\lambda_x = 10\ \mu\text{m}$, $\lambda_y = 100$ to $10\ \mu\text{m}$

length in y direction (λ_y). It can be seen from Fig. 9.2 that the summit radius decreases with an increase in pattern ratio. The reason for decrease in summit radius is fourth order moments which increase with an increase in pattern ratio. More details on topography analysis of anisotropic rough surfaces can be found in Ref. [3]. The summit radius is a critical topography parameter and has been extensively used in modeling of wear [5, 6].

9.2.2 *Fractal Characterization*

Firstly, W-M function has been proposed by Weierstrass-Mandelbrot for simulating the self-affine nature of fractal surfaces [7–9]. Recently, various surfaces such as worn surfaces, fracture surfaces, road-tire surfaces have been successfully characterized by the fractal theory [10–13]. Hurst coefficient (H) which is an important descriptor of fractal nature has also been used to determine the texture direction of the surfaces [14–16].

Topography parameters are considerably used in contact modeling as well as wear modeling of rough surfaces [17]. In this section, surface moments in terms of magnification factors (ξ, ζ), Hurst coefficient (H) and wave vectors (q_r, q_l, q_s) is presented which are further used to determine the topography parameters. The closed form expression for determining the summit density (D_{sum}) in terms of magnification factors (ζ, ξ) and the Hurst coefficient (H) is developed in Ref. [18]. The expression for determining the summit radius and summit deviation developed by Prajapati et al. [1] is given in Eqs. 9.6–9.7.

Table 9.1 Input parameters in numerical simulation

Parameters	Value
Sample length	$l_x = l_y = 10\ \mu m$
Number of sampling points	$n = m = 512$
Sampling interval, Δ	$l_x/N\ \ \mu m$
Lowest wave vector, q_l	$\pi/l_x\ \ \mu m^{-1}$
Upper wave vector, q_s	$\pi/\Delta\ \mu m^{-1}$

$$R(H, \xi, \varsigma) = \frac{0.43310}{\sqrt{Cq_r^{4-2H}\left(\frac{1-\xi^6}{6} + \frac{\varsigma^{4-2H}-1}{4-2H}\right)}} \tag{9.6}$$

$$\sigma_s(H, \xi, \varsigma) = \left(1 - \frac{0.8968}{\alpha}\right)^{1/2} \sqrt{2\pi Cq^{-2H}\left(\frac{1-\xi^2}{2} + \frac{\varsigma^{-2H}-1}{-2H}\right)} \tag{9.7}$$

where, C is the scaling factor, R is the summit radius, σ_s is the standard deviation of summit height, α is the bandwidth parameter [1], $\zeta = q_l/q_r$ and $\xi = q_s/q_r$ (ζ, ξ are magnification factor [1], H is the Hurst coefficient, q_r is the roll-off vector [1], q_s is the highest possible wave vector [1], q_l is the lowest possible vector [1].

Table 9.2 represents the variation of topography parameters with the Hurst coefficient (H). The input parameters used to determine topography parameters are listed in Table 9.1. It can be seen from Table 9.2 that the summit density (D_{sum}) decreases with increase in Hurst coefficient. The summit density is directly related to the number of peaks within the sample area. So, it can be said that the number of peaks decreases with increase in the Hurst coefficient. The summit radius (R) increases with an increase in Hurst coefficient. The standard deviation of summit height (σ_s) also decreases with increase in Hurst coefficient (H). The higher value of summit radius or lower value of summit density indicates more flattening of roughness peaks.

Table 9.2 Variation of topography parameters with hurst coefficient (H)

Hurst coefficient (H)	D_{sum} (μm^{-2})	Summit radius, R (μm)	Summit deviation, σ_s (μm)
$H = 0.1$	1.14×10^3	2.2×10^{-4}	0.43
$H = 0.3$	0.98×10^3	4.7×10^{-4}	0.42
$H = 0.5$	0.81×10^3	1.2×10^{-3}	0.39
$H = 0.7$	0.58×10^3	2.7×10^{-3}	0.35
$H = 0.9$	0.32×10^3	6.7×10^{-3}	0.33

9.3 Topography Parameters Evolution During Dry Sliding Wear

The significance of theory of dry sliding contacts is to simulate the sliding part of non-conformal contacts. The theory of dry sliding wear analysis is also valuable to get an insight as to how thoroughly asperity features change with operational cycles, and surface topography. It is known that most of the asperity features change during the running-in wear. Previously, various studies have been reported on the running-in wear by developing numerical model and experiments [19–21]. However, in those studies only topography parameters related to heights of roughness peaks were reported. It has been stated by Stout et al. [22] that spatial and hybrid parameters also contain useful information of the surface which may also vary during wear. In this chapter it is shown that hybrid and spatial parameters, mean contact pressure and the real contact area also vary significantly during running-in wear. More details on modeling of running-in wear model can be found in Ref. [23]. The flow chart for developing the wear model is shown in Fig. 9.3. More details on flow chart of wear model can be found in Ref. [24].

Figure 9.4 shows the evolution of real contact area for isotropic Gaussian 2*D* profile under different value of normal load. In Fig. 9.4 force-displacement relation is established where normal load is applied on the 2*D* profile and parameters such

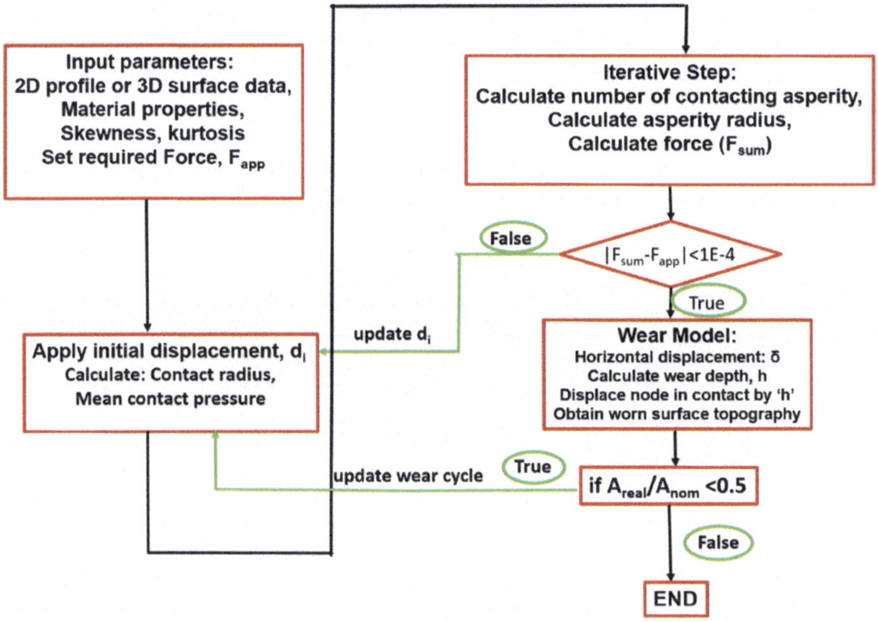

Fig. 9.3 Flow chart for wear model

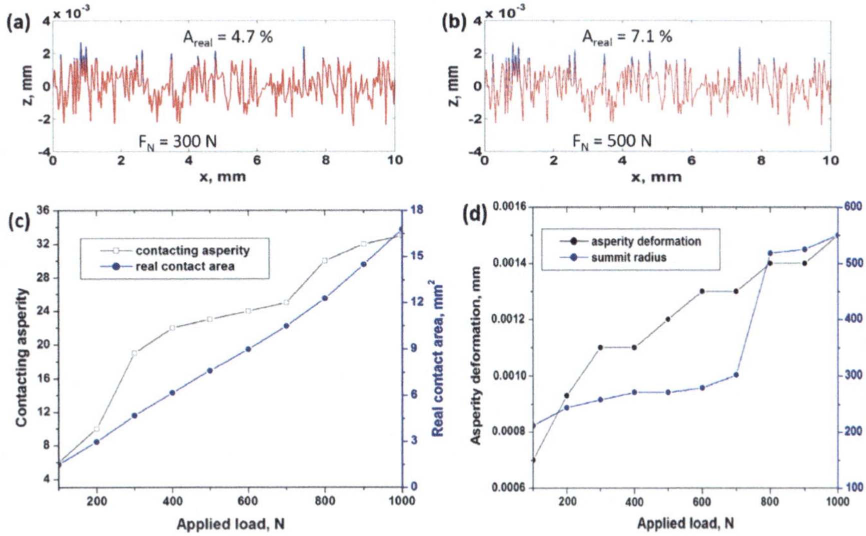

Fig. 9.4 **a** Deformed isotropic gaussian 2D profile under 300 N normal load, **b** deformed isotropic gaussian 2D profile under 500 N normal load, **c** variation of contacting asperity and real contact area with normal load, **d** variation of asperity deformation and summit radius with normal load

as real contact area, asperity deformation and summit radius are calculated corresponding to each value of normal load. It can be seen from Fig. 9.4a, b that the real contact area increases with increase in normal load. As normal load increase, more asperities deform resulting in the real contact area. As illustrated in Fig. 9.4c number of contacting asperity increases with an increase in normal load from 100 N to 1000 N. The real contact area is also found to increase with an increase in normal load. Figure 9.4d shows the variation of asperity deformation and summit radius (R) with an increase in normal load. It can be seen that asperity deformation increases with an increase in normal load.

Figure 9.5 shows evolution of the real contact area during wear process. It can be seen that the real contact area increases with increase in wear cycles. The reason for increasing real contact area is significant removal of asperities. Due to the removal of contact asperities above the mean plane, surface become flattened leading to an increase in the real contact area. As illustrated in Fig. 9.6 the mean contact pressure decreases with an increase in wear cycles. The reason is increase in the real contact area with an increase in wear cycles. Figure 9.7 shows the variation of skewness (R_{sk}) and kurtosis (R_{ku}) with increase in wear cycles. It can be seen from Fig. 9.7 that skewness (R_{sk}) decreases with increase in wear cycles from 0 to 21,000. Figure 9.7 shows the variation of kurtosis (R_{ku}) with an increase in wear cycles. As illustrated in Fig. 9.7 that kurtosis first decreases till 2500 wear cycles and then increases for further increase in wear cycles. Kurtosis decreases initially due to removal of roughness peaks and after some wear cycles starts increasing due to an increase in the real contact area. Figure 9.8a shows the unworn isotropic Gaussian 2D profile

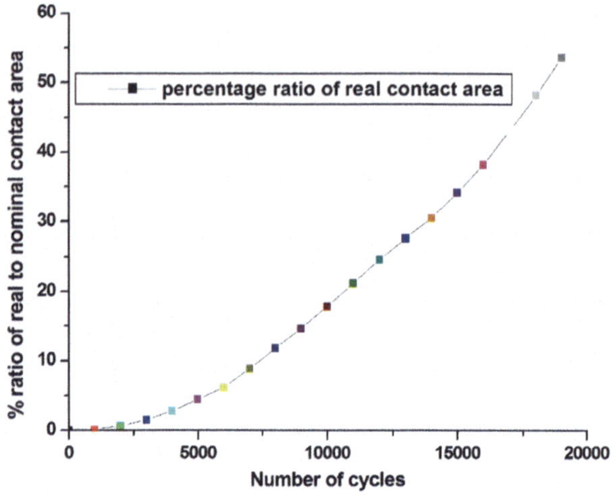

Fig. 9.5 Variation of real contact area with wear cycles ($F_N = 2$ kN, $\delta = 200$ μm)

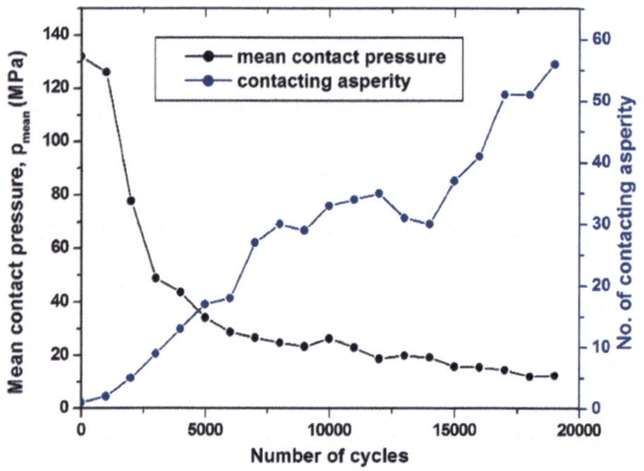

Fig. 9.6 Variation of mean contact pressure and contacting asperity with wear cycles ($F_N = 2$ kN, $\delta = 200$ μm)

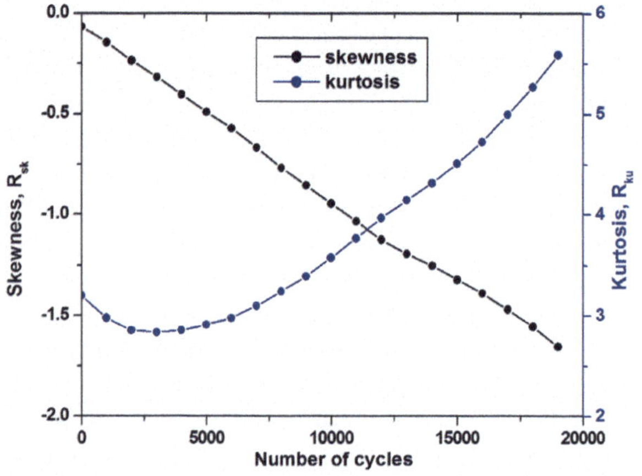

Fig. 9.7 Variation of skewness and kurtosis with wear cycles ($F_N = 2$ kN, $\delta = 200$ μm)

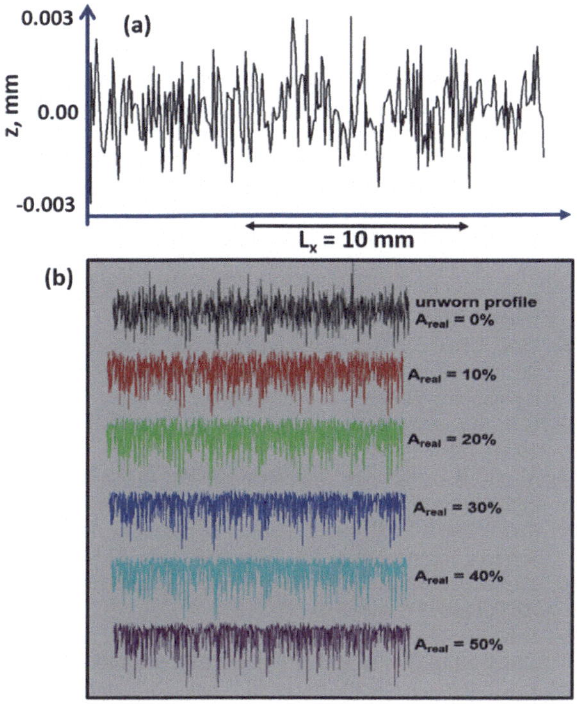

Fig. 9.8 **a** Unworn isotropic gaussian 2D profile, **b** worn profiles at different real contact area ($F_N = 2$ kN, $\delta = 200\ \mu$m)

having the Gaussian property (i.e. mean = 0 μm, $R_q = 1$ μm, $R_{sk} = 0$, $R_{ku} = 3$). Figure 9.8b shows the worn profiles at different value of the real contact area. It can be seen from Fig. 9.8 that roughness peaks height decreases as real contact area increases from 0 to 50%.

9.4 Summary

Characterization of surface roughness can be done using statistical and fractal methods. Solid surface produced by machining process exhibit different topography parameters when calculated in different directions. For directional anisotropic rough surface, mean summit radius is found to decrease with an increase in pattern ratio. It is also shown that for self-affine fractal surfaces, summit radius increases with an increase in Hurst coefficient. Whereas, summit density and standard deviation of summit height decreases with an increase in Hurst coefficient. So, it can be inferred that components which run with higher mean summit radius experience less friction. Evolution of topography parameters during wear is also discussed in detail and it is shown that mean contact pressure, and skewness decreases with an increase in number of cycles. Whereas, the real contact area and kurtosis increase with an increase in number of cycles.

References

1. D.K. Prajapati, Evolution of surface topography parameters for tribological components under rolling and sliding motion. Ph.D. Thesis (IIT, Patna, India, 2020), 1–199
2. D.K. Prajapati, M. Tiwari, The relation between fractal signature and topography parameters: a numerical and experimental. Surf. Topogr.: Metrol. Prop. **6**, 045008 (2018)
3. D.K. Prajapati, M. Tiwari, Topography analysis of random anisotropic gaussian rough surfaces. ASME J. Tribol. **139**, 041402 (2017)
4. R.S. Sayles, T.R. Thomas, Thermal conductance of rough elastic contact. Appl. Energy **2**(4), 249–267 (1976)
5. L. Xiao, B.G. Rosen, N. Amini, P.H. Nilsson, A Study on the effect of surface topography on rough friction in roller contact. Wear **254**(11), 1162–1169 (2003)
6. D.K. Lawrence, R. Shanmugamani, B. Ramamurthy, Evaluation of image based abbott-firestone curve parameters using machine vision for the characterization of cylinder liner surface topography. Wear **55**, 318–334 (2014)
7. B.B. Mandelbrot, *The fractal geometry of nature* (W.H. Freeman and Company, NY, USA, 1983). ISBN 0-7167-1186-9
8. Y. Xu, R.L. Jackson, Statistical models of nearly complete elastic rough surface contact-comparison with numerical solutions. Tribol. Int. **105**, 274–291 (2017)
9. C.Q. Yuan, J. Li, P. Yan, Z. Peng, The use of the fractal description to characterize engineering surfaces and wear particles. Wear **255** (1–6), 315–326 (2003)
10. Z.Q. Mu, C.W. Lung, Studies on fractal dimension and facture toughness of steel. J. Phys. D: Appl. Phys. **21,** 848–850
11. D.K. Jha, D.S. Singh, S. Gupta, A. Ray, Fractal analysis of crack initiation in polycrystalline alloys using surface interferometry. EPL **98,** 44006
12. B.N.J. Persson, On the fractal dimension of rough surfaces. Tribol. Lett. **54**(1), 99–106
13. P. Podsiadlo, G.W. Stachowiak, Analysis of trabecular bone texture by modified Hurst orientation transform method. Med. Phys. **29**(4), 460–474 (2002)
14. M. Wolski, P. Podsiadlo, G.W. Stachowiak, Applications of the variance orientation transform method to the multiscale characterization of surface roughness and anisotropy. Tribol. Int. **43** (11), 2203–2215 (2010)
15. K. Holmberg, A. Laukkanen, H. Ronkainen, R. Waudby, G. Stachowiak, M. Wolski, P. Podsiadlo, M. Gee, J. Nunn, C. Gachot, L. Li, Topographical orientation effects on friction and wear in sliding DLC and steel contacts, part 1: Experimental. Wear (330–331), 3–22 (2015)
16. J.A. Greenwood, J.B.P. Williamson, Contact of nominally flat surfaces. Proc. R. Soc. Lond. A **295**, 300–319 (1966)
17. V.A. Yastrebov, G. Anciaux, J.F. Molinari, From infinitesimal to full contact between rough surfaces: evolution of the contact area. Int. J. Solids Struct. **52**, 83–102 (2015)
18. E.T. George, H. Liang, *Mechanical tribology: materials, characterization, and applications* (CRC Press, Boca Raton, US, 2004)
19. Y. Xie, J.A. Williams, The prediction of friction and wear when a soft surface slides against a harder rough surface. Wear **196** (1–21), 21–34 (1996)
20. D.V. De Pellegrin, A.A. Torrance, E. Haran, Wear mechanisms and scale effects in two-body abrasion. Wear **266**(1– 9), 13–20 (2009)
21. S.K. Roy Chowdhury, H. Kaliszer, G.W. Rowe, An analysis of changes in surface topography during running-in of plain bearings. Wear **57**(2), 331–343 (1979)
22. K.J. Stout, T.G. King, D.J. Whitehouse, Analytical techniques in surface topography and their application to a running-in experiment. Wear **43**(1), 99–115 (1977)
23. D.K. Prajapati, M. Tiwari, 3D numerical wear model for determining the change in surface topography. Surf. Topogr.: Metrol. Prop. **6**, 045006
24. D.K. Prajapati, M. Tiwari, 2D numerical wear model for determining the change in surface topography with number of wear cycles, in *Proceedings of Asia International Conference on Tribology,* vol. 2018, pp. 233–234 (2018)

Chapter 10
Wear Characteristics of LASER Cladded Surface Coating

Manidipto Mukherjee

Abstract Laser cladding by a powder injection technique has been widely used in industrial applications such as rapid manufacturing, parts repair, surface coating, and innovative alloy development. The capability to mix two or more types of powders and to control the feed rate of each powder flow makes laser cladding a flexible process for fabricating heterogeneous components or functionally graded materials. This technology also allows the material gradient to be designed at a microstructure level because of small localized fusion and strong mixing motion in the melt pool of laser cladding. Thus materials can be tailored for a flexible functional performance in particular applications. The inherently rapid heating and cooling rates associated with the laser-cladding process enable extended solid solubility in the metastable or non-equilibrium phases of production, offering the possibility of creating new materials with advanced properties. Laser cladding uses the same concept as arc welding methods, except that a laser is used to melt the surface of the substrate and the additional material, which can be in the form of wire, powder or strip. Laser cladding is commonly performed with CO_2, Nd: YAG, and more recently fibre lasers. Laser cladding typically produces clads having low dilution, low porosity and good surface uniformity. This technique produces minimal heat input on the part, which largely eliminates distortion and the need for post-processing, and avoids the loss of alloying elements or hardening of the base material. The clad material experiences a rapid natural quench when cooling down after deposition, which results in a fine-grained microstructure. Among the different surface treatments used to improve the wear resistance of metallic materials, laser cladding is an attractive alternative to conventional techniques due to the intrinsic properties of laser radiation: high input energy, low distortion, avoidance of undesirable phase transformations and minimum dilution between the substrate and the coating. Furthermore, the advantages of laser cladding include great processing flexibility and the possibility of selectively cladding small areas. These advantages not only result in better quality products but also offer significant economic benefits. This chapter describes in details three major aspects

M. Mukherjee (✉)
CAMM, CSIR-Central Mechanical Engineering Research Institute, Durgapur, West Bengal 713209, India
e-mail: m.mukherjee.ju@gmail.com

J. K. Katiyar et al. (eds.), *Tribology in Materials and Applications*,
Materials Forming, Machining and Tribology,
https://doi.org/10.1007/978-3-030-47451-5_10

of LASER cladding surface coating with respect to wear characteristics. The first subsection will describe different LASER cladding techniques and their effect on surface morphology. And how these surface morphologies are affecting the wear characteristics. The second subsection will provide the details on different aspects of filler metal selection for wear resistance applications. Then, the effect of process parameters on the surface morphology will be discussed thoroughly. The focus will be particularly on the effect of heat input on the wear behaviour of the cladded surface. The overall aim is to cover a broad area of LASER cladding surface coating in terms of wear characteristics and the improvement in wear resistance compared to other techniques.

10.1 Introduction

Laser cladding coupled with powder injection methods is widely used for many industrial applications such as fast manufacturing, repair of parts, surface coating, and the development of modern alloys. It is a versatile method for the production of heterogeneous components or functionally graded materials (FGMs) due to its ability to mix two or more material types together with the regulation of the material flow. The laser cladding technique has small localized fusion and strong mixing motion in the melt pool allowing versatile functional quality in specific applications to be achieved by the gradient material engineered at a microstructure point. The laser-cladding process, combined with rapid heating and cooling speeds, allows for prolonged solid solubility in the development phases of metastability or non-equilibrium, providing the possibility to create new materials with advanced properties.

Laser cladding follows the same principle as arc welding techniques, except that the surface of the substrate and the filler material in the form of metal, powder or strip are melted with a laser source at different energy levels. The CO_2 laser, Nd: YAG laser and more recently the fiber laser are widely used as the energy source for the application of laser cladding. Laser cladding generates typically low dilution clads, low porosity and good surface uniformity. This technique creates low heat feedback on the substrate, which effectively removes distortion and the need for post-processing, and avoids the loss of alloy elements or surface hardening. The clad material experiences a rapid natural quench when cooling down after deposition, which results in a fine-grained microstructure.

Laser cladding is an attractive alternative to traditional techniques due to the intrinsic properties of laser radiation among the various surface treatments used to boost the wear resistance of metallic materials: high input power, low distortion, avoidance of undesirable phase transitions and reduced dilution between the substrate and the coating. In addition, the advantages of laser cladding include great flexibility in processing and the possibility of selectively cladding small areas. Not only do these benefits lead to better quality products, but they also offer significant economic benefits.

This chapter describes in details three major aspects of laser cladding surface coating with respect to wear characteristics. At first, the laser cladding techniques and the effect of process parameters on the surface morphology are briefly discussed. Then the details on different aspects of coating metal selection for wear resistance applications have been presented. Also, an overview on different alloy systems used in laser cladding is provided along with current problems and the development trend in this field are briefly elaborated.

10.2 LASER Cladding Techniques

To understand LASER cladding technology, we must first understand the direct metal deposition (DMD) method which is currently transforming the traditional manufacturing into a twenty first century modern manufacturing.

10.2.1 Direct Metal Deposition

The DMD method is a type of solid free-form manufacturing (SFM) method that allows the production of precise components with a precision of 0.254 mm and similar properties to wrought materials, with a density of almost 100%. DMD production methods are generally classified as additive processes and are commonly referred to as solid free-form manufacturing, layer production or rapid prototyping/manufacturing [1]. The SFM is generally an idea of constructing a 3D object layer by layer from a set of discrete 2D profiles designed and sliced using a dedicated CAM software. In the past two decades, more complex prototyping products were developed with shorter processing time due to rapid advancement of CNC technologies and other CAD-CAM tools [2].

The SFF techniques can be generally classified into four major categories according to the processing methods [1, 2]:

- Photo-curing
- Cutting and gluing
- Melting and solidification or fusing
- Joining or binding.

Most products, such as paper, polymers, nylon, wax, resins, metals, and ceramics, have been manufactured in SFM to date. The SFM technique can be used as input material to fit solid plate, metal, dust, water, droplet and gas/vapor. It technology can be applied widely to (1) product design, (2) technical evaluation and production planning, and (3) software and manufacturing to serve a wide range of industries such as aerospace, automotive, biomedical, industrial, electrical and electronics [3, 4]. The powder-based laser cladding (PBLC) technique achieved a high level of marketing due to its flexibility and accuracy among the various SFM methods. The PBLC

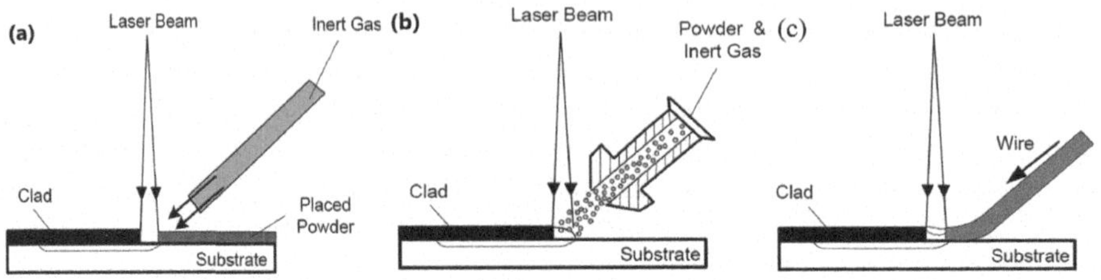

Fig. 10.1 Different methods of laser cladding: two-step cladding i.e. **a** preplaced powder, and one-step cladding i.e. **b** powder injection and **c** wire feeding

technique involves either a powder bed (two-step cladding); for example, selective laser sintering; as shown in Fig. 10.1a or injection of powder flow and wire (one-step cladding); e.g. DMD, net-shape laser engineering; as shown in Fig. 10.1b, c respectively. The PBLC process usually achieved better deposit resolution compared with the powder injection laser cladding (PILC) method. Nevertheless, as a consequence of the powder-paving steps, the PBLC process requires long processing times and delivers small construction envelopes within a specific layer and therefore lacks the flexibility to build heterogeneous materials/structures within one layer [1, 3]. Hence, more emphasis is provided in this chapter on the recent development of one-step laser cladding or PILC process.

The closed-loop DMD is a synthesis of multiple technologies including lasers, sensors, a computer numerical controlled (CNC) work handling stage, CAD/CAM software and cladding metallurgy. The key features of the closed-loop DMD process is the integrated feedback system which ensures close dimensional tolerance. This system actively maintains a uniform layer thickness and saving valuable post processing time. This suggests that by removing many intermediate steps, this method will significantly reduce the lead time from "concept to product." The process can deliver various materials with close dimensional tolerance in different pixels that provide better control of heat transfer in a mould or die depending on the functional requirement. For example, for better wear resistance, the surface of a die may be made of tool steel, but internal parts of the same die may contain conductive materials and cooling channels for improved heat transfer. The closed-loop DMD has already shown the ability to manufacture completely dense metal components with practical designs in three dimensions. As a result, this method can be adapted directly to CAD, magnetic resonance imaging, and X-ray tomography digital data and is capable of producing a wide variety of complex metallic components. This process uses many engineering alloys such as tool steels, Stellite 21 and 6 cobalt alloys, Inconel 718, stainless steel, titanium, etc. for component manufacturing [5–10].

10.2.2 LASER Cladding Process

The closed-loop DMD is an extension of laser-cladding technology and is a multilayer laser-cladding system with feedback control in theory. The alloy material can be incorporated into the laser-melted volume either as pre-deposited layers (Fig. 10.1a) or using a co-depositing technique where the alloy materials are fed in the form of wire, sheet or powder [11]. Cladding with pre-deposited layers is the simplest method where the powder is pre-placed on the substrate and the area is being covered by an inert gas during the melting operation by laser source. On the other hand, the co-depositing technique can be carried out in two different ways i.e. wire or strip feeding method and lateral or coaxial powder injection method [11, 12]. However, the PILC technique, as shown in Fig. 10.2, is widely used in industries because of its explicit capabilities for fabricating heterogeneous components or functionally graded materials [13–16]. In the PILC technique, argon is used to generate a powder stream which is blown under the laser beam while it scans the surface of the substrate to generate a molten pool with a depth corresponding to the thickness of a single clad. Overlapping tracks are made to cover large surface area shielded by argon. The following sections briefly discussed the two one-step laser cladding processes.

10.2.2.1 Laser Cladding with Powder Injection Method

The PILC process is a type of one-step laser-cladding where powders are directly supplied into the molten pool either coaxially around the laser beam or laterally from the side of the laser beam. In both cases, the laser interacts with the powder particles before reaching the substrate. The temperature of powder particles increases by absorbing a part of laser energy, other part of the energy beam is reflected by the particles which reduces the laser intensity and may change its distribution. Gedda et al. [17] developed a semiempirical method of assessing the laser energy redistribution during CO_2 laser cladding with lateral powder injection. The experimental and theoretical studies on laser attenuation and temperature distribution in powder for a

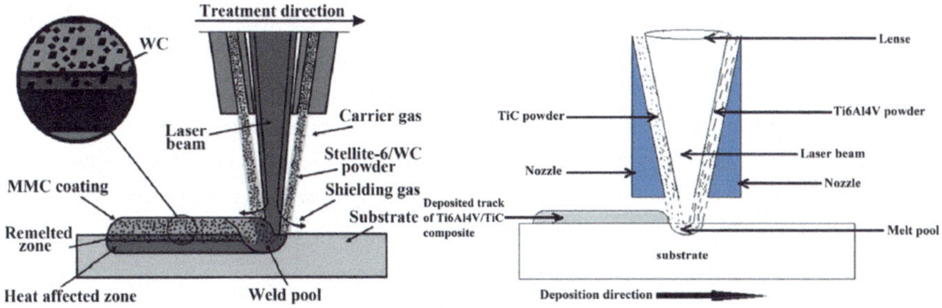

Fig. 10.2 The schematic diagram of coaxial powder injected laser cladding process

focused coaxial powder stream were conducted by Lin et al. [18, 19]. A Gaussian distribution function was used to understand the coaxial powder concentration and later the theoretical data was verified by the actual observation. Many numerical models and methods were developed by Liu and Lin [20] to study the effect of CO_2 laser beam energy coupled with coaxial powder flow on the heating, melting, and evaporation processes of a single spherical powder particle. Similarly, numerous analytical and numerical models of the powder injected cladding process have been developed in past decade to understand the involved mechanisms such as heat flux, dilution, inter-track porosity, continuity, temperature distribution, heat transfer and fluid motion. Some of the common existing mathematical models are: one-dimensional conduction model, finite-difference model, two-dimensional steady-state model, finite-element model, and two-dimensional stationary finite-element model [21–26]. Recently, a three-dimensional transient finite-element model was developed for laser cladding with a powder injection process [27] where the interaction between the powder and melt pool is decoupled to simplify the thermal analysis and melt pool boundary calculation. Such models discussed other aspects of physical phenomena occurring in the system, such as thermal diffusion and dynamics of the melt bath. Most of these simulation models, however, assumed the form of the melt pool as a significant feature or ignored the convection effect, and these models may not be sufficiently accurate to predict the effects of system parameters, such as prediction of dilution. The injection of powder greatly influences the pool dynamics and is crucial to the quality of the product for the following reasons: I the laser power is attenuated by the powder layer, so that the threshold power of the molten pool on the substratum varies with the mass flow rate; and (ii) the injected powder brings additional momentum and energy to the melt pool, which affects the flow pattern and the flow rate.

In powder feeding method, the process requires the control of some parameters such as powder feed rate, process speed, laser power, beam diameter or even melting pool temperature for smooth and continuous flow. The particle critical bonding velocity is also an important factor which is sensitive to spray material, powder quality, the particle size and the impact temperature of the powder. The high process speed would lead to a decrease in the surface temperature and the tracks would not correctly bind to the surface. Inversely, for a slower processing speed, the surface would reach higher temperature, leading to a higher dilution and lower mechanical properties of the clad. Again, optimum particle size is essential to attain acceptable fluidity of powder during feeding. The larger particle size, for instance 140-mesh powder, need more heat for melting, which means that an increasing laser power or decreasing travelling velocity is needed. Smaller particle-sized powders, for instance 300-mesh powder, are easy to melt but when the particle size is smaller than 400-mesh, the fluidity of the powder during powder feeding will be poor, so the feeding will cause difficulties during the laser cladding process [28, 29].

10.2.2.2 Laser Cladding with Wire Feeding Method

Laser wire cladding is another one-step process that combines a wire with a laser beam and offers many benefits over traditional cladding processes. It is ideal for corrosion and wear resistance when a very detailed part or an exact compositional specification need to be met. The wire feeding method has the advantages of high efficiency, low surface roughness and almost 100% of the material usage compare to powder injection method. However, wire feeding claddings are more sensitive to cold cracking resulting from residual thermal stresses produced during rapid cooling and poor ductility of both substrate and cladding materials [30]. The wire is generally fed from the side, and the geometry of the cladding layer is very much related to the wire feeding angle [31]. Due to high laser reflection rate from the wire surface, the process is very inconsistence [32]. Only half of the side feeding wire can be irradiated by the laser, and this unbalanced heat transferring makes the wire easily to bent and jump out of the molten pool during the process. To solve these problems, coaxial feeding methods have been developed to make the wire feed vertically in the middle of the laser to the substrate. Coaxial wire cladding method is very promising because it offers a very high efficiency and a consistent interaction between the laser and wire [33]. Again, by adopting correct wire feeding direction and position, the difficulties in the melting of fed wire and the transferring of molten drop for laser cladding with wire feeding can be solved. The best possible wire direction, in the side feeding method, is to feed the wire from the opposite side of the moving laser source (scan direction) at an angle of 30° [34]. The wire should melt completely during the best wire feeding condition without disturbing the clad layer. If it is not properly melted during the time of laser-wire interaction, the wire will be plunged into the melt pool and will be melted by the high temperature of the melt pool.

10.2.2.3 Effect of Process Parameters on the Surface Characteristics

An important quality aspect of the clad layer is the dilution of the deposited clad layer material, i.e. the extent of mixing between the clad material and the substrate. Even though a certain minimum mixing is necessary to guarantee a good bonding, an excessive mixing is not desirable. The clad layer is characterized by several geometrical quantities. The quantities H_c, W_c, A_c and A_b of Fig. 10.3 are the clad layer height, width and the areas of clad and the molten substrate respectively. The dilution (D_c) of the clad layer can be determined by the concentrations of a specific element in the clad layer, the supplied clad material and the substrate material. Salehi et al. [35] used the iron content to determine the dilution using the following equation for the ferrous based alloy systems:

$$D_C = \frac{L_{Fe} - P_{Fe}}{S_{Fe} - P_{Fe}} \tag{10.1}$$

Fig. 10.3 Schematic cross-sectional representation of a clad layer and dilution abbreviations

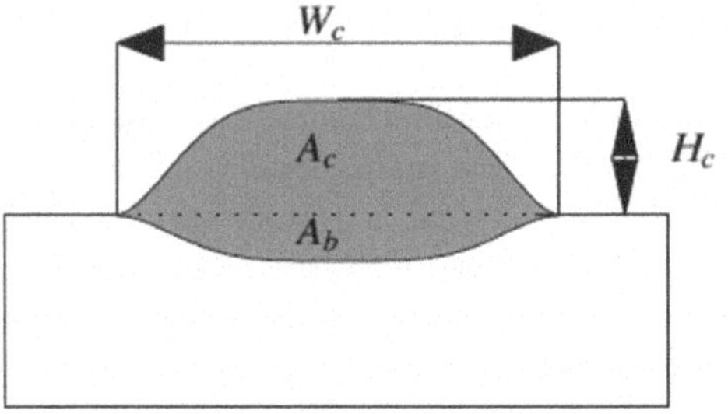

where P_{Fe}, L_{Fe} and S_{Fe} are the iron concentration in the supplied powder, the clad layer and the substrate respectively. However, the general approximation of the dilution level in a clad layer is the ratio between the area of the molten substrate material and the total area of the molten layer.

$$D_C = \frac{A_b}{A_b + A_c} \tag{10.2}$$

The ratio in Eq. 10.2 has to be as small as possible to obtain a surface layer which is hardly diluted by the substrate material. However, there is the risk of no fusion and bonding between cladding material and substrate if the ratio tends to zero. Therefore, a dilution between 2 and 10% is generally accepted. The dilution level depends on power level and the distributions of both the laser beam and the powder jet. It is reported that the laser power and the melt pool temperature have a positive effect and both the cladding speed and clad layer height have a negative effect on the amount of dilution. The melt pool length and melt pool width has a positive effect on the amount of dilution above a critical shape factor [36].

During the process, the initial temperature of the substrate material changes, influencing the optimum level of laser power in relation to the amount of dilution [37]. The depth of the melting pool also depends on the distribution of laser intensity. And a centrally located energy distribution results in a deeper melting pool than a uniform distribution [38]. In order to avoid mixing substrate material in the pool, the depth of the melt pool should be small and close to constant over the beam diameter. This can be accomplished by selecting a higher intensity power density distribution at the edges of the spot [39].

A laser source's scanning velocity significantly affects the diluted region size, microstructural morphology, and hard particle distribution. The interaction time between laser beam and metal surface increases when scanning velocity decreases and both the maximum depth and the molten pool cross-section area increase. Uniform distribution of MC particles or hard particles along with the large number of compounds are obtained in the layered surface with the lower scanning velocity. The uniformity of distributed hard particles decreases the wear constant which is generally results in better wear resistance of the cladding. However, if the particle size decreased beyond a critical value during a longer interaction time it will alter the wear mechanism due to the active participation of metal matrix in the transference process which will lead to more severe wear. Furthermore, the depth of interaction and the adherence of the coating decreased with increasing scanning velocity [40–42].

The laser power density is also based on the microstructure and microhardness of the clad surface. The depth of the alloyed layer increases and with the higher laser power, the non-equilibrium phases decrease. The lowest surface roughness and highest microhardness was shown by the lamellar eutectic structure formed at intermediate power density [43]. However, during laser surfacing of aluminium with WC + Co + NiCr particles, Nath et al. [44] observed that the depth of alloyed zone may be increased initially with increasing applied power due to higher energy absorption but continuously decreases with increasing power level because at the high-power level evaporation of materials started from the surface. The extreme laser power generates a large quenching stress leading to large micro-cracks, whereas very low laser power unable to dissolve the hard particles which incompletely provide higher average surface hardness of the material. Again, the laser surfacing of aluminium alloy with nickel using a pulsed Nd:YAG laser showed that the molten pool depth decreases and the percentage of nickel within the molten pool increases with increasing the laser beam diameter. Maximum absorption of beam energy occurred at the middle of the molten pool when the laser power is lower and the temperature gradient with negative slope created between the middle and the edge of the molten pool. The convective fluid flow is observed on the pool surface when the temperature gradient is closer to zero, starting from the middle to the edges of the molten pool. The stream of convective flow decreases the depth of penetration and increases the size of the molten pool. The shallow depth of the pool creates strong surface tension forces towards the pool's circumferential area which bring the metals to the surface and leading to an increasing nickel content in the cladded area [45].

The laser beam focal position also has significant influence on the cladded coating. Three types of focus modes such as negative defocus, focus, and positive defocus (Fig. 10.4), are generally applied to create laser cladding of non-ferrous alloy systems [46].

The powders are not completely melted in a negative defocus state and the partially melted particle continues to deflect away without depositing for the pool layer as they never went through the beam's highest power area (Fig. 10.4a). The centered condition of the laser beam produces a perfect molten pool where the particles are properly melted and completely submerged in the pool creating deposition of the surface without any transverse deflection which creates almost ideal interfaces for the substrate coating (Fig. 10.4b). The positive defocus condition generates a large melted zone with unnecessary dilution of substrate with the deposited layer which declined the properties of the cladding (Fig. 10.4c).

Additionally, wavelength of the laser beam has important effects on the laser absorption by metals. Different types of lasers with different wavelengths such as CO_2 laser (10.64 μm), YAG laser (1064 nm), diode laser (848 nm), and excimer laser (248 nm), are used to clad coatings on non-ferrous alloys. However, the shorter wavelength is generally advantageous due to higher energy absorption [47]. Besides, the laser absorption rate (η) directly influenced the laser beam and the treated materials. The absorption rate can be calculated by the equation of $\eta = E_a/E_i$. Where,

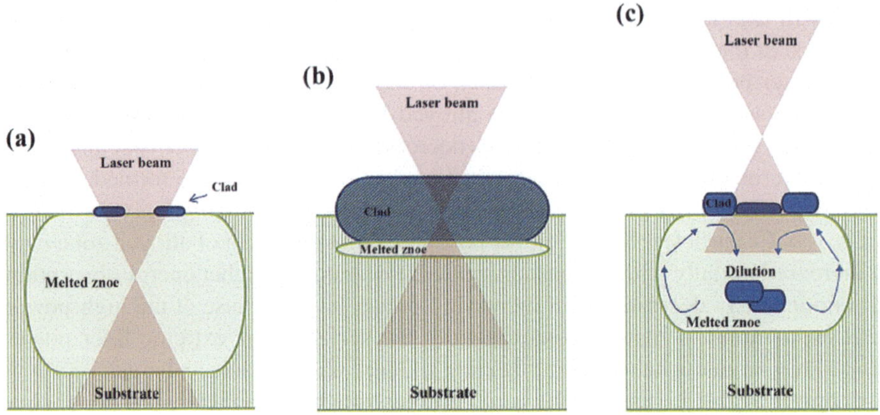

Fig. 10.4 Schematic representation of various laser beam focal positions: **a** negative defocus, **b** focus and **c** positive defocus plane

E_a is the energy absorbed by the irradiated material and E_i is the incident energy of the laser beam. However, the method of determining the laser absorption rate is very sensitive to the stability of the laser source and the pulses while using pulsed source. The laser beam and material may interact better with the pre-treated techniques such as anodising [48].

There are many parameters which can influence the output of the cladding in term of quality. Therefore, optimization of the process parameters such as laser power, scanning speed, etc., is important to achieve better quality. The substrate can collapse or melted down if excessive laser power or very low scanning speed is used to conduct surface modification. On the other hand, low laser power and high scanning speed lead to a low energy density where deposition may not properly react with the substrate leading to poor metallurgical bonding. Furthermore, the depth and width of the molten pool are intensively affected by the laser beam diameter. The surface quality of the cladding also depends on the accurate selection of the wavelength, the beam location, and the absorption rate of the laser.

10.3 Selection of Coating Material

In addition to the parameters of laser cladding, it is important to select proper coating materials to obtain laser cladded coatings with desired properties and surface performance. In general, consideration should be given to both the physicochemical properties of the substrate and the coating materials. On this basis, suitable coating materials would be selected to develop coatings according to the condition of their service. A deeper understanding of the microstructure–composition–synthesis–processing relationships would be established through a large number of experiments and analyzes, helping us further refine the parameters as well as the material system.

Compatibility between the coating materials and the substrate is the most important aspect in which both have similar physical properties, such as melting point (T_m), thermal expansion coefficient (γ) and elasticity modulus (E). Wide variation in these properties may result in undesirable quality between the coating material and the substrate. Wide difference in T_m makes it difficult to achieve a proper metallurgical bonding between the coating and the substrate. Similarly, the residual stress will be very high due to the excessive thermal gradient which generate cracks. If the difference of γ and E between the two materials is too large, then the abscission of the coating can occur. In addition, the crystal structure and the chemical properties of the two should match each other in order to achieve good wettability. Better wettability between the substrate and the coating material creates a good metallurgical bond. And the coating surface tension should be lower than the substrate's essential surface tension. Low-melting metals (e.g. Co, Ni) and some reactive metals (e.g. Ti, Al) as well as the related compounds are therefore typically used as cladding materials [49].

10.4 Wear Characteristics of Cladded Surface

The wear characteristics of any material are generally determined, in laboratory, by the Pin on disk test apparatus where a standard (according to the ASTM G99-17) hemispherical pin of 10 mm diameter is used on rotary disk. The wear rate is calculated from the volume loss of the pin based on the wear diameter. The volume loss and wear rate are calculated by using Eqs. (10.3) and (10.4) respectively [50–52].

$$V = \frac{\pi d^4}{64R} \tag{10.3}$$

$$k = \frac{V}{SF_n} \tag{10.4}$$

whereas V is the volume loss (mm^3) of the pin, d is the wear diameter (mm) of the pin and R is the radius of the pin tip. While, k represents the wear rate (mm^3/N m) and it is defined as the volume loss V (mm^3) per unit of the sliding distance S (mm) and of the applied force (F_n). The wear properties of a laser clad layer are severely dependant on the process parameters (briefly explained in previous section), microstructure (particularly the secondary phases), hardness of the clad layer etc. In this section, the effect of microstructure of cladded layer on the wear characteristics is briefly discussed. Simultaneously, different alloy systems are also discussed to understand the different microstructural behaviour.

10.4.1 *Effect of Clad Composition and Microstructure on Wear Characteristics*

In this section four major alloy systems i.e. High strength steel, Titanium, Magnesium and Aluminium have been briefly discussed to understand the microstructural effect on the wear characteristics.

10.4.1.1 HSS System

The laser clad HSS alloys showed lowest degree of wear because of the highly refined microstructure. At room temperature, wear mechanism of the laser clad HSS alloys is dominated by the abrasion between the interfacial layer due to the presence of hard carbide particles. Several type of metallic carbides can be precipitated in the HSS cladding according to the standard phase diagram as shown in Fig. 10.5 [53].

The wear resistance of the laser clad HSS also depends on the hardness of the matrix, the higher the matrix hardness, the better the wear resistance. However, the high temperature wear characteristics of laser clad HSS are highly complex and governed by the inhomogeneous oxidation process of metallic carbides. The oxidation rate is dependent upon the composition and distribution of carbides. MC type of carbides are highly prone to oxidation whereas M_2C and $M_7C_3/M_{23}C_6$ carbides resist oxidation. Also, the presence of vanadium-tungsten enriched complex nanometer-size carbides reduced the oxidation of matrix. The continuous oxide layer on the surface acts as a protective layer and the matrix experienced marginal wear and vice versa. Therefore, heat treatment process has no significant improvement on the wear characteristics of the laser clad HSS alloys [54–56]. Furthermore, the wear rate of

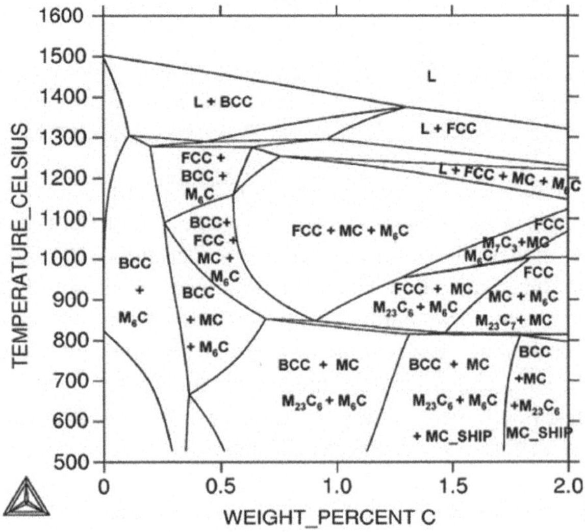

Fig. 10.5 Phase diagram of HSS showing formation of different metallic carbides [53]

laser clad HSS also dependant on the sliding speed. The wear rate of the as clads is generally increased with the increase in sliding speed [57]. The post-deposition heat treatment of the substrate can alter the wear characteristics of the laser clads. The quenching and subsequent tempering of the deposited zone leads to the formation of tempered martensite and austenitic dendrites with fine spherical carbides. The tempering of the deposited clad layer resulted in the formation of tempered martensite within the microstructure, with the retained austenite phase led to a decrease in hardness. The higher wear resistance can be achieved with the non-heat-treated condition. However, the tempered condition can provide reasonably higher hardness than the quench and tempered condition due to the presence of tempered martensite and large size carbides [58].

10.4.1.2 Titanium System

The titanium and its alloys are often cladded with nitride, carbide and oxide, known as ceramics, as laser cladding materials. TiN is used as preplaced material to fabricate coatings on TC4 alloys by laser cladding technique. The hexagonal BN powder is also used as cladding material on the surface of Ti–6Al–4V alloy substrate to create TiN, TiB_2, TiC, Ti_5Si_3 and various alloy phases. These phases are exhibiting exceptional wear resistance when compared with substrate material [59, 60]. The metal matrix composite (MMC) coatings and intermetallic compound (IMC) coatings on the titanium substrate is widely used to enhance the wear resistance [2]. Hard MMC coatings such as TiC, TiB, TiB_2, TiN, SiC, WC etc. always acted as the reinforcements and extremely adherent with the substrate due to the strong interfacial bonding [61]. These adherent coatings can provide outstanding sliding wear resistance and friction improvements. Similarly, the IMCs such as Ti–Al, Ti–Ni, Ti–Co, Ti–Ni–Si etc. have positive impact on the wear resistance. For example, TiC reinforced Ti–Ni–Si intermetallic composite coating with a microstructure of TiC and uniformly distributed phase of Ti_2Ni_3Si–NiTi–Ti_2Ni in the clad matrix on a substrate of TA15 titanium alloy show excellent combination of ductility, hardness and toughness which prevent micro-/macro-cracking or micro-/macro-fracturing and improve wear resistance [62].

10.4.1.3 Magnesium System

Mg-Al alloys are the widely used in several applications and the phase fraction formed during laser cladding operation may be predicted from the equilibrium Al-Mg phase diagram [63] as shown in Fig. 10.6. The Al-Si coatings are fabricated on Mg-Al alloys due to the fact that both have very little difference in the melting points. The defect-free coatings with higher hardness and lower wear susceptibility can be obtained with Al-Si laser cladding consist of fine microstructure [64]. The AZ91D magnesium alloy, cladded with Al-Si powder mainly consist of $Mg_{17}Al_{12}$ + Mg eutectic and Mg_2Si phase, according to Fig. 10.5, which enhance the wear

Fig. 10.6 Al-Mg phase diagram showing formation of secondary precipitation [63]

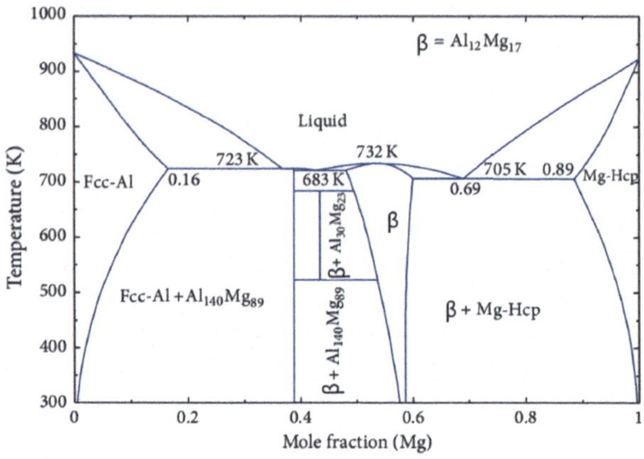

resistance [65]. The phase fraction also depends on the Si content in the system. And when the Si content is 8.0 wt% in the mixed powder, a small amount of Mg_2Si phase formed. But when the Si content in the mixed powder increases above 10%, a slight of dendritic $Mg_{17}Al_{12}$ is synthesized along with the large amount of Mg_2Si. Al-12.5 wt% Si eutectic alloy powder produces a relatively uniform distribution of Mg_2Si phase in the fine dendrite of $Mg_{17}Al_{12}$ which exhibit best wear resistance. Similar to the AZ91D magnesium alloy, the wear resistance of ZE41 magnesium alloy can improved significantly by laser surface alloying with Al-12 wt% Si powders [66]. The formation of Mg_2Al_3, $Mg_{17}Al_{12}$, Mg_2Si and Al rare earth phases at the top surface microstructure are the reason of enhanced wear resistance. The Al-Cu coating is fabricated on AZ91HP magnesium alloy and the clad layer consists of $AlCu_4$ and $Mg_{17}Al_{12}$ grains, embedded in the Al-Mg matrix. The fabrication of multiple intermetallic compounds in the coating may contributed to the enhancement of the wear resistance of AZ91HP magnesium alloy [67].

Some amorphous composite coatings such as $Fe_{38}Ni_{30}Si_{16}B_{14}V_2$ [68] and $Zr_{65}Al_{7.5}Ni_{10}Cu1_{7.5}$ [69], in single layer or multiple layer cladding on magnesium alloy system exhibited graded microstructure consists of amorphous structure, amorphous-nanocrystalline structure and predominant crystalline structure. The presence of amorphous-nanocrystalline composites significantly enhances the wear resistance of the coating, up to 16 times, compared to the substrate. Besides, the addition SiC particles in the amorphous coating on Mg alloy further increased the wear resistance of the same system [70].

10.4.1.4 Aluminium System

Transitional metal aluminides like Fe_3Al, Cu_9Al_4 and Ni_3Al, synthesized with Fe, Cu and Ni powder by laser surface alloy of 2024 aluminum alloy, provide good aluminum wear resistance [71, 72]. Self-fluxing alloys such as Ni-based and Fe-based are common coating materials containing boron and silicon for aluminum alloys that create low-melting fluxes. Boron and silicone decrease the melting temperature of the coating and increase melt liquidity; formation of solid-state chromium boride and complex carbo-borides during coating provide enhanced wear resistance of the alloys [73]. The crack free Ni-Ti coating was made on the AA6061 aluminum alloy using a high-power CW Nd:YAG laser in an argon atmosphere to obtain $TiAl_3$ and Ni_3Al in the clad layer. The presence of $TiAl_3$ and Ni_3Al increase the layer hardness as well as wear resistance by a factor of greater than 5 than that of substrate [74]. The Ni-based alloys are costly which restrict their usage even though they have very good overall properties. Alternatively, the ferro-aluminides can be a new candidate material for high temperature applications due to excellent oxidation resistance, high hardness and relatively low cost. In general, the laser surface alloyed Fe-Al composite layer increased both surface hardness and wear resistance. If the content of Fe increases in the powder, the laser-alloyed layer microstructure changes from a hypo-eutectic structure to a hyper-eutectic structure, forming a needle-like $FeAl_3$, a fine needle-like $FeAl_3$, and finally a lump-like Fe_2Al_5. These IMCs are responsible for further enhancement of the cladded coating's surface durability and wear resistance [75]. The Cu powder alloyed coating on Al substrate formed intermetallic -Al_4Cu_9 phase which increases the hardness of the material by 2.5 times and also significantly improved the wear resistance [76]. In addition, the hypo-eutectic microstructure consisting of primary white (Al) grains and a dark eutectic ((Al) + Al_2Cu) was observed at 24 wt% of added Cu. As the copper ratio increases, the proportion of this fine eutectic microstructure increases slowly as the (Al) phase disappears leading to an improved wear behavior of these coatings compared to Al substratum [77]. Similarly, Al-Mo system has also been studied to improve the surface properties of Al-based light alloys as Mo forms a significant proportion of Al-based hard intermetallic compounds [78]. The IMCs like Al_5Mo, $Al_{22}Mo_5$ and $Al_{17}Mo_4$, formed in Al-Mo systems exhibit acicular morphology at low scanning velocity and an equiaxial, flower-like shape at high scanning velocity of laser source. Such microstructural morphologies will adjust the wear coefficient of the device, followed by specific charging. The wear coefficient of the similar clad microstructure is reported to decrease with an increase in the applied load suggesting that the material is more resistant to wear at the higher loads used than at lower loads [79]. Besides metals and IMCs, due to their high hardness and excellent wear resistance, ceramics such as carbide, oxide and boride are also widely used as laser alloy materials on aluminum and its alloys [80]. The Si doped TiC coating, on Al alloys, was used to enhance the wettability, fluidity, hardness and wear resistance of the cladded coating [81]. The multi-ceramic TiB_2 + TiC coating, with an average coating thickness of 110 μm, enhanced the microhardness by 40% and wear resistance by 8 times than the pure Al substrate [82]. Similar to the multi-ceramic TiB_2 + TiC coating, clad microstructure consists of TiB_2, Ti_3B_4 and Al_3Ti also provide higher hardness and better wear resistance compared with that of the as-received Al substrate [83].

10.5 Current Problems and Evolution of Cladding Process

10.5.1 General Problems

Many problems, related to wear resistance properties of laser cladding, still exists even though the technique has progressed deep into the field. The surface coating quality extremely dependant on the selection of laser parameters, selection of cladding materials and the type of substrates. During laser cladding, the formation of pores is normally occurred due to the inaccurate selection of parameters. For example, high laser energy density may vaporize the substrate and the gases evolved due to such vaporization may not be able to leave the deep molten pool during solidification leading to the formation of pores [84]. Similarly, the surface crater cracks are generally developed due to the presence of high residual stress are the developed stress is directly governed by the laser power [85]. Alloy composition also plays a major contributing factor in the laser cladding process and therefore, selection of proper cladding materials performs a pivotal role in quality control of the cladding. Again, oxide prone materials, such as magnesium and aluminium alloys, should be cladded with a shielding gas environment (such as argon) to prevent oxidation [86]. To overcome the high dilution of the substrate in the cladded layers, the laser parameters should be optimized. The optimum process variable not only reduce dilution effect it would also enhance the metallurgical bonding between the materials which should enhance the service life of the cladding. For example, poor thermal conductivity of the substrate material should retain higher absorbed energy within it which leads to increasing dilution rate [84]. Use of laser source with shorter wavelength [87], pre-heating the substrate [88] and addition of activators [89] on the surface can significantly increase the laser absorptivity for a poorly thermal conductive material.

10.5.2 Evolution of Cladding Process for Better Wear Resistance

Many techniques, methods, systems, materials, etc. have been evolved to enhance the cladding process to achieve better wear resistance properties of the alloy systems and to overcome the current issues involved with cladding. A short description of the current development in the field of cladding process has been discussed in this section. Detail descriptions of these techniques are beyond the scope and mentioned elsewhere [90].

10.5.2.1 Development of Novel Cladding Materials

As discussed earlier that the selection of materials plays an important role in laser cladding process and thus many studies have been carried out, till date, on different materials to overcome the problems with the existing engineering materials. However, selection of a good material or design of a new material for cladding purposes should be governed by certain principles. The first principle says that the cladding should have good wettability with the substrate to achieve superior bonding strength. Secondly, the distribution of alloy segregation [91], amorphous phases [70], and fine nano-meter grain structure [92], across the cladding should be achieved to improve the mechanical properties. Besides, these factor laser absorption rate and proper dilution rate should be maintained in each layer.

10.5.2.2 Development of New Devices

Through constructing a separate shielding gas chamber around the surface, Wang et al. [93] adopted a new technique to prevent the oxidation problem of the highly oxidized alloy framework. The air around the laser beam was expelled by the gas coming from the well-arranged holes on the gas chamber before the laser beam irradiated on the substrate, and the laser melting process was carried out under almost pure shielding gas circumstances. Cui et al. [94] developed a technique of immersion in which the substrate is submerged in a bath of liquid nitrogen. This method is more promising than the method of shielding gas.

10.5.2.3 Laser Cladding Process Combined with Other Methods

Laser plasma hybrid spraying (LPHP) is a newly developed hybrid technology that concurrently uses the combination of deposition of material and laser treatment. Following the subsequent actions of high laser energy, which contributed to the metallurgical bonding between the coatings and substrates, the unfused particles after plasma spraying could be completely melted and a dense coating could be obtained [95–97].

10.5.2.4 Functional Gradient Coating

The functionally graded coating (FGC) is an encouraging method of generating high bonding strength. The FGC has the potential to alleviate localized stresses throughout the material which significantly improve the corrosion and wear resistance resulting in higher component service life [98, 99]. The FGC's greatest advantage is that microstructure gradient change minimizes internal stress and reduces the coating's crack sensitivity. Thus, the FGC achieves the close bonding of excellent mechanical properties between the substrate and the cladding materials. However, the compatibility between different layers of the FGC cladding materials must be considered in the design of the alloys and further study should be made.

10.6 Summary

1. Optimizing process parameters for laser alloy such as laser power, laser beam length, and scanning speed helps to produce high-quality alloy surface. Excessive concentration of laser power can cause the substrate to collapse and melt. The coating and the substrate, on the other hand, cannot adequately react and lead to poor metallurgical bonding.
2. Compatibility and wettability are two key factors in choosing the right materials for alloying. Similar physical properties are required between the substrate and the coatings, such as the melting point, the coefficient of thermal expansion and the modulus of elasticity. It should also be considered a strong matching value for the lattice.
3. In laser surface alloying on different alloys, composite materials, especially metal matrix composite (MMCs) are becoming much more popular. Ceramic phase and metal combinations are more likely to obtain high strength alloyed layer and few defects. Therefore, a theme in future studies will be the creation of practical gradient coatings, high entropy alloys and amorphous coatings.
4. Cracks and pores are two of the most common problems that still occur during the laser surface alloy process. Preheating the substrate before the laser alloy and additional heat treatment of the material can also improve the final properties of the products. Developing new techniques such as gas shielding and immersion techniques can effectively control oxidation rates and defects.
5. Some useful measures to increase the rate of laser absorption should be taken, such as taking the laser with a shorter wavelength, preheating the substrate before laser cladding, inserting activators on the substrate, etc.
6. The combination of laser cladding technique and other techniques (e.g. LPHP) may be effective in refining the microstructure of cladded coatings.

References

1. J. Mazumder, Laser-aided direct metal deposition of metals and alloys, in *Laser Additive Manufacturing* (2017), pp. 21–52
2. C. Chua, K. Leong, C. Lim, *Rapid Prototyping: Principles and Applications*, 2nd edn. (World Scientific, Singapore, 2003)
3. L. Lu, J. Fuh, Y. Wong, *Laser-Induced Materials and Processes for Rapid Prototyping* (Kluwer Academic Publishers, 2001)
4. D.T. Pham, R.S. Gault, A comparison of rapid prototyping technologies. Int. J. Mach. Tools Manuf. **38**(10–11), 1257–1287 (1998)
5. J. Mazumder, J. Choi, K. Nagarathnam, J. Koch, D. Hetzner, The direct metal deposition of H13 tool steel for 3-D components. J. Miner. Metals Mater. Soc. **49**(5), 55–60 (1997)
6. J. Mazumder, D. Dutta, N. Kikuchi, A. Ghosh, Closed loop direct metal deposition: art to part. Opt. Lasers Eng. **34**(4–6), 397–414 (2000)
7. M. Ma, Z. Wang, D. Wang, X. Zeng, Control of Shape and performance for direct laser fabrication of precision large-scale metal parts with 316L Stainless Steel. Opt. Laser Technol. **45**, 209–216 (2013)

8. K. Zhang, S. Wang, W. Liu, X. Shang, Characterization of stainless steel parts by laser metal deposition shaping. Mater. Des. **55**, 104–119 (2014)
9. M. Sachs, J. Cima, P. Williams, D. Brancazio, J. Cornie, Three dimensional printing: rapid tooling and prototypes directly from a CAD model. J. Eng. Ind. Trans. ASME **114**(4), 481–488 (1992)
10. B. Dutta, S. Palaniswamy, J. Choi, L. Song, J. Mazumder, Direct metal deposition. Adv. Mater. Process. **5**, 33–36 (2011)
11. H. Gedda, *Laser Surface Cladding: A Literature Survey*, Technical Report Lulea Tekniska Universitet, 2000. ISSN:1402-1536
12. U. de Oliveira, V. Ocelik, J.T.M. De Hosson, Analysis of coaxial laser cladding processing conditions. Surf. Coat. Technol, **197**(2–3), 127–136 (2005)
13. O. Milewski, G.K. Lewis, J. Fonseca, R.B. Nemec, Laser powder deposition of a near net shape injection mold core e a case study. Mater. Manuf. Process. **15**(2), 247–258 (2000)
14. R. Vilar, Laser alloying and laser cladding, in *Lasers in Materials Science*, vol. 301 (1999), pp. 229–251
15. J. Mazumder, Laser assisted surface coatings, in *Metallurgical and Ceramic Protective Coatings*, ed. by K. Stern (Chapman and Hall, London, 1996), pp. 74–111
16. J. Mazumder, O. Conde, M. Steen, R. Vilar (eds.), *Laser Processing: Surface Treatment and Film Deposition*, NATO ASI Series E, vol. 307 (Kluwer Academic Publishers, 1996), pp. 47–75
17. H. Gedda, J. Powell, G. Wahlstrom, W.B. Li, H. Engstrom, C. Magnusson, Energy redistribution during CO_2 laser cladding. J. Laser Appl. **14**(2), 78–82 (2002)
18. J. Lin, Laser attenuation of the focused powder streams in coaxial laser cladding. J. Laser Appl. **12**(1), 28–33 (2000)
19. J. Lin, Temperature analysis of the powder streams in coaxial laser cladding. Opt. Laser Technol. **31**(8), 565–570 (1999)
20. C.Y. Liu, J. Lin, Thermal processes of a powder particle in coaxial laser cladding. Opt. Laser Technol. **35**(2), 81–86 (2003)
21. A. Kar, J. Mazumder, One-dimensional finite-medium diffusion model for extended solid solution in laser cladding of Hf on nickel. Acta Metall. **36**, 701–712 (1988)
22. V.M. Weerasinghe, W.M. Steen, in *Transport Phenomena in Materials Processing*, ed. by M. Chen, J. Mazumder, C. Tucker (ASME, New York, NY, 1983), pp. 15–23
23. W.M. Steen, V.M. Weerasinghe, P. Monson, in *High Power Lasers and Their Industrial Applications*, vol. 650, ed. by D. Schuöcker (SPIE, Bellingham, WA, 1986), pp. 226–34
24. A.F.A. Hoadley, M. Rappaz, A thermal model of laser cladding by powder injection. Metall. Trans. B **23B**, 631–642 (1992)
25. M. Picasso, M. Rappaz, Laser-powder-material interactions in the laser cladding process. J. Phys. **IV**(C4), 27–33 (1994)
26. M. Picasso, A.F.A. Hoadley, Finite element simulation of laser surface treatments including convection in the melt pool. Int. J. Num. Meth. Heat Fluid Flow **4**, 61–83 (1994)
27. E. Toyserkani, A. Khajepour, S. Corbin, in *ICALEO 2002*, Scottsdale, AZ, 14–17 Oct 2002
28. Y. Yang, H.C. Man, Microstructure evolution of laser clad layers of W-C-Co alloy powders. Surf. Coat. Technol. **132**, 130–136 (2000)
29. W. Ul Haq Syed, A.J. Pinkerton, Z. Liu, L. Li, Coincident wire and powder deposition by laser to form compositionally graded material. Surf. Coat. Technol. **201**: 7083–7091 (2007)
30. B. Borges, L. Quintino, R.M. Miranda, P. Carr, Imperfections in laser cladding with powder and wire fillers. Int. J. Adv. Manuf. Technol. **50**, 175–183 (2010)
31. W. Ul Haq Syed, L. Li, Effects of wire feeding direction and location in multiple layer diode laser direct metal deposition. Appl. Surf. Sci. **248**, 518–524 (2005)
32. R. Vilar, Laser cladding. J. Laser Appl. **11**, 64–79 (1999)
33. S. Jianjun, Z. Ping, F. Geyan, S. Shihong, Geometry characteristics modeling and process optimization in coaxial laser inside wire cladding. Opt. Laser Technol. **101**, 341–348 (2018)
34. J.D. Kim, Y. Peng, Plunging method for Nd:YAG laser cladding with wire feeding. Opt. Lasers Eng. **33**, 299–309 (2000)

35. D.S. Salehi, M. Brandt, Melt pool temperature control using LabVIEW in Nd:YAG laser blown powder cladding process. Int. J. Adv. Manuf. Technol. **29**, 273–278 (2006)
36. J.T. Hofmana, D.F. de Lange, B. Pathiraj, J. Meijer, FEM modelling and experimental verification for dilution control in laser cladding. J. Mater. Process. Technol. **211**, 187–196 (2011)
37. S. Kumar, S. Roy, Development of theoretical process maps to study the role of powder preheating in laser cladding. Comput. Mater. Sci. **37**, 425–433 (2006)
38. D.F. de Lange, J.T. Hofman, J. Meijer, Influence of intensity distribution on the melt pool and clad shape for laser cladding, in *Proceedings of the Third International WLT Conference on Lasers in Manufacturing'05* (2005), pp. 323–327
39. G.R.B.E. Römer, J. Meijer, Inverse calculation of power density for laser surface treatment. Ann. CIRP **49**, 135–138 (2000)
40. M.H. Staia, M. Cruz, N.B. Dahotre, Microstructural and tribological characterization of an A-356 aluminum alloy superficially modified by laser alloying. Thin Solid Films **377**, 665–674 (2000)
41. P. Kadolkar, N.B. Dahotre, Variation of structure with input energy during laser surface engineering of ceramic coatings on aluminum alloys. Appl. Surf. Sci. **199**, 222–233 (2002)
42. P. Kadolkar, N.B. Dahotre, Effect of processing parameters on the cohesive strength of laser surface engineered ceramic coatings on aluminum alloys. Mater. Sci. Eng. A **342**, 183–191 (2003)
43. Y. Yang, J.D. Hu, Effects of laser power density on the microstructure and micro- hardness of Ni–Al alloyed layer by pulsed laser irradiation. Opt. Laser Technol. **43**, 138–142 (2011)
44. S. Nath, S. Pityana, J.D. Majumdar, Laser surface alloying of aluminium with WC + Co + NiCr for improved wear resistance. Surf. Coat. Technol. **206**, 3333–3341 (2012)
45. S.A. Vaziri, H.R. Shahverdi, M.J. Torkamany, S.G. Shabestari, Effect of laser parameters on properties of surface-alloyed Al substrate with Ni. Opt. Laser Eng. **47**, 971–975 (2009)
46. A. Riquelme, P. Rodrigo, M.D. Escalera-Rodríguez, J. Rams, Analysis and optimization of process parameters in Al–SiCp laser cladding. Opt. Lasers Eng. **78**, 165–173 (2016)
47. D. Bergström, J. Powell, A.F.H. Kaplan, The absorptance of steels to Nd:YLF and Nd:YAG laser light at room temperature. Appl. Surf. Sci. **253**, 5017–5028 (2007)
48. S. Ignat, P. Sallamand, D. Grevey, M. Lambertin, Magnesium alloys (WE43 and ZE41) characterisation for laser applications. Appl. Surf. Sci. **233**, 382–391 (2004)
49. X.L. Zhou, X.Z. Li, S.H. Yang, H.Q. Hu, Properties of interface between laser cladding ceramic layer and metal matrix. Laser Technol. **20**(2), 91–94 (1996)
50. J.F. Archard, Contact and rubbing of flat surfaces. J. Appl. Phys. **24**(8), 981–988 (1953)
51. I. Hemmati, V. Ocelik, J.T.M. De Hosson, The effect of cladding speed on phase constitution and properties of AISI 431 stainless steel laser deposited coatings. Surf. Coat. Technol. **205**(21–22), 5235–5239 (2011)
52. K. Rutherford, I. Hutchings, A micro-abrasive wear test, with particular application to coated systems. Surf. Coat. Technol. **79**(1–3), 231–239 (1996)
53. H. Halfa, Thermodynamic calculation for silicon modified AISI M2 high speed tool steel. J. Miner. Mater. Charact. Eng. **5**, 1–14 (2013)
54. N. Ur Rahman, M.B. de Rooij, D.T.A. Matthews, G. Walmag, M. Sinnaeve, G.R.B.E. Römer, Wear characterization of multilayer laser cladded high speed steels. Tribol. Int. **130**, 52–62 (2019)
55. S. Ilo, A. Tomala, E. Badisch, Oxidative wear kinetics in unlubricated steel sliding contact. Tribol. Int. **44**(10), 1208–1215 (2011)
56. Y.P. Ji, S.J. Wu, L.J. Xu, Y. Li, S.Z. Wei, Effect of carbon contents on dry sliding wear behavior of high vanadium high speed steel. Wear **294–295**, 239–245 (2012)
57. N. Hashemi, A. Mertens, H.M. Montrieux, J.T. Tchuindjang, O. Dedry, R. Carrus et al., Oxidative wear behaviour of laser clad High Speed Steel thick deposits: influence of sliding speed, carbide type and morphology. Surf. Coat. Technol. **315**, 519–529 (2017)
58. D.S. Shim, G.Y. Baek, S.B. Lee, J.H. Yu, Y.S. Choi, S.H. Park, Influence of heat treatment on wear behavior and impact toughness of AISI M4 coated by laser melting deposition. Surf. Coat. Technol. **328**, 219–230 (2017)

59. P.A. Molian, L. Hualun, Laser cladding of Ti–6Al–4V with BN for improved wear performance. Wear **130**, 337–352 (1989)
60. Y.S. Tian, C.Z. Chen, Microstructures and wear properties of in situ formed composite coatings produced by laser alloying technique. Mater. Lett. **61**, 635–638 (2007)
61. V. Ocelík, D. Matthews, J.T.M. De Hosson, Sliding wear resistance of metal matrix composite layers prepared by high power laser. Surf. Coat. Technol. **197**, 303–315 (2005)
62. Y.J. Dong, H.M. Wang, Microstructure and dry sliding wear resistance of laser clad TiC reinforced Ti–Ni–Si intermetallic composite coating. Surf. Coat. Technol. **204**, 731–735 (2009)
63. T.B. Massalski, H. Okamoto, P. Subramanian, L. Kacprzak, *Binary Alloys Phase Diagrams 3* (ASM International, Materials Park, OH, 1990)
64. B. Carcel, J. Sampedro, A. Ruescas, X. Toneu, Corrosion and wear resistance improvement of magnesium alloys by laser cladding with Al–Si. Phys. Procedia **12**(Part A), 353–363 (2011)
65. Y. Yang, H. Wu, Improving the wear resistance of AZ91D magnesium alloys by laser cladding with Al–Si powders. Mater. Lett. **63**, 19–21 (2009)
66. A. Fabre, J.E. Masse, Friction behavior of laser cladding magnesium alloy against AISI 52100 steel. Tribol. Int. **46**, 247–253 (2012)
67. Y. Gao, C. Wang, H. Pang, H. Liu, M. Yao, Broad–beam laser cladding of Al–Cu alloy coating on AZ91HP magnesium alloy. Appl. Surf. Sci. **253**, 4917–4922 (2007)
68. Q. Zhu, X. Wang, S. Qu, Z. Zou, Amorphization of $Fe_{38}Ni_{30}Si_{16}B_{14}V_2$ surface layers by laser cladding. Trans. Nonferrous Met. Soc. China **18**, 270–273 (2008)
69. T.M. Yue, Y.P. Su, H.O. Yang, Laser cladding of $Zr_{65}Al_{7.5}Ni_{10}Cu_{17.5}$ amorphous alloy on magnesium. Mater. Lett. **61**, 209–212 (2007)
70. T.M. Yue, Y.P. Su, Laser cladding of SiC reinforced $Zr_{65}Al_{7.5}Ni_{10}Cu_{17.5}$ amorphous coating on magnesium substrate. Appl. Surf. Sci. **255**, 1692–1698 (2008)
71. F.Y. Hu, J.L. Wen, M.C. Wang, Laser surface alloying of cast aluminum alloy with Ni–Cr–Si–Nb–B powder mixture. J. Northeast. Univ. Nat. Sci. **23**, 964–967 (2002)
72. C.S. Liu, B. Zhang, Z. Li, Q.K. Cai, Laser surface alloying of 2024 Aluminum alloy with Ni, Fe and Cu. J. Northeast. Univ. **18**, 34–36 (1997)
73. E. Feldshtein, M. Kardapolava, O. Dyachenko, On the effectiveness of multi-component laser modifying of Fe-based self-fluxing coating with hard particulates. Surf. Coat. Technol. **307**, 254–261 (2016)
74. H.C. Man, S. Zhang, F.T. Cheng, Improving the wear resistance of AA 6061 by laser surface alloying with NiTi. Mater. Lett. **61**, 4058–4061 (2007)
75. S. Tomida, K. Nakata, Fe–Al composite layers on aluminum alloy formed by laser surface alloying with iron powder. Surf. Coat. Technol. **174–175**, 559–563 (2003)
76. Z. Li, C.Y. Ma, Q.K. Cai, G.D. Wang, B.J.A. Aleem, Laser alloying on the 2024 (LY12) surface. Aerosp. Mater. Technol. **29**, 20–24 (1999)
77. L. Dubourg, H. Pelletier, D. Vaissiere, F. Hlawka, A. Cornet, Mechanical characterisation of laser surface alloyed aluminium–copper systems. Wear **253**, 1077–1085 (2002)
78. R.S. Rajamure, H.D. Vora, N. Gupta, S. Karewar, S.G. Srinivasan, N.B. Dahotre, Laser surface alloying of molybdenum on aluminum for enhanced wear resistance. Surf. Coat. Technol. **258**, 337–342 (2014)
79. A. Almeida, F. Carvalho, P.A. Carvalho, R. Vilar, Laser developed Al–Mo surface alloys: microstructure, mechanical and wear behaviour. Surf. Coat. Technol. **200**, 4782–4790 (2006)
80. B.S. Yilbas, C. Karatas, H. Karakoc, B.J.A. Aleem, S. Khan, N. Al-Aqeeli, Laser surface treatment of aluminum based composite mixed with B_4C particles. Opt. Laser Technol. **66**, 129–137 (2015)
81. L.R. Katipelli, A. Agarwal, N.B. Dahotre, Interfacial strength of laser surface engineered TiC coating on 6061 Al using four-point bend test. Mater. Sci. Eng. A **289**, 34–40 (2000)
82. D. Ravnikar, N.B. Dahotre, J. Grum, Laser coating of aluminum alloy EN AW 6082-T651 with TiB_2 and TiC: microstructure and mechanical properties. Appl. Surf. Sci. **282**, 914–922 (2013)
83. J. Xu, W.J. Liu, Y.D. Kan, M.L. Zhong, Microstructure and wear properties of laser cladding Ti-Al-Fe-B coatings on AA2024 aluminum alloy. Mater. Des. **27**, 405–410 (2006)

84. S. Ignat, P. Sallamand, D. Grevey, M. Lambertin, Magnesium alloys laser (Nd: YAG) cladding and alloying with side injection of aluminium powder. Appl. Surf. Sci. **225**, 124–134 (2004)
85. Y. Jun, G.P. Sun, H. Wang, S.Q. Jia, S.S. Jia, Laser (Nd: YAG) cladding of AZ91D magnesium alloys with Al + Si + Al_2O_3. J. Alloy. Compd. **407**, 201–207 (2006)
86. R. Galun, A. Weisheit, B.L. Mordike, Laser surface alloying of magnesium base alloys. J. Laser Appl. **8**, 299–305 (1996)
87. A.F.H. Kaplan, Local absorptivity modulation of a 1 μm–laser beam through surface waviness. Appl. Surf. Sci. **258**, 9732–9736 (2012)
88. J. Chen, Q.L. Zhang, J.H. Yao, J.B. Fu, Study on laser absorptivity of metal material. J. Appl. Opt. **29**, 793–798 (2008)
89. C. Taltavull, B. Torres, A.J. Lopez, P. Rodrigo, E. Otero, A. Atrens, J. Rams, Corrosion behaviour of laser surface melted magnesium alloy AZ91D. Mater. Des. **57**, 40–50 (2014)
90. J. Liu, H. Yu, C. Chen, F. Weng, J. Dai, Research and development status of laser cladding on magnesium alloys: a review. Opt. Lasers Eng. **93**, 195–210 (2017)
91. H. Zhang, Y. Pan, Y. He, H. Jiao, Microstructure and properties of 6FeNiCoSiCrAlTi high–entropy alloy coating prepared by laser cladding. Appl. Surf. Sci. **257**, 2259–2263 (2011)
92. R. Zhu, Z. Li, X. Li, Q. Sun, Microstructure and properties of the low–power–laser clad coatings on magnesium alloy with different amount of rare earth addition. Appl. Surf. Sci. **353**, 405–413 (2015)
93. A.H. Wang, H.B. Xia, W.Y. Wang, Z.K. Bai, X.C. Zhu, C.S. Xie, YAG laser cladding of homogenous coating onto magnesium alloy. Mater. Lett. **60**, 850–853 (2006)
94. Z. Cui, H. Shi, W. Wang, B. Xu, Laser surface melting AZ31B magnesium alloy with liquid nitrogen–assisted cooling. Trans. Nonferrous Met. Soc. China **25**, 1446–1453 (2015)
95. J. Qian, Y. Yin, T. Li, X. Hu, C. Wang, S. Li, Structure, micro–hardness and corrosion behaviour of the Al–Si/Al_2O_3 coatings prepared by laser plasma hybrid spraying on magnesium alloy. Vacuum **117**, 55–59 (2015)
96. N. Serres, F. Hlawka, S. Costil, C. Langlade, F. Machi, Microstructures and mechanical properties of metallic NiCrBSi and composite NiCrBSi–WC layers manufactured via hybrid plasma/laser process. Appl. Surf. Sci. **257**, 5132–5137 (2011)
97. Y. Qian, H. Zhang, G. Wang, Research of rapid and direct thick coatings deposition by hybrid plasma–laser. Appl. Surf. Sci. **252**, 6173–6178 (2006)
98. X. Pei, J. Wang, Q. Wan, L. Kang, M. Xiao, H. Bao, Functionally graded carbon nanotubes/hydroxyapatite composite coating by laser cladding. Surf. Coat. Technol. **205**, 4380–4387 (2011)
99. J.M. Amado, J. Montero, M.J. Tobar, A. Yáñez, Ni–based metal matrix composite functionally graded coatings. Phys. Procedia **39**, 362–367 (2012)

Chapter 11
Analysis of Journal Bearing with Partial Texture Lubricated Using Micropolar and Power-Law Fluids

T. V. V. L. N. Rao, Ahmad Majdi Abdul Rani, Norani Muti Mohamed, Hamdan Haji Ya, Mokhtar Awang, and Fakhruldin Mohd Hashim

Abstract The analysis of micropolar and power-law fluid lubricated journal bearing with partial texture is presented. The pressure and shear stress expressions (nondimensional) for micropolar and power-law fluids are derived based on narrow groove theory (NGT) using Reynolds boundary conditions. The nondimensional partial texture, micropolar and power-law fluid parameters considered in the analysis are: extent of texture region; land to cell extent ratio; depth of recess; coupling number, ratio of characteristic length to film gap and power-law index. The improvement in load capacity and reduction in coefficient of friction of partial texture journal bearings lubricated with micropolar and power-law fluids is evaluated.

11.1 Partial Texture Journal Bearing Lubricated with Micropolar and Power-Law Fluids

Steady state analysis of bearing (journal) with partial texture lubricated using micropolar fluid, power-law fluid are presented. The load enhancement and friction coefficient reduction of partial texture design is analyzed using NGT (narrow-groove-theory) using one-dimensional approximation (long bearing). Texture (partial) at inlet is useful in the case of bearing (journal) with low eccentricity ratio lubricated by micropolar, pseudo-plastic power-law fluid produce beneficial effects of load enhancement and friction coefficient reduction.

T. V. V. L. N. Rao (✉)
Department of Mechanical Engineering, SRM Institute of Science and Technology, Kattankulathur 603203, India
e-mail: tvvlnrao@gmail.com

A. M. A. Rani · H. H. Ya · M. Awang · F. M. Hashim
Department of Mechanical Engineering, Universiti Teknologi PETRONAS, 32610 Bandar Seri Iskandar, Perak Darul Ridzuan, Malaysia

N. M. Mohamed
Department of Fundamental and Applied Sciences, Universiti Teknologi PETRONAS, 32610 Bandar Seri Iskandar, Perak Darul Ridzuan, Malaysia

J. K. Katiyar et al. (eds.), *Tribology in Materials and Applications*, Materials Forming, Machining and Tribology,
https://doi.org/10.1007/978-3-030-47451-5_11

11.1.1 Texture Surfaces for Superior Bearing Performance

Surface texturing has emerged as useful and possible approach to get better tribological properties. Surface texturing has attracted wide attention due its tremendous potential for improving tribological performances. Etsion [1] explored the application of laser surface texturing on mechanical components for improving tribological characteristics. Gropper et al. [2] highlighted potential of textured surfaces on hydrodynamic lubrication. Gachot et al. [3] described a comprehensive review on texturing for improving tribological characteristics of contact surfaces under different conditions of lubrication. Senatore and Rao [4] presented investigations on partial slip texture influence on bearing characteristics. Grützmacher et al. [5, 6] examined the possibilities to reduce friction in lubricated contacts using single and multi-scale patterns. The optimized tribological contact patterns obtained on a ball-on-disk tests were transferred to journal bearing tests. Rosenkranz et al. [7] studied friction reduction performance for texture patterns under thick film lubrication for both low and high convergence ratios.

11.1.2 Partial Texture Bearing Analysis

Hydrodynamic journal bearings with partial texture patterns under low load conditions achieve high load and low friction coefficient. Tønder [8] investigated the effectiveness of bearing pads with roughness locations at bearing inlet. Brizmer and Kligerman [9] analyzed potential of the hydrodynamic journal bearings with partial surface texturing. Fowell et al. [10] and Pascovici et al. [11] assessed influence of textured surfaces (partial) in load enhancing and in friction reducing characteristics of hydrodynamically lubricated bearings. Meng and Khonsari [12] presented on the role of viscosity wedge for predicting the pressure and load of dimple textured parallel surfaces. The effect of viscosity wedge has an important role in generation of pressure in dimple textured surfaces. Tauviqirrahman et al. [13] examined combined texture/slip patterns for significantly improving bearing load and friction performance. Rao et al. [14, 15] investigated couple stress as well as micropolar, power-law fluid with partial texture slip. The pressure distribution and shear stress variation for partial texture slip bearing with non-Newtonian fluids are derived using NGT (narrow-groove-theory) presented by Vohr and Chow [16].

Partial texturing patterns of bearing surfaces under hydrodynamic lubrication is an ideal approach for increasing bearing load and reducing friction coefficient.

11.1.3 Micropolar Fluid Lubricated Journal Bearing

Micropolar fluid lubricated bearings have been widely investigated using the theory of micropolar fluids developed by Eringen [17, 18]. Khonsari and Brewe [19] and Das et al. [20] investigated the performance of micropolar fluid bearing (journal). Wang and Zhu [21] presented numerical study of dynamically loaded micropolar fluid lubricated journal bearing. Nair et al. [22] analyzed the characteristics of a flexible elliptical bearing (journal) operated with micropolar fluid lubricant. Naduvinamani and Santosh [23] studied the characteristics in squeeze film lubricated finite porous journal bearings under the effect of micropolar fluid lubricants. Sharma and Rajput [24] analyzed imperfections in journal geometry on hybrid micropolar fluid lubricated bearing performance. Lin et al. [25] investigated the dynamic analysis of slider bearings with parabolic-film profile under the influence of micropolar fluid lubricants. Khatak and Garg [26] presented a review of characteristics (static and dynamic) of bearings with micropolar fluid lubricants. Bhattacharjee et al. [27] investigated load and stiffness coefficient of porous (double-layered) bearing (micropolar fluid lubricated). Khatri and Sharma [28] analyzed that a textured journal bearing (two-lobe hybrid with micropolar fluid) enhances bearing performance significantly.

11.1.4 Power-Law Fluid Lubricated Journal Bearing

The bearing performance analysis incorporating lubricant additive effects have been investigated employing power-law fluid (non-Newtonian) model. Following perturbation technique, Dien and Elrod [29] derived modified Reynolds equation for a power-law fluid lubricated bearing. Elsharkawy [30] employed power-law (non-Newtonian) fluid in the analysis of bearing lubrication in magnetic recording device applications. Das [31] investigated infinitely wide slider bearings (inclined and parabolic) considering non-Newtonian power-law fluid lubricant. Following slip (Navier) conditions at the bearing surfaces, Li et al. [32] derived the extended modified Reynolds equation. Sharma et al. [33] presented performance improvement of power law fluid lubricated porous journal bearings by incorporating micro dimple (spherical and ellipsoidal) textures. The role of surface texturing and non-Newtonian power-law fluid model in porous bearing designs is investigated. The shear thickening fluid lubricated porous journal bearings with spherical texture show superior performance than plain porous bearings in low load conditions.

11.2 Analysis of Micropolar Fluid Lubricated Journal Bearing with Partial Texture

Analysis of partial texture micropolar fluid lubricated bearing (journal) is presented. The nondimensional expressions of pressure and shear stress based on NGT [16] are derived from Rao et al. [15]. The nondimensional partial texture and micropolar parameters considered are: texture extent (θ_t); ratio of land region in a cell (γ); nondimensional recess depth (H_t); coupling number (N) and characteristic length of micropolar fluid to film gap ratio (Λ).

11.2.1 Modified Reynolds Equation for Partial Texture Journal Bearing with Micropolar Fluid

The bearing (journal) schematic with the partial texture extent θ_t and the extents of cell regions θ_{ts} and θ_{tn} respectively with land and recesses is depicted in Fig. 11.1. The ratio of land region in a cell (γ) is expressed as $\gamma = \theta_{ts}/(\theta_{ts} + \theta_{tn})$. The film thickness (nondimensional) is H ($H = h/C, h = C(1 + \varepsilon \cos\theta)$). The recess film thickness (nondimensional) is $H + H_t$ ($H_t = h_t/C$).

The flow expressions in cell (land and recess) and exit region are expressed as [15]

$$Q = G_{1d} - G_{2d}\frac{dP_d}{d\theta},$$
$$Q = G_{1t} - G_{2t}\frac{dP_t}{d\theta}, \text{ and}$$
$$Q = G_{1r} - G_{2r}\frac{dP}{d\theta} \tag{11.1}$$

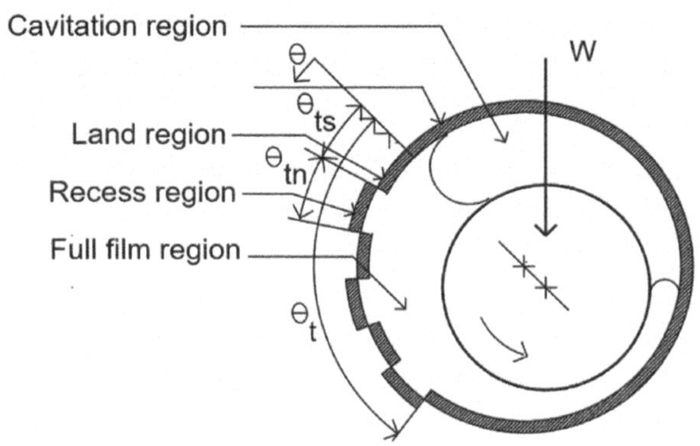

Fig. 11.1 Partial texture journal bearing geometry

The overall pressure gradient is related to pressure gradients in cell (land and recess) based on NGT [16] as

$$\frac{dP}{d\theta} = \gamma \frac{dP_d}{d\theta} + (1-\gamma)\frac{dP_t}{d\theta} \tag{11.2}$$

Simplifying Eqs. (11.1–11.2), the pressure gradient (nondimensional) in partial texture ($0 \le \theta \le \theta_t$) and exit ($\theta_t \le \theta \le \theta_r$) region are

$$\frac{dP}{d\theta} = \gamma \frac{G_{1d} - Q}{G_{2d}} + (1-\gamma)\frac{G_{1t} - Q}{G_{2t}} \text{ and } \frac{dP}{d\theta} = \frac{G_{1r} - Q}{G_{2r}} \tag{11.3}$$

The coefficients G_{1k}, G_{2k} for $k = d, t, r$ respectively for micropolar fluids are determined from Eqs. (11.4–11.5) as follows

$$G_{1d} = G_{1r} = \frac{H}{2} \text{ and } G_{2d} = G_{2r} = \frac{H^3}{12} + \Lambda^2 H - \frac{N\Lambda H^2}{2}\coth\left(\frac{NH}{2\Lambda}\right) \tag{11.4}$$

$$\begin{aligned} G_{1t} = \frac{H+H_t}{2} \text{ and} G_{2t} &= \frac{(H+H_t)^3}{12} + \Lambda^2(H+H_t) \\ &- \frac{N\Lambda(H+H_t)^2}{2}\coth\left(\frac{N(H+H_t)}{2\Lambda}\right) \end{aligned} \tag{11.5}$$

The characteristic length of micropolar fluid is Λ ($\Lambda = \lambda/C$, $\lambda = \sqrt{\eta_m/4\mu_f}$) where η_m is micropolar viscosity coefficient and N is coupling number for micropolar fluids ($N = \sqrt{\kappa/(2\mu_f + \kappa)}$) where κ is micropolar viscosity coefficient and μ_f is fluid viscosity.

11.2.2 Steady State Analysis of Micropolar Fluid Lubricated Partial Texture Journal Bearing

The pressure profile (nondimensional) is P ($P = pC^2/\mu_f u_j R$) for partial texture ($0 \le \theta \le \theta_t$) and exit ($\theta_t \le \theta \le \theta_r$) region. The pressure distribution is obtained from integrating the Eq. (11.3) and substituting the Reynolds boundary conditions at film reformation ($P|_{\theta=0} = 0$), yields as

$$P(0 \le \theta \le \theta_t) = \int_0^\theta \left(\frac{\gamma G_{1d}}{G_{2d}} + \frac{(1-\gamma)G_{1t}}{G_{2t}}\right)d\theta - Q\int_0^\theta \left(\frac{\gamma}{G_{2d}} + \frac{(1-\gamma)}{G_{2t}}\right)d\theta \tag{11.6}$$

$$P(\theta_t \le \theta \le \theta_r) = P|_{\theta=\theta_t} + \int_{\theta_t}^{\theta} \frac{G_{1r}}{G_{2r}}d\theta - Q\int_{\theta_t}^{\theta} \frac{1}{G_{2r}}d\theta \tag{11.7}$$

Based on the Reynolds conditions at film rupture boundary ($P|_{\theta=\theta_r} = 0$ and $dP/d\theta|_{\theta=\theta_r} = 0$), simplifying the pressure in Eqs. (11.6, 11.7) and pressure gradient (nondimensional) at film rupture in Eq. (11.3) results in Q as

$$Q = \frac{\int_0^{\theta_t}\left(\frac{\gamma G_{1d}}{G_{2d}} + \frac{(1-\gamma)G_{1t}}{G_{2t}}\right)d\theta + \int_{\theta_t}^{\theta_r}\frac{G_{1r}}{G_{2r}}d\theta}{\int_0^{\theta_t}\left(\frac{\gamma}{G_{2d}} + \frac{(1-\gamma)}{G_{2t}}\right)d\theta + \int_{\theta_t}^{\theta_r}\frac{1}{G_{2r}}d\theta} \text{ and } Q|_{\theta=\theta_r} = G_{1r}|_{\theta=\theta_r} \tag{11.8}$$

The iterative procedure (Newton-Raphson) is employed to obtain solution of both θ_r and $Q|_{\theta=\theta_r}$ simultaneously using Eq. (11.8).

The load capacity (nondimensional) is W ($W = wC^2/\mu_f u_j R^2 L$) and is evaluated as

$$W = \sqrt{W_\varepsilon^2 + W_\phi^2} \tag{11.9}$$

where $W_\varepsilon = -\int\limits_0^{\theta_r} P\cos\theta d\theta$ and $W_\phi = \int\limits_0^{\theta_r} P\sin\theta d\theta$.

The shear stress (nondimensional) in cell region are

$$\Pi_d|_{Y=0} = \frac{1}{H - 2N\Lambda\tanh\left(\frac{NH}{2\Lambda}\right)} + \frac{H}{2}\frac{dP_d}{d\theta} \tag{11.10}$$

$$\Pi_t|_{Y=0} = \frac{1}{(H+H_t) - 2N\Lambda\tanh\left(\frac{N(H+H_t)}{2\Lambda}\right)} + \frac{(H+H_t)}{2}\frac{dP_t}{d\theta} \tag{11.11}$$

The shear stress (overall) in partial texture region is Π ($\Pi = \tau_{xy}C/\mu_f u_j$) and obtained from the shear stress in cell region based on NGT [16] as

$$\Pi = \gamma\Pi_d + (1-\gamma)\Pi_t \tag{11.12}$$

The shear stress (nondimensional) in partial texture and exit region respectively are

$$\begin{aligned}\Pi|_{Y=0}(0 \le \theta \le \theta_t) = {} & \frac{\gamma}{H - 2N\Lambda\tanh\left(\frac{NH}{2\Lambda}\right)} + \frac{\gamma H}{2}\left(\frac{G_{1d} - Q}{G_{2d}}\right) \\ & + \frac{(1-\gamma)}{(H+H_t) - 2N\Lambda\tanh\left(\frac{N(H+H_t)}{2\Lambda}\right)} \\ & + \frac{(1-\gamma)(H+H_t)}{2}\left(\frac{G_{1t} - Q}{G_{2t}}\right)\end{aligned} \tag{11.13}$$

$$\Pi|_{Y=0}(\theta_t \le \theta \le \theta_r) = \frac{1}{H - 2N\Lambda\tanh\left(\frac{NH}{2\Lambda}\right)} + \frac{H}{2}\left(\frac{G_{1r} - Q}{G_{2r}}\right) \tag{11.14}$$

The friction force (nondimensional) on the journal surface is F ($F = fC/\mu_f u_j RL$) and is evaluated as $F = \int_0^{\theta_r} \Pi|_{Y=0} d\theta$. The friction coefficient (nondimensional) is

$$C_f = \left(\frac{R}{C}\right)\frac{f}{w} = \frac{F}{W} \tag{11.15}$$

11.2.3 *Characteristics of Partial Texture Journal Bearing with Micropolar Fluid*

The partial texture bearing and micropolar fluid parameters used in analysis are presented in Tables 11.1 and 11.2.

The load (nondimensional) (W) of partial texture bearing with micropolar fluid is shown in Fig. 11.2a–d. Figure 11.2a depicts increasing load (nondimensional) (W) with increasing texture extent (θ_t) at $\varepsilon = 0.1$. As depicted in Fig. 11.2a, load (nondimensional) capacity (W) increases with increasing θ_t at $\varepsilon = 0.2$ for lower coupling number (N) of 0.1. The increase in load (nondimensional) (W) increases with increasing (i) $N = 0.1$–0.7 and (ii) $\Lambda = 0.1$–0.2. Figure 11.2b demonstrates increasing load (nondimensional) (W) with increasing recess (texture) [or decreasing ratio of land region in a cell (γ)] at lower eccentricity ratio (ε) of 0.1. As demonstrated in Fig. 11.2b, for $\theta_t = 120°$, load (nondimensional) (W) decreases with increasing recess (texture) [or decrease in ratio of land region in a cell (γ)] for eccentricity ratios (ε) of 0.2–0.3. Figure 11.2c shows increasing load (nondimensional) (W) with increasing N. Figure 11.2d shows a maximum value of load (nondimensional) capacity (W) with nondimensional recess depth (H_t) from 0.2 to 1.0 at ε of 0.1. As shown in Fig. 11.2d, load (nondimensional) (W) decreases with increasing nondimensional recess depth (H_t) from 0.2 to 1.0 for ε of 0.2–0.3.

The friction coefficient (C_f) of partial texture bearing with micropolar fluid is shown in Fig. 11.3a–d. Figure 11.3a depicts that the C_f decreases with increasing (i)

Table 11.1 Partial texture parameters of journal bearing

Eccentricity ratio (ε)	0.1 –0.3
Texture extent (θ_t)	40°–160°
Nondimensional recess depth (H_t)	0.2–1.0
Ratio of land region in a cell (γ)	0.2–0.8

Table 11.2 Micropolar fluid parameters

Coupling number (N)	0.1–0.9
Micropolar fluid characteristic length to film gap ratio (Λ)	0.1–0.3

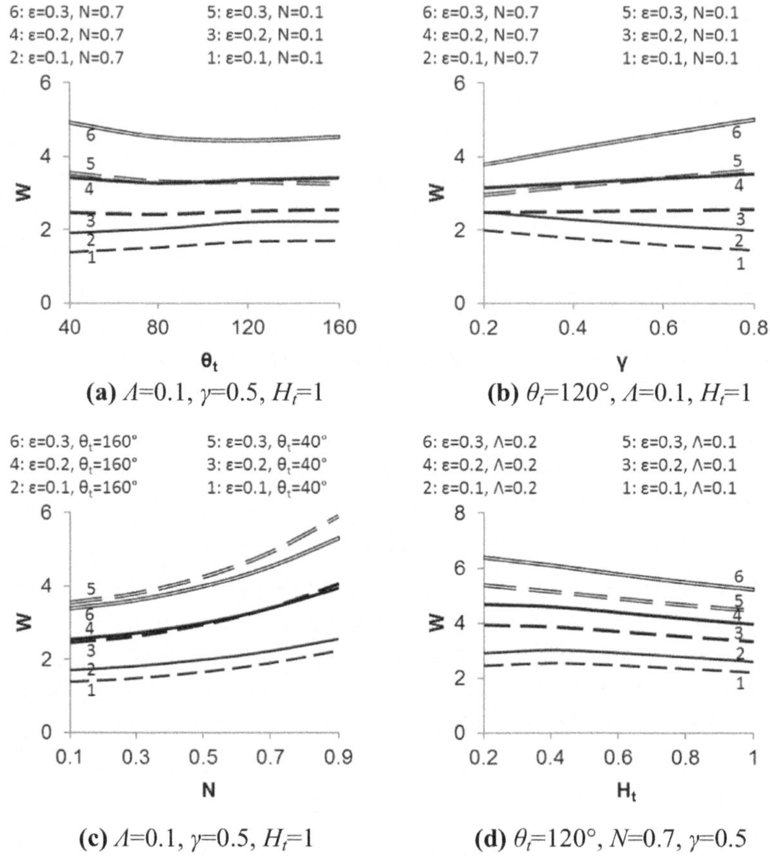

Fig. 11.2 Load (nondimensional) capacity of micropolar fluid lubricated partial texture bearing

$\varepsilon = 0.1$–0.3 and (ii) $N = 0.1$–0.7. Figure 11.3a depicts decreasing C_f with increasing θ_t from 40° to 160° at $\varepsilon = 0.1$. Figure 11.3a also depicts that C_f decreases with increasing θ_t at $\varepsilon = 0.2$ for $N = 0.1$. Figure 11.3b demonstrates that C_f decreases with decrease in γ from 0.8 to 0.2 (recess (texture) region increases) at $\varepsilon = 0.1$. The decrease in C_f (Fig. 11.3c) is shown with increase in θ_t from 40° to 160° at $\varepsilon = 0.1$ for the variation of N from 0.1 to 0.9. With increase in H_t from 0.2 to 1.0, Fig. 11.3d displays (i) minimum value of C_f at ε of 0.1 and (ii) increase in C_f for ε of 0.2–0.3. As demonstrated in Fig. 11.3d, C_f shows no significant influence on the increase in $\Lambda = 0.1$–0.2.

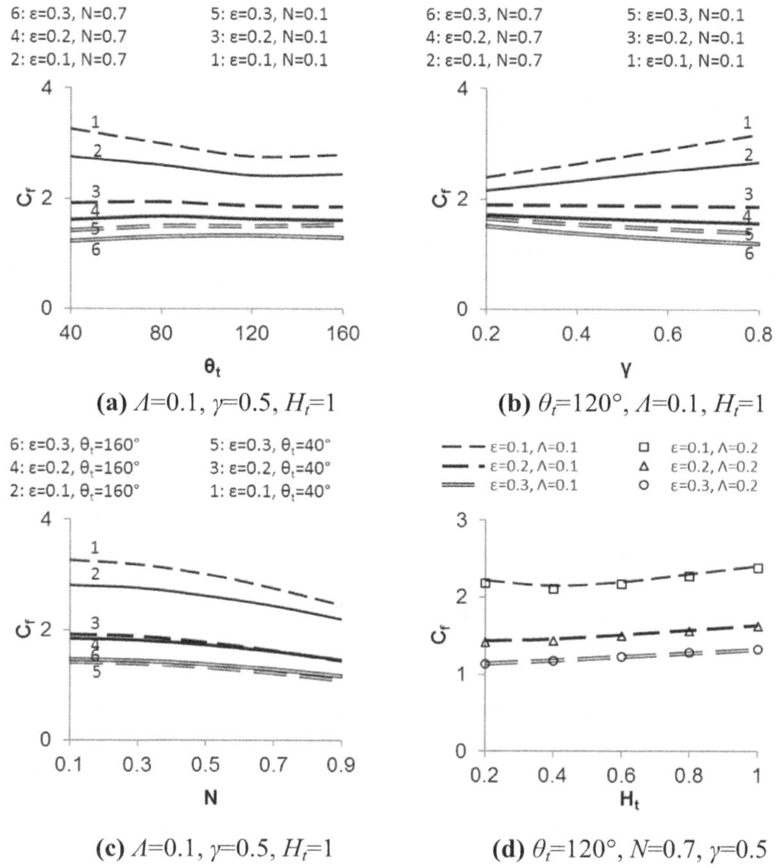

Fig. 11.3 Friction coefficient of micropolar fluid lubricated partial texture bearing

11.3 Analysis of Power-Law Fluid Lubricated Journal Bearing with Partial Texture

The partial texture bearing (journal) analysis with power-law fluids is derived from Rao et al. [15] based on NGT [16]. The nondimensional parameters of partial texture power-law fluid considered are: texture extent (θ_t); ratio of land region in a cell (γ); nondimensional recess depth (H_t) and power-law index (n).

11.3.1 Modified Reynolds Equation for Partial Texture Journal Bearing with Power-Law Fluid

The coefficients G_{1k}, G_{2k} for $k = d, t, r$ respectively for power-law fluids used in the analysis of partial texture ($0 \leq \theta \leq \theta_t$) and exit ($\theta_t \leq \theta \leq \theta_r$) region are

$$G_{1d} = G_{1r} = \frac{H}{2} \text{ and } G_{2d} = G_{2r} = \frac{H^{n+2}}{12n} \quad (11.16)$$

$$G_{1t} = \frac{H + H_t}{2} \text{ and } G_{2t} = \frac{(H + H_t)^{n+2}}{12n} \quad (11.17)$$

The viscosity for reference is η ($\eta = m\left(u_j/C\right)^{n-1}$, $\mu_f = \eta$) where m is flow parameter index.

11.3.2 Steady State Analysis of Partial Texture Journal Bearing with Power-Law Fluid

The pressure (nondimensional) for partial texture ($0 \leq \theta \leq \theta_t$) and exit ($\theta_t \leq \theta \leq \theta_r$) region based on the Reynolds boundary conditions and the nondimensional load capacity are determined from Eqs. (11.6–11.8) and Eq. (11.9) respectively using the coefficients G_{1k}, G_{2k} for $k = d, t, r$ respectively in Eqs. (11.16–11.17) for power-law fluids.

The shear stress (nondimensional) for partial texture region for power-law fluids is

$$\Pi|_{Y=0}(0 \leq \theta \leq \theta_t) = \gamma\left(U_{y,d}^2\right)^{(n-1)/2} U_{y,d} + (1 - \gamma)\left(U_{y,t}^2\right)^{(n-1)/2} U_{y,t} \quad (11.18)$$

where $U_{y,d} = \frac{1}{H} + \frac{H^n}{2n}\left(\frac{G_{1d} - Q}{G_{2d}}\right)$ and $U_{y,t} = \frac{1}{H+H_t} + \frac{(H+H_t)^n}{2n}\left(\frac{G_{1t} - Q}{G_{2t}}\right)$.

The shear stress (nondimensional) for exit region for power-law fluids is

$$\Pi|_{Y=0}(\theta_t \leq \theta \leq \theta_r) = \left(U_y^2\right)^{(n-1)/2} U_y \quad (11.19)$$

where $U_y = \frac{1}{H} + \frac{H^n}{2n}\left(\frac{G_{1r} - Q}{G_{2r}}\right)$.

The friction coefficient (nondimensional) is determined from Eq. (11.15).

Table 11.3 Power-Law fluid parameters

Parameter	
Power-law index (n)	0.7, 1.3

11.3.3 Performance Characteristics of Partial Texture Journal Bearing with Power-Law Fluid

The parameters of power-law fluid partial texture bearing are presented in Table 11.3. The load (nondimensional) capacity (W) and the friction coefficient (C_f) are investigated using parameters of partial texture in Table 11.1.

Figure 11.4a–d show the load (nondimensional) (W) of partial texture bearing with power-law lubricants. Figure 11.4a depicts increasing load (nondimensional) capacity (W) with increasing texture extent (θ_t) for partial texture bearing (journal) of $\varepsilon = 0.1$ (lower bearing eccentricity ratio). Figure 11.4b demonstrates the change

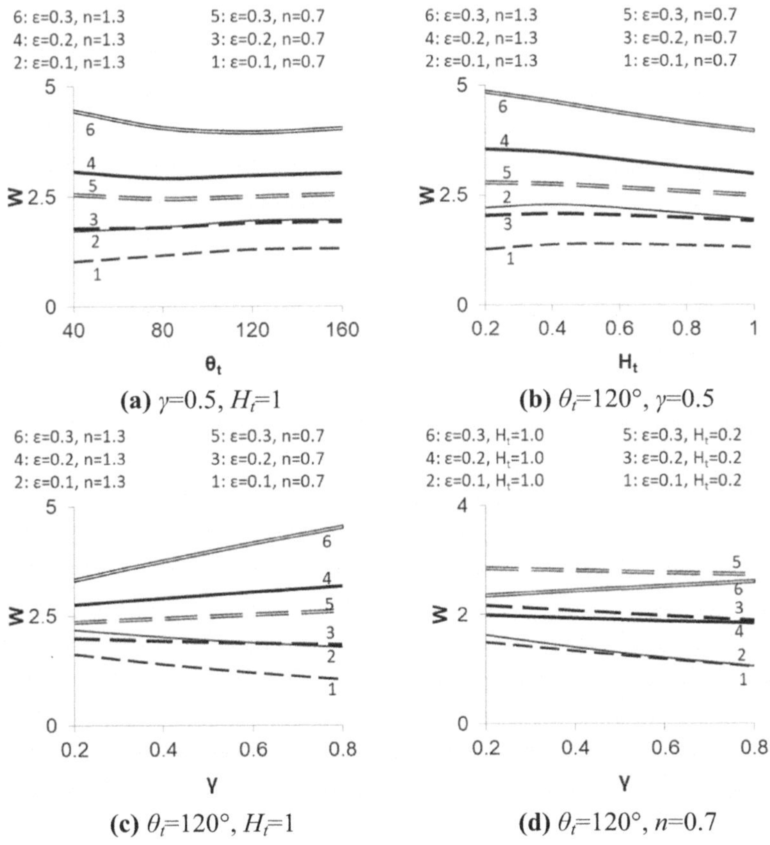

Fig. 11.4 Load (nondimensional) capacity of power-law fluid lubricated partial texture bearing

of load (nondimensional) (W) with H_t. The load (nondimensional) (W) attains a peak value in H_t for $n = 0.7$ and 1.3 at $\varepsilon = 0.1$, but W attains a peak value in H_t only for $n = 0.7$ at $\varepsilon = 0.2$. The load (nondimensional) (W) decreases with increasing H_t for $n = 1.3$ at $\varepsilon = 0.2$. Figure 11.4c depicts increasing load (W) with increasing recess region (or decrease in γ) at $\varepsilon = 0.1$, for both $n = 0.7$ and 1.3, but W increases with increasing texture region only for $n = 0.7$. Figure 11.4d demonstrates increasing load (W) with increasing recess (texture) region (or decrease in γ) at $\varepsilon = 0.1$–0.2 and $n = 0.7$, while W improves with increasing texture at $\varepsilon = 0.3$ only for low depth recess region ($H_t = 0.2$).

Figure 11.5a–d show the friction coefficient (C_f) of partial texture bearing with power-law fluid. Figure 11.5a depicts decrease in C_f with increasing θ_t at $\varepsilon = 0.1$ for partial texture bearing. Figure 11.5b demonstrates that the friction coefficient (C_f) reaches a minimum with H_t at (i) $\varepsilon = 0.1$ for both $n = 0.7$ and 1.3, and (ii) $\varepsilon = 0.2$ for $n = 0.7$. The C_f increases with increasing H_t for $n = 1.3$ at $\varepsilon = 0.2$. Figure 11.5c depicts

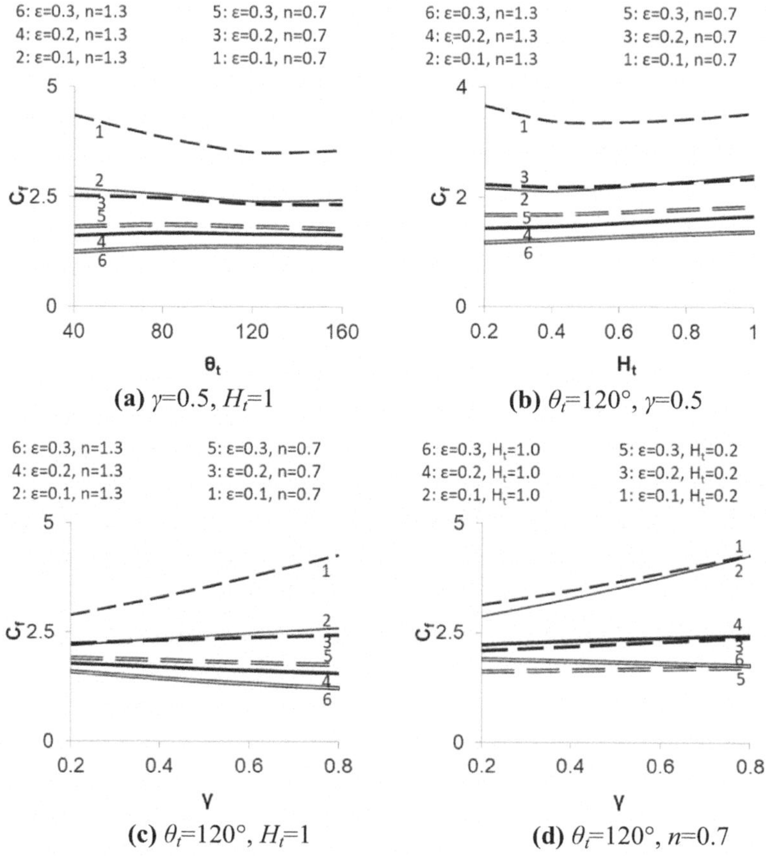

Fig. 11.5 Friction coefficient of power-law fluid lubricated partial texture journal bearing

decreasing friction coefficient (C_f) with increasing recess (texture) (or decrease in γ) (i) at $\varepsilon = 0.1$, for both $n = 0.7$ and 1.3, and (ii) at $\varepsilon = 0.2$, only for $n = 0.7$. Figure 11.5d demonstrates effect of partial texture ($H_t = 0.2$) on decreasing C_f at $\varepsilon = 0.1$–0.3 for $n = 0.7$. The effect of $H_t = 0.2$ and 1.0 are alike at $\varepsilon = 0.1$.

11.4 Conclusion

Micropolar, power-law fluid (non-Newtonian) lubricated partial texture bearing is examined based on NGT. The pressure gradient, pressure distribution and shear stress for partial texture journal bearing based on Sommerfeld bearing analysis are presented.

The load (W) increases and friction coefficient (C_f) decreases for micropolar fluid lubricated partial texture configuration at operating conditions of $\varepsilon = 0.1$. The improvement in partial texture journal bearings at operating conditions of $\varepsilon = 0.1$ using micropolar fluids are influenced by increase in (i) texture measured from maximum film gap position (θ_t), (ii) recess (texture) (or decrease in ratio of land in a cell (γ)), (iii) coupling number (N) and (iv) micropolar fluid characteristic length to film gap ratio (Λ).

The characteristics of partial texture bearing (journal) at $\varepsilon = 0.1$ are improved using power-law fluids. The increase in load (nondimensional) (W) and decrease in friction coefficient (C_f) are obtained with increasing texture extent measured from maximum film gap position (θ_t) and decrease in ratio of land region in a cell (γ). The partial texture bearing (journal) configuration with pseudo-plastic fluids ($n = 0.7$) show superior performance improvement related to dilatant fluids ($n = 1.3$).

References

1. I. Etsion, State of the art in laser surface texturing. J. Tribol. **127**, 248–253 (2005)
2. D. Gropper, L. Wang, T.J. Harvey, Hydrodynamic lubrication of textured surfaces: a review of modeling techniques and key findings. Tribol. Int. **94**, 509–529 (2016)
3. C. Gachot, A. Rosenkranz, S.M. Hsu, H.L. Costa, A critical assessment of surface texturing for friction and wear improvement. Wear **372**, 21–41 (2017)
4. A. Senatore, T.V.V.L.N. Rao, Partial slip texture slider and journal bearing lubricated with Newtonian fluids: a review. J. Tribol. **140**(4), 040801 (2018)
5. P.G. Grützmacher, A. Rosenkranz, A. Szurdak, F. König, G. Jacobs, G. Hirt, F. Mücklich, From lab to application - Improved frictional performance of journal bearings induced by single- and multi-scale surface patterns. Tribol. Int. **127**, 500–508 (2018)
6. P.G. Grützmacher, A. Rosenkranz, A. Szurdak, M. Grüber, C. Gachot, G. Hirt, F. Mücklich, Multi-scale surface patterning—an approach to control friction and lubricant migration in lubricated systems. Industr. Lubr. Tribol. **71**(8), 1007–1016 (2019)
7. A. Rosenkranz, H.L. Costa, F. Profito, C. Gachot, S. Medinae, D. Dini, Influence of surface texturing on hydrodynamic friction in plane converging bearings—An experimental and numerical approach. Tribol. Int. **134**, 190–204 (2019)

8. K. Tønder, Inlet roughness tribodevices: dynamic coefficients and leakage. Tribol. Int. **34**, 847–852 (2001)
9. V. Brizmer, Y. Kligerman, A laser surface textured journal bearing. J. Tribol. **134**, 031702 (2012)
10. M. Fowell, A.V. Olver, A.D. Gosman, H.A. Spikes, I. Pegg, Entrainment and inlet suction: two mechanisms of hydrodynamic lubrication in textured bearings. J. Tribol. **129**, 336–347 (2007)
11. M.D. Pascovici, T. Cicone, M. Fillon, M.B. Dobrica, Analytical investigation of a partially textured parallel slider. Proc. Inst. Mech. Eng. Part J **223**, 151–158 (2009)
12. X. Meng, M.M. Khonsari, On the effect of viscosity wedge in micro-textured parallel surfaces. Tribol. Int. **107**, 116–124 (2017)
13. M. Tauviqirrahman, R. Ismail, J. Jamari, D.J. Schipper, Combined effect of texturing and boundary slippage in lubricated sliding contacts. Tribol. Int. **66**, 274–281 (2013)
14. T.V.V.L.N. Rao, A.M.A. Rani, T. Nagarajan, F.M. Hashim, Analysis of couple stress fluid lubricated partially textured slip slider and journal bearing using narrow groove theory. Tribol. Int. **69**, 1–9 (2014)
15. T.V.V.L.N. Rao, A.M.A. Rani, T. Nagarajan and F.M. Hashim, "Analysis of micropolar and power law fluid–lubricated slider and journal bearing with partial slip–partial slip texture configuration." *Tribology Transactions* 59, no.5, (2016): 896–910
16. J.H. Vohr, C.Y. Chow, Characteristics of herringbone-grooved, gas-lubricated journal bearings. J. Basic Eng. **9**, 568–578 (1965)
17. A. Eringen, Simple microfluids. Int. J. Eng. Sci. **2**, 205–217 (1964)
18. A. Eringen, Theory of micropolar fluids. Journal of Mathematics and Mechanics **16**(1), 1–18 (1967)
19. M.M. Khonsari, D.E. Brewe, On the performance of finite journal bearings lubricated with micropolarfluids. Tribol. Trans. **32**(2), 155–160 (1989)
20. S. Das, S.K. Guha, A.K. Chattopadhyay, On the steady-state performance of misaligned hydrodynamic journal bearings lubricated with micropolar fluids. Tribol. Int. **35**, 201–210 (2002)
21. X.-L. Wang, K.-Q. Zhu, A study of the lubricating effectiveness of micropolar fluids in a dynamically loaded journal bearing. Tribol. Int. **37**, 481–490 (2004)
22. K.P. Nair, V.P.S. Nair, N.H. Jayadas, Static and dynamic analysis of elastohydrodynamic elliptical journal bearing with micropolar lubricant. Tribol. Int. **40**, 297–305 (2007)
23. N.B. Naduvinamani, S. Santosh, Micropolar fluid squeeze film lubrication of finite porous journal bearing. Tribol. Int. **44**, 409–416 (2011)
24. S.C. Sharma, A.K. Rajput, Effect of geometric imperfections of journal on the performance of micropolar lubricated 4-pocket hybrid journal bearing. Tribol. Int. **60**, 156–168 (2013)
25. J.-R. Lin, T.-L. Chou, L.-J. Liang, T.-C. Hung, Non-Newtonian dynamic characteristics of parabolic film slider bearings: micropolar fluid model. Tribol. Int. **48**, 226–231 (2012)
26. P. Khatak, H.C. Garg, Influence of micropolar lubricant on bearings performance—a review. Proc. Inst. Mech. Eng. Part J **226**(9), 775–784 (2012)
27. B. Bhattacharjee, P. Chakraborti, K. Choudhuri, Evaluation of the performance characteristics of double-layered porous micropolar fluid lubricated journal bearing. Tribol. Int. **138**, 415–423 (2019)
28. C.B. Khatri, S.C. Sharma, Behaviour of two-lobe hole-entry hybrid journal bearing system under the combined influence of textured surface and micropolar lubricant. Industr. Lubr. Tribol. **69**(6), 844–862 (2017)
29. I.K. Dien, H.G. Elrod, A generalized steady-state Reynolds equation for non-Newtonian fluids, with application to journal bearings. J. Lubr. Technol. **105**, 385–390 (1983)
30. A.A. Elsharkawy, Magnetic head-rigid disk interface hydrodynamically lubricated with a power-law fluid. Wear **213**, 47–53 (1997)
31. N.C. Das, A study of optimum load capacity of slider bearings lubricated with power law fluids. Tribol. Int. **32**, 435–441 (1999)

32. W.-L. Li, H.-M. Chu, M.-D. Chen, The partially wetted bearing—extended Reynolds equation. Tribol. Int. **39**, 1428–1435 (2006)
33. N. Sharma, S. Kango, R.K. Sharma, Adiabatic analysis of microtextured porous journal bearings functioned with power law fluid model. Proc. Inst. Mech. Eng. Part J **233**(10), 1541–1553 (2019)

Chapter 12
Evaluation of the Effect of Friction in Gear Contact Stresses

Santosh S. Patil and Saravanan Karuppanan

Abstract Stress analysis of gears has become a popular area of research in order to reduce failures and optimize the gear design. Friction between gears is detrimental to the gear surface contacts, but there is no effective method to determine the frictional effects on these contacts. Inclusion of friction in the study of gear contact is sparse. Also, friction depends on various other parameters, which in turn are ambiguous to evaluate. Hence, a quantitative study of frictional effects on gear contact problem is therefore essential. Thus, to solve the present problem, a dimensionless factor needs to be developed which would account for the friction in gear contact stress calculations. A simplified gear contact stress evaluation technique which includes friction needs to be developed. In this work, a 3D frictional Finite Element Method (FEM) was employed for the gear frictional study. Also, an experimental validation was carried out using a customised experimental setup, Gear Dynamic Stress Test Rig (GDSTR). The experimental results provided a good correlation with the results of the developed 3D FE models. The results from the validated FE models showed 15–22% rise in gear contact stresses for increased frictional coefficients, which is significant. The FE analysis was further extended and a parametric study was carried out. A dimensionless friction factor function was developed in this work to estimate the frictional gear contact stresses. The mathematically correlated K_f function was verified and its inclusion provides better frictional contact stress evaluation in gears, thus providing a simplified frictional measure in gear contact stress calculations.

S. S. Patil (✉)
Department of Mechanical Engineering, Manipal University Jaipur, Jaipur 303007, India
e-mail: santosh.patil@jaipur.manipal.edu

S. Karuppanan
Department of Mechanical Engineering, Universiti Teknologi PETRONAS, 32610 Bandar Seri Iskandar, Perak, Malaysia

J. K. Katiyar et al. (eds.), *Tribology in Materials and Applications*,
Materials Forming, Machining and Tribology,
https://doi.org/10.1007/978-3-030-47451-5_12

12.1 Introduction

Contact mechanics is the study of the deformation of solids that touch each other at one or more points. Contact mechanics has been analysed in industry for a long time [1]. The analysis of the contact stresses, surface deformations, contact areas and stiffness generated by the rough surfaces of the contacting bodies needs to be accurately accounted for in a smart design of various engineering components [2]. The reason researchers are interested in contact mechanics is due to the increase in contact related problems, e.g. pitting and scuffing problems on gear pairs [3].

Gearing has been one of the most crucial components in transmitting power and motion. The transmission may be with or without change in its direction or speed [4]. Gears are prominent and would remain so in transmitting power and motion in future machineries due to its high degree of reliability and compactness. Hence, gears have vast applications in industries and various other fields of power transmission. High performance gears are required to operate in various applications. The gear's performance is crucial as the whole transmission system depends on these components. Thus studies of gear performance parameters are a pivotal area for many decades. Amongst these studies, gear teeth contact has been considered as one of the most complicated application. The gear contact complexity is due to their nonlinear behaviour and changing operating conditions [5]. The high stress concentrations at the contacting point and tooth root of a gear are the main reasons for gear tooth failure.

It is well known that friction between gears exerts great effects on surface contact stresses of gears [6, 7]. Frictional variations between the gear contact leads to nonlinear complexities and along with pitting on gear tooth, the complexities in evaluating gear contact stresses increase. Inclusion of friction in the study of contact analysis in gears is sparse and has been a long standing problem [8, 9]. The exact frictional values between gear systems cannot be measured as friction depends on various factors which in turn vary with respect to time and operational conditions. Some of the factors on which friction depends are viscosity of the lubricant, temperature, surface roughness, pressure, surface defects and wear [10]. It is difficult to stabilise these factors, thus making the frictional variation unpredictable. Due to this frictional unpredictability in gear contact stresses, many gear systems have experienced premature failures. So, there is a necessity to include the frictional effects in gear contact stress evaluation. The evaluation of exact frictional behaviour is difficult for the above said reasons.

As discussed, there is no exact measureable technique to evaluate the frictional variations and its effects on the gear contact stresses. Also known, that the empirical equations to evaluate contact stress in gears do not consider frictional variations. The gear contact stresses are only available for frictionless case. AGMA equations are one such examples available for frictionless gear contact stresses, which is based on the Hertz's theory of two contacting cylinders [11]. While, various experiments on gears have also been conducted in the past to obtain more realistic gear contact stresses. But, the measure of friction and the exact contribution of frictional variation in the rise of gear contact stress were missing. Many of the experiments represented the combined contribution of dynamics, surface roughness, friction, misalignments (if any), wear, and defects in materials for the rise in gear contact stress. There are factors predicted to overcome the dynamic effects, in the form of dynamic factor. Similarly, to overcome surface roughness, size and surface factors are available. The misalignments and cracks are studied differently to obtain their effects on gear contact stresses. Meanwhile, the frictional variation prediction or factor to overcome frictional effects is not available. Hence, the analytical gear contact stress evaluation approach does not consider frictional effects in any form.

Frictional contact problems are rather complicated, considering mathematical and experimental point of view. Recently, finite element method (FEM) has been increasingly used to simulate rolling and sliding contact and considerably precise stress results are expected [12]. The numerical analysis is very often used to analyse the equivalent stress distribution of an elastic body with complicated geometry, such as a gear. Hence, FEM for gear stress analysis is widely used and it is also being an important method for the analysis of gear tooth micro geometry. To resolve gear contact problems, finite element analysis (FEA) is a natural tool, and indeed ANSYS contact elements have been used successfully to analyse many problems [13–17]. Recent advancements in the field of computer aided design (CAD) and FEM have encouraged attempts for the gear problem solutions. Also, the frictional effects were either neglected or considered negligible during the gear studies. Numerous investigations are devoted to gear research and analysis, but still a general numerical approach which is capable of predicting the effects of friction on gear contact stresses is not available. Furthermore, few recent studies have been devoted to the frictional study on gears, but mainly on spur gear pairs. Hence, complete understanding and quantified effects of friction on spur and helical gears is required.

In the literatures, various old and recent studies on gear stress analysis have been reviewed. Since the local contact stress along the profile of the tooth is higher than that occurring at the root of the tooth and is a primary source of the surface fatigue failure, many studies were focused on contact stress. The contact stress related literature is extensive and it has been systematically studied. The contact stress analysis with friction has been our main consideration along with other supporting gear analysis. There are literatures where frictional analysis of spur and helical gear has been done and has shown that the gear contact stresses significantly increase with friction [18–23].

Mathematical modelling is based on certain assumptions and also very time consuming. The complicated calculations and evaluation with vast amount of empirical data leads to an abstract outlook to the problem. However, the model gives reliable results in many cases. Experimental testing of contact stress in gears has been done in the past decade. Few experimental test rigs have been developed, such as, twin disc test machine, back-to-back gear test machine and FZG test rig. The twin disc testing is faster, but testing is not done on real gear test conditions [24, 25]. They have been few studies using back-to-back gear test machine to evaluate contact stresses and stiffness [26–28]. The FZG testing is done on real gear conditions, but the results are restricted to a particular set of gear pairs used in the setup [29–34]. Thus both of the approaches have certain amount of limitations in evaluating contact stresses. Many recent works [21, 25, 35–39] on contact stress evaluation show that Finite element analysis is one of the most common methods being used in stress analysis of gear tooth. This is possible due to the development of contact algorithms and their implementation in commercial FEA software in the last few years. Also, the studies showed that the evaluation of gear contact stresses with the inclusion of friction is sparse. Few studies which include friction were mainly on spur gear contact. Hence the frictional gear contact studies have been extended to helical gears in this research.

In the present study, few parameters such as, lubrication type, temperature, gear ratio, surface roughness, number of teeth, normal pressure angle, face width and module have been kept constant so that the variation of results due to the aforementioned parameters were suppressed. Meanwhile the varying parameters were friction coefficient, torque and speed. Due to the limited amount of time and the resources available, it was decided to only focus on developing a function for analysing frictional contact stresses in gears.

12.2 Methodology

The gear contact problem has been solved using three approaches, analytical approach, experimental testing and FEM. Analytical calculations uses the modified AGMA equations represented by Eqs. 12.1 and 12.2 [19].

$$\sigma_c = C_p.\boldsymbol{K}_f\sqrt{\frac{F_n}{b.d.I}K_oK_vK_mK_sZ_R} \tag{12.1}$$

$$\sigma_H = C_p.\boldsymbol{K}_f\sqrt{\frac{F_n}{bdI}\left(\frac{\cos\beta}{0.95CR}\right)K_vK_o(0.93K_m)} \tag{12.2}$$

AGMA contact stress calculations were carried out for the gear module ($m = 5$). The detailed specifications of the module set are shown in Table 12.1. The analytical results of the test gears ($m = 5$) were used for comparison with experimental test and FEM results. In the next step, the models are assembled, constrained and applied with boundary conditions prior to frictional contact analysis via FEM. The experimental testing is done using GDSTR presented in Ref. [20]. The experimental analysis was performed by strain measuring technique, i.e. by the use of carbon slip rings and mounted strain gages on the gear tooth. The flow chart of the study is represented in Fig. 12.1.

Further, the analysis results of the FEM and the experimental testing were compared in order to verify the finite element procedure. Necessary improvements on the mesh refinement and solver settings were performed when the results did not converge and/or results were not in the allowable limits. The mesh refinement was performed and based on the convergence results; a refined mesh was selected for

Table 12.1 Specifications of experimental test gears

Test gears	Spur gear pair		5° helical gear pair		25° helical gear pair	
Parameter	Pinion	Gear	Pinion	Gear	Pinion	Gear
No. of teeth	20	20	20	20	20	20
Normal module	5	5	5	5	5	5
Normal pressure angle (°)	20	20	20	20	20	20
Helix angle (°)	0	0	5	5	25	25
Pitch diameter (mm)	100	100	100	100	110	110
Face width (mm)	8	8	8	8	8	8
Centre distance (mm)	100		100		110	
Contact ratio	1.54		1.52		1.43	
Torque (Nm)	45, 60, 80, 100					
Speed (rpm)	30					

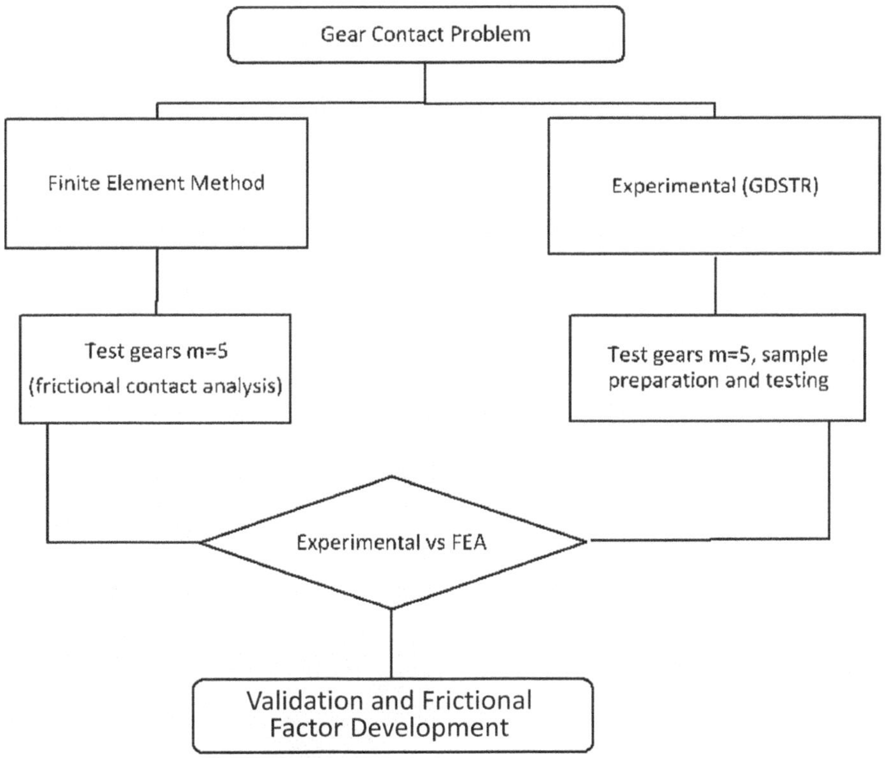

Fig. 12.1 Flow chart of the study

further finite element analyses. Later, the results were compared with the experimental results. The effect of friction, developed in the form of friction factor and the developed friction factor was further validated with a few other gear finite element studies, at different loading conditions.

12.3 Friction Factors for the Gear Pairs

The effect of friction on the contact stresses in gears has been presented in the form of friction factor, shown in Eq. 12.1 [19]. The contact stresses varied with increasing friction coefficient and helix angle of the gears. The effect of friction needs to be accounted in calculating contact stresses and hence, friction factor was developed for spur and different helical gear pairs. Based on this study, friction factors can be generated at different contacting positions, namely first point of tooth contact-FPTC, highest point of single tooth contact-HPSTC, pitch point contact-PPC, lowest point of single tooth contact-LPSTC, and last point of tooth contact-LPTC. However, only the maximum contact stress position i.e. the pitch point position is important and considered in this study.

The combined function of friction factor for contact stress of different gear sets at pitch point contact with varying coefficient of friction was extracted from Fig. 12.2 and is represented in Eq. 12.3. Graph plotting software Origin pro 8.5.1 was used for fitting of the 3rd order polynomial curve for each helical gear friction factor results. Later, the coefficients of the 3rd order polynomial equation of all the helical gear equations were used to plot a function of coefficients. Thus, enabling the development of a combined function of friction factor (Eq. 12.3) which includes the effect of coefficient of friction and helix angle. This is a curve fit equation of the surface plot

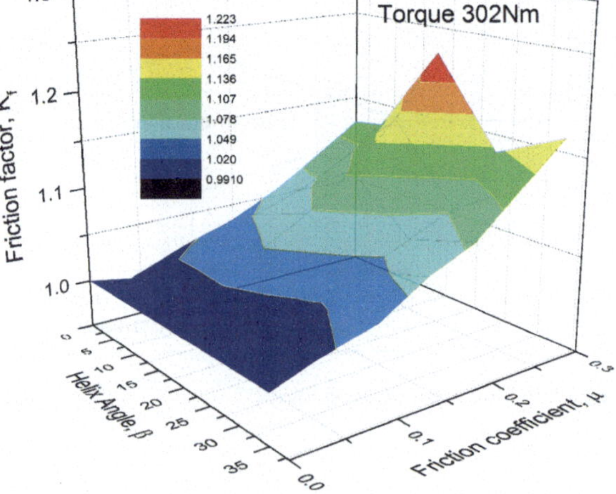

Fig. 12.2 Friction factor for contact stress of 20–20 gear sets with varying coefficient of friction [19]

shown in Fig. 12.2. This defined equation forms the base of the parametric relation and will be a guiding direction for future study.

$$\begin{aligned} \boldsymbol{K}_f =& \left[(-1.1024 \times 10^{-4})\mu^2 + (7.9378 \times 10^{-5})\mu + (1.068 \times 10^{-6})\right]\beta^3 \\ &+ \left[(4.84 \times 10^{-3})\mu^2 - (4.32 \times 10^{-3})\mu - (6.165 \times 10^{-5})\right]\beta^2 \\ &+ \left[-(0.03279)\mu^2 - (0.0581)\mu - (0.00103)\right]\beta \\ &+ \left[(1.01917)\mu^2 + 0.05905\mu + (0.9957)\right] \end{aligned} \quad (12.3)$$

The values of K_f which accounts for different coefficient of friction can be estimated from Eq. 12.3. The value of K_f can be found for any intermediate helical gear pair and for any coefficient of friction value in the range of 0.0–0.3. It can be seen from the surface plot that the K_f values are between 1 and 1.22, hence depicting the frictional rise of up to 22%. The individual study carried out earlier on helical gears also showed a frictional rise of 15–22%. Thus the developed friction factor function represented an appropriate percentage rise.

12.4 Validation of the Dimensionless Friction Factor Function

The K_f values evaluated in the present study are proposed to be incorporated in standard equations (AGMA equations) [40]. The modified AGMA equations including the friction factor are shown in Eqs. 12.1 and 12.2. These equations need to be verified before their application. The friction factor function which has been developed using the polynomial fit also needs to be validated. Thus a few case studies were carried out in support of the friction factor function. The presented function was used to calculate the friction factor for these case studies. Further, using this dimensionless friction factor, frictional contact stresses for the said cases were evaluated analytically (modified AGMA equation). The AGMA calculation now includes frictional effects and was compared with corresponding FE analysis. The finite element gear models for the same cases were established and analyses were carried out. The cases studied are listed in Table 12.2.

The modified AGMA equations including the K_f were employed to calculate the frictional contact stresses for appropriate estimation of contact stresses. Table 12.3 shows Case I supporting comparisons between the newly calculated AGMA contact stresses and FE simulations. The AGMA contact stress for frictionless condition is more than the contact stress estimated by FE simulation and remains constant as it does not consider the frictional parameter. The percentage variation of the mean for Case I studies is 9% and the standard deviation is 4.5%. While the maximum variation for higher frictional coefficient was observed to be in the maximum range of 10–14%. The variations are slightly fluctuating, but are in allowable range, hence it can be said that the inclusion of K_f in the AGMA equation gives a better solution when compared

Table 12.2 Case study details for validation of K_f function

Parameter	Case I	Case II	Case III	Case IV	Case V	Case VI
Gear type	Spur gear	5° Helical gear	15° Helical gear	Spur gear	5° Helical gear	15° Helical gear
N_1/N_2	20/20	20/20	20/20	20/20	20/20	20/20
Face width (mm)	20	20	20	20	20	20
Centre distance (mm)	91.50	91.84	94.72	91.50	91.84	94.72
Module (mm)	4.5	4.5	4.5	4.5	4.5	4.5
Torque (Nm)/speed (rpm)	100/30	100/30	100/30	400/30	400/30	400/30

Table 12.3 Comparison of contact stresses to verify the K_f function (Case I)

Case I: Spur gear ($\beta = 0$) under torque of 100 Nm at speed 30 rpm						
S. No.	Coefficient of friction	AGMA contact stress, MPa (w/o K_f)	Friction factor (K_f)	AGMA contact stress, MPa (with K_f)	FE contact stress, MPa	Percent variation of stress
1.	0.0	697	1	697	693	3.66
2.	0.05		1.006	700	701	0.51
3.	0.1		1.016	708	719	11.13
4.	0.15		1.031	719	730	11.19
5.	0.2		1.053	733	765	10.71
6.	0.25		1.078	751	765	13.68
7.	0.3		1.109	773	785	12.10

to non-frictional AGMA calculations. Therefore, the modified AGMA calculations which is considering the frictional parameter is proposed.

The sample AGMA calculations have been used for the helical gear pair. Thus Case II with coefficient of friction 0, 0.1, 0.2 and 0.3 has been represented as follows;

$$\sigma_H = 191 \times 1 \times \sqrt{\frac{2185.2}{0.02 \times 0.0915 \times 0.0803}\left(\frac{\cos 5}{0.95 \times 1.55}\right) 1.18 \times 1 \times (0.93 \times 1.2)}$$
$$= 695.3 \text{ MPa}$$

$$\sigma_H = 191 \times 1.031 \times \sqrt{\frac{2185.2}{0.02 \times 0.0915 \times 0.0803}\left(\frac{\cos 5}{0.95 \times 1.55}\right) 1.18 \times 1 \times (0.93 \times 1.2)}$$

$$= 717.4\ \text{MPa}$$

$$\sigma_H = 191 \times 1.096 \times \sqrt{\frac{2185.2}{0.02 \times 0.0915 \times 0.0803}\left(\frac{\cos 5}{0.95 \times 1.55}\right) 1.18 \times 1 \times (0.93 \times 1.2)}$$

$$= 762.4\ \text{MPa}$$

$$\sigma_H = 191 \times 1.187 \times \sqrt{\frac{2185.2}{0.02 \times 0.0915 \times 0.0803}\left(\frac{\cos 5}{0.95 \times 1.55}\right) 1.18 \times 1 \times (0.93 \times 1.2)}$$

$$= 825.7\ \text{MPa}$$

The contact stress results obtained by simulation for same case (Case II) are shown in Fig. 12.3. The FEA results and the analytical results (with and without the friction factor) were compared to authenticate the developed K_f function. The detailed comparison between AGMA, modified AGMA and FE simulations for Case study II is presented in Table 12.4.

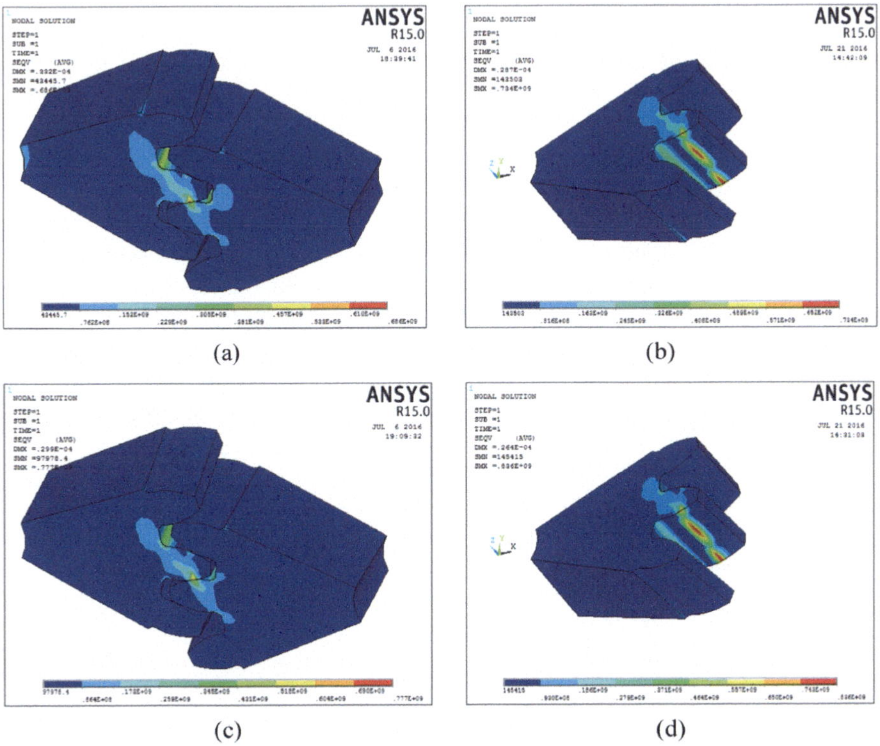

Fig. 12.3 von Mises contact stress distribution for Case II; **a** $\mu = 0$; **b** $\mu = 0.1$; **c** $\mu = 0.2$; **d** $\mu = 0.3$

Table 12.4 Comparison of contact stresses to verify the K_f function (Case II)

Case II: 5° helical gear ($\beta = 5$) under torque of 100 Nm at speed 30 rpm						
S. No.	Coefficient of friction	AGMA contact stress, MPa (w/o K_f)	Friction factor (K_f)	AGMA contact stress, MPa (with K_f)	FE contact stress, MPa	Percent variation of stress
1.	0.0	695	1	695	686	9.33
2.	0.05		1.009	702	715	13.20
3.	0.1		1.032	717	734	16.58
4.	0.15		1.061	738	751	13.38
5.	0.2		1.096	762	777	14.60
6.	0.25		1.139	792	802	10.24
7.	0.3		1.187	826	842	16.30

The mean percentage variation for the Case II studies was 12% and the standard deviation was 2.5%. The maximum variation for two μ values (i.e. 0.1 and 0.3) was observed to be in the range of 10–16%. While for other μ values the maximum variation was well below 15%. The standard deviation is low in Case II condition, thus it can be said that the results are acceptable enough.

Similarly, the detailed comparison of contact stress results for Case study III, between AGMA, modified AGMA and FE analysis is presented in Table 12.5. The percentage variation mean for Case III studies was around 9% and the standard deviation was 3.3%. The maximum variation observed was 12.68% which can be said to be in the acceptable range. The variations slightly fluctuate as the coefficient of friction changes. The standard deviation is lower for Case III too, producing precise results. The case studies showed a good support for the developed K_f function. The K_f function inclusion in AGMA equations gave comparably better stress results. The

Table 12.5 Comparison of contact stresses to verify the K_f function (Case III)

Case III: 15° helical gear ($\beta = 15$) under torque of 100 Nm at speed 30 rpm						
S. No.	Coefficient of friction	AGMA contact stress, MPa (w/o K_f)	Friction factor (K_f)	AGMA contact stress, MPa (with K_f)	FE contact stress, MPa	Percent variation of stress
1.	0.0	685	1	685	672	12.68
2.	0.05		1	686	681	3.68
3.	0.1		1.027	703	707	4.07
4.	0.15		1.075	736	725	11.12
5.	0.2		1.138	779	769	10.48
6.	0.25		1.217	833	822	11.02
7.	0.3		1.31	897	888	8.74

contact stress calculation using the modified frictional AGMA equations gave much closer and reliable results when compared to FE analysis.

The comparison was further extended between AGMA, modified AGMA and FE analysis for different torque value (400 Nm). In Case IV, Case V and Case VI the comparison between the AGMA, modified AGMA and FE analysis also shows that the modified AGMA with friction factor has provided closer and better approximation when compared to AGMA standard equation. It can be seen from Table 12.6 (Case IV) that the calculations with friction factor are compared with the corresponding FE analysis and the percentage variation was not more than 5%. Thus, the evaluation of gear contact stresses with inclusion of friction factor is appreciable. The percentage variation mean for Case IV studies was around 1% and the standard deviation was 2.05%. Also, the maximum variation for this particular case (Case IV) was observed to be less than 5% thus showing a better level of acceptance. This shows that the variations are not much and results fall in acceptable range.

The similar comparison between AGMA, modified AGMA and FE analysis for Case V (listed in Table 12.7) also shows a better compatibility of the newly developed friction factor into the AGMA equation. The stress distribution obtained for this case has been presented in Fig. 12.4. The percentage variation mean for Case V study was around 5%, the standard deviation was 1.68% and maximum variation for all the μ values was less than 8%. Hence, the variations are not much and results are acceptable.

Final comparison of modified AGMA calculations with that of FE results for Case VI is shown in Table 12.8. The percentage variation mean for Case VI studies was around 5%, the standard deviation was 4.38% and maximum variation for all the μ values was less than 12%. This shows that the variations are not much and results fall in acceptable range. The modified AGMA contact stress calculations have shown that they are suitable for torque values of 0–400 Nm and for lower speed range.

The 3D finite element approach was validated and the comparisons have been presented. The FE simulations have also been successfully validated with the available

Table 12.6 Comparison of contact stresses to verify the K_f function (Case IV)

Case IV: Spur gear ($\beta = 0$) under torque of 400 Nm at speed 30 rpm

S. No.	Coefficient of friction	AGMA contact stress, MPa (w/o K_f)	Friction factor (K_f)	AGMA contact stress, MPa (with K_f)	FE contact stress, MPa	Percent variation of stress
1.	0.0	1393	1	1393	1333	4.33
2.	0.05		1.006	1401	1340	4.35
3.	0.1		1.016	1416	1380	2.52
4.	0.15		1.031	1438	1446	0.58
5.	0.2		1.053	1466	1470	0.23
6.	0.25		1.078	1503	1500	0.18
7.	0.3		1.109	1546	1550	0.27

Table 12.7 Comparison of contact stresses to verify the K_f function (Case V)

Case V: 5° helical gear ($\beta = 5$) under torque of 400 Nm at speed 30 rpm						
S. No.	Coefficient of friction	AGMA contact stress, MPa (w/o K_f)	Friction factor (K_f)	AGMA contact stress, MPa (with K_f)	FE contact stress, MPa	Percent variation of stress
1.	0.0	1391	1	1391	1329	4.43
2.	0.05		1.009	1404	1335	4.89
3.	0.1		1.032	1435	1355	5.56
4.	0.15		1.061	1475	1440	2.39
5.	0.2		1.096	1525	1462	4.12
6.	0.25		1.139	1584	1475	6.85
7.	0.3		1.187	1651	1525	7.65

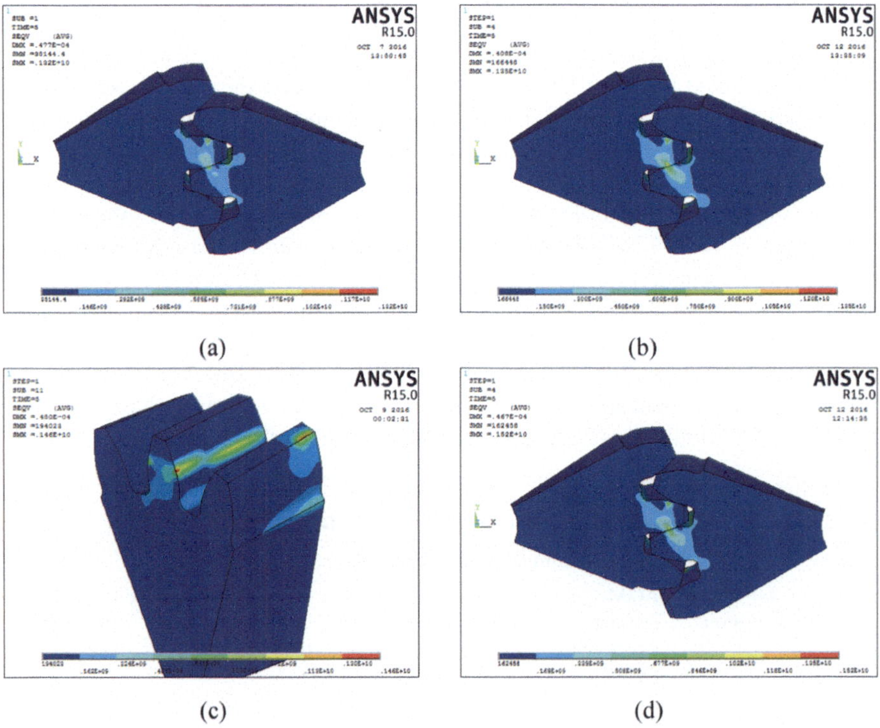

Fig. 12.4 von Mises contact stress distribution for Case V; **a** $\mu = 0$; **b** $\mu = 0.1$; **c** $\mu = 0.2$; **d** $\mu = 0.3$

Table 12.8 Comparison of contact stresses to verify the K_f function (Case VI)

Case VI: 15° helical gear ($\beta = 15$) under torque of 400 Nm at speed 30 rpm						
S. No.	Coefficient of friction	AGMA contact stress, MPa (w/o K_f)	Friction factor (K_f)	AGMA contact stress, MPa (with K_f)	FE contact stress, MPa	Percent variation of stress
1.	0.0	1369	1	1369	1310	4.34
2.	0.05		1	1369	1360	0.68
3.	0.1		1.027	1406	1420	1.01
4.	0.15		1.075	1472	1430	2.87
5.	0.2		1.138	1559	1450	6.99
6.	0.25		1.217	1666	1490	10.57
7.	0.3		1.31	1793	1590	11.35

analytical frictionless gear contact cases. Extensive FE simulations have been done for various cases of gear pairs and the stress distribution along the tooth contact has been represented.

The effect of friction was introduced in the form of a dimensionless factor for spur and helical gears' contact stress calculations. Later, comparison between all the methods has been tabulated and discussed in detail. Lastly, a new friction factor for evaluating the contact stress analysis has been proposed for involute gear pairs. The developed dimensionless friction factor has been supported with the results obtained for various other operating conditions (six cases).

12.5 Conclusion

The FE simulations have been successfully validated, further, the available frictionless gear calculations and experimental validation have been shown for a few cases. The validated FE methodology (with new APDL involute profile program) was then extended for the present parametric studies. The contributions of the simulations and validations carried out have been explained well. Further, gear contact stress analyses, including friction were performed on spur and different helical gears, to understand the effect of friction and helix angle. Later, the effect of friction was introduced in the form of a dimensionless factor for spur and helical gears' contact stress calculations. A new friction factor for evaluating the contact stress analysis has been developed for involute gear pairs. The developed dimensionless friction factor has been supported with the results obtained for various other operating conditions.

Based on the study made on the different sets of gear pairs, the results obtained have been analysed and summarized in the following sections.

1. For spur and different angled helical gear ranging between 0° and 15°, the individual analysis of different set of gears for increased frictional coefficients showed 15–22% rise in gear contact stresses.
2. A dimensionless factor named as friction factor was developed based on the parametric studies carried out on finite element gear models. This friction factor accounted the frictional variations between the gear contacts and was incorporated in AGMA contact stress equations.
3. The function to evaluate the dimensionless friction factor has been presented in this work. The function and the modified AGMA contact stress calculations are suitable for torque values 0–400 Nm and for lower speed range.
4. Also, the friction factor was validated and verified for different torque and speeds. Thus concluding that the developed K_f is suitable for involute parallel axis gears in said environment.

Recommendations for Future Work

This chapter can be of interest to gear researchers working on frictional effects. It provides the frictional characteristics of involute spur and helical gears and stimulates works of various bodies that are involved in gear research and production. Furthermore, this study contributes to a better gear design, assist gear technologist and all those who are interested in involute spur and helical gears. Friction factor is the new idea coined in this particular work. The idea proposed in this study requires improvements in two aspects. Firstly, the gear ratio and the torque applied in the present study are the same for different gear pairs. Hence, more parametric study can be done to understand its effect on the factor. Secondly, more experimental work can be carried out on the Gear Dynamic Stress Test Rig (shown in ref. 20) in order to study extensively the gear tooth behavior. The experimentation can also include proper lubrication to study the elastohydrodynamic nature of the gear contact and its effect to the friction factor. The gear ratio and torque can be varied and similar analysis can be carried out on gears to develop its effects on frictional gear contacts. Finally, the experimentation can be extended to study the effect of tip relief and other profile modifications on gear contact stresses.

References

1. V.L. Popov, in *Introduction*. Contact Mechanics and Friction: Physical Principles and Applications (Springer, Berlin, 2010), pp. 1–7
2. M. Mokhtari, D.J. Schipper, N. Vleugels, J.W.M. Noordermeer, Transversely isotropic viscoelastic materials: contact mechanics and friction. Tribol. Int. **97**, 116–123 (2016)
3. K.L. Johnson, *Contact Mechanics*, vol. 9 (Cambridge University Press, Cambridge, England, 2003)
4. F.L. Litvin, A. Seireg, Theory of gearing. J. Mech. Des. **114**, 212 (1992)
5. M. Amarnath, S. Lee, Assessment of surface contact fatigue failure in a spur geared system based on the tribological and vibration parameter analysis. Measurement **76**, 32–44 (2015)
6. W.J. Qin, C.Y. Guan, Contact fatigue crack initiation prediction of spur gears based on finite element dynamics analysis, in *ICMFF10* (2015), pp. 1–8

7. C.-M. Hsieh, R.L. Huston, F.B. Oswald, Contact stresses in meshing spur gear teeth: use of an incremental finite element procedure. NASA Tech. Rep. **90**, 1–22 (1992)
8. V. Nikolic, I. Atanasovska, The analysis of contact stress on meshed teeth's flanks along the path of contact for a tooth pair. Facta Univ. **3**, 1055–1066 (2003)
9. C. Gosselin, A review of the current contact stress and deformation formulations compared to finite element analysis, in *International Gearing Conference, UK* (1994)
10. M. Björling, Friction in Elastohydrodynamic Lubrication, Doctoral Thesis, Luleå University of Technology, 2014
11. V.L. Popov, in *Rigorous Treatment of Contact Problems—Hertzian Contact*, Contact Mechanics and Friction: Physical Principles and Applications (Springer, Berlin, 2010), pp. 55–70
12. W.J. Qin, C.Y. Guan, An investigation of contact stresses and crack initiation in spur gears based on finite element dynamics analysis. Int. J. Mech. Sci. **83**, 96–103 (2014)
13. D.H. Johnson, Principles of simulating contact between parts using ANSYS, in *International ANSYS Conference* (2002)
14. S. Sezer, An Evaluation of Ansys Contact Elements, Master's Thesis, Louisiana State University, 2005
15. Z. Wei, Stresses and Deformations in Involute Spur Gears By Finite Element Method, Master's Thesis, University of Saskatchewan, 2004
16. N. Alemu, Analysis of Stresses in Helical Gears by Finite Element Method, Master's Thesis, Addis Ababa University, 2007
17. I. Atanasovska, Finite element model for stress analysis and nonlinear contact analysis of helical gears. Sci. Tech. Rev. **64**(1), 61–69 (2009)
18. S.S. Patil, S. Karuppanan, I. Atanasovska, A.A. Wahab, Contact stress analysis of helical gear pairs, including frictional coefficients. Int. J. Mech. Sci. **85**(2014)
19. S.S. Patil, S. Karuppanan, I. Atanasovska, Contact stress evaluation of involute gear pairs, including the effects of friction and helix angle. J. Tribol. **137**(4), 044501 (2015)
20. S.S. Patil, S. Karuppanan, I. Atanasovska, Experimental measurement of strain and stress state at the contacting helical gear pairs. Measurement **82**, 313–322 (2016)
21. S. He, R. Gunda, R. Singh, Inclusion of sliding friction in contact dynamics model for helical gears. J. Mech. Des. **129**(1), 48 (2007)
22. S. He, Effect of Sliding Friction on Spur and Helical gear, Ph.D. Thesis, The Ohio State University, 2008
23. M. Vaishya, R. Singh, Strategies for modeling friction in gear dynamics. J. Mech. Des. **125**(2), 383 (2003)
24. J. Sukumaran et al., Modelling gear contact with twin-disc setup. Tribol. Int. **49**, 1–7 (2012)
25. J. Kleemola, A. Lehtovaara, Experimental simulation of gear contact along the line of action. Tribol. Int. **42**(10), 1453–1459 (2009)
26. M. Ristivojević, T. Lazović, A. Vencl, Studying the load carrying capacity of spur gear tooth flanks. Mech. Mach. Theory 59, 125–137 (2013)
27. R.G. Parker, S.M. Vijayakar, T. Imajo, Non-linear dynamic response of a spur gear pair: modelling and experimental comparisons. J. Sound Vib. **237**, 435–455 (2000)
28. A. Mihailidis, I. Nerantzis, A new system for testing gears under variable torque and speed. Recent Patents Mech. Eng. **2**, 179–192 (2009)
29. M.A. Muraro, U.R. Jr, C. Henrique, The influence of contact stress distribution and specific film thickness on the wear of spur gears during pitting tests. J. Braz. Soc. Mech. Sci. Eng. **34**(2) (2012)
30. K.C. Stoker, *A Finite Element Approach To Spur Gear Response and Wear Under Non-ideal Loading* (University of Florida, 2009)
31. J. Hedlund, A. Lehtovaara, Modeling of helical gear contact with tooth deflection. Tribol. Int. **40**(4), 613–619 (2007)
32. T.T. Petry-Johnson, A. Kahraman, N. Anderson, D.R. Chase, Experimental investigation of spur gear efficiency, in *Proceedings of the ASME 2007 International Design Engineering Technical Conferences & Computers and Information in Engineering Conference IDETC/CIE 2007* (2007), pp. DETC2007-35045, pp. 1–12

33. H. Ding, Dynamic Wear Models for Gear Systems, Ph.D. Thesis, The Ohio State University, 2007
34. M. Slogen, Contact Mechanics in Gears, Master's Thesis, Chalmers University of Technology, Sweden, 2013
35. S.-C. Hwang, J.-H. Lee, D.-H. Lee, S.-H. Han, K.-H. Lee, Contact stress analysis for a pair of mating gears. Math. Comput. Model. **57**(1–2), 40–49 (2013)
36. V.S.N.K. Bommisetty, Finite element analysis of spur gear set, Master's Thesis, Cleveland State University, 2012
37. I. Atanasovska, R. Mitrović, D. Momčilović, A. Subic, Analysis of the nominal load effects on gear load capacity using the finite-element method. Proc. Inst. Mech. Eng. Part C J. Mech. Eng. Sci. **224**(11), 2539–2548 (2010)
38. S.D. Patel, Finite Element Analysis of Stresses in Involute Spur and Helical Gear, Master's Thesis, The University of Texas at Arlington, 2010
39. A.R. Hassan, Contact stress analysis of spur gear teeth pair. World Acad. Sci. Eng. Technol. **58**, 611–616 (2009)
40. ANSI/AGMA, Fundamental rating factors calculation methods for invoule spur and helical gear teeth, in *American National Standard*, vol. 04 (American Gear Manufacturers Association, Alexandria, Virginia, 2001)

Chapter 13
Tribo-mechanical Aspects in Micro-electro Mechanical Systems (MEMS)

Anand Singh Rathaur, Jitendra Kumar Katiyar, and Vinay Kumar Patel

Abstract Micro Electro-Mechanical System (MEMS) devices continue to find new applications in technology. Some of the successful MEMS are micro-reservoir, micro-pumps, cantilever, micro-pillars for holding the mirror devices in projectors, rotors, channels, valves, sensors etc. In majority of the cases, polycrystalline silicon is used as the structural material for MEMS fabrication because of the micro fabrication process knowledge acquired from the semi-conductor industry. Recently, polymer materials have also been used for MEMS fabrication. Some of the polymers used as structural material are acrylic (PMMA), PDMS and the epoxy-based SU-8. SU-8 is a negative photoresist which is UV curable and has excellent mechanical properties over other polymers. However, when compared to silicon, SU-8 is mechanically inferior. SU-8 has excellent thermal stability. Despite many processing advantages, the bulk mechanical and tribological properties of SU-8 are the main limitations in making it is a versatile MEMS material. Therefore, it is important to increase mechanical strength and toughness, and reduce friction and wear of SU-8 without affecting its micro-fabrication efficiency. Tribology of MEMS, in general, is important as these machines do contain moving parts with relative motion in contact with each other. Smaller parts have high surface area to volume ratio which increases the adhesion and friction forces between two surfaces. In MEMS, wear is a result of high friction.

A. S. Rathaur · V. K. Patel
Department of Mechanical Engineering, Govind Ballabh Pant Institute of Engineering and Technology Ghurdauri, Pauri Garhwal, Uttarakhand 246194, India

J. K. Katiyar (✉)
Department of Mechanical Engineering, SRM Institute of Science and Technology, Kattankulathur Campus, Chennai, Tamil Nadu 603203, India
e-mail: jitendrv@srmist.edu.in

J. K. Katiyar et al. (eds.), *Tribology in Materials and Applications*, Materials Forming, Machining and Tribology,
https://doi.org/10.1007/978-3-030-47451-5_13

13.1 Introduction

MEMS have developed from the silicon (Si) semiconductor industry when it was possible to fabricate micro-scale machines (in the range of 100s of μm) that could function within the elastic range of structural deformation. Further developments in the fabrication led to the making of several machines such as engine and actuator which are shown in Fig. 13.1. Though they were capable of functioning, their lives were limited because of surface related issues such as adhesion, friction and wear. Some of the successful MEMS commercially in use are accelerometer, micro mirrors, pressure sensors, etc. Several issues related to fabrication, mechanical properties and tribological properties have been identified in the literature. In this chapter the mechanical and tribological properties for MEMS are addressed for a new structural material, SU-8.

13.2 MEMS Materials and Fabrication of MEMS

Several materials have been used for making MEMS structure. Most used material among them is polysilicon (Si). Si is a widely used semiconductor material for which the micro-fabrication process using the wet-etching method is well established. Though Si has been successfully used for several MEMS, the drawbacks with Si have kept MEMS industry from growing. The micro-fabrication of Si MEMS is a very energy intensive and expensive process. Tribological properties of Si are inferior because of its hydrophilic nature which leads to high adhesion, high friction and high wear against itself or other similar materials. This has led many researchers to look for other materials that can replace Si as the structural material and at the same time provide good mechanical and tribological properties. Among polymers, PMMA (Poly-methyl methacrylate) and SU-8 (an epoxy based negative photoresist polymer) have found application in MEMS structure. Though both polymers are compatible with the micro-fabrication processes, there are serious drawbacks with their mechanical and tribological properties. SU-8 has shown better mechanical and tribological properties when filled with solid particles and lubricant droplets. It has been shown in a previous study [3] that inclusion of lubricant droplets can reduce the coefficient of friction drastically without any effect on the mechanical properties of the polymer. Therefore, it is proposed that an optimized composition of liquid lubricant with appropriate hard phase particle can enhance both the mechanical and tribological properties.

The SU-8 polymer has three basic components (1) bisphenol A novolavglycidyl epoxy resin, (2) organic solvent gamma-butyrolactone and (3) 10 wt% of mixed triarylsulfonium known as photo-acidic generator. The purpose of photo-acidic generator is to accelerate the cross-linking of SU-8 resin on UV exposer [4, 5]. The chemical structure of SU-8 is given in Fig. 13.2 [6]. It has many advantages such

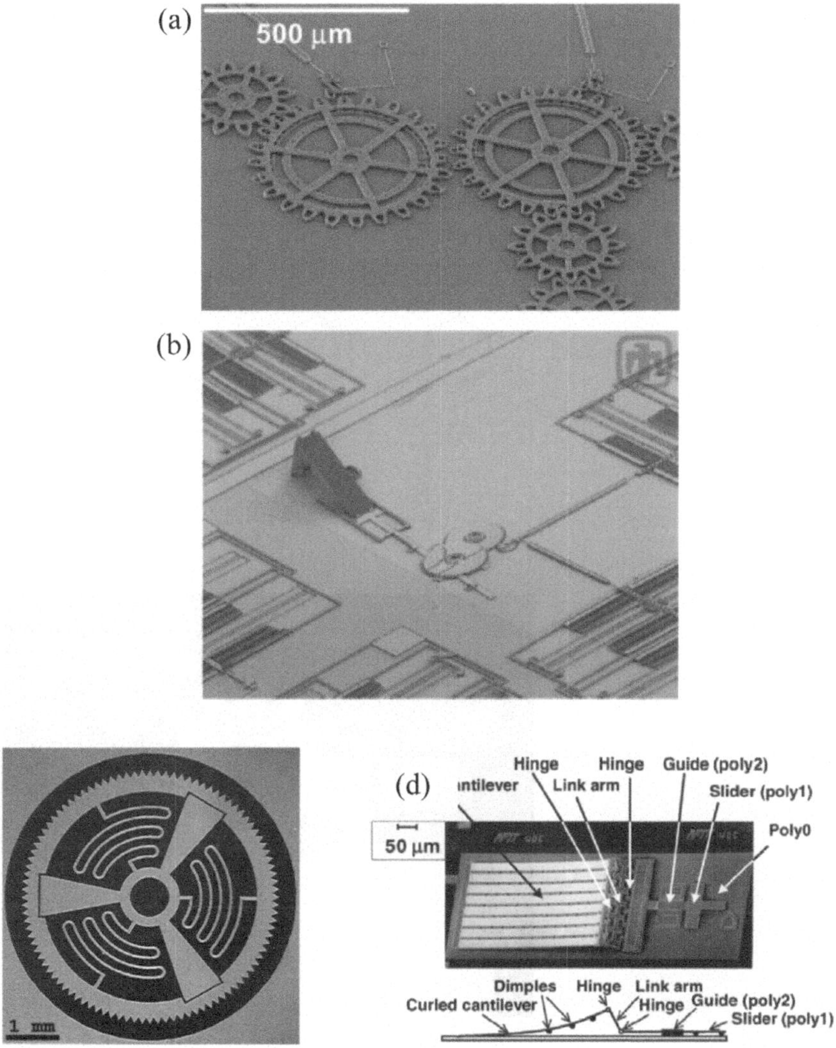

Fig. 13.1 MEMS Structures **a** intermeshing gears and **b** mirror and drive systems produced by Si surface micromachining in Sandia National Laboratories' SUMMiT™ process (Reproduced with the permission of Ref. [1]), **c** rotor gear arrangements (Reproduced with the permission of Ref. [2]) and **d** slider rotor arrangements (Reproduced with the permission of Ref. [1])

as higher thermal decomposition temperature (~330 °C) (compared with other polymers) [7], bio compatibility [8], good optical properties [9], easy fabrication of high aspect ratio components [10] and surface is relatively hydrophobic when compared with silicon [11]. In addition to being used in bulk form, the improved SU-8 composite could also be coated on silicon or other substances such as on other polymers

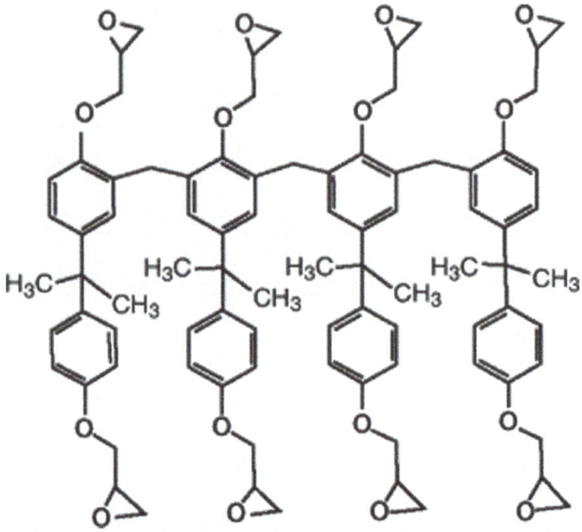

Fig. 13.2 Chemical structure of SU-8 polymer. Reproduced with the permission of Ref. [6]

Fig. 13.3 Digital image of a 200 μm thick gear made of SU-8 + PFPE composite using UV lithographic process. Reproduced with the permission of Ref. [13]

and glass [12] for the fabrication of MEMS structure with improved tribological properties. An example of SU-8 structure is shown in Fig. 13.3 [13].

13.2.1 Tribology in MEMS

The tribological behaviour in microscopic scale is very much different from the macroscopic scale. In macroscopic scale, Amonton's law of friction which states coefficient of friction is independent of contact area and applied load holds good but in the case of microscopic scale this law of friction fails because in microscopic scale surface forces become extremely dominant in comparison to the applied force. One of the important factors that influences the operations of a micro-device is the surface area to volume ratio of its components. The surface area to volume ratio is least for spherical object but increases rapidly as the material is flattened into a thin wafer. Hence, it is expected for MEMS, which are mostly made of thin flat wafer pieces cut into shapes, that the surface area to volume ratio is very high. Hence, surface forces become significant causing very high levels of adhesion and friction. High friction also leads to high wear because of plastic deformation and fracture at the interface where two surfaces meet. High adhesion can also cause stiction between components and thus causing failure of the MEMS. Therefore, it is of utmost importance that we develop SU-8 polymer further to enhance its mechanical and tribological properties for its application as structural material for MEMS and other micro-systems. Figure 13.4 provides some examples of MEMS and the failures by stiction, friction and wear of some components.

Though Si has been the main stay structural material for MEMS, there has been a constant search to find an alternative material. SU-8 has emerged as the material for MEMS because of many advantages it possess in comparison to Si and other competing materials. However, it's mechanical and tribological properties are still of main concern. SU-8 composites provide grate opportunity to tailor-make the required properties. Table 13.1 provide a comparative study of the important properties of SU-8 and other MEMS materials. Pure SU-8 is a brittle type of polymer with tensile fracture strain of 3.2%. The elastic modulus and the hardness are 2.7–3.8 GPa and 0.27 GPa respectively. Its bulk mechanical properties has been enhanced and a literature survey on this is presented below.

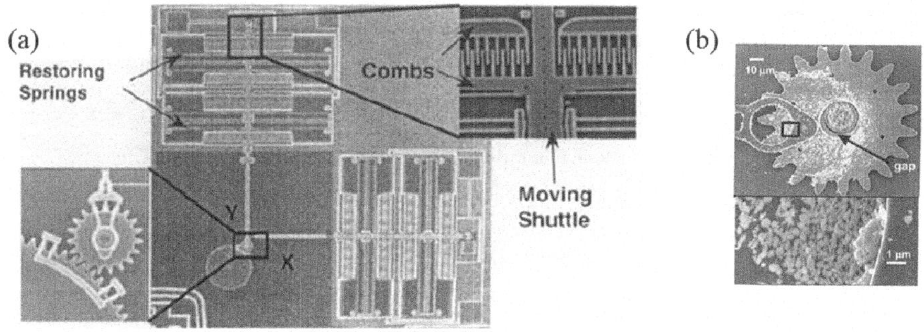

Fig. 13.4 **a** The micro-engine consisting of two orthogonal comb drives, a drive gear, and a load gear and **b** wear debris of the driver gear connected to connecting rod tested to failure. Reproduced with the permission of Ref. [14]

Table 13.1 Comparative study of the important properties of SU-8 and other MEMS materials

Materials	Yield strength (GPa)	Elastic modulus (GPa)	Hardness (GPa)	Poisson's ratio	Density (gm/cm^3)	Sp. Heat (J/gm °C)	References
Si	7.0	190	158	0.23	2.30	0.70	[15]
SiC	21.0	700	24.9	0.14	3.20	0.67	[15]
Si_3N_4	14.0	385	22	0.27	3.10	0.19	[15]
GaAs	2.70	75	6.9	0.31	5.30	0.35	[15]
Quartz	0.5–0.7	76–97	8.8	0.17	2.66	0.82–1.2	[15]
SU-8	0.06	2.7–3.8	0.27	0.22	1.20	–	[16]
PDMS	0.009	0.36–0.47	0.0021	0.5	0.92–0.99	1.46	[17]
PMMA	0.047–0.079	1.6	0.11	0.39	1.15–1.19	1.16–1.47	[18]

13.2.2 Mechanical Properties Concerns in SU-8

From Table 13.2, it is observed that SU-8 has very poor mechanical properties among all the MEMS materials. Therefore, researchers have tried to improve the mechanical properties by adding solid fillers as well as liquid fillers in different weight percentages. Jiguet et al. [2, 19, 20] added 2.5 and 5 wt% of silica nano particles in SU-8 matrix. They fabricated tensile specimen of 10 mm length and performed tensile test. They found that the elastic properties of SU-8 and its composites affect their coefficient of friction. They also performed heat treatment on fabricated samples and observed the mechanical properties before and after the heat treatment. They reported that Young's modulus of SU-8 has increased from 448 to 900 MPa after heat treatment but after adding the fillers, Young's modulus decreases by ~1.5 times to ~1.3 times before and after heat treatment from pure SU-8. Further, they have performed the stress test on the surface of structure of SU-8 and its composite (added 2.5 and

Table 13.2 Comparison of bulk mechanical properties of SU-8 after adding solid fillers

Materials	Percentage (wt%)	Elastic modulus (GPa)	Hardness (GPa)	References
Pure SU-8	0	0.6	–	[19]
SU-8/Silica	2.5	0.68	–	
SU-8/Silica	5	0.28	–	
SU-8/Pure SU-8	0	1.6	–	[21]
SU-8/SWCNT	0.25	1.3	–	
SU-8/Diamondiod	7	1.9	–	
SU-8/MWCNT	5	3.7	–	[23]
Pure SU-8	0	3.8	0.27	[24]
SU-8/NP (CNT, SiO_2, Graphite)	5	3.9	0.17	

5 wt% of silica nano particles) which was made by photolithography on aluminium, quartz and silicon surface. They reported that internal stresses have been reduced by the factor of 5 after adding silicon particles and also coefficient of thermal expansion from 50 to 27 ppm/°C.

Further study was done by Chaimori et al. [6] who mixed nano particles such as diamondoids, SWCNT (single wall carbon nano tube) and gold nanosphere in SU-8 matrix. They fabricated SU-8 composite by using lithography method and conducted ASTM standard tensile test. They reported that the effective modulus of SU-8 was decreased by the addition of nanoparticles at low weight percentage (0.25–1 wt%).

Al-Halhouli et al. [21] have analyzed the hardness property of SU-8 by Oliver-Pharr method on nano mechanical tribometer using Berkovich indenter tip. They performed indentation test on each sample at six different maximum forces ranging from 1500 to 9000 μN with a loading and unloading rate of 300 μN/s and a holding time of 2 s. They have calculated Young's modulus using Eq. 13.1 and reported the value in the range between 5.25 and 6.21 GPa for pure SU-8. They have also reported that elastic modulus, hardness, energy storage modulus and energy loss modulus were reduced by very marginal amount. They have also concluded that SU-8 photoresist material has moderate viscoelastic behaviour.

$$E^* = \left(\frac{1 - \nu_1^2}{E_1}\right) + \left(\frac{1 - \nu_2^2}{E_2}\right) \tag{13.1}$$

where E_1 and E_2 are the elastic modulii of the two interacting materials and ν_1 and ν_2 are their Poisson's ratios. Further, Mionic et al. [22] have studied the mechanical response of SU-8/MWCNT composite by nano indentation test using Berkovich indenter tip. They prepared functionalized and non-functionalized MWCNT and mixed with SU-8 using different solvents. These solvents are gamma-butyrolactone (GBA), propylene glycol methyl-ether acetate (PGMEA), methyl-ethyl ketone (MEK) and acetone. After mixing, solutions were kept in oven for 24 h. Authors observed a significant increase in the elastic modulus when acetone or PGMEA were used as the solvent with maximum increase in the modulus by 104% for acetone.

Table 13.2 provides a comparison of bulk mechanical properties of SU-8 composites using different solid fillers.

Saravanan et al. [23, 24] have taken the approach of adding liquid fillers to SU-8. Though primarily for the improvement of tribological properties, it was found that addition of perfluoropolyether (PFPE), a liquid lubricant, improves the mechanical properties of SU-8. For example an addition of 5 wt% of PFPE in SU-8 increased the elastic modulus and hardness by 5% and 19% respectively. Other lubricants such as base oil and Multiply Alkylated Cyclopentane (MAC) also had similar effects on increasing the bulk mechanical properties of SU-8, however, the increments in the properties was only marginal. Saravanan et al. [25] have explained that the increase in the bulk mechanical properties of SU-8 after adding PFPE is because of an etherification reaction between SU-8 and the OH functional group of PFPE molecules.

The literature survey on the bulk mechanical property improvement of SU-8 shows that the properly improvement has not been enough for MEMS application. For very thin components, it is necessary to have enough stiffness in the structure and hence the elastic modulus must be further improved. Also, for strength purpose, the hardness has to be increased.

13.3 Tribology Properties Improvements in SU-8

For the improvement of tribological properties of SU-8, four methods have been adopted by researchers. These are, bulk composite with solid lubricant, in-situ lubrication, surface (molecular) modification and surface texturing. Detailed descriptions of all the four methods are given in the following sections.

13.3.1 Bulk Composite with Solid Lubricant

In this improvement of tribological properties, Jiguet et al [19, 20] have added silica nano particles in SU-8 with 2.5 and 5 wt% composition. The fabricated SU-8 composites were further tested on linear friction apparatus using carbon steel and polyoxymethylene (POM) balls of 6 mm diameter as counterfaces at 2 N normal load and 33.6 mm/s sliding speed. From these experiments, they have observed the following points, (1) heat treatment of surface reduces wear rate because of better crosslinking of SU-8 molecules, (2) for low wear, the optimal concentration of silica nano particles is required and, (3) the coefficient of friction of composites depends upon the counterface material. In a further study, the same authors have developed SU-8 composite of silica nano particles layer on aluminium, quartz and silicon wafers by photolithography method. They performed experiments on same setup using polyoxymethylene (POM) balls as the counterface. It is reported that nanoparticles reduce coefficient of friction and wear rate by a factor of ~5 in comparison to that for unreinforced SU-8. It was also found that the surface roughness of SU-8 changes from $Ra = 0.115\ \mu m$ before curing to $Ra = 0.012\ \mu m$ after curing as measured by the atomic force microscope (AFM). Further, Saravanan et al. [24] added 5 wt% of nano particles such as graphite, silicon dioxide and carbon nano tube (CNT) in SU-8 matrix and performed the ball-on-disk tribometer test. They observed ~2 times decrement in the coefficient of friction for SU-8/graphite and SU-8/CNT's composites in comparison to that for pure SU-8. It is to be noted that the wear lives of these composites were reported as zero because the COF was $\geq$0.3. They used the criterion of wear failure when the COF $\geq$0.3. Moreover, this work was carried out further by Katiyar et al. [25]. They varied the concentration of solid carbon fillers such as graphite, graphene and multiwall carbon nanotube in SU-8 matrix. They performed physical, mechanical and tribological test on fabricated samples and reported that graphite gives lower coefficient of friction apart from other carbon fillers. Hence, further they optimized the concentration of graphite. From which, they reported that the optimum concentration of graphite is 10–20 wt% at which lower coefficient of friction and higher wear resistance properties were obtained [26]. Katiyar et al. [27] further reported that blending of graphite/talc in SU-8 matrix shows drastic improvement in friction coefficient and higher wear resistance property. This is because of the better bonding between graphite and talc.

13.3.2 Surface (Molecular) Modification

Surface modification is based on the principle of changing only the top surface with attachment of molecules by chemical or physical means. These have been very few works on the surface modification of SU-8. Singh et al. [28] have employed two step method of surface modification. In the first step, the SU-8 surface was given an exposure of oxygen plasma which was followed by dip-coating of PFPE molecules. This surface treatment reduced the initial and steady-state COF by ~ 4–7 times and 2.5–3.5 times respectively, and wear life increased by >1000 times. This method provided a physical means to attach PFPE lubricants molecules onto SU-8 surface. The OH functional groups of PFPE tend to form bonds with the functional groups created on the SU-8 surface during oxygen plasma treatment. Chemical surface modification was also conducted by Singh et al. [29] in which the SU-8 surface was treated with ethanolamine-sodium phosphate buffer solution. It was followed by dip-coated with PFPE in a dilute solution. The chemical reaction scheme in this treatment is explained as follows. The amine group of ethanolamine and epoxy group of SU-8 reacts to form chemical bond. In the next step, the polar hydroxyl (–OH) group on the other end of ethanolamine reacts with the –OH functional group of PFPE [30].

The chemical surface modification also provided similar improvements in the COF and wear life as those obtained from the physical surface modification. This is because the top layer in both cases is made up of the same PFPE molecules. Physical and chemical bonding with SU-8 makes the surface wear durable. The physical surface modification has also been tried on PMMA surface with very improved results [31].

13.3.3 In-Situ Lubrication

In-situ lubrication of SU-8 has been a new concept. In this technique, an optimized quantity (wt%) of a known lubricant is added to the matrix. PFPE was added to SU-8 before curing. PFPE droplets were homogenized thoroughly before the start of curing process for this composite by UV ray exposure [32]. Saravanan et al. [23, 24] found that the micron to nano size droplets of PFPE were trapped in the matrix of SU-8 cross-linked molecules. These trapped droplets act as reservoir of the lubricant at the interface. Thus, initial wear may take place on the surface of SU-8 but this will lead

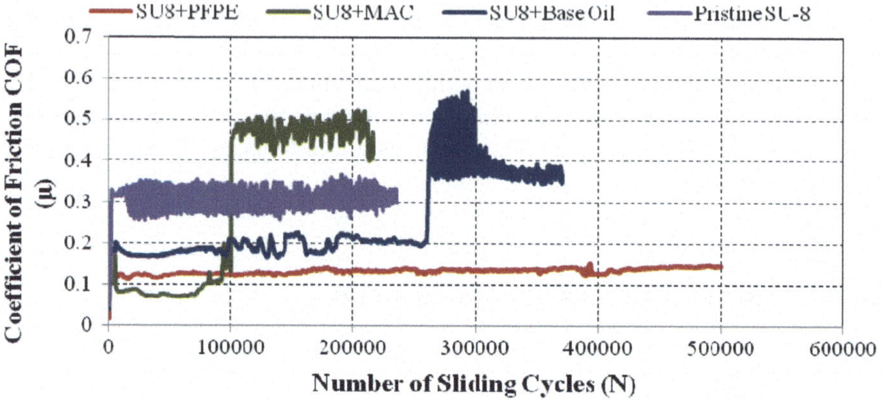

Fig. 13.5 Plot of COF versus sliding cycles. Reproduced with the permission of Ref. [23]

to opening up of the PFPE droplets. The presence of PFPE lubricant at the interface will reduce friction and the wear life is increased by several orders of magnitude. Figure 13.5 provides plots of COF versus sliding cycles for pure SU-8 and compares the result with those of in-situ lubricated SU-8 with PFPE at 2 and 10 wt%. For the highest PFPE wt% of 10 there was a reduction in the COF 0.52–0.07, a ~7 times reduction. There was no failure of the material in any gross manner with wear life more than ~5 times. The mechanism of in-situ lubrication was found to be of mixed lubrication where there is partial formation of fluid film between the solid surfaces. Figure 13.6 shows optical images of the counterface ball surface. The presence of liquid lubricant is seen on the surface along with solid debris particles making a slurry.

Batooli et al. [33] fabricated SU-8 composite with ionic liquid (1-Methyl-3-octylimidazolium hexafluoro-phosphate) as the filler in 4 and 10 wt% composition. Although the ionic liquid (IL) provided same reaction in coefficient of friction over pure SU-8. It was only confined to the first approximately 1000 cycles. COF quickly

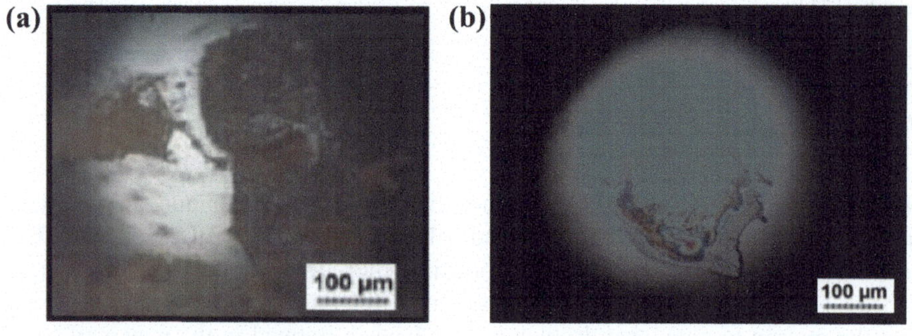

Fig. 13.6 Optical images of counterface ball surface (Reproduced with the permission of Ref. [23]) **a** Pure SU-8, **b** SU-8/PFPE (10 wt%)

rose to higher value in the range of 0.2 and above. Wear was only reduced by half over 1800 cycles of sliding.

The literature survey on the in-situ lubrication of SU-8 shows that it has excellent potential to further reduce friction and increase wear life. Hence, this method of improving tribological properties of SU-8 will be further explored.

13.3.4 Surface Texturing

Surface texturing is very widely used approach for reducing friction and wear. In this approach, the surface contact area is reduced which results in lower adhesion and hence lower friction [34]. One problem with textured surfaces is that the texture tends to get deformed or broken. If the texture material is a hard type then the broken debris may even wear the remaining surfaces. Texturing of Si and polymer surfaces has been tried for reduction in friction. However, wear life were not increased much. The nano-texturing effect provided by the oxygen plasma treatment of PMMA was found to increase the wear life of PMMA surface by many orders.

Tay et al. [35] fabricated rounded micro-dot pattern of SU-8 on silicon wafer of 2 mm × 2 mm dimension. The micro-dots of diameter 108.8 μm and height 1.14 μm were fabricated on Si wafer with different pitch and sliding experiments were conducted against flat Si wafer and Si_3N_4 balls. They observed that the micro-dot pattering can drastically reduce COF and there is an optimum pitch for which the COF is the lowest. This pitch length was found to be 150 μm for this case. It was also confirmed in this case that the wear life cannot be high despite low friction. The initial wear tends to accelerate the wear process in the presence of a texture. However, this effect was mitigated by providing a thin layer of PFPE lubricant on top of SU-8 micro-dots. Thus, texturing with top surface lubrication increased wear life by 2–3 orders of magnitude without failure.

Further, Osborn et al. [36] have fabricated connected and isolated micro structures (pillars) of SU-8, with and without diamond like carbon (DLC) coating as the top layer. They performed reciprocating tribological tests using 7 mm diameter chrome steel ball as the counterface and obtained COF in the range of 0.8–0.4 and very less wear debris (1000 number of cycles).

Further in the field of surface texturing, Myint et al. [37] have fabricated 3D negative fingerprint and honeycomb textured surfaces on SU-8. They performed tribological tests against silicon nitride ball counterface on ball-on-disk tribometer. They was obtained the coefficient of friction of 0.08 for negative figure print as comparison to 0.2 for the untextured SU-8 surface in a rotational test at 100 mN normal load. The wear life of the negative figure print texture on SU-8 was highest compared with untextured and honeycomb texture when PFPE was also used as the lubricant layer.

13.4 Lubrication in MEMS

Lubrication of MEMS have been a challenge because of the small size of MEMS, nature of the structural material Si, need for lubrication process to be compatible to the micro-fabrication processes and no possibility of re-lubrication once the MEMS is in service. Because of these constraints, MEMS lubrication technology is still in the evolving period. A number of solution have been proposed and tried in laboratories. Some of the important ones are gaseous lubrication by 1-pentanol hermetically sealed inside MEMS [38], use of self-assembled monolayers (OTS) [39], localized lubrication of PFPE etc. Though these methods are compatible with MEMS fabrication technique, their efficiency and cost-effectiveness vary. The new concept of in-situ lubrication, in comparison, is very simple and cost effective. It does not require any extra step of lubrication during the micro-fabrication. The method utilizes the changing of the bulk of the material which improves both the Tribological as well as mechanical properties. As mentioned earlier, the in-situ lubrication works in the mixed lubrication regime. The mechanism of mixed lubrication is explained in the next section along with associated model and analysis.

13.4.1 Mechanism of Lubrication

The boundary and mixed lubrication regimes tend to be effective in providing low friction if there is formation and reformation of the boundary film. Formation and reformation of the boundary film is dependent on the role of absorption of additives. The additives are present in the lubricant as floating molecules. Hence, the process of formation and reformation of effective boundary layer is a diffusion controlled phenomenon. Based on this concept, Albertson [40] introduced an equation for the dynamic coefficient of friction, μ_v, at a given sliding velocity, v, as,

$$\mu_v = \mu_d - (\mu_d - \mu_\infty)\left[1 - e^{\left[\frac{-c}{v}\right]}\right] \tag{13.2}$$

where μ_d and μ_∞ are the coefficient of friction of fully damaged boundary film (substrate fully exposed) and fully effective boundary film (possible at low speed), respectively, and c is a diffusion parameter for the rate of diffusion of additives and molecules.

Based on the Eyring [41] activated slip model. The potential barrier to slip can be overcome by the shear stress and thermal fluctuations. Hence, the net rate of passage of the molecules under the influence external stress would a sum of the forward and the backward movements of the molecules. The resultant rate of passage will be given as,

$$k_f - k_b = A\left[e^{-\frac{\left(E-\frac{\tau\varphi}{2}\right)}{kT}} - e^{-\frac{\left(E+\frac{\tau\varphi}{2}\right)}{kT}}\right] \tag{13.3}$$

where k_f and k_b are the rates of passage across the potential barriers, E is the energy barrier between atoms or molecules, τ is shear stress, φ is shear activated volume ($\lambda \times A$), A is the area over which the force act, λ is the distance moved by the molecules from one side to another across a potential barrier, k and T are the Boltzmann constant and absolute temperature, respectively.

The measure of slip is then given as the relative velocity v which is given as

$$v = \lambda\left(k_f - k_b\right)$$

$$v = A\lambda e^{-\frac{E}{kT}}\left[e^{\frac{\tau\varphi}{2kT}} - e^{-\frac{\tau\varphi}{2kT}}\right] \tag{13.4}$$

Since $\sin h(x) = \frac{(e^x - e^{-x})}{2}$.
Equation (13.4) can be re-written as,

$$v = 2A\lambda e^{-\frac{E}{kT}} \sin\ h\left(\frac{\tau\varphi}{2kT}\right) \tag{13.5}$$

For low applied shear stress $kT >>> (\frac{\tau\phi}{2})$ and hence Eq. (13.5) is re-written as

$$v = A\lambda e^{-\frac{E}{kT}}\left(\frac{\tau\varphi}{2kT}\right) \tag{13.6}$$

Equation (13.6), provides a relation of the slipping or sliding speed within the lubricant as the applied shear stress is changed. When the applied shear stress is high

$$e^{\frac{\tau\varphi}{2kT}} >>>> e^{-\frac{\tau\varphi}{2kT}}$$

Hence, the Eq. (13.5) can be modified as:

$$v = A\lambda e^{-\frac{E}{kT}} e^{\frac{\tau\varphi}{2kT}} \tag{13.7}$$

Equation (13.8) shows that the speed increases exponentially with the applied shear stress. Briscoe et al. [42] have shown that the potential barrier E actually changes with the hydrostatic pressure and the actual potential barrier is given as $E + p\Omega$, where p is the pressure and Ω is a pressure coefficient. Hence, Eq. (13.7) is then modified as below,

$$v = v_0 e^{-(E+p\Omega)} e^{\frac{\tau\varphi}{2kT}} \tag{13.8}$$

where $A\lambda = v_{0,}$ a velocity term. Equation (13.8) can be written as

$$\tau = \frac{kT}{\varphi} \ln \frac{v}{v_0} + \frac{1}{\varphi}(E + p\Omega) \tag{13.9}$$

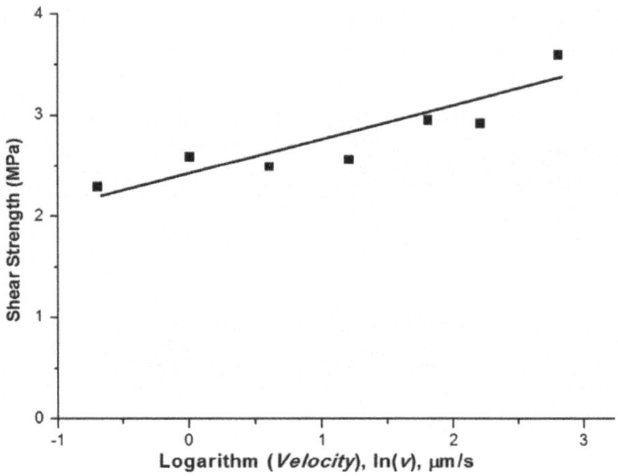

Fig. 13.7 Variation of shear strength with logarithmic velocity for steric acid monolayer

Equation (13.9) can be given in the following form as,

$$\tau = \tau_0 + \theta \ln(v) \tag{13.10}$$

where $\tau_0 = \frac{1}{\phi}(E + p\Omega - kT \ln(v_0))$ and $\theta = \frac{kT}{\phi}$.

In Eq. (13.10), τ_0 and θ are constants for a given monolayer. Equation (13.10) states that there is a logarithmic relation between the shear stress and the sliding speed. Hence, a plot of τ versus $\ln(v)$ will give linear result.

Equation (13.10) was also obtained by Briscoe and Evans [43] in their work on the lubrication by a monolayer of stearic acid. The fatty acid monolayer was deposited on mica sheet using Langmuir-Blodgett method. The results of this classical work is plotted in Fig. 13.7. Saravanan et al [44] have recently conducted tests on in-situ lubricated SU-8 composites and confirmed that Eq. (13.10) hold true for such cases. They observed strong dependence of the coefficient of friction on the sliding speed but very week dependence on the applied normal load.

13.5 Conclusion

SU-8 is a very promising material for making MEMS. SU-8 is compatible with UV-curing and micro-fabrication processes used in making MEMS. The mechanical and tribological properties should be further improved in order to make SU-8 a successful MEMS structural material. Following conclusions can be made from the current state-of-the-art in SU-8 properly improvements.

1. Tribological properties of SU-8 such as coefficient of friction has been improved by ~2 to 3 times by adding solid fillers (graphite, SiO_2, MWCNTs etc) but wear resistance property improved very marginally. When liquid fillers (PFPE, SN150

and MAC, etc.) has blended in SU-8 matrix, the coefficient of friction has been improved by ~6 to 7 times and wear resistance property improved by 4 orders of magnitude as comparison to pure SU-8.
2. Mechanical properties such as elastic modulus and hardness of SU-8 have not been improved much after adding the solid fillers as well as liquid fillers. Saravanan et al reported the value of elastic modulus after adding graphite and PFPE is 5.0 GPa from pure SU-8 (3.8 GPa) and hardness is reduced from 0.27 GPa (SU-8) to 0.16 GPa (SU-8 composite). 15–25% by adding above fillers.
3. The mechanism of lubrication of in-situ liquid-filled SU-8 has been identified as that of mixed lubrication.

References

1. S.H. Kim, D.B. Asay, M.T. Dugger, Nanotribology and MEMS. Nanotoday **2**(5), 22–29 (2007)
2. S. Jiguet, M. Judelewicz, S. Mischler, A. Bertch, P. Renaud, Effect of filler behavior on nanocomposite SU8 photoresist for moving micro-parts. Microelectron. Eng. **83**(4–9), 1273–1276 (2006)
3. P. Saravanan, Tribology of Self Lubricated SU-8 Composite for Mirco-Electro Mechanical Systems (MEMS) Applications, Ph.D. Thesis, NUS Singapore, 2015
4. J.B. Lee, K.H. Choi, K. Yoo, Innovative SU-8 lithography techniques and their applications. Micromachines **6**(1), 1–18 (2015)
5. R. Martinez-Duarte, SU-8 photolithography as a toolbox for carbon MEMS. Micromachines **5**(3), 766–782 (2014)
6. H.C. Chiamori. J.W. Brown, E.V. Adhiprakasha, E.T. Hantsoo, J.B. Straalsund, N.A. Melosh, B.L. Pruitt, Suspension of nanoparticles in SU-8: processing and characterization of nanocomposite polymers. Microelectron. J. **39**, 228–236 (2008)
7. S. Metz, S. Jiguet, A. Bertsch, P. Renaud, Polyimide and SU-8 microfluidic devices manufactured by heat-depolymerizable sacrificial material technique. Lab Chip (Roy. Soc. Chem.) **4**, 114–120 (2004)
8. B.A. Weisenberg, D.L. Mooradian, Hemocompatibility of materials used in microelectromechanical systems: platelet adhesion and morphology in vitro. J. Biomed. Mater. Res. **60**, 283–291 (2002)
9. I.G. Foulds, M. Parameswaran, A planar self-sacrificial multilayer SU-8 based MEMS process utilizing a UV-blocking layer for the creation of freely moving parts. J. Micromech. Microeng. **16**, 2109–2115 (2006)
10. N.C. Labianca, J.D. Gelorme, K.Y. Lee, High aspect ratio optical resist chemistry for MEMS application, in *Proceedings of 4th International Symposium on Magnetic Materials, Processes and Devices, Chicago* (1993), pp. 386–396
11. H.C. Ewan, D.F. Moore, SU-8 thick photoresist processing as a functional material for MEMS applications. J. Micromech. Microeng. **12**, 368–374 (2002)
12. H. Bangert, C. Eisenmenger, Deposition and structural properties of two-component metal coatings for tribological applications. Surf. Coat. Technol. **80**, 162–170 (1996)
13. P. Saravanan, N. Satyanarayana, S.K. Sinha, Self-lubricating SU-8 nanocomposites for microelectromechanical systems applications. Tribol. Lett. **49**(1), 169–178 (2013)
14. D.M. Tanner, N.F. Smith, L.W. Irwin, W.P. Eaton, *"MEMS reliability: infrastructure, test structures, experiments, and failure modes", SAND 2000-0091* (Sandia National Laboratories, Albuquerque, New Mexico, 2000)
15. S.M. Sze, *Physics of Semiconductor Devices* (Wiley Inc., New York, 1981)

16. H. Lorenz, M. Despont, N. Fahrni, N. LaBianca, P. Renaud, P. Vettiger, SU-8: a low-cost negative resist for MEMS. J. Micromech. Microeng. **7**, 121–124 (1997)
17. http://www.mit.edu/~6.777/matprops/pdms.htm
18. C.A. Harper, E.M. Petrie, *Plastics Materials and processes* (Wiley, New York, 2003)
19. S. Jiguet, M. Judelewicz, S. Mischler, H. Hofmann, A. Bertsch, P. Renaud, SU-8 nanocomposite coatings with improved tribological performance for MEMS. Surf. Coat. Technol. **201**, 2289–2295 (2006)
20. S. Jiguet, A. Bertsch, M. Judelewicz, H. Hofmann, P. Renaud, SU-8 nanocomposite photoresist with low stress properties for micro-fabrication applications. Microelectron. Eng. **83**, 1966–1970 (2006)
21. A.T. Al-Halhouli, I. Kampen, T. Krah, S. Büttgenbach, Nanoindentation testing of SU-8 photoresist mechanical properties. Microelectron. Eng. **85**, 942–944 (2008)
22. M. Mionic, S. Jiguet, M. Judelewicz, A. Karimi, L. Forro, A. Magrez, Study of the mechanical response of carbon nanotubes SU-8 composites by nanoindentation. Phys. Status Solidi **B247**(11–12), 3072–3075 (2010)
23. P. Saravanan, N. Satyanarayana, S.K. Sinha, Wear durability study on self-lubricating SU-8 composites with perfluoropolyther, multiply-alkylated cyclopentane and base oil as the fillers. Tribol. Int. **64** (2013)
24. P. Saravanan, N. Satyanarayana, D.H. Minh, S.K. Sinha, An in-situ heating effect study on tribological behavior of SU-8/PFPE composite. Wear **307**, 182–189 (2013)
25. J.K. Katiyar, S.K. Sinha, A. Kumar, Effects of carbon fillers on the tribological and mechanical properties of an epoxy-based polymer (SU-8). Tribol. Mater. Surf. Interf. **10**(1), 33–44 (2016)
26. J.K. Katiyar, S.K. Sinha, A. Kumar, Effect of graphite concentration on the tribological and mechanical properties of filled SU-8. Tribol. Online **11**(2), 152–158 (2016)
27. J.K. Katiyar, S.K. Sinha, A. Kumar, Friction and wear durability study of SU-8 composite coatings with talc and graphite as the fillers. Wear **362–363**, 199–208 (2016)
28. R.A. Singh, N. Satyanarayana, S.K. Sinha, Surface chemical modification for exceptional wear life of MEMS materials. AIP Adv. **1**, 042141 (2011)
29. R.A. Singh N. Satyanarayana, T.S. Kustandi, S.K. Sinha, Tribo-functionalizing Si and SU-8 materials by surface modification for application in MEMS/NEMS actuator-based devices. J. Phys. D: Appl. Phys. **44**, 015301 (2011)
30. P. Saravanan, N. Satyanarayana, P.C. Siong, H.M. Doung, S.K. Sinha, Tribology of self-lubricating SU-8+PFPE composite based lub-tape, in *Procedia Engineering in Malaysian International Tribology Conference* (2013)
31. A.C. Henry, T.J. Tutt, M. Galloway, Y.Y. Davidson, C.S. McWhorter, A.S. Soper, R.L. McCarley, Surface modification of poly(methyl methacrylate) used in the fabrication of microanalytical devices. Anal. Chem. **72**(21), 5331–5337 (2000)
32. K.H. Lau, A. Giridhar, S. Harikrishnan, N. Satyanarayana, S.K. Sinha, Releasing high aspect ratio SU-8 microstructures using AZ photoresist as a sacrificial layer on metallized Si substrates. Microsyst. Technol. **19**(11), 1863–1871 (2013)
33. L. Batooli, S.G. Maldonado, M. Judelewicz, S. Mischler, Novel SU-8/Ionic liquid composite for tribological coatings and MEMS. Micromachines **6**, 611–621 (2015)
34. M. Wakudaa, Y. Yamauchi, S. Kanzaki, Y. Yasuda, Effect of surface texturing on friction reduction between ceramic and steel materials under lubricated sliding contact. Wear **254**(3–4), 356–363 (2003)
35. N.B. Tay, M. Minn, S.K. Sinha, Tribological study of SU-8 micro-dot patterns printed on Si surface in a flat-on-flat reciprocating sliding test. Tribol. Lett. **44**, 167–176 (2011)
36. L. Osborn, Enhanced Tribological Properties of Surfaces Patterned with SU-8/DLC Microstructures, Bachelor of Science Thesis, University of Arkansas, 2012
37. S.M. Myint, M. Minn, R. Yaping, N. Satyanarayana, S.K. Sinha, C.S. Bhatia, Friction and wear durability studies on the 3D negative fingerprint and honeycomb textured SU-8 surfaces. Tribol. Int. **60**, 187–197 (2013)
38. D.B. Asay, M.T. Dugger, S.H. Kim, In-situ vapor-phase lubrication of MEMS. Tribol. Lett. **29**, 67–74 (2008)

39. N. Satyanarayana, S.K. Sinha, Tribology of PFPE overcoated self-assembled monolayers deposited on Si surface. J. Phys. D Appl. Phys. **38**, 3512–3522 (2005)
40. C.E. Albertson, The mechanism of anti-squawk additive behavior in automatic transmission fluids. ASLE Trans. **6**, 300–315 (1963)
41. H. Eyring, Viscosity, plasticity, and diffusion as examples of absolute reaction rates. J. Chem. Phys. **4**(4), 283–291 (1936)
42. B.J. Briscoe, P.S. Thomas, D.R. Williams, Microscopic origins of the interface friction of organic films: the potential of vibrational spectroscopy. Wear **153**(1), 263–275 (1992)
43. B.J. Briscoe, D.C.B. Evans, The shear properties of Langmuir-Blodgett layers. Proc. Roy. Soc. London A: Math., Phys. Eng. Sci. **380**(1779), 389–407 (1982)
44. P. Saravanan, S.K. Sinha, S. Jayaraman, H.M. Duong A comprehensive study on the self–lubrication mechanism of SU-8 composites. Tribol. Int. **95**, 391–405 (2016)

Chapter 14
Analysis of Rotor Stability Supported by Surface Porous Layered Journal Bearing

C. Shravankumar, K. Jegadeesan, and T. V. V. L. N. Rao

Abstract An analysis of modified Reynolds equation for surface porous layer configuration based on short (Ocvirk) bearing approximation is presented under dynamic conditions. The steady state pressure and dynamic pressure gradients are analyzed using Lund's perturbation method. The nondimensional parameters of film thickness ratio of surface porous layers, dynamic viscosity ratio of surface layers, and permeability of porous layers are considered in the analysis. The stiffness and damping coefficients, threshold (critical) speed and whirl ratio (critical) for layered bearings are presented. An improvement in critical (threshold) speed of layer (film) journal bearings is evaluated.

14.1 Stability Analysis of Rotor on Journal Bearing

Stability threshold linearized analysis of layer (film) bearing (journal) is presented for surface/porous/surface-porous film configurations. In a surface layer configuration, surface layer of higher Newtonian viscosity compared to base fluid is considered. The Brinkman equations are used to investigate the influence of porous film structure. The surface porous film structure is composed of a surface film (infinite permeability porous layer) covered with porous film (low permeability layer). Dynamic analysis is presented based on infinitesimally small journal perturbation. A one-dimensional short bearing analysis of Reynolds modified equation is presented for layer (film). Results of critical (threshold) speed and ratio of whirl (critical) are presented based on the coefficients of critical speed and ratio of whirl for layer (film) journal bearing. A bearing (journal) with (i) higher thickness and viscosity of bearing surface film and (ii) higher thickness and lower permeability of porous film results in higher critical (threshold) speed coefficient. A layered bearing (journal) has prospects to enhance critical (threshold) speed limits of bearing.

C. Shravankumar · K. Jegadeesan · T. V. V. L. N. Rao (✉)
Department of Mechanical Engineering, SRM Institute of Science and Technology, Kattankulathur 603203, India
e-mail: tvvlnrao@gmail.com

J. K. Katiyar et al. (eds.), *Tribology in Materials and Applications*, Materials Forming, Machining and Tribology,
https://doi.org/10.1007/978-3-030-47451-5_14

14.1.1 Surface Layer Journal Bearing Analysis

Journal bearing performance characteristics are enhanced with surface layered films of higher viscosity compared to base fluid. Szeri [1] analyzed potential of composite film bearing with fluid film layers of varying viscosities. The coefficients of load bearing capacity and friction coefficient for composite layered film that comprises of two immiscible fluid layers are derived with respect to homogeneous film [1]. Tichy [2] examined a multi (three) film bearing (journal) adsorbent surface films of high viscosity. The numerical predictions show that high viscosity surface adsorbent films considerably improve the bearing performance characteristics. Rao [3] presented stability (linearized analysis) of Newtonian two-layer film bearing (journal) with varying magnitudes of film (layer) thicknesses and viscosity. The load, threshold (critical) speed, and ratio of whirl (critical) coefficients for two-layered film with respect to uniform film are evaluated.

Hydrodynamic journal bearings with surface adsorbent layer of higher viscosity compared to base fluid provides steady and stability performance improvement.

14.1.2 Porous Layer Journal Bearing Analysis

Operating features of bearing (journal) are influenced by lubricant additive properties. Tichy [4] studied the influences of additives in lubricants based on the examination of thin porous film on bearing surfaces. Li [5] derived Reynolds modified equation of a porous thin layer model by Brinkman equations. Li and Chu [6] and Elsharkawy [7] investigated influences of additives in lubricants on the operating features of hydrodynamic bearing (journal) using porous thin layer model and couple stress fluid model respectively. Lin and Hwang [8] examined the stability of porous finite journal bearings by Brinkman equations. The stability threshold limits of porous journal bearings are evaluated by D'Agostino et al. [9] using analytical approach.

Hydrodynamic journal bearings with porous adsorbent layer provides performance improvement with reducing porous film (layer) permeability.

14.1.3 Surface-Porous Layer Journal Bearing Analysis

Saha and Majumdar [10] investigated the enhancement of stability for porous two-layer hydrostatic bearing based on high permeable backing covered by a thin layer surface film of lower permeability. Rao et al. [11] examined a bearing (journal) with a double layer porous film and surface layer porous film. The surface layer simulates the porous film with high permeability. The flow in the porous film region is modelled by Brinkman equations. Rao et al. [12] investigated stability limits of double layer porous film and surface layer porous film bearing (journal) by long and short bearing approximations. The load, threshold (critical) speed and ratio of whirl (critical) coefficients are evaluated for double layer porous film and surface layer porous film patterns. The coefficients are presented in terms of load bearing capacity, threshold (critical) speed and ratio of whirl (critical) of double layer porous film (and surface layer porous film) with uniform film.

Hydrodynamic journal bearings with surface layer (considered as a porous layer of high permeability) covered with low permeable porous adsorbent film layer is an effective approach for performance improvement.

14.2 Analysis of Rotor Stability Supported by Surface Layered Bearing

An analysis (one-dimensional) of bearing (journal) stability taking into account surface film lubricant layer based on short (Ocvirk) bearing approximation is undertaken. The nondimensional stability coefficients based on short bearing approximation are derived from Rao et al. [3]. The nondimensional surface layer film parameters in the stability performance analysis are: nondimensional thickness ratio of surface (bearing) adsorbent film (layer) to film thickness (γ_l) and viscosity (dynamic) ratio of bearing surface adsorbent film (layer) to base homogeneous fluid (β_1). Results of damping coefficients (B_{ij}), threshold (critical) speed ($\omega_{s,l}$) and ratio of whirl (critical) ($\Omega_{s,l}$) of layer film are presented for eccentricity ratios (ε) obtained for the steady state position evaluated with variations in journal speed.

14.2.1 Dynamic Analysis of Surface Film Layer Bearing

The schematic of a surface film layer bearing (journal) with nondimensional thickness ratio of bearing surface adsorbent film (layer) to film thickness (γ_1) ($\gamma_1 = \delta_1/h$) and viscosity (dynamic) ratio of surface (bearing) adsorbent film (layer) to base homogeneous fluid (β_1) ($\beta_1 = \mu_s/\mu$) is shown in Fig. 14.1. The nondimensional film thickness of journal bearing is $H = (1 + \varepsilon \cos\theta)$ ($H = h/C$).

Lund and Thomsen [13] developed numerical model to estimate dynamic bearing fluid film coefficients and stability (linearized) of bearing (journal) based on infinitesimal perturbation method. The dynamic coefficients used in the linear stability analysis are obtained from Reynolds dynamic equation using the pressure perturbation (linearized) with oil film displacements and velocities. Rao [14] evaluated stability short journal bearing using dynamic coefficients obtained from linearized pressure perturbation method. Elrod and Vijayaraghavan [15] analyzed perturbation of Reynolds dynamic equation considering flow within the cavitation region and presented threshold critical stability (linearized) of bearing (journal).

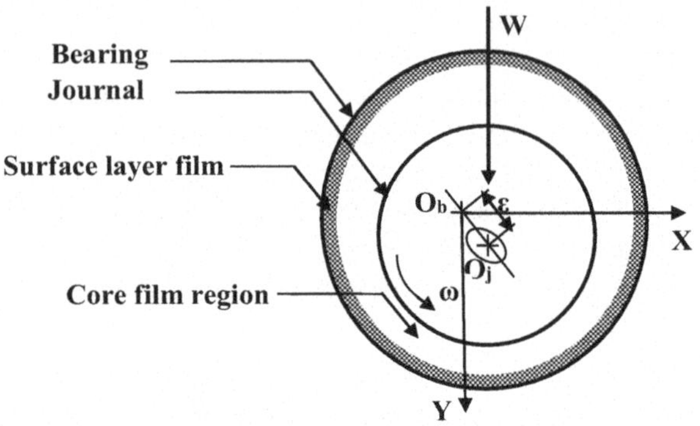

Fig. 14.1 Surface layered journal bearing geometry

The load capacity (nondimensional) (W) and dynamic bearing coefficients (K_{yy}, K_{xy}, K_{yx}, K_{xx}, B_{yy}, B_{xy}, B_{yx} and B_{xx}) calculated ($\bar{K}_{ij} = k_{ij}C^3/\mu_l\omega R^3L$; $K_{ij} = \bar{K}_{ij}/W$ for $i = x, y$ and $\bar{B}_{ij} = b_{ij}C^3/\mu_l R^3L$; $B_{ij} = \bar{B}_{ij}/W$ for $i = x, y$) using nondimensional pressure gradients (P_y, P_x, $P_{\dot{y}}$, and $P_{\dot{x}}$) respectively using short (Ocvirk) bearing estimate are specified in Rao et al. [12].

The load capacity (nondimensional), stiffness and damping coefficients of layer film are derived from homogeneous film as [3, 12]

$$W_l = \frac{\Delta_s}{\Delta_p}W_h,\ \bar{K}_{ij,l} = \frac{\Delta_s}{\Delta_p}\bar{K}_{ij,h} \text{ and } \bar{B}_{ij,l} = \frac{1}{\Delta_p}\bar{B}_{ij,h}\ for\ i = y, x \tag{14.1}$$

$$K_{ij,l} = \frac{\bar{K}_{ij,l}}{W_l} = \frac{\bar{K}_{ij,h}}{W_h} \text{ and } B_{ij,l} = \frac{\bar{B}_{ij,l}}{W_l} = \frac{1}{\Delta_s}\frac{\bar{B}_{ij,h}}{W_h} \text{ for } i = y, x \tag{14.2}$$

The coefficients Δ_s, Δ_p for surface layer journal bearing are expressed as [3, 12]

$$\Delta_s = \left[\frac{1 - Y_1^2 + \beta_1 Y_1^2}{1 - Y_1 + \beta_1 Y_1}\right],$$
$$\Delta_p = \left[Y_1^3 + \frac{(1 - Y_1)^3}{\beta_1} + \frac{3Y_1(1 - Y_1)}{(1 - Y_1 + \beta_1 Y_1)}\right] \text{ and } \gamma_1 = 1 - Y_1 \tag{14.3}$$

Lund [16] presented an overview of stability analysis of fluid film journal bearings. The whirl ratio (critical) and critical (threshold) speed (nondimensional) are evaluated as [16]

$$\Omega_s = \sqrt{\frac{(K_{yy} - \kappa_o)(K_{xx} - \kappa_o) - K_{yx}K_{xy}}{B_{yy}B_{xx} - B_{yx}B_{xy}}} \tag{14.4}$$

$$\omega_s = \sqrt{M} = \frac{\sqrt{\kappa_o}}{\Omega_s} \tag{14.5}$$

where

$$K_{ij} = \bar{K}_{ij}/W \text{ and } B_{ij} = \bar{B}_{ij}/W \text{ for } i = x, y,$$

$$\kappa_o = \frac{B_{yy}K_{xx} + B_{xx}K_{yy} - B_{yx}K_{xy} - B_{xy}K_{yx}}{B_{yy} + B_{xx}} \tag{14.6}$$

The threshold (critical) speed coefficient (C_ω) and ratio of whirl (critical) coefficient (C_Ω) for a layered bearing (journal) are expressed in terms of critical (threshold) speed ($\omega_{s,l}$, $\omega_{s,h}$) and ratio of whirl (critical) ($\Omega_{s,l}$, $\Omega_{s,h}$) of layer film and uniform film respectively as

$$C_\omega = \frac{\omega_{s,l}}{\omega_{s,h}} = \frac{1}{\Delta_s} \text{ and } C_\Omega = \frac{\Omega_{s,l}}{\Omega_{s,h}} = \Delta_s \tag{14.7}$$

The threshold (critical) speed ($\omega_{s,l}$) and ratio of whirl (critical) ($\Omega_{s,l}$) of surface layered journal bearing are influenced by nondimensional damping coefficients as shown in Eq. (14.2). The critical (threshold) speed (ω_{sh}) and whirl ratio (critical) (Ω_{sh}) for lubricant homogenous film are found by substituting $\Delta_s = \Delta_p = 1$.

14.2.2 Stability Characteristics of Surface Layered Bearing

The effect of surface film layer on the bearing (journal) stability is examined by one-dimensional Ocvirk (short) bearing analysis for bearing length-to-diameter ratio (L/D) of 0.5. Results of nondimensional damping coefficients (B_{ij}), threshold (critical) speed ($\omega_{s,l}$) and ratio of whirl (critical) ($\Omega_{s,l}$) of layer film are presented for journal bearing and surface layered film parameters in Tables 14.1 and 14.2 respectively.

Figure 14.2 depicts stiffness (nondimensional) coefficients with bearing eccentricity ratio for base (homogeneous) film of a bearing (journal) obtained using Eq. 14.2. The stiffness (nondimensional) coefficients are unaltered by lubricant layered film considerations as shown in Eq. 14.2.

Table 14.1 Journal bearing parameters

Parameter	
Static load (w)	300 N
Bearing slenderness ratio (L/D)	0.5
Fluid viscosity (μ_f)	0.02 Ns/m^2
Radius of journal (R)	0.025 m
Radial bearing clearance (C)	0.025×10^{-3} m
Angular velocity of journal (ω)	10–300 rad/s

Table 14.2 Surface layered bearing (journal) parameters

Parameter	
Nondimensional thickness (film) ratio of surface film adjacent to bearing surface to film thickness (γ_1)	0.02–0.2
Viscosity (dynamic) ratio of bearing surface film layer to base homogeneous fluid (β_1)	2–5

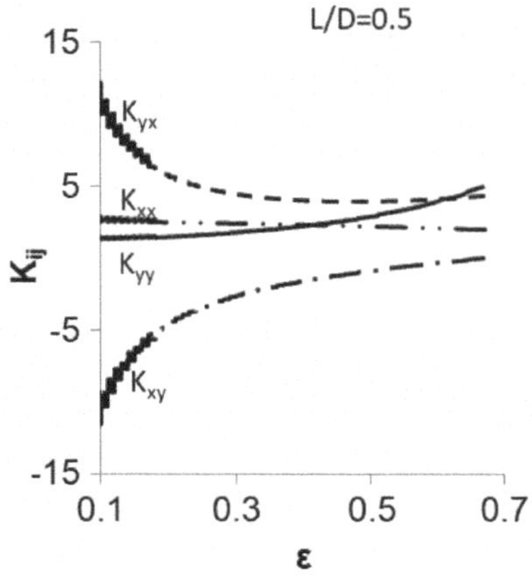

Fig. 14.2 Nondimensional stiffness coefficients of rotor supported on bearing (journal)

Figure 14.3a, b demonstrate the deviation of damping (nondimensional) coefficients with bearing eccentricity ratio for surface layered film of a short (Ocvirk) bearing (journal). The deviation in damping (nondimensional) coefficients with lubricant surface (layer) film is influenced by the parameter ($\frac{1}{\Delta_s}$) as shown in Eq. (14.2).

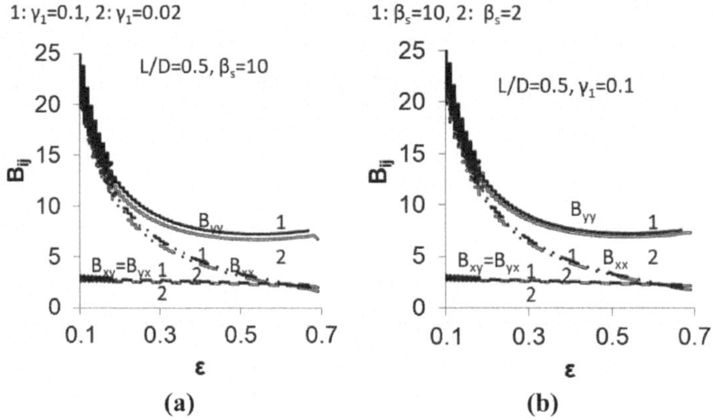

Fig. 14.3 Nondimensional damping coefficients of surface layered bearing (journal)

The nondimensional direct damping coefficient (B_{yy}) rises with increasing nondimensional thickness (film) ratio of surface film adjacent to bearing to film thickness (γ_I) and viscosity (dynamic) ratio of bearing surface film layer to base homogeneous fluid (β_1). Higher γ_I has substantial effect on rise in B_{yy}.

Figure 14.4a, b indicate the change in eccentricity ratio of surface layered bearing (journal) with angular journal velocity (ω). Figure 14.4 shows that eccentricity ratio decreases with increase in both nondimensional thickness (film) ratio of film (surface) adjacent to bearing to film thickness (γ_I) and viscosity (dynamic) ratio of bearing film (surface) to base homogeneous fluid (β_1). Higher γ_I has substantial influence on decrease in eccentricity ratio in comparison with increase in β_1.

Figure 14.5a, b illustrate the deviation in critical (threshold) speed (ω_s) of surface layered journal bearing with eccentricity ratio. Figure 14.5 shows that critical (threshold) speed (ω_s) increases with increase in both nondimensional thickness (film) ratio of film (surface) adjacent to bearing surface to film thickness (γ_I) and viscosity (dynamic) ratio of bearing film (surface) to base uniform fluid (β_1). Both higher γ_I

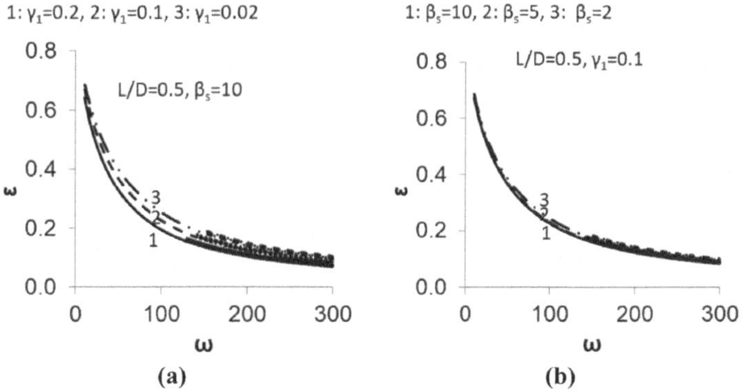

Fig. 14.4 Eccentricity ratio of surface layered bearing (journal)

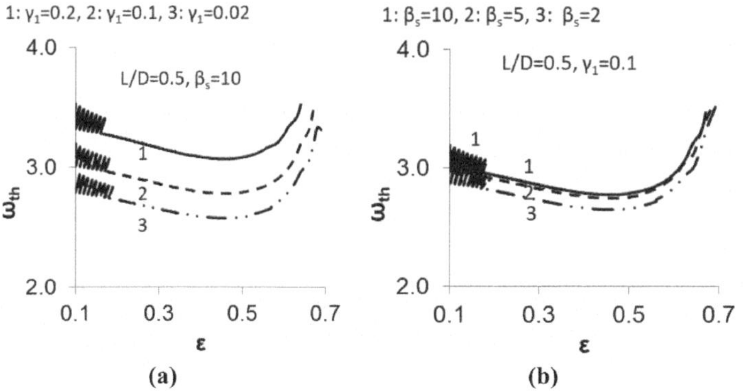

Fig. 14.5 Threshold speed of surface layered bearing (journal)

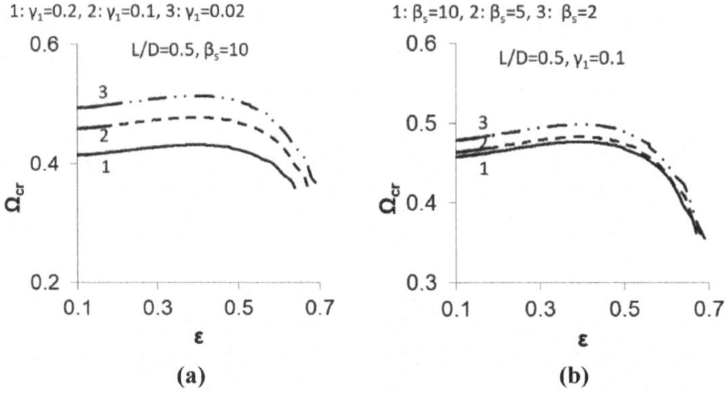

Fig. 14.6 Critical whirl frequency ratio of surface layered bearing (journal)

and β_1 has superior effect on increase in critical (threshold) speed (ω_s). The fluid flow resistance increases with increasing viscosity and thickness of surface (layer) film of bearing.

Figure 14.6a, b display the change in whirl ratio (critical) (Ω_s) of surface layered (journal) bearing with eccentricity ratio. Figure 14.6 shows that ratio of whirl (critical) (Ω_s) decreases with increase in both γ_I and β_1. Lower whirl ratio (critical) (Ω_s) and higher stability limits are obtained with both higher γ_I and β_1. Both stability limit and whirl ratio (critical) are influenced by the variation of whirl ratio (critical) coefficient (C_Ω).

14.3 Analysis of Rotor Stability Supported by Porous Layered Bearing

A porous film (layer) bearing (journal) one-dimensional stability analysis considering using short (Ocvirk) bearing approximations is investigated. The short (Ocvirk) bearing approximation based nondimensional stability coefficients are derived from Rao et al. [11, 12]. The nondimensional parameters of porous layer film considered in the stability performance analysis are: nondimensional thickness (film) ratio of film (porous) adjacent to bearing to film thickness (γ_I) and permeability (nondimensional) of porous film (K_1) ($K_1 = k_1/h^2$). Results of damping coefficients (B_{ij}), threshold (critical) speed ($\omega_{s,l}$) and ratio of whirl (critical) ($\Omega_{s,l}$) of layer film are presented with steady state eccentricity ratios (ε) evaluated based on journal speed variations.

14.3.1 Dynamic Analysis of Porous Film (Layer) Bearing

The schematic of a porous film (layer) bearing (journal) is similar to surface film (layer) bearing (journal) depicted in Fig. 14.1 with nondimensional thickness (film) ratio of film (porous) adjacent to bearing to film thickness (γ_1) and permeability (nondimensional) of porous film (K_1).

The coefficients Δ_s, Δ_p for porous layer journal bearing are expressed as

$$\Delta_s = 2F_1 H_1^* + (F_1 + 1)(1 - \gamma_1) \tag{14.8}$$

$$\Delta_p = 12F_2 H_1^* + 6F_2(1 - \gamma_1) + (1 - \gamma_1)^3 + 12K_1\left(\gamma_1 - 2H_1^*\right) \tag{14.9}$$

where

$$F_1 = \frac{1}{(1-\gamma_1)\left[\frac{1}{\sqrt{K_1}}\coth\left(\frac{\gamma_1}{\sqrt{K_1}}\right) + \frac{1}{(1-\gamma_1)}\right]},\ F_2 = \left[\frac{H_1^* + \frac{1}{2}(1-\gamma_1)}{\frac{1}{\sqrt{K_1}}\coth\left(\frac{\gamma_1}{\sqrt{K_1}}\right) + \frac{1}{(1-\gamma_1)}}\right],$$

$$H_1^* = \sqrt{K_1}\left[\coth\left(\frac{\gamma_1}{\sqrt{K_1}}\right) - \operatorname{csch}\left(\frac{\gamma_1}{\sqrt{K_1}}\right)\right] \tag{14.10}$$

14.3.2 Stability Characteristics of Porous Film (Layer) Bearing

The instability of bearing (journal) considering the influence of porous layer film is examined by one-dimensional Ocvirk (short) bearing technique for bearing L/D (length to diameter) ratio of 0.5. Results of nondimensional damping coefficients (B_{ij}), threshold (critical) speed ($\omega_{s,l}$) and ratio of whirl (critical) ($\Omega_{s,l}$) of layer film are presented for journal bearing and porous layered film parameters in Tables 14.1 and 14.3 respectively.

Figure 14.7a, b demonstrate the deviation of damping (nondimensional) coefficients for porous layered film of a short bearing (journal) with eccentricity ratio. The direct damping (nondimensional) coefficient (B_{yy}) increases with increase in nondimensional thickness (film) ratio of film (porous) adjoining to bearing to film

Table 14.3 Porous layered bearing (journal) parameters

Parameter	
Nondimensional thickness (film) ratio of porous film adjacent to bearing surface to film thickness (γ_l)	0.02–0.2
Permeability (nondimensional) of porous film (K_1)	10^{-1}–10^{-5}

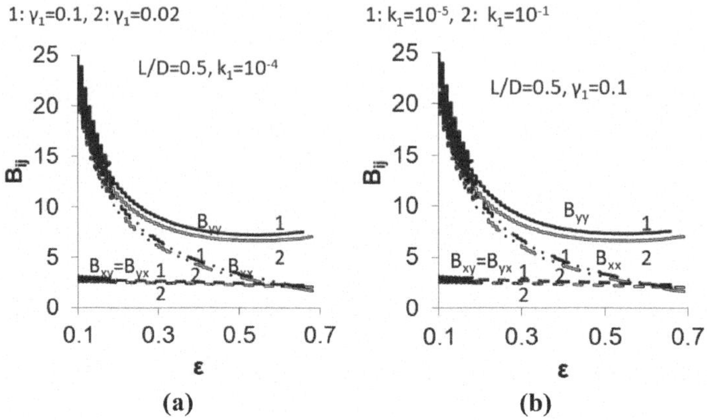

Fig. 14.7 Nondimensional damping coefficients of porous layered bearing (journal)

thickness (γ_I) and decrease in permeability (nondimensional) of porous film (K_1). Higher γ_I and lower K_1 results in significant increase in B_{yy}. The increasing fluid flow resistance in porous film (layer) gives to increasing direct damping coefficient. The influence of γ_1 on the increase in B_{yy} considering permeability of porous film (layer) in bearing (journal) is similar to bearing (journal) with high viscosity ratio adsorbent surface film (layer).

Figure 14.8a, b indicate change in eccentricity ratio of with angular velocity of journal (ω) of porous layered (journal) bearing. Figure 14.8 shows that eccentricity ratio decreases with increase in nondimensional thickness (film) ratio of film (porous) adjoining to bearing to film thickness (γ_I) and decrease in permeability (nondimensional) of porous film (K_1). Lower K_1 has notable effect on decreasing eccentricity ratio with higher γ_I.

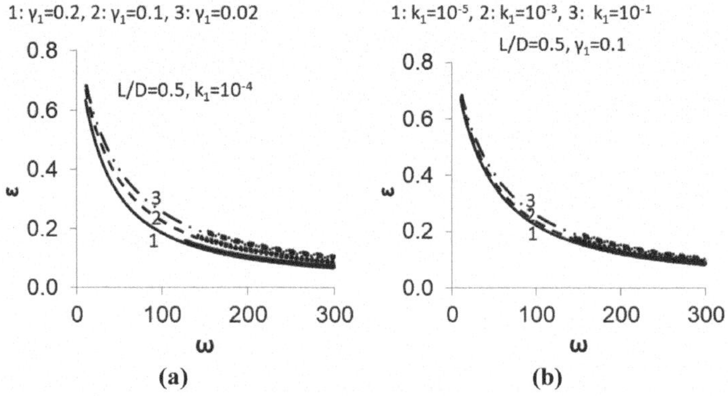

Fig. 14.8 Eccentricity ratio of porous layered bearing (journal)

Figure 14.9a, b illustrate porous layered journal bearing critical (threshold) speed (ω_s) deviation with eccentricity ratio. As shown in Fig. 14.9, critical (threshold) speed (ω_s) increases with increasing nondimensional thickness (film) ratio of film (porous) adjoining to bearing to film thickness (γ_l) and decrease in permeability (nondimensional) of porous film (K_1). Lower K_1 and higher γ_l have significant effect on increasing critical (threshold) speed (ω_s). The increase in critical (threshold) speed results due to opposition to flow in porous film (layer) with low permeability and high thickness.

Figure 14.10a, b display change in whirl ratio (critical) (Ω_s) with eccentricity ratio of porous (journal) bearing. As shown in Fig. 14.10, increasing nondimensional thickness (film) ratio of film (porous) adjoining bearing to film thickness (γ_l) and decreasing permeability (nondimensional) of porous film (K_1) results in decrease in critical whirl frequency ratio (Ω_s). Lower whirl ratio (critical) (Ω_s) and higher stability limits are obtained with lower K_1 and higher γ_l.

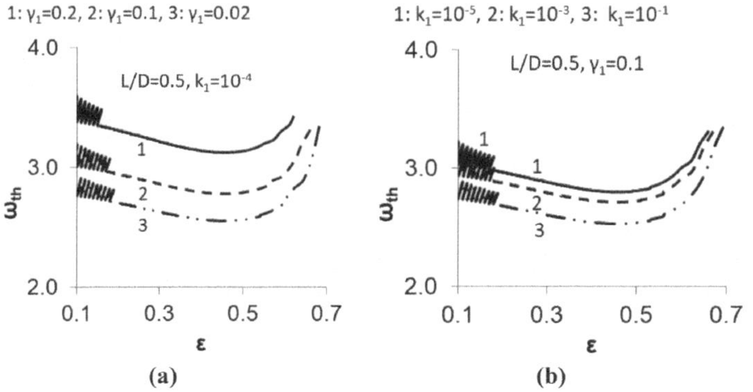

Fig. 14.9 Threshold speed of porous layered bearing (journal)

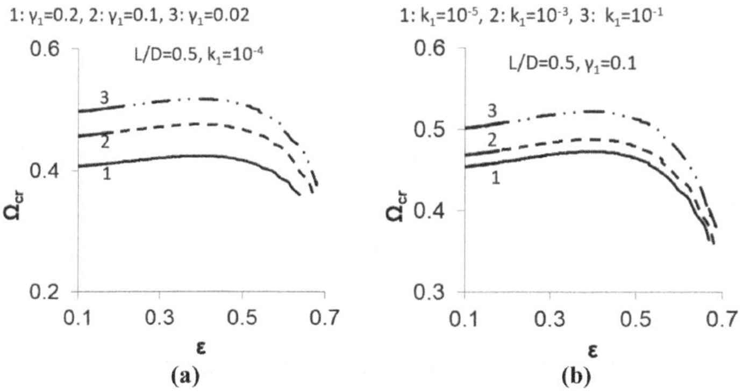

Fig. 14.10 Critical whirl frequency ratio of porous layered bearing (journal)

14.4 Analysis of Rotor Stability Supported by Surface Porous Layered Bearing

A stability analysis of surface porous film (layer) bearing (journal) taking into account one-dimensional short (Ocvirk) bearing approximation is presented. The nondimensional stability coefficients of surface porous film are derived from Rao et al. [11, 12]. The nondimensional parameters of surface porous layer film considered are: nondimensional thickness (film) ratio of film (surface) adjacent to bearing to film thickness (γ_I); nondimensional thickness (film) ratio of film (porous) adjacent to bearing to film thickness (γ_2) ($\gamma_2 = \delta_2/h$) and permeability (nondimensional) of film (porous) adjacent to bearing film (surface) (K_2) ($K_2 = k_2/h^2$). The variation of damping coefficients (B_{ij}), critical (threshold) speed ($\omega_{s,l}$) and whirl ratio (critical) ($\Omega_{s,l}$) of layer film are presented with steady state eccentricity ratios (ε).

14.4.1 Dynamic Analysis of Surface Porous Layered Bearing

The schematic of a surface porous film bearing (journal) with nondimensional thickness (film) ratio of surface film adjoining to bearing to film thickness (γ_1), nondimensional thickness (film) ratio of porous film adjoining to bearing film (surface) to film thickness (γ_2) and permeability (nondimensional) of porous film adjoining to bearing film (surface) (K_2) is shown in Fig. 14.11. In surface porous film bearing (journal) configuration, surface (layer) film (may also be treated as porous layer film-I of infinite permeability) is covered with porous (layer) film-II.

The coefficients Δ_s, Δ_p for surface porous (layer) film bearing (journal) are expressed as

$$\Delta_s = 2F_1 H_1^* + 2(F_1 + F_3)H_2^* + (F_3 + 1)(1 - \gamma_1 - \gamma_2) \tag{14.11}$$

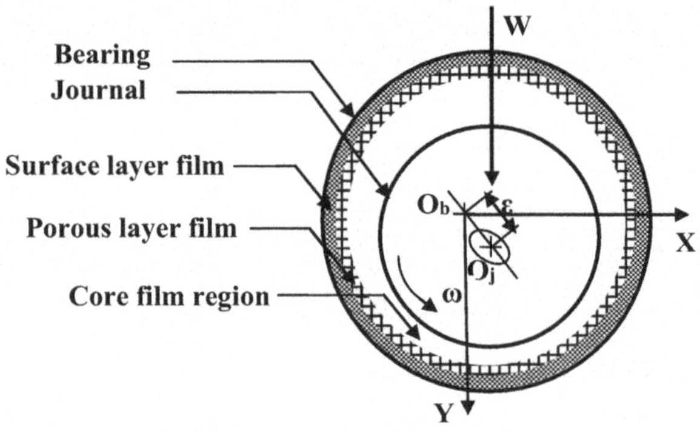

Fig. 14.11 Surface porous layered bearing (journal) geometry

$$\Delta_p = 12F_2H_1^* + 12(F_2 + F_4)H_2^* + \left[6F_4 + (1 - \gamma_1 - \gamma_2)^2\right](1 - \gamma_1 - \gamma_2) + 12K_2\left(\gamma_2 - 2H_2^*\right) + \gamma_1^3 \quad (14.12)$$

where

$$F_1 = \frac{-E_{12}E_{231}}{E_{11}E_{22} - E_{21}E_{12}},\ F_2 = \frac{E_{22}E_{13} - E_{12}E_{232}}{E_{11}E_{22} - E_{21}E_{12}}$$
$$F_3 = \frac{E_{11}E_{231}}{E_{11}E_{22} - E_{21}E_{12}},\ F_4 = \frac{-E_{21}E_{13} + E_{11}E_{232}}{E_{11}E_{22} - E_{21}E_{12}} \quad (14.13)$$

$$E_{11} = \frac{1}{\gamma_1} + \frac{1}{\sqrt{K_2}}\coth\left(\frac{\gamma_2}{\sqrt{K_2}}\right),\ E_{22} = \frac{1}{\sqrt{K_2}}\coth\left(\frac{\gamma_2}{\sqrt{K_2}}\right) + \frac{1}{(1 - \gamma_1 - \gamma_2)}$$
$$E_{12} = E_{21} = -\frac{1}{\sqrt{K_2}}\operatorname{csch}\left(\frac{\gamma_2}{\sqrt{K_2}}\right) \quad (14.14)$$

$$E_{13} = H_1^* + H_2^*,\ E_{231} = \frac{1}{(1 - \gamma_1 - \gamma_2)},\ E_{232} = H_2^* + \frac{1}{2}(1 - \gamma_1 - \gamma_2) \quad (14.15)$$

$$H_1^* = \frac{\gamma_1}{2},\ H_2^* = \sqrt{K_2}\left[\coth\left(\frac{\gamma_2}{\sqrt{K_2}}\right) - \operatorname{csch}\left(\frac{\gamma_2}{\sqrt{K_2}}\right)\right],\ H = 1 + \varepsilon\cos\theta \quad (14.16)$$

14.4.2 Stability Characteristics of Surface Porous Layered Bearing

Stability analysis of surface-porous film bearing is presented. Results of nondimensional damping coefficients (B_{ij}), threshold (critical) speed ($\omega_{s,l}$) and ratio of whirl (critical) ($\Omega_{s,l}$) of film are presented based on one-dimensional short bearing analysis for bearing L/D (length to diameter ratio) of 0.5. The journal bearing and surface porous layered film parameters used in analysis are presented in Tables 14.1 and 14.4 respectively.

Table 14.4 Surface porous layered journal bearing parameters

Parameter	
Nondimensional thickness (film) ratio of surface film adjoining to bearing to film thickness (γ_1)	0.02–0.2
Nondimensional thickness (film) ratio of porous film adjoining to bearing film (surface) to film thickness (γ_2)	0.02
Permeability (nondimensional) of porous film adjoining to bearing film (surface) (K_2)	10^{-1}–10^{-5}

Figure 14.12a, b demonstrate the deviation of short journal bearing nondimensional damping coefficients for surface porous layered film. The nondimensional direct damping coefficient (B_{yy}) increases with increase in nondimensional thickness (film) ratio of surface film adjoining to bearing to film thickness (γ_1) and decrease in permeability (nondimensional) of porous film adjoining to bearing film (surface) (K_2). Higher γ_1 and lower K_2 results in significant increase in direct damping coefficient (B_{yy}). The increasing flow resistance in layer (porous) adjoining to bearing film (surface layer) gives to increasing direct damping coefficient. The influence of γ_1 is higher compared to γ_2 on the increase in direct damping coefficient (B_{yy}).

Figure 14.13a, b indicate eccentricity ratio change with angular velocity of journal for surface porous layered journal bearing. Figure 14.13 shows that low permeability of film (porous) adjacent to bearing film (surface) (K_2) has superior influence on decrease in eccentricity ratio compared to nondimensional thickness (film) ratio of surface film adjoining to bearing to film thickness (γ_1).

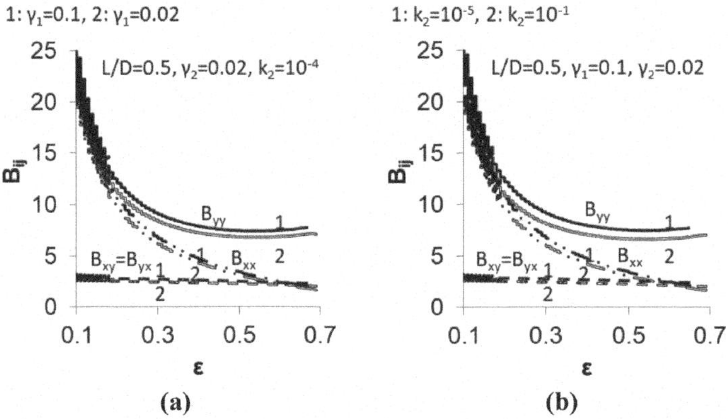

Fig. 14.12 Nondimensional damping coefficients of surface porous layered bearing (journal)

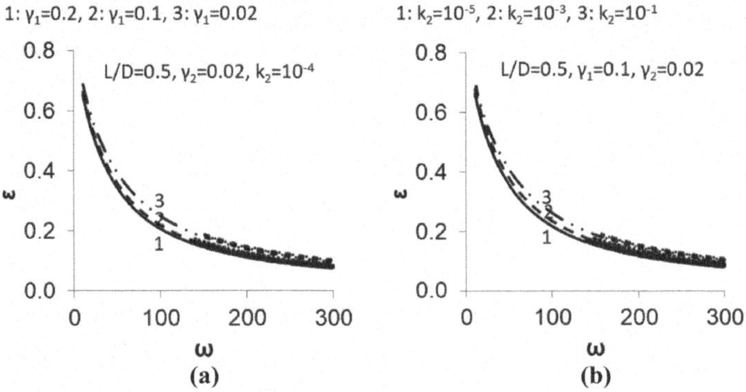

Fig. 14.13 Eccentricity ratio of surface porous layered bearing (journal)

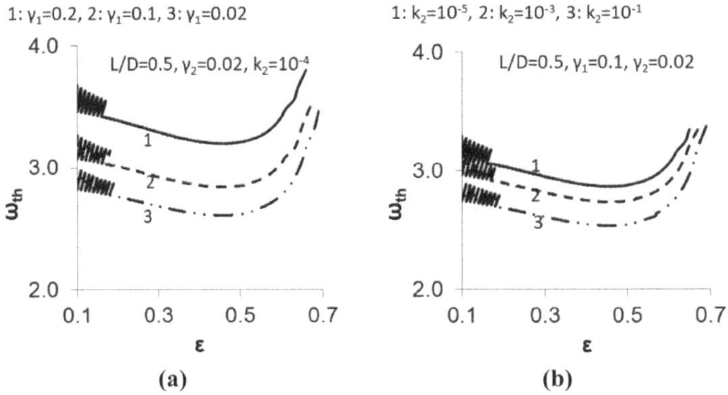

Fig. 14.14 Threshold speed of surface porous layered bearing (journal)

Figure 14.14a, b illustrate critical (threshold) speed (ω_s) deviation with eccentricity ratio of surface porous layered bearing. Figure 14.14 shows that critical (threshold) speed (ω_s) increases with increase in nondimensional thickness (film) ratio of surface film adjoining to bearing to film thickness (γ_1) and decrease in permeability (nondimensional) of porous film adjoining to bearing film (surface) (K_2). Similar to porous layered configuration, lower permeability (nondimensional) of porous film adjoining to bearing film (surface) (K_2) and higher nondimensional thickness (film) ratio of surface film adjoining to bearing to film thickness (γ_1) has superior influence on increase in threshold speed. Lower K_2 and higher γ_1 have significant effect on increase in critical (threshold) speed (ω_s) compared to nondimensional thickness (film) ratio of porous film adjoining to bearing film (surface) to film thickness (γ_2). The increase in threshold (critical) speed results due to higher opposition to flow in layer (porous film) with lower permeability of porous film (layer) and higher thickness of surface layer.

Figure 14.15a, b display change of whirl ratio (critical) (Ω_s) with eccentricity ratio for surface porous film bearing. Lower K_2 and higher γ_1 results in lower whirl ratio (critical) (Ω_s). Lower whirl ratio (critical) (Ω_s) and higher stability limits are obtained with lower K_2 and higher γ_1 similar to porous layered configuration.

The significance of lower K_2 and higher γ_1 is superior compared to nondimensional thickness (film) ratio of porous film adjoining to bearing film (surface) to film thickness (γ_2) on decrease in whirl ratio (critical) (Ω_s) of surface porous layer journal bearing.

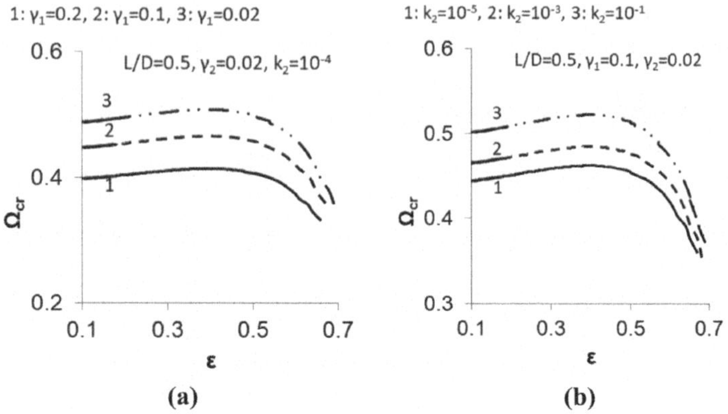

Fig. 14.15 Critical whirl frequency ratio of surface porous layered bearing

14.5 Conclusion

The stability of surface/porous/surface-porous film bearing is presented. Dynamic coefficients, threshold (critical) speed and whirl ratio (critical) for layered bearing (journal) design are obtained using short (Ocvirk) bearing approximation (one-dimensional analysis).

For a surface layer bearing (journal), stability limits are improved for both higher nondimensional thickness (film) ratio of film (surface) adjoining to bearing to film thickness (γ_1) and viscosity (dynamic) ratio of bearing film (surface) to base uniform fluid (β_1). The stability improvements for a porous film bearing are obtained for higher thickness (film) ratio of film (porous) adjoining to bearing to film thickness (γ_1) and lower permeability (nondimensional) of porous film (K_1). The stability improvements of surface-porous layer bearing are analogous to those attained for surface/porous film (layer) bearing (journal) such as higher nondimensional thickness (film) ratio of surface film adjoining to bearing to film thickness (γ_1) and lower permeability (nondimensional) of porous film adjoining to bearing film (surface) (K_2). In case of surface-porous layer configuration, higher nondimensional thickness (film) ratio of surface film adjoining to bearing to film thickness (γ_1) has superior influence compared to nondimensional thickness (film) ratio of porous film adjoining to bearing film (surface) to film thickness (γ_2) on the stability enhancement of bearing (journal).

Surface-porous film (film (adsorbent to bearing) covered with low permeability layer (porous film)) causes in increase in the stability.

References

1. A.Z. Szeri, Composite-film hydrodynamic bearings. Int. J. Eng. Sci. **48**(11), 1622–1632 (2010)
2. J.A. Tichy, A surface layer model for thin film lubrication. Tribol. Trans. **38**, 577–582 (1995)
3. T.V.V.L.N. Rao, Stabilization of journal bearing using two-layered film lubrication. J. Tribol. **134**, 014504 (2012)
4. J.A. Tichy, A porous media model for thin film lubrication. J. Tribol. **117**, 16–21 (1995)
5. W-L. Li, Derivation of modified reynolds equation—a porous media model. J. Tribol. 121, 823–828 (1999)
6. W.-L. Li, H.-M. Chu, Modified reynolds equation for couple stress fluids—a porous media model. Acta Mech. **171**, 189–202 (2004)
7. A.A. Elsharkawy, Effects of lubricant additives on the performance of hydrodynamically lubricated journal bearings. Tribol. Lett. **18**, 63–73 (2005)
8. J.-R. Lin, C.-C. Hwang, Linear stability analysis of finite porous journal bearings—use of the brinkman-extended darcy model. Int. J. Mech. Sci. **36**, 645–658 (1994)
9. V. D'Agostino, A. Ruggiero, A. Senatore, Unsteady oil film forces in porous bearings: analysis of permeability effect on the rotor linear stability. Meccanica **44**, 207–214 (2009)
10. N. Saha, B.C. Majumdar, Steady state and stability characteristics of hydrostatic two-layered porous oil journal bearings. Proc. Inst. Mech. Eng. Part J **218**, 99–108 (2004)
11. T.V.V.L.N. Rao, A.M.A. Rani, T. Nagarajan, F.M. Hashim, Analysis of journal bearing with double-layer porous lubricant film: influence of surface porous layer configuration. Tribol. Trans. **56**(5), 841–847 (2013)
12. T.V.V.L.N. Rao, A.M.A. Rani, M. Awang, T. Nagarajan, F.M. Hashim, Stability analysis of double porous and surface porous layer journal bearing. Tribol. Mater. Surf. Interf. **10**(1), 19–25 (2016)
13. J.W. Lund, K.K. Thomsen, A calculation method and data for the dynamic coefficients of oil-lubricated journal bearings, in *Topics in 'Fluid Film Bearing and Rotor Bearing System Design and Optimization* (ASME, New York, 1978), pp. 1–28
14. J.S. Rao, *Rotor Dynamics* (New Age International (P) Limited, New Delhi, 2009)
15. H.G. Elrod, D.J. Vijayaraghavan, A stability analysis for liquid-lubricated bearings incorporating the effects of cavity flow: part i: classical one-dimensional journal bearing. J. Tribol. **116**(2), 330–335 (1994)
16. J.W. Lund, Review of the concept of dynamic coefficients for fluid film journal bearings. J. Tribol. **109**, 37–41 (1987)

Chapter 15
Tribological Effects of Diesel Engine Oil Contamination on Steel and Hybrid Sliding Contacts

Ramkumar Penchaliah

Abstract In advanced automotive engines, especially in diesel engines, consumer demand for ever increasing service intervals for vehicles has led to longer oil drain periods. Consequently this has increased contamination levels in lubricating oils that will in turn reduce engine efficiency and increase the possibility of system failure due to increases in viscosity and the potential of oil starvation leading to scuffing and catastrophic failure of the engine. Therefore it is necessary to understand the effects of contaminants in diesel engine oil on the tribological performance of tribo-contacts and also the possible interaction between the contaminants. The paper aims to investigate the influence of contaminants and their interactions on diesel engine oil using Electro sensing (ES) monitoring. Using pin-on-disc (PoD) tribometer, all tests were carried out under ambient conditions at 5 m/s sliding speed and contact stress of 1.5–2.05 GPa to simulate a valve-train in a diesel engine with fully formulated heavy-duty diesel engine oil used as lubricant. In the first phase, using a parametric study examining the effect of four contaminants (soot, oxidation, moisture, and sulphuric acid) at varying levels (four for each) on steel-on-steel sliding contact. It was observed that all contaminants and contaminant levels reduce the conductivity of the oil. Oxidation and soot contaminants produced large increases in viscosity. The wear rate was mainly influenced by acid and soot additions, while the coefficient of friction was increased by all contaminants and contaminant levels. The steady-state charge levels changed for some contaminants. The best correlation of steady-state charge with the other measured tribological parameters of wear rate, friction, and temperature is seen for the series of oxidized oils. The multi-contaminated oil (L4 × 4) shows remarkably little degradation in tribological performance. Analysis of the wear mechanisms shows that soot and oxidation produced abrasion and polishing wear, respectively, while sulphuric acid and moisture produced corrosive wear. In the second phase, investigates the effects of diesel contaminants and their interaction on tribological properties for bearing steel (En31) and ceramic (Si_3N_4) sliding contacts using a factorial study. The contaminants are soot, sulphuric acid, moisture

R. Penchaliah (✉)
Machine Design Section, Department of Mechanical Engineering, Indian Institute of Technology Madras, Chennai, India
e-mail: ramkumar@iitm.ac.in

J. K. Katiyar et al. (eds.), *Tribology in Materials and Applications*,
Materials Forming, Machining and Tribology,
https://doi.org/10.1007/978-3-030-47451-5_15

and oxidation, and each contaminant has three different level of concentration (low, medium and high) in the test matrix. The factorial test matrix consisted of 20 tests, constructed from a quarter fractional factorial test matrix with four points at the medium values for the contaminants. Results from this matrix required six further tests to elucidate aliased pairs of interactions using Bayesian model selection. A pin-on-disc tribometer was used to carry out all the experiments. The factorial study showed that charge was influenced by tribo-couple material; the silicon nitride discs produced higher charge than steel discs. However, it was opposite for friction; the silicon nitride disc gave lower friction and the pins showed higher friction than their steel counterparts. For wear scar and temperature, soot contaminant was found to be important. The two important interactions were found for the charge response, with the interaction between sulphuric acid and pin material being more important than sulphuric acid–oxidation interaction. Similarly to charge, an interaction between sulphuric acid and pin material interaction was found for friction. To conclude, the ES monitoring was sensitive to the presence and levels of contaminants in diesel lubricating oil, particularly diesel soot. The change in charge levels indicated the concentration of soot level present in the contact, which was directly related to wear. ES monitoring also detected interactions between the contaminants through statistical analysis. ES monitoring has shown that monitoring lubricant performance and the effects of contamination are feasible under laboratory conditions.

15.1 Introduction

Vehicle emissions accounted for about 25% of all energy-related CO_2 discharges and is a prime source of pollutants influencing quality of air. In addition, internal combustion engines are responsible of other pollutants, as particulate matter, nitrogen oxides and hydrocarbons. Even though these pollutants are not considered as a universal problem, they pose threat for plant and animal life in a regional scale. Along with this depleting fossil fuel reserves also demands for another source of energy for transportation.

Electric vehicles are going to have a vital role moving forward, but as of now it is very difficult to evaluate how big role they will finally play. Various projections say by 2030 around 10–25% of vehicles will be electrified, which leaves a lion's share of vehicles to continue with I.C engines. Also, large trucks and ships may be still running on diesel. Besides, having domestic recharger for electric vehicles is problematic and expensive. This is why it is still important to work on IC engines and make them as efficient and clean as we can.

The main drawbacks of the IC engine are low thermal and mechanical efficiencies with a significant proportion of the fuel energy are dissipated as heat. Only around 15% of the total energy generated from the combustion of fuel filled in a vehicle is converted to useful power such as moving a car forward or running different accessories. The remaining portion of the energy is wasted as engine and driveline losses and during idling. Therefore, there is enormous potential to improve efficiency of engine with advanced technologies.

15.1.1 Energy Losses in a Vehicle

(a) Losses in Engine

In gasoline fueled automobiles, around 60% of the energy from fuel is dissipated in the engine itself. IC engines possess very poor efficiency at converting the fuel's chemical energy to useful mechanical energy. Energy is wasted in an engine as friction, pumping air into and out of the engine, and wasted heat. These losses can be minimized with the use of latest engine technologies such as turbocharging, direct fuel injection, variable valve timing and lift, and cylinder deactivation. In addition, diesel engines are having 30–35% more efficiency than petrol engines, and new advances in diesel engine technologies and fuels are making these vehicles more appealing to customers.

(b) Accessories

Around 2% of energy produced in the engine are consumed by accessories such as power steering, air conditioning etc. More efficient accessories can contribute up to 1% increase in overall fuel economy in the case of automobiles.

(c) Driveline Losses

Transmission and other driveline parts create a loss of around 5% energy produced by fuel. Technologies, such as continuously variable transmission and automated manual transmission are capable of reducing these losses.

(d) Aerodynamic Drag

A vehicle has to do some work to push air out of the way as it moves forward and energy needed to perform this work increases as speed of vehicle increases. Around 2–3% of energy is used to overcome aerodynamic drag in vehicles. Drag mainly depends on the vehicle's shape. Modern automobiles are designed considering aerodynamics, but further reductions of 20–30% are possible in the case of most vehicles.

(e) Rolling Resistance

Rolling resistance is the force that is needed to move the tire forward. It is directly proportional to the load that is acting on the tire supports. Some of the techniques used to minimize rolling resistance includes using better performing tire thread and shoulder designs, and selecting advanced materials in the tire belt and traction surfaces. A 5–7% reduction in rolling resistance is reported to increase the fuel efficiency by 1% in automobiles.

(f) Inertia and Braking Losses

While moving along a road, a vehicle has to spend some energy to overcome inertia forces which comes from its own weight. For reducing inertia losses, overall weight of the vehicle should be reduced by using lightweight materials and optimized design. Also, while braking, the energy initially used to overcome inertia is lost.

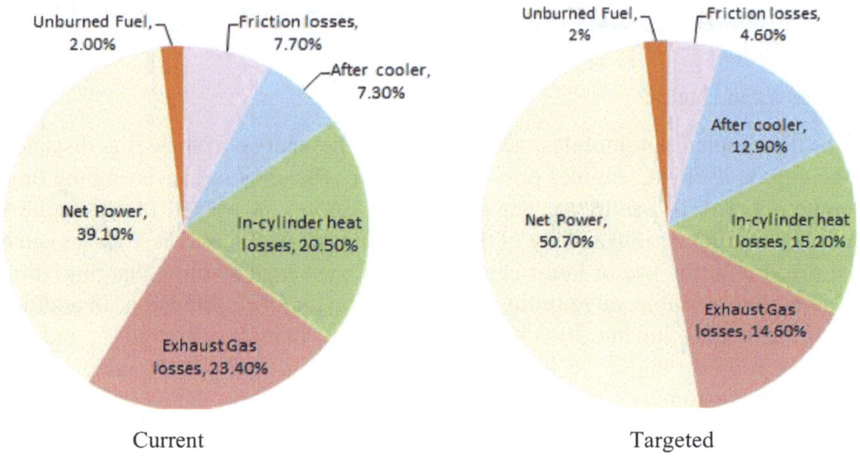

Fig. 15.1 Energy distribution diagram showing frictional losses in an I.C engine

15.1.2 Energy Losses in Engine

Around 60% of energy losses in a vehicle occur in the engine itself. The figure below indicates the distribution of different losses in engine. The overall power output in an engine can be increased by 10% by minimising major losses such as frictional losses, exhaust gas losses and heat losses [1] (Fig. 15.1).

About 8% of overall energy loss occurring in an automobile engine is contributed by friction. The prime subsystems of automobile engine contributing to frictional losses include piston ring-cylinder liner system, bearings in crank shaft system and valve drives. The spread of loss among these sub-systems are shown in the above diagram, nevertheless the exact values depend on the engine type and conditions of use.

15.1.3 Engine Systems Causing Friction

The Stribeck curve shown below indicates various parts in an engine and the lubrication regime they go through during each stroke of their operation (Fig. 15.2).

(a) **Piston Assembly System**

The major components of a piston assembly system are piston, piston rings, piston pin, connecting rods and bearings. The friction generated from piston assembly system can be classified as contributions from three major groups (a) piston skirt reciprocating against liner, (b) ring pack moving up and down against cylinder liner, (c) bearings located in connecting rods and wrist pins.

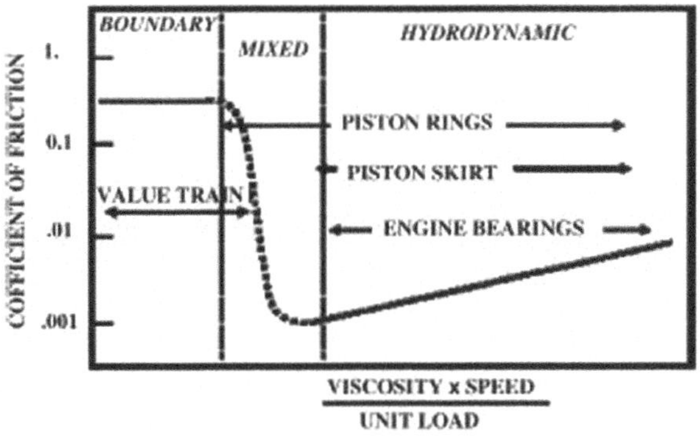

Fig. 15.2 Engine components with lubrication regime they operate. Reproduced with the permission of Ref. [3]

The piston ring pack is considered as one of the most complex frictional elements in an automobile engine. It is exposed to highly alternating load, temperature, speed and oil supply. In a single stroke piston itself, the piston ring cylinder liner (PRCL) system go through boundary, mixed and hydrodynamic lubrication regimes [2].

It is generally assumed that during middle portion of every stroke of I.C engine, piston ring-cylinder liner contact is in hydrodynamic lubrication regime. However, when the piston motion reaches very slow speeds near top and bottom dead centers, the piston speed approaches near zero values and is not sufficient to develop hydrodynamic lubrication. The tribofilm become very thin and results in excessive cylinder liner wear near TDC and BDC. This severe wear on the surface could affect the sealing action of ring-liner system which may finally result in the formation of excessive blow-by gas and increased consumption of fuel.

(b) **The Crankshaft and Connecting-rod Bearing Systems**

The main bearings in crank shaft, connecting rod crank shaft interface, connecting rod and piston pin etc. are operating under journal bearing lubrication. The primary source of friction in crank shaft assembly is the main bearings that support crank shaft in its motion. The main bearings generally operate in hydrodynamic lubrication regime. The continuous supply of oil to the bearings is achieved through oil feeds along the crank shaft. This adequate lubrication is necessary to make sure that the wear in the contact is low after initial run in period. Sometimes lubricants containing particulate matter or a wrong alignment of shaft can cause excessive wear at the contact surface. Another predominant cause of bearing failures is corrosion [3].

(c) **Valve Train System**

The valve train system includes of a number of mechanical components that helps in controlling the intake and exhaust valves. Valve train system undergo all lubrication regimes during its operation from boundary to hydrodynamic. The major culprits creating friction in valve train system are camshaft bearings, interface between cam and follower, and other linear oscillatory components such as valves and lifters, guides, valve seal etc. The cam- follower interface runs mostly in boundary lubrication regime, where the action of additives is critical. [4].

15.1.4 Lubricants

Using the appropriate lubricant makes an engine run easily with low levels of friction and wear. It produces a thin lubricant film between contact surfaces, which prevents the metal surfaces from rubbing against each other and makes them less prone to wear and other surface damages. Lubricant also delay corrosion, remove excess heat away from the engine and assist in keeping the engine surfaces clean and free from contaminants.

Automobile lubricant generally consists of approximately 80% base oil and remaining part is a mixture of different additives. The additive package consists of anti-wear and extreme pressure additives, friction modifiers, antioxidants, dispersants and detergents to clean the engine, and chemicals that helps in maintaining the oil viscosity within the specified range throughout the engine's operating conditions. The base oil take these chemicals to the locations where they are required, and carry the heat away from engine components, There are both mineral and synthetic base oils available in market, and lubricants can have complete mineral base oil, complete synthetic base oil, or a mixture of mineral and synthetic oils. Engine oils consisting of such mixture are generally referred as part-synthetic oils.

Synthetic oils are produced by chemical methods to fulfil the pressing demands of modern automobile engines. As they are specifically synthesised for each application, they provide superior performance, protection and fuel efficiency than mineral oils. They are stable at elevated temperatures and remain in liquid state even at substantially lower temperatures.

Part-synthetic oils are produced by combining mineral oil and synthetic oil. They are more fuel economical and provide improved performance and protection compared to mineral oils, but their performance is inferior compared to fully synthesised oils. Mineral oils are produced from natural crude oil which is refined and treated to remove waxes and other contaminants, and used in higher viscous lubricants.

15.1.5 Causes of Lubricant Degradation

Oils will eventually degrade or become contaminated after a finite period of use need to be changed. There are multiple reasons for degradation of oil in an engine among which most common are oxidation and thermal decomposition of the oil, depletion of additives and contamination.

Oxidation: Oxidation of lubricant is a result of reaction between oil and oxygen molecules. It can lead to an increase in viscosity and eventually the formation of sludge and sediment. Additive depletion and a breakdown in the base oil can also happen due to oxidation. An increase in acid number can be used as an indicator for oxidation of oil. Additionally, oxidation can result in rust and corrosion in the equipment.

Thermal breakdown: Increase in temperature of lubricant will result in a decrease in thermal stability. The oxidation rate of oil is doubled for every 10 °C raise in oil operating temperature and, as a result the life time of the oil is halved. This phenomenon is not as severe as it sounds since lubricants have quite a long life time. Oil temperature generally becomes a critical parameter only over 65 °C and lubricants that are exposed to high temperatures for long durations are combined with additives that retard oxidation behaviour [5].

Additive depletion: Additives pack in an engine oil typically constitutes of more than twenty components that are blended with the base oil to enhance existing properties or impart new properties. Nearly in all cases, additives in the lubricant follow sacrificial mechanism, that is they get used up during the lifetime of the oil. After a finite period, the additives become depleted and inactive and eventually the lubricant will need to be replaced before the friction and wear at the contact become too high.

Contamination: Contamination such as carbon, moisture, air, etc., can substantially affect the rate of lubricant decomposition. Fine metal particles sometimes act as a catalyst that speeds up the breakdown process of the lubricant. Contaminants such as air and water supplies oxygen that reacts with the lubricant and results in oxidation of oil. Oil analysis helps in monitoring contamination levels of the lubricant.

15.2 Effect of Lubricant Contaminants in Friction and Wear

In this chapter, the influence of different lubricant contaminants on tribological characteristics of the contact surfaces is discussed. In the first study, concentrations of different contaminants like soot, sulphuric acid, moisture and oxidation level were varied on a steel on steel contact in a pin-on-disc experiment. The second study was carried out by varying the pin material with different levels of soot contamination.

15.2.1 The Effects of Diesel Contaminants on Friction and Wear Characteristics on Steel Contacts During Sliding Contact

This study will examine the effect of four of these contaminants: soot, water, oxidation, and sulphuric acid. Soot and water were chosen as these are the most common contaminants, while oxidation and sulphuric acid representation long-term degradation mechanisms. Soot is a major contaminant in diesel engines with levels being increased dramatically by the introduction of exhaust gas recirculation, EGR [6]. Major contaminants along with their sources and major problems are listed in Table 15.1.

Table 15.1 Lubricant contaminants types and their origins [5]

Contamination	Primary sources	Major problems
Soot	Blow-by gases	Interfere with additives, abrasive wear, heavy deposits, oil thickening/gelation
Water	Combustion blow-by, leakage of coolant	Metal corrosion, Promotes lubricant breakdown
Oxidation of oil	thermal degradation/ contact with atmospheric air	Oil thickening
Acids	Combustion blow-by, Lubricant breakdown	Metal corrosion, autocatalysis of lubricant breakdown
Metallic particles	Component wear	Abrasion, surface roughening leading to adhesion, catalysis of lubricant breakdown
Metal oxides	Component wear, Oxidation of metallic particles	Abrasion, surface roughening leading to adhesion
Minerals (i.e. silica sand) and dirt	Induction air	Abrasion, surface roughening leading to adhesion
Exhaust gases	Combustion blow-by	Acids promoting lubricant breakdown
Glycol	Leakage of coolant	Breakdown of lubricant
Fuel	Blow-by-rich mixture	Breakdown of lubricant

EGR is an effective means of reducing NOx in light-duty diesel engines, the process involves the recirculation of a fraction of the engine exhaust gas to the engine intake to dilute the fresh mixture reducing peak burned gas temperatures, and hence, reducing NOx formation. Though EGR has proven to be effective for light-duty diesel engines, it results in a sharp increase in soot [7]. Elemental analysis of soot particulate matter from diesel oils indicates a high carbon content (88.3%) with other species [8, 9] (Fig. 15.3).

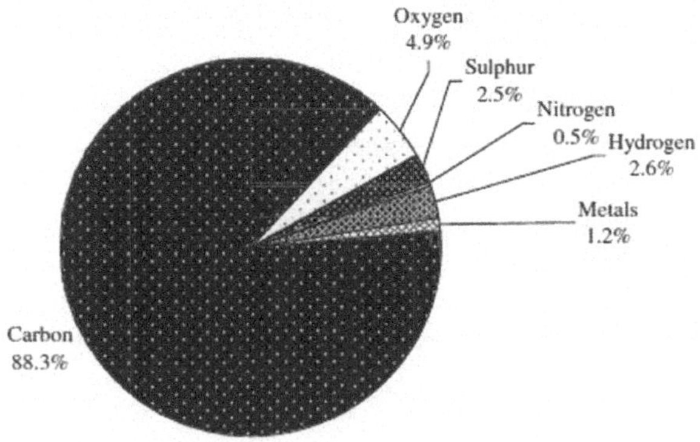

Fig. 15.3 Composition of the typical diesel contaminants by mass

Water can contaminate the lubricant by leakage through weak seals and from moisture from ambient sources including combustion and condensation. Free water may deplete polar additives from the oil [10]. The amount of water contamination has been found to vary between 0.2 and 10% by weight [11, 12].

Corrosive species found in lubricating oils include naturally occurring sulphur compounds, acidic combustion products (oxy-acids of nitrogen and sulphur), acidic oil oxidation products, water, and carbon dioxide. The use of EGR is also known to increase corrosion due to the formation of increased levels of sulphuric acid, by reaction of sulphur oxides (formed during combustion) with condensed water on the cylinder surfaces. When the acid reaches the oil sump, it reduces the total base number (TBN) of the lubricating oil and thereby reduces lubricant life [13]. Sulphuric acid is also known to break tribofilms down on cam material surfaces and cause extreme corrosive wear [14]. Oxidation is a natural phenomenon and considered the leading indicator of oil degradation [15, 16]. Oxidation leads to the formation of deposits, sludge, and corrosive by-products [17]. In an engine, oxidation products would combine in propagation reactions to form large molecules that would eventually precipitate from the oil to form varnish, sludge, acids, and salts.

Results of post-test analysis with different contaminants with varying concentrations are discussed below.

Uncontaminated oil: In the case of wear scar of the test conducted with uncontaminated oil, there is no evidence of scoring or grooving caused by abrasion processes. However, within the inlet/ entrainment region of the contact, some deeper scratches and indentations are seen and these are thought to be caused by the wear debris of the pin re-entering the contact. Thus, the predominant mechanism is plastic deformation and this agrees with work by Green et al. [18] (Fig. 15.4).

Soot: The primary wear mechanism in the case of soot contamination is abrasion irrespective of soot concentration. Different wear mechanisms proposed in the case of soot contamination in lubricant is discussed below.

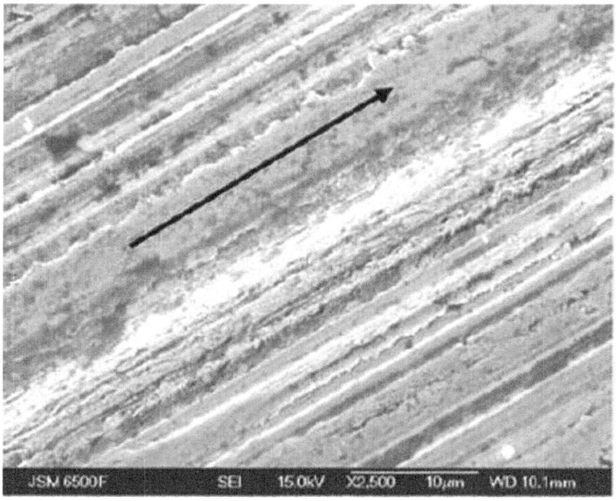

Fig. 15.4 Wear scar with uncontaminated oil [32]

Competitive adsorption: Some studies suggest that soot competes with different additives in lubricants. Soot is being adsorbed on the contact surface and prevent the adsorption of additives and cause subsequent decomposition of anti-wear films. Also, the adsorption of carbon particles on surfaces can restrict the quantity of oxygen reaching the surface resulting the formation of less wear resistant FeO instead of Fe_3O_4. This can eventually result in an increase in wear between the metallic surfaces in contact [19].

Oil starvation: In some of the researches it has been pointed out that the carbon particles in the lubricant accumulates at the contact entry region and hinders the entry of lubricant to the contact. This restricts the formation of proper lubricant layer on the rubbing interface which in turn increases the wear due to metal on metal contact [20].

Abrasion: Abrasion is believed to be the most prominent wear mechanism in the case of soot contamination. Soot particles increase the wear either by directly abrading the metal particles or by penetrating through the lubricant film. Removal of lubricant film by soot particles exposes the underlying metal surfaces to wear [21, 22].

Corrosive-Abrasive wear: According to some studies, wear rate is increased when soot contaminants interact with Sulphur or phosphorous containing additives in the lubricants. For example, ZDDP and carbon black in lubricant are having antagonistic behavior in terms of wear. That is, presence of ZDDP and carbon black together produces much higher wear than ones containing ZDDP without carbon black or carbon black oil without ZDDP [23]. Corrosive-abrasive wear mechanism proposes that the abrasive soot particles continuously remove the phosphorus containing tribofilms formed on the steel surface by ZDDP. Since creation of tribofilms involves formation of bonds between substrate with the additive, the removal of these lubricant films also results in the removal material from the substrate surface. Again, a new lubricant layer forms on the newly created surface and is then removed by soot particles as described above. This repeated cycles of tribofilm formation by the additives and its removal by carbon black contaminants leads to severe wear on the tribo-contact [22, 24].

Severity of abrasive wear is increased with increase in soot percentage in oil. For example, in an experiment with 2% soot in oil, mild abrasive wear with widely spaced groves were seen, whereas an increased wear rate, more debris formation and deeper groves were seen with 4% soot level. Also, high abrasive wear and some adhesive plucking occurred at an increased soot level of 10% (Fig. 15.5).

At the high levels of soot, dispersants can become depleted and this allows the soot particles to rapidly agglomerate to form particles with diameters well over a micron. Adhesive material removal from the worn pin surface due to agglomerated soot particles blocking oil entrainment leading to partial oil starvation. The oil film thickness is reduced and leads to boundary lubrication and in the absence of any additive tribofilms. This is followed by direct asperities contact between the contacting surfaces which increases friction and wear rate. The soot which acts as abrasive

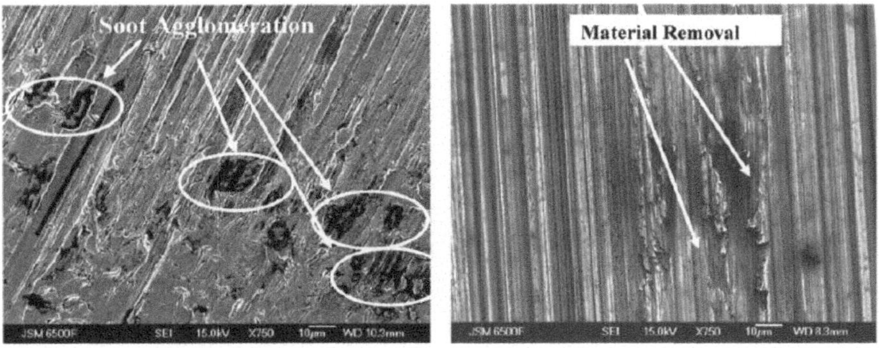

Fig. 15.5 Soot agglomeration and material removal due to adhesive wear with soot contamination [32]

particles breaks the tribofilm while rubbing against it. As tribofilms are softer compared to metals, they are more susceptible to abrasion by soot particles. Along with this, adsorption of additives by soot particles enhanced wear rate drastically [25].

Sulphuric acid: Concentration of sulphuric acid also plays a critical role in determining the wear mechanism. While testing with oil having 1.25 mM sulphuric acid, the test pin shows mild abrasive wear with a few cracks. For 2.5 mM acid level, the pin wear scar shows pitting with large perpendicular cracks and when the acid level was increased to 25 mM, it reacted with the pin surface producing corrosion and pitting wear (Figs. 15.6 and 15.7).

This could be because the detergents present in the oil might not be sufficient to neutralize the acid. Also the water in the diluted acid might have hydrolyzed ZDDP in the lubricant and further increased the amount of sulphuric acid.

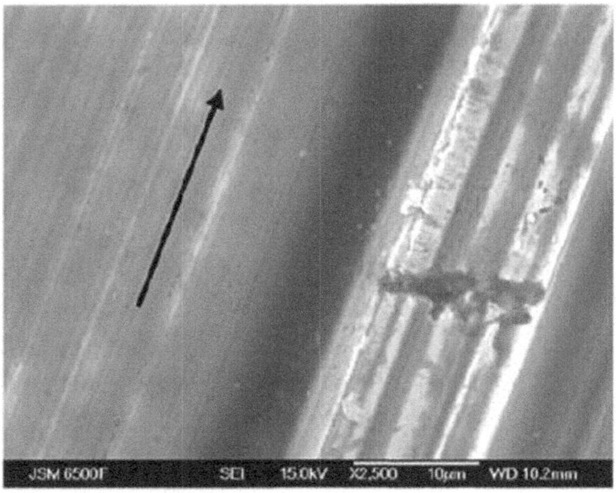

Fig. 15.6 Wear scar for 1.25 mM sulphuric acid

Fig. 15.7 Corrosive wear in 10 mM sulphuric acid [32]

Moisture content: At high levels of moisture content, predominant wear mechanism was polishing wear. Low- level moisture content shows a combination of pitting and localized corrosive wear. At low moisture concentration, water existed as droplets on the interface which caused localized corrosion. At this condition wear was even lower than uncontaminated condition (Figs. 15.8 and 15.9).

Oxidation level: Oxidation of lubricant is having is less influence on wear rate compared to other factors considered before. As oxidation increases, viscosity of lubricant also increases which in turn reduces wear rate. However, when the oxidation crossed some level, wear rate started increasing due to degradation of lubricant and absence of continuous tribo film.

Combination of contaminants: When all the contaminants (soot, sulphuric acid, moisture, oxidation level) are combined together, it did not produce the expected

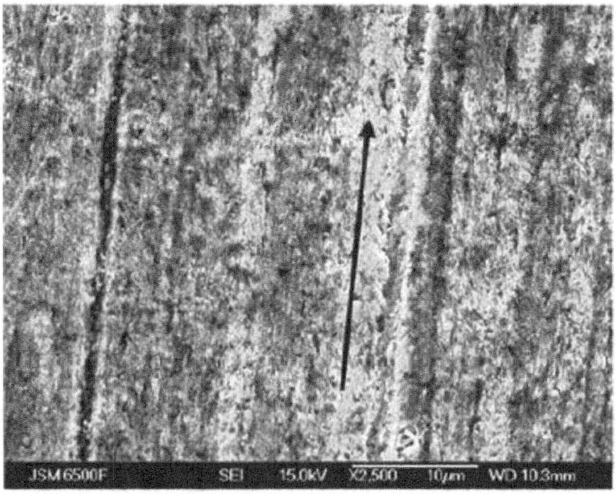

Fig. 15.8 Wear scar for 2.5% moisture content

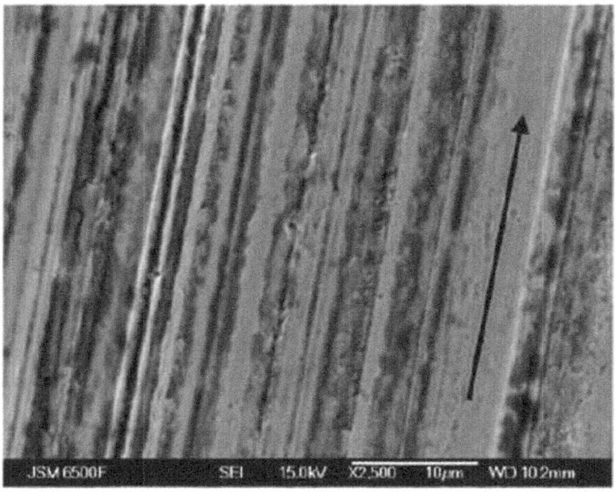

Fig. 15.9 Wear scar for 4 h oxidation level [32]

accumulated wear or friction behavior. The results seem to suggest that there is a negative synergy when all the contaminants are combined together which makes friction and wear low. But it should be noted that the value of friction and wear in this case is above that of uncontaminated condition, which indicates that the interaction are not cancelled out completely. Moisture and oxidation have minimal influence on wear. Sulphuric acid contamination exhibits more influence, especially at the higher concentrations, which is believed to be a tribo-chemical attack due to the corrosive nature of the contaminants. However, the greatest influence is observed for soot contamination, which is believed to be related to its abrasive nature [17, 18]. It is also believed that at increasing levels of soot contamination, large soot agglomerates are produced, as the dispersant is less able to effectively separate (disperse) the particles. Soot particles were found to entrain into the elasto-hydrodynamic contact by a mechanical entrapment mechanism adhering to surfaces, accumulating and forming an inhomogeneous solid-like boundary film. This boundary film is likely to influence the wear rate particularly when the soot particles are larger than the film thickness.

15.2.2 Effect of Soot on Viscosity of Oil

Viscosity of lubricant was increased linearly with increase in soot concentration. Soot particles and dispersant molecules chemically react to form long chains of hydrocarbon molecules. As a result, the average size and weight of the lubricant molecules increases. It leads to thickening of oil and ultimately in the viscosity of the oil increases. Pumping more viscous oil requires more power and also leads to more wear [26, 27].

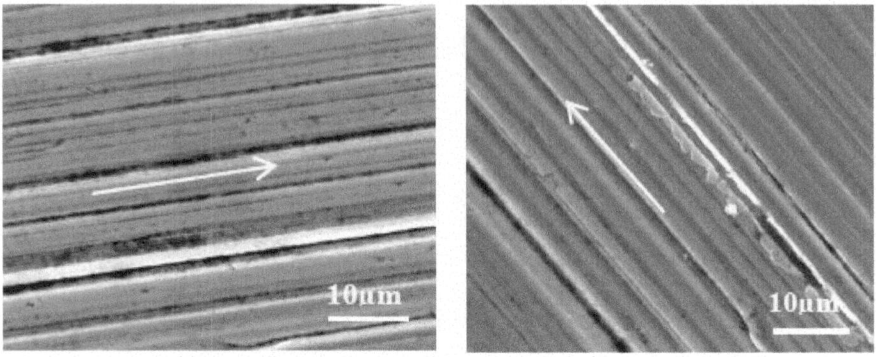

2% soot contamination 6% soot contamination

Fig. 15.10 SEM micrographs of pin wear scars for different concentration of soot contamination: steel on steel. Reproduced from my master student [33]

15.3 Tribological Behaviour of Steel and Ceramic Contacts in Presence of Soot Contamination

In this chapter, friction and wear characteristics of different materials sliding against steel is discussed. Abrasive wear was identified as the primary wear mechanism for all soot contaminated samples and as the percentage of soot in lubricant increases, severity of wear also increases. The influence of soot contamination on tribological properties of various contact surfaces is discussed below.

15.3.1 Steel on Steel Contact

Friction and wear of steel on steel contact showed only a slight increase up to 4% soot concentration, since the dispersant additives in the lubricants were able to disperse the abrasive soot particles. However, at 6% soot, which is highest concentration, dispersant was not sufficient enough to stop the formation of large clusters of soot particles. These soot particles start build-up around the contact and reduce lubricant flow into the contact zone and cause starvation. It leads to lubrication starvation and results in metal to metal contact [28, 29] (Fig. 15.10).

15.3.2 Si_3N_4 on Steel

For Si_3N_4 on steel also, similar trend was observed as in steel on steel contact. The influence of soot on friction and wear was less significant up to 4% of soot level and at 6% soot level values of friction and wear became high (Fig. 15.11).

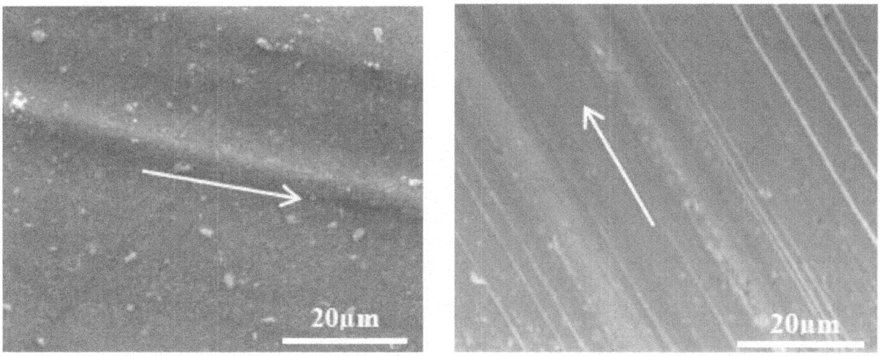

Fig. 15.11 SEM micrographs of pin wear scars for different concentration of soot contamination: Si_3N_4 on steel. Reproduced from my master student [33]

Abrasive wear and polishing wear were seen in the case of Si_3N_4 pins rubbing against steel with soot contamination. Impingement of soot particles in the surface caused rougher surface. At higher soot level, agglomeration of particles caused oil starvation in the contact which resulted in deeper grooves.

15.3.3 ZrO_2 on Steel

For ZrO_2 on steel contact, the rate of increment of coefficient of friction was higher compare to Si_3N_4 tribo-couple and a clear difference in coefficient of friction can be observed at each level of soot. Zirconia surface was more affected by soot particles compare to steel and silicon nitride. Due to the continuous penetration of soot particles, crack initiation occur which propagates and results in the release of a wear particle. Soot particles caused rougher surface due to the pits formed and resulted in increased friction [30, 31] (Fig. 15.12).

High contact pressure and penetration of soot particles promote micro cracks in pores and grain boundaries. These sub surface cracks propagate and intersect on the contact surface and thus material removal happens. This is like delamination theory of wear.

Friction was found less in silicon nitride and zirconia due to the interatomic bonding which leads to narrow dislocation. Lowest coefficient of friction was found in the case of silicon nitride in the presence of soot contamination. In hybrid contacts, wear rate was found much less compared to metallic contact (Figs. 15.13 and 15.14).

Due to the interatomic bonding and higher hardness of silicon nitride and zirconia, it acts as wear resistant and influence of soot particles becomes less effective and

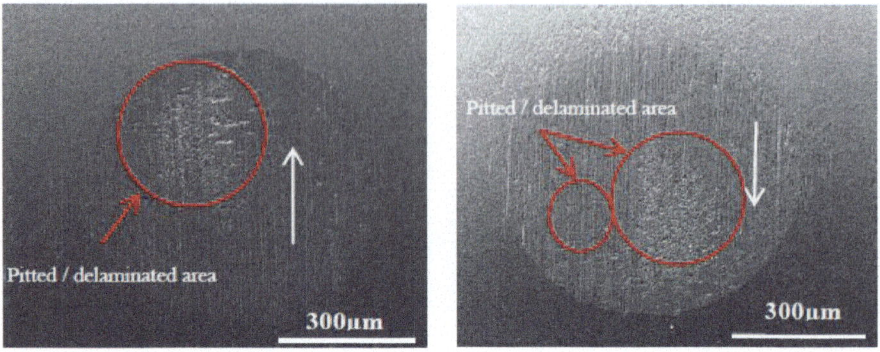

2% soot contamination 6% soot contamination

Fig. 15.12 SEM micrographs of pin wear scars for different concentration of soot contamination: ZrO_2 on steel. Reproduced from my master student [33]

Fig. 15.13 Comparison of steady state coefficient of friction at varying level of soot contamination. Reproduced from my master student [33]

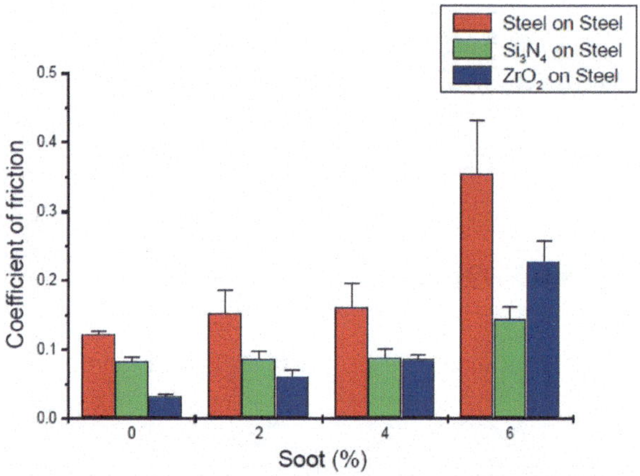

Fig. 15.14 Comparison of wear rate calculated from pin wear scar at different concentration of diesel soot. Reproduced from my master student [33]

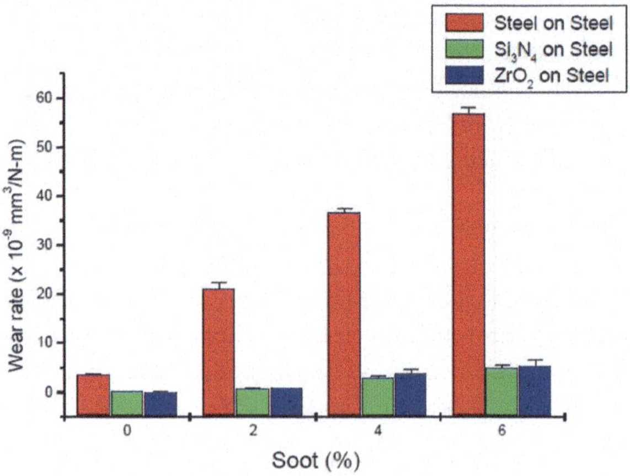

causes less wear compare to steel sliding contacts. As the soot level increases in the case of hybrid contacts, wear rate is marginally affected and increases slightly with soot levels.

15.4 Conclusion

Improving friction and wear properties of engine components is one most beneficial challenge in tribology. Contaminants such as soot, sulphuric acid, moisture and oxidation are influencing tribological properties of engine parts. Moisture and oxidation are having less influence on wear whereas sulphuric acid increases wear rate especially at high concentrations due to corrosive attack. However, effect of soot is highest on friction and wear among the four contaminants considered. The primary wear mechanism of soot is abrasion, but at higher concentrations soot particles agglomerates on surface producing adhesive wear. Soot also causes increase in viscosity, which makes pumping of lubricant and finally results in a higher wear rate.

Friction and wear rate were found less in the case of ceramic contacts compared to steel in the presence of soot contaminants. This is due to the higher inter atomic forces and narrow dislocations in ceramics compared to steel.

References

1. M. McNeely, ARES—Gas Engines for Today & Beyond. Diesel Gas Turbine Worldwide (2003)
2. D. Dowson, Piston assemblies; background and lubrication analysis, in *Engine Tribology, Tribology Series*, ed. by Taylor, (Elsevier, Amsterdam, vol. 26, 1993, Chap. 9), pp. 213–240
3. S.C. Tung, M.L. McMillan, Automotive tribology overview of current advances and challenges for the future. Tribol. Int. **37**(2004), 517–536 (2004)
4. C.M. Taylor, Automobile engine tribology—design considerations for efficiency and durability. Wear **221**(1), 1–8 (1998)
5. W.M. Needelman, P.V. Madhavan, Review of lubricant contamination and diesel engine wear. No. 881827. SAE technical paper (1988)
6. S. Aldajah et al., Effect of exhaust gas recirculation (EGR) contamination of diesel engine oil on wear. Wear **263**(1–6), 93–98 (2007)
7. M. Gautam et al., Effect of diesel soot contaminated oil on engine wear—investigation of novel oil formulations. Tribol. Int. **32**(12), 687–699 (1999)
8. D.R. Snelling et al., Particulate matter measurements in a Diesel engine exhaust by laser-induced incandescence and the standard gravimetric procedure. No. 1999-01-3653. SAE Technical paper (1999)
9. A.V. Oliver, Gear lubrication—a review. Proc. IMechE, Part J: J. Eng. Tribol. **216**(5), 255–267 (2002)
10. M.F. Smiechowski, V.F. Lvovich, Electrochemical monitoring of water-surfactant interactions in industrial lubricants. J. Electroanal. Chem. **534**(2), 171–180 (2002)
11. Y. Murakami, Analysis of corrosive wear of diesel engines: relationship to sulfate ion concentrations in blow by and crankcase oil. JSAE Rev. **16**(1), 43–48 (1995)

12. S.K. Singh, A.K. Agarwal, D.K. Srivastava, M. Sharma, Experimental investigation of the effect of exhaust gas recirculation on lubricating oil degradation and wear of a compression ignition engine. ASME J. Eng. Gas Turbines Power **128**, 921–927 (2006)
13. K. Akiyama, K. Manunaga, K. Kado, T. Yoshioka, Cylinder wear mechanism in an EGR equipped diesel engine and wear protection by engine oil. SAE paper 872158 (1987)
14. W.F. Bowman, G.W. Stachowiak, Determining the oxidation stability of lubricating oils using sealed capsule differential scanning calorimetry (SCDSC). Tribol. Int. **29**(1), 27–34 (1996)
15. N. Gracia, S. Thomas, P. Bazin, L. Duponchel, F. Thibault-Starzyk, O. Lerasle, Combination of mid-infrared spectroscopy and chemometric factorization tools to study the oxidation of lubricating base oils. Catal. Today **155**(3–4), 255–260 (2010)
16. B.K. Sharma, A.J. Stipanovic, Development of a new oxidation stability test method for lubricating oils using high-pressure differential scanning calorimetry. Thermochim. Acta **402**, 1–18 (2003)
17. D.A. Green, R. Lewis, R.S. Dwyer-Joyce, Wear effects and mechanisms of soot-contaminated automotive lubricants. Proc. Inst. Mech. Eng. Part J: J. Eng. Tribol. **220**(3), 159–169 (2006)
18. M. Masuko, A. Suzuki, T. Ueno, Influence of chemical and physical contaminants on the antiwear performance of model automotive engine oil. Proc. IMechE, Part J: J. Eng. Tribol. **220**(5), 455–462 (2006)
19. F.G. Rounds, Soots from used diesel-engine oils: their effects on wear as measured in 4-ball wear tests. SAE technical paper 810499 (1981)
20. T. Skurai, K. Yoshida, Tribological behaviour of dispersed phase systems, in: International Tribology Conference 1987, Melbourne, 2–4 Dec 1987: Preprints of Papers. Institution of Engineers, Australia (1987)
21. P.R. Ryason, I.Y. Chan, J.T. Gilmore, Polishing wear by soot. Wear **137**(1), 15–24 (1990)
22. M. Ratoi et al., The influence of soot and dispersant on ZDDP film thickness and friction. Lubr. Sci. **17**(1), 25–43 (2004)
23. Y. Olomolehin, R. Kapadia, H. Spikes, Antagonistic interaction of antiwear additives and carbon black. Tribol. Lett. **37**(1), 49 (2010)
24. F.M. Salehi et al., Corrosive–abrasive wear induced by soot in boundary lubrication regime. Tribol. Lett. **63**(2), 19 (2016)
25. W.M. Needleman, P.V. Madhavan, Review of lubricant contamination and Diesel engine wear, SAE paper 881827 (1988)
26. M.F. Smiechowski, V.F. Lvovich, Characterization of non-aqueous dispersions of carbon black nanoparticles by electrochemical impedance spectroscopy. J. Electroanal. Chem. **577**(1), 67–78 (2005)
27. S. George et al., Effect of diesel soot on lubricant oil viscosity. Tribol. Int. **40**(5), 809–818 (2007)
28. M. Kaneta et al., Effects of soot on wear in elastohydrodynamic lubrication contacts. Proc. Inst. Mecha. Eng. Part J J. Eng. Tribol. **220**(3), 307–317 (2006)
29. D.A. Green, R. Lewis, R.S. Dwyer-Joyce, Wear effects and mechanisms of soot-contaminated automotive lubricants. Proc. Inst. Mech. Eng. Part J J. Eng. Tribol. **220**(3), 159–169 (2006)
30. K. Kitamura, Y. Imada, K. Nakajima, Effect of soot introduced between sliding ceramic surfaces. Lubr. Eng. **49**, 185–190 (1993)
31. Y. He et al., Grain-size dependence of sliding wear in tetragonal zirconia polycrystals. J. Am. Ceram. Soc. **79**(12), 3090–3096 (1996)
32. R. Penchaliah et al., The effects of diesel contaminants on tribological performance on sliding steel on steel contacts. Proc. Inst. Mech. Eng. Part J J. Eng. Tribol. **225**(8), 779–797 (2011)
33. K. Yadvendra, Master Thesis, IIT Madras (2018)

Index

J. K. Katiyar et al. (eds.), *Tribology in Materials and Applications*, Materials Forming, Machining and Tribology,
https://doi.org/10.1007/978-3-030-47451-5

Index

J. K. Katiyar et al. (eds.), *Tribology in Materials and Applications*, Materials Forming, Machining and Tribology,
https://doi.org/10.1007/978-3-030-47451-5

The manufacturer's authorised representative in the EU is Springer Nature Customer Service Centre GmbH, Europaplatz 3, 69115 Heidelberg, Germany. If you have any concerns regarding our products, please contact ProductSafety@springernature.com

Printed and bound by CPI Group (UK) Ltd, Croydon, CR0 4YY
30/09/2026
02225646-0014

Strafensystem und Gefängniswesen

in

England.

Von

Dr. P. F. Aschrott,

Amtsrichter.

Berlin und **Leipzig.**

Verlag von J. Guttentag.

(D. Collin).

1887.

Vorwort.

Die Notwendigkeit einer auf gleichmäßigen Grundsätzen beruhenden Neugestaltung des Gefängniswesens in Deutschland und des Erlasses des hierzu erforderlichen Strafvollzugsgesetzes für das Deutsche Reich wird allgemein anerkannt. Es ist dies bereits bei der Beratung des Norddeutschen Strafgesetzbuchs zum Ausdruck gekommen, indem der Bundeskanzler durch einen Beschluß des Reichstags des Norddeutschen Bundes vom 4. März 1870 aufgefordert wurde, ein Strafvollzugsgesetz vorzulegen.

Die auf andern Gebieten so überaus thätige Reichsgesetzgebung ist jedoch hier bis jetzt im Rückstande geblieben. Im März 1879 ist zwar der Entwurf eines Strafvollzugsgesetzes dem Bundesrate vorgelegt worden, ohne jedoch hier eine Erledigung zu finden. Die Gründe für das Scheitern dieses Entwurfs sind offiziell nicht bekannt gegeben; man wird jedoch kaum fehlgehen, wenn man einen nicht unerheblichen Anteil daran der im Jahre 1879 erschienenen Schrift Mittelstädts „Gegen die Freiheitsstrafen" und dem ungeheuren Aufsehen zuschreibt, welches diese Schrift damals erregte.

Mittelstädt hatte nicht nur in energischer Weise den Kampf gegen die im Strafvollzuge herrschenden Besserungstendenzen eröffnet, sondern es war durch ihn die Berechtigung der Stellung der Freiheitsstrafen überhaupt im Rahmen des deutschen Strafensystems in Frage gezogen worden. Seit dem Erscheinen der

Schrift Mittelstädts hat sich denn auch die wissenschaftliche Litteratur weit mehr auf dem Gebiete der Strafmittel bewegt, als auf demjenigen der eigentlichen Gefängniswissenschaft. Wie immer man über die Ansichten und Vorschläge Mittelstädts im einzelnen denken mag, über den einen Punkt besteht fast allgemeine Übereinstimmung: mit der gesetzlichen Regelung des Strafvollzugs wird eine Abänderung des in dem deutschen Strafgesetzbuche niedergelegten Strafensystems Hand in Hand gehen müssen.

Es sind demgemäß umfassende Reformen, welche der Lösung durch die Gesetzgebung harren. Bei der schwierigen Aufgabe, welche den gesetzgebenden Faktoren hier bevorsteht, muß eine Kenntnis der bezüglichen Institutionen des Auslandes und der dort gesammelten Erfahrungen von Wert sein.

An eine direkte Übertragung fremder Einrichtungen wird auf den in Frage kommenden Gebieten kein Sachverständiger denken. Es gibt wenige Zweige staatlicher Thätigkeit, bei welchen auf die Eigentümlichkeiten des Landes und seiner Bevölkerung, auf die bereits bestehenden und ohne übermäßige finanzielle Opfer nicht auf einmal zu beseitigenden Einrichtungen, sowie endlich auf den Umfang der gesellschaftlichen Mithilfe in so umfassendem Maße Rücksicht genommen werden muß, als bei dem Kampfe gegen das Verbrechertum. Aber, so irrationell eine Nachahmung fremder Einrichtungen sein würde, so unklug wäre es, wenn den im Auslande bestehenden Verhältnissen nicht die größte Beachtung geschenkt und aus den von andern Nationen gemachten Erfahrungen nicht, soweit es irgend möglich ist, Schlußfolgerungen für unser eigenes Land zu ziehen versucht würde. Es handelt sich um Fragen von so großer sozialer und finanzieller Bedeutung, daß ein Experimentieren auf Grund von Ideen, wie sie in der Studierstube gewonnen oder von einem menschenfreundlichen Herzen eingegeben sein mögen, nicht am Platze ist. „Die Strafe in der Hand des Staates ist ein zweischneidiges Schwert; bei verkehrtem Gebrauch kehrt sie ihre Spitze gegen ihn selbst, schädigt mit dem Verbrecher zugleich ihn selber" (v. Ihering, Zweck im Recht, Band I Seite 375).

Wenn hiernach jede Darstellung der in einem andern Kulturstaate mit Bezug auf den Kampf gegen das Verbrechertum getroffenen Einrichtungen ein allgemeines Interesse in Deutschland erwarten kann, so dürfte eine Bearbeitung des Strafensystems und Gefängniswesens in England besonders zeitgemäß sein. Nicht nur die heute in England bestehenden Einrichtungen verdienen in Deutschland die vollste Beachtung, sondern es sind vor allem die reichen Erfahrungen überaus lehrreich, welche in England im Laufe der Zeit mit mehrfach wechselnden Einrichtungen und unter verschiedenartiger Handhabung derselben gemacht worden sind.

Es erscheint daher in der That auffallend, daß eine deutsche Bearbeitung der bezüglichen englischen Verhältnisse bis jetzt völlig gefehlt hat. Während wir aus der Feder des Geheimen Oberjustizrats Starke eine vorzügliche Darstellung des belgischen Gefängniswesens besitzen und auch über die Verhältnisse in Frankreich und andern Ländern leidlich unterrichtet sind, ist die Litteratur über England hier eine mehr als dürftige.

Eine deutsche Bearbeitung des gesamten Strafensystems in England ist mir überhaupt nicht bekannt; einzelne Strafmittel, wie Deportation, Todesstrafe und Friedensbürgschaft, werden in Spezialarbeiten behandelt. Ausführlichere Nachrichten über das Gefängniswesen in England besitzen wir nur aus den fünfziger Jahren, insbesondere von Mittermaier; die bezüglichen Arbeiten, so wertvoll sie an sich sind, sind jedoch heutigentags vollkommen veraltet. Nur über das bei der Strafe der penal servitude zur Anwendung gebrachte Progressivsystem existieren beachtenswerte deutsche Arbeiten. Allein auch diese Schriften — es kommen in erster Linie diejenigen von Professor v. Holtzendorff in Betracht — liegen zunächst ein volles Vierteljahrhundert zurück, in welchem Zeitraume zahlreiche Veränderungen und Verbesserungen eingeführt worden sind. Ferner aber beschränken sich diese Schriften auf eine Darstellung der von Sir Walter Crofton in Irland getroffenen Einrichtungen, wobei zu berücksichtigen ist, daß einmal die Verhältnisse in Irland ganz besonders geartete sind und daß anderseits Sir Walter Crofton ein exzeptionell fähiger Gefängnisleiter war. Es erscheinen daher allgemeine Schlußfolgerungen für andre Länder, in denen die Verhältnisse nicht so wie in Irland liegen und wo man nicht mit Bestimmtheit auf eine gleich tüchtige Leitung rechnen kann, mindestens recht gewagt. Man kann also sagen, daß es für die heutige Gestaltung des Gefängniswesens in England an zuverlässigem Material in deutscher Sprache fehlt.

Auf die hiernach bestehende Lücke in der deutschen Litteratur dürfte es auch zurückzuführen sein, daß in den Motiven und Beilagen zum deutschen Strafgesetzbuche, in welchen im allgemeinen den ausländischen Verhältnissen eingehende Beachtung geschenkt wird, England, abgesehen von dem Abschnitte über die Todesstrafe, kaum Erwähnung findet. Es ist dies um so leichter erklärlich, als zu der damaligen Zeit auch in englischer Sprache halbwegs zuverlässige und dabei für den Ausländer verständliche Bücher über den Gegenstand nicht existierten. Erst in den allerletzten Jahren sind aus der Feder von Sir Edmund Du Cane und Sir James Stephen englische Bücher erschienen, in welchen der Deutsche zuverlässige Auskunft zu finden sicher sein kann.

Die Schriften dieser beiden Autoren waren denn auch fast das einzige, was an litterarischen Hilfsmitteln bei der vorliegenden Arbeit zu Rate gezogen werden konnte. Die umfangreichen Arbeiten von Sir James Stephen boten für die Darstellung der historischen Entwickelung des Strafensystems in England ein reiches Material. Die beiden Schriften von Sir Edmund Du Cane fanden, wie sich dies bei der hervorragenden Stellung des Autors von selbst verstehen mußte, bei der Behandlung des heutigen englischen Gefängniswesens die aufmerksamste Beachtung. Eine direkte Verwendung dieser Schriften war jedoch für die vorliegende Arbeit nur in beschränktem Maße möglich. Sir Edmund Du Cane hat sich in seinen beiden Schriften — die bedeutendste: „The Punishment and Prevention of Crime“ ist in der von Macmillan herausgegebenen Serie „The English Citizen, his rights and responsibilities“ erschienen — die Aufgabe gestellt, in populärer Weise den englischen Leser über die Art und Weise der Vollstreckung der Strafen in England aufzuklären. Von einer systematischen, streng wissenschaftlich gegliederten Darstellung ist dabei keine Rede; das Ganze ist mehr in erzählender Form gehalten. Der Zweck der vorliegenden Arbeit erheischte nicht nur eine ganz andre Art der Behandlung, sondern vor allen Dingen ein viel größeres Eingehen auf theoretische Fragen und auf Einzelheiten.

Hierfür habe ich in sehr umfassendem Maße auf das Urmaterial, die englischen Blaubücher, zurückgehen müssen. In den Jahresberichten der Gefängnisbehörden und in den Berichten der zahlreichen im Laufe dieses Jahrhunderts niedergesetzten Kommissionen zur Untersuchung der Gefängniszustände ist ein ungeheures Material angesammelt, welches bis jetzt auch in England noch keine wissenschaftliche Verwertung gefunden hat. Ich kann wohl sagen, daß die vorliegende Arbeit der erste Versuch einer wissenschaftlichen Bearbeitung des englischen Gefängniswesens ist.

Über die Schwierigkeit eines derartigen Versuchs habe ich mich von Anfang an keiner Täuschung hingegeben. Ich bin mir auch weiter bewußt gewesen, wie schwer es ist, in einem Buche von relativ beschränktem Umfange dem deutschen Leser ein zutreffendes Bild von zwei so umfangreichen Materien, wie Strafensystem und Gefängniswesen in England, zu geben. Eine Reihe von Anschauungen und Begriffen, welche dem Engländer ganz geläufig sind, erscheinen dem deutschen Leser fremd und bedürfen näherer Erörterung. Eine weitere Schwierigkeit bestand darin, daß ich nicht an irgend einem andern Werke ein Muster für die Art und Weise der Behandlung besaß. Strafensystem und Gefängniswesen sind bisher vollkommen getrennt bearbeitet worden; ein Buch, welches diese beiden Materien zu gleicher Zeit und einheitlich behandelt, ist mir weder in englischer noch in deutscher Sprache bekannt.

Auch hierin enthält somit das vorliegende Buch einen ersten Versuch; einen Versuch, welchen ich machen mußte, da nach meiner persönlichen Überzeugung ein richtiges Urteil über die Art und Weise des Vollzuges von Strafen unmöglich ist ohne eine Kenntnis des zu Grunde liegenden Strafensystems, welches jedem einzelnen Strafmittel seine ganz bestimmte Stellung in dem Gefüge der vom Staate gegenüber den Verbrechen ausgeübten Repression anweist. Ich glaube, daß es — im wesentlichen als Resultat der schon angeführten Mittelstädtschen Schrift — mehr und mehr anerkannt werden wird, daß Strafensystem und Gefängniswesen zusammen zu behandeln sind, und daß es nicht länger zulässig bleibt, daß der

Kriminalist sich ausschließlich mit dem Strafensysteme beschäftigt und die Behandlung des Strafvollzugs, die sogenannte Gefängniswissenschaft, den Praktikern d. h. den Gefängnisbeamten überläßt.

Bei den Schwierigkeiten, welche hiernach die vorliegende Arbeit dargeboten hat, darf ich wohl gütige Nachsicht erhoffen. Manchem Leser mag die Art und Weise der Behandlung als eine gar zu juristische erscheinen, andre mögen ein näheres Eingehen auf die philosophischen Probleme des Strafrechts oder auf Detailfragen der Gefängnisverwaltung vermissen. Es ist aber beim besten Willen kaum möglich, hier von vornherein die juste milieu zu treffen und alle zu befriedigen.

Das Eine kann ich versichern: ich bin von Anfang bis zu Ende der Arbeit bestrebt gewesen, die in der deutschen Litteratur bestehende Lücke in möglichst objektiver Weise auszufüllen. Ich habe es auf das sorgfältigste vermieden, irgendwo rein subjektive Anschauungen zum Ausdrucke zu bringen. Zweck des Buches sollte es sein, das Material zu einer richtigen Beurteilung des Strafensystems und Gefängniswesens in England zu liefern. Der Leser mag daraus selbst seine Schlüsse ziehen, und ich zweifle nicht, daß dieselben je nach dem Parteistandpunkte, von dem ausgegangen wird, in mancher Beziehung voneinander abweichende sein werden.

Mit voller Absicht habe ich es unterlassen, in dem Buche selbst Schlußfolgerungen zu geben, inwieweit die eine oder andre Einrichtung ein Muster für Deutschland abgeben könnte oder sich in irgend einer Weise bei den in Deutschland bevorstehenden Reformen verwerten ließe. Durch derartige, auf subjektiven Anschauungen beruhende Schlüsse würde die Arbeit unwillkürlich mehr oder weniger ihren objektiven Charakter verloren haben. Meine persönlichen Ansichten darüber, was wir aus den englischen Einrichtungen entnehmen könnten und sollten, werde ich in einer besonderen Arbeit, welche ich über diese Frage jetzt nach Vollendung des Buches zu schreiben beginne, zusammenfassen und in einem der nächsten Hefte der „Zeitschrift für die gesamte Strafrechtswissenschaft" veröffentlichen.

Was die Disposition des vorliegenden Buches betrifft, so zerfällt dasselbe in zwei Hauptabschnitte, zu welchen kurze Vorbemerkungen und ein ebenfalls kurzer Schlußabschnitt hinzutreten.

In den Vorbemerkungen (Seite 1—22) soll in möglichst knapper Form dem deutschen Leser dasjenige aus der englischen Gerichtsverfassung, dem Strafrechte und Strafprozesse vorgeführt werden, was zum Verständnisse der ganzen Arbeit erforderlich erscheint. Es ist unmöglich, ein richtiges Urteil über das Strafensystem eines fremden Staates zu gewinnen ohne Kenntnis von der Art und Weise, in welcher die staatliche Repression gegenüber dem Verbrechen ausgeübt wird. Wenn man sich ein Bild von der Verwendung der einzelnen Strafmittel machen will, so muß man zunächst wissen, in welcher Weise die Polizei eingreift; ob ein öffentlicher Ankläger die Strafverfolgung übernimmt oder ob resp. in welchem Umfange die letztere dem Verletzten überlassen bleibt; wie die Gerichte zusammengesetzt sind, welche über die Bestrafung zu erkennen haben; in welcher Weise das Strafverfahren vor sich geht u. s. w. Es erschien ferner angezeigt, eine Reihe technischer englischer Ausdrücke und Einteilungen, welche im Verlaufe der Arbeit häufig wiederkehren, vorweg zu erläutern.

Der erste Abschnitt (Seite 23—128) behandelt sodann die Strafmittel des englischen Rechts. Bei jedem einzelnen Strafmittel ist die historische Entwickelung bis auf den heutigen Tag verfolgt, und es sind statistische Notizen über den Umfang, in welchem von dem Strafmittel Gebrauch gemacht wird, gegeben. Auf die Art und Weise der Vollstreckung der Freiheitsstrafen ist in dem ersten Abschnitte nur insoweit eingegangen, als es zum Verständnisse der Entwickelung erforderlich erschien. Mit Rücksicht auf die hervorragende Bedeutung, welche die Transportationsstrafe für die Gestaltung des englischen Strafensystems gehabt hat, ist dieses Strafmittel hier behandelt worden, obwohl auf dasselbe heutigentags nicht mehr erkannt werden kann.

In dem zweiten Abschnitte (Seite 129—292) wird das heutige englische Gefängniswesen in drei Kapiteln zur Darstellung gebracht: Kapitel I (Seite 133—161) handelt von der Organisation des Gefängniswesens, Kapitel II (Seite 162—180) von den Gefängniseinrichtungen sowie den Kosten des Gefängniswesens, Kapitel III (Seite 180—292) von der Behandlung der Gefangenen. Anhangsweise (Seite 267—274) ist die Thätigkeit der Fürsorgevereine für entlassene Sträflinge erörtert worden. Auch bei diesem zweiten Abschnitte werden statistische Notizen in reichlichem Maße gegeben.

Im Schlußabschnitte (Seite 293—311) endlich wird noch kurz auf das sogenannte irische Gefängnissystem eingegangen. Dieser Abschnitt ist lediglich in der Erwägung hinzugefügt worden, daß das irische System in der deutschen Litteratur mehrfache Behandlung erfahren hat, und daß es daher angezeigt ist, kurz hervorzuheben, ob und inwieweit dasselbe Abweichungen von den im Abschnitte II ausführlich dargestellten englischen Einrichtungen enthält.

Zum Schlusse möchte ich noch dankbarst anerkennen, welche Förderung ich während meines Aufenthaltes in England und Irland von allen Seiten, von den Beamten des Gefängnisressorts wie von Vereinen, bei meinen Studien gefunden habe. Mit der größten Bereitwilligkeit sind mir überall Auskünfte erteilt und Materialien zur Verfügung gestellt worden. Zu ganz besonderem Danke fühle ich mich aber dem jetzigen Leiter des Gefängniswesens in England, Sir Edmund Du Cane, verpflichtet. Infolge seiner liebenswürdigen Empfehlungen habe ich die am meisten sehenswerten Anstalten eingehend besichtigen können und habe über alles, was mir von Interesse war, mit der größten Offenheit Bescheid erhalten.

Berlin, im März 1887.

Dr. P. F. Aschrott.

Inhalt.

Abschnitt II.

Das heutige englische Gefängniswesen. 129—292

Kapitel I.

Die Organisation des Gefängniswesens. 133—161

Kapitel II.

Die Gefängniseinrichtungen. 162—180

Titel II.

Seite

Titel III.

Abschnitt III.

Litteraturübersicht.*)

A. Englische Schriften.

Crofton (Sir Walter), I. A few remarks on the Convict Question. Dublin 1857.

II. Memoranda relative to the intermediate convict prisons in Ireland. Dublin 1857.

III. Convict System and Transportation. Bristol 1862.

IV. A brief description of the Irish Convict System. London 1862.

Du Cane (Sir Edmund), I. An Account of the manner in which sentences of penal servitude are carried out in England. London 1882.

II. The Punishment and Prevention of Crime. London 1885.

Howard (John), The State of the Prisons in England and Wales with preliminary observations and an account of some foreign prisons. London 1777.

Leone Levi, A survey of indictable and summary jurisdiction offences in England and Wales from 1857 to 1876; Journal of the London Statistical Society Band XLIII Seite 423 ff.

Maitland, Justice and Police. London 1885.

Monro, Report on the history of the Metropolitan Convict Supervision Office. London 1887.

Shirley, I. Digest of the Criminal Law. London 1880.

II. Elementary Treatise on Magisterial Law. London 1881.

Stephen (Sir James), I. A History of the criminal law of England. 3 vols. London 1883.

II. A Digest of the criminal law. (3. ed.). London 1883.

III. A Digest of the law of criminal procedure in indictable offences. London 1883.

---

*) Die hier angeführten Schriften werden im nachfolgenden regelmäßig nur mit dem Namen des Autors citiert. Soweit von demselben Autor mehrere Schriften vorliegen, wird dem Namen des Autors die römische Ziffer hinzugefügt, unter welcher die betreffende Schrift in der vorstehenden Litteraturübersicht angeführt steht. — Die englischen Blaubücher, welche für die vorliegende Arbeit die hauptsächlichste Quelle gebildet haben, sind stets mit der Jahreszahl ihrer Publikation citiert; es bezeichnet also beispielsweise „Report of the Directors of Convict Prisons 1886“ den im Jahre 1886 erschienenen Bericht über das Jahr 1885/6.

Stephen (Sir James), IV. Über den gegenwärtigen Zustand des englischen Strafrechts und die Pläne zu dessen Kodifikation. Zeitschrift für die gesamte Strafrechtswissenschaft Band I Seite 439 ff.

Stone, Justices' Manual. London 1884.

Wilkinson, The law of prisons. London 1878.

Wines, The state of prisons and of child-saving institutions in the civilized world. Cambridge 1880.

B. Deutsche Schriften.

Behrend, Geschichte der Gefängnisreform. Berlin 1859.

Eberty, Das Gefängniswesen. Berlin 1858.

Föhring, Die Reformatory und Industrial Schools der vereinigten Königreiche Großbritannien und Irland im 14. Vereinshefte des nordwestdeutschen Vereins für Gefängniswesen Seite 118 ff.

Glaser, Das englisch-schottische Strafverfahren. Wien 1850.

Gneist, I. Das englische Verwaltungsrecht der Gegenwart. 2 Bände. Berlin 1883, 1884.

II. Selfgovernment, Kommunalverfassung und Verwaltungsgerichte in England. Berlin 1871.

Götting, Strafrechtspflege und Gefängniswesen in England und Irland. Hildesheim 1876.

v. Grosz, Wanderung durch irische Gefängnisse. Berlin 1868.

v. Holtzendorff, I. Die Deportation als Strafmittel. Leipzig 1859.

II. Das irische Gefängnissystem. Leipzig 1859.

III. Die Kürzungsfähigkeit der Freiheitsstrafen. Berlin 1861.

IV. Untersuchungen über die Ergebnisse des irischen Strafvollzugs Berlin. 1865.

John, Über Strafanstalten Berlin 1865.

Liepmann, Das Vorverfahren im englischen Strafprozesse; Zeitschrift für die gesamte Strafrechtswissenschaft Band VI Seite 413 ff.

Marquardsen, Die Todesstrafe in England; Kritische Zeitschrift für Rechtswissenschaft Band 22 Seite 478 ff., Band 23 Seite 202 ff., 311 ff.

Mittermaier, I. Gefängniseinrichtungen in England. Heidelberg 1850.

II. Die Gefängnisverbesserung. Erlangen 1858.

III. Der gegenwärtige Zustand der Gefängnisfrage. Erlangen 1860.

Röder, I. Der Strafvollzug im Geiste des Rechts. Leipzig 1863.

II. Die herrschenden Grundlehren von Verbrechen und Strafe in ihrem innern Widerspruch. Wiesbaden 1864.

Schierlinger, Die Friedensbürgschaft. Erlangen 1877.

Tauffer, I. Erfolge des progressiven Strafvollzugs. Berlin 1885.

II. Beiträge zur neuesten Geschichte des Gefängniswesens. Stuttgart 1885.

Tellkampf, Die Besserungsgefängnisse in Nordamerika und England. Berlin 1844.

Vorbemerkungen.*)

§ 1. Das englische Strafrecht.

England besitzt kein Strafgesetzbuch. Bis zum Jahre 1827 war ein erheblicher Teil des englischen Strafrechts in schriftlicher Form auf autoritative Weise überhaupt nicht festgestellt, und im übrigen befanden sich die strafrechtlichen Normen in zahlreichen, über mehrere Jahrhunderte hinausreichenden Gesetzen vor.

In den Jahren 1827—1832 wurde dann ein großer Teil dieses Statutarrechtes unter Beseitigung von Hunderten früherer Parlamentsakte in eine konsolidierte Form gebracht, wobei gleichzeitig die Strafen für die meisten, in den konsolidierten Statuten behandelten Delikte erheblich gemildert wurden. Es wurde ferner eine Kommission eingesetzt, welche über den Zustand des englischen Strafrechts weiter berichten und Vorschläge zu fernerer Konsolidierung desselben machen sollte.

Nachdem in den Jahren 1833—1861 eine große Anzahl von Berichten dieser Kommission erschienen waren (Reports of the Criminal Law Commissioners), kamen endlich 1861 5 große Gesetze [1]) zustande, in welchen die Materien des Diebstahls, der Sachbeschädigung, der Fälschung, der Münzverbrechen und der Angriffe auf Leib und Leben in umfassender Weise geregelt wurden, wobei

*) Quellen: In erster Linie die Arbeiten von Sir James Stephen, sodann die Bücher von Maitland, Shirley und Stone, die Judicial Statistics 1884 und der Report of the Commissioner of Police of the Metropolis 1884; conf. auch Gneist I Seite 780 ff. und Seite 981 ff., II Seite 227 ff. und Seite 430 ff.

[1]) 24 a. 25 Vict. c. 96—100, das Diebstahlsgesetz (c. 96) allein umfaßt 123 Artikel.

wiederum vielfach eine Milderung der Strafbestimmungen stattfand.

Diese 5 Gesetze aus dem Jahre 1861 stellen gegenüber dem bisherigen Zustande des englischen Strafrechts einen entschiedenen Fortschritt dar, allein sie entsprechen den an ein Strafgesetzbuch zu stellenden Erfordernissen in keinerlei Weise. Abgesehen davon, daß eine erhebliche Zahl wichtiger Delikte — so insbesondere die Delikte gegen die öffentliche Ordnung und gegen die Sittlichkeit — hier gar nicht berücksichtigt werden, fehlt es in diesen Gesetzen auch bezüglich der darin behandelten Materien an den allerwichtigsten Verbrechensdefinitionen. Es werden zum Beispiel sehr ausführliche Bestimmungen über die Bestrafung der verschiedenen Arten des Diebstahls gegeben; aber was juristisch unter Diebstahl zu verstehen ist, erfährt man aus dem Gesetze nicht.[1] Für alle Definitionen wie für die wichtigsten Materien des allgemeinen Teils eines Strafgesetzbuchs — beispielsweise für die ganze Lehre von den Strafausschließungsgründen — muß man auf alte Gesetze und in nicht unerheblichem Maße auf ungeschriebenes Recht zurückgreifen.

Es ist klar, daß man auf dem zuerst im Jahre 1827 beschrittenen Wege der Kodifikation des Strafrechts weiter vorwärts schreiten mußte. Und so liegt denn auch ein Entwurf eines vollständigen Strafgesetzbuchs bereits vor. Nachdem Sir James Stephen, wohl der hervorragendste jetzt lebende englische Jurist, im Jahre 1877 in seinem Buche A Digest of the Criminal Law das gesamte, zur Zeit in England geltende Strafrecht in etwa 400 Artikeln zusammengefaßt und damit die Möglichkeit einer Kodifikation des gesamten Strafrechts dargelegt hatte, erhielt er von dem damaligen Lordkanzler, Earl Cairns, den Auftrag, den Entwurf eines Strafgesetzbuchs auszuarbeiten. Der fertiggestellte Entwurf, welchem ein Entwurf einer Strafprozeßordnung beigefügt war, wurde 1878 dem Parlamente vorgelegt. Das Parlament verwies denselben zunächst an eine Kommission von 4 hervorragenden Juristen, welche ihren Bericht über den Entwurf bereits 1879 dem Parlamente erstatteten.

---

[1] Sir James Stephen IV Seite 477 bezeichnet dieses Diebstahlsgesetz als „eins der verwickeltsten, schwerfälligsten und auf den ersten Anblick trostlos unverständlichsten Erzeugnisse gesetzgeberischer Thätigkeit, welchem ich je begegnet bin.“

Seitdem ist leider die Sache nicht vorwärts gediehen, [1]) doch dürfte die Ansicht von Sir James Stephen, dessen bereis in dritter Auflage erschienener Digest of the Criminal Law inzwischen eine geradezu autoritative Bedeutung erhalten hat, eine gar zu skeptische sein, wenn er ausführt, [2]) daß die Aufgabe der Kodifikation „eine solche ist, welche eine volkstümliche Versammlung, wie das britische Parlament, weder selbst zu erfüllen geeignet, noch irgend jemand anderm anzuvertrauen gewillt ist". „Eine gründliche Revision und Umarbeitung der ganzen Masse der Strafrechtssätze zu verbesserter Gestalt würde so viele Fragen der mannigfaltigsten Art, über welche große Meinungsverschiedenheit herrscht, hervorrufen, daß meiner Meinung nach kaum irgend ein Ministerium geneigt sein könnte, mit einer so weitaussehenden Arbeit sich zu befassen, und kaum ein Parlament willens sein dürfte, das Gesetz durchzuberaten." — Wir denken, das englische Parlament wird, wenn die Notwendigkeit einer Kodifikation des Strafrechts erst aus dem Kreise der Juristen heraus zum allgemeinen Volksbewußtsein geworden ist, sich zu der That schon aufraffen. Eine That wird es allerdings für den Engländer sein, all die antiquarischen Raritäten, welche sich in dem heutigen englischen Strafrecht vorfinden, über Bord zu werfen! Wie groß die Zahl derselben ist, das wird sich selbst in dem beschränkten Rahmen unserer Arbeit zeigen.

§ 2. Die Einteilung der Delikte.

Das englische Strafrecht kennt, wie das deutsche und französische, eine Dreiteilung der Delikte: die Delikte sind entweder treason oder felony oder misdemeanour.

Felonies sind die Kapitalverbrechen, für welche ursprünglich Todesstrafe und Vermögenskonfiskation angedroht war. Treasons — der Begriff ist umfassender, als der unsers Hoch- und Landes-Verrates — bilden eine besonders qualifizierte Art der felonies; hier bleibt unter anderm das noch näher zu behandelnde benefit of clergy ausgeschlossen. Misdemeanours sind diejenigen Delikte,

---

[1]) Der Entwurf der Strafprozeßordnung wurde in der Thronrede von 1882 als Regierungsvorlage erwähnt, ist jedoch nicht zur Beratung gelangt.

[2]) Sir James Stephen IV Seite 478.

bei welchen ursprünglich nur auf Geldstrafe oder Gefängnis oder körperliche Strafen erkannt werden konnte.

Diese, noch heute in dem englischen Strafrechte aufrecht erhaltene Dreiteilung der Delikte hat infolge der im Laufe der Zeit vorgenommenen Veränderung in den Strafbestimmungen jede Berechtigung verloren: die Strafe der Vermögenskonfiskation ist im Jahre 1870 — 33 a. 34 Vict. c. 23 s. 1 — überhaupt abgeschafft worden; die Todesstrafe ist jetzt nur noch für 4 Delikte zulässig; im allgemeinen ist ferner dem Richter bei Erkennung der Strafe für eine felony fast ein ebenso weiter Spielraum gelassen, als für die misdemeanours. Im großen und ganzen ist es nicht unrichtig zu sagen, daß unter felonies die schweren, unter misdemeanours die leichteren Straffälle verstanden werden, doch trifft dies nicht überall zu: z. B. ist jede Art Diebstahl eine felony, während der Meineid nur als misdemeanour gilt.

So wenig gerechtfertigt hiernach die Dreiteilung erscheint, so hat dieselbe doch noch vielfache praktische Bedeutung behalten. Um nur einiges anzuführen, so ist eine Verhaftung ohne Haftbefehl im allgemeinen nur bei felonies, nicht bei misdemeanours zulässig; der wegen felony Angeklagte kann ferner in Untersuchungshaft behalten werden, während der wegen misdemeanour Angeklagte im allgemeinen ein Recht auf Freilassung gegen Sicherheitsstellung hat u. s. w.

Von viel größerer Wichtigkeit jedoch, als diese Dreiteilung der Delikte, ist die Einteilung der Delikte [1]) in solche, welche

[1]) In den Judicial Statistics, und regelmäßig auch in den englischen Lehrbüchern, findet sich folgende Klassifikation der Delikte:

1) Delikte gegen die Person (offences against the person); hierunter fällt auch die bei uns gebräuchliche Kategorie der Delikte gegen die Sittlichkeit;
2) Delikte gegen das Eigentum unter Anwendung von Gewalt (offences against property with violence;
3) Delikte gegen das Eigentum ohne Anwendung von Gewalt (offences against property without violence): Diebstahl, Unterschlagung, Hehlerei u. s. w.;
4) böswillige Delikte gegen das Eigentum (malicious offences against property); es ist dies eine recht bunte Musterkarte von Delikten; das wichtigste darunter dürfte die Brandstiftung sein;
5) Fälschung und Münzverbrechen (forgery and offences against the currency);

nach erhobenem Anklagebeschlusse vor einer Jury verhandelt werden müssen — indictable offences —, und in solche, welche im summarischen Verfahren abgeurteilt werden können — offences punishable upon summary conviction —. Hier wird in der That auf die Schwere der Strafthat Rücksicht genommen, und es bleibt dabei nur zu beachten, daß im Falle der Jugend oder des Geständnisses des Angeschuldigten auch an sich schwerere Delikte (indictable offences) summarisch behandelt werden können.

§ 3. Die Strafgerichte.

Wenn man von der, eine wirklich praktische Bedeutung kaum mehr beanspruchenden Strafgerichtsbarkeit des House of Lords, des Privy Council und der Queen's Bench Division des High Court of Justice [1]) absieht, so kommen als ordentliche Strafgerichte in Betracht:

1) die Assisenhöfe und der Central Criminal Court in London;
2) die Gerichtshöfe der Quartalssitzungen der Friedensrichter (County oder Borough Quarter Sessions);
3) die Gerichte mit summarischer Gerichtsbarkeit (Petty Sessional Courts und Courts of Police Magistrates).

Die sub 1) angeführten Gerichte sind für die eigentlich schweren Straffälle kompetent. Die Gerichte sub 2) sind Mittelgerichte, deren Kompetenz in neuerer Zeit erheblich erweitert worden ist; es sind jetzt nur einige wenige Delikte aus ihrem Geschäftskreise ganz ausgeschlossen, insbesondere diejenigen Delikte, bei denen auf Todesstrafe oder lebenslängliche Strafknechtschaft erkannt werden kann. In noch höherem Maße als die Gerichte sub 2) haben die Gerichte mit summarischer Gerichtsbarkeit in neuerer Zeit an Bedeutung ge-

---

6) andere, in den obigen Klassen nicht untergebrachte Delikte (other offences not included in the above classes); z. B. Meineid, Bankrott, Delikte wider die öffentliche Ordnung.

Sir James Stephen verwirft mit Recht diese Einteilung als jedes Prinzips entbehrend. Er selbst teilt II Seite 40 ff. die Delikte ein in solche gegen die öffentliche Ordnung, gegen die staatliche Autorität, gegen das Publikum im allgemeinen (?! hier die Delikte gegen die Sittlichkeit, gegen die Religion u. s. w.), gegen Rechte der Person und gegen Vermögensrechte.

[1]) Über diese letztere conf. unten Seite 7, 11.

wonnen. Im allgemeinen können diese letzteren Gerichtshöfe auf Strafen bis zu 3 Monaten, ausnahmsweise bis zu 6 Monaten Gefängnis erkennen. Um einen ungefähren Begriff von dem Umfange der Thätigkeit der verschiedenen Gerichte zu geben, seien folgende Durchschnittszahlen der letzten Jahre angeführt. Von den Gerichtshöfen unterster Instanz wurden jährlich rund 725 000, von denen mittlerer Instanz knapp 10 000, von denen höchster Instanz rund 5000 Fälle behandelt. Man ersieht aus diesen Zahlen, welch große Bedeutung heute das summarische Verfahren in England gewonnen hat.

Gehen wir nun auf die Zusammensetzung der verschiedenen Gerichte kurz ein.

I. Die Assisenhöfe oder, wie der technische Ausdruck lautet, Courts of Commissioners of Oyer and Terminer and of General Gaol Delivery sind Gerichte vom höchsten Alter. Ihre heutige Verfassung datiert aus der Mitte des 12. Jahrhunderts, wo unter Heinrich II. ganz England in 6 Gerichtsbezirke (circuits) eingeteilt wurde. Für jeden dieser Bezirke wird alljährlich eine bestimmte Zahl von Richtern des höchsten Gerichtshofs bestimmt und denselben 3 Vollmachten erteilt, nämlich die Vollmacht, alle Klagen aus dem betreffenden Bezirke zu verhandeln und zu entscheiden (commission of Oyer and Terminer), sodann die Vollmacht, die Gefängnisse zu räumen (commission of Gaol Delivery), und endlich die im wesentlichen auf Zivilsachen bezügliche commission of the peace. Neben den Richtern des höchsten Gerichtshofs werden zur Unterstützung derselben eine Anzahl Anwälte (Queen's Counsels) und jetzt auch Kreisrichter (County Court Judges) in die Vollmacht eingeschlossen. Auf Grund der Vollmachten haben die Ernannten alljährlich 4 mal — früher nur 2 mal — den betreffenden Bezirk zu bereisen und in den besonders bestimmten Assisenstädten unter Zuziehung der, aus dem Bezirke gewählten Jury die anstehenden Zivil- wie Strafsachen zu erledigen; gewöhnlich übernimmt dabei einer der Richter durchweg die Zivil-, der andere die Strafsachen.

London und seine Umgebung sind nicht in einen dieser Gerichtsbezirke eingeschlossen, sondern es besteht hier seit 1834 ein besonderer Gerichtshof (Central Criminal Court), welcher jeden Monat seine Sitzungen hält und diejenigen Strafsachen zu erledigen hat, welche in andern Teilen des Landes vor den Assisen verhandelt werden. Den Vorsitz führt hier bei den wichtigsten Straffällen ein Richter des höchsten Gerichtshofs, im übrigen einer der 3, von der City of London ernannten Richter (Recorder, Common Sergeant und Judge of the City of London Court).

II. Die Quartalssitzungen der Friedensrichter stammen aus dem 14. Jahrhundert aus der Regierungszeit Eduard III. Die County Quarter Sessions werden 4 mal im Jahre von den Friedensrichtern jeder Grafschaft abgehalten. Das Verfahren ist im wesentlichen dasselbe, wie bei den Assisenhöfen; der Unterschied besteht nur darin, daß an Stelle eines gelehrten Richters hier eine Richterbank von sämtlichen anwesenden Friedensrichtern der Grafschaft gebildet wird. Die bekanntlich unbesoldeten Friedensrichter wählen aus ihrer Mitte den Vorsitzenden, und es ist dies regelmäßig ein studierter Jurist. Außer diesem Vorsitzenden genügt zur Zusammensetzung der Richterbank die Anwesenheit eines weiteren Friedensrichters, wenngleich gewöhnlich die Zahl der Anwesenden eine viel größere ist.

Von der Jurisdiktion der County Quarter Sessions sind auf Grund besonderer Verleihung eine Anzahl größerer inkorporierter Städte (boroughs) eximiert worden. In diesen Städten finden die Quartalssitzungen (Borough Quarter Sessions) unter dem Vorsitze eines von der Königin ernannten Recorder statt, welcher aus der Zahl der Advokaten von wenigstens 5 jähriger Praxis gewählt und von der Stadtgemeinde besoldet wird.

Die Kompetenz der Quartalssitzungen ist im Laufe der Zeit vielfachen Veränderungen, bald einschränkender, bald ausdehnender Natur, unterworfen gewesen; in neuerer Zeit ist dieselbe, wie schon erwähnt wurde, wieder erweitert worden. Es besteht jedoch die Bestimmung, daß die Assisenhöfe — resp. deren Präsident — und ebenso die Queen's Bench Division jede Strafsache mittels writ of certiorari von den Quartalssitzungen an sich ziehen können.[1])

[1]) Im Jahre 1884 wurden von den bisher behandelten Gerichten im ganzen 14 407 Personen abgeurteilt, davon entfallen auf:

die Assisenhöfe	25,2 %
den Central Criminal Court	9,5 %
die County Quarter Sessions	46,8 %
die Borough Quarter Sessions	18,5 %.

Unter den County Quarter Sessions verdienen diejenigen der Grafschaft

III. Viel jüngeren Datums, als die bisher behandelten Gerichte, ist die Gerichtsbarkeit der Courts of Summary Jurisdiction. Seit dem 17. Jahrhundert wurde in einer großen Anzahl von Spezialgesetzen den Friedensrichtern die Befugnis gegeben, ohne Zuziehung einer Jury wegen Übertretung bestimmter, in den Gesetzen enthaltener Vorschriften auf Geldstrafen und kurze Gefängnisstrafen zu erkennen. Im wesentlichen handelte es sich dabei um Polizeiübertretungen, nicht um eigentliche Delikte. Erst im Laufe dieses Jahrhunderts werden in immer weiterem Maße eigentliche Delikte in den Kreis der summarischen Gerichtsbarkeit hineingezogen.

Im Jahre 1847 wird 2 Friedensrichtern (justices in petty sessions) die Befugnis zugesprochen, gegen Personen unter 14 Jahren, welche wegen einfachen Diebstahls angeklagt sind, zu erkennen; im Jahre 1855 wird den Friedensrichtern in petty sessions die Kompetenz zur Entscheidung aller Diebstahlsfälle im Werte von nicht mehr als 5 Schilling beigelegt. Seitdem ist die Kompetenz von Jahr zu Jahr ausgedehnt worden.

Gleichzeitig mit dieser Bewegung sind denn auch in allen größeren Städten an die Stelle der 2 Friedensrichter (petty sessional court) besoldete Polizeirichter, (Stipendiary Police Magistrates) getreten. Die Metropolis ist in dieser Bewegung vorangegangen: es bestehen hier jetzt 13 Polizeigerichte (police courts) mit 23 besoldeten Richtern. Die Polizeirichter werden von dem Lordkanzler aus der Zahl der Advokaten mit wenigstens 5 resp. 7 jähriger Praxis ernannt und erhalten zumeist ein Gehalt von £ 1500 (30 000 M.). Ein Court of Summary Jurisdiction wird hiernach in den großen Städten von einem besoldeten Polizeirichter, in den übrigen Teilen des Landes von 2 Friedensrichtern gebildet. [1])

Die Kompetenz dieser Courts of Summary Jurisdiction und das Verfahren vor denselben hat im Jahre 1879 durch den

---

Middlesex besondere Erwähnung. Hier führt ein von der Königin ernannter besoldeter Richter (Assistant Judge) den Vorsitz. Von den oben angeführten 46,8 % entfallen auf die Middlesex County Sessions 13 %. Conf. Judicial Statistics 1884 Seite XXVII.

[1]) Neben diesem Gerichtshof mit summarischer Gerichtsbarkeit besteht die Befugnis des einzelnen Friedensrichters, wegen einzelner Polizeistraffälle geringfügige Strafen zu erkennen, fort.

Summary Jurisdiction Act — 42 a. 43 Vict. c. 49 — umfassende Veränderungen erfahren. Es sei hier nur hervorgehoben, daß fortab im Wege des summarischen Prozesses verhandelt werden kann:

1) gegen Kinder unter 12 Jahren wegen aller Delikte mit Ausnahme des Mordes, vorausgesetzt, daß nicht der Vater oder Vormund Widerspruch gegen diese Verhandlungsweise erhebt; die Strafe darf jedoch nicht höher bemessen werden, als auf 1 Monat Gefängnis oder 40 Schilling Geldstrafe;

2) gegen jugendliche Personen, d. h. Personen zwischen 12 und 16 Jahren, wegen der meisten Eigentumsdelikte, insbesondere wegen Diebstahls, Unterschlagung u. s. w., vorausgesetzt, daß sich der Angeschuldigte mit dieser Verhandlungsweise einverstanden erklärt. Die höchste Strafe, auf welche erkannt werden kann, ist 3 Monate Gefängnis mit harter Arbeit oder 10 £ Geldstrafe;

3) gegen Erwachsene wegen Diebstahls, wenn das Wertobjekt nicht mehr als 40 Schillinge beträgt oder wenn bei einem höheren Objekte der Angeklagte die That einräumt. Auch in diesen Fällen ist die Zustimmung des Angeklagten zu der summarischen Verhandlungsweise erforderlich. Der höchste Betrag der Strafe beträgt hier 6 Monate Gefängnis mit harter Arbeit.

In all diesen Fällen hat sich also der Court of Summary Jurisdiction zunächst darüber schlüssig zu machen, ob eine Strafe bis zum Höchstbetrage der ihm zustehenden Kompetenz für eine genügende Sühne zu halten ist, andernfalls hat er sich der Entscheidung der Sache zu enthalten und dieselbe an die Quarter Sessions abzugeben. Will der Court of Summary Jurisdiction selbst entscheiden, so hat er in den oben erwähnten Fällen den Angeklagten darauf aufmerksam zu machen, daß er Aburteilung vor den Quarter Sessions verlangen könne. Thatsächlich macht übrigens der Angeklagte von dieser Befugnis nur selten Gebrauch, besonders wohl mit Rücksicht darauf, daß die Quarter Sessions ja nur vierteljährlich zusammentreten und durch eine Verweisung an dieselben die Aburteilung regelmäßig längere Zeit hinausgeschoben wird. Bei einer Anzahl bestimmt bezeichneter Delikte, insbesondere bei thätlicher Beleidigung, steht übrigens dem Angeklagten das Wahlrecht überhaupt nicht zu; hier kann der Court of Summary Jurisdiction ohne weiteres bis zu 3 Monaten Gefängnis, in einzelnen Fällen sogar noch zu höherer Strafe verurteilen. [1])

Eine Berufung gegen die Entscheidung des Court of Summary Jurisdiction an die Quarter Sessions steht dem Verurteilten nur in denjenigen Fällen zu, wo die Berufung ausdrücklich im Gesetze zugelassen worden ist. Die Zahl dieser Berufungen ist eine minimale. Im Jahre 1884 betrug gegenüber einer Gesamtzahl von 583 330 im summarischen Verfahren Verurteilten die Zahl aller Berufungen 167. [2])

Nach diesem Überblicke über die Gerichtsorganisation soll zunächst die Gestaltung der Strafverfolgung kurz besprochen werden, ehe die Hauptmomente des Strafverfahrens selbst skizziert werden.

§ 4. Die Strafverfolgung und die Polizei.

Wohl die Haupteigentümlichkeit des englischen Strafverfahrens, diejenige, durch welche sich das englische Strafverfahren von dem aller anderen Länder unterscheidet, besteht darin, daß in England die Strafverfolgung im wesentlichen der privaten Initiative überlassen ist. Nur in einigen wenigen — das System an sich kaum berührenden — Ausnahmefällen geschieht die Verfolgung begangener Verbrechen durch den Staat resp. durch staatlich bestellte Personen.

Um diese Ausnahmefälle vorweg zu nehmen, so hat zunächst der Attorney General, der höchste Justizbeamte der

---

[1]) Um einen Begriff von der Bedeutung des summarischen Verfahrens in der englischen Strafgerichtsorganisation zu geben, seien hier folgende Ziffern angeführt. Während sich die Gesamtzahl der Personen, gegen welche vor den sub I. und II. behandelten Gerichtshöfen im Jahre 1884 verhandelt wurde, auf 14 407 belief, wurde in diesem Jahre im Wege des summarischen Verfahrens allein wegen Diebstahls und Diebstahlsversuchs gegen 48 899 Personen und wegen thätlicher Beleidigung gegen 87 691 Personen verhandelt. Dasjenige Delikt, wegen dessen die Courts of Summary Jurisdiction am meisten in Anspruch genommen wurden, war Trunkenheit; es wurde gegen 198 274 drunk and disorderly persons verhandelt. Cf. Judicial Statistics 1884 Seite 25.

[2]) Judicial Statistics 1884 Seite 37. Bei den Berufungen wurde in 93 Fällen die Entscheidung vollständig aufrecht erhalten, in 20 Fällen wurde das Strafmaß geändert, in 54 Fällen wurde die frühere Entscheidung aufgehoben.

Krone, die Befugnis, wegen gewisser misdemeanours bei dem höchsten Gerichtshofe direkt unter Umgehung der Vorinstanzen die Anklage zu erheben („filing a criminal information“). Von diesem Rechte wurde früher zuweilen bei politischen Schmähschriften oder wegen Verbreitung von Schriften aufrührerischen Inhalts Gebrauch gemacht; die Befugnis ist in neuerer Zeit überhaupt nur selten ausgeübt worden.

Von etwas größerer Bedeutung ist schon die Verfolgung von Delikten, durch welche das Finanzministerium oder andere öffentliche Behörden direkt verletzt sind, durch den Solicitor to the Treasury, einen im Finanzministerium als ständigen Beamten angestellten Anwalt. Es sind vorzugsweise Münzverbrechen, Bankrottfälle und schwere Gewaltthätigkeiten gegen Polizeibeamte, bei welchen die Strafverfolgung durch den Treasury Solicitor eingeleitet wird.

Endlich ist noch für eine Strafverfolgung bei gewaltsamen Todesfällen gesorgt. Ein hierfür besonders angestellter Beamter, der Coroner, hat bei verdächtigen Todesfällen eine Untersuchung vorzunehmen, den Thatbestand festzustellen, und, wenn die Jury, welche er hierbei zuzuziehen hat, gegen eine bestimmte Person ein Verdikt fällt, so liegt ihm die weitere gerichtliche Verfolgung derselben ob. [1])

Abgesehen von diesen Ausnahmefällen ist es im wesentlichen der verletzten Privatperson anheimgestellt, ob sie ein begangenes Delikt zur Anzeige bringen und weiterhin — allerdings in erheblich beschränkterem Maße —, ob sie die weitere Strafverfolgung übernehmen will. Ein gewisser indirekter Zwang wird dabei durch die Bestimmung des englischen Rechts ausgeübt, daß niemand wegen ihm zugefügter felony auf Entschädigung im Zivilwege klagen kann, bevor die strafrechtliche Verfolgung vor sich gegangen ist. Ferner ist zu berücksichtigen, daß bei gemeingefährlichen Verbrechen gesetzlich jedem, der zur Bestrafung des Thäters wirksame Hilfe geleistet hat, eine Belohnung aus öffentlichen Mitteln zugesichert ist. Endlich haben noch die Polizeibeamten, die Constables, die Pflicht, gewisse Delikte bei dem Friedensrichter zur Anzeige zu bringen, und sie werden von dem letzteren dann eventuell verpflichtet, für die weitere Strafverfolgung zu sorgen.

---

[1]) Im Jahre 1884 betrug die Gesamtzahl der Coroners Inquests 28603; das Verdikt der Jury lautete in 192 Fällen auf murder. Conf. Judicial Statistics 1884 Seite XXIII.

Die Mängel, welche diesem Systeme der Popularklage anhaften, sind seit lange in England gefühlt worden. Es ist klar, daß es bei diesem Systeme häufig an einem Interessenten fehlte, welcher geneigt und im stande war, die nicht unerheblichen Mühen und Beschwerden einer Strafverfolgung zu übernehmen; speziell, wenn der Verletzte zu den ärmeren Volksklassen gehört, muß es unbillig erscheinen, ihm diese Mühen zuzumuten. [1]) Auch sind in England vielfach direkte Mißbräuche vorgekommen: einem Verbrecher, welcher vermögend ist, gelingt es häufig genug, die Sache aus der Welt zu schaffen.

Diese und andre Mängel haben seit dem Jahre 1856 zu vielfachen Vorschlägen, parlamentarischen Verhandlungen und Gesetzentwürfen über die Einführung einer Staatsanwaltschaft geführt. Endlich im Jahre 1879 ist ein Gesetz, der Prosecution of Offences Act — 42 a. 43 Vict. c. 22 —, durchgegangen, durch welches eine öffentliche Anklagebehörde allerdings in äußerst beschränktem Umfange geschaffen worden ist. Nach diesem Gesetze hat der Staatssekretär des Innern (Home Secretary) einen erfahrenen Juristen gegen ein Gehalt von 2000 £ zum Director of Public Prosecutions zu ernennen, demselben werden zu seiner Unterstützung noch eine Anzahl — höchstens 6 — Juristen als Assistants zur Seite gestellt. Es soll zunächst die Pflicht dieses Director of Public Prosecutions sein, Polizeibeamte und Privatpersonen bei der Verfolgung von Verbrechen zu unterstützen. Sodann soll er nach besonderen, von dem Attorney General, dessen Aufsicht er unterstellt ist, zu erlassenden Regulativen selbst Anklage erheben in Fällen von Wichtigkeit oder Schwierigkeit oder in besonders gearteten Fällen, in denen die verletzte Privatperson die Strafverfolgung verweigert oder mangelhaft führt. Nichts in dem Gesetze soll aber das Recht einer jeden Privatperson, Anklagen zu erheben und durchzuführen, einschränken.

---

[1]) Sir Edmund Du Cane I S. 15 sagt mit Recht: „The law by which the person who has already suffered the wrong has further to take on himself the burthen and expense of carrying on a prosecution in which he has no more interest than any other member of society is both a grievous wrong to him and may give an undue advantage to the criminal. Probably many cases occur in which a man prefers to suffer in silence the first loss caused by the crime rather than add to it the trouble and loss he will suffer if he has to prosecute."

Soweit das Gesetz! Die praktische Bedeutung desselben war eine so geringfügige, die Zahl der von dem Director of Public Prosecutions betriebenen Strafverfolgungen eine so kleine, daß man im August 1884 das Amt überhaupt wieder aufhob, resp. das Amt eines Public Prosecutor mit demjenigen des Treasury Solicitor vereinigte.[1]) Fortab soll der Treasury Solicitor neben seiner bisherigen oben geschilderten Thätigkeit auch diejenigen Aufgaben erfüllen, welche das Gesetz von 1879 dem Public Prosecutor auferlegte.

Die Gesamtzahl der im Jahre 1884 von dem Public Prosecutor resp. dem Treasury Solicitor in seiner Eigenschaft als Public Prosecutor betriebenen Strafverfolgungen belief sich auf nur 247, die Zahl der dabei angeklagten Personen betrug 377. Daneben leitete der Treasury Solicitor in seiner Eigenschaft als Vertreter der verschiedenen Regierungsdepartements die Strafverfolgung ein wegen 258 Münzverbrechen und 436 andern Delikten (vorzugsweise Bankrottfälle, Ungesetzlichkeiten bei öffentlichen Wahlen und thätliche Angriffe gegen Beamte).[2])

In all diesen Fällen führt der Treasury Solicitor die Strafverfolgung nicht in eigner Person, sondern er instruiert einen Advokaten (Queen's Counsel), im Namen der Krone den Strafprozeß zu betreiben. Der öffentliche Ankläger hat dabei keinerlei Vorrechte vor einer Privatperson, welche die Bestrafung eines Verbrechers verfolgt.[3])

Ein Punkt dürfte hier noch der Erwähnung wert sein: nämlich die Frage nach der Tragung der Kosten der Strafverfolgung.

In früheren Zeiten mußte der Ankläger die gesamten Kosten der Strafverfolgung selbst übernehmen. Seit 1752 begann jedoch

---

[1]) Man ist in England fast allgemein der Ansicht, daß das Gesetz von 1879 ein Fehlgriff gewesen ist. Conf. Alaric Rumsey „The Prosecution of Offences Act" im Law Magazine VIII Seite 389 ff.

[2]) Judicial Statistics 1884 Seite XXXII.

[3]) Es ist hier nicht möglich, auf die in England bestehenden Anschauungen über Staatsanwaltschaft und Anklagemonopol näher einzugehen, doch existiert darüber eine solche Fülle interessanten Materials, daß nicht unterlassen werden soll, hier wenigstens die Hauptquellen anzugeben: Parliamentary Papers 1854/5 Band XII, 1856 Band VII (Select Committee on Public Prosecutors), 1874 Band XXIV (5 Reports of Judicature Commission), 1884 C. 4016 (Committee on Office of Public Prosecutor).

die Praxis, dem Ankläger bei felonies die notwendigen Kosten und Auslagen aus der Grafschaftskasse zu ersetzen. Allmählich bildete sich das Recht dahin aus, daß nicht nur bei felonies, sondern auch bei misdemeanours der Gerichtshof bei seiner Entscheidung Bestimmung darüber traf, ob dem Ankläger die Kosten und Auslagen vergütet werden sollen. Die Vergütung kann auch dann zugebilligt werden, wenn eine Verurteilung nicht erfolgt.

Im Jahre 1836 wurden dann die Hälfte dieser Kosten und 1846 — 14 a. 15 Vict. c. 45 — die gesamten Kosten auf die Staatskasse übernommen. Im Jahre 1884 betrug die hierdurch für den Staat entstehende Belastung insgesamt £ 140 354.[1])

Um ein Urteil über die in England gegenüber den Verbrechen ausgeübte Repression zu ermöglichen, sollen hier schließlich noch einige Angaben über die Polizei gemacht werden.

In England ist seit dem Jahre 1856 das System der besoldeten Polizei durchweg eingeführt worden. Zuerst hatte die Metropolis im Jahre 1829 an die Stelle der alten Friedensbewahrung eine besoldete Polizeimacht geschaffen, im Jahre 1839 wurde den Grafschaften das fakultative Recht zur Bildung einer neuen Polizeimacht gegeben, 1856 wurde diese Neubildung obligatorisch gemacht. Seitdem hat sich die Polizeimacht von Jahr zu Jahr vergrößert;[2]) sie beziffert sich jetzt insgesamt auf 34 999 Mann, d. h. es kommt in ganz England 1 constable auf eine Bevölkerungsziffer von 775, in der City von London schon auf 56, in den übrigen Teilen der Metropolis auf 404.[3])

---

[1]) Cf. Judicial Statistics 1884 Seite XXXII. Insgesamt hatte der Staat bei 38 319 Strafsachen die Kosten zu erstatten. Die durchschnittlich auf die einzelne Strafsache entfallenden Kosten waren bei den einzelnen Gerichtshöfen sehr verschieden; sie beliefen sich

bei den Assisenhöfen	auf 14 £	19 s	9 d
bei dem Central Criminal Court	„ 8 „	10 „	8 „
bei den County Quarter Sessions	„ 8 „	11 „	5 „
bei den Borough Quarter Sessions	„ 6 „	17 „	0 „
bei den Middlesex County Sessions	„ 5 „	0 „	1 „
bei den Courts of Summary Jurisdiction	„ 0 „	18 „	1 „

[2]) Die Vermehrung beträgt für die Zeit von 1872 bis 1884; 21,2 %.

[3]) Cf. für die hier gegebenen Ziffern Judicial Statistics 1884 Seite V—XIII. Die Gesamtausgaben für die Polizei beliefen sich 1884 auf £ 3 476 000, wovon etwa 40 % (nämlich £ 1 353 145) aus dem allgemeinen

Die Polizei führt in jedem Bezirke ein Register der notorischen Diebe, Hehler und sonstiger verdächtiger Personen, wobei die Bestimmung gilt, daß eine Person, welche sich nachweislich 1 Jahr lang in ehrlicher Weise den Lebensunterhalt erworben hat, aus dem Register zu streichen ist. Die Gesamtzahl der in diesen Registern 1884 eingetragenen Personen betrug 35 757, hiervon entfielen auf die Metropolis allein 2203. Es sei dabei bemerkt, daß die Zahl der in den Registern Eingetragenen seit längerer Zeit in starker Abnahme begriffen ist. Im Jahre 1873/4 betrug die Gesamtzahl 43 555; sie hat also innerhalb 10 Jahre um mehr als $^1/_6$ abgenommen.

Die Polizei führt weiterhin in jedem Bezirke ein Register verdächtiger oder als Zufluchtsstätten von Verbrechern bekannter Häuser. Auch hier ist eine ständige Abnahme zu konstatieren. Die Gesamtzahl der 1884 in diesem Register aufgeführten Häuser war 3842, davon entfielen auf die Metropolis 286.

Jede Polizeibehörde besitzt ferner ein für ganz England alljährlich zusammengestelltes Zentralregister der wegen schwerer Verbrechen zum zweiten Male Bestraften.

Dieses Zentralregister verdankt seine Entstehung dem Habitual Criminals Act 1869 — 32 a. 33 Vict. c. 99 —, welcher anordnete, daß alle mehrfach Bestraften in ein besonderes Zentralregister eingetragen werden sollten.[1]) Die Zahl dieser Eintragungen wurde jedoch so groß und das Register dadurch so unhandlich, daß man sich 1876 — 39 a. 40 Vict. c. 23 s. 2 — entschloß, die Eintragungen auf die Fälle schwerer Verbrechen, wie sie unter Anlehnung an s. 20 des Prevention of Crimes Act 1871 von dem Staatssekretär des Innern durch besonderes Regulativ vom 15. März 1877 näher bezeichnet wurden, zu beschränken.

Das erste hiernach im Jahre 1877 auf der Basis der von 1869—1876 ergangenen Verurteilungen zusammengestellte Zentralregister umfaßt die Namen von 12 164 Personen.[2]) Alljährlich erscheint seitdem eine, in den Strafanstalten gedruckte Fortsetzung

---

Staatsfonds gezahlt wurden. Durchschnittlich erhält ein Polizeibeamter an Gehalt jährlich 73 £ 14 s. 1 d.

[1]) Cf. auch Prevention of Crimes Act 1871 — 34 a. 35 Vict. c. 112 s. 6 und 20.

[2]) Cf. Du Cane II Seite 195.

dieses Registers, welche allen Polizeibehörden zugesandt wird. Das Register enthält bei jeder Person eine ganze Anzahl von Notizen, welche in den Strafanstalten aufgenommen worden sind und für eine Identifizierung der betreffenden Person außerordentlich wertvolle Anhaltspunkte geben.

Zur Ergänzung ist in den letzten Jahren noch ein neues Register, **Distinctive Marks Register**, hinzugekommen; dasselbe enthält eine genaue Beschreibung der bei den meisten schweren Verbrechern sich vorfindenden Tättowierungen. [1])

Es ist leicht einleuchtend, daß die Möglichkeit der Ergreifung eines Verbrechers durch diese Register in außerordentlicher Weise gefördert wird. [2])

Die Gesamtzahl der von der Polizei im Jahre 1884 vorgenommenen Ergreifungen wegen indictable offences — betreffs der von den Gerichtshöfen mit summarischer Jurisdiktion abgeurteilten Delikte existiert hier keine Ziffernangabe — betrug 20 366 gegenüber einer Gesamtzahl von 47 089 zur Anzeige gebrachten indictable offences.

§ 5. Das Strafverfahren.

I. Jeder Privatmann im Besitze der bürgerlichen Ehrenrechte kann die Einleitung eines Strafverfahrens veranlassen. Zu diesem Zwecke hat er in Person vor dem Friedensrichter — in den Städten vor dem Police Magistrate — eine Anzeige von dem begangenen

---

[1]) Bei meinen Besuchen englischer Strafanstalten habe ich den Eindruck gehabt, als ob diese Tättowierungen dort noch häufiger vorkämen als in den deutschen Anstalten. Auf meine Fragen bei den Gefängnisbeamten, welches wohl der Grund für diese Tättowierungen sein mag, habe ich sehr verschiedene Antworten erhalten. Manche meinten, der Gefangene habe es wohl lediglich aus Langeweile oder aus Nachahmungssucht gethan, während andere der Ansicht waren, der Gefangene wolle damit prahlen, daß er sich vor dem Schmerze der 1000 Nadelstiche nicht fürchte.

[2]) In dieser Richtung ist jetzt ein weiterer Fortschritt gemacht worden; es erscheinen nämlich neuerdings als periodische Beilagen der amtlichen Polizei-Zeitung (Police Gazette) Abbildungen von Gewohnheitsverbrechern. Jeder zu Strafknechtschaft oder längerer Gefängnisstrafe Verurteilte wird bei Antritt der Strafverbüßung und bei seiner Entlassung photographiert — cf. 34 a. 35 Vict. c. 112 s. 6 und die sehr eingehende Verfügung über die Art des Photographierens vom 3. Juli 1872. — Diese Photographieen werden dem in-

Delikte zu Protokoll zu geben (information). Wenn auf die Anzeige hin ein Haftbefehl (warrant of apprehension) erlassen werden soll, so hat der informer seine Anzeige zu beeidigen. Betrifft die Anzeige ein leichteres Delikt, oder ist eine Gefahr, daß sich der Beschuldigte der Verantwortung entziehen würde, nicht vorhanden, so erläßt der Friedensrichter einen einfachen Erscheinungsbefehl (summons). Wenn der Beschuldigte diesem Erscheinungsbefehle nicht nachkommt, so erfolgt seine Verhaftung.

II. Nachdem der Beschuldigte infolge Erscheinungs- oder Haftbefehls oder infolge seiner Verhaftung durch einen constable vor den Friedensrichter gebracht worden ist, wird er zunächst gefragt, ob er den ihm mitgeteilten Inhalt der information als richtig einräume und sich schuldig bekenne. In den meisten Fällen gibt der Beschuldigte eine Antwort hierauf überhaupt nicht ab, und es beginnt nun sofort, ohne daß irgend ein weiteres Verhör mit dem Angeschuldigten angestellt wird, die eidliche Vernehmung der Belastungszeugen, insbesondere des informer. Dem Angeschuldigten ist es dabei gestattet, persönlich oder durch seinen Verteidiger Fragen an die Zeugen zu stellen, die Zeugen dem Kreuzverhöre (cross-examination) zu unterwerfen.

Nach Beendigung des Belastungsbeweises wird der Angeschuldigte gefragt, ob er behufs seiner Verteidigung Zeugen benennen wolle. Diese Entlastungszeugen sind dann zu vernehmen. Erforderlichenfalls wird zu diesem Zwecke die Verhandlung vertagt.

Auch sonst kann der Friedensrichter, wenn es ihm ratsam erscheint, die Verhandlung vertagen. Während der Vertagung kann der Angeschuldigte auf schriftlichen Befehl des Friedensrichters in Haft gehalten werden

III. Wenn der Fall zur Kompetenz der Courts of Summary Jurisdiction gehört, so erläßt der Police Magistrate — resp. der petty sessional court von 2 Friedensrichtern — am Schlusse der Verhandlung das Urteil auf Verurteilung oder Freisprechung. Die Durchschnittsziffer der Verurteilungen ist dabei etwa 80 %.[1])

London auf dem Büreau des Chief Superintendent of the Criminal Investigation Department aufbewahrten Zentralregister beigefügt, und eine Anzahl derselben wird in der angegebenen Weise vervielfältigt.

[1]) Judicial Statistics 1884 Seite XVII: im Jahre 1884 betrug die Durchschnittsziffer: 80,4 %; im Jahre 1883: 81,1 %; im Jahre 1882: 80,3 %;

Bei den indictable offences hat der Friedensrichter auf Grund der Beweisaufnahme lediglich zu entscheiden, ob ein prima facie Beweis gegen den Angeschuldigten erbracht ist. Wenn der Friedensrichter den Belastungsbeweis nicht für hinreichend hält, um ein Hauptverfahren gegen den Angeschuldigten zu eröffnen, so hat er den letzteren zu entlassen (discharge); wenn er dagegen den Beweis in dieser Richtung für hinreichend hält oder doch wenigstens eine starke oder wahrscheinliche Vermutung der Schuld für begründet erachtet, so hat er den Angeschuldigten zum Hauptverfahren zu überweisen (commit).

Es ergeht in diesem Falle regelmäßig ein Untersuchungshaftbefehl (warrant of commitment). Doch kann der Friedensrichter den Beschuldigten auch freilassen, falls derselbe unter Sicherheitsleistung eine Verpflichtung zum Erscheinen in der Hauptverhandlung eingeht (bailed to appear for trial). Wird die von dem Friedensrichter verlangte Sicherheitsleistung nicht gestellt, so erfolgt die nachträgliche Verhaftung (committed for want of sureties). Das Verhältnis zwischen diesen verschiedenen möglichen Entscheidungen hat sich in den letzten Jahren folgendermaßen gestellt: [1])

	1884	1883	1882
a) discharged	26,1%	25,4%	26,0%
b) committed for trial	65,8 „	67,2 „	67,3 „
c) committed for want of sureties	0,3 „	0,2 „	0,3 „
d) bailed to appear for trial . . .	7,8 „	7,2 „	6,4 „

In allen Fällen, in welchen der Friedensrichter den Angeschuldigten zum Hauptverfahren überweist, hat er von den für das Hauptverfahren nötig erscheinenden Zeugen eine dahingehende Verpflichtung (recognizance) [2]) aufzunehmen (binding over to give evidence). Zur Eingehung dieser Verpflichtung können die Zeugen mittels Haft, welche bis zur Beendigung des Hauptverfahrens ausgedehnt werden kann, angehalten werden.

In gleicher Weise kann der Friedensrichter den informer

im Jahre 1881: 79,3 %. Die Gesamtzahl der im Jahre 1884 von den Courts of Summary Jurisdiction Verurteilten belief sich auf 583 330, diejenige der Freigesprochenen auf 141 378.

[1]) Judicial Statistics 1884 Seite XIII; die Zahl der Personen betrug unter den obigen Rubriken a) 5460, b) 13 410, c) 59, d) 1437.

[2]) Wegen der Form der recognizance cf. Abschnitt I § 7 Anm. 1.

zwingen, die weitere Strafverfolgung zu übernehmen (binding over to prosecute). Doch ist es in neuerer Zeit Sitte geworden, anstatt eines informer, welcher die weitere Strafverfolgung nicht zu übernehmen wünscht, einen Polizeibeamten oder den Sekretär (clerk) des Friedensrichters oder eine sonst geeignete Persönlichkeit damit zu beauftragen. Die eigentlich juristische Durchführung der Sache übertragen diese Personen — ebenso wie es der informer zu thun pflegt — einem Anwalte.

IV. Nachdem in dieser Weise das friedensrichterliche Vorverfahren bei den schwereren, vor der Jury der Assisen oder der Quartalssitzungen zu verhandelnden Strafsachen beendigt ist, hat der die Strafverfolgung Betreibende auf Grund der information und der Protokolle des Friedensrichters eine Anklageschrift (bill of indictment) auszuarbeiten. Diese Anklageschrift hat unter Einhaltung bestimmter, streng vorgeschriebener Formeln den Thatbestand unter Angabe der Hauptbelastungszeugen zu enthalten und ist der bei jeder Schwurgerichtsperiode[1]) zu bildenden Grand Jury zur Beschlußfassung über die Versetzung des Angeschuldigten in den Anklagestand zu unterbreiten.

Die Grand Jury besteht aus wenigstens 12 und höchstens 23 Geschwornen. Bei Beginn jeder Schwurgerichtsperiode werden die Geschwornen zunächst beeidigt und dann von dem Richter, welcher in der betreffenden Periode den Vorsitz führt, mit der rechtlichen Natur der sämtlichen in der Periode anstehenden Strafsachen bekannt gemacht.

Daraufhin ziehen sich die Geschwornen zur geheimen Verhandlung und Beschlußfassung zurück. Der Angeschuldigte wird dabei nicht zugezogen, die Entscheidung beruht vielmehr ausschließlich auf der Anklageschrift und den Aussagen der erforderlichenfalls hierüber nochmals vernommenen Belastungszeugen.[2]) Wenn die Geschwornen auf

[1]) Es ist oben schon hervorgehoben worden, daß die Assisenhöfe und die Quartalssitzungen alle Vierteljahre zusammentreten; der Central Criminal Court hat in jedem Monate eine Schwurgerichtsperiode. Die längste Zeit, welche zwischen dem Abschlusse des Vorverfahrens und der Eröffnung des Hauptverfahrens liegen kann, beträgt somit 3 Monate, bei den Sachen, welche vor den Central Criminal Court gelangen, 1 Monat. Wenn der Angeschuldigte verhaftet worden ist, so befindet er sich während dieser Zeit in einem local prison als Untersuchungsgefangener (unconvicted prisoner).

[2]) Die Grand Jury ist nicht verpflichtet, die Belastungszeugen zu vernehmen.

Grund dieses Materials mit mindestens 12 Stimmen der Ansicht sind, daß der Grund zur Erhebung der Anklage ein genügender ist, so setzt der Obmann der Geschwornen auf die Rückseite der Anklageschrift die Worte „a true bill". Hiermit erst ist die Anklage als definitiv erhoben anzusehen. Der Angeklagte gelangt nunmehr zur Aburteilung vor die Petty Jury. Wenn sich die Geschwornen nicht dahin einigen können, das Verdikt „a true bill" auszusprechen, so setzt der Obmann auf die Anklageschrift die Worte „no bill"; der in Haft befindliche Angeschuldigte ist alsdann ohne weiteres aus der Haft zu entlassen.

Das ganze Verfahren vor der Grand Jury ist ein außerordentlich summarisches. Es handelt sich um eine bloß oberflächliche Vorprüfung, ob ein „prima facie case" gegen den Angeschuldigten vorliegt; diese Frage wird nur relativ selten verneint. Im Jahre 1884 gab die Grand Jury gegenüber einer Zahl von 14 407 Angeschuldigten nur 454 mal das Verdikt „no bill" ab. [1])

Der Vollständigkeit halber sei mit wenigen Worten noch auf einige zulässige Abweichungen von dem bisher beschriebenen regelmäßigen Gang des Verfahrens hingewiesen.

a) Die Anklageschrift kann, ohne daß ein friedensrichterliches Vorverfahren stattgefunden hat, direkt der Grand Jury unterbreitet werden. Nachdem jedoch durch den Vexatious Indictment Act — 22 a. 23 Vict. c. 17 — dies Verfahren für eine größere Anzahl von Delikten als unzulässig erklärt worden ist, geschieht dies nur noch in wenigen, vereinzelten Fällen.

b) Die Grand Jury kann, ohne daß ein friedensrichterliches Vorverfahren stattgefunden hat, auf Grund eigner Wissenschaft von Amts wegen die Anklage (in diesem Falle presentment genannt) erheben. Dies ist jedoch jetzt fast ganz außer Gebrauch gekommen. [2])

---

[1]) Cf. Judicial Statistics 1884 Seite XXVII. In letzter Zeit sind vielfach Stimmen a.f Abschaffung der Grand Jury laut geworden. Cf. Kinghorn, „Ought Grand Juries to be abolished?" im Law Magazine VII Seite 12 ff.

[2]) In früherer Zeit war gerade diese Art des Verfahrens die regelmäßige. Zu einem richtigen Verständnisse des ganzen Instituts der Grand Jury ist die Kenntnis ihrer früheren Bedeutung als Anklagebehörde unbedingt erforderlich.

c) Unter Umgehung der Grand Jury kann die Strafsache direkt zur Entscheidung an die Petty Jury gebracht werden, falls die Jury des Coroner bei gewaltsamen Todesfällen die Anklage gegen eine bestimmte Person erhebt (conf. oben Seite 11). In der Praxis geschieht auch dies sehr selten; meistenteils wird auch hier zur genauen Feststellung des Sachverhaltes die Eröffnung eines Vorverfahrens durch den Friedensrichter vorgezogen.

d) Die Queen's Bench Division kann auf Anklage (in diesem Falle information genannt) des Attorney General oder des Master of the Crown Office ohne weiteres in die Verhandlung und Entscheidung einer Strafsache eintreten. Es ist dies ein Ausnahmeverfahren, welches nur im öffentlichen Interesse und höchst selten angewandt wird (conf. oben Seite 11). [1])

V. Über die Verhandlung vor der Petty Jury, welche aus 12 Geschwornen besteht, zum Schlusse noch wenige Worte.

Die Verhandlung beginnt damit, daß der Vertreter der Anklage (Counsel for the Crown) [2]) kurz die Thatsachen anführt, welche er durch seine Zeugen gegen den Angeklagten beweisen will. Im Anschlusse hieran vernimmt er die Belastungszeugen, welche alsdann von dem Anwalte des Angeklagten in Kreuzverhör (cross-examination) genommen werden können. Hieraufhin bespricht der Counsel for the Crown das Resultat des Anklagebeweises, und der Anwalt des Angeklagten hält seine Verteidigungsrede. Hat der Angeklagte seinerseits Zeugen benannt, so werden diese jetzt von seinem Anwalte verhört und dem Kreuzverhöre des Counsel for the Crown unterworfen. Der Verteidiger erhält dann noch einmal das Wort und der Counsel for the Crown darf darauf replizieren. Die Verhandlung schließt mit dem Resümee des vorsitzenden Richters, in welchem derselbe unter Darlegung der rechtlichen Fragen und unter Eingehen auf die Beweisführung den Geschwornen die Hauptgesichtspunkte an die Hand gibt, auf Grund deren die Entscheidung zu fällen ist.

Die Geschwornen ziehen sich sodann zur Beratung zurück. Sie

---

[1]) Eine andere, ebenfalls sehr selten angewandte Befugnis des Attorney General mag hier beiläufig erwähnt werden. Der Attorney General kann jede Strafverfolgung bis zum Erlasse des Urteils durch ein Nolle prosequi niederschlagen.

[2]) Dies ist, wie oben erörtert wurde, der Anwalt derjenigen Privatperson, welche die Strafverfolgung in dem betreffenden Falle betreibt.

müssen ihren Wahrspruch einstimmig abgeben. Erklären die Geschwornen nach einer, von dem Richter für angemessen erachteten Zeit der Beratung, daß sie sich nicht einigen können, so werden sie entlassen (discharged) und die Sache kommt vor einer andern Jury zur neuen Verhandlung.

Wenn der Spruch der Geschwornen auf Freisprechung lautet, so wird der Angeklagte sofort entlassen; lautet der Spruch auf schuldig, so hat der Richter das Strafmaß festzusetzen.[1])

Mit dieser Festsetzung des Strafmaßes treten wir in den eigentlichen Rahmen dieses Buches ein. Zum Schlusse dieser Vorbemerkungen sei noch der Prozentsatz der von den Geschwornen ausgesprochenen Verurteilungen für die letzten 5 Jahre angegeben; derselbe betrug:[2])

1884	1883	1882	1881	1880
77,2	77,4	76,6	76,7	75,9 %.

[1]) Ausnahmsweise kann der Richter, nachdem die Geschwornen das Schuldig gefunden haben, die Vollstreckung des Urteilsspruchs aussetzen, indem er bestimmt zu formulierende Rechtsfragen, welche im Laufe des Verfahrens hervorgetreten sind, zur endgültigen Entscheidung an den Gerichtshof für vorbehaltene Straffälle (Court for Crown Cases Reserved) verweist. Dieser Gerichtshof besteht aus 5 Richtern der Queen's Bench Division unter dem Vorsitze des Lord Chief Justice of England. Wenn dieser Gerichtshof zu Gunsten des Verurteilten entscheidet, so ist die Verurteilung aufzuheben. Die Gesamtzahl der Crown Cases Reserved betrug im Jahre 1884: 19. Judicial Statistics Seite XXXI. Cf. über den vor dem Court for Crown Cases Reserved verhandelten Mignonette-Fall Zeitschr. f. d. ges. Strafrechtswissensch. Band V S. 367 ff.

Es mag noch hervorgehoben werden, daß es eine eigentliche Berufung gegen Urteile der Schwurgerichtshöfe in England nicht gibt. Im Jahre 1882 legte die Regierung einen Gesetzentwurf vor, wonach bei Verurteilung zu Todesstrafe fortab die Berufung zulässig sein sollte. Das Committee, welchem der Entwurf zur Vorberatung überwiesen wurde, wünschte das Berufungsrecht auch auf Verurteilungen zu längerer Strafknechtschaft ausgedehnt zu haben. Die Regierung verweigerte jedoch diese Ausdehnung und zog schließlich den Gesetzentwurf zurück. Bei der Beratung dieses Entwurfs zeigte sich vielfach, und besonders in richterlichen Kreisen, eine entschiedene Abneigung gegen die Einführung der Berufung überhaupt. Man führte von dieser Seite an, daß die Gerichtsverhandlungen, sobald sie nicht endgültige wären, weniger sorgfältig geführt werden würden, und daß das bestehende Begnadigungsrecht der Königin vollständig genügte, um zu verhindern, daß zweifelhafte Urteile vollstreckt würden.

[2]) Judicial Statistics 1884 Seite XXVIII.

Abschnitt I.*)

Die Strafmittel des englischen Rechts.

*) Als Quellen für diesen ersten Abschnitt sind die Arbeiten von Sir James Stephen I und II, sowie das Buch von Sir Edmund Du Cane II zu nennen. Insoweit bei der Darstellung der einzelnen Strafen anderweitige Litteratur benutzt worden ist, ist dieselbe am Eingange der betreffenden Paragraphen namhaft gemacht.

§ 1. Einleitung.

Die nach dem heutigen englischen Strafrechte zulässigen Strafmittel sind: Todesstrafe, Strafknechtschaft (Penal Servitude), Gefängnis mit oder ohne harte Arbeit (Imprisonment with or without hard labour), Geldstrafe (Fine), Friedensbürgschaft (Putting under Recognizances), körperliche Züchtigung (Whipping), Überweisung an eine Besserungsanstalt (Confinement in a Reformatory School) und Polizei-Aufsicht (Police Supervision).

Außer diesen Strafen ist hier noch des nähern zu behandeln die Transportation (Transportation), eine Strafe, auf welche jetzt nicht mehr erkannt werden kann, in deren Verbüßung sich aber immer noch eine kleine Anzahl von Personen[1]) befindet und welche außerdem zum richtigen Verständnisse der historischen Entwickelung des englischen Strafensystems von maßgebender Bedeutung ist.

Diese historische Entwickelung läßt sich kurz dahin charakterisieren: die Zahl der ursprünglich — nach common law — mit Strafe bedrohten Handlungen ist eine sehr kleine gewesen. Diese Delikte zerfielen in felonies und misdemeanours. Für die felonies bestand die Strafe in der Todesstrafe, für die misdemea-

---

[1]) Der Report of the Directors of Convict Prisons 1884/5 gibt Seite XXIV die Ziffer approximativ auf 50 an.

nours in Gefängnis, körperlicher Züchtigung, Geldstrafe und Friedensbürgschaft. Allmählich wurde die Anwendung der Todesstrafe mehr und mehr eingeschränkt; an ihre Stelle trat zunächst die Transportationsstrafe und nach deren Aufhebung die Strafknechtschaft. Die Gesetzgebung — statute law — hat sodann im Laufe der Zeit, insbesondere seit dem Beginne des 18. Jahrhunderts, die Zahl der strafbaren Handlungen erheblich vermehrt und eine große Anzahl neuer felonies und misdemeanours geschaffen. [1]) Dabei ist man immer mehr und mehr von dem Systeme der absoluten Strafen abgewichen und hat dem freien richterlichen Ermessen sowohl hinsichtlich der Auswahl der Strafmittel wie der Höhe der im Einzelfalle zu erkennenden Strafen einen immer größeren Spielraum gewährt. Das Resultat dieses Entwickelungsprozesses ist, daß der Richter heutzutage — von einigen Ausnahmen [2]) abgesehen — wegen desselben Deliktes je nach Lage des Falles auf Gefängnis von 1 Tage bis zu 2 Jahren [3]) oder auf Strafknechtschaft von 5 Jahren und mehr erkennen kann.

Dieser, dem Richter gelassene weite Spielraum bei Zuerkennung der Strafe ist eine besonders charakteristische Erscheinung des heutigen englischen Strafrechts. [4]) Es ist die ganze konstante Richtung der neueren Strafgesetzgebung gewesen, von der gesetzlichen Fixierung eines Strafmaximum oder Strafminimum nur bei wenigen, schweren Delikten Gebrauch zu machen.

Eine andere charakteristische Erscheinung, welche hier Erwähnung verdient, ist die geringe Beachtung, welche in England dem sogenannten philosophischen Teile des Strafrechts — der Frage nach dem Rechte zu strafen und nach dem Zwecke der Strafe — geschenkt wird. Gerade uns Deutschen

---

[1]) Diese Vermehrung der mit Strafe bedrohten Handlungen findet sich übersichtlich zusammengestellt bei Stephen I Band II Seite 197 ff.

[2]) So ist für widernatürliche Verbrechen das Strafminimum von 10 Jahr Strafknechtschaft bestehen geblieben.

[3]) Es gibt Ausnahmefälle, in denen auf Gefängnis über 2 Jahre erkannt werden kann; z. B. ist wegen Meineids eine Gefängnisstrafe bis zu 7 Jahren zulässig.

[4]) Das Gesetz, durch welches vorzugsweise die diskretionäre Befugnis des Richters eingeführt resp. erweitert worden ist, ist der Act 9 a. 10 Vict. c. 24 aus dem Jahre 1846.

muß es auffallen, in englischen Strafrechtsbüchern diese Fragen kaum berührt zu finden: in dem ausgezeichneten, auf der Höhe der Wissenschaft stehenden dreibändigen Werke von Sir James Stephen, History of the criminal law, sind diesen Fragen nur gelegentlich einige Worte gewidmet, und, soweit man daraus die Ansicht des gelehrten Verfassers entnehmen kann, dürfte dieselbe lediglich dahin gehen: wenn der Gesetzgeber, und im Einzelfalle der Richter, es für angemessen hält, eine gewisse Handlung in einer gewissen Weise zu bestrafen, so würde er sich einer Schwäche schuldig machen, wenn er diese Strafe nicht eintreten ließ.[1]) Der einzige Punkt, welchen Stephen etwas ausführlicher behandelt, ist das Verhältnis zwischen Strafrecht und Moral.[2])

Diese geringe Beachtung der philosophischen Fragen des Strafrechts bei den englischen Kriminalisten steht im auffallenden Gegensatze zu der eingehenden Erörterung, welche allen strafprozessualen Fragen gewidmet wird. Die englische Strafrechtswissenschaft sieht ihre Aufgabe und ihr Ziel nicht sowohl in der philosophischen Behandlung von Strafbarkeitsprinzipien und Strafrechtstheorieen, als in der Ausbildung strafprozessualer Formen, durch welche genügende Garantieen zur Verhütung des Mißbrauchs der richterlichen Gewalt gewonnen werden.

Soweit es bei dieser Lage der Sache überhaupt möglich ist, von dem Vorherrschen einer bestimmten Strafrechtstheorie in England zu sprechen, kann man wohl sagen, daß die juristischen Kreise und die Kreise der Gefängnisverwaltung im wesentlichen der Abschreckungstheorie huldigen.[3]) Dagegen ist die

---

[1]) Cf. übereinstimmend hiermit Oliver Smith in der Zeitschr. für die gesamte Strafrechtswiss. Band IV Seite 519.

[2]) Stephen I Band II Seite 79 ff. In folgenden Sätzen ist das Wesentlichste der Ansicht von Stephen enthalten: „The infliction of punishment by law gives definite expression and a solemn ratification and justification to the hatred, which is excited by the commission of the offence.“ „The criminal law proceeds upon the principle that it is morally right to hate criminals and it confirms and justifies that sentiment by inflicting upon criminals punishments which express it.“

[3]) Cf. Du Cane II S. 1 f. „the object of the penal element is more to deter others than the effect on the individual subjected to the punishment“ „if by punishing those who have an incurable tendency to crime we can deter fresh recruits from joining the ranks of the criminal class, the object of punishment is effected.“ Cf. auch Du Cane in der

Howard Association, welche sich eines großen praktischen Einflusses auf alle Fragen der Gefängnisverwaltung erfreut,[1]) den Anhängern der Besserungstheorie zuzuzählen. Doch sei hier schon erwähnt, daß diese Gesellschaft es geradezu ängstlich vermeidet, von einem „Bessern" (improving) der Verbrecher zu sprechen, sie bezeichnet vielmehr als ihr Streben, „to reform the criminals", die Verbrecher „umzubilden". Es wird hierauf später noch näher eingegangen werden.

Wir wenden uns nunmehr zu den einzelnen Strafmitteln[2]) und der Darlegung ihrer geschichtlichen Entwickelung.

§ 2. Die Todesstrafe.

Litteratur: Report of the Capital Punishment Commission 1866. — Marquardsen, die Todesstrafe in England in der kritischen Zeitschr. für Rechtswissenschaft Band 22 Seite 478 ff., Band 23 Seite 202 ff., 311 ff. — Anlage zu den Motiven des Strafgesetzentwurfs für den Norddeutschen Bund Seite 52 ff.

Von der ersten Hälfte des 13. Jahrhunderts ab ist nachweislich die Todesstrafe die regelmäßige Strafe für alle nach gemeinem Rechte (common law) strafbaren treasons und felonies — diese letztere Rubrik umfaßt: Mord und Totschlag, Notzucht, Raub, Falsch-

---

Law Quarterly Review 1886 Seite 231 „the amount of punishment should be such as will deter others." — Cf. ferner Sir Joshua Jebb, Report of the Directors of Convict Prisons 1860 Seite XVIII: „means of reformation should be united with punishment, but the deterring features of a sentence should predominate."

[1]) Die Howard Association wurde unter dem Protektorate von Lord Brougham gegründet „for the promotion of the best methods of crime prevention and penal treatment." „It labours to indoctrinate the public mind with the importance of a reformatory and radically preventive treatment." William Tallack entwickelt als Sekretär der Gesellschaft insbesondere auf litterarischem Gebiete eine sehr umfangreiche und segensreiche Thätigkeit. Die Gesellschaft gibt alljährlich einen ausführlichen Jahresbericht heraus, in welchem alle aktuellen Fragen auf dem Gebiete des Strafrechts und des Gefängniswesens Erörterung finden.

[2]) Bei der Aufzählung der nach heutigem Rechte zulässigen Strafen hätte noch angeführt werden können, daß seit einem Gesetze von 1870 — 33 a. 34 Vict. c. 23 s. 2 — bei einer Verurteilung wegen einer felony zu einer Strafe

münzerei und Diebstahl — gewesen, und sie ist gesetzlich — mit alleiniger Ausnahme des sogenannten kleinen Diebstahls (petty larcency) — für diese Delikte die regelmäßige Strafe geblieben bis zum Jahre 1827. In der Praxis ist jedoch im Laufe dieser Zeit der Zustand durch die Entwickelung der sogenannten Rechtswohlthat des Klerus (benefit of clergy) erheblichen Veränderungen unterworfen gewesen; thatsächlich ist die Verhängung der Todesstrafe infolge dieser Rechtswohlthat für die Mehrzahl der genannten Delikte fast zur Ausnahme geworden, und es sind andere Strafmittel an deren Stelle getreten.[1])

Das benefit of clergy bestand ursprünglich in dem Rechte des Klerus, von der Jurisdiktion der staatlichen Gerichtsbarkeit eximiert und lediglich den kirchlichen Gerichten unterworfen zu sein. Es sei dabei bemerkt, daß diese Exemtion in England niemals vollständig anerkannt worden ist, und daß sich im Laufe der Zeit eine größere Reihe von Einschränkungen herausgebildet hat. Hervorzuheben ist vor allem ein Gesetz vom Jahre 1275 — 3 Edward I —, wonach auch gegen einen Geistlichen die Anklage vor den staatlichen Gerichtshöfen zu erheben war und er erst nach Erhebung der Anklage von dem Rechte der Exemtion Gebrauch machen konnte. Unter Heinrich VI. bildete sich daraufhin sogar die Praxis, daß der staatliche Gerichtshof das „Schuldig" oder „Nichtschuldig" gegen den Geistlichen aussprach und erst im Falle des Schuldigspruches der Geistliche von der Rechtswohlthat Gebrauch machte. Das benefit of clergy wurde so von einer Exemtion von der staatlichen Gerichtsbarkeit zu einer Exemtion von der staatlichen Strafe.

In dieser letzteren Gestalt ist das benefit of clergy dann auf einen immer weiteren Kreis von Personen ausgedehnt worden. Bereits ein Gesetz von 1350 — 25 Edw. III c. 3 — billigte das benefit of clergy allen „secular as religious

---

von 1 Jahr Gefängnis mit harter Arbeit oder zu einer härteren Strafe der Verlust aller öffentlichen Ämter und Würden sowie der aus öffentlichen Ämtern herrührenden Pensionen u. s. w. eo ipso eintritt. Eine Strafe der Aberkennung der bürgerlichen Ehrenrechte kennt das englische Recht nicht.

[1]) Blackstone, Commentaries on the laws of England Band IV (2. Ausgabe) Seite 364, sagt von dieser Entwickelung: „the English legislature, in the co rse of a long and laborious process, extracted by noble alchemy rich medicines out of poisonous ingredients."

clerks" zu. Die Praxis dehnte dann diesen Begriff der „secular clerks" allmählich dahin aus, daß darunter alle männlichen Personen verstanden wurden, welche lesen und schreiben konnten. Weibliche Personen waren lange Zeit hindurch — mit Ausnahme der Nonnen — von dem Privileg ganz ausgeschlossen; im Jahre 1622 — 22 James I c. 6 — erhielten auch sie das Privileg für kleinere Diebstahlsfälle (bis zu einem Wertobjekte von 10 Schilling), im Jahre 1692 — 4 Will. a. Mary c. 9 — wurden endlich die Frauen den Männern ganz gleich gestellt. Im Jahre 1705 hörte das benefit of clergy überhaupt auf, ein Privileg bestimmter Klassen zu sein, indem das Erfordernis der Fähigkeit des Lesens und Schreibens gesetzlich — 5 Anne c. 6 — aufgehoben wurde.

Mit dieser Richtung der Ausdehnung des benefit of clergy auf immer weitere Klassen der Bevölkerung geht nun aber ganz naturgemäß eine andere Richtung, nämlich die Beschränkung des benefit auf einen kleineren Kreis von Delikten, einher. Von vornherein erscheint das benefit ausgeschlossen bei Hochverrat, bei Brandstiftung und bei Straßenraub. Diese Exemtionen bestimmter Delikte von der Rechtswohlthat des Klerus wurden durch eine sehr detaillierte Gesetzgebung immer weiter ausgedehnt: 1547 werden alle Arten des Mordes und des nächtlichen Diebstahls mit Einbruch davon ausgeschlossen; die nachfolgende Gesetzgebung hebt dann weitere Arten des Diebstahls hervor, auf welche das benefit keine Anwendung finden soll. Das Resultat dieser Bewegung ist eine scharfe Trennung zwischen felonies clergyable, bei denen das benefit bestand, und felonies not clergyable.

Für die clergyable felonies war ursprünglich jedwelche Bestrafung ausgeschlossen. Diejenigen, welche eines derartigen Deliktes schuldig befunden waren, wurden auf Grund eines Gesetzes von 1487 — 4 Henry VII c. 13 — am Daumen mit der Brandmarke M (murder) im Falle des Mordes und mit der Brandmarke T (theft) im Falle des Diebstahls versehen. Wenn eine so gebrandmarkte Person zum zweiten Male eine felony beging, so blieb sie — wenn es nicht ein ordinierter Geistlicher war — von dem benefit of clergy ausgeschlossen. Im Jahre 1576 — 18 Eliz. c. 7 s. 3 — wurde den Richtern die Befugnis gegeben, Personen, welche einer clergyable felony schuldig befunden waren, bis zur Dauer von 1 Jahr im Gefängnis zu behalten.

Bei den *felonies without benefit of clergy* — hierzu gehörte während des 18. Jahrhunderts Hochverrat, Seeräuberei, Mord, nächtlicher Diebstahl mit Einbruch (burglary), Mordbrennerei, Brandstiftung, Straßenraub, Kirchenraub, Entführung mit der Absicht zu heiraten, Pferdediebstahl, Diebstahl im höheren Wertobjekte [1]) — war die *Todesstrafe zu verhängen*. Allein diese Strafe wurde, soweit nach den vorhandenen Nachrichten beurteilt werden kann, *nur relativ selten vollstreckt*. Für einige wenige Delikte wurde der Todesstrafe körperliche Verstümmelung — besonders beliebt war dabei das Ohrstutzen (cropping the ears) — substituiert, meistenteils erfolgte *Begnadigung unter der Bedingung der Transportation* nach Amerika oder später nach Australien, sei es für Lebenszeit, sei es für eine längere Reihe von Jahren (das Nähere hierüber im nächsten Paragraphen).

Wenn in dieser Weise die Härte des englischen Strafrechts thatsächlich erheblich gemildert wurde, [2]) so blieb doch der grelle Widerspruch der Gesetzgebung mit dem im übrigen vorgeschrittenen Kulturzustande Englands offenkundig bestehen. Seit dem Anfange dieses Jahrhunderts beginnen denn auch die Reformbestrebungen sich mit Macht zu regen. Unter den Männern, welche an der Spitze dieser Bewegung standen, sind in erster Linie Bentham, Samuel Romilly und Makintosh zu nennen.

Zunächst erfolgte im Jahre 1827 ein Gesetz Georgs IV. — 7 a. 8 George IV c. 28 —, durch welches *das ganze Institut des benefit of clergy aufgehoben wird*. Todesstrafe soll fortab nur erkannt werden bei denjenigen Delikten, für welche dieselbe ausdrücklich im Gesetze festgesetzt ist. Für andere felonies wird Transportation oder auch Gefängnisstrafe angeordnet. Es folgen sodann unter Wilhelm IV. und besonders seit dem Regierungsantritte der Königin Viktoria eine lange Reihe von Gesetzen, durch

[1]) Die Zahl dieser Delikte ist im Laufe des 18. Jahrhunderts dadurch, daß durch besondere Gesetze (statute law) neue felonies geschaffen wurden, noch vermehrt worden.

[2]) Immerhin wurden doch in dem Zeitraume von 1820—1826 in England 528 Personen hingerichtet; davon waren zur Todesstrafe verurteilt wegen nächtlichen Diebstahls mit Einbruch (burglary): 116, wegen Falschmünzerei: 53, wegen Pferdediebstahls: 50 Personen.

welche der Kreis der mit Todesstrafe bedrohten Delikte immer mehr und mehr eingeschränkt wird. [1])

Nach heutigem englischen Rechte sind nur noch folgende Delikte mit Todesstrafe bedroht: Hochverrat (high treason), Mord (murder), Seeräuberei unter Anwendung von Gewalt (piracy with violence) und Brandstiftung an Schiffsdocks und Arsenalen (setting fire to dockyards and arsenals).

Thatsächlich sind seit dem Jahre 1861, in welchem der Kreis der todeswürdigen Verbrechen in besonders erheblichem Maße eingeschränkt worden war, bis zum Jahre 1884 überhaupt nur 2 Personen wegen eines andern Verbrechens, als Mord, zum Tode verurteilt worden. [2]) Die Zahl der Verurteilungen wegen Mordes betrug im Jahre 1884: 38; es ist das die höchste Ziffer, welche sich überhaupt in den seit 1836 vorliegenden offiziellen Statistiken findet. Die niedrigste Ziffer weisen die Jahre 1837, 1850, 1854 und 1855 auf mit je 11 wegen Mordes zum Tode verurteilten Personen. Die Ziffer beträgt jährlich durchschnittlich in der Periode 1836—1884: 23; die Schwankungen zwischen den einzelnen Jahren sind recht erheblich, so betrugen die Verurteilten 1883 nur 23 gegenüber 38 im Jahre 1884 wegen Mordes zum Tode Verurteilten.

Die Zahl der Hinrichtungen beträgt seit 1861 durchschnittlich etwas über die Hälfte der Zahl der Verurteilungen, [3]) doch variiert dies Verhältnis in den einzelnen Jahren ziemlich erheblich. Im Jahre 1884 wurden 16 Personen hingerichtet. Die höchste Ziffer der Hinrichtungen zeigt das Jahr 1877 mit 23, die niedrigste das Jahr 1871 mit 4.

Über die Begnadigung zum Tode Verurteilter hat in

---

[1]) 2 a. 3 Will. IV c. 62, 5 a. 6 Will. IV c. 81, 1 Vict. c. 84—91, 4 a. 5 Vict. c. 38, 24 a. 25 Vict. c. 96—100.

[2]) Cf. für die hier gegebenen Ziffern Report of the Directors of Convict Prisons für 1884/5 Seite XXII. Im Jahre 1836 betrug die Zahl der wegen andrer Verbrechen als Mord zum Tode Verurteilten: 474, im Jahre 1838 nur noch 91, seitdem war die Zahl fortdauernd bis 1861, wo sie 24 betrug, gesunken.

[3]) Bis zum Jahre 1861 ist ein sehr viel geringerer Prozentsatz der Todesstrafen wirklich vollzogen worden; z. B. 1836 gegenüber 494 Todesurteilen nur 17 Hinrichtungen; 1838 gegenüber 116 Todesurteilen 6 Hinrichtungen. Während die absolute Zahl der Todesurteile seit 1861 sehr erheblich abgenommen hat, hat die absolute Zahl der Hinrichtungen entschieden zugenommen.

England der Staatssekretär des Innern (Home Secretary) zu entscheiden, welcher dem Parlamente gegenüber die volle Verantwortung dafür zu übernehmen hat. Es ist durchaus nicht selten, daß der Staatssekretär im Parlamente sowohl wegen der Vollstreckung von Todesurteilen wie wegen der Begnadigung zur Rede gestellt wird. Eine Bestätigung der Todesurteile durch die Königin findet nicht statt;[1]) die Königin unterzeichnet aber, nachdem ihr der Staatssekretär den regelmäßig nach vorgängiger Anhörung des erkennenden Richters angefertigten Begnadigungsbericht vorgelegt hat, den Gnadenerlaß unter Gegenzeichnung des Staatssekretärs.

Was die Vollstreckungsart der Todesstrafe anbetrifft, so geschieht dieselbe in England von alters her durch Hängen. Eine Zeitlang bestand die Vollstreckungsart besonders bei Frauen sowie im Falle der Verurteilung wegen Ketzerei im Verbrennen. Die Hinrichtung geschah bis 1868 öffentlich; in London war der regelmäßige Platz für Hinrichtungen Tyburn und seit 1785 Newgate. Die von englischen Schriftstellern vielfach geschilderten Szenen bei diesen Hinrichtungen entsprachen wenig dem Ernste des Aktes und riefen eine förmliche Bewegung auf Einführung der Intramuran-Hinrichtung hervor. Durch ein Gesetz von 1868 — 31 a. 32 Vict. c. 24 — ist dieselbe denn auch eingeführt worden.

Seitdem geschieht die Hinrichtung innerhalb der Baulichkeiten des Gefängnisses, in welchem der Verurteilte aufbewahrt wird. Regelmäßig findet dieselbe um 8 Uhr morgens am dritten Montage nach der Verurteilung statt. Während der Exekution wird eine schwarze Flagge auf dem Gefängnisgebäude aufgehißt, die Glocken des Gefängnisses und einer benachbarten Kirche werden 15 Minuten vor und 15 Minuten nach der Exekution geläutet. Bei der Hinrichtung, deren Vollstreckung zu den Pflichten des Sheriff der betreffenden Grafschaft gehört, haben außer dem Sheriff oder dessen Stellvertreter der Direktor, der Arzt und der Geistliche des Gefängnisses zugegen zu sein. Alle Friedensrichter der Grafschaft haben

[1]) Bis zum Regierungsantritt der Königin Victoria mußten sämtliche in London selbst (nicht aber in andern Teilen des Landes) gefällten Todesurteile dem King in Council — d. i. dem Könige unter Zuziehung des Privy Council — zur Bestätigung unterbreitet werden. Diese Einrichtung wurde jedoch durch ein besonderes, bald nach dem Regierungsantritte der jugendlichen Königin erlassenes Gesetz aufgehoben.

ein Recht zugegen zu sein; andern Personen kann dies von dem Sheriff gestattet werden. Unmittelbar nach der Exekution hat der Coroner mit einer Jury die Leichenschau abzuhalten und ein Attest auszustellen, daß die Strafe an der betreffenden Person vollstreckt ist. Dieses Attest wird auf die Dauer von 24 Stunden an dem Gefängnisthore angeschlagen. Die Leiche wird innerhalb der Gefangenanstalt begraben.

Es erübrigt noch kurz auf die in England bestehende Bewegung auf gänzliche Abschaffung der Todesstrafe einzugehen. An der Spitze dieser Bewegung steht seit ihrer Gründung die Howard Association. Diese Gesellschaft hat die Frage fortdauernd auf der Tagesordnung erhalten und wiederholentliche Erörterungen im Parlamente veranlaßt.

Nachdem im Jahre 1840 ein Antrag auf Abschaffung der Todesstrafe im Unterhause von einem Mr. Ewart eingebracht, aber abgelehnt worden war, wurde dieser Antrag im Jahre 1847 erfolglos wiederholt. In demselben Jahre setzte das Oberhaus eine Kommission nieder, welche unter anderm auch über die Wirksamkeit der Todesstrafe berichten sollte. Die Meinungen der von der Kommission vernommenen Sachverständigen waren sehr geteilt, besonders über die Frage, ob die Todesstrafe in der That eine abschreckende Wirkung ausübe. In den Jahren 1848, 1849 und 1850 wurden wiederum Anträge auf Abschaffung der Todesstrafe gestellt und abgelehnt. Im Jahre 1864 beriet eine Königliche Kommission (Royal Commission) über die Frage der Abschaffung der Todesstrafe; 4 von den 12 Mitgliedern der Kommission erklärten sich dafür. Die Kommission selbst schlug in ihrem, im Januar 1866 erstatteten Berichte vor, die Todesstrafe nur für Hochverrat und Mord beizubehalten, dabei aber 2 Grade von Mord zu unterscheiden und nur für den ersten Grad die Todesstrafe anzudrohen.

Es muß hierbei bemerkt werden, daß die Definition des Mordes und die Unterscheidung zwischen Mord und Totschlag im englischen Rechte eine recht wenig bestimmte ist. Als Mord (murder) wird bezeichnet die rechtswidrige Tötung mit vorbedachter Böswilligkeit (unlawful homicide with malice aforethought), als Totschlag (manslaughter) die rechtswidrige Tötung ohne vorbedachte Böswilligkeit (unlawful homicide without malice aforethought). [1]

---

[1]) Stephen II Seite 158.

Was dabei unter „Vorbedacht“ und was unter „Böswilligkeit“ zu begreifen ist, ruft juristisch vielen Zweifel hervor. Es sei hier nur erwähnt, daß Stephen „malice“ als vorhanden annimmt 1) bei der Absicht zu töten, 2) bei der Absicht, einer Person einen schweren körperlichen Schaden zuzufügen, 3) bei der Absicht, irgend eine felony zu begehen, 4) bei der Kenntnis, daß die beabsichtigte That wahrscheinlich den Tod oder die schwere körperliche Verletzung einer Person verursachen wird, 5) bei der Absicht, einem in Ausübung seines Berufs begriffenen Beamten mit Gewalt Widerstand zu leisten. Das englische Recht begreift somit unter Mord eine Reihe von Fällen, welche wir als Totschlag, Kindesmord oder als Körperverletzung mit tödlichem Erfolge charakterisieren würden.

Die Royal Commission schlug nun in ihrem Berichte vom Jahre 1866 vor, diese verschiedenartigen Fälle des Mordes in 2 Klassen zu teilen und die Todesstrafe nur für die erste Klasse beizubehalten. Das Oberhaus verwarf jedoch im Jahre 1866 diesen Antrag. Es sind dann in den späteren Jahren noch mehrfach im Parlamente Anträge gestellt worden, welche unter Abweichungen im einzelnen alle darauf hinzielen, 2 Kategorieen bei den Mordfällen zu unterscheiden. Ein Gesetzentwurf vom Jahre 1882 wollte es, ohne Aufstellung einer gesetzlichen Definition, den Geschwornen überlassen, ob sie im Einzelfalle Mord im ersten Grade, strafbar mit dem Tode, oder Mord im zweiten Grade, strafbar mit Strafknechtschaft, annehmen wollten. Hiergegen machte der damalige Attorney General, Sir Henry James, mit Recht geltend, daß hier, allen Grundsätzen zuwider, den Geschwornen die Befugnis gegeben werden sollte, Strafen zu bestimmen, anstatt über „Schuldig“ oder „Nichtschuldig“ ihr Verdikt abzugeben. [1])

Gegenüber dieser Bewegung auf Abschaffung oder Einschränkung der Todesstrafe existiert aber auch in England eine andre Bewegung,

---

[1]) Am 11. Mai 1886 brachte Sir Joseph W. Pease, im Unterhause einen Antrag auf Abschaffung der Todesstrafe ein; der Antrag wurde, wie alle frühern diesbezüglichen Anträge, abgelehnt. In der Debatte erklärte der damalige Staatssekretär des Innern, Mr. Childers, man könne den 1866 von der Royal Commission gemachten Vorschlag wegen der Schwierigkeit einer richtigen Abgrenzung von 2 Mordklassen nicht durchführen. Sir John Gorst und Sir Richard A. Croß, zwei hervorragende Juristen, erwiderten dem Minister, daß sie diese Schwierigkeit durchaus nicht für unüberwindlich hielten. Cf. auch Howard Association Report 1886 Seite 13 ff.

welche auf eine Erweiterung des Kreises der todeswürdigen Verbrechen hinwirkt. Diese Richtung darf, wenn die Zahl ihrer Anhänger auch keine große ist, hier nicht unerwähnt bleiben, weil derselben kein Geringerer angehört, als Sir James Stephen, Englands größter lebender Kriminalist. Er gibt sein Urteil offen dahin ab: „Meine Meinung ist, daß wir mit der Aufhebung derselben (scil. der Todesstrafe) zu weit gegangen sind, und daß sie in vielen Fällen verhängt werden sollte, welche jetzt nicht todeswürdig sind." [1])

§ 3. Die Transportationsstrafe.

Litteratur: Report of the Committee of the House of Commons on the system of Transportation 1838. — Report of the Committee of Inquiry on Transportation 1856. — Sir Edmund Du Cane's Artikel über das Transportationssystem im November-Hefte der Nineteenth Century 1879. — von Holtzendorff, Die Deportation als Strafmittel. Leipzig 1859.

Die erste Spur von der Strafe der Transportation ist in einem Gesetze der Königin Elisabeth aus dem Jahre 1598 (39 Eliz. c. 4) zu finden. Die Richter werden hier ermächtigt, Vagabunden und Landstreicher nach den Kolonieen transportieren zu lassen. Wieweit hiervon Gebrauch gemacht worden ist, läßt sich nicht mit Sicherheit feststellen. Im Jahre 1619 schickte Jakob I. hundert liederliche Dirnen nach der Kolonie Virginia, um dem dort bestehenden Mangel an Frauen abzuhelfen. Wirklich in das Strafensystem aufgenommen wurde die Transportationsstrafe jedoch erst unter Karl II.

Durch 2 Gesetze aus den Jahren 1678 und 1682 (18 Charles II c. 3 und 22 Charles II c. 5) wurde den Richtern die Befugnis gegeben, Personen, welche zur Todesstrafe verurteilt worden sind,

[1]) Stephen I Band I Seite 478; es werden alsdann eine Anzahl bestimmter Fälle angeführt, und dabei heißt es u. a.: „People should be made to understand that to attack the existing state of society is equivalent to risking their own lives", „they should be destroyed partly in order to gratify the indignation which such crimes produce and which it is desirable that they should produce and partly in order to make the world wholesomer than it would otherwise be by ridding it of people as much misplaced in civilized society as wolves or tigers would be in a populous country." „What is the use of keeping such a wretch alive at the public expense for, say, half a century?"

unter Aufhebung dieser Strafe nach Amerika zu transportieren unter der Bedingung, daß im Falle ihrer Rückkehr aus Amerika die Todesstrafe nachträglich vollstreckt werden würde. Für die Ausführung der Transportation hatte der Sheriff zu sorgen. Der Sheriff pflegte mit einem Unternehmer einen Vertrag dahin abzuschließen, daß der letztere die Transportation nach Amerika unentgeltlich unternahm und dafür das Recht erhielt, über die Arbeitskraft der Transportierten während einer bestimmten Zeitdauer zu seinem Nutzen zu verfügen.

In etwas bestimmterer Form wurde dann die Transportationsstrafe und ihre Ausführung geregelt durch ein Gesetz vom Jahre 1717 — 4 Geo. I c. 2 —. Interessant ist die Einleitung dieses Gesetzes. Es wird hier ausgeführt, daß viele Verbrecher, welche unter der Bedingung der Transportation nach Amerika und des Verbleibens dort begnadigt worden seien, diese Bedingung nicht erfüllt, sondern in England neue Verbrechen begangen hätten, daß anderseits aber in den englischen Kolonieen in Amerika ein großer Mangel an Arbeitskräften herrsche.

Diese offiziellen Ausführungen bezeichnen in deutlicher Weise den Zweck, welchen man bei der Transportationsstrafe überhaupt im Auge hatte: man suchte durch die Transportation in entfernte Kolonieen einerseits den Verbrecher für das heimische Gemeinwesen unschädlich zu machen, anderseits die Arbeitskräfte des Verbrechers im Interesse der Kolonisation und des Anbaues bis dahin unkultiviert gebliebener Länder zu verwerten.

Das Gesetz von 1717 setzte nun fest, daß die Richter bei den felonies without benefit of clergy die Todesstrafe unter der Bedingung der Transportation nach Amerika zu 14 jährigem Aufenthalte erlassen und bei den clergyable felonies statt Brandmarkung auf 7 jährige Transportation erkennen könnten. Wenn die Verbrecher vor Ablauf der Zeit nach England zurückkehrten, so sollten sie der Todesstrafe verfallen. Zur Ausführung der Transportation sollten die Gerichte mit Unternehmern, welche Sicherheit dafür zu geben hatten, daß sie den Verbrecher an den Gouverneur der Kolonie ablieferten, Verträge abschließen. Die Unternehmer sollten als Entschädigung für die Kosten der Überfahrt Eigentum und Interesse an den Arbeitsleistungen des Transportierten für die ganze Dauer der Transportationszeit erhalten („shall have pro perty and interest in the service of such person for such term of years").

Welchen Gebrauch der Unternehmer von den ihm hiernach zugesprochenen Arbeitskräften des Sträflings machte, ob er dieselben verkaufte oder im eignen Interesse verwendete, darum kümmerte sich der Staat nicht.[1]) Der Sträfling war innerhalb der ihm auferlegten Strafzeit nicht anders gestellt als ein Sklave, und wie ein Sklave wurde er thatsächlich in Amerika verkauft. Viele Kolonieen — insbesondere Maryland, wohin die Mehrzahl transportiert wurde — erhoben zwar Widerspruch gegen diese Einführung von Verbrechern, welche die öffentliche Sicherheit und Sittlichkeit bedrohten, allein diese Proteste blieben in England, wo man froh war, sich in so billiger Weise der schädlichen Subjekte entledigen zu können, um so mehr unbeachtet, als viele Parlamentsmitglieder und hochgestellte Personen, welche in den Kolonieen Grundeigentum besaßen, ein direktes Interesse an der Transportation hatten, weil ihnen durch dieselbe für ihre ausgedehnten Ländereien eine genügende Anzahl Arbeitskräfte gegen billigen Lohn sichergestellt waren.

Erst durch den Aufstand der amerikanischen Kolonieen wurde diesem System ein Ende gemacht: mit dem Unabhängigkeitskriege hörten die Transportationen nach Amerika auf. Für England entstand damit die schwierige Frage, was mit den Verbrechern, deren man sich bisher durch Transportation entledigt hatte,[2]) fortab geschehen sollte. Versuche, an der Westküste von Afrika Strafkolonieen anzulegen, mißglückten.

Man entschloß sich dazu, die zur Transportation verurteilten Verbrecher zunächst auf einigen abgetakelten Schiffen (hulks) in Woolwich und an andern Orten unterzubringen. Dieses durch das Gesetz 16 Geo. III c. 43 eingeführte hulk system sollte jedoch

---

[1]) Mit der Ankunft in der Kolonie hörte jede Kontrolle über den Sträfling auf.

[2]) Es ist dabei hervorzuheben, daß die Zahl der nach Amerika Transportierten ständig zugenommen hatte. Die englische Gesetzgebung des 18. Jahrhunderts hatte eine erhebliche Zahl neuer felonies geschaffen, für welche direkt gesetzlich die Transportationsstrafe angeordnet war. Außerdem war die Begnadigung von zum Tode Verurteilten unter der Bedingung der Transportation in erheblichem Maße durch ein Gesetz von 1768 — 8 Geo. III c. 15 — erleichtert worden.

nach einem 3 Jahre später (1779) erlassenen Gesetze nur eine provisorische Maßregel sein.[1])

Dieses letztere, unter dem Einflusse des bekannten Gefängnis-Reformers John Howard[2]) erlassene Gesetz sah nämlich als definitive

---

[1]) Das ursprünglich nur als provisorische Maßregel gedachte **hulk system** hat sich thatsächlich fast ein Jahrhundert lang in Geltung erhalten. In England selbst hörte es im Jahre 1857, wo die hulks in Woolwich abbrannten, auf; es bestand in wenig veränderter Form jedoch bis zum Jahre 1875 in der nach dem hulk system im Jahre 1842 errichteten Gefangenanstalt auf Gibraltar fort. Durch 19 Geo. III c. 74 ward Einschließung in die hulks zu einem besondern Strafmittel gemacht, auf welches an Stelle der Transportationsstrafe wegen bestimmter Delikte erkannt werden konnte; dabei wurde Einschließung in die hulks auf 1—5 Jahre einer 7 jährigen Transportationsstrafe und Einschließung auf 7 Jahre einer 14 jährigen Transportationsstrafe gleich geachtet.

Die in den hulks untergebrachten Sträflinge wurden zu Hafen- und Dockarbeiten verwandt. Die Zustände in den hulks werden als geradezu entsetzliche geschildert. Eine als hulk fever bezeichnete Krankheit raffte eine große Zahl der Sträflinge hinweg; für das Jahr 1779 wird die Sterblichkeitsziffer auf 30 % angegeben. Im Jahre 1802 wurde ein Inspektor ernannt, um fortlaufend über die Zustände in den hulks an den obersten Gerichtshof zu berichten. Im Jahre 1815 wurde einem ständigen Beamten im Ministerium des Innern — dem Superintendent of the hulks — die Beaufsichtigung der hulks übertragen. Im Jahre 1850 endlich gingen die damals noch bestehenden hulks in die Aufsicht der Directors of Convict Prisons über.

Die Einschließung in die hulks wurde während der ganzen Zeit des Bestehens des Systems als Ergänzung und Ersatz der Transportationsstrafe behandelt; soweit die letztere Strafe in den einzelnen Jahren aus irgendwelchen Gründen nicht an der Gesamtzahl der dazu Verurteilten vollstreckt werden konnte, wurden dieselben in den hulks untergebracht. Eine ausführliche Darstellung des hulk system und seiner Mängel ist in dem Report of the Committee of the House of Commons on secondary punishment vom Jahre 1832 gegeben.

[2]) **John Howard**, geboren 1737, war 1773 Sheriff der Grafschaft Bedford geworden. In dieser Eigenschaft hatte er Gelegenheit, die Mängel des damaligen englischen Gefängniswesens zu sehen, und erhielt dadurch die Anregung zu seinen Studien über das Gefängniswesen überhaupt. Nachdem er eine große Anzahl englischer und kontinentaler Gefängnisse eingehend besichtigt hatte, gab er im Jahre 1777 sein berühmtes Buch The State of the Prisons in England and Wales with preliminary observations and an account of some foreign prisons heraus, welches ihm wegen der darin enthaltenen scharfen Kritik der Gefängniszustände in fast allen europäischen Staaten den Namen eines europäischen censor morum verschaffte. Das Buch, welches 1779 in zweiter und 1784 in dritter Auflage erschien und von welchem Gottl. Ludw. Wilhelm Köster einen Auszug in deutscher Sprache (Leipzig 1780) veröffentlicht hat, bildete den

Maßregel die Errichtung von Strafanstalten (penitentiaries) vor, in welchen die Verbrecher in einzelnen Zellen getrennt gehalten werden sollten. Allein, ehe es zur Ausführung dieses Planes kam,[1]) führte ein zufälliger Umstand, die Auffindung der Botany-Bai durch Cook, zur Rückkehr zu der Tranportationsstrafe.

Ein Gesetz von 1784 — 24 Geo. III c. 56 — ermächtigte den König, geeignet erscheinende Orte zur Vollstreckung der Transportationsstrafe zu wählen. Am 13. Mai 1787 segelte der erste Transport von 184 männlichen und 100 weiblichen Sträflingen nach Botany-Bai ab. New South Wales, Van Diemen's Land und Norfolk Island bildeten fortab die Orte, wohin die zur Transportationsstrafe Verurteilten oder Begnadigten geschafft wurden.

Von Anfang an war freilich die Ausführung der Transportation hierher größeren Schwierigkeiten unterworfen, als es früher die Transportation nach Amerika gewesen war. Man fand nicht ein geordnetes Gemeinwesen und freie Ansiedler, denen man die Arbeitskräfte der Sträflinge überlassen konnte, vor, sondern die Regierung hatte zumeist selbst die Überwachung, die Verpflegung und die Beschäftigung der Transportierten zu übernehmen. Die Regierung that alles mögliche, um freie Ansiedler heranzuziehen; sie zahlte bis 1818 die Überfahrtskosten und war mit Landüberweisungen sehr freigebig. Allein bis zum Jahre 1822 blieb die Ziffer der freien Einwanderung nur eine sehr geringe, und die Einwanderer waren durchweg Leute ohne alles Kapitalvermögen, welche mit Hilfe der ihnen von der Regierung gewährten Unterstützung, insbesondere der unentgeltlichen Überlassung der Arbeitskräfte der Sträflinge, zu einigem Vermögen zu kommen suchten.

All dies änderte sich jedoch seit dem Jahre 1822. In der Zeit von 1787 bis 1822 war es einer Reihe von thatkräftigen englischen Gouverneuren gelungen, den ökonomischen Zustand der neuen Kolonieen in außerordentlicher Weise zu entwickeln, insbesondere Ackerbau und Viehzucht zu vorher nicht geahnter Höhe zu bringen.

Ausgangspunkt für die Gefängnisreformen in England (in § 5 wird hierauf näher eingegangen werden). John Howard starb im Januar 1791.

[1]) Die erste penitentiary, Millbank Prison, wurde erst im Jahre 1821 vollendet; im Dezember 1842 ward die zweite penitentiary, Pentonville Prison, eröffnet. Das Nähere hierüber im § 4.

Nachdem dies erreicht war, begann in England das Kapital und der Unternehmungsgeist auf dem neugewonnenen Felde Beschäftigung zu suchen. Es ergoß sich ein immer größer werdender Strom freier Einwanderer nach den australischen Kolonieen. Gleichzeitig damit begannen nun aber die *Proteste der neuen Ansiedler gegen weitere Zuführung von verbrecherischen Elementen.*

Die *Art und Weise*, wie in *Australien mit den Transportierten verfahren* wurde, war je nach der Person des Gouverneurs, welcher fast uneingeschränkte Machtbefugnis hatte, und nach der Fortentwickelung der einzelnen Kolonie ständigen Veränderungen unterworfen gewesen. Anfangs hatte natürlich überall die Regierung die Arbeitskräfte selbst notwendig, um die erforderlichen Gebäulichkeiten herzustellen, Wege zu bauen und sonstige öffentliche Arbeiten zu vollenden. Als dann die Zahl der freien Einwanderer zugenommen hatte, unternahm die Regierung die Ausführung gewisser Arbeiten für die neuen Ansiedler, insbesondere die Rodung von Land u. s. w., durch die Sträflinge in sogenannten Rodungskompanieen (clearing gangs), welche von Wächtern beaufsichtigt wurden. Noch später ging man dazu über, den neuen Ansiedlern eine Anzahl Sträflinge im Wege der *Assignation* zuzuteilen: der assignee erhielt gegen die Verpflichtung zum Unterhalt und zur Verpflegung die freie Verfügung über die ihm zugeteilten Sträflinge. Bei weiterer Entwickelung der Kolonie begann man dann die Arbeitskräfte von geschickten Sträflingen, welche ein Handwerk verstanden, besser zu verwerten. Die Regierung behielt diese wertvolleren Kräfte für sich und vermietete dieselben nur zu einzelnen Arbeitsleistungen gegen entsprechendes Entgelt.

Von der Assignation blieben ferner diejenigen ausgeschlossen, welche man für besonders gefährliche Verbrecher hielt oder welche sich in der Kolonie selbst Verbrechen hatten zu schulden kommen lassen. Für diese Klasse schuf man besondere Strafkolonieen (penal settlements), deren wichtigste auf Norfolk Island errichtet wurde. Die Sträflinge hatten hier in Ketten schwere Arbeiten zu verrichten; unerbittliche Strenge galt als einziges Prinzip.

Für Unterricht der Sträflinge war in den australischen Kolonieen überhaupt nicht gesorgt; für geistlichen Zuspruch nur ganz minimal. Bis 1807 war es ein einziger Geistlicher, welchem die Seelsorge für alle australischen Kolonieen oblag. Die Zahl der Disziplinarbestrafungen, denen auch die an die freien Ansiedler überwiesenen

Sträflinge unterworfen blieben, war eine ungeheuer große. Aufstände der Gefangenen gehörten nicht zu den Seltenheiten.

Es war natürlich, daß dieser Zustand für das Sicherheitsgefühl der sich allmählich vermehrenden Zahl von freien Einwanderern ein wenig erfreulicher war. Einzelne Gouverneure suchten durch eine mehr systematische Behandlung der Transportierten Besserung zu schaffen. Der Neuankömmling wurde zuerst in eine Strafkolonie gebracht, um zu strenger Arbeit angehalten zu werden. Wenn er sich dort zur Zufriedenheit betragen hatte, so verwendete man ihn in den clearing gangs, und erst, wenn er sich hier, wo ihm schon mehr Freiheit gelassen war, gut geführt hatte, wurde er einem freien Ansiedler assigniert. Hatte der Assignierte dann während bestimmter Zeit zu keinem Tadel Anlaß gegeben, so wurde er unter gewissen, streng einzuhaltenden Bedingungen für den Rest seiner Strafzeit freigelassen; er erhielt ein sogenanntes ticket of leave, einen Urlaubsschein, welcher die Bedingung enthielt, daß der Betreffende den ihm angewiesenen Distrikt nicht verlassen und sich periodisch melden mußte.

Die Erteilung dieser Urlaubsscheine erfolgte anfangs seitens der Gouverneure, ohne daß eine besondere gesetzliche Ermächtigung dafür vorlag; erst ein Gesetz von 1829 — 9 Geo. IV c. 83 — führte die Urlaubsscheine als gesetzliche Institution ein. Ein Gesetz von 1832 — 2 a. 3 William IV c. 62 — gab dann nähere Regulative über die Erteilung der Urlaubsscheine, indem es bestimmte, daß ein zu lebenslänglicher Strafe Verurteilter erst nach Ablauf von 8 Jahren, ein Sträfling von 14 Jahren nach 6 Jahren, ein Sträfling von 7 Jahren nach 4 Jahren einen Urlaubsschein erhalten dürfte.

Inzwischen hatte die Bewegung gegen weitere Transportationen immer mehr an Umfang gewonnen. In England selbst machte sich die Meinung geltend, daß die Strafe der Transportation von den Verbrechern nicht mehr gefürchtet würde: unentgeltliche Überfahrt nach dem im Aufblühen begriffenen Australien mit der sichern Aussicht, bei einigermaßen gutem Verhalten sich dort nach wenigen Jahren eine behagliche, selbständige Existenz gründen zu können, das würde von der Verbrecherklasse nicht als Strafe empfunden.

Unter diesen Umständen beschloß das Unterhaus im Jahre 1837 ein spezielles Komitee von 15 Mitgliedern einzusetzen, um eine umfassende Untersuchung anzustellen über das System der Transportation, seine Wirksamkeit als Strafmittel und seinen Einfluß auf den Sittenzustand in den Kolonieen. Der von diesem Komitee, welchem Lord John Russel, Sir Robert Peel, Sir William Molesworth, Sir George Grey und andre hervorragende Staatsmänner angehörten, am 3. August 1838 erstattete Bericht faßt das Resultat der Untersuchung dahin zusammen, daß die Transportationsstrafe der Abschreckung für die heimische Bevölkerung in England entbehre, für die Besserung des Sträflings wirkungslos sei, daß das System in sich selbst mangelhaft sei, weil die Natur und der Grad der Strafe im wesentlichen unsicher und ungleich erscheine und von der Laune des Herrn, dem der Sträfling assigniert werde, abhänge, daß das System ferner sich als sehr kostspielig herausgestellt habe und für den Sittenzustand der Kolonie verderblich wirke. Demgemäß wird möglichst baldige Aufhebung der Transportationsstrafe beantragt.

Auf Grund dieses Berichtes wird denn auch im Jahre 1840 die Transportation nach New South Wales als derjenigen Kolonie, welche in ihrer Entwickelung am meisten vorgeschritten war und bei welcher sich die erwähnten Mißstände am schärfsten gezeigt hatten, vorläufig eingestellt. In den übrigen australischen Kolonieen wird das Assignationsverfahren gänzlich aufgegeben[1]) und ein neues System der Behandlung der Transportierten eingeführt.

Die Ausbildung dieses neuen, sogenannten Probesystems (probation system) ist das Verdienst von Lord Stanley als damaligen Staatssekretärs für die Kolonieen (Colonial Secretary). Die wesentlichsten Grundzüge des Systems in der Gestalt, welche dasselbe in der im Jahre 1842 von Lord Stanley und Sir James Graham — der letztere war damals Staatssekretär des Innern — erlassenen Instruktion erhielt, sind folgende. Es werden unter den Sträflingen fortab 3 Klassen streng geschieden:

a) Sträflinge, welche zu mehr als 15 Jahre oder auf Lebenszeit verurteilt worden sind oder welche sich besonders gefährlicher Verbrechen schuldig gemacht haben;

b) Sträflinge mit nicht mehr als 7 Jahr Strafe, bei welchen in hervorragendem Maße Aussicht auf Besserung vorliegt;

c) alle übrigen Sträflinge.

---

[1]) Der Bericht hatte besonders eingehend das Assignationsverfahren behandelt und die auf der Hand liegenden Nachteile und Ungerechtigkeiten dieser neuen Art Sklaverei mit Ernst hervorgehoben.

Betreffs dieser letzteren, weitaus umfangreichsten Klasse zunächst wird bestimmt, daß dieselben folgende 4 Stadien durchzumachen haben:

1) Öffentliche Arbeiten in Van Diemen's Land. Die Sträflinge werden in Rotten (probation gangs) eingeteilt und arbeiten unter Aufsicht eines Gefangenwärters. Jeder Rotte wird ein Geistlicher oder ein Schullehrer beigegeben.

2) Der Sträfling arbeitet bei Privaten gegen Lohn, bleibt aber der Gefängnisaufsicht und Disziplin unterworfen. Von dem Lohne hat er bestimmte Prozentsätze an die Regierung abzuführen; die betreffende Summe dient der Regierung als Sicherheit; sie verfällt, sobald der Sträfling die ihm auferlegten Beschränkungen übertritt. Die in diesem Stadium — dem sogen. stage of probation pass — befindlichen Sträflinge werden in 3 Unterabteilungen eingeteilt; mit dem Vorrücken in eine höhere Unterabteilung werden die dem Sträflinge auferlegten Beschränkungen erleichtert und es verringert sich der an die Regierung abzuführende Teil des Arbeitsverdienstes.

3) Der Sträfling erhält einen Urlaubsschein — ticket of leave — er kann fortab über seine Arbeitskraft frei verfügen und darf sich nur ohne besondere Erlaubnis nicht außerhalb des ihm zugewiesenen Bezirks begeben; er muß sich ferner periodisch melden.

4) Der Sträfling erhält conditional pardon, er wird unter der Bedingung begnadigt, daß er innerhalb der Strafzeit nicht nach England zurückkehrt.

Für die unter a) angeführten schwersten Verbrecher erhält dies System in sofern eine Änderung, als dieselben zunächst nach Norfolk Island gebracht werden, wo sie einer sehr strengen Disziplin unterworfen werden und in mehr angestrengter Weise arbeiten müssen. Erst wenn sich die Sträflinge dieser Klasse während geraumer Zeit in Norfolk Island gut geführt haben, so daß man zu ihnen Vertrauen haben kann,[1]) werden dieselben nach Van Diemen's Land gebracht, um dort die angeführten 4 Strafstadien durchzumachen.

Was endlich die unter b) angeführte Klasse der leichteren und besserungsfähigen Sträflinge anbetrifft, so sollen diese sogen.

[1]) Von einem der Gouverneure von Norfolk Island, Kapitän Maconochie, ist zur Kontrolle der guten Führung das in dem heutigen englischen Gefängniswesen so wichtige Markensystem erfunden worden.

selected men die erste Zeit ihrer Strafe bis zur Dauer von 18 Monaten in England selbst in einer der neu errichteten Strafanstalten Millbank oder Pentonville verbüßen und sodann erst nach Australien geschafft werden; wenn sie sich bisher gut geführt haben, so erhalten sie in Australien sofort einen Urlaubsschein.

Dies ist das neue, auf dem Prinzip der Klassifikation der Sträflinge und des allmählichen Fortschreitens in der Behandlung derselben basierte System.

Leider zeigten sich die Umstände zu einer ruhigen Entwickelung desselben wenig günstig.[1]) Es stellte sich zunächst heraus, daß es unmöglich war, in Van Diemen's Land für die probationary pass men genügend Arbeit bei Privatpersonen zu finden. Die Regierung wurde dadurch genötigt, auch diese Klasse zu öffentlichen Arbeiten in sogen. hiring gangs zu verwenden; da es aber bei diesen Arbeiten an dem erforderlichen Aufsichtspersonale fehlte, so ergaben sich schwere Mißstände. Auch die große Zahl der als ticket of leave men gar keiner eigentlichen Aufsicht unterworfenen Sträflinge erweckte den Unwillen der freien Ansiedler.

Dem fortwährenden Drängen der letzteren gab endlich Gladstone als Colonial Secretary im Jahre 1846 nach, indem er die vorläufige Einstellung der Transporte von Sträflingen nach Van Diemen's Land anordnete. Infolgedessen sah sich denn der damalige Staatssekretär des Innern, Sir George Grey, zu einer Änderung des probation system genötigt. Im Einverständnisse mit dem neuen Colonial Secretary, Earl Grey, ordnete er im Jahre 1847 an, daß alle zur Transportationsstrafe Verurteilten — mit Ausnahme der schwersten Verbrecher, für welche Norfolk Island reserviert blieb — das erste Stadium ihrer Strafe in den Anstalten von Pentonville oder Millbank verbüßen, dann zu öffentlichen Arbeiten in England, Gibraltar oder Bermuda[2]) verwendet werden und erst im dritten Strafstadium als ticket of leave men nach den Kolonieen gesandt werden sollten.

Während des zweiten Strafstadiums, d. i. während der Be-

---

[1]) Über das probation system und seine Ausführung geben vorzugsweise eine Anzahl von Depeschen zwischen den Staatssekretären der Kolonieen und den Staatssekretären des Innern Auskunft. Dieselben finden sich abgedruckt im Anhang 25 zum Report of the Directors of Convict Prison 1880/81.

[2]) Auf den Bermudas-Inseln war im Jahre 1824 ein hulk establishment errichtet worden.

schäftigung der Sträflinge bei öffentlichen Arbeiten, wurden dieselben Anfangs in England in den hulks sowie in den hulks establishments in Gibraltar und auf den Bermudas-Inseln untergebracht. Da diese hulks establishments jedoch in jeglicher Beziehung mangelhaft waren, so ging man nunmehr dazu über, in England selbst neue Strafanstalten zu errichten, wo die Sträflinge während des Stadiums der Beschäftigung bei öffentlichen Arbeiten verwahrt werden konnten (Public Work Prisons). Die erste derartige Anstalt wurde 1848 in Portland eröffnet, andre Anstalten folgten rasch nach: 1850 Dartmoor, 1852 Portsmouth, 1856 Chatham.

Die Transportation bildete bei diesem neuen System nur noch den Schluß der Strafvollstreckung; sie hatte aufgehört, der wesentliche Bestandteil der Strafe zu sein, und auch in dieser abgeschwächten Bedeutung war ihr nur noch ein kurzes Leben beschieden.

In Australien hatte sich ein Bund „Australian League" gebildet, um sich mit allen zu Gebote stehenden Kräften der weiteren Einführung englischer Sträflinge zu widersetzen. England mußte nachgeben und im Jahre 1852 die Transportation nach Van Diemen's Land definitiv einstellen, nach New South Wales waren bereits seit 1849 keine Sträflinge mehr transportiert worden. [1]) Als die einzige australische Kolonie, welche sich zur weiteren Aufnahme englischer Sträflinge noch bereit erklärte, blieb Western Australia übrig, aber bei der geringen Anzahl freier Ansiedler (im Jahre 1853 ca. 6000) konnte die Zahl der dorthin zu sendenden Sträflinge nur eine beschränkte sein. Man versuchte in andern Weltteilen Strafkolonieen zu gründen, aber vergebens: als man 1850 ein Schiff mit Sträflingen nach dem Kap der guten Hoffnung sandte, leisteten die dortigen Ansiedler der Landung thätlichen Widerstand.

Ein Gesetz von 1853 — 16 a. 17 Vict. c. 99 — trug diesen veränderten Verhältnissen Rechnung. Das Strafmittel der Transportation wurde für Strafen unter 14 Jahren gänzlich abgeschafft und durch das neue Strafmittel der Strafknechtschaft (penal servitude) ersetzt: an die Stelle langjähriger Transportationsstrafe sollte Strafknechtschaft von kürzerer Dauer treten.

---

[1]) Auch die Transportationen nach Norfolk Island hörten 1853 auf.

Prinzipiell wurde damit eigentlich nicht viel Neues geschaffen. Das probation system, wie es im Jahre 1847 festgestellt war, blieb mit der einzigen Modifikation erhalten, daß die ticket of leave men nicht mehr nach den Kolonieen geschickt wurden, sondern im Lande verblieben. Dem so modifizierten probation system wurde der neue Name penal servitude beigelegt.

Im Jahre 1856 setzte sowohl das Ober- wie das Unterhaus besondere Komitees ein, um endgültig über die Beibehaltung der Transportationsstrafe zu beraten. Beide Komitees erklärten sich im Widerspruche mit den Beschlüssen des Komitees von 1838 für die Beibehaltung. Freilich verliert dieser Widerspruch seine Bedeutung, wenn man bedenkt, welche Veränderungen in der Zwischenzeit in der Vollstreckung der Transportationsstrafe eingetreten waren: es handelte sich jetzt nicht mehr, wie im Jahre 1838, darum, ob England seine Verbrecher zur Verbüßung der gegen sie erkannten Strafzeit nach den Kolonieen senden sollte oder nicht; es stand fest, daß die eigentliche Strafvollstreckung in England selbst zu erfolgen habe, und es war nur die Frage zu entscheiden, ob man durch die Transportation nach einer entfernten Kolonie dem Sträflinge, nachdem er den größten Teil seiner Strafe verbüßt hatte, den Wiedereintritt in die bürgerliche Gesellschaft mittels redlicher Beschäftigung erleichtern könnte und erleichtern sollte?[1]) Und wenn man die Frage in dieser Weise stellt, so ist es in der That nicht zu verwundern, wenn die beiden Parlamentskomitees zu einer Bejahung gelangten. Sie sahen in der Transportation ein rationelles Übergangsstadium zwischen Gefangenschaft und Freiheit, ein Mittel, den Sträfling vor einem Rückfalle in die frühere schlechte Lebensweise zu bewahren und ihn in den Stand zu setzen, leichter als in der alten Heimat ein neues Leben zu beginnen. In den Kolonieen hatte es sich, insbesondere seit Einführung einer rationellen Behandlung der Trans-

---

[1]) Dafür, daß diese Auffassung der Transportation die maßgebende geworden war, gibt eine im Jahre 1847 getroffene Bestimmung einen treffenden Beleg: es wurde angeordnet, daß den mit Urlaubsscheinen nach Australien Transportierten erst dann conditional pardon bewilligt werden sollte, wenn sie 15 £ als Entschädigung für die Kosten der Überfahrt an die Regierung gezahlt hätten, und daß die so zusammengebrachte Summe, zu welcher noch besondere Zuschüsse vom Parlamente bewilligt wurden, dazu verwendet werden sollte, die Frauen und Kinder der Transportierten nach Australien zu schaffen.

portierten, gezeigt, „wie die für unbrauchbar gehaltenen Granitmassen verbrecherischer Bevölkerungsbestandteile so weit verwittern können, daß eine reiche Kultur auf ihnen Wurzel schlägt" (Holtzendorff).[1])

So erklärlich aber auch hiernach das Urteil der beiden Komitees auf Beibehaltung der Transportationsstrafe erscheinen muß, so wenig konnte dasselbe doch gegenüber dem energischen Widerstande der selbständig gewordenen Kolonieen zur praktischen Durchführung gelangen.[2]) Das englische Parlament mußte diesem Widerstande weichen, und es geschah dies, indem durch Gesetz vom Jahre 1857 — 20 a. 21 Vict. c. 3 — die Transportation als richterlich zu erkennende Strafe endgültig abgeschafft wurde.

Das Gesetz erklärte es dabei für zulässig, Personen, welche zu der an Stelle der Transportation tretenden Strafe der Strafknechtschaft verurteilt sind, im Verwaltungswege während der Strafperiode nach den Kolonieen zu transportieren. Von dieser Befugnis ist wenig Gebrauch gemacht worden,[3]) und auch sie ist seit 1867 beseitigt. An ihre Stelle ist die Unterstützung der freiwilligen Auswanderung entlassener Sträflinge getreten,[4]) auf welche später (Abschnitt II § 24) noch zurückzukommen ist.[5])

---

[1]) Napoleon I. soll sich einmal ähnlich dahin ausgesprochen haben, er halte für das beste Präventivsystem dasjenige, welches die alte Welt reinigt und die neue bevölkert.

[2]) Dieser Widerstand verschärfte sich im Anfange der 50er Jahre, als sich infolge der Entdeckung der Goldfelder in New South Wales und Victoria ein neuer Strom von Auswanderern nach Australien ergoß, in erheblicher Weise.

[3]) Die Grundzüge, nach welchen bei der administrativen Wegsendung von Sträflingen verfahren wurde, sind in einem Rundschreiben von Sir George Grey vom 27. Juni 1857 enthalten.

[4]) Es mag hier bemerkt werden, daß eine derartige Maßregel schon im Jahre 1838 von dem Erzbischofe von Dublin, Whatley, empfohlen worden war. Er bezeichnete es vor dem Komitee von 1837/8 als wünschenswert, wenn unter Abschaffung der Transportationsstrafe die entlassenen Sträflinge vor einem Rückfalle in ihre frühere schlechte Lebensweise dadurch bewahrt würden, daß nach Ablauf der Strafzeit diejenigen, welche eine Neigung zur Besserung an den Tag gelegt hätten, auf ihren eignen Wunsch mit den Mitteln versehen würden, um nach den britischen Kolonieen auszuwandern.

[5]) Einige wenige statistische Notizen über die Transportationsstrafe mögen hier noch Platz finden. Die Gesamtzahl der überhaupt nach Australien Transportierten wird schätzungsweise auf 134 300 Personen angegeben. In der ersten Zeit bis zum Jahre 1816 war die Zahl der Transportierten nur eine sehr kleine; sie wird auf durchschnittlich jährlich 474 Personen berechnet. Von

Mit dem Jahre 1857 hat somit die Transportation aufgehört, ein Strafmittel zu sein. Während der langen Zeit, in welcher sie ein wichtiges Glied in dem englischen Strafensysteme bildete, ist, wie hier nachzuweisen versucht wurde, ihre Gestalt eine vielfach wechselnde gewesen. Die Transportation wird ursprünglich als ein bequemes Mittel für England, sich seiner verbrecherischen Bevölkerung zu entledigen, gehandhabt: man trägt Sorge dafür, daß der Verurteilte nach der Kolonie geschafft wird und daß er von dort nicht nach England zurückkehrt; was aus dem Transportierten in der Kolonie wird, darum kümmert sich das Mutterland nicht. Erst später sieht sich England in die Notwendigkeit versetzt, die Beaufsichtigung, Beschäftigung und Verpflegung der Transportierten ganz oder doch teilweise selbst in die Hand zu nehmen. In dem folgenden Entwickelungsstadium wird es dann überhaupt unmöglich, die zur Transportation Verurteilten ohne weiteres nach den Kolonieen zu entsenden, das Mutterland muß den ersten Teil der Strafvollstreckung bei sich zu Hause übernehmen und so die Verurteilten erst für die Transportation vorbereiten. Schon damit hörte die Transportation eigentlich auf, ein Strafmittel zu sein, sie wird zu einem Stadium des Strafvollzuges. Und aus einem wünschenswerten Abschlusse des Strafvollzuges wird sie dann schließlich ein Gnadenakt, mittels dessen

---

1817 bis 1834 ist die Zahl dann im ständigen Wachsen begriffen; sie erreicht im Jahre 1834 mit 4920 Transportierten ihren Höhepunkt. Die Zahl der Transportierten sinkt dann bis zum Jahre 1839 auf 2732; seit 1839 überschreitet sie die Ziffer 3000 nur noch 2 mal (im Jahre 1842 mit 4166, im Jahre 1844 mit 3279 Transportierten); im Jahre 1853 sinkt sie auf 600 und bleibt von da ab auf kleine Ziffern beschränkt. Im Jahre 1857 werden noch 532 Personen nach Australien transportiert; im Jahre 1867 endlich geht der letzte Transport von 451 Personen nach Australien ab. Cf. Du Cane I Seite 109, II Seite 111 und 123. Im Jahre 1886 hat ein Abkommen zwischen der englischen Regierung und der Kolonialregierung in Western Australia stattgefunden, nach welchem die letztere fortab die noch in der Strafverbüßung befindlichen Transportierten auf eigne Kosten gegen eine Entschädigung von 42—45 £ per Kopf übernimmt; dabei ist festgestellt, daß sich in Strafanstalten noch 38, in Lunatic Asylums und Invalid Depots noch 87 Personen befinden, während als Gesamtzahl der nach Western Australia Transportierten die Ziffer 9669 ermittelt ist. Cf. Rep. of Directors of Conv. Prisons 1885/6 S. XII f. (Es wird hier eine Übersicht über das allmähliche Aufblühen dieser Kolonie gegeben unter Hinweis darauf, daß dieses Aufblühen größtenteils auf die Arbeitsleistungen der Transportierten zurückzuführen sei.)

dem Sträflinge der Wiedereintritt in die bürgerliche Gesellschaft erleichtert werden soll.

Die hier geschilderte Entwickelung ist nicht hervorgegangen aus einer veränderten strafrechtlichen Auffassung, sondern aus Veränderungen in den kolonialen Verhältnissen, von deren Gestalt die Transportationsstrafe abhängig erscheint. Aber unbewußt haben sich aus den Versuchen, die Transportation mit den kolonialpolitischen Interessen in Einklang zu bringen, die Elemente herausgebildet, auf denen die Gestaltung der neuen Freiheitsstrafe basiert werden konnte und basiert wurde. Und hierin besteht die bleibende Bedeutung der Transportationsstrafe für das englische Strafensystem.

§ 4. Die Strafe der Strafknechtschaft (penal servitude).

Litteratur: Report of the Select Committee of the House of Commons 1856 to inquire into the working of the Penal Servitude Act 1853. — Report of the Royal Commission 1863 to inquire into the operation of the Penal Servitude Acts 1853 and 1857. — Report of the Commissioners appointed 1878 to inquire into the working of the Penal Servitude Acts.

Als das Gesetz von 1853 das neue Strafmittel der Strafknechtschaft schuf, waren alle Elemente für die Ausführung dieser Strafe bereits durch die bisherige Entwickelung der Transportationsstrafe gegeben. Betrachten wir zunächst diese Elemente der neuen Strafe.

Im Jahre 1779, kurze Zeit nachdem die bisherigen Transportationen nach Amerika infolge des Unabhängigkeitskrieges hatten eingestellt werden müssen, hatte das schon erwähnte, im wesentlichen auf Anregung John Howards und seiner Anhänger (Blackstone und William Eden) ergangene Gesetz — 19 Geo. III c. 74 — die Erbauung von Strafanstalten (Penitentiaries) angeordnet, in welchen die Verbrecher in einzelnen Zellen getrennt gehalten werden sollten. Das Gesetz bezweckte, an Stelle der Transportationen in diesen Strafanstalten ein System einzuführen, welches neben dem Abschreckungszwecke sich auch die Besserung der Sträflinge angelegen sein ließ. [1]) Zur Ausführung dieses Ge-

[1]) In sectio 5 des Gesetzes wird ausgeführt: „that if many offenders convicted of crimes for which transportation has been usually inflicted

setzes waren 3 Kommissare, unter denen sich auch Howard befand, eingesetzt worden. Es wurde oben schon ausgeführt, daß die Verwirklichung des Planes durch die Auffindung neuer Kolonieen in Australien und die dadurch geschaffene Möglichkeit, die Transportationen wieder aufzunehmen, einen langen Aufschub erlitt: es vergingen fast 40 Jahre, bis in Millbank Prison im Jahre 1816 die erste Penitentiary eröffnet wurde. Die näheren Bestimmungen über die Verwaltung dieser Strafanstalt wurden in einem Gesetze vom Jahre 1816 — 56 Geo. III c. 63 — niedergelegt und durch eine große Anzahl späterer Gesetze [1]) ergänzt und modifiziert. In die Anstalt sollten alle zur Transportationsstrafe Verurteilten gebracht werden und dort so lange bleiben, bis die Transportation nach Australien erfolgen konnte. Während dieser Zeit, welche bald von größerer, bald von geringerer Dauer war, wurden die Sträflinge in strenger Isolierhaft gehalten, sie erhielten in ihrer Zelle geistlichen Zuspruch und Ermahnung und wurden in angestrengter Weise mit mehr oder weniger nutzloser Arbeit (Wergzupfen, Tretmühle u. s. w.) beschäftigt. Im Jahre 1842 trat zu der Anstalt von Millbank eine zweite in dem unter Leitung von Sir Joshua Jebb neuerrichteten Pentonville Prison. [2]) In demselben Jahre war das schon erwähnte probation system durch Lord Stanley eingeführt worden. Nach diesem Systeme sollten die leichteren und besserungsfähigen Sträflinge, die sog. selected men von nicht mehr als 7jähriger Strafzeit, die ersten 18 Monate ihrer Strafe in einer der beiden Penitentiaries verbüßen, ehe sie nach Australien gesandt wurden. Die hiermit zuerst eingeführte systematische Benutzung der Isolierhaft als erste Periode der Strafverbüßung bewährte sich sehr

---

were ordered to solitary imprisonment accompanied by well regulated labour and religious instruction, it might be the means under providence not only of deterring others, but also of reforming the individuals and inuring them to habits of industry."

[1]) 59 Geo. III c. 136, 7 a. 8. Geo. IV c. 33, 7. William IV c. 13, 6 a. 7 Vict. c. 26, 11 a. 12 Vict. c. 104. Die Verwaltung wurde ursprünglich von einem Verwaltungskuratorium (Committee of Management) geführt, dessen Mitglieder keine Besoldung erhielten; später wurden 2 besoldete Inspektoren ernannt.

[2]) Die Verwaltung von Pentonville wurde, getrennt von derjenigen von Millbank, durch ein Kollegium unbesoldeter Commissioners geführt. 5 a. 6 Vict. c. 29.

gut.[1]) Dies Resultat bewog Sir George Grey bei der im Jahre 1847 vorgenommenen Abänderung des probation system, die 18 monatliche Isolierhaft auf alle zur Transportationsstrafe Verurteilten auszudehnen.

Das zweite Element bei der Strafe der penal servitude besteht in der Verwendung der Sträflinge zu öffentlichen Arbeiten. Seit Beginn der Transportationen nach Australien war die Zwangsarbeit ein wesentlicher Bestandteil der Transportationsstrafe geworden und das 1842 eingeführte probation system hatte die Anordnung getroffen, daß alle Transportierten nach ihrer Ankunft in Australien zunächst zu öffentlichen Arbeiten — insbes. zum Wegebau — verwendet werden sollten. Als die Ausführung weiterer Transportationen nach Australien Schwierigkeiten bereitete, ordnete dann Sir George Grey 1847 an, daß die Sträflinge dieses Stadium der öffentlichen Zwangsarbeit in England selbst durchmachen sollten. Zu diesem Zwecke wurde im Laufe des Jahres 1848 Portland Prison errichtet, wo die Sträflinge unter der Leitung von Kapitän Whitty und Sir Joshua Jebb zu Hafenbauten verwendet wurden, während das 1850[2]) eröffnete Dartmoor Prison für landwirtschaftliche Arbeiten bestimmt wurde. Die zur Transportationsstrafe Verurteilten wurden nunmehr, nachdem sie die 18 monatliche Isolierhaft in Pentonville oder Millbank durchgemacht hatten, zu einem dieser Public Works Prisons geschafft,[3]) wo dieselben während der Nacht- und der Essenszeit in Zellen getrennt gehalten wurden, während der übrigen Zeit aber in Gemeinschaft im Freien arbeiteten.

---

[1]) Die von Sir James Graham 1842 eingesetzten Commissioners zur Verwaltung von Pentonville Prison — darunter Sir Joshua Jebb, William Crawford und Rev. Whitworth Russell — konstatierten in ihrem Berichte vom Jahre 1847 „moral results without parallel in the history of penal discipline".

[2]) Die Zahl der Public Works Prisons vermehrte sich schnell: 1852 wurde ein neues zu Porthmouth errichtet, 1856 folgte Chatham, später Parkhurst, Woking, Borstal, Wormwood Scrubs (zu gleicher Zeit als Penitentiary benutzt); das neueste Public Works Prison ist im August 1885 in Dover eröffnet worden.

[3]) Im Jahre 1850 wurden alle die Anstalten in England, welche für die Aufnahme der zur Transportationsstrafe Verurteilten dienen (Penitentiaries, Public Works Prisons, hulks establishments) unter die einheitliche Aufsicht eines Kollegiums — der Directors of Convict Prisons — gestellt. 13 a. 14 Vict. c. 39.

Diese Zeit der Beschäftigung bei öffentlichen Arbeiten dauerte so lange, bis sich der Sträfling durch sein gutes Verhalten [1]) der Erteilung eines Urlaubsscheines — eines ticket of leave — würdig gezeigt hatte.

Das ticket of leave bildet das dritte Element, welches aus der Transportationsstrafe in das neue Strafmittel der penal servitude übergegangen ist. Die Erteilung von Urlaubsscheinen war ursprünglich von den Gouverneuren der australischen Kolonieen ohne eine besondere gesetzliche Ermächtigung erfolgt. Für die Einführung des Instituts war wohl wesentlich der Kostenpunkt maßgebend: die Gouverneure wollten der Sorge und der Kosten für den Unterhalt der Transportierten möglichst bald überhoben sein. Daneben kamen auch Rücksichten der Kolonialpolitik in Betracht, welche es zweckmäßig erscheinen ließen, die Transportierten zur Ansiedelung und zum dauernden Verbleiben in den Kolonieen zu veranlassen: da die ticket of leave men Aufenthaltsbeschränkungen für längere Zeit unterworfen waren, so war anzunehmen, daß sich dieselben in den betreffenden Bezirken dauernd niederlassen würden. Nachdem dann durch das Gesetz von 1829 — 9 Geo. IV c. 83 — diese Institution gesetzlich anerkannt und durch ein weiteres Gesetz von 1832 — 2 a. 3 William IV c. 62 — näher reguliert war, wurde dieselbe in dem probation system von 1842 als dritte Stufe in der Verbüßung der Transportationsstrafe aufgenommen. Als vierte Stufe schloß sich hier die völlige Begnadigung unter der Bedingung an, daß der Betreffende nicht nach England zurückkehre (conditional pardon). Dieses letztere Stadium mußte naturgemäß fortfallen, als man durch den Widerstand der Kolonieen, die zur Transportationsstrafe Verurteilten weiter aufzunehmen, genötigt wurde, die ticket of leave men in England zu behalten. Die Institution des ticket of leave, ursprünglich ein Stadium der Strafvollstreckung, näherte sich damit einem Begnadigungsakte, [2]) und dementsprechend wurde denn auch in dem Gesetze von 1853 — 16 a. 17 Vict. c. 99 — bestimmt, daß die „Urlaubsscheine" oder, wie nunmehr der Name

---

[1]) Zu dessen Feststellung das in Abschnitt II § 13 näher zu besprechende Markensystem diente.

[2]) Die juristisch wichtige Unterscheidung zwischen diesen beiden Auffassungen des Urlaubsscheins ist von v. Holtzendorff III Seite 23 ff. ausführlich behandelt worden; ich komme darauf noch später zurück.

lautete, die „Erlaubnisscheine zum ungehinderten Aufenthalte" „licenses to be at large in the United Kingdom and the Channel Islands" von einem Staatssekretär im Namen der Königin ausgestellt werden sollten.

Hiermit sind die 3 Bestandteile des im Jahre 1853 neu eingeführten Strafmittels der penal servitude gegeben: Isolierhaft in einer der Penitentiaries, dann gemeinschaftliche Zwangsarbeit in einem Public Works Prison und endlich Gewährung von Urlaubsscheinen.

Wenngleich die nähere Würdigung dieses Strafmittels einem späteren Abschnitte vorbehalten bleibt, so muß doch schon hier der von Nicht-Engländern häufig aufgestellten Behauptung widersprochen werden, daß man bei der Bestimmung des Inhalts der neuen Strafe prinziplos verfahren und, ohne einen innern Zusammenhang herzustellen, Gegensätzliches nebeneinander aufgenommen habe. Der Zusammenhang zwischen den 3 Stadien der penal servitude, wie derselbe bereits bei Einführung dieses neuen Strafmittels aufgefaßt wurde, ist folgender: Durch die Isolierhaft soll der Sträfling zur Selbsterkenntnis, zur Reue über seine Vergangenheit und zu guten Entschlüssen für seine Zukunft umgestimmt werden; er soll dadurch vorbereitet werden für das zweite Stadium der gemeinschaftlichen Zwangsarbeit, deren Zweck darin besteht, den Sträfling zu der Gewohnheit regelmäßiger Arbeit und voller Ausnützung seiner körperlichen Kräfte zu erziehen und ihn auf diese Weise für einen erfolgreichen Rücktritt in die bürgerliche Gesellschaft fähig zu machen.[1]) Es wird dabei nicht verkannt, daß die Gemeinschaft mit andern Sträflingen jeden Einzelnen vielfachen Ver-

[1]) Cf. den Brief des Home Secretary Sir George Grey an den Colonial Secretary Earl Grey vom 20. Januar 1847 (abgedruckt bei Du Cane I Seite 107); hier heißt es: „they will have passed through a course of separate imprisonment, accompanied by a system of moral and religious instruction and of industrial training, which, it is hoped will in most cases be attended with a beneficial effect on their character, and have prepared them for that intercourse with their fellow-prisoners which is inseparable from any plan for the employment of convicts on public works; and which, under certain restrictions, may in itself be made conducive to the progress of their reformation, and to their preparation for a return to society on the expiration of their sentence."

suchungen aussetzt, aber das erfolgreiche Bestehen dieser Versuchungen wird als die unerläßliche Probe dafür angesehen, daß der Sträfling die innere Kraft besitzt, die in der Isolierhaft gefaßten guten Vorsätze wirklich zur Ausführung zu bringen. Um ihn in diesem Bestreben zu stärken, wird ihm dann weiter für den Fall seines guten Verhaltens die Aussicht eröffnet, seine Strafzeit abzukürzen, indem er sich durch Arbeitsamkeit und Gehorsam während eines bestimmten Teils der Strafzeit den Urlaubsschein erwerben kann.[1]) Die gemeinschaftliche Zwangsarbeit bildet den Mittelpunkt der neuen Strafe, die Isolierhaft soll darauf vorbereiten, die Aussicht auf die Abkürzung der Strafzeit soll den bei der gemeinschaftlichen Arbeit bestehenden Gefahren nach Möglichkeit entgegenwirken.[2])

---

[1]) Sir Joshua Jebb konstatiert in dem Report of Directors of Convict Prisons 1852 Seite 337 den guten Erfolg der Beschäftigung der Sträflinge in den Public Works Prisons und fügt hinzu: „it is due to the principle of encouragement held out in the regulations".

[2]) In Deutschland wird die innere Bedeutung, welche in England dem Strafmittel der penal servitude von vornherein beigelegt worden ist, durchaus nicht gewürdigt. Man findet durchweg die Auffassung verbreitet, als ob Sir Walter Crofton in dem sog. irischen Strafsysteme etwas ganz Neues geschaffen hätte. Dem gegenüber muß hervorgehoben werden, daß in England bereits vor Sir Walter Crofton das System eines progressiven Strafvollzugs bestanden hat. Das Verdienst von Sir Walter Crofton, einem sehr hervorragenden Gefängnispraktiker, ist nicht die Auffindung eines neuen Systems des Strafvollzugs gewesen, sondern die eminent geschickte praktische Durchführung des bereits bestehenden Systems, welches erst durch Croftons Handhabung zur vollen Würdigung gelangt ist. Cf. Law Quarterly Review 1886 Seite 235. Das einzig Neue, was Crofton dabei geschaffen hat, war die Einführung des Stadiums der Zwischenanstalt zwischen dem Public Works Prison und der Erteilung des Urlaubsscheins. Diese Neuschaffung, deren Wert recht verschieden beurteilt wird, war vollständig im Sinne des bestehenden Systems gelegen: durch verschiedene Strafstadien hindurch auf den Wiedereintritt in die bürgerliche Gesellschaft vorzubereiten. — Eine nähere Darstellung und Würdigung des irischen Systems und seiner Verschiedenheit von dem englischen wird im Abschnitte III gegeben werden. Doch erschien es angemessen, das Verhältnis des sog. irischen oder Croftonschen Strafsystems zu dem im Texte behandelten Strafmittel der penal servitude schon hier in das richtige Licht zu setzen. Das irische Strafsystem ist durch die ausgezeichneten Schriften von v. Holtzendorff II—IV in Deutschland wohl bekannt gemacht, während die deutsche Litteratur über das Strafmittel der penal servitude eine mehr als dürftige ist. Dieser rein äußere Umstand dürfte die Beurteilung der Sache selbst beeinflußt haben.

Gehen wir nunmehr zur Darstellung der weiteren Entwickelung des neuen Strafmittels über.

Der erste Penal Servitude Act vom Jahre 1853 — 16 a. 17 Vict. c. 99 — hatte bestimmt:

1) daß an Stelle einer Transportationsstrafe unter 14 Jahren fortab auf penal servitude erkannt werden sollte;

2) daß auch in allen übrigen Fällen statt auf Transportation auf penal servitude erkannt werden könnte;

3) daß zwischen dem neuen Strafmittel der penal servitude und der Transportationsstrafe folgendes Verhältnis bestehen sollte: 4 Jahre penal servitude = 7 Jahre Transportation; 6 Jahre penal servitude = 10 Jahre Transportation; 8 Jahre penal servitude = 15 Jahre Transportation; an die Stelle von mehr als 15 jähriger Transportationsstrafe sollte 10 Jahre penal servitude treten;

4) daß sowohl den zu Transportationsstrafe wie den zu penal servitude Verurteilten bei guter Führung während der Periode der gemeinschaftlichen Zwangsarbeit Urlaubsscheine (licenses to be at large) durch den Staatssekretär des Innern gewährt werden könnten.

Thatsächlich wurde jedoch von dieser letzteren Befugnis bei den zur penal servitude Verurteilten kein Gebrauch gemacht, weil man der Ansicht war, die neue Strafe sei infolge der sub 3 angeführten Strafumwandlungs-Grundsätze schon so erheblich milder, als die früheren Strafen, daß eine weitere Strafkürzung nicht angebracht erscheine.

Im Jahre 1856 setzte dann das Unterhaus eine Kommission nieder, um über die Handhabung und Wirksamkeit des Penal Servitude Act 1853 Bericht zu erstatten. Der Bericht fiel im allgemein engünstig aus. Es wurde jedoch getadelt, daß die Periode der Isolierhaft zu lang bemessen sei; es hätte sich gezeigt, daß eine 18 monatliche Isolierhaft für den körperlichen und noch mehr für den geistigen Zustand der Sträflinge nachteilig sei. Es wurde ferner getadelt, daß Strafkürzung bei penal servitude thatsächlich fast gar nicht zur Anwendung gebracht würde, und es wurde vorgeschlagen, daß sich der Sträfling durch sein gutes Verhalten nach Ablauf einer bestimmten Strafzeit ein Recht auf den Urlaubsschein erwerben sollte, so daß die Erteilung des Urlaubsscheins der Willkür des Staatssekretärs entzogen sein würde. Endlich sprach man sich dafür aus, daß dem Richter ein weiterer Spielraum in der Bemessung der Zeitdauer der Strafe der penal servitude gewährt würde.

Auf der Basis dieser Vorschläge beruhte der zweite Penal Servitude Act vom Jahre 1857 — 20 a. 21 Vict. c. 3 —. Durch dieses Gesetz — welches, wie in § 3 ausgeführt wurde, die Transportation als Strafmittel gänzlich beseitigte und der Regierung nur die Ermächtigung erteilte, die zur penal servitude Verurteilten zur Strafverbüßung nach den Kolonieen zu schaffen — sowie durch die daraufhin von Sir George Grey erlassene Ausführungsverordnung wurden betreffs des Strafmittels der penal servitude folgende Veränderungen getroffen:

1) Der Richter kann an Stelle der bisherigen Transportationsstrafe auf penal servitude von gleicher Zeitdauer erkennen; die in dem Gesetze von 1853 gegebenen Bestimmungen über die Strafumwandlung werden aufgehoben. Weiter wird das bisherige Strafminimum der penal servitude von 4 Jahren auf 3 Jahre herabgesetzt;[1])

2) die Isolierhaft soll fortab nur 9 Monate dauern;

3) der Sträfling erhält durch sein gutes Verhalten in dem Public Works Prison einen Anspruch auf die Erteilung eines Urlaubsscheins vor Ablauf seiner Strafzeit. Die dadurch herbeigeführte Strafkürzung beträgt je nach der Dauer der Strafzeit $^1/_6$—$^1/_3$ (bei 3 Jahren penal servitude: 6 Monate; bei 15 Jahren penal servitude: 5 Jahre);

4) die mit der Erteilung eines Urlaubsscheines zu verbindenden Einschränkungen und Verpflichtungen sollen streng innegehalten werden. Der Beurlaubte soll sich sofort bei der Polizei melden; er soll ohne besondere Erlaubnis den betreffenden Ort nicht verlassen dürfen, und er soll sich periodisch bei der Polizei wieder melden, damit diese eine fortlaufende Aufsicht über sein gutes Verhalten ausüben kann.

Auch hier war es wiederum die letztere Bestimmung, deren praktische Ausführung am wenigsten befriedigte. Die über die Entlassenen ausgeübte polizeiliche Aufsicht erwies sich durchweg als zum Schutze der bürgerlichen Gesellschaft unzureichend und an vielen Orten als direkt schädlich für die Entlassenen, welche durch die Art,

[1]) Das Maximum der Strafe beträgt 25 Jahre, daneben ist aber auch lebenslängliche penal servitude zulässig.

wie die Aufsicht ausgeübt wurde, häufig an dem Auffinden einer Beschäftigung und damit an einem redlichen Lebenserwerbe gehindert wurden. Die Zahl der Verbrechen nahm im Anfange der 60er Jahre, insbesondere in London selbst, in geradezu erschreckender Weise zu, und die Zahl der dabei beteiligten ticket of leave men — oder license holders, wie nunmehr die gebräuchlichere Bezeichnung ist — war eine so große, daß erhebliche Zweifel an der Richtigkeit des ganzen Systems laut wurden.

Unter diesen Umständen berief die Königin im Jahre 1863 eine Royal Commission mit dem Earl Grey als Vorsitzenden, um über die praktische Wirksamkeit der Penal Servitude Acts 1853 und 1857 Bericht zu erstatten. Die Kommission verhörte eine große Anzahl Zeugen; alle auf dem Gebiete des Gefängniswesens in England hervorragenden Persönlichkeiten gaben ihr Urteil ab.[1]) Aus dem, dem Parlamente erstatteten umfangreichen Berichte, auf welchen im Abschnitte II mehrfach zurückzukommen ist, kann hier nur einiges wenige hervorgehoben werden.

Das Gesamturteil der Kommission ging dahin, daß eine zu große Milde bestehe, sowohl was die Höhe der erkannten Strafen als was die Art und Weise ihrer Vollstreckung anbetrifft. Im einzelnen wird ausgeführt: die Minimalstrafe der penal servitude sei von 3 Jahren auf 7 Jahre zu erhöhen, bei Rückfälligen sei das freie richterliche Ermessen in der Bestimmung der Strafhöhe einzuschränken und es sei hier durchweg auf höhere Strafen zu erkennen. Während des Aufenthaltes in den Public Works Prisons seien strengere Disziplinarstrafen, insbesondere die häufigere Anwendung von körperlichen Strafen, einzuführen. Die Höhe der dem Sträflinge bei seiner Entlassung gezahlten Arbeitsbelohnung (gratuity) sei erheblich zu beschränken, insbesondere für die Rückfälligen, welche an die geforderten Arbeiten gewöhnt wären und sich dadurch im Vorteile gegenüber den zum ersten Male Bestraften befänden. Die Strafkürzung sei für die längeren Freiheitsstrafen, also für die schweren Verbrecher, zu verringern. Das Markensystem, auf welchem ja die Erteilung des Urlaubsscheines beruht, sei umzugestalten; es sei unrichtig, dem Sträflinge für sein gutes Verhalten, zu welchem er ja

---

[1]) Insbesondere wurde Sir Walter Crofton, der Leiter des irischen Gefängniswesens, einer eingehenden Vernehmung unterzogen.

schon an sich verpflichtet sei und das sich somit von selbst verstehen müsse, eine Belohnung zu gewähren. Die Marken und damit zusammenhängend die Strafkürzung sollten dem Sträflinge fortab nur gewährt werden für die von ihm bewiesene Arbeitsamkeit; die so verdienten Marken sollten aber bei etwaigem schlechten Verhalten wieder entzogen werden. Für die mit einem Urlaubsscheine Entlassenen sei fortab eine strenge Kontrolle einzuführen: es sollen zur Beaufsichtigung derselben besondere Beamte ernannt werden, welchen die Polizeiorgane Hilfe zu leisten hätten, während ein direktes Eingreifen der letzteren unterbleiben sollte. Das dem Entlassenen auferlegte gute Verhalten sei näher zu spezifizieren. Man solle nicht warten, bis der Entlassene eine neue strafbare Handlung begangen habe, sondern solle den Urlaubsschein schon dann für verfallen erklären, wenn der Entlassene ein träges, unregelmäßiges Leben führe und einen redlichen Unterhaltserwerb nicht nachweisen könne, oder wenn er mit Leuten von notorisch schlechtem Charakter Umgang pflege. Für den Fall, daß der Urlaubsschein für verfallen erklärt würde, solle der Betreffende nicht nur für die noch nicht abgelaufene Strafzeit in die Anstalt zurückgeschafft werden, sondern er solle die ganze Strafzeit, welche ihm bei Erteilung des Urlaubsscheines erlassen worden sei, ohne Rücksicht auf die in der Freiheit verbrachte Zwischenzeit, abbüßen; und er solle dazu noch eine neue Strafe für die Nichtinnehaltung der ihm bei seiner Entlassung auferlegten Bedingungen erhalten.

Man sieht, die Kommission glaubte durchweg gegenüber dem bisherigen Zustande recht erhebliche Verschärfungen empfehlen zu sollen. Anderseits aber hielt sie — und dies verdient besonders hervorgehoben zu werden — an dem wesentlichen und charakteristischen Inhalte der penal servitude, an der Aufeinanderfolge von Isolierhaft und gemeinschaftlicher Zwangsarbeit und an der in die Macht des Sträflings gestellten Strafkürzung, fest.

Auf Grund der Vorschläge der Kommission erging im Jahre 1864 der dritte Penal Servitude Act — 27 a. 28 Vict. c. 47 —. Die wesentlichsten Änderungen, welche durch dieses Gesetz resp. die später ergangenen Ausführungsbestimmungen eingeführt wurden, sind folgende:

1) Die Minimalstrafzeit der penal servitude wird auf 5 Jahre festgesetzt; für Rückfällige, d. h. diejenigen, welche

schon einmal wegen einer felony bestraft worden sind, beträgt die Minimalstrafzeit 7 Jahre;

2) Während der Zwangsarbeit in den Public Works Prisons werden die Sträflinge in 5 Klassen geteilt: probation, third, second, first und special class. Um aus einer niederen Klasse in eine höhere aufzurücken, muß der Sträfling eine bestimmte Anzahl Marken verdient haben. Diese Marken werden fortab ledig-lich für Arbeitsamkeit gegeben; im Falle schlechter Führung kann ein gewisser Teil derselben wieder entzogen werden. Mit jeder höheren Klasse sind besondere Vergünstigungen verbunden, insbesondere steigt der Arbeitsverdienst, dessen Maximalhöhe im allgemeinen auf 3 £ begrenzt wird, von Klasse zu Klasse. Der Sträfling muß die ersten 4 Klassen durchgemacht haben, ehe er eine license to be at large erhält. Die Strafkürzung beträgt in maximo $^1/_4$ derjenigen Zeit, welche der Sträfling in dem Public Works Prison zuzubringen hat [1]) (also bei 5 Jahren penal servitude 1 Jahr 3 Wochen, da nach Abzug der 9 Monate Einzelhaft 4 Jahre 3 Monate auf das Public Works Prison fallen);

3) Die license, welche dem Sträflinge bei seiner Entlassung vor Ablauf der Strafzeit gegeben wird, enthält eine lange Reihe von Verhaltungsmaßregeln, welche im allgemeinen den Vorschlägen der Kommission entsprechen. Jede Übertretung dieser Verhaltungsmaßregeln soll fortab als ein selbständiges Delikt angesehen werden, welches im summarischen Verfahren mit Gefängnis bis zu 3 Monaten strafbar ist. Die license holders können von der Polizei wegen jeder derartigen Übertretung, wie auch wegen jedes andern Deliktes ohne Haftbefehl verhaftet werden. Wenn ein license holder die Verhaltungsmaßregeln übertritt oder irgend ein Delikt begeht, so kann der Staatssekretär die license widerrufen. Im Falle des Widerrufs hat der Betreffende die ganze Strafzeit, welche ihm erlassen worden war, nachträglich zu verbüßen.

Diese letzteren Bestimmungen sind noch etwas verschärft worden durch ein Gesetz von 1871: Prevention of Crimes Act — 34 a. 35 Vict. c. 112). [2]) Hier wird der Polizei

---

[1]) Bei weiblichen Sträflingen, für welche auch sonst spezielle Bestimmungen gelten, $^1/_3$ der Strafzeit.

[2]) Dieses Gesetz, auf dessen nähere Bestimmungen noch unten im § 10 eingegangen werden wird, ist an die Stelle des Habitual Criminal Act vom Jahre 1869 — 32 a. 33 Vict. c. 99 — getreten. Das Gesetz hat durch den

das Recht gegeben, jeden license holder, welcher verdächtig ist, seinen Lebensunterhalt auf unredliche Weise zu erwerben, ohne weiteres zu verhaften und denselben einem Gerichtshofe mit summarischer Jurisdiktion vorzuführen; findet der letztere den Verdacht begründet, so erkennt derselbe auf Verwirkung der license. Es wird ferner bestimmt, daß ein license holder, welcher seine Verpflichtung, jeden Wohnungswechsel binnen 48 Stunden der Polizei anzuzeigen und sich monatlich 1 mal persönlich oder schriftlich auf der Polizei zu melden, nicht erfüllt, im summarischen Prozesse nicht nur zur Verwirkung seiner license, sondern auch zu einer Gefängnisstrafe bis zu 1 Jahre verurteilt werden kann.

Das auf Grund dieser Gesetze gestaltete Strafmittel der penal servitude hat sich nach dem in England allgemein bestehenden Urteile sehr gut bewährt. Die im Anfange der 60er Jahre vielfach laut gewordenen Klagen sind mehr und mehr verstummt. Erst im Jahre 1877 wurde die öffentliche Meinung durch mehrere, schnell nacheinander erscheinende Schriften und durch eine Anzahl von Zeitungsartikeln wieder auf den Gegenstand hingelenkt.[1]) Obwohl man diesen litterarischen Erzeugnissen, welche sämtlich von entlassenen Sträflingen verfaßt waren, an sich keine allzugroße Bedeutung beilegte, erschien es doch angemessen, nachdem eine lange Reihe von Jahren seit der letzten gründlichen Untersuchung verflossen war, sich über den gegenwärtigen Zustand durch eine besondere Kommission Aufklärung zu verschaffen.

Im Januar 1878 ernannte die Königin 6 Commissioners unter dem Vorsitze des Earl of Kimberley, um über die praktische Wirksamkeit der Penal Servitude Acts Bericht zu erstatten. Die Commissioners, welche ihre Untersuchung auf ganz Großbritannien[2]) ausdehnten, überzeugten sich an Ort und Stelle von der Ausführung der gesetzlichen Bestimmungen und dem Zustande der Anstalten, und ver-

---

Prevention of Crimes Act vom Jahre 1879 — 39 a. 40 Vict. c. 23 — noch einige Ergänzungen erfahren.

[1]) Besonderes Aufsehen erregte eine Schrift: Five Years Penal Servitude by one who has endured it. Der anonyme Verfasser, dessen Persönlichkeit übrigens ermittelt worden ist, schildert darin in recht grell aufgetragenen Farben seine eigenen Erlebnisse während der Strafverbüßung.

[2]) Es sei hier bemerkt, daß die Gesetze über die penal servitude in allen wesentlichen Punkten für die 3 Königreiche übereinstimmende sind.

nahmen dann eine große Anzahl von Zeugen, darunter auch mehrere Personen, welche selbst Strafknechtschaft erlitten hatten und so aus eigner Wahrnehmung sprechen konnten.

Der Bericht, welchen die Commissioners daraufhin im Jahre 1879 erstatteten, spricht sich, soweit England in Frage kommt, über die bestehenden Verhältnisse sehr günstig aus.[1]) Es werden eine kleine Anzahl Verbesserungsvorschläge gemacht, wovon jedoch nur ein einziger eine gesetzliche Änderung erforderte, und diese Änderung ward denn auch sofort im Jahre 1879 durch das Gesetz 42 a. 43 Vict. c. 55 vorgenommen.

Es handelt sich dabei um die in dem Gesetze von 1864 getroffene Bestimmung, daß bei Rückfälligen die Minimalstrafzeit der penal servitude 7 Jahre betragen solle. Hervorragende Juristen hatten bekundet, daß die Richter häufig bei Beurteilung von Rückfälligen eine Strafe von 7 Jahren nach der Lage des konkreten Falles für zu hoch hielten und, da eine Bestrafung mit 5 Jahren penal servitude durch das Gesetz ausgeschlossen wäre, überhaupt nicht penal servitude, sondern Gefängnis verhängten; daß somit die gesetzliche Beschränkung in der richterlichen Strafabmessung ganz gegen den Willen des Gesetzes zu einer thatsächlichen Strafmilderung führte.[2]) Da die Bestimmung in dem Gesetze von 1864 mit der allgemeinen Richtung des heutigen englischen Strafrechts, wonach dem Richter bei der Strafabmessung fast völlig freie Hand gelassen wird, im Widerspruch stand, so wurde die Aufhebung derselben ohne alles Bedenken vom Parlamente beschlossen.

Bei dieser Gelegenheit kam auch die in dem Berichte der Commissioners eingehend erörterte und auch sonst in England viel behandelte Frage zur Sprache, ob es sich nicht empfehlen würde, die Kluft, welche zwischen der regelmäßigen Maximalstrafzeit der Gefängnisstrafe von 2 Jahren

---

[1]) Weit weniger günstig betreffs der Einrichtungen in Irland. Es wird darauf im Abschnitte III noch näher eingegangen werden.

[2]) Von Bedeutung ist besonders das Gutachten der Criminal Code Commissioners gewesen, welches dahin ging „that the 7 years minimum is too high, that it is very injurious, and that many sentences of imprisonment, which are felt to be inadequate, are passed in order to avoid that which would be an excess“.

und der Minimalstrafzeit der penal servitude von 5 Jahren liegt, in irgend einer Weise auszufüllen.

Die Commissioners haben diese Frage nur in sofern erörtert, als die Ausdehnung der Maximalstrafzeit der Gefängnisstrafe auf 3 Jahre in Betracht kommt. Ihre Ansicht ging gegen eine derartige Ausdehnung, und sie haben hierbei zweifellos die Majorität in England auf ihrer Seite. Man führt an, daß eine Isolierhaft von mehr als 2 Jahren — die Gefängnisstrafe wird, was hier vorausgenommen werden muß, in England regelmäßig in Einzelhaft verbüßt — der geistigen wie auch der körperlichen Gesundheit der Gefangenen nachteilig sein würde, und daß ferner eine geeignete Beschäftigung von Gefangenen während so langzeitiger Einzelhaft Schwierigkeiten bereiten müßte.

Die andre Frage, ob man nicht die Minimalstrafzeit der penal servitude auf 3 Jahre herabsetzen sollte, ist von den Commissioners nicht erörtert worden. Es macht sich hierfür augenblicklich in England, besonders in juristischen Kreisen, eine lebhafte Bewegung geltend. Es wird dabei von sachkundigster Seite versichert, daß die Anwendung des jetzigen Strafvollzugsmodus der penal servitude auf eine dreijährige Strafzeit keinerlei Schwierigkeiten oder Bedenken unterliegen würde. *Die Einführung einer Minimalstrafzeit der penal servitude von 3 Jahren dürfte hiernach nur eine Frage der Zeit sein.*[1])

Was die übrigen, im Verwaltungswege einzuführenden Verbesserungsvorschläge der Commissioners anbetrifft, welche im Abschnitte II eingehendere Erörterung finden werden, so seien davon hier nur zwei kurz hervorgehoben.

Die Commissioners wiederholten den bereits im Jahre 1864 von der Royal Commission gemachten *Vorschlag, zur Beaufsichtigung der license holders besondere Beamte zu ernennen.* Wie es schon damals geschehen ist, so wird auch jetzt

---

[1]) Ich habe mir gerade über diese Frage während meines Aufenthaltes in England durch persönliche Besprechungen mit hervorragenden Männern der Strafrechtswissenschaft wie des Gefängniswesens Aufklärung zu verschaffen gesucht und glaube, auf Grund hiervon die obige Ansicht aussprechen zu können. Ich berufe mich dabei speziell auf das Urteil von Sir Edmund Du Cane dafür, daß die jetzige Art und Weise der Vollstreckung der penal servitude auch bei kürzerer, etwa 3 jähriger Strafzeit anwendbar sein würde.

scharf hervorgehoben, daß die Aufsicht über die Entlassenen resp. Beurlaubten ihren Zweck nicht erfüllen könne, wenn sie in die Hand der gewöhnlichen Polizeiorgane gelegt würde. Dieselben seien vor allem in den großen Städten, welche bei der Frage ja vorzugsweise in Betracht kämen, viel zu sehr mit anderweitigen Geschäften belastet, um eine wirksame Aufsicht führen zu können. Die gewöhnlichen Polizeiorgane seien aber ferner auch an sich schon meistenteils für diese Aufgabe nicht qualifiziert; es seien Leute, welche an eine routinenmäßige Erledigung der Geschäfte gewöhnt seien, während es doch bei der Aufsicht über die Entlassenen darauf ankomme, zu spezialisieren, je nach dem Einzelfalle verschieden zu handeln, wenn anders nicht gerade der auf gutem Wege Befindliche durch ungerechtfertigte Einmischung wieder in die Verbrecherlaufbahn zurückgestoßen werden solle. Es sei zwar anzuerkennen, daß sich diese Verhältnisse gegenüber dem Zustande im Jahre 1863 infolge der segensreichen Thätigkeit der Vereine für entlassene Sträflinge (Discharged Prisoners Aid Societies) erheblich gebessert hätten, allein es sei notwendig, um das Ziel vollständig zu erreichen, daß besondere, und zwar speziell für diese Aufgabe qualifizierte Organe bestellt würden.

Dem Vorschlage der Commissioners ist denn auch nunmehr wenigstens für die großen Städte Folge gegeben worden.[1]) Speziell in London ist im Jahre 1879 ein Convict Supervision Office eingerichtet worden, welches seine Aufgabe in mustergültiger Weise erfüllt. (Auf die Thätigkeit desselben wird im § 10 noch eingegangen werden).

Der zweite, hier zu erwähnende Vorschlag der Commissioners betrifft die Klassifizierung der Sträflinge und der Strafanstalten. Eine derartige Klassifizierung hatte bereits in sofern stattgefunden, als diejenigen Sträflinge, welche infolge ihres geistigen oder körperlichen Zustandes einer speziellen Behandlung bedürftig erschienen oder welche infolge von Alter und Gebrechlich-

---

[1]) Es ist vor allem der unermüdlichen Thätigkeit des um die Reform des englischen Gefängniswesens hochverdienten Friedensrichters in Gloucester, Mr. Barwick Baker, zuzuschreiben, daß die Frage einer genügenden und gut eingerichteten Aufsicht in fortwährender Diskussion erhalten worden ist. Ein Lebensbild dieses ausgezeichneten Mannes, welcher Anfang Dezember 1886 leider verstorben ist, hat v. Holtzendorff in seiner Erzählung „Ein englischer Landsquire" gegeben.

keiten zu den schweren, in den Public Works Prisons vorgenommenen Arbeiten nicht herangezogen werden konnten, in die besonders eingerichteten Anstalten zu Woking, Parkhurst und Dartmoor gebracht wurden. Die Commissioners sprachen zunächst ihre Billigung dieses Systemes aus und wünschten, das dasselbe streng eingehalten würde: wenn man diese Klassen von Sträflingen nicht in besondere Anstalten brächte, so würde die Verwaltung und die Beobachtung einer strengen Disziplin in den Anstalten erheblich erschwert werden. Die Commissioners verlangten sodann aber noch eine weitere Klassifizierung der Sträflinge: es sollten einerseits die Gewohnheits- oder sonst gefährlichen Verbrecher und anderseits die zum ersten Male bestraften Gelegenheitsverbrecher von den übrigen Sträflingen gesondert werden.

Was die Gewohnheitsverbrecher anbetrifft, so war von mehreren Zeugen der Vorschlag gemacht worden, dieselben einer besonders strengen Disziplin zu unterwerfen und sie vor Ablauf ihrer Strafzeit eine nochmalige Periode der Isolierhaft durchmachen zu lassen. Die Commissioners sprachen sich gegen diesen Vorschlag aus: eine derartige ungleichmäßige Behandlung würde unbillig und ungerecht sein, weil der Richter schon bei der Ausmessung der Strafe den Umstand in Betracht ziehe, daß der Betreffende ein Gewohnheitsverbrecher sei, und weil somit eine strengere Behandlung in der Strafanstalt einer nochmaligen Bestrafung gleichkommen würde. Außerdem wäre es im allgemeinen nicht ratsam, den Sträfling gerade die letzte Periode seiner Strafzeit in Isolierhaft zubringen zu lassen, weil erwiesenermaßen durch die Isolierhaft die körperlichen Kräfte des Sträflings und seine Fähigkeit zu einer wirklich anstrengenden Beschäftigung vermindert würden, und weil es unbillig wäre, den Sträfling mit dieser verminderten Arbeitsfähigkeit in die Freiheit zu setzen und ihm so den redlichen Erwerb seines Lebensunterhaltes noch mehr zu erschweren. Eine Zurückversetzung in die Isolierhaft erschiene jedoch stets gerechtfertigt im Falle von Trägheit und schlechtem Verhalten des Sträflings während der gemeinschaftlichen Zwangsarbeit. Im übrigen sollten Sträflinge, von denen ein übler Einfluß auf die andern Insassen der Anstalt zu befürchten wäre, während der gemeinschaftlichen Zwangsarbeit möglichst abgesondert und in besondere Arbeitsabteilungen gebracht werden.

Gemäß diesen Vorschlägen wird seitdem verfahren. Eine allgemeine Verfügung vom 12. März 1878 [1]) bestimmt, daß ein Sträfling, welcher während seines Aufenthaltes in einem Public Works Prison durch sein Verhalten derartig zu Unzufriedenheit des Anstaltsdirektors Anlaß gibt, daß ein übler Einfluß auf die andern Sträflinge befürchtet werden muß, in die Isolierhaft auf 6—9 Monate zurückversetzt werden kann und alsdann, nach seiner Rückkehr zu dem Public Works Prison, in einer zu bildenden besonderen Arbeitsabteilung zu beschäftigen ist. In diese Abteilung (separate working party) sind ferner diejenigen einzurangieren, von welchen nach dem Charakter des von ihnen begangenen Verbrechens oder aus sonstigen Gründen ein übler Einfluß auf andre Sträflinge befürchtet werden kann. [2])

Was die Neulinge im Verbrechen betrifft, so verkannten die Commissioners durchaus nicht, daß die Thatsache der erstmaligen Bestrafung an sich noch keinen Beweis dafür liefere, daß der Betreffende im Verbrechen nicht bewandert sei und daß von ihm ein übler Einfluß auf andre Sträflinge nicht zu befürchten sei. Sie glaubten jedoch anderseits, daß es möglich sei, durch eingehende Erkundigungen nach den Antezedenzien und durch scharfe Beobachtung während der Zeit der Isolierhaft den Charakter des Sträflings mit einiger Sicherheit festzustellen. Die Commissioners schlugen vor, diejenigen zum ersten Male Bestraften, welche sich auf Grund dieser Ermittelungen als „not versed in crime“ charakterisieren ließen, von allen übrigen Sträflingen scharf zu trennen und, wenn möglich, in eine besondere Strafanstalt zu bringen.

Auch dieser Vorschlag ist verwirklicht worden. Der Direktor und der Geistliche derjenigen Strafanstalt, in welcher der Sträfling die Isolierhaft verbüßt, ist jetzt verpflichtet, über jeden Fall, welcher nach seiner Ansicht in die Rubrik der Neulinge im Verbrechen gerechnet werden kann, an die Zentralbehörde eingehend zu berichten; vorher hat er nach einem festen, von der Zentralbehörde aufgestellten Formulare [3]) Erkundigungen über die Antezedenzien des betreffenden Sträflings einzuziehen, insbesondere bei der Ortsbehörde, dem

---

[1]) Abgedruckt bei Du Cane I Seite 113 f.

[2]) cf. Report of the Royal Commissioners 1879 Seite XXX „notorious thieves, burglars and receiviers of stolen goods, men who have made crime their profession and are eager to initiate others in the mysteries of their nefarious craft.“

[3]) Ein solches Formular findet sich abgedruckt bei Du Cane I Seite 115 f.

früheren Arbeitgeber, dem Geistlichen u. s. w. Auf Grund all dieser Ermittelungen erfolgt dann die event. Zuteilung des betreffenden Sträflings zu der sog. star class. Die zu dieser Klasse Gehörigen sind bisher in einer abgesonderten Abteilung der Public Works Prisons gehalten und getrennt von den übrigen Sträflingen beschäftigt worden. Jetzt ist das neuerrichtete Public Works Prison in Dover [1]) zur ausschließlichen Aufnahme von star class prisoners bestimmt.

Diese neue Einrichtung hat sich nach dem ganz allgemeinen Urteile vorzüglich bewährt. Sämtliche Strafanstaltsdirektoren stimmen darin überein, daß die Arbeitsleistungen der star class prisoners weit den Durchschnitt derjenigen der übrigen Sträflinge übertreffen und daß Disziplinarstrafen in der star class Abteilung fast als Seltenheiten angesehen werden können. Das glänzendste Zeugnis für die neue Einrichtung und zugleich den besten Beweis, daß man bei der Zuteilung zur star class den richtigen Maßstab anwendet, liefert die Ziffer der Rückfälligen unter den star class prisoners.

Vom Jahre 1879 bis 1885 wurden der star class insgesamt 1332 Sträflinge zugeteilt (im Jahre 1885 betrug die Ziffer nur 107, sie war niedriger, als in irgend einem der Vorjahre). Davon sind vom Jahre 1882 ab zur Entlassung gekommen: 455. Von diesen 455 ist ein Einziger als rückfällig zur Bestrafung gelangt. [2]) Das ist ein geradezu staunenswertes Ergebnis, hauptsächlich wenn man dabei berücksichtigt, daß die Entlassenen einer strengen Aufsicht unterworfen sind und daß Rückfälle während der ersten Jahre nach der Entlassung wohl nur selten unentdeckt bleiben. [3])

Im Anschlusse hieran seien schließlich noch einige Ziffern über die Anwendung des Strafmittels der penal servitude angeführt. [4])

[1]) Dies Gefängnis ist am 31. August 1885 eröffnet worden und enthält jetzt 188 star class prisoners. Cf. Report of Directors 1886 Seite VII.

[2]) Cf. Rep. of Directors 1886 Seite XI.

[3]) Um ein ganz sicheres Urteil zu gewinnen, müßte allerdings die Ziffer derjenigen ermittelt werden, welche nach ihrer Entlassung ausgewandert sind. Aber auch wenn man diese Ziffer recht hoch annimmt, bleibt das Ergebnis ein glänzendes.

[4]) Diese Ziffern sind entnommen den Reports of Directors 1885 Seite XVI ff. und 1886 Seite XVIII ff. sowie den Judicial Statistics 1884 Seite XXIX.

Im Jahre 1885/6[1]) wurden 1027 Personen zur penal servitude verurteilt; es ist dies die niedrigste Ziffer, welche jemals vorgekommen ist. Im Jahre 1855 betrug die Zahl der zur penal servitude Verurteilten 2590; im Jahre 1857: 2841; die Zahl stieg im Jahre 1862 auf 3369 und sank dann bis zum Jahre 1869 auf 2000. Von 1870—1881 bewegten sich die Ziffern in kleinen Schwankungen zwischen 1493 und 1788. Im Jahre 1882 sank die Ziffer auf 1364, und erhielt sich während der nächsten 2 Jahre in etwa gleicher Höhe (1883: 1378; 1884: 1349), um dann im Jahre 1885 auf 1027 zu fallen. Es ist also seit dem Jahre 1863 eine konstante Abnahme bemerkbar.[2])

Die Delikte, wegen welcher auf penal servitude erkannt wurde, sind vorzugsweise: Diebstahl, Hehlerei, Einbruch und Raub. In den Jahren 1884 und 1885 betrug die Zahl der Personen, welche zur penal servitude verurteilt wurden:[3])

---

[1]) Seit 1879 wird die Statistik von April zu April aufgestellt. Der Einfachheit halber ist die zweite Jahresziffer im folgenden regelmäßig fortgelassen. Die für 1885 angeführten Ziffern beziehen sich somit auf die Periode vom 31. März 1885 bis 31. März 1886.

[2]) Es dürfte von Interesse sein, die Zahl der zur penal servitude Verurteilten der Bevölkerungsziffer gegenüberzustellen. In der folgenden Tabelle ist dies für 5 jährige Perioden geschehen.

Während der folgenden 5jährigen Perioden betrug:	a) die Durchschnittszahl der zur penal serv. Verurteilten:	b) die Durchschnittsziffer der Bevölkerung in Tausenden:
1855—1859	2589	19,257
1860—1864	2800	20,370
1865—1869	1978	21,681
1870—1874	1622	23,088
1875—1879	1633	24,700
1880—1884	1427	26,399
Im Jahre 1885 betrug	die Zahl der Verurteilten: 1027	die Bevölkerungsziffer: 27,499,(041).

[3]) Die hier für die Jahre 1884 und 1885 angeführten Ziffern der Bestrafungen wegen der 3 Delikte erklären vollständig die schon hervorgehobene große Abnahme der Bestrafungen im Jahre 1885/6; diese Abnahme ist ausschließlich zurückzuführen auf die geringere Zahl der zur Bestrafung gelangten Eigentumsverbrechen, insbesondere des Diebstahls.

	1884	1885
a) wegen Diebstahls und Hehlerei:	638	299
b) wegen Einbruchs (burglary):	190	136
c) wegen Raubes und Raubversuchs:	119	38

Unter Zugrundelegung der oben Seite 4 Anm. 1 mitgeteilten Einteilung der Delikte in den Judicial Statistics verteilten sich die Bestrafungen zur penal servitude im Jahre 1884

1) auf offences against property:	
a) with violence	19,4 %
b) without violence	56,6 %
c) malicious offences	2,2 %
zusammen:	78,2 %
2) auf offences against the person:	15,9 %
3) auf alle übrigen Rubriken zusammen:	5,9 %.

Man sieht, weit über 3/4 der Bestrafungen zur penal servitude erfolgten im Jahre 1884 wegen Eigentumsverbrechen. Dieser Prozentsatz nimmt von Jahr zu Jahr ab, derselbe betrug noch für das Jahr 1883: 80,9 %. Im Jahre 1885 hat er dann weiter in erheblichem Maße abgenommen.

Unter den im Jahre 1885 mit penal servitude Bestraften befanden sich 330, also etwas weniger als 1/3 im Rückfalle, im Jahre 1884 betrug die Zahl der Rückfälligen 480, also etwas über 1/3 der Gesamtzahl der Bestraften.

Was endlich die Höhe der Strafen betrifft, so wurde im Jahre 1885 erkannt auf penal servitude

von	5	Jahren	gegen	646	Personen
"	6	"	"	37	"
"	7	"	"	181	"
"	8	"	"	19	"
"	9	"	"	1	"
"	10	"	"	94	"
"	11—15	"	"	33	"
"	16—20	"	"	7	"
"	21—25	"	"	1	"

Dazu treten 8 Urteile auf penal servitude für Lebenszeit. Der Schwerpunkt der Strafe liegt also in dem Strafmaße von 5 und 7 Jahren.

Die Höhe des Strafmaßes hat von Jahr zu Jahr abgenommen. Die Ziffern für die Zeit von 1879 sind dabei allerdings wegen der in diesem Jahre vorgenommenen, oben Seite 62 erwähnten gesetzlichen Änderung nicht ohne weiteres vergleichbar. Es seien deshalb bloß folgende Ziffern angeführt: die Verurteilungen zur penal servitude verteilten sich auf folgende Strafmaße:

	1880	1881	1882	1883	1884
5 und 6 Jahre	57,3 %	59,5 %	51,8 %	64,2 %	64,6 %
7 Jahre	23,1 „	21,1 „	21,6 „	19,8 „	20,0 „
8—10 Jahre	14,5 „	13,8 „	11,2 „	11,3 „	10,4 „
11—15 Jahre	2,8 „	3,6 „	3,5 „	2,0 „	3,0 „
über 15 Jahre	1,8 „	0,9 „	1,4 „	2,0 „	1,1 „
Lebenszeit	0,5 „	1,1 „	0,5 „	0,7 „	0,9 „

Man sieht, das Strafmaß von 5 und 6 Jahren — das letztere wird übrigens nur selten gegeben — ist von Jahr zu Jahr mehr zur Anwendung gebracht worden, während in entsprechender Weise die Strafen von 7—10 Jahren abgenommen haben.

§ 5. Die Gefängnisstrafe.

Litteratur: Report of the Committee of the House of Lords on the state of Gaols and Houses of Correction 1835. — Report of the Committee of the House of Commons on the rules and discipline in Gaols and Houses of Correction 1850. — Report of the Committee of the House of Lords on Gaols and Houses of Correction 1863. — Wines, the state of Prisons and of Child-saving Institutions in the civilized world· Cambridge 1880. — Wilkinson, the law of Prisons in England and Wales. London 1878.

Die Gefängnisstrafe besteht seit alters her als Strafmittel für misdemeanours. Zur Vollstreckung derselben dienten die unter der Verwaltungsaufsicht der Sheriffs stehenden Grafschaftsgefängnisse (common gaols), welche als im Eigentume des Königs stehend betrachtet wurden. Der König verlieh ferner gelegentlich an reiche Grundherren zugleich mit der Gerichtsgewalt ein Gefängnisrecht; die Grundherren erbauten daraufhin Gefängnisse, welche unter ihrer eignen Verwaltung verblieben.

Auf Grund eines Gesetzes der Königin Elisabeth wurden sodann in Verbindung mit der neuerlassenen Armengesetzgebung Korrektionshäuser (Houses of Correction oder auch Bridewells genannt) errichtet, in welche arbeitsscheue Personen, Bettler, Vagabunden und liederliche Dirnen gebracht werden sollten. Im Jahre

1609 wurde gesetzlich — 9 James I c. 4 — bestimmt, daß in jeder Grafschaft ein derartiges Korrektionshaus bestehen müßte, über welches die Friedensrichter die Verwaltungsaufsicht auszuüben hätten.

Da nun die Zahl der Gefängnisse eine sehr kleine war — manche Grafschaften waren überhaupt nicht im Besitze eines Gefängnisses —, so bildete sich mehr und mehr die Praxis, die Korrektionshäuser zugleich als Gefängnisse zu benutzen, eine Praxis, welche später auch die gesetzliche Sanktion erhielt: ein Gesetz von 1719 — 6 Geo. I c. 19 — erklärte es ausdrücklich für zulässig, daß Untersuchungsgefangene in die Korrektionshäuser gebracht würden; in einer Reihe von Strafgesetzen wurde es ferner dem Richter freigestellt, ob er zu Gefängnis oder zu Korrektionshaus verurteilen wollte. Die beiden Arten von Strafanstalten wurden so allmählich fast völlig gleichberechtigt; es durften nur einerseits Schuldgefangene (debtors) nicht in die Korrektionshäuser, anderseits Bettler und Vagabunden nicht in die Gefängnisse gebracht werden.

Die Zustände in beiderlei Anstalten waren während des 17. und 18. Jahrhunderts überaus mangelhafte und traurige.[1])

In den Korrektionshäusern führten die Gefangenen trotz der gesetzlichen Bestimmung Karls II. vom Jahre 1668, wonach die Friedensrichter dafür sorgen sollten, daß die Gefangenen zur Arbeit gesetzt würden, durchweg ein müßiges Leben. Untersuchungsgefangene, bloße Übertreter und schwere Verbrecher wurden in demselben Raume ohne irgendwelche Scheidung und ohne gehörige Aufsicht zusammengebracht; diejenigen, welche noch unverderbt in die Anstalt gekommen waren, wurden in dieser „Schule des Lasters" sehr bald den verderbtesten Subjekten gleich.

Nicht besser stand es in den Gefängnissen. Die ganze Verwaltung war hier fast vollständig der Willkür des Gefängnisvorstehers (gaoler) anheimgegeben. Derselbe wurde nicht besoldet, sondern bezahlte vielmehr häufig seinerseits große Summen für seine Anstellung.[2]) Er erhielt nämlich von den Gefangenen,

---

[1]) Eine sehr ausführliche Schilderung dieser Zustände, welcher die im Texte gegebene Darstellung teilweise entnommen ist, findet sich bei Du Cane II Seite 30 ff. Cf. auch State Trials Band 17 Seite 297—618.

[2]) Ein gewisser Huggins zahlte an den Lordkanzler Clarendon, welcher das Gefängnisrecht für Fleet Prison in London besaß, 5000 £ für die Über-

welche die Kosten ihres Aufenthaltes in dem Gefängnisse stets selbst zu tragen hatten, bestimmte Gebühren und hatte das Recht, die vermögenslosen Gefangenen so lange im Gefängnisse zu behalten, bis sie durch Arbeitsleistungen diese Gebühren abverdient hatten; dieses Recht erstreckte sich sogar auf Untersuchungsgefangene, welche häufig nach ihrer Freisprechung noch lange Zeit hindurch im Gefängnisse verbleiben mußten, weil sie die Gebühren des gaoler nicht bezahlen konnten.

Der gaoler betrachtete seine Stellung vollständig geschäftsmäßig: er suchte so wenig wie möglich für die Gefangenen auszugeben und so viel wie möglich für sich selbst zu erwerben.[1]) Die Stellen von Gefängniswärtern wurden denjenigen Gefangenen überlassen, welche dem gaoler dafür das meiste Geld boten. Wer das nötige Geld besaß, konnte sich alle möglichen Erleichterungen von dem gaoler erkaufen; er wurde von den Gefangenen, welche arm waren und die Gefängnisgebühren abverdienen mußten, vollständig bedient. Die Arbeitskraft dieser armen Gefangenen wurde auf das Äußerste ausgenutzt, die Zeit ihrer Detention möglichst ausgedehnt.

Infolgedessen waren die meisten Gefängnisse in geradezu grauenhafter Weise überfüllt[2]) und es entstanden die schlimmsten Krankheiten. Das sog. gaol fever raffte unzählige Gefangene hinweg. Fast noch schlimmer, als diese körperlichen Krankheiten, war jedoch die moralische Pest, welche in den Gefängnissen herrschte: Münz- und Banknotenfälschungen wurden in den Gefängnissen selbst vorgenommen; es kam vor, daß Gefangene von dem gaoler selbst zu Diebstahl und Plünderung in der Nacht aus dem Gefängnisse herausgelassen wurden und dann die Beute mit dem ehrenwerten Gefängnisvorsteher teilten; das Gefängnis war der beliebteste Schlupfwinkel, um unrechtmäßig erworbenes Gut zu hehlen. All das spielte sich vor den Augen der Gefangenen

---

lassung der Stelle des gaoler für sich und seinen Sohn auf Lebenszeit. Cf. Du Cane II Seite 35.

[1]) Das jährliche Einkommen des gaoler des Schuldgefängnisses Marshalsea wurde auf 3000—4000 £ angegeben. Cf. Du Cane II Seite 36.

[2]) Im Mai 1750 befanden sich in einem Gefängnisse in London, welches 2 Räume von je 7 Fuß Höhe und 14 × 7 Fuß Grundfläche hatte, 100 Gefangene. Cf. Du Cane II Seite 44.

ab, und die noch nicht im Verbrechen Bewanderten wurden hier dazu erzogen.

Es fehlte nicht an Leuten, welche diese unwürdigen Zustände zur Sprache brachten, und es fehlte auch nicht an Verbesserungsvorschlägen; doch dauerte es lange, bis man sich zu umfassenden Reformen entschloß.

Bereits im Jahre 1618 erschien eine Schrift „Essays and Character of a Prison and Prisoners" von Geffrey Mynshull in welcher von dem Verfasser, der selbst als Schuldgefangener die Zustände in den Gefängnissen kennen gelernt hatte, in grellen Farben alle Mißstände dargestellt und die englischen Gefängnisse bezeichnet wurden als „a school of villany, a moral pest-house, a sink of debauchery and a place where the penniless might starve to death".

Im Jahre 1699 wurde die Christian Knowledge Society gegründet, welche es zu einer ihrer Hauptaufgaben machte, durch ein besonderes Komitee die verschiedenen Gefängnisse in London besuchen zu lassen, um die dabei vorgefundenen Mißstände an das Tageslicht zu bringen. Der Vorsitzende dieses Komitees, Dr. Thomas Bray, gab in seinem Buche „Essay towards the Reformation of Newgate and the other Prison in and about London 1699" ein trauriges Bild von dem Zustande der Gefängnisse und macht eine Reihe von Verbesserungsvorschlägen. Darunter befindet sich neben Verteilung guter religiöser Schriften an die Gefangenen auch das Verlangen, daß jeder Gefangene in einer gesonderten Zelle gehalten werden solle „to keep every prisoner in a distinct cell". Von den Vorschlägen ist jedoch nur die Verteilung religiöser Schriften in den Gefängnissen zur Ausführung gelangt.

Im Jahre 1728 setzte das Parlamentsmitglied General Oglethorpe die Einsetzung eines Parlaments-Komitees durch, um die Zustände der Gefängnisse zu untersuchen „to inquire into the state of jails [1]) in this kingdom". Der Erfolg dieser Untersuchung war, daß wenigstens diejenigen gaolers, in deren Gefängnissen die schlimmsten Zustände aufgedeckt worden waren (Acton, Bambridge, Barnes, Huggins), zur strafrechtlichen Verantwortung gezogen wurden. [2]) Aber ein neues, im Jahre

[1]) „jail" ist eine andere, auch jetzt noch gebräuchliche Schreibart statt „gaol" = Gefängnis.

[2]) State Trials Band 17 Seite 297—618.

1735 eingesetztes Parlamentskomitee fand schon wieder ähnliche Mißstände vor.

Erst die Bemühungen John Howard's[1]) führten zu gesetzlichen Reformen. Er war der erste Sheriff, welcher die ihm gesetzlich obliegende Verpflichtung zur Beaufsichtigung der Gefängnisse ernst nahm. Im Jahre 1773 zum Sheriff der Grafschaft Bedford ernannt, erkannte er sofort die Mißstände wie ihre Ursachen. Er schlug den Friedensrichtern seiner Grafschaft vor, die Gebühren des gaoler zu beseitigen und besoldete gaolers anzustellen. Allein er erhielt die Antwort, er solle erst einmal zeigen, daß eine derartige Einrichtung irgendwo in England bestehe, dann sei man bereit, dieselbe auch in Bedfordshire einzuführen. Howard selbst gibt diese Antwort als die eigentliche Veranlassung seiner Reisen und seiner Besuche englischer[2]) wie kontinentaler Gefängnisse an. Howard's berühmtes Buch „The State of the Prisons", das Ergebnis dieser Reisen, ist oben im § 3 schon erwähnt worden. Dasselbe enthält eine genaue Beschreibung der englischen Gefängnisse, denen einige von Howard besuchte kontinentale Gefängnisse gegenübergestellt werden,[3]) und sodann auf 25 Quartseiten Verbesserungsvorschläge.

Der wichtigste Vorschlag ist neben der Einführung besoldeter Gefängnisbeamten die Herstellung von Einzelzellen für jeden Gefangenen[4]) mit einem geordneten Arbeitszwange und regelmäßigem geistlichen Zuspruche. Abschreckung dürfe nicht das einzige Ziel des Gefängnisses sein, man müsse auch darauf bedacht sein, Leben und Gesundheit des Gefangenen zu erhalten und denselben

---

[1]) Cf. über Howard's Persönlichkeit die Anmerkung 2 auf Seite 39.

[2]) Howard besuchte im ganzen 244 Gefängnisse in England; die Zahl der Insassen dieser Gefängnisse gibt er auf 4379 an.

[3]) Die ganz besondere Bewunderung Howard's erregte das neue maison de force in Gent, wo die Sträflinge des Nachts in Isolierzellen getrennt waren, dagegen bei Tage in gemeinschaftlicher Arbeit beschäftigt wurden.

[4]) Im Jahre 1772 hatte ein Geistlicher, Namens Denne, in einem offenen Briefe an Sir Robert Ladbroke die Einrichtung von Einzelzellen anempfohlen als a means of reforming criminals and thereby checking crime. Dieser Brief hatte in der damaligen Zeit, wo durch die in englischer Sprache erschienene Übersetzung von Beccarias Werk über Verbrechen und deren Bestrafung das Interesse für die Frage der Gefängnisreform in England angeregt war, großes Aufsehen gemacht. Howard nimmt bei seinem Vorschlage ausdrücklich hierauf Bezug.

mit Eindrücken der Religion und des sozialen Pflichtbewußtseins der Welt zurückzugeben („restore him to the world with impressions of religion and social love in his mind").

Howard's Ideen fanden in einer Reihe hervorragender englischer Parlamentsmitglieder, wie Blackstone, Popham, Eden und Romilly, warme und energische Anhänger. Es gelang denselben, rasch hintereinander wichtige Reformgesetze durchzubringen.

Im Jahre 1774 — 14 Geo. III c. 59 — wurde das Recht der gaolers auf den Empfang von Gebühren von den Gefangenen beseitigt und zugleich den Friedensrichtern anbefohlen, für eine genügende Zahl gesunder Gefängnisräume und besonderer Krankenabteilungen (infirmaries) zu sorgen, um so der Überfüllung der Anstalten und der Ausbreitung von Krankheiten in denselben vorzubeugen.

Im Jahre 1779 führte ein Gesetz — 19 Geo. III c. 74 — eine Strafe der Einzelhaft mit harter Arbeit (solitary confinement with hard labour) ein, welche insbesondere auch für einzelne Fälle an die Stelle der Transportationsstrafe treten sollte.

Im Jahre 1782 — 22 Geo. III c. 64 — wurden genaue Bestimmungen über die in den Korrektionshäusern zu handhabende Disziplin getroffen. Im Jahre 1784 — 24 Geo. III c. 54 — ordnete ein Gesetz an, daß in den Gefängnissen die Schuldgefangenen und die Untersuchungsgefangenen von den Strafgefangenen getrennt gehalten und daß unter den Strafgefangenen besondere Abteilungen für die wegen felony und für die wegen misdemeanour Bestraften gebildet werden sollten. Es wurden gleichzeitig besondere Bestimmungen über den Bau und die Einrichtung neuer Gefängnisse getroffen.

Endlich im Jahre 1791, dem Todesjahre Howard's, wird gesetzlich den Friedensrichtern anbefohlen, in jedem Quartal wenigstens dreimal die in ihrem Bezirke gelegenen Gefängnisse zu besichtigen und darüber an die Quarter Sessions zu berichten.

Man sieht, die Gesetzgebung ist in kurzer Zeit auf dem Gebiete der Reform des Gefängniswesens äußerst thätig gewesen. Leider entsprachen nur die praktischen Erfolge den Bemühungen der Gefängnisreformer recht wenig. Es waren nur einzelne Orte, wie vor allem Gloucester,[1]) wo pflichtbewußte Friedensrichter auf die wirklich ernste Durchführung der Gesetze Bedacht nahmen, während man an anderen Orten es sogar ruhig geschehen ließ, daß die gaolers Gebühren, trotz der gesetzlichen Aufhebung derselben, von den Gefangenen einzogen.[2]) Wenn der Zustand der Gefängnisse bisher überall gleichmäßig schlecht gewesen war, so entstand jetzt in der Ausführung der Gefängnisstrafe eine große Verschiedenheit in den einzelnen Bezirken.

In London selbst begann seit dem Anfange dieses Jahrhunderts das Publikum sich lebhafter für den Zustand der Gefängnisse zu interessieren. Es ist dies vor allem der Thätigkeit der Mrs. Elizabeth Fry zuzuschreiben. Sie begann im Jahre 1813 die weiblichen Gefangenen in Newgate zu besuchen und suchte, durch Eingehen auf die Verhältnisse jeder Einzelnen, dieselbe besseren Einflüssen zugänglich zu machen. Ihr Beispiel regte unter den höheren Klassen Londons zur Nachahmung an, und Mrs. Fry errichtete im Jahre 1817 eine Association for the improvement of female prisoners in Newgate. Die Besuche in dem Gefängnisse wurden systematisch geregelt: die Damen lasen mit den Gefangenen die Bibel, erteilten Unterricht, leiteten die Gefangenen zu bestimmten Arbeiten an, verkauften dann die hergestellten Arbeiten und besorgten für den Erlös den Gefangenen Kleidungsstücke und sonst Nothwendiges.[3])

Ziemlich gleichzeitig mit dieser Gesellschaft entstand die Society for the improvement of prison discipline. Dieselbe verfolgte vor allem das Ziel, die Fragen der Gefängnisreform im

---

[1]) Hier errichtete der Friedensrichter Sir George Paul ein Gefängnis genau nach dem von Howard in seinem Buche angegebenen Plane.

[2]) Die Gesetzgebung wiederholte infolgedessen im Jahre 1815 das Verbot des Bezuges von Gefängnisgebühren 55 Geo. III c. 50 und 116.

[3]) Der Einfluß von Mrs. Fry auf die Verbesserung des Gefängniswesens in England ist ein ganz außerordentlich großer gewesen, und sie dürfte in dieser Hinsicht John Howard wohl als ebenbürtig an die Seite zu stellen sein. Mrs. Fry war es, deren unermüdlicher Agitation es im wesentlichen zu danken war, daß allmählig in den englischen Gefängnissen weibliche Wärter für die weiblichen Gefangenen angestellt wurden und daß mehr und mehr für eine Beschäftigung der Gefangenen und für Erteilung von Unterricht in den Gefängnissen gesorgt wurde.

Gange zu halten.[1]) Alljährlich wurde ein Bericht über den Zustand der englischen Gefängnisse und die in denselben vorgefundenen Mängel veröffentlicht. Aus dem Berichte vom Jahre 1818 ist zu entnehmen, daß es damals in den Vereinigten Königreichen 518 Gefängnisse gab, von denen in 445 die Gefangenen in keinerlei Weise beschäftigt wurden, in 100 eine Überfüllung herrschte, in 59 Männer und Frauen in denselben Räumen zusammen gehalten wurden und nur in 23 eine Trennung zwischen Schuld-, Untersuchungs- und Strafgefangenen stattfand.

Diese Enthüllungen veranlaßten die Einsetzung einer Parlamentskommission, als deren Resultat sich Sir Robert Peel's Consolidated Gaol Acts — 4 Geo. IV c. 64 und 5 Geo. IV c. 85 — aus den Jahren 1823 und 1824 darstellen.

In diesen beiden Gesetzen wurden unter Aufhebung von 22 älteren Gesetzen alle gesetzlichen Bestimmungen über das Gefängniswesen zusammengefaßt und dabei eine Reihe von Neuerungen eingeführt: es sollten fortab für alle Anstalten, in denen sich weibliche Gefangene befinden, weibliche Gefangenwärter ernannt werden, es sollte für Unterricht im Lesen und Schreiben gesorgt werden, der Gefängnisgeistliche[2]) sollte täglich Gottesdienst halten, es war eine ständige Beschäftigung der Gefangenen einzuführen. Daneben wurde die schon durch das Gesetz von 1784 angeordnete Trennung der verschiedenen Klassen von Gefangenen noch einmal besonders eingeschärft; für die beiden Geschlechter sollten unbedingt ganz gesonderte Abteilungen eingerichtet werden.

Leider bezogen sich jedoch diese Gesetze nur auf die Grafschaftsgefängnisse; die meist sehr kleinen Gefängnisse der Marktflecken (boroughs), in welchen sich gerade die schlimmsten Übelstände vorfanden, wurden nicht von dem Gesetze betroffen. Sir Robert Peel war mit Recht der Ansicht, daß sich in diesen kleinen Gefängnissen all die angeordneten kostspieligen Maßregeln nicht würden durch-

[1]) Während Mrs. Fry ihre Bemühungen wesentlich der Verbesserung der Lage der weiblichen Gefangenen zugewandt hatte, suchte die Gesellschaft diese Bestrebungen zu verallgemeinern und noch andere, für das Gefängniswesen wichtige Fragen, wie z. B. die nach der geeignetsten Bauart der Anstalten, in den Kreis der Diskussion zu ziehen.

[2]) Seit 1814 war jede Gefängnisbehörde verpflichtet, einen besonderen Geistlichen anzustellen.

führen lassen, aber vor einer Schließung dieser Anstalten, welche die einzig richtige Konsequenz dieser Ansicht gewesen wäre, scheute er zurück.[1])

Die Art und Weise, wie die Gefängnisstrafe ausgeführt wurde, blieb so nach wie vor in den einzelnen Anstalten eine vielfach verschiedene; es hing nach wie vor im wesentlichen von den Lokalbehörden ab, inwieweit sie auf eine strenge Innehaltung der gesetzlichen Vorschriften hinwirkten. Es kamen noch immer an einzelnen Orten die größten Mißstände vor, welche aber im Gegensatze zu der früheren Zeit jetzt durch die Thätigkeit der Gesellschaft für Gefängnisverbesserung bald in die Öffentlichkeit gebracht wurden. Die Frage weiterer Reformen, insbesondere die Frage einer staatlichen Beaufsichtigung der Gefängnisse, wurde infolgedessen immer lebhafter diskutiert.

Endlich im Jahre 1835 geschah auf Empfehlung eines von dem Oberhause eingesetzten Komitees, welches einen überaus sorgfältigen und eingehenden Bericht erstattete,[2]) in dem Gesetze 5 a. 6 William IV. c. 38 ein entscheidender Schritt vorwärts.

Das Gesetz von 1835[3]) bestimmte zunächst, daß die Lokalbehörden alle von ihnen erlassenen Gefängnisordnungen dem Staatssekretär des Innern zur Genehmigung vorzulegen hätten. In diesen Gefängnisordnungen sollten vor allem genaue Bestimmungen über die zu verabreichende Gefängniskost und über die Arbeitszeit der Gefangenen getroffen werden. Der Staatssekretär

---

[1]) Das Einzige, was bezüglich dieser kleinen Gefängnisse geschah, bestand darin, daß den Gemeinderäten der boroughs u. s. w. gestattet wurde, Abkommen mit andren Orten oder mit der Grafschaft behufs gemeinschaftlicher Anlegung oder gemeinschaftlicher Benutzung von Gefängnissen zu treffen. 5 Geo. IV c. 12, c. 85; cf. auch 6 a. 7 William IV c. 105, 16 a. 17 Vict. c. 43, wo die Befugnisse noch erweitert wurden.

[2]) Dem Komitee, welches auf Antrag des Herzogs von Richmond ernannt worden war, lag der Bericht des von Lord Melbourne nach Amerika zum Studium der dortigen Gefängnisse gesandten Mr. Crawford über das Mustergefängnis Cherry Hill in Philadelphia vor. Das Komitee trat in eine eingehende Würdigung dieses Berichtes, wie überhaupt aller Fragen des Gefängniswesens.

[3]) Der vollständige Titel dieses wichtigen Gesetzes lautet: An Act for effecting greater uniformity of practice in the government of the several prisons in England and Wales, and for appointing Inspectors of prisons in Great Britain.

hatte die Befugnis, diese Gefängnisordnungen zu ändern und seinerseits eine anderweitige Gefängnisordnung zu erlassen. Von diesem Rechte ist denn auch mehrfach Gebrauch gemacht worden, und die Gefängnisordnungen sind dadurch im wesentlichen einheitlich gestaltet worden. Damit diese Einheitlichkeit aber auch thatsächlich zur Durchführung gelangte, bestimmte das Gesetz weiter, daß Gefängnisinspektoren ernannt werden sollten, welche die Gefängnisse regelmäßig zu besuchen und alljährlich in einem dem Parlamente vorzulegenden Berichte sich über den Zustand der Gefängnisse zu äußern hätten [1]).

Mit dem Jahre 1836 beginnen nunmehr die wertvollen Gefängnisberichte, welche besonders in den ersten Jahren vielfach allgemeine Fragen des Gefängniswesens zu einer eingehenden Erörterung ziehen. [2])

Dem mit dem Gesetze vom Jahre 1835 gethanen ersten Schritte auf dem Wege einer Zentralisation folgten rasch hintereinander weitere nach.

Ein Gesetz vom Jahre 1839 — 2 a. 3 Vict. c. 56 — ordnete an, daß fortab die Pläne für Neubau oder Umbau von Gefängnissen dem Staatssekretär zur vorherigen Genehmigung vorgelegt werden sollten. Zur besseren Durchführung dieser Bestimmung wurde dann im Jahre 1845 — 7 a. 8 Vict. c. 50 — ein besonderer Beamter, der Surveyor General of Prisons, ernannt, welcher unter

---

[1]) Unter den für England und Wales ernannten Inspektoren befanden sich 2 um das englische Gefängniswesen hochverdiente Männer: Crawford und Russel. Auch für Schottland und Irland, für welche im Übrigen spezielle Gefängnisgesetze erlassen worden sind, wurden auf Grund des Gesetzes von 1835 Gefängnisinspektoren ernannt. Auch der für Schottland ernannte Inspektor, Frederic Hill, hat sich in der Geschichte der Gefängnisreformen einen hervorragenden Namen erworben.

[2]) Die Gefängnisinspektoren gaben in ihrem ersten Jahresberichte die Gesamtzahl der englischen Gefängnisse auf 256 an. In besonders ausführlicher Weise werden in den Berichten die Verhältnisse in dem Londoner Gefängnisse Margate behandelt und dabei werden mannigfache Mißstände hervorgehoben, zu denen die Besuche der von Mrs. Fry gegründeten Gesellschaft geführt hatten. Die segensreiche Wirksamkeit von Mrs. Fry selbst wird voll anerkannt, aber hinzugefügt, daß die Besuche von Gefängnissen seitens der Damen der höheren Stände zu einer Art Sport geworden wären, daß viele dieser Damen für den Zweck durchaus nicht qualifiziert wären und daß eine geordnete Gefängnisverwaltung durch derartige gelegentliche Besuche Außenstehender erheblich erschwert würde.

dem Staatssekretär des Innern als Dezernent für alle gefängnisbautechnischen Fragen fungiert.

Auf die Fragen der Gefängnisbautechnik begann man, in der damaligen Zeit besonders großen Wert zu legen. Die Erörterung der diesbezüglichen Fragen nahm in den Jahresberichten der Gefängnisinspektoren einen breiten Raum ein, und dabei wurde die Frage, ob Einzelhaft oder Gemeinschaftshaft?, im unmittelbaren Zusammenhange mit den bautechnischen Fragen behandelt. Die beiden englischen Gefängnisinspektoren waren entschiedene Anhänger der Einzelhaft, und sie betonten besonders in ihrem 3. Jahresberichte, daß es wünschenswert wäre, wenn der Staat selbst ein Mustergefängnis auf der Basis des Einzelhaftsystems errichten würde, damit sich die Lokalbehörden zunächst ein Bild von diesem Systeme und von den dazu erforderlichen baulichen Einrichtungen machen könnten.

Der damalige Staatssekretär des Innern, Lord John Russel, ging auf diesen Vorschlag ein und beantragte beim Parlamente die Bewilligung einer Summe von 90000 £ zum Bau eines model prison. Diese Summe wurde bewilligt, und im Jahre 1840 begann unter Leitung der beiden Gefängnisinspektoren und des bisherigen Kapitäns im Ingenieurkorps Sir Joshua Jebb[1]) der Bau von Pentonville Prison. Diese, nach dem Vorbilde des Gefängnisses Cherry Hill in Philadelphia nach dem Radialsystem erbaute und für 520 Gefangene bestimmte Anstalt wurde im Jahre 1842 eröffnet. Die englische Regierung benutzte, wie im § 3 ausgeführt wurde, die neue Anstalt zur Aufnahme eines Teiles der zur Transportationsstrafe Verurteilten, welche hier das erste Stadium des im Jahre 1842 eingeführten probation system durchzumachen hatten. Nach dem Muster von Pentonville wurden in den nächstfolgenden Jahren von den Lokalbehörden eine große Anzahl von Gefängnissen[2]) errichtet. Innerhalb 6 Jahren entstanden 54 neue Gefängnisbauten mit zusammen 11000 Einzelzellen.

---

[1]) Sir Joshua Jebb wurde im Jahre 1845 zum Surveyor General of Prisons ernannt.

[2]) Für diese Anstalten kommt von dieser Zeit ab anstatt des Namens gaol die Bezeichnung local prison, Lokalgefängnis, in Gebrauch, zum Unterschiede von den beiden Staatsanstalten zu Millbank und Pentonville. Für diese Staatsanstalten, welche lediglich für die zur Transportationsstrafe

Die englische Gesetzgebung hielt sich bezüglich der Frage der Einführung der Einzelhaft zunächst sehr zurückhaltend. Es wurde im Jahre 1837 — 7 William IV a. 1 Vict. c. 90 — dem Richter die Befugnis gegeben, in dem Urteile auszusprechen, daß der zur Gefängnisstrafe Verurteilte in Einzelhaft (solitary confinement) gehalten werden sollte; [1]) dabei wurde jedoch die Einschränkung hinzugefügt, daß diese Einzelhaft auf einmal nicht die Dauer eines Monats und innerhalb eines Jahres nicht die Dauer von 3 Monaten überschreiten dürfte. [2]) Das Gesetz von 1839 gestattete ferner den Lokalbehörden, einen Strafgefangenen in Isolierhaft zu halten; die dazu benutzte Haftzelle muß jedoch seitens der Regierung vorher untersucht worden sein, und bei dieser Untersuchung ist ein Attest darüber auszustellen, ob die betreffende Zelle zur Aufbewahrung eines Gefangenen während der Nachtzeit oder bei Tag und Nacht geeignet erscheint, und ferner in beiden Fällen, einen wie langen Zeitraum der Gefangene in der betreffenden Zelle gehalten werden darf.

Die Zustände in den englischen Gefängnissen hatten sich seit dem Jahre 1835 auf Grund der verschiedenen hier erwähnten Gesetze erheblich gebessert. Immerhin aber ließen dieselben doch noch viel zu wünschen übrig: in den kleinen Gefängnissen war, wie das vorausgesehen werden mußte, die Durchführung von Verbesserungen und

---

Verurteilten bestimmt waren, wird neben der Bezeichnung penitentiaries der Name convict prisons gebräuchlich.

[1]) Noch nach einer andern Richtung hin wurde dem Richter die Befugnis beigelegt, über die Art und Weise der Strafvollstreckung Bestimmung zu treffen: der Richter kann fortab den Verurteilten auf Antrag von der Gefängnisordnung hinsichtlich der Gefängniskost, der Gefängniskleidung und der Beschäftigung dispensieren, indem er den Verurteilten als first class misdemeanant bezeichnet. Der Verurteilte erhält dadurch das Recht, seine eigne Kleidung im Gefängnisse weiter zu tragen, sich auf seine Kosten unter Einhalten gewisser Schranken zu verpflegen und die Art und Weise seiner Beschäftigung zu bestimmen. Die Gefängnisstrafe besteht hier im wesentlichen in der Freiheitsentziehung. Die nähere Regelung dieser neueingeführten Art der Gefängnisstrafe ist später erst getroffen worden; es wird davon noch im Texte gehandelt werden. Cf. speziell unten Abschnitt II § 11.

[2]) Diese Einschränkung ist zurückzuführen auf die Besorgnis, welche man in England, insbesondere auf Grund des Berichtes von Crawford über die Zustände in den amerikanischen Gefängnissen, hegte, daß eine Einzelhaft von längerer Dauer ein physisches und moralisches Siechtum der Gefangenen herbeiführen würde.

Neuerungen an sich schon unmöglich; aber auch in den größeren Gefängnissen erwiesen sich die der Zentralgewalt beigelegten Befugnisse als nicht ausreichend.

Die Gefängnisinspektoren konnten zwar die von ihnen in den einzelnen Anstalten vorgefundenen Mißstände in ihren Jahresberichten zur Sprache bringen, und, da diese Berichte dem Parlamente vorgelegt und publiziert wurden, so war die dadurch gegebene Möglichkeit eines öffentlichen Tadels ein praktisch wirksames Mittel, die Lokalbehörden zur Erfüllung ihrer Pflichten anzuhalten. Ein weiteres Mittel, auf die Lokalbehörden einzuwirken, bestand darin, daß dem Staatssekretär des Innern seit dem Ende der 40er Jahre alljährlich von dem Parlamente bestimmte Summen bewilligt wurden, um dieselben den Lokalbehörden als Zuschüsse zu den Gefängniskosten zuzuweisen.[1]) Wenn nun eine Lokalbehörde die von den Gefängnisinspektoren gerügten Mißstände nicht abstellte oder sonst einer Anordnung des Staatssekretärs nicht Folge leistete, so konnte der letztere die Zuschüsse an die betreffende Lokalbehörde suspendieren event. gänzlich aufheben. Es fehlen leider alle Angaben, inwieweit von dieser Befugnis thatsächlich Gebrauch gemacht worden ist. Jedenfalls aber vermochte dieselbe ebensowenig wie die Aufsicht und der Tadel der Inspektoren alle Übelstände[2]) zu beseitigen und eine einheitliche Behandlung der Gefangenen in der Praxis sicherzustellen.

In dem Berichte eines im Jahre 1850 zur Untersuchung des Zustandes der Gefängnisse vom Unterhause ein-

---

[1]) Du Cane I Seite 77 gibt an, daß sich diese Zuschüsse durchschnittlich auf 18,5 % der gesamten Gefängniskosten belaufen hätten.

[2]) Es verdient hier hervorgehoben zu werden, daß sich diese Übelstände zu der damaligen Zeit nicht nur nach der Richtung hin zeigten, daß in manchen Anstalten für die Gefangenen nicht genügend gesorgt wurde, sondern wohl in noch höherem Maße nach der Richtung, daß für die Gefangenen zuviel geschah, daß in einzelnen Anstalten infolge der Bestrebungen, den Gefangenen zu bessern, die Strafe ihren eigentlichen Charakter verlor. Es sei hier nur ein Beispiel dieser gefährlichen Humanitätsduselei angeführt. In dem Gefängnisse zu Reading war die Bestimmung getroffen, daß der größere Teil des Tages durch Unterricht resp. Vorbereitung zum Unterrichte ausgefüllt werden sollte, während der übrigen Tageszeit durften sich die Gefangenen mit der Lektüre der ihnen aus der reichhaltigen Gefängnisbibliothek gelieferten Bücher beschäftigen, und nur, wenn sie des Lesens müde waren, sollten sie etwas Werg zupfen („when tired of reading pick a little oakum"). Infolge dieser Bestimmung wurde diese „Criminal University" witzig „Read-Reading Gaol" genannt. Cf. Du Cane II Seite 57.

gesetzten Komitees wurde dieser Mangel denn auch offen anerkannt. Es wurde hier ausdrücklich ausgesprochen, daß die noch vorhandenen Übelstände in vielen Gefängnissen nur dadurch beseitigt werden könnten, daß einer Zentralbehörde umfassende Befugnisse beigelegt würden, um gegenüber den Lokalbehörden eine gleichmäßige und rationelle Behandlung der Gefangenen zu erzwingen. Das Komitee empfahl dann weiter die allgemeine Einführung einer beschränkten Einzelhaft: alle Gefangenen sollten mit Ausnahme der Stunden des Gottesdienstes, des Unterrichts und der Arbeit isoliert gehalten werden; bei Gefangenen unter 3 Monaten sollte während der ganzen Strafzeit vollständige Einzelhaft stattfinden, bei den anderen Gefangenen sollte vollständige Einzelhaft bis zur Dauer von 12 Monaten gestattet sein.

Die Vorschläge des Komitees führten zunächst praktische Folgen nicht herbei. Erst auf den Bericht eines im Jahre 1863 eingesetzten Komitees des Oberhauses erfolgten umfassende gesetzliche Änderungen. Das Komitee des Oberhauses hatte als Hauptmängel des bestehenden Zustandes hervorgehoben, daß der Willkür der Lokalbehörden in allzugroßem Umfange die Gestaltung des Vollzugs der Gefängnisstrafe überlassen sei, daß es an einem eigentlichen Zwangsmittel gegenüber den Lokalbehörden fehle, wodurch die letzteren zum genauen Befolgen der Gefängnisordnungen gezwungen werden könnten, und endlich, daß eine zu große Zahl kleiner Gefängnisse vorhanden sei, [1]) in welchen ein rationeller Strafvollzug überhaupt nicht durchzuführen sei.

Zur Beseitigung dieser Mißstände erging nun im Jahre 1865 der Prison Act 28 a. 29 Vict. c. 126. Dieses Gesetz [2]) gibt unter Aufhebung einer großen Zahl älterer Gesetze eine vollständige Ordnung des Gefängniswesens bis in die Einzelheiten der Verwaltung hinein. Das Gesetz selbst besteht aus 82 zum Teil sehr umfangreichen Paragraphen, denen als Schedule I ein Gefängnisregulativ in 104 Paragraphen angehängt ist. Jede Änderung oder

---

[1]) Die Zahl der Gefängnisse hatte übrigens seit 1837, in welchem Jahre dieselbe 256 betrug, erheblich abgenommen. Es gab im Jahre 1862 in England und Wales 193 Gefängnisse, von welchen die Durchschnittsbelegung bei 27: weniger als 6, bei 14: zwischen 6 und 10, bei 22: zwischen 11 und 24, bei 63: circa 25 betrug.

[2]) Der vollständige Titel lautet: An Act to consolidate and amend the laws relating to prisons.

Ergänzung des Regulativs seitens einer Lokalbehörde bedarf der Zustimmung des Staatssekretärs des Innern. Der letztere wird in dem Gesetze zu weiteren Ausführungsverordnungen ermächtigt, welche dem Parlamente zur Genehmigung vorzulegen sind. In dem Gesetze wird die bisherige Unterscheidung zwischen gaols und houses of correction — eine Unterscheidung, welche schon seit lange ohne größere praktische Bedeutung gewesen war — gänzlich aufgehoben.

Im einzelnen bestimmte das Gesetz folgendes: Die Behörde, welche für die Vollstreckung der Gefängnisstrafe zu sorgen hat, sind die Quartalssitzungen der Friedensrichter. Denselben liegt insbesondere ob, eine genügende Zahl von geeigneten Zellen herzustellen [1]) und die erforderlichen Gefängnisbeamten [2]) zu ernennen. Die laufende Verwaltung wird von einem Ausschusse von zwei ad hoc zu wählenden Friedensrichtern geführt, welche die Bezeichnung Visiting Justices erhalten. Das Institüt der Gefängnisinspektoren bleibt in der bisherigen Weise erhalten. Die Quartalssitzungen haben alljährlich dem Staatssekretär des Innern einen Generalbericht über den Zustand und die Verwaltung ihres Gefängnisses einzureichen. Der Staatssekretär kann, wie bisher, denjenigen Lokalbehörden, welche bestimmten, als wesentlich im Gesetze bezeichneten Anordnungen nicht nachkommen, den Staatszuschuß entziehen und außerdem, falls die Zuwiderhandlung innerhalb 4 Jahren nicht abgestellt ist, das betreffende Gefängnis schließen. Vierzehn ganz kleine Gefängnisse werden durch das Gesetz selbst aufgehoben.

Betreffs der Behandlung der Gefangenen wird bestimmt, daß für Männer und Frauen völlig getrennte Abteilungen herzustellen und daß innerhalb dieser Abteilungen 4 Klassen streng zu sondern sind: 1. Untersuchungsgefangene, 2. Schuldgefangene (debtors), 3. first class misdemeanants, welche nicht als eigentliche Strafgefangene (criminal prisoners) angesehen werden sollen, sondern im allgemeinen

---

[1]) Wie bisher muß jede Zelle vor ihrer Benutzung durch den Surveyor General besichtigt und ein Attest darüber ausgestellt sein, ob dieselbe lediglich als Schlafzelle geeignet ist oder zum Aufenthalte eines Gefangenen bei Tag und Nacht benutzt werden darf. Das Attest ist an der Zellenthür anzuschlagen.

[2]) Jedes Gefängnis muß einen Aufseher (Gaoler), einen Geistlichen (Chaplain), einen Arzt und die erforderliche Zahl von Aufsehern haben. Die Befugnisse und Pflichten all dieser Personen werden in dem Gesetze und dem beigegebenen Regulative auf das genaueste festgestellt.

den Schuldgefangenen gleich zu behandeln sind, 4. Strafgefangene. Hinsichtlich der 3 ersten Klassen wird dem Staatssekretär noch der Erlaß besonderer Anordnungen vorbehalten. Jeder Gefangene ist in einer abgesonderten Zelle zu halten; bei den Strafgefangenen hängt es von dem Ermessen der Lokalbehörde ab, ob dieselben nur bei Nacht oder bei Tag und Nacht in der Zelle gelassen werden sollen. Für den Fall, daß die Strafgefangenen am Tage in gemeinsamer Arbeit beschäftigt werden, sind Maßregeln zu treffen, daß keinerlei Verkehr zwischen den Gefangenen möglich ist. Alle Strafgefangenen sind während bestimmter Tagesstunden zu beschäftigen. Durch Unterrichtsstunden, welche auf Lesen, Schreiben und Rechnen zu beschränken sind, darf die Arbeitszeit in keinerlei Weise verkürzt werden. Diejenigen Strafgefangenen, welche zu harter Arbeit (hard labour) verurteilt worden sind, müssen während wenigstens 3 Monaten täglich 9—10 Stunden zu bestimmt vorgeschriebener Arbeit (Tretmühle, Steineklopfen u. s. w.), sog. first class hard labour, angehalten werden. Für den Rest der Strafzeit sowie bei allen übrigen Strafgefangenen kann die Lokalbehörde die Art und Weise der Beschäftigung, sog. second class hard labour, bestimmen. Zu Arbeiten in der Anstaltsverwaltung sollen die Gefangenen jedoch nur bei gutem Verhalten, gleichsam als Belohnung, herangezogen werden. Die Lokalbehörden sind ferner befugt, dem Sträflinge bei gutem Verhalten und fleißiger Arbeit eine Arbeitsbelohnung (gratuity) bei seiner Entlassung auszuzahlen.

Dies die wesentlichsten Bestimmungen des Gesetzes von 1865, auf dessen Einzelheiten, soweit dieselben noch heute Geltung haben, im zweiten Abschnitte näher eingegangen werden wird.

Das Gesetz führte, wie gar nicht bestritten werden kann, in vielfacher Beziehung eine Besserung der bisherigen Zustände herbei. Aber seinem ganzen Inhalte nach stellte sich dasselbe doch nur als eine Stufe auf dem Wege der Reform dar, nicht aber als ein Abschluß derselben. All die getroffenen Anordnungen mögen als Übergangsbestimmungen aus dem bisherigen Zustande zu einem ganz neuen eine gewisse Berechtigung gehabt haben, an sich waren es aber doch nur halbe Maßregeln, welche die von dem Oberhauskomitee von 1863 gerügten Mängel verminderten, aber nicht beseitigten: eine Anzahl kleiner Gefängnisse wird geschlossen, aber eine viel größere Zahl für eine rationelle Behandlung der Gefangenen unbrauchbarer Gefängnisse bleibt bestehen; die

Machtbefugnisse der Zentralbehörde gegenüber den Lokalbehörden werden erweitert, aber die Zentralbehörde muß 4 Jahre lang Mißstände in einer Anstalt ruhig mit ansehen, ehe sie zur Schließung derselben schreiten kann, und selbst dann ist die Schließung praktisch unausführbar, wenn nicht die Möglichkeit eines anderweitigen Unterbringens der Gefangenen vorhanden ist; in die Behandlung der Gefangenen endlich wird ein gewisses System hineingebracht, indem die verschiedenen Klassen von Gefangenen voneinander getrennt, die Isolierung während der Nachtzeit allgemein angeordnet und über die Art und Weise der Beschäftigung einiges festgesetzt wird. Aber auch hier wird der Willkür der Lokalbehörden noch ein weiter Spielraum gelassen, welcher naturgemäß von den einzelnen Behörden je nach der individuellen Auffassung und nach der größeren oder geringeren Energie in der Durchführung und in der Überwachung der Anstalt recht verschieden ausgefüllt wurde.

Wenn das Ziel, welches seit 1835 der englischen Gesetzgebung vorgeschwebt hatte: — einheitliche Gestaltung des Gefängniswesens und Durchführung der für rationell gehaltenen Behandlungsweise der Gefangenen in einheitlicher Weise im ganzen Lande — wirklich erreicht werden sollte, so gab es dafür nur ein Mittel: Übertragung der Vollstreckung der Gefängnisstrafe von den Lokalbehörden auf die Zentralgewalt. Freilich war dies ein gewichtiger, folgenschwerer Schritt, und es ist zweifelhaft, ob sich dafür im englischen Parlamente so bald eine Majorität gefunden haben würde, wenn nicht zwei Momente hinzugekommen wären.

Dies war einmal die allgemeine Anerkennung, deren sich die Verwaltung der convict prisons durch die Zentralbehörde mit Recht zu erfreuen hatte, und zweitens der Wunsch, die Lokalbesteuerung zu erleichtern. Nun erhielt gerade der Aufwand für die Erhaltung der Gefängnisse eine schwere Belastung der Lokalsteuer, und der Vorschlag der Regierung, diese Kosten auf die allgemeine Staatskasse zu übernehmen, wurde deshalb nicht ungünstig aufgenommen. Es verstand sich aber von selbst, daß die Regierung, wenn sie einmal die Kosten der Verwaltung der Gefängnisse übernahm, die Verwaltung selbst nicht fremden Händen überlassen konnte, und dies um so weniger, als die Regierung überzeugend nachwies, daß sie bei Übernahme der Gefängnisse im stande sein würde, diese Kosten erheblich einzuschränken. Die Verwaltungsausgaben für die große Anzahl der kleinen Gefängnisse waren, wie das ja selbstverständlich sein mußte, ganz unverhältnismäßig hohe; die Regierung erklärte nun, sie würde alle diese kleinen Gefängnisse eingehen lassen und damit natürlicherweise die Gesamtkosten für das Gefängniswesen erheblich vermindern.

So gelang es denn im Jahre 1877, den Prison Act 40 a. 41 Vict. c. 21 durchzusetzen, welcher die beiden Prinzipien enthielt: der Staat übernimmt fortab die Verwaltung der Gefängnisse und trägt die Kosten des Vollzugs der Gefängnisstrafe.[1]) Im übrigen werden die in dem Gesetze von 1865 aufgestellten Grundsätze über die Behandlung der Gefangenen im allgemeinen aufrecht erhalten.

Aus dem Inhalte des 61 Paragraphen umfassenden Gesetzes, auf welches bei der Darstellung des heutigen englischen Gefängniswesens in Abschnitt II des nähern zurückzukommen ist, sei hier nur so viel hervorgehoben, als zum Abschlusse der hier versuchten Klarlegung der Entwickelung der Gefängnisstrafe und ihrer Vollstreckung erforderlich erscheint.

Das Gesetz enthält zunächst eine Anzahl Übergangsbestimmungen: jede Gefängnisbehörde ist verpflichtet, dem Staate so viele attestierte Zellen zu übergeben, als die durchschnittliche tägliche Belegungsziffer während der letzten 5 Jahre (1872—1876) betrug. Bleibt die Zahl der thatsächlich vorhandenen Zellen hinter dieser Ziffer zurück, so hat die Lokalbehörde für jede fehlende Zelle eine Pauschalsumme von 120 £ an den Staat abzuführen; beträgt anderseits die Zahl der vorhandenen Zellen mehr als die verlangte Ziffer, so hat der Staat für jede weitere Zelle der Lokalbehörde 120 £ zu vergüten. Der Staat ist unter gewissen Einschränkungen berechtigt, ein Gefängnis, dessen er zur Vollstreckung der Gefängnisstrafe nicht mehr bedarf, an die Lokalbehörde zurückzuübertragen, wobei dann wieder der Satz von 120 £ pro Zelle zu Grunde gelegt wird.[2])

---

[1]) Für Schottland und Irland ergingen im wesentlichen gleichlautende Gesetze, so daß jetzt ein irgendwie erheblicher Unterschied in dem Gefängniswesen der 3 Königreiche nicht mehr besteht.

[2]) Der Satz von 120 £ blieb hinter den thatsächlichen Herstellungskosten durchweg zurück. Die Herstellungskosten der in den letzten Jahren neu errichteten Gefängnisse betrugen pro Zelle zwischen 140 £ (Manchester) und 209 £ (London, Holloway).

Bei Inkrafttreten des Gesetzes waren in England und Wales 113 Gefängnisse vorhanden, davon wurden 38 sofort geschlossen; weitere Schließungen fanden dann in den folgenden Jahren statt, nachdem die erforderlichen Erweiterungsbauten bei den beibehaltenen Gefängnissen vollendet waren. Zur Zeit giebt es 60 Anstalten zur Vollstreckung der Gefängnisstrafe.[1])

Die Verwaltung der Gefängnisse wird fortab von einer Anzahl (3—4) Commissioners of Prisons geführt. Die Befugnisse der Quartalssitzungen der Friedensrichter und der Visiting Justices werden aufgehoben. Eine gewisse Beaufsichtigung der Gefängnisverwaltung durch die Lokalbehörden wird jedoch insofern beibehalten, als aus der Zahl der Friedensrichter für jedes Gefängnis ein Visiting Committee von regelmäßig 12 Friedensrichtern bestellt wird mit der Befugnis, das betr. Gefängnis jederzeit zu besuchen, Beschwerden von Gefangenen anzuhören und über etwaige Mißbräuche an den Staatssekretär zu berichten.[2]) Den Friedensrichtern wird ferner die Befugnis übertragen, Disziplinarstraffälle, welche über die Kompetenz des Anstaltsvorstehers (Governor) hinausgehen, zu entscheiden.

Betreffs der Behandlung der Gefangenen wird gegenüber dem Gesetze von 1865 die Änderung getroffen, daß sog. first class hard labour (Tretmühle, Steineklopfen) auf die Dauer von 1 Monat (bisher 3 Monate) und auf die Zeit von täglich 6 Stunden (bisher bis zu 10 Stunden) eingeschränkt wird.[3]) Im übrigen wird der Staatssekretär angewiesen, besondere Regulative über die Behandlung der 3 Klassen: Untersuchungsgefangene, Schuldgefangene und first class misdemeanants zu erlassen.[4]) Der Staatssekretär ist des

---

[1]) Die Gesamtzahl der Zellen beträgt jetzt 23 494 gegenüber 27 151 im Jahre 1876/7 (also Abnahme 13,4 %); darunter befinden sich certified separate cells 22 841 gegenüber 24 614 im Jahre 1876/7 (also Abnahme 7,3 %). Cf. Judicial Statistics 1884 Seite XXXIX.

[2]) Die näheren Bestimmungen über die Befugnisse des Visiting Committee sind von dem Staatssekretär unter dem 19. Februar 1878 erlassen worden.

[3]) Gerade bezüglich der Bestimmungen über hard labour war das Gesetz von 1865 von den Lokalbehörden vielfach nicht befolgt worden, und man glaubte den Grund hierfür in einer zu großen Härte dieser Bestimmungen zu sehen, welche man deshalb jetzt wesentlich milderte.

[4]) Diese Regulative sind unter dem 19. Februar 1878 ergangen. Es mag hierbei bemerkt werden, daß die Zahl der Schuldgefangenen nach der jetzigen Gesetzgebung nur eine kleine ist, da Schuldhaft nur noch in Ausnahme-

weitern berechtigt, über die Behandlung der eigentlichen Strafgefangenen (criminal prisoners) durch besondere Regulative (rules) Bestimmung zu treffen. Die Regulative treten jedoch erst in Kraft, nachdem dieselben 28 Tage im Parlamente ausgelegen haben und ein Widerspruch dagegen nicht erfolgt ist.

Auf Grund dieser Befugnis hat der Staatssekretär vor allem besondere Regulative über die Verpflegung und den Unterricht in den Anstalten erlassen.[1]) Er hat ferner das Markensystem, welches sich in den convict prisons so gut bewährt hatte, auch in den local prisons eingeführt und damit auch hier ein System des stufenweisen Fortschreitens im Strafvollzuge (system of progressive stages) verbunden: der Gefangene rückt während der Strafverbüßung fortlaufend von einer niederen zu einer höheren Klasse vor, nachdem er sich jeweilig durch fleißige Arbeit bei gutem Verhalten eine bestimmte Anzahl Marken erworben hat; mit jeder höheren Klasse tritt eine gewisse Erleichterung im Strafzwange und gleichzeitig eine Erhöhung der Arbeitsvergütung ein.[2])

Der Staatssekretär hat ferner umfassende Anordnungen über die Benutzung der einzelnen Strafanstalten gegeben; einige Anstalten sind ausschließlich für Frauen, andere ausschließlich für Männer bestimmt, mehrere Anstalten dienen speziell zur Aufnahme von Gefangenen mit längerer Strafzeit; in einigen Anstalten werden bestimmte Handwerke und Industrieen vorzugsweise betrieben, und es werden in dieselben besonders Gefangene aufgenommen, welche sich für den betreffenden Arbeitszweig qualifizieren. Mit dieser Spezialisierung der Strafanstalten, welche für eine geordnete und nach Möglichkeit billig arbeitende Verwaltung von der höchsten Wichtigkeit ist, werden von Jahr zu Jahr weitere Fortschritte gemacht.

---

fällen zulässig ist, cf. 32 a. 33 Vict. c. 62, 33 a. 34 Vict. c. 76. Dem Schuldgefangenen gleich werden übrigens diejenigen behandelt, welche wegen Ungehorsams gegen einen richterlichen Befehl (contempt of court) in Haft genommen worden sind.

[1]) Die bisher erlassenen Regulative finden sich zusammengestellt in einem Heftchen The Prison Code, containing the rules made by the Secretary of State. London 1885 (nicht im Buchhandel erschienen).

[2]) Die nähere Darstellung der heutigen Behandlung der Gefangenen folgt im Abschnitte II Kapitel 3.

Schon nach dem wenigen, was hier über die auf Grund des Gesetzes von 1877 eingeführten Verbesserungen und Neuerungen mitgeteilt worden ist, wird man es begreiflich finden, daß man in England mit dem neuen System allgemein zufrieden ist. Es ist das allgemeine Urteil, daß jetzt die lang erstrebte Einheitlichkeit im Strafvollzuge wirklich erreicht ist, daß der Strafvollzug selbst als ein rationeller bezeichnet werden kann, daß infolge der Zentralisation des Gefängniswesens ein besseres, kompetenteres Beamtenpersonal, eine geeignete Ausnutzung der Arbeitskräfte der Gefangenen und damit zusammenhängend eine erhebliche Ersparnis von Kosten ermöglicht worden ist. Für dies letztere seien wenige Ziffern angeführt. [1])

Die Gesamtkosten für die Gefängnisse beliefen sich im Jahre 1876/7 auf 501 348 £, sie sind jetzt auf 372 433 £ im Jahre 1884/5 herabgegangen. Die Kosten betrugen im Jahre 1876/7 durchschnittlich auf den Kopf des Gefangenen 24 £ 12 s 5 d und jetzt (1884/5) 22 £ 8 s 2 d. [2]) Dabei ist die Gesamtsumme der Gehälter der Gefängnisbeamten von 222 487 £ in 1876/7 auf 212 024 £ in 1884/5 herabgegangen, und diese Kostenersparnis — im wesentlichen zurückzuführen auf die Schließung der kleinen Gefängnisse — ist Hand in Hand gegangen mit einer erheblichen Vermehrung der Zahl der Gefängnisärzte und der Gefängnisschullehrer.

Neben diesen Angaben über die Resultate des neuen Gesetzes in ökonomischer Hinsicht sei angeführt, daß sich die Zahl der Selbstmorde (während der 6 Jahre vor Inkrafttreten des Gesetzes jährlich durchschnittlich 17,6, während der 6 Jahre nach

---

[1]) Die nachfolgenden Ziffern sind entnommen den Judicial Statistics 1884 Seite XXXIII ff. und den Reports of Commissioners of Prisons 1884/5 Seite 15, 30, 78 ff., 1885/6 Seite 5, 11, 23 ff., 28, 74 ff.

[2]) Auf den Kopf des Gefangenen entfielen in den 6 Jahren vor Inkrafttreten des neuen Gesetzes (1871/2—1876/7): rund 26 £; in den 6 Jahren nach Inkrafttreten des neuen Gesetzes (1879/80—1884/5): rund 23 £ jährliche Kosten. Bei all diesen Ziffern sind die Erträgnisse der Arbeit der Gefangenen nicht abgerechnet; dieselben sind vor Inkrafttreten des neuen Gesetzes von den einzelnen Gefängnisbehörden in vielfach abweichender, und zwar meistenteils unzuverlässiger Weise berechnet worden, so daß Vergleiche mit den jetzigen, auf einheitlicher und sicherer Unterlage beruhenden Angaben nicht möglich sind.

Inkrafttreten 13,8), der Todesfälle (während der 6 Jahre vor Inkrafttreten des Gesetzes jährlich durchschnittlich 1,08 %, während der 6 Jahre nach Inkrafttreten 0,82 %) und der Disziplinarstrafen (im Jahre 1877 auf 1000 Gefangene täglich 7,7, im Jahre 1885/6 dagegen nur 6,2) erheblich gebessert hat.

Es mögen hier schließlich noch einige statistische Angaben über die Anwendung der Gefängnisstrafe ihren Platz finden. Es ist dabei an 2 schon früher erwähnte Momente zu erinnern: nämlich erstens, daß durch die Gesetzgebung im letzten Menschenalter eine erhebliche Anzahl neuer mit Strafe bedrohter Handlungen geschaffen worden ist, und zweitens, daß die Zahl der Verurteilungen zur penal servitude seit dem Jahre 1863 erheblich abgenommen hat. Trotz dieser Momente ist gegenüber der fortdauernden Bevölkerungszunahme die Gesamtziffer der Verurteilungen zur Gefängnisstrafe nur wenig gestiegen. Was die Zahl der von den Assisengerichtshöfen zur Gefängnisstrafe Verurteilten betrifft, so war dieselbe im Jahre 1885/6 niedriger als in irgend einem Vorjahre seit 1836.[1]) Dagegen mußte sich die Zahl der von den Gerichtshöfen mit summarischer Jurisdiktion zur Gefängnisstrafe Verurteilten erhöhen, einmal wegen der erweiterten Kompetenz dieser Gerichtshöfe, anderseits weil die neugeschaffenen mit Strafe bedrohten Handlungen in ihrer großen Mehrheit — Delikte gegen die Schul- und Wegegesetze u. s. w. — zu den, vor den Gerichtshöfen mit summarischer Jurisdiktion abzuurteilenden Straffällen gehören.

Es ist nicht leicht, aus den amtlichen Publikationen sich ein Bild über die Gesamtheit der Bestrafungen zur Gefängnisstrafe zu machen, weil diejenigen, welche zu einer Geldstrafe oder zur Friedensbürgschaft und nur subsidiär zu Gefängnis verurteilt worden sind, im Falle des Eintretens dieser eventuellen Strafe denjenigen ohne weiteres beigezählt werden, welche von vornherein zu Gefängnis verurteilt wurden. Mit dieser Maßgabe seien folgende Ziffern angeführt. Es betrug die Zahl der Personen, welche zur Gefängnisstrafe verurteilt wurden:

---

[1]) Nämlich 8891. Die Zahl war im Jahre 1836: 10 125, stieg bis 1854 auf: 20 388, fiel bis 1860 auf 9656, schwankte zwischen 1861 und 1871 zwischen 10 083 und 12 358; von 1872—1884 ist die Ziffer 10 000 nur noch dreimal überschritten worden (1876: 10 020; 1878: 10 218; 1879: 10 440).

In den Jahren	von den Assisengerichtshöfen	von den Gerichtshöfen mit summarischer Jurisdiktion	Totalsumme	Bevölkerungsziffer
1857	12,507	89,558	102,065	19,256,516
1858	10,834	86,739	97,573	19,471,291
1859	10,060	78,432	88,492	19,686,701
1860	9,656	73,460	83,116	19,902,713
1861	11,233	81,754	92,987	20,066,224
1862	11,944	95,933	107,877	20,371,013
1863	12,251	97,647	109,898	20,625,855
1864	11,948	97,534	109,482	20,883,889
1865	12,358	95,738	108,096	21,145,151
1866	11,582	95,114	106,696	21,409,684
1867	11,801	101,476	113,277	21,677,525
1868	12,546	113,915	126,461	21,948,713
1869	12,002	126,647	138,649	22,223,299
1870	10,908	128,972	139,880	22,501,316
1871	10,083	122,716	132,799	22,712,266
1872	9,318	121,728	131,046	23,095,819
1873	9,141	129,481	137,622	23,407,317
1874	9,793	131,956	141,789	23,723,017
1875	9,282	135,897	145,179	24,042,974
1876	10,020	140,415	150,435	24,367,247
1877	9,793	147,941	157,734	24,695,894
1878	10,218	16,037 [1]	86,255	25,028,973
1879	10,440	146,450	156,890	25,366,544
1880	9,663	152,347	162,010	25,708,666
1881	9,266	141,913	151,179	25,974,439
1882	9,715	153,267	162,982	26,406,820
1883	9,455	156,215	165,670	26,770,744
1884	9,135	152,628	161,763	27,132,449
1885	8,891	140,213	149,104	27,499,041

Ein besseres Bild, bei welchem gleichzeitig hervortritt, daß nur die ganz kleinen Straffälle sich vermehrt haben können, gibt die Ziffer der Durchschnittsbelegung der Gefängnisse. Dieselbe betrug:

im Jahre	1867/8:	18,677
" "	1868/9:	20,080
" "	1869/70:	19,830
" "	1870/1:	18,565
" "	1871/2:	17,505
" "	1872/3:	17,680
" "	1873/4:	17,896
" "	1874/5:	18,487
" "	1875/6:	18,986

[1]) Die Ziffer für 1878 ist infolge der in diesem Jahre stattgehabten Verlegung des Rechnungsjahres unvollständig.

im Jahre	1876/7:	20,361
„ „	1877/8:	20,833
„ „	1878/9:	19,818
„ „	1879/80:	19,835
„ „	1880/1:	18,027
„ „	1881/2:	17,798
„ „	1882/3:	17,876
„ „	1883/4:	17,194
„ „	1884/5:	16,619
„ „	1885/6:	15,375

Bemerkenswert ist die erhebliche Abnahme in den letzten Jahren, mit welcher die schon hervorgehobene Abnahme der zur penal servitude Verurteilten korrespondiert.[1])

Über die Höhe des Strafmaßes bei Verurteilung zur Gefängnisstrafe dürfte die folgende Zusammenstellung orientieren. Unter den am 31. März 1886 in den local prisons befindlichen 12 365 Strafgefangenen hatten zu verbüßen eine Strafe:

von über 2 Jahren	6	Personen,
„ 2 Jahren	222	„
„ 18 Monaten bis 2 Jahre exkl.	999	„
„ 1 Jahre bis 18 Monaten exkl.	1859	„
„ 9 Monaten bis 1 Jahre exkl..	676	„
„ 6 bis 9 Monaten	1561	„
„ 3 „ 6 Monaten	1946	„
„ 2 „ 3 Monaten	844	„
„ 1 „ 2 Monaten . . .	1978	„
unter 1 Monat	2274	„

Es ist dabei daran zu erinnern, daß eine Strafe von mehr als 2 Jahre Gefängnis nur ausnahmsweise zulässig ist.

Bemerkenswert ist in der Zusammenstellung die geringe Zahl der zu einer Gefängnisstrafe über 18 Monate Verurteilten. Ja, auch die Zahl der zu einer Strafe über 1 Jahr Verurteilten ist eine relativ kleine; denn unter den in der vierten Reihe der Zusammenstellung aufgeführten 1859 Personen betrug bei 1414 die Strafe genau

---

[1]) Es sei dabei hervorgehoben, daß auch die Zahl der zu Gefängnis mit harter Arbeit Verurteilten in den letzten Jahren abgenommen hat; dieselbe betrug 1884/5: 120 651, 1883/4: 122 633, 1882/3: 125 476. Eine Angabe über die Zahl der first class misdemeanants ist in den offiziellen Statistiken nicht enthalten.

1 Jahr, so daß sich die Gesamtsumme aller zu Gefängnis über 1 Jahre Verurteilten auf 2641 stellt gegenüber einer Gesamtziffer von 12365. Bemerkenswert ist ferner die relativ kleine Zahl der zu Gefängnis unter 1 Monat Verurteilten. Dabei dürfte die Angabe nicht ohne Interesse sein, daß zu einer Strafe unter 1 Woche überhaupt nur 80 Personen verurteilt worden sind; bei leichten Straffällen wird regelmäßig auf die in den beiden nächsten Paragraphen zu besprechenden Strafmittel erkannt, nicht auf Gefängnisstrafe. Der Schwerpunkt der Gefängnisstrafe liegt innerhalb des Strafrahmens von 1 Monat bis zu 1 Jahre; innerhalb dieses Strafrahmens fallen von den 12365 Verurteilungen 8419, also weit über $^2/_3$ aller Gefängnisstrafen.

§ 6. Die Geldstrafe.

Litteratur: Shirley, Elementary Treatise on Magisterial Law. London 1881.

Das für misdemeanours am häufigsten angewandte Strafmittel ist die Geldstrafe. Bereits in der ältesten Zeit war dieselbe sowohl als Hauptstrafe wie als Nebenstrafe, insbesondere neben einer Gefängnisstrafe, in Gebrauch. In historischer Hinsicht ist hier wenig zu bemerken. Es gibt nur wenige allgemeine Bestimmungen über die Geldstrafe: regelmäßig wird in jedem einzelnen Spezialgesetze über die Maximalhöhe der Geldstrafe und der event. an ihre Stelle tretenden Gefängnisstrafe das Nähere bestimmt.

In der Magna Charta [1]) wie in der Bill of Rights [2]) wird angeordnet, daß die Geldstrafe sich in rationellen Grenzen halten, nicht übermäßig hoch sein soll. Eine Anzahl Bestimmungen über die Vollstreckung der Geldstrafen sind enthalten in dem Hauptgesetze über die summarische Gerichtsbarkeit — 11 a. 12 Vict. c. 43 s. 21—31. Der Small Penalties Act vom Jahre 1865 — 28 a. 29 Vict.

---

[1]) c. 20 „liber homo non amercietur pro parvo delicto, nisi secundum modum delicti, et pro magno delicto amercietur secundum magnitudinem delicti salvo contenemento suo; et mercator eodem modo salva mercandisa sua; et villanus eodem modo amercietur salvo wainagio suo."

[2]) 1 William a Mary 2, c 2 „that excessive bail ought not to be required, nor excessive fines imposed."

c. 127 — endlich trifft einige allgemeine Anordnungen über die Umwandlung einer nicht beizutreibenden Geldstrafe in Gefängnisstrafe. Diese letzteren Anordnungen sind dann durch den Summary Jurisdiction Act vom Jahre 1879 — 42 a. 43 Vict. c. 49 — in einigen Punkten abgeändert worden.

Das letztere Gesetz hat zu gleicher Zeit die Anwendung der Geldstrafe erheblich ausgedehnt: es wird nämlich den Gerichtshöfen mit summarischer Jurisdiktion die Befugnis gegeben, in denjenigen Strafsachen, welche zu ihrer Kompetenz gehören, im Falle einer erstmaligen Verurteilung auch dann auf Geldstrafe — von in maximo 25 £ — zu erkennen, wenn das zur Anwendung kommende Strafgesetz seinerseits eine Geldstrafe nicht vorsieht. Es ist auch dies wiederum ein Beleg für die schon mehrfach erwähnte Tendenz, dem Ermessen des Richters bei der Bestrafung einen möglichst weiten Spielraum zu geben: der Polizeistrafrichter kann jetzt in allen zu seiner Kompetenz gehörigen, d. h. in allen vom objektiven Standpunkte aus leichten Straffällen bei einer erstmaligen Bestrafung auf Geldstrafe erkennen.

Von den allgemeinen, jetzt gültigen Bestimmungen über die Geldstrafe dürften folgende von Interesse sein.

Der Richter kann in dem auf Geldstrafe lautenden Urteile eine Frist festsetzen, bis zu welcher die betreffende Summe bezahlt sein muß. Er kann ferner dem Verurteilten Ratenzahlungen unter der Bedingung gestatten, daß bei nicht pünktlicher Zahlung einer einzelnen Rate der gesamte Restbetrag als nicht gezahlt angesehen und dementsprechend die event. Gefängnisstrafe vollstreckt werden solle. Bei der Bewilligung von Ratenzahlungen ist der Richter befugt, eine Sicherstellung durch Bürgen zu verlangen.

Der Richter kann weiter in dem Strafurteile bestimmen, daß im Falle der Nichtzahlung der Verurteilte ohne weiteres verhaftet werde. Eine derartige Bestimmung ist jedoch eine Ausnahme, welche nur behufs Abkürzung des Verfahrens da eintritt, wo der Richter von vornherein befürchtet, daß der Verurteilte die Geldstrafe nicht zahlen werde. Das regelmäßige Verfahren besteht darin, daß zunächst Zwangsvollstreckung (warrant of distress) wegen der nicht gezahlten Geldstrafe eintritt, und daß erst bei fruchtlosem Verlaufe der Zwangsvollstreckung ein Verhaftungsbefehl (warrant of commitment) erlassen wird.

Die im Falle der Nichtbeitreibung der Geldstrafe eintretende Gefängnisstrafe braucht nicht schon im Urteile festgesetzt zu sein; diese Festsetzung kann auch nachträglich noch erfolgen. Der Richter hat bei der Bestimmung der Höhe der Geldstrafe wie der event. eintretenden Gefängnisstrafe die, in dem zur Anwendung zu bringenden Gesetze festgesetzte Maximalhöhe zu beachten, ist aber innerhalb dieser Schranke in der Strafabmessung völlig frei. Wenn z. B. das Gesetz für leichte Körperverletzung Geldstrafe bis 5 £ event. Gefängnis bis 2 Monate androht,[1]) so kann der Richter je nach den Vermögensverhältnissen des Angeklagten auf jede Geldstrafe zwischen 1 Schilling und 5 £ erkennen,[2]) und er kann der erkannten Geldstrafe von 5 £ wiederum je nach Lage der Sache eine beliebig hohe Gefängnisstrafe bis zur Höhe von 2 Monaten substituieren.

Der Summary Jurisdiction Act 1879 bestimmt folgende Skala bei der Umwandlung von Geldstrafen in Gefängniß.

Geldstrafe bis 10 Schilling	:	Gefängnis bis	7	Tage,
" " 1 £	:	" "	14	" ,
" " 5 £	:	" "	1	Monat,
" " 20 £	:	" "	2	Monate,
" über 20 £	:	" "	3	" .

Es ist dabei aber zu bemerken, daß, insoweit ein Spezialgesetz ein höheres Maximum der zu substituierenden Gefängnisstrafe zuläßt — wie dies z. B. bei leichter Körperverletzung der Fall ist —, hieran durch die neue Bestimmung nichts geändert werden soll. Die Umwandlungsskala soll nur einen Anhalt da geben, wo gesetzliche Bestimmungen fehlen.

Die charakteristische Erscheinung bei all diesen Bestimmungen besteht darin, daß dem Richter in umfassendem Maße die Möglichkeit gegeben ist, je nach der Lage des Einzelfalles die Höhe der Geldstrafe, die Art und Weise ihrer Beitreibung wie

---

[1]) Stephen II Seite 183 art. 252.

[2]) Aus der Praxis eines Londoner Gerichtshofes ist mir ein Fall bekannt, wo gegen eine größere Anzahl von jungen Burschen, welche in der Neujahrsnacht 1884/5 auf der Straße allerhand Unfug verübt hatten, ohne daß gegen irgend einen einzelnen erschwerende Momente festgestellt wurden, lediglich mit Rücksicht auf die Vermögensverhältnisse auf ganz verschiedene Strafen von 5 Schilling bis zu 3 £ erkannt wurde.

die im Nichtzahlungsfalle eintretenden Konsequenzen zweckentsprechend zu gestalten. Die thatsächlich vorhandene umfassende Anwendung der Geldstrafe, wie sie in England auch allgemein für wünschenswert gehalten wird,[1]) ist eben hierdurch ermöglicht worden.

Statistische Angaben über die Anwendung der Geldstrafe finden sich in den amtlichen Publikationen nur in sehr dürftigem Maße. Angaben über die Anwendung der Geldstrafe als Nebenstrafe fehlen ganz; es fehlen ferner alle Angaben über die Höhe der Geldstrafen. Die Judicial Statistics geben lediglich die Zahl der alljährlich von den Gerichtshöfen mit summarischer Jurisdiktion erkannten Geldstrafen, soweit dieselben die alleinige Strafe bildeten. Diese Zahl betrug[2]) im Jahre 1883/4: 428619 gegenüber einer Gesamtzahl von 583330 Verurteilungen überhaupt. Etwa $^3/_4$ aller von den Gerichtshöfen mit summarischer Jurisdiktion erkannten Strafen bestanden also ausschließlich in Geldstrafen.

Zu den Delikten, bei denen vorzugsweise auf Geldstrafen erkannt wurde, gehören: Trunkenheit und ungebührliches Verhalten (drunk and disorderly persons):[3]) 161752 Geldstrafen bei 177190 Strafen überhaupt; Übertretung der Schulgesetze: 60075 Geldstrafen unter 74688 Bestrafungen; Übertretung von Ortspolizei-Verordnungen u. s. w. (Local Acts and Byelaws): 34500 Geldstrafen bei 37616 Strafen überhaupt; leichte Körperverletzung: 27216 Geldstrafen unter 42518 Bestrafungen; einfacher kleiner Diebstahl: 6741 Geldstrafen gegenüber 27103 Strafen überhaupt, endlich Nahrungs- und Genußmittel-Diebstahl: 2216 Geldstrafen unter 3105 Bestrafungen.

---

[1]) Ein englischer Jurist, Mayhew, bemerkte einmal sehr richtig: „the very first principle of enlightened penology is to endeavour to keep people out of prison as long as possible rather than thrust them into it for the most trivial offences".

[2]) Judicial Statistics 1884 Seite XVII, 32.

[3]) Es wird unter diesen Begriff so ziemlich alles untergebracht, was in Deutschland als grober Unfug zu fassen gesucht wird.

§ 7. Die Friedensbürgschaft.

Litteratur: Schierlinger, die Friedensbürgschaft. Erlangen 1877.

Die Institution der Friedensbürgschaft, schon dem ältesten englischen Rechte bekannt, ist erst in neuester Zeit zu einem eigentlichen Strafmittel ausgebildet worden.

Ursprünglich charakterisiert sich die Friedensbürgschaft als die dem Friedensrichter, als dem Hüter des allgemeinen Friedens und der allgemeinen Ordnung, zustehende Befugnis, von Personen, welchen eine Gefährdung des Friedens oder der Ordnung zuzutrauen ist, ein besonderes verpflichtendes Versprechen für ihr Verhalten zu verlangen. Als derartige präventive polizeiliche Schutzmaßregel besteht die Friedensbürgschaft in zweierlei Gestalt, nämlich als surety oder recognizances for the peace und for good behaviour.

In der ersteren Gestalt als Bürgschaft für Erhaltung des Friedens kommt die Institution zur Anwendung a) wenn jemand in Gegenwart des Friedensrichters einen andern gewaltsam angreift oder bedroht oder mit einem andern Händel beginnt, b) wenn ein Polizeibeamter (constable) eine Person auf frischer That bei einem Friedensbruche ertappt und vor den Richter bringt, c) wenn irgend eine Privatperson gerechten Grund zu der Befürchtung hat, daß ein anderer ihr das Haus anstecken oder ihr körperlichen Schaden zufügen oder solches durch dritte Personen thun lassen werde, und von dieser Befürchtung unter eidlicher Versicherung der näheren Umstände für den Verdacht dem Friedensrichter Anzeige macht (swearing the peace against another). Der Friedensrichter ladet in diesem Falle zunächst den Angeschuldigten und erfordert die Bürgschaft erst dann, wenn er sich nach summarischer Anhörung des Beschuldigten überzeugt hat, daß die Befürchtung als eine gerechtfertigte anzusehen ist.

Die zweite Gestalt der Friedensbürgschaft, die Bürgschaft für gutes Verhalten, ist umfassender als die erstere und schließt dieselbe vollständig in sich. Auf Grund des Gesetzes 34 Edward III c. 1 ist jeder Friedensrichter ermächtigt, von allen Personen, welche nicht von gutem Rufe sind („that be not of good fame"), Bürgschaft für ihr gutes Verhalten gegenüber dem Könige und seinem Volke zu verlangen („good behaviour towards the king and his people“), „damit das Volk nicht beunruhigt oder beschädigt, noch der Frieden vermindert werde, noch Kaufleute und andere auf den Landstraßen gestört und gefährdet werden“. In der Praxis hat diese Bestimmung eine weite Interpretation gefunden: der Ausdruck not of good fame ist sowohl von einem Verhalten contra bonos mores wie contra pacem verstanden und demgemäß die Friedensbürgschaft in dieser Gestalt zur Anwendung gebracht worden gegen Trunkenbolde, Kuppler, Nachtschwärmer, Vagabunden, Diebe, Gauner wie sonstige einer strafbaren Handlung verdächtige Personen.

Die Friedensbürgschaft selbst nun besteht darin, daß der Beschuldigte sowie eine vom Richter bestimmte Anzahl von Bürgen sich als Schuldner des Königs (resp. der Königin) auf einen vom Richter festgesetzten Geldbetrag bekennen (recognizance) [1]) für den Fall, daß der Beschuldigte die ihm gestellte Bedingung (condition) nicht erfüllt. Diese Bedingung enthält die nach dem Einzelfalle verschieden gestaltete Verpflichtung, in der und der Weise Frieden zu halten resp. in der und der Weise ein gutes Verhalten zu beobachten. [2]) Der Friedensrichter hat dabei den Zeitraum anzugeben, für welchen diese Verpflichtung fortbestehen soll. Je nach Lage des Falles ist diese Zeitdauer wie die Zahl der verlangten

---

[1]) Die Form der recognizance, wie sie auch bei Prozeßbürgschaften (cf. oben Seite 18) vorkommt, ist etwa folgende:

„Es wird beurkundet, daß vor mir, dem Friedensrichter der Grafschaft X, in Person erschienen sind A, B und C und daß sich ein jeder besonders zum Schuldner der Frau Königin in Höhe der nachfolgenden Beträge bekannt hat; nämlich A . . . £, B . . . £, C . . . £, zahlbar und einziehbar aus deren Gütern und beweglicher Habe, Ländereien und Pachtungen, zur Verfügung der Frau Königin, Ihrer Erben und Thronfolger für den Fall, daß obengenannter A gegen umstehende Klausel verstoßen sollte.“

Hieran schließt sich die Klausel, welche je nach Lage des Falles verschieden gestaltet ist; bei der gewöhnlichen recognizance for the peace lautet dieselbe dahin, daß A verpflichtet sein soll, Frieden zu halten gegenüber der Königin und allen ihren Unterthanen — insbesondere gegenüber dem Z — für den Zeitraum von 12 Monaten.

[2]) Der Friedensrichter kann sich auch auf die Aufnahme der Verpflichtung (recognizance) beschränken, „daß der Beschuldigte in der nächsten Quartalssitzung erscheinen werde“, wodurch der letzteren die definitive Entscheidung über die Beschuldigung vorbehalten wird.

Bürgen — regelmäßig sind dies 2 — wie endlich die Höhe der verlangten Sicherheiten eine verschiedene.

Die Friedensbürgschaft ist verwirkt (forfeited) in jedem Falle des Zuwiderhandelns gegen die auferlegte Verpflichtung (condition), wobei der Friedensrichter den Thatbestand detailliert festzustellen hat. Die Einziehung der betreffenden Kautionsbeträge von dem Beschuldigten und den Bürgen kann alsdann ohne weiteres bewirkt werden. — Die Friedensbürgschaft ist erledigt (discharged) nach Ablauf der in der condition festgesetzten Zeit resp. mit dem Tode des Beschuldigten.

Wenn sich der Beschuldigte weigert, die Friedensbürgschaft in der verlangten Form zu geben, oder wenn er keine dem Richter genügende Bürgen zu finden vermag, so kann ihn der Friedensrichter ohne weiteres for want of sureties verhaften lassen und ihn bis zur Dauer eines Jahres in Haft behalten.

Dem Betreffenden steht sowohl gegen diese Verhaftung wie gegen die Auferlegung der Friedensbürgschaft überhaupt und gegen das event. Urtheil über die Verwirkung der Bürgschaft das Recht der Berufung zu.

Soviel über die Friedensbürgschaft, soweit sich dieselbe als Ausfluß der dem Friedensrichter obliegenden Friedensbewahrung darstellt. Die Friedensbürgschaft dient in dieser Gestalt in umfassendem Maße den Bedürfnissen der Rechtssicherheit, sie ist ein sehr wirkungsvolles Präventivmittel, um beabsichtigten Rechtsverletzungen rechtzeitig entgegenzutreten.[1]) Wie sie ein Ausfluß der polizeilichen Gewalt ist,

[1]) Eine Art der Anwendung der Friedensbürgschaft dürfte vom Standpunkte der Strafpolitik besonderes Interesse bieten und sei deshalb hier noch hervorgehoben. Wenn es zweifelhaft ist, ob eine beabsichtigte Handlung sich als eine rechtswidrige und strafbare charakterisiert, so kann die an der Unterlassung und Verhinderung der Handlung interessierte Partei dadurch einen richterlichen Ausspruch über die Strafbarkeit der Handlung herbeiführen, daß sie den Friedensrichter in der oben angeführten Weise angeht, von dem Betreffenden die Stellung einer Friedensbürgschaft zu erfordern. Der Friedensrichter hat dann, vorausgesetzt daß das Vorhaben selbst als erwiesen anzunehmen ist, sich darüber schlüssig zu machen, ob die Handlung unter ein bestehendes Strafgesetz fallen würde. Wenn er dies annimmt, so verlangt er von dem Beschuldigten die Stellung einer Friedensbürgschaft. Hiermit ist es dem letzteren zum Be-

so trägt sie auch einen wesentlich polizeilichen Charakter. Als ein Strafmittel kann sie in dieser Gestalt schon deshalb nicht angesehen werden, weil bei ihrer Verhängung eine Rechtsverletzung noch gar nicht vorzuliegen braucht.

Die Thatsache, daß die Friedensbürgschaft als Mittel zur Verhütung von Verbrechen sich vorzüglich bewährt hat, ist nun aber für die englische Gesetzgebung der Anlaß gewesen, die Institution auch als Strafmittel zu verwenden. Dabei waren zwei Gesichtspunkte maßgebend.

Zunächst stellte sich die Friedensbürgschaft als ein praktischer Ausweg dar, um die Gesellschaft gegen die ihr bei Entlassung von Verbrechern nach verbüßter Strafe drohenden Gefahren der Rückfälligkeit möglichst zu schützen, ohne daß auf der andern Seite der Strafentlassene irgendwie in seinem Bestreben, sich fortab auf redliche Weise seinen Lebenserwerb zu verschaffen, gehindert würde. Die Auferlegung der Friedensbürgschaft (putting under recognizance) erschien als eine wirkungsvollere und dabei gleichzeitig weniger

---

wußtsein gebracht, daß die von ihm beabsichtigte Handlung rechtlich als eine strafbare aufgefaßt wird, er mag daraus Anlaß nehmen, von der Ausführung seines Vorhabens abzustehen. —

In jüngster Zeit ist in einer sehr wichtigen politischen Frage dieses Verfahren zur Anwendung gebracht worden. Der irische Abgeordnete Dillon hatte im Herbst 1886 den Plan aufgestellt, eine Herabsetzung der Pachtzinsen in Irland dadurch zu erzwingen, daß alle Pächter, denen die verlangte Herabsetzung der Zinsen nicht bewilligt würde, an den Gutsherrn überhaupt keinerlei Pachtzins abführen sollten; vielmehr sollte der herabgesetzte Zinsbetrag an eine, von einem Komitee bestimmte Vertrauensperson gezahlt und die so zusammengebrachte Summe im wesentlichen dazu verwendet werden, die Herabsetzung der Pachtzinsen durchzusetzen. Die öffentliche Meinung war zweifelhaft, ob die Ausführung dieses Planes eine strafbare Handlung in sich schließen würde. Die Regierung, welche davon eine Gefährdung des öffentlichen Friedens in Irland befürchtete, beantragte durch ihren Treasury Solicitor (cf. oben Seite 11 ff.) dem Mr. Dillon die Stellung einer Friedensbürgschaft aufzugeben. Mr. Dillon wurde vorgeladen und gab zu, daß er den obigen Plan aufgestellt und auch schon zur Ausführung desselben eine Organisation in das Leben gerufen hätte. Der Richter erklärte darauf zunächst, daß eine derartige Organisation eine gesetzwidrige wäre und daß sich die Ausführung des Planes mittels einer derartigen Organisation als Verschwörung charakterisieren würde, und ordnete sodann an, daß Mr. Dillon binnen 12 Tagen für sein gutes Verhalten eine Verpflichtung (recognizance) in Höhe von 1000 £ einzugehen und 2 Bürgen mit einer recognizance von je 1000 £ zu stellen habe, widrigenfalls er Gefängnis von 6 Monaten zu gewärtigen haben würde.

drückende und nachteilige Nebenstrafe, als die Polizeiaufsicht, welche besonders in früherer Zeit dem Zwecke der Sicherung der Gesellschaft nur wenig gerecht wurde und dem Entlassenen die Rückkehr in die bürgerliche Gesellschaft erheblich erschwerte.

Die Friedensbürgschaft war aber nicht nur ein geeignetes Mittel der Strafschärfung neben einer Freiheitsstrafe, sondern sie schien auch geeignet, für manche Straffälle eine Freiheitsstrafe überflüssig erscheinen zu lassen. In Fällen, in welchen die Strafthat an sich durch eine Geldstrafe genügend geahndet war, in welchen der Richter aber wegen der von dem Thäter nach der subjektiven oder objektiven Richtung hin bewiesenen Gemeingefährlichkeit eine härtere Strafe für angezeigt halten mußte, bot die Auferlegung der Friedensbürgschaft neben einer Geldstrafe ein der Sachlage mehr entsprechendes Strafmittel als die Verurteilung zu einer Freiheitsstrafe.

Nachdem in Berücksichtigung der beiden hier angeführten Momente einzelne Strafgesetze die Auferlegung der Friedensbürgschaft als Nebenstrafe bereits zugelassen hatten, bestimmten die im Jahre 1861 ergangenen 5 großen Gesetze, durch welche das Strafrecht bezüglich der Mehrzahl der Delikte konsolidiert wurde (cf. oben Seite 1) — 24 und 25 Vict. c. 96—100 — ganz allgemein, daß der Richter bei all den hier behandelten Delikten die Befugnis haben solle, soweit es sich um eine felony handelt, neben der Freiheitsstrafe eine Friedensbürgschaft aufzuerlegen,[1]) und soweit es sich um eine misdemeanour handelt, neben oder statt der Freiheitsstrafe auf Geldstrafe und Stellung einer Friedensbürgschaft zu erkennen.[2])

Hiermit ist die Friedensbürgschaft als ein außerordentlich wichtiges Strafmittel in das englische Strafensystem eingefügt worden. Nicht bloß der Friedensrichter, sondern jeder Gerichtshof kann jetzt auf Stellung einer Friedensbürgschaft erkennen. Die Friedensbürgschaft ist bei der Mehrzahl aller Delikte als Nebenstrafe zulässig, bei den schwereren Deliktsfällen neben einer Freiheitsstrafe, bei den leichteren Fällen neben einer Geldstrafe.

---

[1]) Dem Richter wird hier bei Auferlegung der Friedensbürgschaft die Wahl gelassen, ob er von dem Verurteilten und von Bürgen oder bloß von dem Verurteilten oder bloß von Bürgen die Eingehung einer recognizance verlangen will.

[2]) 24 a. 25 Vict. c. 96 s. 117, c. 97 s. 73, c. 98 s. 51, c. 99 s. 38, c. 100 s. 71.

Daneben bleibt die Friedensbürgschaft als polizeiliches Präventivmittel bestehen.

Die Friedensbürgschaft hat somit jetzt teils den Charakter einer polizeilichen Maßregel, teils denjenigen einer gerichtlichen Strafe. Dabei sind alle wesentlichen Bestimmungen der polizeilichen Friedensbürgschaft auch für die gerichtliche Friedensbürgschaft beibehalten worden.[1]) Aus dieser Übernahme älterer Bestimmungen ist es zu erklären, daß auch die gerichtliche Friedensbürgschaft sich mit dem Eingehen einer schuldnerischen Verpflichtung begnügt, obwohl dies doch nicht als eine eigentliche Strafe angesehen werden kann.[2])

Über den Umfang der Anwendung der Friedensbürgschaft enthalten die Judicial Statistics folgende Angaben.[3]) Im Jahre 1883/4 wurde von den Gerichtshöfen mit summarischer Gerichtsbarkeit Friedensbürgschaft von 16 571 Personen verlangt, von den andern Gerichtshöfen wurden 454 Personen zu einer Geldstrafe resp. Friedensbürgschaft verurteilt.[4]) Diese Angaben sind offenbar unvollständig; es fehlen ganz die Fälle, in denen neben einer Freiheitsstrafe auf Friedensbürgschaft erkannt worden ist.

§ 8. Körperliche Züchtigung.

Körperliche Züchtigung, in früherer Zeit bei der Bestrafung fast aller Delikte häufig zur Anwendung gebracht,[5]) besteht auch jetzt noch

---

[1]) In einzelnen Fällen der gerichtlichen Friedensbürgschaft ist die Maximalzeitdauer derselben eine gesetzlich beschränkte; so kann z. B. bei Körperverletzung nur auf Stellung einer Friedensbürgschaft von 6 Monaten erkannt werden. Cf. Stephen II Seite 182 art. 150.

[2]) Um den Charakter einer Strafe zum Ausdruck zu bringen, hätte wenigstens Kautionsstellung oder sonstige Realsicherheit verlangt werden müssen.

[3]) Judicial Statistics 1884 Seite XVII, XXVII.

[4]) Die Zahl der im Jahre 1883/4 for want of sureties Verhafteten betrug 2532 (Judicial Statistics 1884 Seite XXXIII); wie viele hiervon wegen Nichtleistung einer verlangten Friedensbürgschaft verhaftet worden sind und wie viele wegen Nichtleistung der für das Erscheinen im Haupttermin verlangten Sicherheit (cf. oben Seite 18) ist nicht angegeben.

[5]) und zwar sowohl bei Männern wie bei Frauen. Bei den Frauen fand die Züchtigung meistenteils nicht öffentlich statt, dagegen bei den Männern regelmäßig öffentlich.

als allgemeines Strafmittel dem Gesetze nach fort, wenngleich davon in der Praxis nur in denjenigen Fällen Gebrauch gemacht wird, für welche die Prügelstrafe durch neuere Gesetze angeordnet worden ist.[1])

Rechtlich ist noch heute körperliche Züchtigung als Nebenstrafe neben einer Freiheitsstrafe bei allen felonies zulässig, bei denen nicht durch spezielles Gesetz die Strafe anderweitig bestimmt ist.[2]) Dabei besteht über die Art und Weise sowie über das Maß der Züchtigung keine allgemeine gesetzliche Bestimmung, abgesehen von der Anordnung der Bill of Rights, daß keine grausamen und ungewöhnlichen Strafen auferlegt werden sollen.[3])

Für die Gerichtshöfe mit summarischer Gerichtsbarkeit wird in einem Gesetze vom Jahre 1861 — 25 Vict. c. 18 s. 1 — angeordnet, daß bei Verurteilung zu einer körperlicher Züchtigung der Richter die Zahl der Hiebe und das Instrument, mit welchem die Hiebe beigebracht werden, in dem Urteile selbst genau anzugeben hat und daß bei einer Person unter 14 Jahren die Zahl der Hiebe nicht mehr als 12 betragen und die Hiebe mittels einer Rute (birch) beigebracht werden sollen.[4])

In dreien der großen Konsolidationsgesetze aus dem Jahre 1861 — nämlich in dem Gesetze gegen Diebstahl, gegen böswillige Eigentumsbeschädigung und gegen Vergehen an der Person — ist die körperliche Züchtigung als Strafmittel bei Personen männlichen Ge-

---

[1]) Stephen I Band 1 Seite 490, Du Cane II Seite 15.

[2]) Stephen II Seite 14 art. 18, Seite 15 art. 19.

[3]) 1 William a. Mary s. 2, c. 2 „that excessive bail ought not to be required nor excessive fines imposed nor cruel and unusual punishments inflicted.“

[4]) Ein früheres Gesetz, welches die Gerichtshöfe mit summarischer Jurisdiktion für kompetent zur Aburteilung von kleinen Diebstahlsfällen bei Personen unter 14 Jahren erklärte, hatte körperliche Züchtigung sowohl als allein zu erkennende Strafe wie als Nebenstrafe zugelassen. 10 a. 11 Vict. c. 82. Zu erwähnen ist hier ferner die Bestimmung des Summary Jurisdiction Act 1879 — 42 a. 43 Vict. c. 49 s. 10 —, wonach bei Kindern unter 12 Jahren ein Gerichtshof mit summarischer Jurisdiktion auf nicht mehr als 6 Rutenschläge erkennen soll. Es wird hier weiter angeordnet, daß bei der durch einen Polizeibeamten (constable) vorzunehmenden Vollstreckung der Züchtigung stets ein höherer Polizeibeamter zugegen sein soll und daß der Vater oder Vormund des Kindes auf sein Verlangen zuzuziehen ist.

schlechts ausdrücklich beibehalten worden,[1]) wobei wiederum bestimmt ist, daß der Gerichtshof die Zahl der Schläge und das Instrument, mittels dessen dieselben beigebracht werden, in dem Urteile bestimmt anzugeben hat, und ferner, daß die Züchtigung eine nicht öffentliche sein soll.

Ein Gesetz vom Jahre 1863 „zum besseren Schutze der Unterthanen gegen persönliche Gewalt" — 26 a. 27 Vict. c. 44 p. 1 — bestimmt, daß im Falle des Raubes unter Anwendung von Gewalt gegen die Person, sowie im Falle des Versuchs, eine Person zu erdrosseln oder zu ersticken (to choke, suffocate or strangle) neben der anderweitigen Strafe bei männlichen Angeklagten auf 1—3 malige körperliche Züchtigung erkannt werden kann. Auch hier soll der Richter in dem Urteile die Zahl der Schläge und das dabei anzuwendende Instrument genau angeben und, falls der Angeklagte unter 16 Jahren alt ist, auf nicht mehr als 25 Rutenschläge erkennen, bei erwachsenen Angeklagten soll die Zahl der Schläge bei der jedesmaligen Züchtigung nicht mehr als 50 — zusammen also in maximo 150 — betragen. Die Züchtigung selbst soll nicht öffentlich ausgeführt werden und muß binnen 6 Monaten nach der Urteilsverkündung vollstreckt werden.

In der Praxis wird die Prügelstrafe heutzutage bei weiblichen Personen gar nicht mehr, bei erwachsenen männlichen Personen nur in den im Gesetze von 1863 angeführten 2 Fällen zur Anwendung gebracht.[2]) Bei jugendlichen Personen männlichen Geschlechts dagegen wird die Prügelstrafe noch heute häufig angewendet.[3]) Die Gesamtzahl der im Jahre 1883/4 von den Gerichtshöfen mit summarischer Jurisdiktion ergangenen

---

[1]) 24 a. 25 Vict. c. 96 s. 119, c. 97 s. 75, c. 100 s. 70.

[2]) Auch als Disziplinarstrafe kommt die körperliche Züchtigung jetzt nur noch bei Männern zur Anwendung. Im Jahre 1885/6 wurde in den local prisons an 138, in den convict prisons an 69 Personen körperliche Züchtigung als Disziplinarstrafe vollzogen. Report of Commissioners 1886 Seite 47, Report of Directors 1886 Seite XXXII.

[3]) Im Jahre 1882 wurde im englischen Unterhause die Einbringung eines Gesetzentwurfes beantragt, wonach auch gegen erwachsene Personen bei körperlichen Angriffen (assaults) auf Prügelstrafe erkannt werden sollte. Der damalige Staatssekretär des Innern, Sir W. Vernon Harcourt, lehnte es jedoch ab, einen derartigen Entwurf einzubringen.

Urteile auf körperliche Züchtigung belief sich auf 3284.[1]) Weitere statistische Angaben, insbesondere über die Zahl der von den Assisengerichtshöfen erkannten Prügelstrafen, fehlen gänzlich.

§ 9. Überweisung an eine Besserungs- oder Erziehungs-Anstalt (Reformatory oder Industrial School) und die Bestrafung jugendlicher Personen überhaupt.

Litteratur: Report of the Royal Commission on Reformatory, Industrial and Day Industrial Schools 1883. — Die jährlichen Reports of the Inspector appointed to visit the Certified Reformatory and Industrial Schools of Great Britain. — Föhring, die Reformatory und die Industrial Schools der vereinigten Königreiche Großbritannien und Irland im 14. Vereinshefte des nordwestdeutschen Vereins für Gefängniswesen Seite 118 ff. — Blätter für Gefängniskunde Band 19 S. 243 ff. über den Bericht der Royal Commission von 1883.

Die Strafmündigkeit tritt nach englischem Rechte mit dem vollendeten 7. Lebensjahre ein.[2]) Bei Angeschuldigten zwischen 7 und 14 Jahren darf eine Verurteilung nur dann erfolgen,[3]) wenn der Richter positiv feststellt, daß der Angeschuldigte bei Begehung der That die zu ihrer Strafbarkeit erforderliche Einsicht besaß („unless it be shewn affirmatively that such person had sufficient capacity to know that the act was wrong").

Diese Feststellung ist offenbar in früherer Zeit wenig ernst genommen worden: die Zahl der zu Ende des vorigen und zu Anfang dieses Jahrhunderts in den Strafanstalten befindlichen Kinder wird als eine sehr hohe angegeben.[4]) Die Art der Bestrafung verurteilter Kinder war zu der damaligen Zeit von derjenigen der Erwachsenen in keinerlei Weise verschieden.

Schon Howard wies jedoch in seinem Buche: The State of the Prisons auf die Mißstände hin, welche dadurch hervorgerufen würden,

---

[1]) Judicial Statistics 1884 Seite XVII.

[2]) Stephen I Band II Seite 98 plaidiert dafür, die Strafmündigkeit erst mit dem vollendeten 12. Lebensjahre eintreten zu lassen; man könnte jedoch von dieser Regel bei gewissen schweren Verbrechen Ausnahmen machen.

[3]) Über die ausgedehnte Kompetenz der Courts of Summary Jurisdiction zur Bestrafung Jugendlicher cf. oben Seite 9.

[4]) cf. Du Cane II Seite 199 ff., Stephen I Band II Seite 98.

daß Kinder mit erwachsenen Gefangenen in denselben Räumen zusammengesperrt und so dem nachteiligen Einflusse der im Verbrechen Ergrauten ausgesetzt wären. Er verlangte abgesonderte Anstalten nicht nur für verurteilte Kinder, sondern für jugendliche Verbrecher überhaupt.

Auch von andrer Seite wurde anerkannt, daß es unrichtig wäre, jugendliche Personen denselben Strafen zu unterwerfen wie Erwachsene, da einerseits bei den ersteren zumeist eine viel größere Schuld den Eltern zuzuschreiben wäre, welche die Erziehung ihrer Kinder vernachlässigt oder die Kinder gar direkt zum Verbrechen angeleitet hätten, als den Bestraften selbst, und da es anderseits bei jugendlichen Sträflingen möglich und im allgemeinen Interesse geboten wäre, dieselben noch von der Bahn des Verbrechens abzulenken und auf richtige Lebenswege hinzuführen.

Es fiel zunächst privaten Unternehmungen zu, diese an sich sicherlich richtigen Gedanken zu verwirklichen. Die im Jahre 1788 in London gegründete Philantropic Society machte es sich unter anderm zur Aufgabe, für die Unterbringung bestrafter Kinder in einer besonderen Anstalt zu sorgen.[1]) Andere Wohlthätigkeitsgesellschaften folgten diesem Beispiele nach, und die Regierung unterstützte diese Unternehmungen in der Weise, daß sie denjenigen zu einer Strafe verurteilten Kindern und jugendlichen Personen, zu deren Aufnahme sich die Wohlthätigkeitsgesellschaften bereit erklärten, die Verbüßung der erkannten Strafe im Gnadenwege erließ.

Dieser Zustand wurde in einem Gesetze vom Jahre 1838 — 1 a. 2 Vict. c. 82 — sanktioniert und dabei bestimmt, daß Flucht oder Fluchtversuch aus diesen Anstalten, sowie Übertretung der fortab staatlich zu genehmigenden Anstaltsordnungen als selbständiges Delikt bestraft werden solle. Durch dasselbe Gesetz wurde ferner eine besondere Strafanstalt für die Aufnahme jugendlicher Verbrecher eingerichtet. Ein Gebäude in Parkhurst auf der Insel Wight, welches bisher als Militärhospital be-

[1]) Die erste von der Philantropic Society eröffnete Anstalt in St. George's Field war nach dem Kollektivsystem eingerichtet, während die auf besonderes Betreiben von Gladstone errichtete und berühmt gewordene neue Anstalt „Farm School for the Reformation of Criminal Boys at Redhill", welche im Jahre 1849 eröffnet wurde, das englische Familiensystem annahm.

nutzt worden war, sollte fortab der Strafverbüßung jugendlicher Verbrecher ausschließlich dienen. [1])

Nachdem so praktische Einrichtungen zur Aufnahme jugendlicher Verbrecher geschaffen waren, ging die Gesetzgebung einen Schritt weiter und erließ nunmehr besondere strafrechtliche Normen für die Bestrafung jugendlicher Angeschuldigten. Der im Jahre 1854 ergangene Reformatory School Act [2]) — 17 a. 18 Vict. c. 86 — bestimmte, daß gegen jugendliche d. h. unter 16 Jahre alte Angeschuldigte zusätzlich zu einer Gefängnisstrafe von nicht weniger als 14 Tagen auf Unterbringung in eine Besserungsanstalt (Reformatory School) auf die Dauer von 2—5 Jahren erkannt werden könnte.

Die Gründung von Reformatory Schools wird der privaten Initiative überlassen, welche sich ja schon bisher die Unterbringung jugendlicher Verbrecher hatte angelegen sein lassen. Die Anstalten müssen jedoch, ehe jugendliche Verbrecher aufgenommen werden dürfen, staatlich genehmigt (certified) sein und unterliegen einer fortdauernden staatlichen Aufsicht. Die Ausübung der staatlichen Rechte gegenüber den Anstalten wird dem Staatssekretär des Innern übertragen, welcher dieselben durch einen für die Reformatory Schools ernannten Inspektor wahrnehmen läßt. Dem Inspektor sind vor Erteilung des Zertifikats die Baupläne, Statuten, Hausordnungen u. s. w. für die Anstalt vorzulegen. Entsprechen dieselben den an die Anstalt zu stellenden Anordnungen nicht, so ist wegen ihrer Abänderung zu verhandeln. Erweist sich die Anstalt, nachdem sie das Zertifikat erhalten hat, späterhin als ungenügend, so kann ihr jederzeit das Zertifikat wieder entzogen werden. Der Inspektor hat behufs Beaufsichtigung der

---

[1]) In der Vorrede des Gesetzes wird ausgeführt „it might be of great public advantage, that a prison be provided, in which young offenders may be detained and corrected, and receive such instruction and be subject to such discipline as shall appear most conducive to their reformation and the repression of crime“. Die Anstalt zu Parkhurst ist im Jahre 1864, nachdem sie durch die inzwischen errichteten Reformatory Schools überflüssig geworden war, geschlossen worden.

[2]) Das Gesetz bezieht sich auf England und Schottland; auch die spätere Gesetzgebung in dieser Materie ist mit unwesentlichen Abweichungen für beide Länder dieselbe. Für Irland sind besondere Gesetze ergangen, welche zwar im einzelnen mannigfache Abweichungen enthalten, aber in der Grundidee mit den englisch-schottischen Gesetzen übereinstimmen.

Anstalt dieselbe jährlich wenigstens einmal zu besichtigen und hierüber an den Staatssekretär zu berichten. Alljährlich ist ein Generalbericht über die Reformatory Schools dem Parlamente vorzulegen. Durch das Zertifikat erhält die Anstalt die Berechtigung der Aufnahme eines bestraften Jugendlichen; dagegen ist sie zur Aufnahme nicht verpflichtet.[1]) Der Staatssekretär kann einen in die Anstalt Aufgenommenen jederzeit wieder daraus entfernen und einer andern Anstalt zuweisen. Der Staat leistet zur Unterhaltung der Anstalten per Kopf und Woche festzustellende Zuschüsse, außerdem können von den Eltern der aufgenommenen jugendlichen Verbrecher Verpflegungsgelder bis zur Höhe von 5 Schilling per Woche beigetrieben werden.

Soviel über das Gesetz von 1854. Dasselbe hat in den folgenden Jahren einige Zusätze und Abänderungen erfahren, welche dann im Jahre 1866 in dem Reformatory School Consolidating and Amending Act — 29 a. 30 Vict. c. 117 — zu einer Neukodifikation geführt haben; auch hier sind durch spätere Gesetze noch einige Abänderungen hinzugefügt worden. Der heutige Stand der Gesetzgebung zeigt gegenüber demjenigen von 1854 folgende Neuerungen.

Es kann schon bei einer Verurteilung zu 10 Tagen Gefängnis (1854: 14 Tage) auf Unterbringung in eine Besserungsanstalt als Zusatzstrafe erkannt werden. Anderseits soll gegen ein Kind unter 10 Jahren nur dann auf Unterbringung erkannt werden, wenn das Kind bereits vorbestraft ist oder wenn es sich um ein schwereres, von den Quartalssitzungen abzuurteilendes Delikt handelt. Die Bestimmung der Anstalt erfolgt durch den erkennenden Richter unter möglichster Berücksichtigung des Glaubensbekenntnisses des Jugendlichen. Der Staatssekretär kann jederzeit die Versetzung in eine andere Anstalt oder auch die definitive Entlassung verfügen. Die Anstaltsverwaltung selbst kann den Jugendlichen nach Verlauf von 18 Monaten vorläufig entlassen; diese vorläufige Entlassung wird immer nur auf 3 Monate ausgesprochen und kann bei schlechter Führung widerrufen werden.

Von seiten des Staatssekretärs sind für alle

---

[1]) Von praktischer Bedeutung dürfte übrigens das gesetzliche Recht zur Ablehnung der Aufnahme mit Rücksicht auf den staatlichen Zuschuß nicht sein.

Anstalten verbindliche Hausordnungen, General Rules and Regulations, erlassen. Die einzelne Anstalt kann mit Genehmigung des Staatssekretärs noch ergänzende Bestimmungen (byelaws) geben, im allgemeinen ist jedoch die Verwaltung und Hausordnung in allen Anstalten dieselbe. [1]) Wegen bestimmter Vergehen gegen die Hausordnung, wie fortdauernde Unbotmäßigkeit, Versuch der Entweichung, kann von einem Friedensrichter auf Gefängnisstrafe mit oder ohne harte Arbeit bis zur Dauer von 3 Monaten erkannt werden.

Die Vierteljahrssitzungen der Friedensrichter werden ermächtigt, Zuschüsse zu den Unterhaltungskosten wie zum Bau von Besserungsanstalten oder zum Landankaufe für dieselben zu leisten oder auch bei vorhandenem Mangel an Anstalten für eigne Kosten Anstalten anzulegen und zu verwalten. [2]) Der Staat zahlt zu den Unterhaltungskosten einen wöchentlichen Zuschuß von 4—6 Schilling per Kopf.

Bezüglich der praktischen Verwaltung der Anstalten schreiben die General Rules vor, daß 6 Stunden per Tag der Ausbildung in der Arbeit (industrial training) und 3 Stunden dem Unterrichte gewidmet sein sollen, und daß der letztere regelmäßig nur Lesen, Schreiben und Rechnen und nur bei besonderer Genehmigung noch Singen, Geschichte und Geographie umfassen solle. Daneben ist für Religionsunterricht und tägliche Hausandacht zu sorgen. Die Arbeitsausbildung umfaßt bei den Mädchen: Nähen, Waschen und Handarbeit, bei den Knaben die Erlernung eines Handwerks; dazu tritt Beschäftigung in Feld- und Gartenarbeit. Ein Teil der Knaben, welcher sich der seemännischen Laufbahn widmen will, erhält eine seemännische Ausbildung auf einem Training Ship. [3])

Die Direktoren (managers) der Anstalten bemühen sich, für die zur Entlassung Kommenden eine geeignete Stellung ausfindig zu machen. Ein großer Teil der Entlassenen tritt in das Heer oder

---

[1]) Insbesondere ist das disziplinare Strafrecht in den Anstalten einheitlich geregelt. Körperliche Züchtigung ist nur bei Knaben zulässig und darf nicht über 18 Rutenschläge hinausgehen.

[2]) Von dieser Befugnis haben bis jetzt nur 2 Grafschaften (Kent und Staffordshire) Gebrauch gemacht.

[3]) Ein Training Ship liegt in der Nähe von London (in Purfleet), zwei andre bei Liverpool.

die Marine ein. Die Anstalten stehen ferner mit Auswanderungsgesellschaften in Beziehung und veranlassen vielfach die zur Entlassung Kommenden, nach den Kolonieen überzusiedeln. Die Zahl der so nach den Kolonieen Ausgewanderten wird bis zum Jahre 1881 auf 2241 angegeben.[1]) Über die in England Verbleibenden pflegen die Anstalten während 3 Jahre nach der Entlassung eine Aufsicht auszuüben. Die Verbindung mit der Anstalt wird jedoch von den Entlassenen vielfach auch noch nach dieser Zeit aufrecht erhalten. Die Anstalten stehen auch später noch den Betreffenden mit Rat und That bei, um ihnen eine ehrliche Existenz zu gründen oder zu sichern.

Es gibt zur Zeit in England 49 Certified Reformatory Schools, in denen sich Ende 1884: 3639 Knaben und 804 Mädchen, zusammen also 4443 Kinder befanden.[2]) Im Jahre 1883/4 wurden einer Besserungsanstalt überwiesen: 1410 Knaben und 221 Mädchen. Bei 79,6 % derselben erfolgte die Verurteilung wegen Diebstahls. Von den 1041 Knaben waren 539, also über die Hälfte, schon früher einmal bestraft worden, 5 waren bereits: 5 mal, 2: 6—7 mal vorbestraft.[3]) Die überwiesenen Kinder verteilen sich auf folgende Altersklassen

	a) noch nicht bestrafte Kinder	b) bereits vorbestrafte Kinder
1) 10 Jahre und darunter	3,1 %	3,0 %
2) 10—11 Jahre	7,8 „	7,7 „
3) 11—12 „	11,5 „	3,3 „
4) 12—13 „	19,9 „	18,5 „
5) 13—14 „	24,1 „	24,2 „
6) 14—15 „	23,9 „	23,8 „
7) 15—16 „	9,7 „	9,5 „

Die Altersklassen 13—15 Jahre stehen im Vordergrunde. Von den 1262 im Jahre 1883/4 neu überwiesenen Kindern konnten 312 weder lesen noch schreiben, bei 795 war die Kenntnis im Lesen und Schreiben nur eine unvollkommene. Nur bei 619 dieser Kinder lebten beide Eltern; bei 307 lebte Vater oder Mutter.

[1]) cf. Du Cane II Seite 224.

[2]) cf. für die folg. statistischen Angaben Judicial Statistics 1884 S. XLV ff.

[3]) Unter den Mädchen ist die Zahl der Vorbestraften eine viel geringere: von den 221 im Jahre 1883/4 überwiesenen Mädchen waren nur 18 vorbestraft.

Von den 1262 Kindern hatten 496 vor ihrer Aufnahme in die Anstalt eine Gefängnisstrafe von 10 Tagen, 758 eine solche von 10 Tagen bis zu 1 Monate und 8 eine noch längere Gefängnisstrafe zu verbüßen.

Die Totalausgaben für die Certified Reformatory Schools betrugen im Jahre 1882/3: 134204 £,[1]) davon entfielen auf die Staatsbeiträge 87241 £, während von den Eltern der Kinder 5918 £ beigetrieben wurden.[2])

Neben den Reformatory Schools besteht nun aber in England heutigen Tages noch eine andere Institution zur Behandlung jugendlicher Verbrecher, nämlich die Industrial Schools. Dieselben, ursprünglich mehr für verwahrloste Kinder bestimmt, dienen nach den jetzt gültigen Gesetzen auch zur Aufnahme von 2 Klassen von Kindern, welche bereits mit dem Strafgesetze in Kollision gekommen sind.[3]) Es kann nämlich von dem Richter die Unterbringung in eine Industrial School angeordnet werden:

a) bei Kindern unter 12 Jahren, welche wegen einer mit Gefängnis oder geringerer Strafe bedrohten Handlung angeklagt und noch nicht vorbestraft sind, falls der Richter von einer eigentlichen Bestrafung mit Rücksicht auf das Alter oder sonstige Umstände noch Abstand zu nehmen wünscht;

b) bei Kindern unter 14 Jahren, welche bettelnd oder vagabundierend aufgegriffen werden.

Das Verhältnis zwischen den Reformatory und den Industrial Schools läßt sich kurz dahin charakterisieren: die Überweisung an eine Reformatory School ist ein wirkliches Strafmittel, welches zusätzlich zu einer Gefängnisstrafe eintritt, die Industrial Schools[4]) dagegen sind für

---

[1]) Die Kosten werden per Kopf auf wöchentlich 7 Schilling 6 Pence bis 8 Schilling angegeben, cf. Du Cane II Seite 221.

[2]) Im Jahre 1883/4 wurden von den Eltern beigetrieben 5704 £, die Staatsbeiträge beliefen sich auf 69565 £. — Über die Zuschüsse der Lokalbehörden fehlen genauere Angaben. Jedenfalls muß ein recht erheblicher Teil der Ausgaben durch freiwillige Spenden aufgebracht werden.

[3]) cf. betreffs der übrigen Klassen von Kindern, welche in die Industrial Schools Aufnahme finden: Aschrott, das englische Armenwesen. Leipzig 1886. Seite 337, 338.

[4]) Man wird dieselben deutsch am richtigsten „Erziehungsanstalten“ benennen.

solche jugendliche Verbrecher bestimmt, welche der Richter zwar noch straffrei lassen will, bei denen er jedoch eine Zwangserziehung, wie sie verwahrlosten Kindern zu teil wird, für angebracht hält.

Die Industrial Schools wurden zuerst in Schottland im Jahre 1854 — 17 a. 18 Vict. c. 74 — eingeführt, im Jahre 1857 sind die wesentlichsten Bestimmungen dieses Gesetzes auf England ausgedehnt worden — 20 a. 21 Vict. c. 48 —. Das Hauptgesetz für die Industrial Schools ist vom Jahre 1866 — 29 a. 30 Vict. c. 118 —, dazu sind noch eine größere Anzahl Zusatzgesetze, insbesondere mit Rücksicht auf die inzwischen eingetretene staatliche Regelung des Elementarschulwesens, ergangen.[1])

Die Einrichtung der Industrial Schools stimmt im allgemeinen mit derjenigen der Reformatory Schools überein; es ist nur eine mildere Form dieser Institution: die Disziplinarvorschriften sind gelindere, die tägliche Arbeitsausbildung (industrial training) ist von 6 auf 4 Stunden herabgesetzt; die Kinder sind regelmäßig mit vollendetem 16. Lebensjahre aus der Anstalt zu entlassen. Im übrigen gelten durchweg die gleichen Vorschriften wie bei den Reformatory Schools: die Anstalten unterstehen der durch einen besonderen Inspektor ausgeübten staatlichen Aufsicht; sie müssen vor ihrer Eröffnung ein Zertifikat des Inspektors erhalten haben;[2]) sie basieren zunächst auf freiwilligen Beiträgen. Der Staat leistet bestimmte Zuschüsse, die Lokalbehörden (insbesondere die School Boards) können Zuschüsse gewähren; von den Eltern des Kindes kann ein wöchentlicher Beitrag bis zu 5 Schilling beigetrieben werden. Die Gesamtzahl der Certified Industrial Schools betrug 1883/4: 107. In dieselben

[1]) Besondere Arten der Industrial Schools bilden die Day Industrial Schools und die Truant Schools, cf. Aschrott l. c. Seite 337 Anm. 3 und Seite 338 Anm. 3.

[2]) Neben den Certified Industrial Schools bestehen noch eine große Anzahl Uncertified Schools; es sind dies reine Privatunternehmungen, bei denen die staatliche Aufsicht wie der staatliche Zuschuß fortfällt. Die meisten derselben stehen unter der Leitung der Reformatory and Refuge Union, einer Institution, welche die Zentralisation des gesamten Besserungs- und Erziehungswesens erstrebt. Im Jahre 1881 standen in Verbindung mit der Union: 99 Uncertified Schools in London und 78 Anstalten in den Provinzen, welche zur Aufnahme von insgesamt 5197 resp. 2806, also zusammen 8003 Kindern bestimmt waren. Cf. Du Cane I Seite 96.

wurden während des Jahres überwiesen: 3347 Knaben und 627 Mädchen, zusammen 3974 Kinder; davon waren 42 % zwischen 10 und 12 Jahre, 38,8 % zwischen 12 und 14 Jahre alt; nur 19,2 % waren unter 10 Jahre alt. Die Gesamtzahl der in den Anstalten am Ende des Jahres befindlichen Kinder belief sich auf 9756 Knaben und 2289 Mädchen, zusammen also 12045 Kinder. Die Gesamtkosten betrugen 240198 £, hiervon wurde ein Betrag von 11435 £ von den Eltern der Kinder beigetrieben.[1])

Eine eigentümliche Zwischenstellung zwischen einer Reformatory und einer Industrial School nimmt die Middlesex Industrial School ein, welche ausschließlich für die Grafschaft Middlesex, wozu das Londoner Stadtgebiet mit Ausnahme der City gehört, bestimmt ist. Die Anstalt verdankt ihre Entstehung einem besonderen Gesetze vom Jahre 1854 — 17 a. 18 Vict. c. 169 — und wird unter Beitragspflicht der Eltern aus der Grafschaftssteuer erhalten. Ursprünglich war sie für jugendliche Verbrecher im Alter von 7—14 Jahren bestimmt, welche auf 1—3 Jahre der Anstalt überwiesen werden konnten, ohne daß der Überweisung eine Gefängnisstrafe vorherzugehen brauchte. Im Jahre 1868 erhielt die Anstalt daneben die Rechte einer Industrial School. Im Jahre 1875 — 38 a. 38 Vict. c. 87 — wurde in Abänderung des Gesetzes von 1854 bestimmt, daß der Anstalt nur jugendliche Verbrecher im Alter von 10—14 Jahren überwiesen werden sollten, daß die Dauer der Überweisung 2—5 Jahre zu betragen hätte und daß die Entlassung spätestens mit vollendetem 16. Lebensjahre erfolgen sollte. Von den im Jahre 1883/4 der Anstalt überwiesenen 68 Knaben[2]) hatten 80,1 % sich eines Diebstahls oder Diebstahlsversuchs schuldig gemacht.[3]) In die Middlesex Industrial

[1]) Judicial Statistics 1884 Seite L u. LI. Für das Jahr 1882 wird der Staatszuschuß auf 170473 £ angegeben; der Staat leistet für jedes Kind wöchentlich einen Zuschuß von 3 Schilling 6 Pence bis 5 Schilling. Die durchschnittlichen Kosten in den Anstalten werden per Kopf auf wöchentlich 7 Schilling bis 7 Schilling 6 Pence berechnet. Du Cane II Seite 221 ff.

[2]) Die Anstalt war ursprünglich ausschließlich für Knaben bestimmt erst im Jahre 1881 wurde eine Nebenanstalt zu Bedfont behufs Aufnahme von Mädchen eröffnet.

[3]) Im Jahre 1883/4 wurden 68 Knaben der Anstalt überwiesen. Am Ende des Jahres befanden sich in der Anstalt 143 Knaben. Die jährlichen Kosten beliefen sich per Kopf auf 27 £ 18 s. 4 d. Von den Eltern der Knaben

School findet ein Teil der anderswo zu einer Reformatory School gesandten Kinder neben den auch anderswo für eine Industrial School bestimmten Kindern Aufnahme, während man im allgemeinen bisher in England streng auf eine Scheidung der beiden Arten von Anstalten und der dafür bestimmten Kinder gehalten hat.[1])

Über die günstigen Erfolge der Reformatory und der Industrial Schools herrscht in England nur eine Stimme:[2]) die in den letzten Jahrzehnten thatsächlich eingetretene erhebliche Abnahme der Delikte wird von allen Sachverständigen, soweit dieselben auch sonst in den von ihnen für diese Besserung angeführten Gründen voneinander abweichen, überein-

---

wurde ein Betrag von 181 £ 9 s. 6 d. eingezogen. Judicial Statistics 1884 Seite XLIX u. L.

[1]) Lord Norton tritt in einem Artikel „Schools as Prisons and Prisons as Schools" im Januarhefte der Nineteenth Century 1887 für eine Verschmelzung der Industrial und der Reformatory Schools ein. Lord Norton schlägt ferner vor, die Schulen in Zukunft nicht mehr dem Home Secretary, sondern dem Education Department zu unterstellen.

[2]) Wines l. c. Seite 223 sagt mit Recht: „The English Reformatory School as corrective of criminality, and the English Industrial School as preventive of crime, furnish to the world the best model, upon the whole, of which I have any knowledge". Um die Ausbildung dieser Institutionen haben sich u. a. verdient gemacht: Lord Shaftesbury, Miß Mary Carpenter und Sydney Turner. Den eifrigen Bemühungen des ebenso einflußreichen wie persönlich wohlthätigen Lord Shaftesbury ist es im wesentlichen zuzuschreiben, daß auf diesem Gebiete — wie auch bei der Lösung andrer sozialer Probleme — ein thatkräftiges Ineinandergreifen freiwilliger und staatlicher Thätigkeit stattgefunden hat. Miß Mary Carpenter ist in Schriften (cf. besonders die Schrift Reformatory Prison Discipline, London 1872) und Reden unermüdlich thätig gewesen, um das öffentliche Interesse für Reformatories für das weibliche Geschlecht zu gewinnen und wachzuerhalten. Sydney Turner endlich hat während einer 20jährigen Thätigkeit als Inspektor der Reformatory and Industrial Schools auf die praktische Gestaltung und Ausbildung dieser Institutionen den wesentlichsten Einfluß ausgeübt. — Auch der Name von Lord Norton verdient hervorgehoben zu werden. Lord Norton ist es gewesen, welcher das erste Gesetz über die Reformatory Schools im Parlamente einbrachte, und er hat auch an der weiteren gesetzlichen Regelung dieser Materie fortdauernd einen hervorragenden Anteil gehabt. Für das Zustandekommen des Gesetzes von 1854 war auch der damalige Home Secretary, Lord Palmerston, sowie Lord John Russell — damals Minister ohne Portefeuille — energisch eingetreten.

stimmend zu einem wesentlichen Teile der segensreichen Thätigkeit der beiden Arten von Anstalten zugeschrieben.[1]) Es ist gelungen, eine große Anzahl zum Verbrechen geneigter oder schon im frühen Kindesalter dem Verbrechen anheimgefallener jugendlicher Personen zu einer geordneten redlichen Lebensthätigkeit zu erziehen und damit der Armee der Verbrecher ein erhebliches Rekrutierungskontingent abzuschneiden.[2])

Dies ist das Gesamturteil der im Jahre 1883 berufenen Königlichen Kommission zur Untersuchung der Reformatory und Industrial Schools der vereinigten Königreiche.[3]) Diese aus 14 Mitgliedern bestehende Kommission

---

[1]) Cf. die Rede von Leone Levi vor der London Statistical Society und die sich an diese Rede anschließende Debatte. Journal of the London Statistical Society 1880 Seite 423—461.

[2]) Folgende Ziffern mögen als Beleg hierfür dienen. Unter den zu einer Gefängnisstrafe Verurteilten befanden sich:

	1879/80	1880/1	1881/2	1882/3	1883/4	1884/5
a) Personen unter 12 Jahren	866	528	397	415	387	270
b) Personen zwischen 12 und 16 Jahren	6550	5051	5086	5285	4888	4609

Am 31. März 1886 war die Gesamtzahl aller Gefangenen unter 12 Jahren in den local prisons: 6, und aller Gefangenen zwischen 12 und 16 Jahren: 217. In den convict prisons befanden sich:

	Sträflinge überhaupt	Sträflinge unter 24 Jahren
a) am 31. Dez. 1871	10 160	2949
b) am 31. Dez. 1878	10 225	2197
c) am 31. März 1880	10 268	1935
d) am 31. März 1886	7973	1196

Judicial Statistics 1884 Seite XXXVI; Report of Commissioners 1886 Seite 27; Du Cane I Seite 93; Report of Directors 1886 Seite XXXV. Die Zahlen über die convict prisons sind besonders beachtenswert. Hier nimmt die Zahl der Sträflinge unter 24 Jahren seit 1871 von Jahr zu Jahr in starken Progressionen ab, während die Gesamtzahl der Sträflinge lange Zeit hindurch eine stabile bleibt. Erst in den letzten Jahren — dann aber in sehr ausgesprochener Weise — zeigt sich eine Rückwirkung der bei den jüngeren Altersklassen eingetretenen Besserung auf die allgemeine Verbrecherziffer.

[3]) Es heißt in dem Berichte dieser Kommission „the work already done may be credited with putting an end to the training of boys as professional thieves and with rescuing children fallen into crime from becoming habitual hardened offenders, while it has undoubtedly had the effect of

hat, nachdem sie eine große Anzahl von Anstalten in England, Schottland und Irland besucht hatte, einen außerordentlich ausführlichen Bericht — 86 Folioseiten nebst 795 Seiten Anlagen — über den heutigen Zustand der Reformatory und Industrial Schools veröffentlicht und daran eine Reihe von Verbesserungsvorschlägen geknüpft.[1]) Es seien nur die wichtigsten derselben hier noch hervorgehoben.[2])

Die Kommission empfiehlt: die Minimalzeit der Überweisung an eine Reformatory School auf 3 Jahre festzusetzen und dem Richter die Befugnis zu geben, nicht nur neben Gefängnisstrafe auf Überweisung an eine Reformatory zu erkennen, sondern auch neben einer Strafe der körperlichen Züchtigung, für deren vermehrte Anwendung bei Knaben lebhaft plaidiert wird.

Es empfehle sich ferner, mit den Reformatory Schools Einzelzellen zu verbinden, in denen die Jugendlichen die erkannte Freiheitsstrafe verbüßen könnten, anstatt daß dieselben wie bisher zu diesem Zwecke für kurze Zeit in ein Gefängnis gebracht würden.

Es wird sodann zwar anerkannt, daß größere Anstalten eine bessere Gelegenheit für eine vielseitige Arbeitsausbildung und die Gewinnung geeigneter Lehrkräfte gewährten, allein kleinere Anstalten schienen im allgemeinen doch mehr angezeigt, weil hier den Kindern die mit Rücksicht auf ihre bisherige Vernachlässigung besonders wünschenswerte eingehende persönliche Fürsorge in höherem Maße zu teil werden könnte. Im übrigen empfehle es sich, ein gewisses Klassifikationssystem der Anstalten eintreten zu lassen unter Berücksichtigung des Alters der Insassen, sowie des Grades der Verderbtheit.[3]) Vor allem müßten besondere Anstalten zur Aufnahme von Jugendlichen bestehen, welche bereits früher in einer Anstalt

---

preventing large numbers of children from entering on a career of crime".

[1]) Praktische Folgen haben diese Verbesserungsvorschläge bis jetzt noch nicht gehabt.

[2]) Da ich im Abschnitte II auf die Reformatory und Industrial Schools nicht wieder zurückkomme, so glaubte ich, die für ein richtiges Verständnis der gesetzlichen Bestimmungen wichtigsten Fragen aus der praktischen Verwaltung der Institutionen hier ausführlicher behandeln zu sollen, als es sonst im Abschnitte I geschehen ist.

[3]) Auch die Howard Association tritt in ihrem letzten Jahresberichte warm für eine derartige Klassifikation der Anstalten ein, cf. Report 1886 Seite 9, 10.

erfolglos gewesen oder aus einer Anstalt entwichen wären oder welche sich in einer Anstalt fortgesetzt unbotmäßig aufführten.

Endlich sei es in Überlegung zu ziehen, ob den Eltern der an eine Reformatory überwiesenen Kinder nicht die elterlichen Rechte abgesprochen werden sollten. [1]) Die geeignete Unterbringung der aus einer Reformatory Entlassenen, insbesondere die Sendung derselben nach den Kolonieen wäre häufig dadurch unmöglich gemacht worden, daß die Eltern den dazu erforderlichen Konsens verweigert und die Kinder wieder zu sich genommen hätten, wodurch mit einem Schlage all die Erziehungserfolge der Anstalt zu nichte gemacht würden.

§ 10. Die Polizei-Aufsicht und die Bestrafung Rückfälliger überhaupt.

Litteratur: Monro, Report on the History of the Metropolitan Convict Supervision Office. London 1887.

A. Polizei-Aufsicht tritt nach heutigem englischen Rechte in zwei voneinander gesondert zu haltenden Fällen ein:

I. bei denjenigen zur penal servitude Verurteilten, welche vor vollständiger Verbüßung ihrer Strafe zur vorläufigen Entlassung gekommen sind, bei den sogenannten ticket of leave men oder license holders;

II. bei Vorbestraften, bei welchen auf Polizei-Aufsicht als Nebenstrafe erkannt worden ist.

I. Betreffs der vorläufig Entlassenen ist bereits in den §§ 3 und 4 die historische Entwickelung des Instituts der Polizei-Aufsicht als eines wesentlichen Bestandteils der vorläufigen Entlassung gegeben worden. [2]) Es wurde dabei hervorgehoben, daß das System der vorläufigen Entlassung, mit welcher Aufenthaltsbeschränkungen und die Pflicht zu periodischen Meldungen verbunden waren, zuerst bei den nach den australischen

---

[1]) Cf. über die jetzt in England bestehende Bewegung, die elterlichen Rechte denjenigen, welche für ihre Kinder nicht gehörig sorgen, zu nehmen oder wenigstens zu beschränken: Aschrott, l. c. Seite 336 besonders Anmerk. 2.

[2]) Sehr reiches Material über die Entwickelung des Instituts ist enthalten in dem Report of the Royal Commission 1863 to inquire into the working of the Penal Servitude Acts 1853 and 1857.

Kolonieen Transportierten zur Anwendung gebracht worden ist, daß dann bei Einführung der Strafe der penal servitude im Jahre 1853 diese Institution nach England übernommen worden ist, ohne daß jedoch damals nähere Bestimmungen über die den vorläufig Entlassenen aufzuerlegenden Beschränkungen und über die Beaufsichtigung der Innehaltung derselben ergingen, [1]) und daß erst durch die Penal Servitude Acts von 1857 und 1864 die Beaufsichtigung der vorläufig Entlassenen näher geregelt worden ist.

Nachdem durch die Prevention of Crimes Acts von 1871 und 1876 — 34 a. 35 Vict. c. 112 und 39 a. 40 Vict. c. 23 — noch einige Neuerungen in der Beaufsichtigung der vorläufig Entlassenen eingeführt worden sind, gelten jetzt bezüglich der Polizei-Aufsicht über diese Klasse von Personen folgende Bestimmungen:

1) Der vorläufig Entlassene hat sich binnen 48 Stunden nach Ankunft an dem von ihm gewählten Bestimmungsorte [2]) bei der Polizeibehörde persönlich zu melden und seine Wohnung anzugeben. Er hat von jedem Wohnungswechsel binnen 48 Stunden der Polizei Meldung zu machen. Wenn er an einen andern Ort Großbritanniens [3]) verziehen will, so hat er sich an seinem bisherigen Wohnorte abzumelden und an seinem neuen Wohnorte binnen 48 Stunden nach Ankunft anzumelden. Er darf sich während der Dauer der license an keinem Orte länger als 48 Stunden ohne Meldung aufhalten. An einem von der Polizei bestimmten Tage hat er sich allmonatlich zu melden.

All diese Meldungen haben prinzipiell persönlich zu erfolgen; die Polizei kann dem Entlassenen jedoch gestatten, daß nach seiner

---

[1]) In Irland ist eine gehörige Polizei-Aufsicht über die vorläufig Entlassenen früher eingerichtet worden, als in England. Die in dieser Richtung durch Sir Walter Crofton in Irland getroffenen und unter der Leitung von Mr. Organ in vorzüglicher Weise ausgeführten Einrichtungen haben den späteren englischen Bestimmungen zum Muster gedient.

[2]) Dem Sträflinge steht bei seiner Entlassung aus dem convict prison die freie Wahl zu, nach welchem Orte er befördert sein will. Die Gefängnisverwaltung bestreitet die Kosten dieser Beförderung, soweit der gewählte Ort von der Strafanstalt nicht weiter entfernt ist, als der Ort, wo der Sträfling verurteilt worden ist. Ist die Entfernung eine größere, so wird der Mehrbetrag der Fahrkosten aus der Arbeitsbelohnung des Betreffenden genommen.

[3]) Die hier angeführten Bestimmungen gelten gleichmäßig für England, Schottland und Irland.

ersten Meldung an dem Orte die weiteren Meldungen entweder von ihm selbst oder durch eine dritte Person schriftlich durch die Post eingesendet werden. Wenn sich der Entlassene der Aufsicht eines Vereins zur Unterstützung Strafentlassener (Discharged Prisoners Aid Society) unterwirft, [1]) so wird regelmäßig das Abkommen getroffen, daß alle Meldungen zunächst an den Verein zu richten sind und von dem letzteren dann an die Polizeibehörde weitergegeben werden. Wenn der Arbeitgeber des Entlassenen sich bereit erklärt, die Aufsicht über denselben zu übernehmen, [2]) so kann die Polizei den Entlassenen selbst von allen weiteren Meldungen befreien, vorausgesetzt, daß sich der Arbeitgeber verpflichtet, periodisch über das Verhalten des Betreffenden zu berichten und von einer Dienstentlassung sofort Anzeige zu machen.

In den beiden hier erwähnten Fällen unterläßt es die Polizei, den Strafentlassenen durch Polizeibeamte aufsuchen zu lassen. Im übrigen sind die Polizeibeamten, welchen die Beaufsichtigung der Entlassenen obliegt, bei Strafe verpflichtet, alles zu vermeiden, was geeignet sein würde, fremden Leuten den Zweck ihrer Besuche kund zu thun.

Unterläßt ein license holder die ihm obliegenden Meldungen, so kann er, ohne daß es eines besonderen

---

[1]) In diesem Falle wird regelmäßig ein Teil der dem Entlassenen gebührenden Arbeitsbelohnung an den Verein abgeführt, damit derselbe in der für die betreffende Persönlichkeit geeignet erscheinenden Weise darüber verfüge; cf. über die Thätigkeit der Discharged Prisoners Aid Societies überhaupt unten Abschnitt II § 24 Anhang.

[2]) Sowohl die Polizei wie die in Verbindung mit der Polizei arbeitenden Vereine für Strafentlassene haben mit einzelnen Arbeitgebern Beziehungen angeknüpft, um den Strafentlassenen eine Beschäftigung zu verschaffen. Während früher in England die Ansichten darüber geteilt waren, ob man den Arbeitgebern die Vorbestrafungen mitteilen sollte oder nicht, erhalten jetzt diejenigen Arbeitgeber, welche sich zur Beschäftigung geeigneter Strafentlassener bereit erklären und damit ihre Absicht, ihrerseits an der Fürsorge für Strafentlassene thätigen Anteil zu nehmen, kund thuen, stets eine vollständige Auskunft über die Antezedenzien des von ihnen Beschäftigten. Man ist allgemein der Ansicht, daß nur auf diese Weise eine genügende Anzahl geeigneter Arbeitgeber zur Beschäftigung von Strafentlassenen veranlaßt werden kann. Die Arbeitgeber, welchen in geeigneten Fällen von der Polizei die Aufsicht über die Betreffenden übertragen wird, erlangen durch die den Strafentlassenen auferlegten Beschränkungen (cf. im Texte sub 2 besonders das Verbot, keinen Verkehr mit Personen von notorisch schlechtem Charakter zu pflegen) eine besondere Sicherung für ein gutes Verhalten dieser Leute.

Haftbefehls bedarf, verhaftet und einem Gerichte vorgeführt werden. Das Gericht hat alsdann die Befugnis, die license für verwirkt zu erklären, wodurch der Betreffende verpflichtet wird, die gesamte Strafzeit, welche bei seiner vorläufigen Entlassung noch unverbüßt war, nachträglich zu verbüßen. [1]) Beträgt diese noch nicht verbüßte Strafzeit weniger als 1 Jahr, so kann das Gericht den Betreffenden fernerhin zu Gefängnis mit oder ohne harte Arbeit verurteilen; die Gesamtzeitdauer der noch zu verbüßenden penal servitude und der zusätzlich erkannten Gefängnisstrafe soll jedoch die Zeitdauer 1 Jahres nicht überschreiten.

2) Für die Dauer der license übernimmt der Entlassene die besondere, auf der license selbst abgedruckte Verpflichtung: a) sich jeder Gesetzesübertretung zu enthalten; b) keinerlei Verkehr mit Personen von notorisch schlechtem Charakter (Dieben, Prostituierten u. s. w.) zu pflegen; c) nicht ein faules und unordentliches Leben zu führen ohne sichtbare Quellen eines redlichen Lebenserwerbs (not lead an idle and dissolute life without visible means of obtaining an honest livelihood); d) die license stets bei sich zu führen und dieselbe jedem Polizeibeamten oder Friedensrichter auf Verlangen vorzuzeigen. Auch hier kann wegen jeder Verletzung einer dieser Verpflichtungen der Betreffende ohne Haftbefehl verhaftet werden, und es kann von jedem Gerichte mit summarischer Jurisdiktion die license für verwirkt erklärt und außerdem — soweit nicht wegen einer Gesetzesübertretung eine härtere Strafe eintritt — auf Gefängnisstrafe bis zu 3 Monaten erkannt werden.

Man sieht, die Einschränkungen, welchen der vorläufig Ent-

---

[1]) Früher war es zweifelhaft, ob bei Verwirkung der license der Betreffende für die ganze, bei seiner vorläufigen Entlassung noch unverbüßte Strafzeit in die Strafanstalt zurückzubringen wäre, oder nur für denjenigen Zeitraum, welcher im Momente der ausgesprochenen Verwirkung von der Strafzeit noch übrig blieb. Die jetzige Form der license läßt darüber keinen Zweifel mehr; es wird für den Fall des Widerrufs und der Verwirkung der license ausdrücklich gesagt: „he will be liable to undergo a term of penal servitude equal to the portion of his term which remained unexpired when his license was granted“. Cf. die jetzt gültige Form einer license Du Cane I Seite 135 ff.

lassene unterworfen ist, sind sehr erhebliche [1]) und die der Polizei gegen diese Klasse von Personen gegebenen Machtbefugnisse sind außerordentlich weitgehende. Die Gefahr, daß durch diese Bestimmungen dem Entlassenen die Auffindung eines redlichen Lebenserwerbs wesentlich erschwert, ja daß derselbe durch eine ungebührliche Einmischung der Polizei aus einer erlangten Beschäftigung wieder herausgerissen werde, erscheint in der That sehr naheliegend.

Nichtsdestoweniger haben die im Jahre 1878 ernannten Commissioners, nachdem sie eine große Anzahl Zeugen über diese Frage verhört hatten, ihr Urteil dahin abgegeben, daß derartige Vorkommnisse zu den Ausnahmen zu rechnen seien. [2]) Es gibt dies der englischen Polizei ein glänzendes Zeugnis. Daneben ist freilich zu berücksichtigen, daß die Polizei in ihrer Thätigkeit eine gewichtige Unterstützung in den Vereinen für entlassene Sträflinge gefunden hat. [3])

Zur Abhilfe der relativ wenigen Fälle, in welchen die Commissioners eine nachteilige Einwirkung der Polizei-Aufsicht feststellten, wurde in dem Berichte von 1879 empfohlen, in den großen Städten, vor allem in London, eine besondere Polizei-Abteilung mit für diesen Zweck ausgesuchten geeigneten Beamten für die Beaufsichtigung der Strafentlassenen einzurichten. [4]) Eine derartige

---

[1]) Es ist dabei hier noch daran zu erinnern (cf. oben Seite 60), daß neben den Fällen, wo das Gericht die Verwirkung einer license ausspricht, dem Staatssekretär das Recht des jederzeitig freien Widerrufs der license zusteht („if it shall please Her Majesty to revoke such license".)

[2]) In dem Report of the Commissioners appointed to inquire into the working of the Penal Servitude Acts 1879 heißt es Seite XXXV: „Upon the whole we are satisfied that such interference as has taken place with the honest employment of discharged convicts has been of comparatively rare occurrence".

[3]) Mit Recht weist v. Holtzendorff II Seite 113 darauf hin, daß für den Erfolg der Polizei-Aufsicht von maßgebender Bedeutung nicht nur die Thätigkeit und Wirksamkeit der Polizei selbst ist, sondern auch das Vertrauen, welches die Polizei in der öffentlichen Meinung genießt, und die Unterstützung, welche der Polizei vom Publikum zu teil wird. Es ist hinlänglich bekannt, wie abweichend die diesbezüglichen Auffassungen über die Polizei in den verschiedenen Ländern sind und wieweit wir in Preußen speziell hinter England hier zurückstehen.

[4]) In dem Report wird auf Seite XXXVI die Befürchtung ausgesprochen, „that supervision, if left in the hands of ordinary police constables in the

Spezialabteilung ist denn auch noch im Jahre 1879 in London gebildet worden, und der vor kurzem von dem Dirigenten dieser Abteilung, Mr. Monro, herausgegebene Bericht über die bisherige Thätigkeit der ihm unterstellten Abteilung gibt ein lehrreiches Bild von der vorzüglichen Handhabung dieses äußerst schwierigen Gebietes polizeilicher Aufgaben.

Folgende Ziffern über das Verhältnis der verwirkten licenses zu den in dem betreffenden Zeitraum neuerteilten licenses, nach fünfjährigen Perioden berechnet, dürfte von Interesse sein. Die verwirkten licenses betrugen:

1855—1859:	14,9 %	der erteilten licenses,
1860—1864:	7,7 %	" " "
1865—1869:	8,5 %	" " "
1870—1874:	8,5 %	" " "
1875—1879:	8,7 %	" " "
1880—1884:	9,4 %	" " "

Im Jahre 1885 wurden 267 licenses für verwirkt erklärt gegenüber 1656 neuerteilten licenses. [1])

II. Polizei-Aufsicht als richterlich erkannte Nebenstrafe neben einer Freiheitsstrafe (Gefängnis oder Strafknechtschaft) ist im Jahre 1869 durch den Habitual Criminal Act — 32 a. 33 Vict. c. 99 — eingeführt worden. Die bezüglichen Bestimmungen wurden durch die beiden Prevention of Crimes Acts 1871 und 1876 — 34 a. 35 Vict. c. 112 und 39 a. 40 Vict. c. 23 — etwas modifiziert und sind jetzt folgende: [2])

Wenn eine Person wegen irgend einer felony oder wegen einer der im Gesetze angegebenen misdemeanours — die wichtigsten sind Münzvergehen, Betrug und eine Reihe von zur Nachtzeit ausgeführten Eigentumsdelikten [3]) — bestraft worden ist und später-

---

metropolis, will tend more and more to become a mere matter of routine, harassing to the men who are subjected to it and affording no real security to society against the criminal classes".

[1]) Report of Directors 1886 Seite XXII.

[2]) Auch diese Bestimmungen gelten gleichmäßig für England, Schottland und Irland.

[3]) Durch 24 a. 25 Vict. c. 96 s. 58 sind eine Reihe von zur Nachtzeit vorgenommenen Handlungen, welche geeignet sind, den Verdacht zu erwecken, daß der Betreffende eine strafbare Handlung auszuführen beabsichtigte, besonders

hin wegen einer felony oder einer der genannten misdemeanours zur nochmaligen Bestrafung gelangt, so kann der Richter neben der anderweitig zu erkennenden Strafe auf Polizei-Aufsicht bis zu 7 Jahren erkennen. Es ist ganz in das freie Ermessen des Richters gestellt, ob er im Einzelfalle von dieser Befugnis Gebrauch machen, und event. für welchen Zeitraum — bis zur Maximalgrenze von 7 Jahren — er die Polizei-Aufsicht eintreten lassen will. Die Polizei-Aufsicht beginnt mit dem Tage der Verbüßung der Hauptstrafe.

Derjenige, gegen welchen auf Polizei-Aufsicht erkannt worden ist, wird dadurch zu genau denselben Meldungen, wie ein license holder, verpflichtet (cf. oben I, 1). Für jeden Fall der Nichtmeldung kann er ohne Haftbefehl verhaftet und von einem Gerichtshofe mit summarischer Gerichtsbarkeit zu Gefängnis mit oder ohne harte Arbeit bis zur Dauer 1 Jahres verurteilt werden.

Die Zahl der Personen, gegen welche im Jahre 1884 auf Polizei-Aufsicht erkannt wurde, belief sich auf 1250. Diese Ziffer ist erheblich geringer, als diejenige der Vorjahre, wo dieselbe betrug 1883: 1448; 1882: 1478; 1881: 1454; 1880: 1433. In all diesen Jahren waren es fast ausschließlich Eigentumsdelikte, bei denen auf Polizei-Aufsicht erkannt wurde. Im Jahre 1884 waren von den zur Polizei-Aufsicht verurteilten Personen bestraft: wegen Delikte gegen das Eigentum: 1228; gegen die Person: 9; wegen Münzverbrechens: 11; wegen anderer Delikte: 2. [1])

B. Gleichzeitig mit der Einführung dieser Polizei-Aufsicht hat die englische Gesetzgebung noch weitere verschärfende Maßregeln gegen vorbestrafte Subjekte getroffen.

Durch die schon angeführten Gesetze wird nämlich weiter bestimmt: Wenn eine Person bereits zweimal wegen irgend einer felony oder wegen einer der im Gesetze genannten misdemeanours bestraft worden ist, so kann dieselbe innerhalb eines Zeitraums von 7 Jahren nach Verbüßung der letzten Strafe in folgenden Fällen von einem

---

unter Strafe gestellt. — Dazu gehört z. B., wenn jemand zur Nachtzeit mit geschwärztem Gesichte oder sonst verkleidet in einem fremden Hause ohne hinreichenden Grund — dessen Nachweis dem Angeklagten obliegt — getroffen wird.

[1]) Judicial Statistics 1884 Seite XXVIII.

Gerichtshofe mit summarischer Gerichtsbarkeit mit Gefängnis bis zu 1 Jahre verurteilt werden:[1])

a) wenn ein Polizeibeamter den Betreffenden beschuldigt, daß derselbe sich seinen Lebensunterhalt auf unredliche Weise beschaffe (getting his livelihood by dishonest means), und der Richter diese Beschuldigung als vernünftigerweise annehmbar erachtet (if there are reasonable grounds for believing);

b) wenn der wegen irgend einer strafbaren Handlung Angeschuldigte seinen richtigen Namen und seine richtige Wohnung vor Gericht nicht angibt;

c) wenn der Betreffende an einem öffentlichen oder privaten Orte unter Umständen getroffen wird, welche dem Richter die Überzeugung geben, daß derselbe im Begriffe war („was about to"), irgend eine strafbare Handlung zu begehen oder Beihilfe zu einer strafbaren Handlung zu leisten, oder auf eine Gelegenheit hierfür wartete („was waiting for an opportunity");

d) wenn der Betreffende in einem Wohn- oder Geschäftsraume, Garten, Lagerplatze u. s. w. getroffen wird, ohne seinen Aufenthalt dort genügend rechtfertigen zu können („without being able to account for his being found on such premises").

Am weitgehendsten unter diesen Bestimmungen ist die sub c) angeführte: hier wird *schon der Verdacht des Wartens auf eine Gelegenheit zur Begehung einer strafbaren Handlung als genügender Thatbestand für eine Bestrafung erklärt*, eine Bestimmung, welche sich mit allgemeinen strafrechtlichen Grundbegriffen kaum in Einklang bringen läßt.[2])

---

[1]) Die ratio legis ist, daß bei Personen, welche schon zweimal mit dem Strafgesetze in ernsten Konflikt gekommen sind, in den ersten Jahren nach ihrer Entlassung aus der Strafanstalt eine besonders große Gefahr besteht, daß sie den Weg des Verbrechens wieder beschreiten werden. Deshalb soll diesen Personen *während dieser kritischen Zeit* schon auf den bloßen Verdacht der Rückkehr zum Verbrechen die Freiheit wieder genommen werden.

[1]) Du Cane II Seite 194 faßt den *juristischen Charakter der Bestimmungen* richtig dahin, daß *eine zweimal bestrafte Person für die Dauer von 7 Jahren bei gewissen Handlungen von der sonst bei Angeklagten geltenden Vermutung des Nichtschuldig ausgeschlossen* werde. „A person convicted on indictment after a previous conviction is considered not to be entitled for seven years to the presumption of innocence under certain circumstances of suspicion, which is the right of every member of the community whose character is untainted."

Von den sonstigen Bestimmungen über die Bestrafung Rückfälliger ist bereits im § 4 erwähnt worden, daß im Jahre 1864 die Minimalstrafdauer der penal servitude für Rückfällige auf 7 Jahre festgesetzt wurde,[1] daß diese Einschränkung des freien richterlichen Ermessens jedoch auf Vorschlag der Royal Commissioners von 1878 seit dem Jahre 1879 beseitigt ist — 27 a. 28 Vict. c. 47 s. 2 und 42 a. 43 Vict. c. 54 s. 1 —.

Dagegen bestehen heute noch für einzelne Delikte Spezialbestimmungen, wonach die Maximalzeitdauer der penal servitude bei Rückfälligen höher bemessen ist, als bei erstmals Bestraften; so ist z. B. für den einfachen Diebstahl die Maximalstrafe im allgemeinen 5 Jahre penal servitude, wenn der Betreffende jedoch vorbestraft war, so erhöht sich diese Maximalstrafe je nach dem Charakter des früher begangenen Deliktes auf 7 resp. 10 Jahre penal servitude.[2])

Die allgemeine Meinung geht in England auf eine sehr strenge Behandlung der Rückfälligen. Man glaubt, wie dies ja die oben angeführten Bestimmungen genügend darthun, daß hier der wesentliche, wenn nicht der einzige Zweck der Strafe darin bestehen müsse, die Gesellschaft vor neuen Angriffen seitens derjenigen zu schützen, welche sich durch ihr bisheriges Verhalten als Feinde der gesellschaftlichen Ordnung gezeigt haben. Selbst ein Mann von so gemäßigten Anschauungen wie Sir Edm. Du Cane drückt sich in dieser Richtung folgendermaßen aus:[3] „Nach meiner Meinung sollten diejenigen, welche wohlüberlegt ein verbrecherisches Leben führen, nachdem ihnen ein oder zweimal die Gelegenheit zur Besserung gegeben war, für sehr lange Zeit eingeschlossen werden, da dies der einzige Weg ist, die Gesellschaft gegen sie zu schützen und sie zu verhindern, eine ebenso schlechte Klasse von Verbrechern in die Welt zu setzen, als sie selbst sind".

---

[1]) Eine andre, auch im Jahre 1864 eingeführte Bestimmung besteht noch fort: Personen, welche zum dritten Male mit penal servitude bestraft werden, erhalten keine license; dieselben müssen also, einerlei wie ihr Verhalten in der Strafanstalt ist, die volle Strafzeit ohne Kürzung abbüßen.

[2]) Stephen II Seite 15 art. 19 und Seite 251 art. 321.

[3]) Du Cane I Seite 94; cf. auch II Seite 7: „for incorrigibles the only mode of protecting society against them is that they should be

Was die Ziffer der Rückfälligen betrifft,[1]) so waren im Jahre 1885/6 von 1027 zur penal servitude Verurteilten 330 bereits vorbestraft.[2]) Die höchste Ziffer fällt dabei auf die Delikte: Diebstahl und Hehlerei mit 180, dann folgt Einbruch mit 53, Münzverbrechen mit 16 bereits vorbestraften Subjekten.

Von den im Jahre 1885/6 zur Gefängnisstrafe Verurteilten, deren Gesamtziffer oben Seite 92 auf 149 104 berechnet worden ist, waren 77 068 Personen bereits vorbestraft,[3]) darunter 14 055 über zehnmal! Sehr merkwürdig ist dabei das Verhältnis der beiden Geschlechter: von den Vorbestraften waren 50 042: Männer, 27 026: Frauen; von den bereits mehr als zehnmal Vorbestraften waren 5074: Männer und 8981: Frauen. Während also die Gesamtzahl der vorbestraften Frauen

---

entirely removed from temptation which they cannot withstand, and be made use of as examples to others."

[1]) Cf. für die im folgenden angeführten Ziffern Judicial Statistics 1884 Seite XXXV, Report of Directors 1886 Seite XVIII und XXII; Report of Commissioners 1886 Seite 72; und Du Cane I Seite 8.

[2]) Die Zahl dieser Vorbestraften variiert in den einzelnen Jahren ziemlich erheblich, und es ist nicht möglich, hier eine bestimmte Richtung der Bewegung zu erkennen. Im Jahre 1884 belief sich die Zahl auf 480; die Durchschnittsziffern für 5 jährige Perioden betrugen

1855—1859: 357	1870—1878: 363
1860—1864: 668	1875—1879: 398
1865—1869: 523	

Mit Recht macht Du Cane I S. 8, 9 darauf aufmerksam, daß aus diesen Ziffern keinerlei Schlüsse auf die Wirksamkeit des Strafen- und Gefängnissystems zulässig sind: zunächst seien die statistischen Angaben für die verschiedenen Perioden nicht in gleicher Weise zuverlässig. Es sei ferner zu berücksichtigen, daß die Zahl der nach Verbüßung der Strafe Ausgewanderten in den einzelnen Perioden eine verschiedene gewesen sei, was auf die in England zur Zählung gelangenden Rückfälligen natürlich von Einfluß sein müsse. Endlich könne ein günstiger oder ungünstiger Erfolg der Behandlung der Verbrecher aus der Zahl der Rückfälligen überhaupt kaum entnommen werden, weil dabei sehr viele andre Momente mit in Betracht kämen; ein Erfolg würde sich aber auch, abgesehen hiervon, immer erst nach längeren Zeiträumen — entsprechend der langen Dauer der Strafen selbst — ermitteln lassen.

[3]) Von den zu Gefängnis Verurteilten ist nur eine sehr kleine Zahl schon einmal zur penal servitude verurteilt gewesen. Im Jahre 1884/5 hatte 1,7 % aller Vorbestraften bereits einmal die Strafe der penal servitude erlitten (1883/4 war die Ziffer: 1,5 %; 1882/3: 1,3 %).

etwa halb so groß ist als diejenige der Männer, übersteigt bei den mehr als zehnmal Vorbestraften die absolute Zahl der Frauen sehr erheblich diejenige der Männer. Und dieselbe Erscheinung findet sich in allen Jahren: so betrugen die mehr als zehnmal Vorbestraften

im Jahre 1884/5: bei den Männern: 10,0 %, bei den Frauen: 31,6 % aller Vorbestraften,

im Jahre 1883/4: bei den Männern: 9,4 %, bei den Frauen: 30,2 % aller Vorbestraften.

Es ist dies ein interessanter Beleg für die schon anderwärts[1]) festgestellte Zähigkeit der Weiber im Verbrechen. Überhaupt ist das Verhältnis des weiblichen Geschlechts unter den Rückfälligen ein sehr ungünstiges. Es befanden sich unter den zu einer Gefängnisstrafe verurteilten Männern:

im Jahre 1884/5: 40,3 % Vorbestrafte,
" " 1883/4: 39,6 % " ;

dagegen unter den zu einer Gefängnisstrafe verurteilten Frauen:

im Jahre 1884/5: 61,5 % Vorbestrafte,
" " 1883/4: 60,4 % " .

---

[1]) Cf. v. Oettingen, Moralstatistik Seite 527.

Abschnitt II.*)

Das heutige englische Gefängniswesen.

*) Quellen: Die jährlichen Berichte der Directors of Convict Prisons und der Commissioners of Prisons (citiert: Report of Directors resp. of Commissioners). — Der Report of the Royal Commissioners appointed 1878 to inquire into the working of the Penal Servitude Acts: (citiert Report of the Royal Commissioners 1879). — The Local Prison Rules 1885 und The Convict Prison Rules 1886 (diese beiden, lediglich als Manuskript gedruckten offiziellen Schriften sind mir durch die Liebenswürdigkeit von Sir Edm. Du Cane zugänglich gewesen). — Die beiden Bücher von Sir Edmund Du Cane.

Einleitung.

Das englische Gefängniswesen umfaßt zwei heute noch formell scharf getrennte Gebiete: nämlich einerseits die Einrichtungen zur Vollstreckung der Gefängnisstrafe, anderseits diejenigen zur Vollstreckung der Strafknechtschaft (penal servitude).

Die historische Entwickelung dieser beiden Zweige des Gefängniswesens ist im Abschnitte I in den §§ 4 und 5 gegeben worden. Es wurde dabei hervorgehoben, daß der Staat die Vollstreckung der penal servitude von vornherein selbst in die Hand nahm, während die Vollstreckung der Gefängnisstrafe ursprünglich den Lokalbehörden ganz überlassen blieb. Es wurde weiter gezeigt, welche Mißstände in den lokalen Gefängnisanstalten zu Tage traten und wie der Staat sich zum Zwecke der Beseitigung dieser Mißstände zu einem immer weiter gehenden Eingreifen in diesen Zweig der Lokalverwaltung veranlaßt sah, bis schließlich durch das Gesetz vom Jahre 1877 auch die Vollstreckung der Gefängnisstrafe ganz in die Hand des Staates gelegt wurde.

Erst seit dieser Zeit besteht eine wirkliche Einheitlichkeit im englischen Strafvollzuge. Formell ist allerdings diese Einheitlichkeit auch heute noch nicht hergestellt: auch heute noch bestehen getrennte Behörden und getrennte Einrichtungen für die Vollstreckung der Gefängnisstrafe einerseits und der penal servitude anderseits; allein diese Trennung ist heute nur noch eine formelle: es haben sich einheitliche Grundsätze sowohl hinsichtlich der Organisation wie hinsichtlich der Behandlung der Gefangenen bei beiden Strafarten Bahn gebrochen.

Unter diesen Umständen dürfte ein klareres Bild von dem heutigen englischen Gefängniswesen gewonnen werden, wenn bei der Behandlung desselben von der thatsächlich im wesentlichen vorhandenen Einheitlichkeit der Grundsätze des Strafvollzugs ausgegangen wird,[1]) als wenn die beiden Zweige des Strafvollzugs der formell noch bestehenden Trennung entsprechend abgesondert behandelt werden. Im folgenden wird daher das gesamte englische Gefängniswesen lediglich nach allgemein sachlichen Rubriken geordnet dargestellt.

Es werden dabei 3 Hauptabschnitte gemacht werden:

1) die Organisation des Gefängniswesens (Kap. I),

2) die Gefängniseinrichtungen (Kap. II),

3) die Behandlung der Gefangenen (Kap. III).

Im Kap. I soll eine Darstellung der Gefängnisbehörden, im Kap. II der Gefängnisanstalten und sonstiger Gefängniseinrichtungen gegeben werden, an welche sich eine Übersicht über die Kosten des Gefängniswesens anschließt. Es wird dabei lediglich in den einzelnen Paragraphen hervorgehoben, inwieweit eine Verschiedenheit zwischen den convict prisons, in denen die penal servitude zur Vollstreckung gelangt, und den zum Vollzuge der Gefängnisstrafe bestimmten local prisons besteht. Im Kap. III soll im Titel I die Behandlung der Gefangenen nach allgemeinen Gesichtspunkten in ihren Hauptmomenten zur Darstellung gebracht werden; in den Titeln II und III werden dann die Spezialbestimmungen über die Behandlung der zur Gefängnisstrafe Verurteilten einerseits und der zur Strafe der penal servitude Verurteilten anderseits noch einmal kurz unter Hinweis auf die im Titel I gegebenen ausführlichen Erörterungen zusammengestellt werden.

[1]) Dies Verfahren erscheint insbesondere bei einer deutschen Bearbeitung des englischen Gefängniswesens das allein zulässige, da andernfalls infolge der ständigen Wiederholung im wesentlichen gleicher Bestimmungen bei den beiden Strafarten der an sich schon spröde Stoff völlig ungenießbar zu werden droht.

Kapitel I.

Die Organisation des Gefängniswesens.

§ 1. Die Zentralisation des Gefängniswesens und die Zentralbehörden.

Das gesamte Gefängniswesen untersteht dem Staatssekretär des Innern (Home Secretary).[1]) Er hat den Etat der Gefängnisse im Parlamente zu vertreten; die Ausführungsverordnungen zu den gesetzlichen Bestimmungen und die Gefängnisregulative gehen von ihm aus; er ernennt die höheren Gefängnisbeamten und beaufsichtigt in höchster Instanz das gesamte Gefängniswesen.

Zur Bearbeitung der Gefängnisangelegenheiten besteht im Ministerium des Innern (Home Office) eine besondere Abteilung, das Prison Department. Dasselbe zerfällt in 2 Sektionen: 1) das Convict Prison Board, die Zentralbehörde für den Strafvollzug der penal servitude, 2) die Prison Commission, die Zentralbehörde für den Vollzug der Gefängnisstrafe.

1) Das Convict Prison Board wurde im Jahre 1850 eingerichtet.[2]) Bis dahin hatten die beiden damals bestehenden

[1]) Gneist I Seite 802 bezeichnet den Home Secretary treffend als zweiten Justizminister für das Kriminaldepartement, während der Lord Chancellor in beschränktem Sinne wie ein erster Justizminister für das Zivildepartement erscheine. Es sei dabei daran erinnert, daß in der Hand des Home Secretary auch das Begnadigungsrecht liegt. Cf. oben Seite 33.

[2]) 13 a. 14 Vict. c. 39.

Penitentiaries von Millbank und Pentonville unter besonderen Verwaltungskomitees (Committees of management) gestanden. Im Jahre 1850 wurden diese beiden Anstalten sowie gleichzeitig die hulks establishments [1]) — somit alle zur Vollstreckung der Transportationsstrafe in England bestimmten Anstalten — unter die einheitliche Leitung des Board gestellt. Als dann im Jahre 1853 das Strafmittel der penal servitude eingeführt wurde, erfolgte die Bestimmung, daß alle zur Vollstreckung dieser Strafe bestimmten Anstalten, die convict prisons, dem Board unterstellt sein sollten.

Das Board besteht aus einem Vorsitzenden (Chairman) und zwei bis drei Gefängnisdirektoren (Directors of Convict Prisons). [2])

Es ist die Aufgabe der Direktoren, sämtliche convict prisons periodisch zu inspizieren und bei diesen Inspektionen etwaige Beschwerden von Sträflingen gegen die Gefängnisverwaltung anzuhören und zu entscheiden, sowie das Disziplinarstrafrecht, insoweit für dasselbe nicht der einzelne Gefängnisvorsteher kompetent ist, auszuüben. Jedes convict prison soll monatlich wenigstens einmal inspiziert werden. Dabei ist die Einrichtung getroffen, daß dem einzelnen Direktor bestimmte Anstalten speziell unterstellt werden, daß jedoch daneben diese Anstalten auch von jedem der anderen Direktoren einmal im Jahre inspiziert werden. Auf diese Weise glaubt man, einerseits eine gewisse Ständigkeit in der Beaufsichtigung der einzelnen Anstalt herbeizuführen, anderseits jedem Direktor einen Einblick in die Verwaltung sämtlicher Anstalten zu gewähren. [3])

Der Chairman nimmt an den regelmäßigen Inspizierungen nicht teil, sucht sich aber durch gelegentliche Besuche der einzelnen Anstalten über den Zustand derselben zu informieren. Die Haupt-

[1]) cf. oben Seite 39 Anmerkung 1.

[2]) Die Zahl der Directors war früher regelmäßig zwei; seit einer Reihe von Jahren beträgt sie jedoch ständig drei. Der erste Chairman des Board war Colonel Sir Joshua Jebb, ihm folgte im Jahre 1863 Lieutenant-Colonel Sir E. Y. W. Henderson und diesem im Jahre 1869 der jetzige Chairman Colonel Sir Edmund Du Cane. Alle diese Herren hatten früher dem Ingenieurkorps angehört.

[3]) In dem Report of the Royal Commissioners 1879 Seite LX wird das Ziel dieser übrigens erst in den letzten Jahren getroffenen Einrichtung dahin angegeben „each prison might have the advantage of a fresh eye from time to time and the Directors themselves might have a thorough personal knowledge of all the establishments.“

aufgabe des Chairman besteht darin, die von dem Staatssekretär zu treffenden Verordnungen,[1]) Entscheidungen und Ernennungen bezüglich der convict prisons vorzubereiten.

Das Board tritt mehrere Male im Jahre unter dem Vorsitze des Chairman in London zusammen, um über den Zustand der Anstalten und etwa zu treffende neue Verordnungen zu beraten, sowie die dem Board zustehenden Ernennungen der Gefängnisunterbeamten vorzunehmen. Alljährlich hat das Board einen dem Parlamente vorzulegenden Generalbericht über den Zustand der Anstalten festzustellen.[2])

2) Die Prison Commission wurde im Jahre 1878, als der Staat die Vollstreckung der Gefängnisstrafe selbst in die Hand nahm, als Zentralinstanz für die local prisons eingesetzt.[3])

Die neugeschaffene Behörde soll aus höchstens fünf Commissioners bestehen,[4]) von denen einer als Chairman zu ernennen ist. Den Commissioners liegen im allgemeinen für die local prisons dieselben Befugnisse ob, wie den Directors bezüglich der convict prisons. Nur sind den Commissioners mit Rücksicht auf die viel größere Anzahl der local prisons besondere Beamte, die Inspectors, für die regelmäßige Besichtigung der Anstalten zur Seite gestellt.

Die Inspectors, deren Zahl zur Zeit sieben beträgt, sollen als „Augen und Ohren" (eyes and ears) der Commissioners dienen; sie sollen in den ihnen zugewiesenen Bezirken alle local prisons periodisch in bestimmten Zeiträumen besichtigen, um sich zu überzeugen, daß die Gesetze und Verordnungen innegehalten werden und daß keine Mißstände bestehen, und sie sollen über den vorgefundenen Zustand an die Commissioners berichten.[5]) Ein selbständiges

---

[1]) Das Verordnungsrecht des Staatssekretärs ist bezüglich der convict prisons ein unbeschränktes und lediglich an die Bedingung geknüpft, daß gesetzliche Bestimmungen dabei nicht verletzt werden.

[2]) Diese Generalberichte (Reports of the Directors of Convict Prisons) beginnen mit dem Jahre 1851; in denselben ist ein außerordentlich wertvolles Material zur Kenntnis des Gefängniswesens enthalten.

[3]) 40 a. 41 Vict. c. 21.

[4]) Bis jetzt ist die Zahl der Commissioners nie mehr als vier gewesen.

[5]) Die Commissioners nehmen übrigens auch in eigner Person von Zeit zu Zeit Besichtigungen von local prisons vor.

Eingreifen oder irgendwelche Entscheidung steht den Inspectors nicht zu.

Der Erlaß von Anordnungen zur Beseitigung etwaiger Mißstände wie die Entscheidung von Beschwerden der Gefangenen gebührt vielmehr ausschließlich den Commissioners. Die letzteren haben ferner, in gleicher Weise wie die Directors, die Gefängnisunterbeamten zu ernennen, über etwa erforderliche neue Gefängnisvorschriften zu beraten und ebenfalls alljährlich einen dem Parlamente vorzulegenden Bericht über den Zustand der Anstalten festzustellen. [1]) Nur in einer Richtung sind ihre Befugnisse geringere als diejenigen der Directors: die Commissioners sind nämlich nicht befugt, Disziplinarstrafen gegen Gefangene zu erkennen; vielmehr ist das höhere Disziplinarstrafrecht bei den local prisons einem Komitee von Friedensrichtern (Visiting Committee) übertragen (cf. unten § 5).

Auch die Machtbefugnisse des Staatssekretärs sind bezüglich der local prisons in sofern etwas eingeschränkt, als derselbe verpflichtet ist, die von ihm erlassenen allgemeinen Verordnungen [2]) zunächst 40 Tage auf den Tisch der beiden Häuser des Parlaments niederzulegen; die Verordnungen treten erst in Kraft, wenn im Parlamente während dieser Zeit ein Widerspruch dagegen nicht erhoben worden ist. [3])

Im übrigen ist es bei den local prisons wie bei den convict prisons der Chairman, welcher die von dem Staatssekretär zu treffenden Verordnungen, Ernennungen u. s. w. vorbereitet.

Da nun aber das Amt eines Chairman der Prison Commission dem Chairman des Board

---

[1]) Die Jahresberichte der Commissioners, von denen bis jetzt neun erschienen sind, enthalten noch detaillierte Angaben als die Berichte der Directors. Insbesondere geben dieselben wertvolles kriminalstatistisches Material.

[2]) Der Prisons Act 1865 — 28 a. 29 Vict. c. 126 — hatte dem Gesetze als Anlage eine Gefängnisordnung in 104 Artikeln einverleibt. Dieselbe besteht noch heute zu Recht. Allgemeine Abänderungen oder Zusätze zu dieser Gefängnisordnung haben in der Form von Additional Rules zu erfolgen.

[3]) Da das Parlament regelmäßig während sechs Monaten im Jahre nicht versammelt ist, so enthält diese Bestimmung eine große Erschwerung für Einführung von Verbesserungen. Du Cane I Seite 78 plaidiert daher mit Recht für eine Abänderung der Bestimmung.

of Directors übertragen worden ist, so ist faktisch diese Person der eigentliche Leiter des gesamten Gefängniswesens. Er hat die Stellung, welche in andern Zweigen der englischen Zentralverwaltung der Permanent Under-Secretary einnimmt; das heißt: er ist der ständige technische Leiter des Gefängnisressorts, während dem mit der Parlamentsmajorität wechselnden Minister die Vertretung des Ressorts im Parlamente zufällt.

Es macht einen eigentümlichen Eindruck, wenn man die sämtlichen Titel des derzeitigen Leiters des Gefängniswesens Sir Edmund Du Cane liest; derselbe ist nämlich nicht nur Chairman of the Directors of Convict Prisons und Chairman of the Prison Commissioners, sondern auch Inspector General of Military Prisons und Surveyor General of Prisons. In seiner Eigenschaft als Inspector General of Military Prisons führt er die Aufsicht über die Militärgefängnisse, wobei ihn die Directors of Convict Prisons in der Eigenschaft als Inspectors zu unterstützen haben.[1]) In seiner Eigenschaft als Surveyor General untersteht ihm das gesamte Gefängnisbauwesen, wofür ihm ein besonderer bautechnischer Rat (Surveyor) beigegeben ist.

Die gesamte Leitung des heutigen englischen Gefängniswesens ist zentralisiert in der Person von Sir Edmund Du Cane, und es ist bloß eine Frage der Zeit, wann die augenblicklich lediglich durch Personalunion der wichtigsten Ämter hergestellte Zentralisation zu einer organischen Einrichtung erhoben werden wird. Vorläufig hält man noch an dem althergebrachten Schema fest, während thatsächlich die Zentralleitung bereits hergestellt ist.

Das nachfolgende, unter Anwendung der entsprechenden preußischen Titulaturen hergestellte Schema dürfte ein klareres Bild der englischen Zentralbehörde geben.

---

[1]) Seit 40 Vict. c. 7 und c. 8 werden Militärgefangene regelmäßig in den local resp. convict prisons untergebracht, so daß die Hervorhebung der Militärgefängnisse jetzt ohne weitere Bedeutung ist.

Ministerialdirektor (Chairman of the Directors of Convict Prisons and of the Prison Commissioners, Inspector General of Military Prisons, Surveyor General of Prisons) Gehalt: 2000 £.

Vorsteher des Zentralbüreaus (Private Secretary), Gehalt: 535 £.

Abteilung I für Convict Prisons u. Militärgefängnisse: 3 vortragende Räte (Directors of Convict Prisons, Inspectors of Military Prisons) Gehalt: 800—1000 £; Büreaudirektor (Secretary): 500 £.	Abteilung II für Local Prisons: 3 vortragende Räte (Commissioners of Prisons), Gehalt: 1000 £; 7 Hilfsarbeiter (Inspectors of Prisons), Gehalt: 700—800 £; Büreaudirektor (Secretary); 600 £.	Abteilung III für technische Angelegenheiten: 1 vortragender Rat für das gesamte Gefängnisbauwesen (Surveyor of Prisons), Gehalt: 800 £; 1 vortragender Rat für das gesamte Gefängnismedizinalwesen (Medical Inspector of Prisons) Gehalt: 800 £.

Der zuletzt genannte Posten eines Medical Inspector of Prisons ist im Jahre 1879 auf Grund der Empfehlung der Royal Commissioners[1]) kreiert worden. Die Commissioners betonten die Notwendigkeit eines medizinaltechnischen Rates in der Zentralbehörde, um die gesundheitlichen Einrichtungen in den Anstalten zu überwachen, der Zentralbehörde über Fragen, welche ärztliche Kenntnisse voraussetzen, Aufschluß zu geben und der Zentralbehörde ferner bezüglich der Ernennung, Pensionierung u. s. w. von Anstaltsärzten Vorschläge zu machen.

Die sämtlichen höheren Beamten der Zentralbehörde werden auf Vorschlag des Staatssekretärs des Innern von der Königin ernannt und unterliegen den allgemeinen Bestimmungen über die permanenten Staatsbeamten (sie können insbesondere nicht in das Parlament gewählt werden). Die Büreaubeamten (Clerks), einschließlich der Secretaries, werden auf Vorschlag des Ministerialdirektors von dem Staatssekretär ernannt.

Unter den Büreaubeamten sind noch besonders hervorzuheben die beiden Rechnungsrevisoren (Store Accountants), von denen der eine der ersten, der andre der zweiten Abteilung der Zentralbehörde angehört. Denselben untersteht das gesamte Rechnungswesen der Anstalten; sie haben die von den einzelnen Anstalten einzureichenden Abrechnungen in Einnahmen

[1]) Report of the Royal Commissioners 1879 Seite XL; cf. über die günstigen Erfolge dieser Einrichtung Report of the Royal Commission on Prisons in Ireland 1884 Seite 16, 17.

und Ausgaben zu prüfen, insbesondere dabei auch auf eine richtige Einsetzung der Erträgnisse der Gefangenarbeit zu halten,[1]) und sie sollen sich durch periodische Visitationen der Anstalten davon überzeugen, daß die Magazine in guter Ordnung und die Vorräte in gutem Zustande sind.

Was die Anstellung der höheren Beamten der Zentralbehörde anbetrifft, so ist ein besonderer Vorbereitungsdienst nicht vorgeschrieben. Die überwiegende Mehrzahl derselben hat früher der Armee oder Marine angehört. Die Stellen der Commissioners und der Directors werden regelmäßig aus der Zahl der Gefängnisinspektoren und der Gefängnisvorsteher besetzt.

Es sei zum Schlusse an einem praktischen Falle ein Bild von der Thätigkeit der Zentralbehörde gegeben.

Wir nehmen an, es solle eine neue Anstalt errichtet werden. Je nachdem es sich um ein convict prison oder um ein local prison handelt, geht der Beschluß von dem Kollegium der Directors oder der Commissioners aus. Der Beschluß wird dem Staatssekretär durch den Ministerialdirektor (Chairman) vorgetragen. Genehmigt der Staatssekretär den Beschluß, so läßt der Ministerialdirektor durch den bautechnischen Rat (Surveyor) den Bauplan entwerfen, damit der Staatssekretär daraufhin im Parlamente die nötige Etatsforderung stelle. Wird der betreffende Posten angenommen, so beauftragt der Ministerialdirektor einen der vortragenden Räte, je nach Lage der Sache Director oder Commissioner, die Ausführung des Baues in die Hand zu nehmen. Derselbe hat den erforderlichen Grund und Boden zu erwerben,[2]) die Baukontrakte abzuschließen und deren Innehaltung zu überwachen. Bevor der Bau abgenommen wird, ist er von dem Surveyor zu besichtigen und darüber an den Ministerialdirektor zu berichten. Es tritt sodann das betreffende Kollegium zusammen, um

[1]) Erst seit dieser, in den letzten Jahren getroffenen Einrichtung können die Angaben über die Arbeitserträgnisse als zuverlässige und wertvolle angesehen werden. Die in den früheren Jahren über den Wert der Gefangenarbeit gemachten Angaben waren vielfach willkürliche und die einzelnen Anstalten gingen dabei von ganz verschiedenen Grundsätzen aus. Cf. über die früheren Einrichtungen einen Aufsatz von Francis Peck in der Contemporary Review vom Juli 1884 „Official Optimism“.

[2]) Der Erwerb geschieht auf den Namen des Convict Prison Board resp. der Prison Commission.

1) über die dem Staatssekretär durch den Ministerialdirektor zu machenden Vorschläge bezüglich der Ernennung der höheren Anstaltsbeamten (Vorsteher, Arzt und Geistlicher) Beschluß zu fassen, 2) die Ernennungen der übrigen Anstaltsbeamten vorzunehmen, 3) die für die neue Anstalt zu erlassende Hausordnung festzustellen. Die Hausordnungen stimmen für alle Anstalten derselben Kategorie im wesentlichen überein;[1]) es können also nur Spezialbestimmungen für den Fall, daß die neue Anstalt zur Aufnahme einer speziellen Klasse von Gefangenen bestimmt sein soll, in besondere Erwägung kommen, 4) Bestimmung darüber zu treffen, aus welchen Bezirken Sträflinge in die neue Anstalt gebracht werden sollen, eventuell ob die Anstalt nur zur Aufnahme einer bestimmten Klasse von Sträflingen dienen soll.

Nachdem so die erste Einrichtung für die neue Anstalt getroffen ist, bestimmt der Ministerialdirektor denjenigen vortragenden Rat, welchem die Aufsicht und das Dezernat über die Anstalt übertragen wird. Wenn es sich um ein local prison handelt, so muß gleichzeitig Bestimmung getroffen werden, welcher Inspektor die Anstalt besichtigen soll und wie häufig — je nach der Größe der Anstalt ist dies verschieden — die Besichtigungen vorgenommen werden sollen.

Der betreffende Rat ist nunmehr für den Zustand der Anstalt in erster Linie verantwortlich. Er hat darauf zu sehen, daß die Anstaltsbeamten die ihnen obliegenden Pflichten erfüllen. Kommen Pflichtverletzungen vor, so hat er über die event. Bestrafung oder Entlassung dem betreffenden Kollegium der Zentralbehörde Vortrag zu halten. Wenn von seiten der Gefangenen Beschwerden einlaufen, so hat er darüber zu entscheiden.[2]) Dem betreffenden Rate liegt es ferner ob, die finanziellen Verhältnisse der Anstalt zu überwachen. Alle für die Anstalt abzuschließenden Verträge hat er entweder selbst zu vollziehen oder, insoweit dem Anstaltsvorsteher ein Recht zum selbständigen Abschlusse bestimmter Verträge gegeben ist, zu kontrollieren.

---

[1]) Das Schema ist in den Local resp. Convict Prison Rules gegeben.

[2]) Wenn es sich um ein local prison handelt, so werden erforderlichen Falls zunächst von dem Inspektor Ermittelungen an Ort und Stelle angestellt und die in Betracht kommenden Personen vernommen.

§ 2. Die Anstaltsbeamten.

Die Anstaltsbeamten zerfallen in zwei Kategorieen: die höheren, deren Anstellung durch den Staatssekretär des Innern erfolgt; und die Subalternen, welche von der Zentralgefängnisbehörde ernannt werden.

Jede Anstalt muß drei höhere Beamten haben: einen Vorsteher (Governor), einen Geistlichen (Chaplain) und einen Arzt (Medical Officer oder Surgeon). In den größeren Anstalten sind zumeist noch Hilfskräfte angestellt, so neben dem Governor ein Deputy Governor, neben dem Chaplain ein Assistant Chaplain oder ein Roman Catholic Priest, neben dem Arzte ein Assistant Surgeon.

Der Governor leitet die gesamte Anstaltsverwaltung und ihm unterstehen die subalternen Beamten, doch sind die Lehrer (Schoolmasters) zunächst dem Chaplain und das in den größeren Anstalten vorhandene besondere Krankenpflegerpersonal (infirmary staff) zunächst dem Arzte untergeordnet.

Bezüglich der Anstellung aller Beamten der Gefängnisverwaltung hat, seitdem der Staat dieselbe in die Hand genommen hat, das Prinzip geherrscht, daß lediglich die persönliche Fähigkeit ausschlaggebend ist. Das Patronatswesen, welches bis vor gar nicht langer Zeit in den meisten Zweigen der englischen Staatsverwaltung allgemein bestand,[1]) hat in der staatlichen Gefängnisverwaltung niemals Eingang gefunden. Daß dies der Fall gewesen ist, ist dem energischen Einschreiten von Sir George Grey und Lord Palmerston zuzuschreiben. Der letztere erließ unter dem 10. September 1853 eine allgemeine Verfügung, in welcher ausgesprochen wurde, daß nur Verdienst und nicht Gunst („merit and not favour") bei der Anstellung und bei der Beförderung maßgebend sein sollte, und daß ein Gefängnisbeamter, welcher versuchte, durch persönliche Beziehungen einen Einfluß auszuüben, schon hierdurch als unqualifiziert für eine Beförderung zu betrachten wäre.[2])

[1]) Cf. darüber Gneist I Seite 240 ff.

[2]) Die Verfügung ist abgedruckt im Report of Commissioners of Prisons 1878 Seite 51 „any officer who may attempt to bring private interest to bear, for the purpose of influencing the Directors to promote

Als dann im Jahre 1878 die local prisons in die Staatsverwaltung übergingen, war es der erste Schritt des damaligen Staatssekretärs Sir Richard Croß, daß er die Verfügung von Lord Palmerston mit dem Bemerken bekannt machte, daß dieselbe fortab auch in den local prisons, wo bisher das Patronatswesen in hoher Blüte gestanden hatte, zur Anwendung gebracht werden würde.

Was die hiernach für die Anstellung maßgebenden Grundsätze betrifft, so gilt zunächst allgemein, daß nur solche Personen im Gefängnisdienste angestellt werden, welche ähnliche Stellungen bereits mit Erfolg bekleidet haben.[1]) Es kommt dabei vor allem in Betracht, daß die Betreffenden einerseits an Ordnung, Disziplin und strikten Gehorsam gewöhnt sind und anderseits bereits ihre Fähigkeit, mit einer großen Anzahl von Menschen fertig zu werden („dealing with large bodies of men"), dargethan haben.

Es ist einleuchtend, daß diesen Erfordernissen in erster Linie Personen entsprechen, welche der Armee oder Marine angehört haben. Und so rekrutiert sich in der That das gesamte Gefängnispersonal zum weit überwiegenden Teile aus der Armee und Marine.[2]) Die Mehrzahl der Governors besteht aus früheren Offizieren, die Mehrzahl des Subalternpersonals aus früheren Unteroffizieren. Die Governors der Public Works Prisons werden mit Rücksicht auf die von ihnen zu leitenden Gefangenarbeiten, bei denen technische Kenntnisse erforderlich sind, fast ausschließlich dem Ingenieurkorps entnommen.

Eine besondere Vorbereitungsschule für den Gefängnisdienst besteht nicht. Der frühere Leiter des irischen Gefängniswesens, Sir Walter Crofton, hat wiederholentlich den Wunsch nach Einrichtung einer solchen Schule ausgesprochen.[3]) Es ist ihm jedoch darauf immer erwidert worden, daß für eine der-

---

him, will be considered as having disqualified himself for the promotion which he may thus have sought to obtain".

[1]) „who have already filled with credit situations in which similiar qualifications are requisite." Du Cane II Seite 188.

[2]) Natürlicherweise wird nicht selten gegen eine derartige Bevorzugung Widerspruch erhoben und darüber die Zentralbehörde oder der Staatssekretär in der Presse angegriffen, cf. Report of the Royal Commissioners 1879 Seite LIII. Mit Recht bemerkt Du Cane II Seite 188, daß es aus der Natur der Sache sich ergebe, daß die Armee ein den gestellten Anforderungen entsprechendes Material in größerer Zahl liefere, als die andern Berufsstände.

[3]) Cf. Report of the Royal Commissioners 1879 Seite LIV.

artige theoretische Vorbereitung ein Bedürfnis nicht bestände, da durch die bestehenden Einrichtungen, nach welchen die Anstaltsvorsteher zunächst als Deputy Governors bei den großen, unter besonders tüchtigen Governors stehenden Anstalten beschäftigt und weiter die Subalternbeamten zunächst nur probeweise auf 3 Monate als Assistant Warders ebenfalls in den großen Anstalten angestellt werden, für eine gute Schulung der Beamten genügend gesorgt wäre.

Über die höheren Gefängnisbeamten übt die Zentralgefängnisbehörde die Disziplinargewalt aus. Bei den Subalternbeamten steht dem Gefängnisvorsteher das Recht der Auferlegung von Geldstrafen zu; im Falle gröberer Disziplinarvergehen, bei welchen auf Entlassung, Degradation oder Zurücksetzung im Avancement und bei der Gehaltserhöhung erkannt werden kann, entscheidet die Zentralbehörde.[1]) Bezüglich der letzteren Strafe, welche von allen Disziplinarstrafen am häufigsten zur Anwendung gelangt, ist zu bemerken, daß das Gehalt der Subalternbeamten alljährlich in bestimmt vorgeschriebenen Proportionen steigt; die Strafe besteht in der Einstellung dieser regelmäßigen Gehaltserhöhung („stoppage of the increment of pay") für eine, in dem Urteile festzusetzende Anzahl von Jahren.

Die Gefängnisbeamten sind nach Ablauf von zehn Dienstjahren pensionsberechtigt.[2]) Die Pension beträgt $^{10}/_{60}$ des Gehaltes und steigt mit jedem ferneren Dienstjahre um $^{1}/_{60}$ bis zur Maximalhöhe von $^{2}/_{3}$ des Gehaltes. Bei weniger als zehn Dienstjahren darf nur ausnahmsweise Pension gewährt werden; gewöhnlich wird hier dem Betreffenden eine nach der Zahl der Dienstjahre verschiedene einmalige Gratifikation (allowance) gewährt.[3])

Die Rechte und Pflichten der einzelnen Klassen von Gefängnisbeamten werden in den beiden nächsten Paragraphen, von

---

[1]) Einige Vergehen von Gefängnisbeamten sind nicht im Disziplinarwege, sondern durch die Gerichte zu bestrafen, so z. B. Beihilfe zur Flucht eines Gefangenen, Hineinschaffung von verbotenen Artikeln in die Anstalt u. s. w.

[2]) Wenn Gefängnisbeamte aus früheren Dienstverhältnissen, besonders in der Armee oder Marine, beim Eintritt in den Gefängnisdienst im Genusse von Pensionen sind, so verlieren die höheren Beamten mit ihrer Anstellung im Gefängnisdienste eo ipso diese Pensionen, während die Subalternbeamten die Pensionen weiter beziehen. Du Cane I Seite 84.

[3]) 22 Vict. c. 26. Cf. Du Cane I Seite 167.

denen § 3 die höheren, § 4 die subalternen Beamten behandelt, angegeben werden.

Eine Reihe allgemeiner Bestimmungen für Gefängnisbeamte sind von dem Staatssekretär unter dem 19. Februar 1878 zunächst für die local prisons erlassen,[1]) finden jedoch auch in den convict prisons Anwendung. Es wird hier den Beamten zur Pflicht gemacht, die Gefangenen mit Güte und Humanität („with kindness and humanity") zu behandeln, aber gleichzeitig auch streng in der Aufrechterhaltung der Ordnung und Disziplin und in der strikten Anwendung der Gefängnisordnung zu sein. Die Subalternbeamten werden ausdrücklich davor verwarnt, etwa aus Gunst oder mißverstandener Milde („through favour or mistaken notions of kindness") einen Gefangenen, welcher sich irgend einer Übertretung der Gefängnisordnung schuldig gemacht hat, nicht zur Anzeige zu bringen. Kein Beamter soll in irgendwelche geschäftliche Beziehungen („any pecuniary dealing") zu einem Gefangenen treten[2]) noch einem Gefangenen Vertraulichkeiten erlauben („nor allow any familiarity"). Der Gefangene soll niemals von einem Beamten zu dessen privaten Angelegenheiten, noch direkt oder indirekt zu irgend einem privaten Zwecke oder Nutzen verwendet werden.

Am Schlusse der hier gegebenen Darstellung der allgemeinen Verhältnisse der Anstaltsbeamten mögen noch einige statistische Notizen ihren Platz finden.[3])

Die Gesamtzahl der Anstaltsbeamten belief sich am 31. März 1886 in den 60 local prisons auf 1895, in den 12 convict prisons auf 1500. Die Anstaltsbeamten verteilen sich auf folgende Kategorieen:

---

[1]) Local Prison Rules s. 64—78.

[2]) In dieser Richtung ist noch besonders der Verkauf von Tabak oder Spirituosen an Gefangene verboten. In den local prisons wurde bis zu deren Übernahme durch den Staat ein derartiger Handel seitens der Wärter vielfach betrieben. Derselbe kam aber auch in den convict prisons vor. Es ist jetzt bestimmt, daß kein Beamter während des Dienstes Tabak oder Spirituosen bei sich führen darf. Cf. Report of the Royal Commissioners 1879 Seite LIV.

[3]) Cf. für die nachfolgenden Ziffern Report of Commissioners 1886 Seite 38, 39 und Report of Directors 1886 Seite XXVII, XXVIII; cf. auch Judicial Statistics 1884 Seite LI und LIV.

	in den local prisons	in den convict prisons	Totalsumme
a) Gefängnisvorsteher	70	22	92
b) Gefängnisgeistliche	65	22	87
c) Gefängnisärzte	67	21	88
d) Gefängnisoberbeamte	236	276	512
e) Gefängniswärter	1457	1159	2616

Zu den Gefängnisoberbeamten sind dabei gezählt: Lehrer und Lehrerinnen, Anstaltssekretäre (clerks), Gefängnisoberaufseher und Oberaufseherinnen (chief oder principal warders und matrons).[1])

Wenn man die Durchschnittsbelegung im Jahre 1885/6 zu Grunde legt, so kommt in den local prisons 1 Anstaltsbeamter auf 8,1 und 1 Gefängniswärter auf 10,5 Gefangene, in den convict prisons 1 Anstaltsbeamter auf 5,5 und 1 Gefängniswärter auf 7,2 Sträflinge.

Die Kosten für die Anstaltsbeamten inkl. Gratifikationen, Ausgaben für Uniformen u. s. w. beliefen sich im Jahre 1885/6 in den local prisons auf 208 792 £, in den convict prisons auf 163 379 £; wenn man die Durchschnittsbelegung zu Grunde legt, so betragen diese Kosten auf den Kopf des Gefangenen in den local prisons jährlich 13 £ 11 sh 7 d, in den convict prisons 19 £ 11 sh 10 d; in deutscher Währung also etwa 272 resp. 392 Mark.[2])

---

[1]) Interessant ist es, den obigen Ziffern für die local prisons diejenigen gegenüberzustellen, welche am 1. April 1878 bei Übernahme der local prisons in die Staatsverwaltung bestanden. Es gab damals

133	Gefängnisvorsteher
132	Gefängnisgeistliche
109	Gefängnisärzte
295	Gefängnisoberbeamte
1904	Gefängniswärter
zusammen 2573	Anstaltsbeamte

(cf. Report of Commissioners 1878 Seite 30). Die Zahl der Gefängnisbeamten hat sich also seit Übernahme der Anstalten in die Staatsverwaltung in sehr erheblichem Maße verringert; anderseits sind aber seit jener Zeit gewisse Kategorieen von Beamten, so insbesondere das Lehrerpersonal erheblich verstärkt worden. Es gab in den local prisons

am 1. April 1878: 93 Lehrer und Lehrerinnen
am 31. März 1886: 178 „ „ „

[2]) Die Kosten für die Anstaltsbeamten haben sich in den local prisons gegenüber dem Jahre 1878 zwar vermindert, aber diese Verminderung ist hier

§ 3. Rechte und Pflichten der höheren Anstaltsbeamten.

I. Der Vorsteher einer Anstalt führt den Titel Governor, bei den kleineren local prisons wird derselbe auch Gaoler genannt. In allen großen Anstalten — insbesondere in den convict prisons — hat der Vorsteher in dem Deputy Governor einen ständigen Stellvertreter.

Wie schon hervorgehoben wurde, ist die Stellung als Deputy Governor besonders dazu geeignet, den in den Gefängnisdienst eintretenden Personen eine praktische Schulung zu geben: der Deputy Governor hat die ihm von dem Governor übertragenen Geschäfte an dessen Stelle wahrzunehmen und hat auf diese Weise Gelegenheit, den praktischen Dienst unter sachverständiger Leitung kennen zu lernen, bevor er die selbständige Leitung einer Anstalt übertragen erhält.

In den 3 lediglich zur Aufnahme von weiblichen Gefangenen bestimmten Anstalten — dem local prison von Westminster und den convict prisons zu Fulham und Woking — führt eine Lady Superintendent die Anstaltsverwaltung. In den anderen Anstalten, in welchen sich männliche wie weibliche Gefangene befinden, steht eine Matron der Abteilung für weibliche Gefangene selbständig vor, doch verbleibt dem Governor die Oberleitung.

Der Governor — resp. die Lady Superintendent — ist für die gesamte Anstaltsverwaltung allein verantwortlich. Er muß in der Anstalt wohnen und darf ohne besonderen Urlaub während der Nachtzeit nicht aus der Anstalt abwesend sein. Wenn er sich am Tage aus der Anstalt entfernt, so hat er — soweit er nicht in dem Deputy Governor einen ständigen Vertreter hat — einen anderen Beamten mit seiner Vertretung zu beauftragen.

---

eine geringere, als bei der Zahl der Anstaltsbeamten; die Kosten betrugen 1878: 239 247 £. Dabei haben sich die Ausgaben für den Unterricht 1885 gegenüber 1878 um 2230 £, für das Ärztepersonal um 1207 £, für die sonstigen Gefängnisoberbeamten um circa 9000 £ erhöht. Cf. Du Cane II Seite 100, Report of Commissioners 1885 Seite 3/4. Das Prinzip, von welchem die Zentralverwaltung ausging, war: weniger, aber besser bezahlte und tüchtigere Beamte!

Es ist die Pflicht des Governor, täglich einmal die ganze Anstalt zu durchwandern und alle männlichen Gefangenen zu sehen; in der Abteilung für weibliche Gefangene hat dies durch die Matron zu geschehen. Einmal in der Woche hat er zu einer vorher nicht bestimmten Zeit nachts durch das ganze Gefängnis zu gehen, bei dem Besuche der weiblichen Abteilung wird er dabei von der Matron begleitet. Alle diese Besuche sind unter genauer Zeitangabe in das von dem Governor zu führende Journal einzutragen.

Der Governor hat darüber zu wachen, daß alle unter ihm stehenden Gefängnisbeamten ihren Pflichten nachkommen. Er kann über dieselben leichtere Disziplinarstrafen, wie Verweis und Geldstrafe bis zu 3 Schilling verhängen; in Vergehensfällen, deren Entscheidung der Zentralbehörde zu überlassen ist, kann er den betreffenden Beamten zunächst ohne weiteres vom Dienste suspendieren. Alle subalternen Beamten haben den Anweisungen des Governor unbedingt Folge zu leisten unter Vorbehalt eines Beschwerderechts an die Zentralbehörde.

Alle Meldungen über die Gefangenen sind an den Governor zu richten. Er übt in erster Instanz das Disziplinarstrafrecht gegen die Gefangenen aus.[1]) Wenn ein Gefangener den Governor zu sprechen wünscht, so hat er dies dem Gefängniswärter zu sagen, welcher alsbald davon Meldung zu machen verpflichtet ist. Alle von dem Governor auf die Anträge von Gefangenen hin getroffenen Entscheidungen sind in das Journal einzutragen.

Wenn ein Gefängniswärter glaubt, daß ein Gefangener körperlich oder geistig nicht wohl ist, so hat er sofort davon dem Governor Meldung zu machen; der letztere ordnet dann den Besuch des Arztes an. Der Governor soll in regelmäßigen Perioden mit dem Arzte zusammen alle hygienischen Einrichtungen der Anstalt, wie Heizung, Wasserleitung, Ventilation u. s. w., inspizieren.

Der Verkehr zwischen der einzelnen Anstalt und der Zentral-Gefängnisbehörde wird aus-

[1]) Cf. das Nähere über das Disziplinarstrafrecht und die Kompetenz des Governor unten § 23.

schließlich durch den Governor geführt. Der Governor hat alljährlich einen Generalbericht an die Zentralbehörde einzureichen. In demselben hat sich der Governor über den allgemeinen Zustand der Anstalt, sowie das Verhalten der Gefängnisbeamten zu äußern und detaillierte Angaben über die Art und Weise der Beschäftigung und den Gesundheitszustand der Gefangenen zu machen, ferner ziffermäßig die Ergebnisse des Schulunterrichts und die im Laufe des Jahres erkannten Disziplinarstrafen anzugeben. [1])

Der Governor hat eine große Anzahl von Büchern zu führen. Die wichtigsten sind das Journal, in welchem alle wesentlichen Vorkommnisse des Tages einzutragen sind; das Record Book, in welchem für jeden Gefangenen ein Blatt bestimmt ist, auf das alle bezüglich des einzelnen Gefangenen getroffenen Anordnungen zu vermerken sind, [2]) ein Punishment Book, in welchem alle Disziplinarstrafen einzutragen sind, endlich eine Reihe von Rechnungs- und Inventarisations-Büchern (Account books).

Das Gehalt der Governors variiert erheblich; in den convict prisons bezieht Kapitän Harvey [3]) das höchste Gehalt mit 752 £ und die Lady Superintendent von Fulham Prison das niedrigste mit 300 £; das Gehalt der Mehrzahl der Governors beträgt hier 650 £. In den local prisons variiert das Gehalt je nach der Größe der Anstalt zwischen 200 und 750 £. Das Gehalt der Deputy Governors beträgt in den convict prisons 300—400 £, in den local prisons 200—250 £.

II. In jedem Gefängnisse muß ein Geistlicher der englischen Hochkirche als Chaplain angestellt sein. In größeren Anstalten steht ihm ein Assistant Chaplain zur Seite. Innerhalb eines

---

[1]) Diese Berichte werden auszugsweise in einem Anhange zu den jährlichen Reports der Directors resp. Commissioners publiziert und dem Parlamente vorgelegt.

[2]) In den convict prisons ist hiermit ein Character Book verbunden zur Aufnahme von Bemerkungen über Charakter- und sonstige Eigenschaften jedes einzelnen Gefangenen.

[3]) Kapitän Harvey gilt neben Sir Edmund Du Cane augenblicklich als die größte Autorität auf dem Gebiete des Gefängniswesens in England. Ich habe diesem ausgezeichneten Manne, welchem das zur Zeit größte convict prison zu Wormwood Scrubs untersteht, wertvolle Aufschlüsse über die englischen Einrichtungen zu verdanken.

Monats ist dem Bischofe der Diözese von der erfolgten Ernennung Mitteilung zu machen, und der Geistliche darf erst nach erhaltenem Konsense amtieren. Neben dem Chaplain werden, soweit ein Bedürfnis dafür vorliegt, Geistliche anderer Konfessionen bestellt zur Wahrnehmung der Seelsorge bei denjenigen Sträflingen, welche nicht der anglikanischen Kirche angehören. [1])

Der Chaplain wohnt regelmäßig nicht in der Anstalt. Seine Pflichten bestehen in der Abhaltung des Gottesdienstes, in der Erteilung von Religionsunterricht und in der Beaufsichtigung des Schulunterrichts. Die Lehrer unterstehen, soweit Fragen des Unterrichts in Betracht kommen, ausschließlich dem Chaplain, während sie sich in ihrer Stellung als Gefängnisbeamte unter der Disziplinargewalt des Governor befinden. Dem Chaplain untersteht ferner die Einrichtung der Gefängnisbibliothek, auf welche in England ein großer Wert gelegt wird. [2])

Der Chaplain hat jeden Gefangenen, welcher nach ihm verlangt, zu besuchen; er soll ferner Gefangene, welche sich in der Verbüßung einer Disziplinarstrafe befinden, täglich wenigstens einmal aufsuchen. Im übrigen hat er den Weisungen des Governor betreffs der von ihm abzustattenden Besuche Folge zu leisten, und er soll den Governor in seinen Bemühungen, den Charakter des einzelnen Gefangenen festzustellen, nach Kräften unterstützen. [3])

Das Gehalt der Chaplains variiert je nach der Größe des Gefängnisses zwischen 150 und 450 £; der Assistant Chaplain resp. der Roman Catholic Priest oder der Jewish Minister erhält zwischen 150 und 250 £.

III. Als Gefängnisarzt darf nur angestellt werden, wer ein den Bestimmungen des Gesetzes 21 a. 22 Vict. c. 90 entsprechendes Zertifikat erhalten hat. In den großen Gefängnis-

---

[1]) Es ist hierüber ein besonderes Gesetz, der Prison's Ministers Act 1863 — 26 a. 27 Vict. c. 79 — ergangen.

[2]) Cf. Report of Commissioners 1885 Seite 11. In den local prisons sind erst in den letzten Jahren überall Gefängnisbibliotheken eingerichtet worden; bei Übernahme der local prisons in die Staatsverwaltung im Jahre 1878 waren viele Anstalten noch ohne Bibliotheken.

[3]) Auch sonst wird dem Chaplain ein einträchtiges Zusammenwirken mit dem Governor zur Pflicht gemacht.

anstalten wohnt der Arzt in der Anstalt selbst und darf alsdann Privatpraxis nicht annehmen.

Es ist die Pflicht des Arztes, jeden Gefangenen bei der Aufnahme in die Anstalt auf seinen Gesundheitszustand zu untersuchen und darüber schriftlich an den Governor zu berichten. Er hat täglich die in ärztlicher Behandlung befindlichen Gefangenen zu besuchen;[1]) alle übrigen Gefangenen hat er wöchentlich wenigstens einmal zu sehen. Wenn er dabei die Beobachtung macht, daß ein Gefangener, sei es infolge der von ihm verlangten Arbeit oder der strengen Gefängnisdisziplin oder aus anderen Gründen sich körperlich oder geistig leidend befindet, so hat er davon sofort dem Governor ausführlichen schriftlichen Bericht zu erstatten. Er kann dabei auch die Gewährung einer Extradiät in Vorschlag bringen.

Die verschriebenen Arzneimittel und sonstigen ärztlichen Anordnungen sind in ein besonderes Buch einzutragen, und jeder einzelne Krankheitsfall ist darin besonders zu vermerken (Medical Register).

Jeden Monat hat der Gefängnisarzt dem Governor einen Gesamtbericht über den Gesundheitszustand in der Anstalt, die Zahl der vorgekommenen Erkrankungen, Todesfälle u. s. w. einzureichen.

Wenn ein Gefangener körperlich gezüchtigt werden soll, so wird der Gefängnisarzt zuvor gefragt, ob der Gesundheitszustand des Betreffenden die Züchtigung zuläßt. In gleicher Weise wird von dem Arzte Auskunft erfordert, bevor ein Gefangener zu harter Arbeit gesetzt wird. Gefangene, welche sich unter Disziplinarstrafe befinden, sind von dem Arzte täglich zu besuchen.

Wenn bei dem Gesundheitszustande eines Gefangenen dessen Verbleiben in der Anstalt nicht angezeigt erscheint, so hat der Arzt darüber ausführlich an den Governor zu berichten.[2]) Besteht

---

[1]) In den größeren Anstalten sind für die Pflege der Erkrankten besondere Hospital Warders angestellt, welche den vom Arzte bezügl. der Krankenpflege getroffenen Anordnungen Folge zu leisten haben.

[2]) Der Staatssekretär hat in dieser Richtung im Dezember 1884 eine besondere Verfügung erlassen, in welcher ausgesprochen wird „that a prison is a place for penal discipline and that it is not under ordinary circumstances the proper scene for a death-bed and ought not to be converted into a hospital for incurables". Cf. Report of Directors 1885 S. IX.

Todesgefahr, so ist zu gleicher Zeit der Chaplain davon in Kenntnis zu setzen. Bei schwereren Krankheitsfällen kann der Gefängnisarzt nach vorgängigem Berichte an den Governor einen zweiten Arzt zuziehen.

Der Gefängnisarzt hat ferner periodisch die ganze Anstalt mit Rücksicht auf die Vorkehrungen in gesundheitlicher Richtung, Heizung, Ventilation u. s. w. zu besichtigen und darüber dem Governor zu berichten, [1]) auch sonst Vorschläge zu machen, wenn er aus hygieinischen Rücksichten irgendwelche Änderungen in der Anstaltsverwaltung für wünschenswert hält.

Das Gehalt der Gefängnisärzte [2]) beträgt zwischen 200 und 500 £; wo neben dem Medical Officer noch ein Assistant Surgeon angestellt ist, erhält der letztere 200—300 £.

§ 4. Rechte und Pflichten der subalternen Anstaltsbeamten.

Die subalternen Beamten zerfallen in die Gefängnisoberbeamten und die Gefängniswärter (Warders).

Zu den Gefängnisoberbeamten sind zu zählen: die Gefängnisaufseher und Oberaufseher (Principal resp. Chief Warders, Matrons), die Anstaltsverwaltungsbeamten (Clerks,

---

[1]) Für diese Sanitary Reports, welche in jedem Vierteljahre mindestens einmal zu erstatten sind, ist von der Zentralbehörde ein bestimmtes Formular mit 8 Fragen vorgeschrieben. Die Fragen sind sehr detailliert und betreffen u. a. die Temperatur in den Zellen, die Beschaffenheit des Wassers und der Gefangenenkost, die allgemeine Reinlichkeit in der Anstalt, den Zustand der water closets u. s. w. Ein derartiges Formular findet sich abgedruckt bei Du Cane I Seite 169. Auch für die mindestens einmal monatlich zu erstattenden Berichte über den Gesundheitszustand der Gefangenen ist ein Formular vorgeschrieben. Du Cane I Seite 168.

[2]) Die Gefängnisärzte haben ferner die Pflicht, alle sich zum Eintritt in den Gefängnisdienst meldenden Personen zu untersuchen und ein ärztliches Attest über den Gesundheitszustand der Betreffenden auszustellen. Regelmäßig liegt es den Gefängnisärzten auch ob, die Gefängnisbeamten — zuweilen auch deren Familien — unentgeltlich ärztlich zu behandeln. Bei den kleineren Gefängnissen haben sie auch die Herstellung der Arzneimittel selbst zu übernehmen, während bei den größeren Anstalten hierfür ein besonderer Apotheker angestellt ist.

Stewards, Storekeepers, Foreman of Works) und die Lehrer (Schoolmasters, Schoolmistresses).

Die Zahl dieser Oberbeamten ist naturgemäß je nach der Größe der Anstalt eine verschiedene. In den kleineren Anstalten sind mehrere Ämter in einer Person vereinigt. Ja, in den beiden kleinsten local prisons ist überhaupt ein Oberbeamter nicht angestellt; der gaoler hat hier alle diesbezüglichen Pflichten selbst zu übernehmen. In allen andern Anstalten ist wenigstens ein Lehrer angestellt, welchem dann event. die Funktionen eines Clerk und Storekeeper mitübertragen werden. [1])

Ausdrücklich ist es untersagt, einen Gefangenen mit der Wahrnehmung irgend eines Amtes in der Anstalt zu beauftragen; es soll damit den früher in den local prisons häufigen Mißbräuchen vorgebeugt werden. Es ist ferner ausdrücklich vorgeschrieben, daß in jedem Gefängnisse, in welches weibliche Gefangene eingeliefert werden, eine Gefangenaufseherin (Matron) oder wenigstens eine Gefangenwärterin (Female Warder) angestellt sein muß.

Die Pflichten der Gefängnisoberbeamten ergeben sich zumeist schon aus ihrem Amtscharakter. Dem Chief Warder untersteht unter dem Governor der innere Dienst in der Anstalt; er vertritt den Governor, wenn derselbe abwesend ist; er erstattet demselben täglichen Rapport; er beaufsichtigt die Gefängniswärter; alle Meldungen der Gefängniswärter gehen durch ihn an den Governor. In den großen Anstalten stehen zwischen dem Chief Warder und den Gefängniswärtern noch die Principal Warders, welchen die Leitung des inneren Dienstes in den einzelnen Gefängnisabteilungen oder Gefängnisflügeln übertragen ist. Soweit Frauen diese Stellungen wahrnehmen, führen sie den Titel „Matron“.

Dem Clerk liegt die gesamte Buch- und Rechnungsführung in der Anstalt ob. In den größeren Anstalten steht ihm dabei ein Storekeeper zur Seite, welcher die Aufsicht über alle Vorräte der Anstalt, den Einkauf der Lebensmittel u. s. w. führt. In den großen convict prisons gibt es mehrere Storekeepers, neben denen

[1]) Cf. die Zusammenstellung der Beamten in den local prisons: Report of Commissioners 1886 Seite 38 und in den convict prisons: Report of Directors 1886 Seite XXVII.

dann ein Steward als Oberaufseher über das gesamte Proviantierungs- und Magazin-Wesen der Anstalt angestellt ist.

Es sei dabei bemerkt, daß die gesamte Rechnungsführung in den Anstalten eine sehr sorgfältige und eingehende ist und bis in das kleinste Detail hinein von dem in der Zentralbehörde angestellten Store Accountant (cf. oben Seite 138) kontrolliert wird. [1])

In den convict prisons, wo die Gefangenenarbeiten einen großen Umfang einnehmen, ist ferner noch ein besonderer Foreman of Works angestellt, welcher die Beschäftigung der Sträflinge zu beaufsichtigen und darüber in der genau vorgeschriebenen Weise Abrechnung zu liefern hat. In den andern Anstalten fungieren für die Beaufsichtigung der Gefangenen in den einzelnen Zweigen der Beschäftigung bestimmte Gefängniswärter als Trade Warders. [2])

Die Lehrer endlich haben den Unterricht in der Anstalt zu erteilen, worüber unten § 20 noch näher gehandelt werden wird. Es sind ihnen ferner unter Leitung des Chaplain die Bibliotheksgeschäfte übertragen.

Die Gefängnisoberbeamten dürfen bei ihrer Anstellung nicht unter 25 und nicht über 40 Jahre alt sein. Für jede Klasse derselben wird eine bestimmte Bildungsstufe vorausgesetzt, über welche sich die Kandidaten durch ein Attest der in London bestehenden Civil Service Commission auszuweisen haben. [3])

Über die Gehaltsverhältnisse der Gefängnisoberbeamten sei folgendes bemerkt: die Chief Warders erhalten jährlich zwischen 100 und 150 £, die Principal Warders zwischen 85 und 90 £, die Clerks zwischen 80 und 190 £, die Storekeepers zwischen 200 und 300 £, die Stewards zwischen 225 und 325 £. [4])

---

[1]) Die Einführung eines geordneten und gleichmäßigen Rechnungswesens war eine der ersten Maßregeln bei der Übernahme der local prisons in die Staatsverwaltung. Cf. Report of Commissioners 1878 Seite 12 u. Seite 48 ff.

[2]) Das Nähere über die bezüglich der Beschäftigung der Gefangenen getroffenen Bestimmungen cf. unten § 21 ff.

[3]) Eine genaue Zusammenstellung der Vorbedingungen für die Anstellung ist abgedruckt bei Du Cane I Seite 164 ff.

[4]) Das Finanzministerium hat unter dem 5. Juli 1878 eine Skala der Gehälter der subalternen Gefängnißbeamten aufgestellt, welcher die obigen Ziffern entnommen sind. Die Anstalten werden dabei nach der Durchschnittsbelegung in Klassen eingetheilt, für welche Minimal- und Maximal-

Was nun die Gefängniswärter betrifft, so zerfallen dieselben in zwei Klassen: die Warders und die Assistant Warders. Jeder Gefängniswärter wird zunächst probeweise auf drei Monate als Assistant Warder angestellt; während dieser Zeit kann er mit einwöchentlicher Kündigung entlassen werden. Auch nach seiner definitiven Übernahme in den Gefängnisdienst hat er noch geraume Zeit in einer größeren Anstalt als Assistant Warder zu fungieren, ehe er zum Warder befördert wird.

Den Warders wird eine bestimmte Anzahl von Gefangenen zur ausschließlichen Beaufsichtigung übergeben, über welche dieselben an jedem Morgen an ihren nächsten Vorgesetzten (Principal oder Chief Warder) Rapport zu erstatten haben. Einige Warders, welche persönlich in einem bestimmten Handwerke bewandert sind, werden als Trade Warders mit dem gewerblichen Unterrichte und der Arbeitsbeaufsichtigung der in dem betreffenden Handwerke beschäftigten Gefangenen beauftragt. Von der Mehrzahl der Warders werden bei ihrer Anstellung genügende praktische Vorkenntnisse in einem bestimmten Handwerke verlangt. [1])

Im übrigen sind die Vorbedingungen der Anstellung folgende: der Kandidat muß zwischen 24 und 42 Jahre alt sein, er muß lesen, schreiben und rechnen können und er muß sich endlich in gutem Gesundheitszustande befinden. Bei der Meldung ist ein bestimmtes Formular, in welchem eingehende Fragen über das Vorleben gestellt sind, eigenhändig auszufüllen [2]) und dabei zwei Referenzen anzugeben, wenn thunlich von früheren Arbeitgebern. Der Kandidat wird sodann von einem Gefängnisarzte auf seinen Gesundheitszustand untersucht, wobei der Kandidat wiederum auf bestimmt vorgeschriebene Fragen eine detaillierte schriftliche Auskunft über seinen Gesundheitszustand, etwaige frühere Krankheiten, ja sogar über den Gesundheits-

---

Gehaltssätze aufgestellt werden, cf. Report of Commissioners 1878 Seite 30 ff. Für die Lehrer fehlen wohl mit Rücksicht auf die ganz verschiedene Geschäftsbelastung derselben in den einzelnen Anstalten bestimmte allgemeine Sätze.

[1]) In neuerer Zeit werden ferner in den größeren Anstalten besondere Hospital Warders angestellt, bei welchen eine Kenntnis der Krankenpflege vorausgesetzt wird.

[2]) Ein derartiges Formular ist abgedruckt bei Du Cane I Seite 159 ff. Darin befinden sich u. a. Fragen nach der bisherigen Beschäftigung, dem Grunde der Entlassung, dem bisher bezogenen Gehalte u. s. w.

zustand seiner Eltern resp. die Ursache ihres Todes zu geben hat.[1]) Der Arzt hat darauf einen genauen Bericht über den Gesundheitszustand des Betreffenden einzureichen und zwölf in dieser Richtung vorgeschriebene Fragen zu beantworten.[2])

Erst nachdem auf diese Weise festgestellt ist, daß der Kandidat nach seinem Gesundheitszustande für den Gefängnisdienst geeignet ist, wird von der in London bestehenden Civil Service Commission an die aufgegebenen Referenzen sowie an sonstige zur Auskunftserteilung geeignet erscheinende Persönlichkeiten ein Fragebogen zur Ausfüllung mit dem Bemerken gesandt, daß die Auskunft auf Wunsch als streng vertraulich behandelt werde. Der Fragebogen schließt nach zwölf detaillierten Fragen über die Charaktereigenschaften des Kandidaten folgendermaßen: „Würden Sie Selbst dem Kandidaten eine Stellung anvertrauen, welche unzweifelhafte Ehrenhaftigkeit erfordert, und würden Sie denselben für eine derartige Stellung Ihren persönlichen Freunden empfehlen?“[3])

Erst wenn in dieser Weise nach allen Seiten hin befriedigende Auskunft über den Kandidaten eingegangen ist,[4]) erfolgt dessen probeweise Anstellung im Gefängnisdienste. Man wird in der That zugeben müssen, daß bei der Anstellung mit der allergrößten Umsicht vorgegangen wird.

In England besteht die allgemeine Überzeugung, daß ein gutes Beamtenpersonal die erste und wichtigste Bedingung für einen wirksamen Strafvollzug ist, daß auch die besten Einrichtungen und Verordnungen und die strengste

[1]) Cf. Du Cane I Seite 162, wo ein Formular abgedruckt ist. Es wird im einzelnen gefragt, ob der Kandidat bestimmte Krankheiten (Rheumatismus, Asthma, Herzkrankheit u. s. w. jemals gehabt hat, ob er oder irgendwelche Verwandte von ihm jemals an wiederum genau angegebenen Krankheiten (Auszehrung, Geisteskrankheit, Epilepsie u. s. w.) gelitten haben? wie alt seine Eltern geworden sind u. s. w.

[2]) Dabei ist speziell die Frage zu beantworten, ob der Gesundheitszustand ein solcher ist, um den betreffenden geeignet erscheinen zu lassen für ein Amt, mit welchem fortdauernde Beschwerlichkeiten und Exponierung bei jeglicher Witterung verbunden sind (involving constant fatigue and exposure to weather).

[3]) Du Cane I Seite 166, 167. Auch hier kehrt neben den Fragen nach der Ehrlichkeit, Nüchternheit, Fleiß, Intelligenz u. s. w. die Frage nach dem Gesundheitszustande wieder.

[4]) Bei all den angeführten Formularen ist die Fragestellung eine ebenso eingehende wie präzise. Die Aufstellung guter Fragebogen ist überhaupt eine den Engländern speziell eigne Fertigkeit.

Aufsicht wirkungslos bleibt, wenn nicht die Behandlung der Gefangenen in die richtigen Hände gelegt wird.[1]) Von diesem Standpunkte ist es von der allergrößten Wichtigkeit, Vorsorge zu treffen für ein tüchtiges Wärterpersonal, aus welchem sich dann auch regelmäßig die Klasse der Aufseher rekrutiert.

Und wer englische Gefängnisse besichtigt hat, wird sich davon haben überzeugen können, daß das angestrebte Ziel vollauf erreicht ist. Es ist in der That merkwürdig, daß es trotz der hohen Anforderungen und trotz der strengen Bedingungen bei der Anstellung niemals an der genügenden Zahl geeigneter Bewerber fehlt.[2]) Wie schon hervorgehoben wurde, liefert dabei das Heer und die Marine das größte Kontingent.

Die Gefängniswärter beginnen als Assistant Warders mit einem jährlichen Gehalte von 60 £, welches sich alljährlich um 1 £ bis auf 65 £ erhöht. Das Anfangsgehalt der Warders ist 70 £; auch hier tritt eine jährliche Steigerung um 1 £ bis zur Maximalhöhe von 75 £ ein,[3]) daran schließen sich dann die oben angegebenen Sätze von 85—90 £ bei den Principal Warders und schließlich von 100—150 £ bei den Chief Warders an. Tüchtigen Wärtern ist somit die Gelegenheit zu einem guten Avancement gegeben. Bei allen Sätzen ist die Uniform, welche diesen Beamten von der Gefängnisverwaltung geliefert wird, inbegriffen.

Außerdem erhalten die Mehrzahl von Wärtern, soweit dieselben in den Gefängnissen wohnen, für ihre Person freie Dienstwohnung inkl. Licht und Feuerung. In neuerer Zeit hat man damit begonnen, in unmittelbarer Nähe der Anstalten Dienstwohnungen auch für die verheirateten Beamten herzustellen. Nach allgemeinem Urteile hat sich diese Einrichtung in jeglicher Beziehung

---

[1]) Cf. Du Cane II Seite 187 „forming a good staff of officers — a matter of the first consequence.“

[2]) Doch gilt dies nur für England und Schottland, nicht aber für Irland, wo die Gefängnisverwaltung Schwierigkeiten hat, ein genügendes Personal zu finden, cf. Report of the Royal Commission on Prisons in Ireland 1884 Seite 19 ff.

[3]) Bei den weiblichen Wärtern sind die Gehaltssätze niedriger; sie betragen ür die Warders 55—70 £, für die Assistant Warders 45—50 £ Cf. Report of Commissi oners 1878 Seite 31.

vorzüglich bewährt[1]) und dürfte in absehbarer Zeit bei allen Anstalten zur Durchführung gelangen.

§ 5. Die Kontrolle über die Anstaltsverwaltung durch außerhalb derselben stehende Personen, insbesondere das Visiting Committee.

Als sich im Jahre 1877 der Staat zu dem eingreifenden Schritte entschloß, die Verwaltung der local prisons durch die Lokalbehörden aufzuheben und in die eigne Hand zu nehmen, waren vielfach Bedenken laut geworden, ob es dem Staate möglich sein würde, von einer Zentralstelle aus die Verwaltung der zahlreichen und so weit zerstreut liegenden local prisons zu überwachen.

Zwar hat die Zentralbehörde in den Inspektoren ihre „Augen und Ohren", allein dieselben besuchen doch nur periodisch die einzelne Anstalt, und da ein erheblicher Teil der Gefangenen die local prisons schon nach kurzer Zeit wieder verläßt, so kann es vorkommen, daß Gefangene von seiten der Gefängnisbeamten eine unbillige Behandlung erfahren, ohne die Gelegenheit zu haben, sich darüber bei den Inspektoren zu beklagen.

Diese Überlegung führte dazu, in dem Gesetze von 1877 neben den Inspektoren noch andern Personen eine gewisse Kontrolle zu übertragen. Das Gesetz von 1877 bestimmte, daß für jedes local prison aus der Zahl der Friedensrichter des betreffenden Bezirkes ein Visiting Committee gewählt werden sollte.[2]) Über die Wahl selbst wurde dem Staatssekretär des Innern der Erlaß der näheren Anordnungen übertragen. Es sind hierüber unter dem 19. Februar 1878, 16. Juni 1879, 23. Juni 1880 eine Anzahl von Verordnungen ergangen.[3]) Der Staatssekretär hat in den-

[1]) Cf. hierüber Report of the Royal Commission on Prisons in Ireland 1884 Seite 20 und Du Cane II Seite 97. Es wird hier ausgeführt, daß besonders in den großen Städten die Errichtung von Beamtenwohnungen in der Nähe der Anstalten wünschenswert sei, weil sonst bei den hohen Mietspreisen die Wärter in Berührung mit einer Klasse der Bevölkerung kämen, von welcher man dieselben möglichst fern zu halten suchen müßte.

[2]) 40 a. 41. Vict. c. 21 s. 13—15.

[3]) Cf. Local Prison Rules Seite 9—16.

selben für jede einzelne Anstalt genau bezeichnet, welche Anzahl von Mitgliedern jedes einzelne Kollegium von Friedensrichtern des Bezirkes zu dem Visiting Committee zu entsenden habe. Die Wahl erfolgt stets auf ein Jahr. Doch findet in der Praxis regelmäßig eine Wiederwahl derselben Personen statt. Die Zahl der Mitglieder ist bei den meisten Visiting Committees: 12. Dieselben bestimmen unter sich, in welcher Reihenfolge sie die Anstalt besuchen wollen.

Das Visiting Committee hat die Pflicht darüber zu wachen, daß die Anstaltsverwaltung den gesetzlichen und reglementarischen Bestimmungen gemäß geführt werde und daß die Gefangenen nicht unnötigerweise mit Härte oder gar Grausamkeit behandelt werden. Zu diesem Zwecke steht jedem Mitgliede des Committee zu jeder Zeit der freie Zutritt zu der Anstalt, der freie Verkehr mit den Gefangenen und die freie Einsicht aller Anstaltsbücher zu. Die Gefangenen werden durch einen in jeder Zelle angebrachten Anschlag ausdrücklich darauf aufmerksam gemacht, daß sie etwaige Beschwerden dem Visiting Committee vortragen können.

Zu diesem Zwecke wird der Name des betreffenden Gefangenen in ein in der Anstalt ausliegendes Minute Book eingetragen. Dasjenige Mitglied des Committee, welches dann zunächst die Anstalt besichtigt, hört den Gefangenen an, wobei auf Wunsch desselben die Gegenwart von Gefängnisbeamten ausgeschlossen wird.[1]) Es können auch andre Personen über den Fall vernommen werden. Das Mitglied des Committee trägt schließlich seine Ansicht über den Fall sowie etwaige Vorschläge zur Abhilfe in das Minute Book ein, welches dem Inspektor jedesmal bei seinem Besuche vorzulegen ist.

In gleicher Weise erfolgen Eintragungen in das Minute Book, wenn ein Mitglied des Visiting Committee von selbst auf irgendwelche Mißstände in der Anstalt aufmerksam wird. In der von dem Staatssekretär über die Thätigkeit der Visiting Committees erlassenen Instruktion wird den letzteren ein einträchtiges Zusammenwirken mit den Inspektoren besonders an

---

[1]) Mit Recht tadelt Du Cane II Seite 72 diese Bestimmung, welche leicht dahin führen könne, daß der Ruf eines Beamten durch ungerechtfertigte, aber ohne Anhören des Betreffenden für wahr gehaltene Anschuldigungen eines Gefangenen ernstlich gefährdet werde.

das Herz gelegt. Und es besteht auch in der That zwischen diesen beiden Kontrollbehörden allgemein das beste Einvernehmen.

Dem Visiting Committee steht nun aber weiter die Befugnis zu, anstatt die vorgefundenen Mißstände dem ebenfalls nur kontrollierenden Inspektor mitzuteilen, direkt die Entscheidung der Zentralbehörde oder des Staatssekretärs anzurufen. Hierzu ist nicht nur das Committee als solches, sondern jedes einzelne Mitglied desselben berechtigt.

Das Visiting Committee soll alljährlich dem Staatssekretär einen allgemeinen Bericht über den Zustand der betreffenden Anstalt einreichen und dabei etwa wünschenswerte Verbesserungsvorschläge machen, welchen von seiten der Zentralbehörde die eingehendste Beachtung zu teil wird.

Neben dieser kontrollierenden Thätigkeit steht dem Visiting Committee nun aber ferner noch eine Disziplinarstrafgewalt gegen die Gefangenen zu. In den local prisons kann der Governor nur auf die geringeren Disziplinarstrafen erkennen; hält er für ein Vergehen eine höhere Strafe für angezeigt, so hat er sich an ein Mitglied des Visiting Committee zu wenden; für einzelne Disziplinarstrafen müssen sogar zwei Mitglieder des Committee zusammenwirken (cf. das Nähere unten § 23).

Endlich hat jedes Mitglied des Visiting Committee das Recht auf Antrag eines Untersuchungsgefangenen, eines Schuldgefangenen oder eines first class misdemeanant (cf. unten § 11) demselben bestimmte Vergünstigungen insbesondere rücksichtlich der Kost, der Bewegung im Freien, der Einrichtung der Zelle u. s. w. zu gewähren.[1])

Außer dem Visiting Committee gewährt das Gesetz von 1877 noch ferner jedem Friedensrichter des Bezirkes die Befugnis,[2]) das betreffende local prison zu besichtigen und etwaige, ihm dabei entgegentretende Mißstände in der Anstalts-

---

[1]) Es soll hierdurch die Möglichkeit gegeben werden, bei diesen Klassen von Gefangenen — welche sämtlich nicht als eigentliche Strafgefangene angesehen werden — der Stellung und den Lebensverhältnissen der Betreffenden in angemessener Weise Rücksicht zu tragen.

[2]) 40 a. 41 Vict. c. 21 s. 15; bis 1877 waren die Befugnisse der Friedensrichter gegenüber den local prisons viel weitergehende, cf. 28 a. 29 Vict. c. 126 s. 55.

verwaltung oder in der Behandlung der Gefangenen in dem, in jeder Anstalt aufzulegenden Visitor's book zu vermerken.

Nachdem so für die local prisons eine umfassende Kontrolle durch außerhalb der Verwaltung stehende Personen hergestellt worden war, und nachdem sich gezeigt hatte, daß durch diese Maßregel die Autorität der für die Gefängnisverwaltung zunächst verantwortlichen Beamten in keinerlei Weise beeinträchtigt, und daß anderseits das Vertrauen des Publikums in eine geordnete Anstaltsverwaltung durch diese Teilnahme der Friedensrichter erheblich gestärkt wurde, machten die Royal Commissioners von 1879 den Vorschlag, eine ähnliche Einrichtung auch in den convict prisons einzuführen.[1]) Die Zentralbehörde ging sofort darauf ein, indem sie für jedes convict prison eine Anzahl Visitors zumeist aus der Klasse der Friedensrichter ernannte.

Die regelmäßig für ein Jahr ernannten Visitors haben in gleicher Weise, wie die Mitglieder eines Visiting Committee bei den local prisons, das Recht,[2]) jederzeit die Anstalt zu besichtigen, sich über alle Einzelheiten der Verwaltung zu informieren, Beschwerden von Sträflingen anzuhören und ihre Beobachtungen und Verbesserungsvorschläge in ein für diesen Zweck ausliegendes Buch einzutragen oder auch direkt der Zentralbehörde mitzuteilen.[3]) Auf der andern Seite dürfen sie, ebensowenig wie die Mitglieder eines Visiting Committee, selbständig irgendwelche Anordnungen treffen. Ihre Aufgabe ist im wesentlichen eine öffentliche

[1]) Report of the Royal Commissioners 1879 Seite LXI, LXII: „it seems to us inconsistant that such jealous precautions should have been taken as regards the local prisons in which prisoners are only confined for comparatively short sentences, and that no similiar safeguard should be provided in the case of the convict prisons, in which prisoners are confined for long periods".

[2]) In den an die zuerst ernannten Visitors im Jahre 1879 von der Zentralbehörde gerichteten Zirkularbriefen werden die Berechtigungen der Visitors des nähern angegeben.

[3]) Ein Disziplinarstrafrecht ist den Visitors nicht gegeben. Bei den convict prisons, in denen alle Insassen lange Zeit zu verweilen haben, war hierfür ein Bedürfnis nicht vorhanden. Die höhere Disziplinarstrafgewalt wird hier von der Zentralbehörde, den Directors, ausgeübt.

Kontrolle der Anstaltsverwaltung herzustellen und dabei zugleich der von dem Sitze der Anstaltsverwaltung entfernten Zentralbehörde als lokalkundige Ratgeber zu dienen.

Augenblicklich besteht eine, besonders eifrig von der Howard Association betriebene Bewegung,[1]) diese öffentliche Kontrolle noch weiter auszudehnen. Man behauptet, die Zulassung einer so beschränkten Anzahl von Personen in die Anstalten sei nicht genügend. Es müsse jedem, welcher Interesse für die Besserung der Gefangenen habe, gestattet sein, die Anstalten zu besuchen und dort mit den Gefangenen zu verkehren.

Mit Recht lehnt die Zentralbehörde dies als mit der Aufrechterhaltung der Disziplin in den Anstalten unvereinbar ab. Es ist etwas ganz anderes, ob einer Anzahl sachverständiger Männer, wie es die Friedensrichter sind, eine derartige Befugnis gegeben wird, oder jedem beliebigen gutherzigen — und vielleicht auch schwachherzigen — Menschen.[2]) Einzelnen Personen, wie insbesondere den Vorständen der Vereine für entlassene Sträflinge, sowie den Mitgliedern gewisser religiöser Orden, gestattet die Zentralbehörde in durchaus liberaler Weise den Besuch der Anstalten und den Verkehr mit den Gefangenen.[3])

[1]) Cf. Reports of the Howard Association 1885 Seite 9, 10 und 1886 Seite 4, 5.

[2]) Schon vor den Royal Commissioners haben Sir Edm. Du Cane und Dr. Guy auf die Gefahren hingewiesen, welche die Gewährung eines Besichtigungsrechtes an Personen, welche der Verwaltung fern stehen, mit sich bringen muß, wenn nicht bei der Auswahl dieser Personen die größte Vorsicht angewandt wird. Cf. Report of the Royal Commissioners 1879 Seite LXI und Seite LXV, LXVI (Memorandum von Dr. Guy, welcher sich auch gegen die Einführung von Visitors erklärte).

[3]) Bei den Vorständen der Vereine für entlassene Sträflinge erscheint diese Vergünstigung in der That gerechtfertigt und wünschenswert, damit dieselben in die Lage gesetzt werden, rechtzeitig in Verkehr mit den zur Entlassung Kommenden zu treten und von deren Eigenschaften, Fähigkeiten und Wünschen ein richtiges Bild zu erhalten.

Kapitel II.

Die Gefängniseinrichtungen.

§ 6. Übersicht über die Gefängnisanstalten.

Es gibt zur Zeit in England 60 local prisons und 12 convict prisons.

Betrachten wir zunächst die local prisons! Die Zahl derselben hat seit der Übernahme dieser Anstalten in die Staatsverwaltung im Jahre 1878 erheblich abgenommen: im Jahre 1878 gab es 113 Anstalten zur Verbüßung der Gefängnisstrafe, [1]) also 53 mehr als jetzt. Die Zentralbehörde hatte es von vornherein als ihr Ziel bezeichnet, eine größere Zentralisation der Anstalten herbeizuführen, da nur auf diese Weise eine rationelle Behandlung der Gefangenen unter gleichzeitiger Kostenersparnis erreicht werden könnte. Man ist Schritt für Schritt auf diesem Wege vorwärts gegangen, indem man die Schließung der kleinen Anstalten erst eintreten ließ, nachdem durch Neubauten bei den größeren Anstalten für die Unterbringung der Gefangenen genügend gesorgt war. Man ist so auch heute noch nicht vollständig aus dem Übergangsstadium heraus; es gibt auch heute noch eine allerdings nur kleine Anzahl von Anstalten, bei denen die Tage des Fortbestehens gezählt sind.

Der letzte Jahresbericht der Zentralbehörde führt noch 61 local

---

[1]) Im Jahre 1837 betrug die Zahl sogar 256, sie war jedoch schon 1864 auf 146 gefallen. Cf. Report of Commissioners 1878 Seite 38.

prisons auf,[1]) bei denen sich die Durchschnittsbelegung folgendermaßen stellte:

unter		50	Gefangene	: 5	Anstalten
zwischen	50 und	100	"	: 12	"
"	100 "	200	"	: 21	"
"	200 "	500	"	: 15	"
"	500 "	1000	"	: 6	"
über		1000	"	: 2	"

Anstalten unter 50 Gefangene Durchschnittsbelegung wird es voraussichtlich schon sehr bald nicht mehr geben. Die allgemeine Meinung ist aber auch gegen das Fortbestehen von Anstalten unter 100 Gefangene.[2]) Eine Durchschnittsbelegung von 100 bis 500 Gefangene wird — von den großen Städten (London, Liverpool, Manchester), wo sich die Ziffern noch höher stellen — als das Normale angesehen.

Daß eine derartige Konzentration der Gefangenen in England möglich ist, dürfte auf mehrere Momente zurückzuführen sein. Es kommt zunächst in Betracht, daß die Zahl der Gerichtssitze eine relativ kleine ist, sodann weiter, daß — ebenfalls relativ — nur selten auf ganz kurze Gefängnisstrafen erkannt wird (cf. oben Seite 94). Endlich ermöglicht aber auch die Entwickelung des Verkehrswesens in England eine leichte Beförderung der Verurteilten zu den Orten, wo die Strafen zu verbüßen sind.[3])

Was die convict prisons betrifft, so ist im Jahre 1885 die Musteranstalt von Pentonville als convict prison geschlossen worden — sie wird jetzt als local prison benutzt —, auf der andern Seite ist in Dover eine neue Anstalt errichtet worden, welche jedoch erst teilweise vollendet ist.

Für das erste Stadium der penal servitude, die Einzelhaft,

---

[1]) Report of Commissioners 1886 Seite 15.

[2]) Cf. Report of the Royal Commission on Prisons in Ireland Seite 5 ff. „the larger the number of prisoners in any prison within certain limits, the greater should be both the efficiency and the economy of their maintenance and treatment" „the nearer the numbers approach to 800 or 900 probably the more efficient will it be" „the daily average population should in no case be less than 100."

[3]) Die Kosten der Beförderung sind übrigens nicht unerheblich, sie werden für 1885 auf rund 26000 £ angegeben. Report of Commissioners 1885 Seite 5. Der Staat zahlt bei der Entlassung aus der Anstalt die Kosten der Rückfahrt bis zum Orte der Aburteilung.

dient jetzt bei männlichen Sträflingen die Anstalt zu Wormwood Scrubs in der Nähe von London, welche 1874 eröffnet worden ist. Diese Anstalt enthält aber auch eine Anzahl Sträflinge, welche sich im zweiten Stadium der Strafverbüßung — den öffentlichen Arbeiten (Public Works) — befinden und mit der Vollendung der Gefängnisbauten zu Wormwood Scrubs beschäftigt sind. Die Gesamtzahl der Sträflinge in Wormwood Scrubs belief sich im Jahre 1886 [1]) auf 1043.

Millbank, die älteste Penitentiary in England, wurde bisher als convict prison für weibliche Sträflinge benutzt. Dieselben sind jedoch vor Kurzem nach Fulham und Woking Prison geschafft worden, um Platz für 290 männliche Sträflinge zu machen, welche beim Umbau von Millbank, das in Zukunft als local prison dienen soll, [2]) Verwendung finden.

Die männlichen Sträflinge machen im übrigen das zweite Stadium der penal servitude in den Public Works Prisons von Portsmouth (1125 Insassen), Portland (1111 Insassen), Chatham (994 Insassen), Dartmoor (913 Insassen), Parkhurst (618 Insassen), Woking (536 Insassen), Borstal (505 Insassen) und der neuen noch nicht vollendeten Anstalt von Dover (198 Insassen) durch.

Zur Aufnahme von weiblichen Sträflingen dienen die beiden convict prisons von Woking und Fulham, das erste mit 457, das zweite mit 183 Insassen.

Da die Anstalt von Fulham voraussichtlich demnächst geschlossen werden wird [3]) und die Anstalt von Millbank nur vorübergehend noch für convict prisoners benutzt wird, so bleiben zehn Anstalten zur Verbüßung der penal servitude übrig. Wenn man von Dover, welches noch nicht vollendet ist, absieht, [4]) so hat die kleinste dieser Anstalten 457, [5]) die größte 1125 Insassen.

---

[1]) Cf. für die im folgenden angeführten Ziffern Report of Directors 1886 Seite XXV.

[2]) Mehrere Flügel der Anstalt Millbank wurden schon im letzten Jahre als local prison benutzt.

[3]) Cf. Report of the Royal Commission on the Working Classes 1885 Seite 680—682.

[4]) Die Anstalt in Dover soll, wenn vollendet, circa 900 Sträflinge aufnehmen können.

[5]) Diese niedrigste Ziffer ist übrigens auch nur im beschränkten Sinne zutreffend, da die beiden Anstalten Woking Female Prison (457 Insassen) und

§ 7. Spezialisierung der Anstalten.

Gleich in ihrem ersten Jahresberichte gab die Zentralbehörde der local prisons ihre Absicht kund, einzelne Anstalten möglichst ausschließlich für bestimmte Klassen von Gefangenen zu benutzen: es sollte danach gestrebt werden, soweit als möglich, Spezialanstalten für weibliche Gefangene, ferner für Personen, welche zu hard labour verurteilt wären, und endlich Spezialanstalten für einzelne Industriezweige zu bilden, wohin vorzugsweise Verurteilte, deren Beschäftigung in der betreffenden Industrie angemessen erschiene, geschafft würden. [1])

Eine Trennung der Anstalten nach dem Geschlechte der Gefangenen ist bis jetzt vor allem in London zur Durchführung gelangt: hier nehmen Holloway und Pentonville ausschließlich männliche und Westminster — sowie in Zukunft Millbank — ausschließlich weibliche Gefangene auf. [2])

Eine Klassifizierung der Anstalten nach der Beschäftigungsart der Mehrzahl der Gefangenen ist bis jetzt nur wenig durchgeführt: in Northampton wird viel Schuhmacherei, in Wakefield, Manchester und Kirkdale: Weberei, in Wandsworth und Pentonville: Papiersortieren, in Stafford und Portsmouth: Mehlmahlen (grinding wheat), in Holloway: Fabrikation von Postsäcken betrieben. Es handelt sich im wesentlichen darum, daß in den betreffenden Anstalten stets für einen gewissen Stamm von Arbeitern des fraglichen Industriezweiges gesorgt ist, damit die übernommenen Aufträge ausgeführt werden können. Da die Beschäftigungsarten regelmäßig nach der in dem betreffenden Bezirke vorzugsweise betriebenen Industrie gewählt sind, so bedarf es zu dem Zwecke nur selten der Hinsendung von Gefangenen aus andern Bezirken, und eine solche Hinsendung findet natürlicherweise nur bei Gefangenen mit längerer Strafdauer statt.

Woking Male Prison (536 Insassen) so nahe zusammenliegen und in der Verwaltung so eng verbunden sind, daß man sie auch als eine Anstalt mit zwei selbständig geleiteten Abteilungen ansehen kann.

[1]) Report of Commissioners 1878 Seite 11, 49 ff.

[2]) In den Provinzen sind die beiden Anstalten von Wandsworth und Kirkdale ausschließlich zur Aufnahme von männlichen Gefangenen bestimmt.

Eine größere Spezialisierung der Anstalten ist in den Public Works Prisons durchgeführt. Da hier ausschließlich langjährige Strafen vollstreckt werden, so kommen die Kosten, welche durch die Transportierung der Sträflinge nach bestimmten Anstalten hervorgerufen werden, nur wenig in Betracht und die Durchführung einer Spezialisierung der Anstalten ist daher bei den convict prisons eine wesentlich leichtere als bei den local prisons. Es bleibt aber bemerkenswert, daß auch bei den letzteren Anstalten möglichste Spezialisierung erstrebt wird. [1])

In den Public Works Prisons ist zunächst die Scheidung zwischen Anstalten für männliche und für weibliche Sträflinge vollständig durchgeführt. Weibliche Sträflinge kommen ausschließlich in die Anstalten von Woking und Fulham.

Es sind ferner besondere Anstalten zur Aufnahme von Sträflingen bestimmt, welche zum ersten Male bestraft sind und besserungsfähig erscheinen. Es wurde bereits oben Seite 67 erwähnt, daß diese Personen als star class prisoners ausgesondert werden. [2]) In die neuerrichtete Anstalt zu Dover kommen ausschließlich Sträflinge dieser Klasse. Ferner befinden sich in der Anstalt von Borstal fast nur Sträflinge, welche zu relativ kürzeren Strafen verurteilt worden sind; ein großer Teil der Insassen gehört auch hier der star class an. Auf der andern Seite dienen die' Anstalten zu Chatham und Portland vorzugsweise zur Aufnahme von schweren Verbrechern, welche langjährige Strafen zu verbüßen haben. [3])

Während die Sträflinge in Chatham, Portland und Portsmouth fast ausschließlich zu Arbeiten verwendet werden, welche —

---

[1]) Es hätte noch hervorgehoben werden können, daß einzelne local prisons — u. a. Portsmouth — fast ausschließlich zur Aufnahme von Militär- resp. Marine-Gefangenen dienen. Personen des Zivilstandes kommen nur ausnahmsweise — wenn die Strafe zu kurz ist, um eine Beförderung nach einer andern Anstalt zu rechtfertigen — in diese Gefängnisse.

[2]) Im Report of Directors 1885 Seite X werden dieselben charakterisiert als „convicts who do not belong to the criminal class".

[3]) In Chatham befinden sich 55 Sträflinge, welche zu 15 Jahren; 39, welche zu 20 Jahren; 4, welche zu 25 Jahren und 74, welche auf Lebenszeit zur penal servitude verurteilt sind. In Portland sind 48 zu 15; 40 zu 20; 5 zu 25 Jahren; 47 auf Lebenszeit Verurteilte. Report of Directors 1886 Seite XXXIII.

wie Hafen- und Festungsbauarbeiten — große Körperkräfte und volle Rüstigkeit erfordern, besteht die Beschäftigung in Dartmoor vorzugsweise in Urbarmachung von Land und leichterer landwirtschaftlicher Arbeit. Hier befindet sich dementsprechend eine größere Anzahl von Sträflingen in höherem Lebensalter: während in Chatham von 994 Sträflingen 32, in Portland von 1111 Sträflingen 74, in Portsmouth von 1125 Sträflingen 75 Personen über 55 Jahre alt waren, fallen in Dartmoor auf diese Altersklasse von 913 Sträflingen 147.[1])

Die Anstalten von Parkhurst und Woking endlich sind ausschließlich für solche Sträflinge bestimmt, welche infolge von Alter, Gebrechen oder Krankheit überhaupt nur noch im geringen Maße arbeitsfähig sind oder welche einer ständigen ärztlichen Behandlung bedürfen.[2]) In Parkhurst waren im Jahre 1886 von 618 Insassen 109, in Woking von 536 Insassen 133 über 55 Jahre alt. In beiden Anstalten ist die Beschäftigung von den ärztlichen Anordnungen abhängig. In Woking konnten durchschnittlich täglich 174 Sträflinge krankheitshalber überhaupt nicht beschäftigt werden, die andern Sträflinge wurden vorzugsweise mit Schneiderei, Schuhmacherei und Buchbinderei beschäftigt.[3])

Am 4. April 1881 hat eine besondere statistische Aufnahme über die an diesem Tage in den convict prisons befindlichen 10261 Sträflinge stattgefunden, wobei festgestellt wurde, daß 3,6 % für jede Arbeit unfähig, 21,5 % nur für leichte Arbeit verwendbar waren, so daß also nur 74,9 % die schwere Arbeit verrichten konnten, zu welchen die convict prisoners prinzipiell in den Public Works Prisons angehalten werden sollen.[4])

---

[1]) Report of Directors 1886 Seite XXXV.

[2]) In Woking war bisher ein Flügel der Anstalt ausschließlich zur Aufnahme von Geisteskranken bestimmt, von 1886 ab sollen jedoch die geisteskranken convict prisoners nach der für criminal lunatics bestimmten Anstalt zu Broadmoor gebracht werden, über welche unten im § 18 noch näher berichtet werden wird.

[3]) Report of Directors 1886 Teil II Seite 175 ff. Unter den Insassen von Parkhurst befindet sich eine große Anzahl schwachsinniger Sträflinge, welche dort vorzugsweise mit landwirtschaftlichen Arbeiten beschäftigt werden.

[4]) Cf. Du Cane I Seite 145 ff. Daselbst sind auch die Ergebnisse von zwei in den Jahren 1862 und 1873 vorgenommenen Untersuchungen angeführt,

Mit Recht glaubt man in England, daß es im Interesse der Aufrechterhaltung der Disziplin liege, die hiernach nicht unerhebliche Zahl von Sträflingen, welche mit Rücksicht auf ihren Gesundheitszustand in der allgemein vorgeschriebenen Weise nicht beschäftigt werden können, in besonderen Anstalten unterzubringen.

§ 8. Bauliche Einrichtung und Anlage der Anstalten.

Die Gefängnisbaukunde ist zu einem selbständigen Wissenschaftszweige herangewachsen, so daß sich derjenige, welcher nicht Bausachverständiger ist, auf diesem Gebiete kaum mehr ein Urteil erlauben darf. In voller Erkenntnis dieser Thatsache sollen denn auch hier lediglich einige Einrichtungen auf dem Gebiete des Gefängnisbauwesens in England hervorgehoben werden, welche von allgemeinerem Interesse sein dürften.

Die größeren local prisons sowie das neue convict prison zu Wormwood Scrubs sind fast sämtlich nach dem Muster der bekannten Anstalt von Pentonville, welche in den Jahren 1840—1842 gebaut wurde, errichtet. Dies heißt: es ist das sogenannte panoptische oder Radial-System zur Anwendung gebracht. Die Anstalt besteht aus mehreren (zumeist 4—6) Flügeln, welche in einer Zentralhalle zusammenlaufen, von der aus der gesamte Aufsichtsdienst geleitet wird. Der einzelne Flügel hat 2—3 Stockwerke. Die Zahl der Zellen in jedem Stockwerke ist je nach der Größe der Anstalt eine verschiedene.

Die Zellen selbst zerfallen in solche, in denen sich die Insassen Tag und Nacht aufzuhalten haben, und in solche, welche nur

welche sich ausschließlich auf die neu eingelieferten männlichen Sträflinge bezogen. Dieselben werden charakterisiert:

	1862	1873
a) kräftig	477	484
b) gesund aber nicht stark	243	286
c) nicht gesund aber zu leichterer Arbeit verwendbar	228	166
d) krank und arbeitsunfähig	52	64

Die Klasse d) wird in Woking, die Klasse c) in Woking und Parkhurst, die Klasse b) in Dartmoor, die Klasse a) in den andern Anstalten untergebracht.

zum Aufenthalte während der Nachtzeit bestimmt sind. Die ersteren sind regelmäßig 13 Fuß lang, 6—7 Fuß breit und 9 Fuß hoch; die letzteren sind erheblich kleiner. Gemeinsame Schlafräume existieren nur in den Krankenabteilungen. Aber auch für die Kranken sucht man jetzt das Prinzip der Einzelzelle nach Möglichkeit durchzuführen und nur diejenigen Gefangenen in gemeinschaftlichen Sälen unterzubringen, bei denen dies aus sanitären Rücksichten wünschenswert ist.

In den Public Works Prisons, wo sämtliche Sträflinge nur die Nachtzeit in den Zellen zubringen, beträgt die Größe der Zelle regelmäßig 7 Fuß in der Länge, 4 Fuß in der Breite und 7 Fuß in der Höhe. In diesen Anstalten ist vielfach von den sogenannten iron cells Gebrauch gemacht, eisernen Hütten für etwa 50 Personen, welche leicht von einer Stelle zur andern zu schaffen sind. [1]) Diese Einrichtung eignet sich vorzüglich für bloß temporäre Bedürfnisse; gegen den ständigen Gebrauch derselben wird der Einwand erhoben, daß bei den dünnen Wänden, welche die einzelnen Zellen voneinander trennen, Unterredungen zwischen den Gefangenen nicht zu verhindern sind.

In den Zellen befindet sich das erforderliche Mobiliar: Tisch, Stuhl, Wandschränkchen zur Aufbewahrung der Eßgeschirre, Bücher u. s. w. und eine Bettstelle mit Matratze. [2]) Vielfach ist die Bettstelle so konstruiert, daß sie am Tage zusammengelegt werden kann und dann den Tisch bildet. Die früher in den Zellen befindlichen Klosets sind fast durchweg beseitigt und an deren Stelle Latrinen am Ende jedes Ganges eingerichtet worden. In der Zelle ist ein Klingelzug angebracht, durch dessen Anziehen eine Klappe an der äußeren Seite der Zellenwand aufspringt. Die Zellenthüren haben ferner eine Klappe, durch welche den Gefangenen Speisen oder sonst Erforderliches hineingereicht wird. In der Mitte der Thür ist ein Loch angebracht, durch welches der Gefängniswärter den Insassen beobachten kann.

Für eine genügende Ventilation ist man auf das sorg-

---

[1]) Diese iron cells sind nach dem Muster der zuerst in der Krim und jetzt im Lager von Aldershot gebrauchten Militärbaracken hergestellt.

[2]) Ein Teil der Gefangenen in den local prisons erhält keine Matratze, sondern schläft auf harter Lagerstätte („plank beds"), cf. darüber unten § 25.

fältigste bedacht. In allen größeren Anstalten besteht Zentralheizung und Gasbeleuchtung. [1])

Eine Anzahl Zellen sind als Strafzellen eingerichtet; dieselben sind kleiner als die andern und ein Teil derselben ist dunkel. In jedem Gefängnisflügel befinden sich ferner eine Anzahl Badezellen.

Mit jeder Anstalt ist ein großer freier Spazierhof verbunden, wo sich die Gefangenen unter Aufsicht eines Wärters einzeln in langen Reihen mit Zwischenräumen von 12 Fuß im sogenannten Gänsemarsche bewegen. [2]) Von den früher gebräuchlichen Isolierspazierhöfen[3]) ist man ganz zurückgekommen; dieselben werden nur noch ausnahmsweise benutzt.

Desgleichen hat man in der Kirche von Anbringung von stalls — Verschlägen, welche nur nach vorne offen sind, so daß die Gefangenen zwar sämtlich den Geistlichen sehen können, aber gegeneinander abgesperrt sind — jetzt überall abgesehen und da, wo dieselben in früherer Zeit eingerichtet waren, dieselben entfernt, da man der Ansicht ist, daß dadurch der feierliche Eindruck des Gottesdienstes beeinträchtigt werde.

Auch in den Schulräumen sind keine stalls angebracht. Der Unterricht wird übrigens — worauf unten § 20 noch eingegangen werden wird — jetzt regelmäßig in der Zelle selbst gegeben. [4])

---

[1]) Man kann von der Einrichtung der englischen Anstalten wohl sagen, daß durchweg nicht nur der Zweckmäßigkeit, sondern auch dem Kostenpunkte die gebührende Berücksichtigung geschenkt wird.

[2]) In früheren Jahren gingen die Gefangenen in Gruppen von je zwei. Da sich jedoch hierbei das Gebot des Schweigens als undurchführbar erwies, so läßt man jetzt die Gefangenen einzeln marschieren cf. Report of the Royal Commissioners 1879 Seite XVI.

[3]) Die Einrichtung ist bekanntlich die, daß ein kreisförmiger Platz durch Zwischenmauern, welche von der Mitte aus nach der Peripherie gehen, in Abschnitte geteilt wird, deren jeder einen Isolierspazierhof für einen Gefangenen bildet.

[4]) Eine andere Übertreibung der Isolierung, der Gebrauch von Masken beim Verlassen der Zelle seitens eines Gefangenen, besteht jetzt in keiner englischen Anstalt mehr. — Dagegen habe ich in dem Gefängnisse von Holloway eine eigentümliche Art von Sprechzimmer für die Gefangenen gefunden. Dieselbe besteht darin, daß zwei Reihen vergitterter stalls in einer Entfernung von etwa zwei Fuß sich parallel gegenüber stehen; in die eine Reihe von stalls werden die Gefangenen, in die andre deren Besucher

Bei jeder Anstalt befinden sich natürlicherweise die erforderlichen Wirtschafts- und Geschäftsräume. Auch sind überall Wohnungen für eine Anzahl von Beamten vorhanden. Wie bereits hervorgehoben wurde, ist man jetzt überall bemüht, die Zahl dieser Beamtenwohnungen zu vermehren.

Zwei andre Neuerungen, welche nach und nach in allen größeren Anstalten vorgenommen werden, bestehen in der Errichtung von Lazarethen (infirmaries), getrennt von der übrigen Anstalt, sowie in der Herstellung von Aufnahmeräumen (reception wards), wohin die Neueingelieferten kommen, um dort zunächst von dem Gefängnisarzte untersucht zu werden.

Das gesamte Gefängnisbauwesen ist einem technischen Rate der Zentralbehörde, dem Surveyor, unterstellt (cf. oben Seite 138). Demselben sind nicht nur vor der Einrichtung einer Anstalt die genauen Baupläne zur Begutachtung vorzulegen, sondern es ist ihm auch von allen beabsichtigten Neubauten oder Umbauten von größerem Umfange vorher Kenntnis zu geben. Der Surveyor hat ferner für jede einzelne Zelle, bevor dieselbe in Benutzung genommen werden darf, zu attestieren, in welcher Weise dieselbe verwendet werden soll, ob zum Aufenthalte eines Gefangenen bei Tag und Nacht, oder lediglich als Schlafzelle, ob als Strafzelle und alsdann für welchen Maximalzeitraum.

Von besonderem Interesse dürfte es sein, daß sämtliche in den letzten Jahren neugebauten Gefängnisse wie überhaupt alle Gefängnisbauarbeiten ausschließlich von Gefangenen hergestellt worden sind. Diese Bauarbeiten bilden eine der Hauptbeschäftigungen der Gefangenen. In den local prisons sind seit Übernahme derselben in die Staatsverwaltung fast in jeder Anstalt umfassende bauliche Veränderungen nötig gewesen

von hinten eingeführt. In dem zwischen den vergitterten stalls liegenden Gange hält sich ein Wärter auf. In den Sprechstunden werden so viel Gefangene und Besucher gleichzeitig eingeführt, als stalls vorhanden sind. Es kann daher nicht nur der Wärter, sondern auch der in dem nebenliegenden stall befindliche Gefangene das geführte Gespräch mit anhören. Ob diese Einrichtung, von deren Nützlichkeit ich mich nicht habe überzeugen können, auch in andern Gefängnissen besteht, ist mir nicht bekannt, cf. übrigens Starke, das belgische Gefängniswesen, Berlin 1877 Seite 60, welcher genau dieselbe Einrichtung in Belgien schildert.

und dieselben sind ausschließlich von Gefangenen ausgeführt worden.[1]) In den convict prisons sind seit 1863 Zellen für 4771 Sträflinge ausschließlich durch Gefangenenarbeit hergestellt. Die thatsächlichen Kosten für diese neuhergestellten Anstalten beliefen sich auf 358 700 £, während für den Bau dieser Anstalten, wenn derselbe Privatunternehmern übertragen worden wäre, 636 400 £ hätten verausgabt werden müssen, so daß sich hier ein Erträgnis der Gefangenenarbeit von 277 700 £ ergibt.[2])

Ein interessantes Beispiel davon, wie bei dem Bau der Anstalten vorgegangen wird, gibt das neue convict prison von Wormwood Scrubs. In dieser für 1050 Sträflinge bestimmten Anstalt ist alles mit einziger Ausnahme der Thürschlösser durch Gefangenenarbeit hergestellt worden. Man hatte zunächst 100 convict prisoners, welche in Public Works Prisons bereits Bauarbeiten gethan hatten, dorthin versetzt, indem man iron cells zu ihrer Unterbringung benutzte. Diese 100 Gefangenen begannen den Bau, und erst allmählich wurden zur Fortsetzung desselben weitere Gefangenenkräfte herbeigezogen. Der Bau, welcher im Jahre 1874 begonnen wurde, ist heute noch nicht vollendet. Man arbeitet jetzt an der Fertigstellung der Kapelle. Mit dem Wohnhause für den Governor hat man noch gar nicht begonnen; derselbe behilft sich vorläufig mit einigen provisorisch für ihn hergestellten Räumen. — Auf diese Weise werden dem Staate enorme Summen erspart, und es werden gleichzeitig die Gefangenen in nützlicher Weise beschäftigt.

Was die Lage der Anstalten betrifft, so ist man bemüht, die local prisons nicht allzuweit von den Gerichtshöfen entfernt zu legen, auf der andern Seite aber zu vermeiden, daß die Gefängnisse in dichtbevölkerten Stadtteilen liegen. Die letztere Erwägung ist die Veranlassung gewesen, daß die Zentralbehörde jüngst mehrere Gefängnisse in London — besonders Clerkenwell und Coldbath Fields — geschlossen hat.[3])

---

[1]) Der Surveyor veröffentlicht alljährlich eine Zusammenstellung dieser baulichen Arbeiten. In dem Berichte für 1885/6 (Report of Commissioners 1886 Seite 48, 49) erscheint eine besonders große Zahl von neuerrichteten reception und infirmary wards sowie von quarters for officers and married warders.

[2]) Cf. Du Cane II Seite 180, 181.

[3]) An einer nützlichen anderweitigen Verwendung dieser Anstalten wird

§ 9. Die Kosten des Gefängniswesens.

Um ein richtiges Bild des englischen Gefängniswesens zu gewähren, muß auch die Kostenfrage ihre Berücksichtigung finden.

Die Kosten der local prisons erscheinen in den alljährlichen Reports der Commissioners of Prisons, diejenigen der convict prisons in den alljährlichen Reports der Directors of Prisons. Zu diesen beiden Posten müßten streng genommen noch die Kosten der Zentralbehörde hinzugerechnet werden, welche den Civil Services zu entnehmen sind. Dieser Posten kann jedoch hier übergangen werden, nachdem die Hauptausgaben desselben — nämlich die Gehälter der Mitglieder der Zentralbehörde — bereits oben im Kap. I § 1 erörtert worden sind.

Die Ausgaben der local wie der convict prisons erscheinen unter drei Überschriften:

1) Kosten des Beamtenpersonals (cost of staff).
2) Kosten der Unterhaltung der Gefangenen (Maintenance of convicts resp. of prisoners).
3) Diverse Ausgaben (other expenses).

Unter der ersten Rubrik sind mitenthalten die Ausgaben für Uniformen und gewährte Wohnungsgelder sowie einmalige Vergütungen (allowances); dagegen sind nicht berücksichtigt die Pensionszahlungen. Was die dritte Rubrik anbetrifft, so sind aus derselben in dem letzten Jahresberichte einige Posten in die zweite Rubrik gebracht worden, wodurch genaue Vergleichungen mit früheren Jahrgängen unmöglich gemacht sind. So erscheinen beispielsweise die Ausgaben für Mobiliar, welche früher als besonderer Posten in der dritten Rubrik aufgeführt waren, jetzt in der zweiten Rubrik zusammen mit den Ausgaben für Bekleidung der Gefangenen.

Die Ausgaben für Transportierung der Gefangenen werden zwar in der Jahresrechnung der convict

---

es nicht fehlen; man hat insbesondere in Aussicht genommen, dieselben zu Wohnungen für die arbeitenden Klassen umzubauen, wozu sie sich ihrer Lage nach vorzüglich eignen. Cf. Report of the Royal Commission on the Housing of the Working Classes 1885 Seite 680—682.

prisons, nicht aber in derjenigen der local prisons aufgeführt.[1]) Es hängt dies wohl damit zusammen, daß bisher die Ausgaben für den Transport von Gefangenen nach und von den Gerichtshöfen auf dem Etat der Polizei standen.

Im übrigen finden bei Aufstellung der Jahresrechnung für die convict prisons und für die local prisons dieselben Grundsätze Anwendung, und es können daher die beiden Etats nebeneinandergestellt werden.[2])

Es betrugen in dem Rechnungsjahre 1885/6:
die Gesamtausgaben für die convict prisons:
299 867 £ 1 s 6 d;
die Gesamtausgaben für die local prisons:
343 697 £ 4 s 10 d;
die Kosten des Gefängniswesens überhaupt somit:
643 564 £ 6 s 4 d,
d. h. in deutscher Währung 12 871 286 Mark.

Die Ausgaben verteilen sich folgendermaßen:[3])

Titel der Ausgaben	convict prisons	local prisons
I. Kosten für das Beamtenpersonal	163 379 £	208 792 £
II. Unterhaltung der Gefangenen:		
a) Beköstigung	69 845 „	58 008 „
b) Bekleidung, Mobiliar, Geschirr 2c.[4])	22 297 „	18 359 „

[1]) Die zur penal servitude Verurteilten kommen zunächst in das nächstgelegene local prison — in welchem sie auch gewöhnlich schon in Untersuchungshaft gewesen waren — und werden erst, nachdem sich eine gewisse Zahl angesammelt hat, nach dem convict prison transportiert. Cf. Du Cane II Seite 103.

[2]) Cf. für die im folgenden gegebenen Ziffern Reports of Directors 1885 Seite XXVI, XXVII. 1886 Seite XXVIII, XXIX und Reports of Commissioners 1885 Seite 41—47, 78—80; 1886 Seite 39—45, 74-77.

[3]) Der Übersichtlichkeit halber wird von Anführung der Schillinge und Pence hier abgesehen.

[4]) Im Jahre 1884/5 wird dieser Posten unter drei gesonderten Rubriken aufgeführt:

aa) Bekleidung	in den convict prisons:	22 719;	in den local prisons:	13 867 £
bb) Betten	„ „	2305;	„ „	1790 „
cc) Sonstiges Mobiliar	„	2033;	„ „	3432 „

Titel der Ausgaben	convict prisons	local prisons
c) Heizung, Licht, Seife und sonstige Reinigungsartikel [1])	20 348 £	32 555 £
d) Medizin und sonstige Heilmittel	2 041 „	1 438 „
III. Diverse Ausgaben:		
a) Gewöhnliche Anstaltsreparaturen	5 515 „	12 381 „
b) Steuern und Abgaben	612 „	146 „
c) Arbeitsgeschenke (gratuities)[2])	7 123 „	6 914 „
d) Transport der Gefangenen	5 172 „	
e) Andre gelegentliche Ausgaben	3 539 „	5 099 „

Da sich im Jahre 1885/6 in den convict prisons durchschnittlich per Tag 9247, in den local prisons 15 375 Gefangene befanden, so entfallen per Jahr auf den Kopf des Gefangenen:

	in den convict prisons	in den local prisons
a) an Kosten für das Beamtenpersonal	19 £ 11 s 10 d	13 £ 11 s 7 d
b) für die Unterhaltung der Gefangenen	13 „ 14 „ 8 „	7 „ 3 „ 7 „
c) Kosten überhaupt	35 „ 19 „ 3 „	22 „ 7 „ 1 „

Ein etwas andres Bild ergibt sich, wenn man von den Ausgaben die durch die Gefangenenarbeit erzielten Einnahmen in Abzug bringt. Die Arbeiten der convict prisoners lieferten im Jahre 1885/6 ein Gesamterträgnis von

165 413 £ 8 s 9 d.

Diese Summe von der oben angegebenen Ausgabensumme abgezogen, ergibt für die convict prisons einen Nettoausgabenbetrag von:

134 462 £ 12 s 9 d,

---

[1]) Im Jahre 1884/5 wird dieser Posten unter zwei gesonderten Rubriken aufgeführt:
aa) Heizung u. Licht in den conv. pris.: 19 204; in den loc. pris. 28 691 £
bb) Seife und sonstige Reinigungsartikel 2481; „ „ 3841 „

[2]) Das Nähere hierüber cf. unten § 22.

d. h. pro Kopf des Sträflings eine jährliche Ausgabe von:

16 £ 2 s 6 d.

Betreffs der Berechnung des Wertes der Gefangenenarbeit wird im § 21 das Nähere angegeben werden. Hier sei nur bemerkt, daß für die Beschäftigung von Gefangenen im Anstaltsdienste [1]) bei der Berechnung kein Betrag in Ansatz gebracht worden ist.

Über die Erträgnisse der Gefangenenarbeit in den local prisons ist für 1885/6 eine Gesamtsumme noch nicht berechnet; man muß hier auf das vorhergehende Jahr zurückgehen. Im Jahre 1884/5 betrugen die Gesamtausgaben in den local prisons:

359 461 £ 11 s 1 d,

d. h. pro Kopf der sich in diesem Jahre auf durchschnittlich 16 619 belaufenden Gefängnisbevölkerung:

21 £ 12 s 7 d.

Das Gesamterträgnis der Gefängnisarbeit betrug:

66 235 £ 16 s 1 d, [2])

Es ergibt sich also ein Nettoausgabenbetrag von:

293 235 £ 19 s,

d. h. pro Kopf des Gefangenen

17 £ 12 s 10 d.

Hiernach stellen sich die Nettoausgaben pro Kopf des Gefangenen in den local prisons etwas höher als in den convict prisons, obwohl die einzelnen Ausgabeposten in den letzteren Anstalten wesentlich höhere sind als in den erstern. Die Arbeitskräfte der Gefangenen können in den convict prisons, wo der Einzelne für lange Jahre detiniert ist, in viel wirksamerer Weise ausgenutzt werden, als in den local prisons mit ihrer ständig wechselnden Bevölkerung. Auf der andern Seite ist in den convict prisons ein relativ viel größeres Beamtenpersonal erforderlich, und auch die Ausgaben für die Unter-

---

[1]) „employment in the ordinary service of the prison, cooking, cleaning etc.

[2]) Auch hier ist für die Beschäftigung von Gefangenen im Anstaltsdienste kein Betrag in Ansatz gebracht worden. Der Wert der Arbeiten im Anstaltsdienste wird für die local prisons im Jahre 1884/5 auf 57 335 £ 12 s 7 d angegeben. Judicial Statistics 1884 Seite 70.

haltung der Gefangenen stellen sich hier höher, weil mit Rücksicht auf die Art der Beschäftigung die Beköstigung eine bessere sein muß.

Gehen wir nun etwas näher zunächst auf die Ausgaben für die convict prisons ein. Es liegen bei diesen Anstalten für eine lange Reihe von Jahren Ziffern vor, welche miteinander vergleichbar sind, weil sie im wesentlichen stets nach denselben Grundsätzen aufgestellt wurden. [1]) Im Jahre 1864, wo zuerst die neuen Bestimmungen über die Behandlung der convict prisoners in Geltung traten, beliefen sich die Gesamtausgaben auf:

247 227 £;

dieselben stiegen in langsamer Progression bis zum Jahre 1877, wo sie:

360 281 £

betrugen. Da sich während dieser Zeit aber auch die Zahl der convict prisoners von 7418 im Jahre 1864 auf 10 735 im Jahre 1877 erhöht hatte, so war der auf den Kopf des Sträflings entfallende jährliche Ausgabenbetrag ziemlich derselbe geblieben: im Jahre 1864: 33 £ 6 s 8 d, im Jahre 1877: 33 £ 11 s 2 d. Vom Jahre 1877 ab fallen die Gesamtausgaben mit einer Unterbrechung im Jahre 1879 langsam bis auf 299 876 £ 1 s 6 d im Jahre 1885/6. Die Zahl der convict prisoners hält sich von 1877 bis 1880/1 noch ziemlich auf gleicher Höhe (10 735 im Jahre 1877 und 10 745 im Jahre 1880/1) und beginnt erst im Jahre 1881/2 in langsamem und erst 1884/5 in beschleunigtem Tempo zu fallen. Da ein großer Teil der Ausgaben, insbesondere diejenigen für Beamtengehälter, auf ziemlich gleicher Höhe geblieben sind, so stellt sich naturgemäß der auf den Kopf des Sträflings entfallende Ausgabenbetrag für die letzten Jahre erheblich höher. Die Beträge von 35 £ 2 s 4 d im Jahre 1884/5 und von 35 £ 19 s 3 d im Jahre 1885/6 pro Kopf des Sträflings sind die höchsten, welche überhaupt in der ganzen Periode vorgekommen sind. Läßt man diese beiden Jahre außer Betracht und sieht man ferner von den Jahren 1870—1872 ab, wo die Kosten pro Kopf sich ausnahmsweise niedrig auf 30 £ 14 s 9 d bis 31 £ 4 s 9 d beliefen, so sind die Schwankungen in den Ausgaben nur geringfügige und bewegen sich

[1]) Cf. Du Cane I Seite 90.

innerhalb der Grenzen von 32 £ bis 34 £ pro Kopf des Sträflings.

Interessant dürfte es noch sein, auf die Verschiedenheit der jährlichen Nettoausgaben — d. h. Ausgaben abzüglich Erträgnis der Sträflingsarbeiten — pro Kopf des Sträflings in den einzelnen convict prisons hinzuweisen. Dieselben beliefen sich im Jahre 1885/6 in Portsmouth, wo nur die kräftigsten Männer und meistenteils Leute mit langer Strafdauer hingeschickt werden, auf: 2 s 7 d — hier werden also die Anstaltsausgaben so gut wie vollständig durch das Erträgnis der Arbeit gedeckt — dagegen in Parkhurst, wohin nur Sträflinge geschickt werden, welche infolge ihres körperlichen Zustandes eine verminderte Arbeitsfähigkeit besitzen, auf: 23 £ 18 s 3 d. In Pentonville, wo sich nur Sträflinge im Stadium der Einzelhaft befinden, beliefen sich die Ausgaben auf: 27 £ 13 s 6 d. In Fulham, wo nur weibliche Sträflinge sind, erreichten die Ausgaben die Höhe von 51 £ 14 s 4 d.

Übrigens darf bei diesen Ziffernangaben nicht außer Acht gelassen werden, daß die Etats der einzelnen Anstalten nicht ohne weiteres miteinander vergleichbar sind, weil besonders der Umfang der Beschäftigung für den Gefängnisdienst ein ganz verschiedener ist: die Anstalt Fulham z. B. besorgt die Wäscherei, welche noch besondere Ausgaben für Seife erfordert, für mehrere andere convict prisons. In den Anstalten von Woking und Parkhurst ferner werden die neuen Gefangenenkleider für alle convict prisons angefertigt; diese Anstalten sind daher bei dem Ausgabenposten für Bekleidung unverhältnismäßig belastet.

Was die Ausgaben für die local prisons betrifft, so sind hier Vergleiche mit früheren Jahrgängen nur im beschränkten Maße möglich. Bis zur Übernahme dieser Anstalten in die Staatsverwaltung im Jahre 1878 war die Buch- und Rechnungsführung durch die einzelnen Lokalbehörden eine vielfach mangelhafte; es bestand keinerlei Garantie, daß die einzelnen Lokalbehörden beispielsweise unter „gewöhnlichen Reparaturen“ — im Gegensatze zu Neu- und Umbauten — überall dasselbe verstanden, oder daß sie die Erträgnisse der Gefangenenarbeit in gleichmäßiger Weise berechneten. Es werden ferner einzelne Posten, wie z. B. die Kosten für den Transport der Gefangenen, erst seit 1878 den Anstalten belastet; anderseits sind jetzt von dem Etat der

Anstalten die Beamtenpensionen verschwunden, welche früher darauf gestanden haben.

Aber auch die Ausgaben seit 1878 können als normale kaum angesehen werden. Die Verwaltung befindet sich noch in einem Übergangsstadium, welches besondere Kosten verursacht. Beispielsweise ist noch eine Anzahl höherer Gefängnisbeamter mit dem früheren — die jetzigen Normen weit überschreitenden — Gehalte im Dienste belassen worden.

Nur mit allem Vorbehalt mag daher angeführt werden, daß die jährlichen Ausgaben pro Kopf des Gefangenen seit Übernahme der Anstalten in die Staatsverwaltung auf durchschnittlich jährlich 23 £ angegeben werden — im Jahre 1885/6 betrugen sie 22 £ 7 s 1 d —, während dieselben für die Periode vor 1878 auf durchschnittlich 26 £ geschätzt werden. [1])

Eine Verringerung der Ausgaben ergibt sich ganz naturgemäß aus der Thatsache, daß eine große Anzahl ganz kleiner Anstalten, bei denen sich die Generalkosten unverhältnismäßig hoch stellen, aufgehoben worden sind. Dies hat es beispielsweise ermöglicht, daß sich die Kosten für die höheren Beamten gegenüber dem Jahre 1878 um 40871 £ pro Jahr ermäßigt haben, obwohl die Gehälter für die Anstaltsärzte wesentlich erhöht worden sind. Die Ersparnisse an Gas werden auf jährlich 300 £, die Ersparnisse an Heizungsmaterial auf jährlich 240 £ berechnet u. s. w. [2])

---

[1]) Report of Commissioners 1885 Seite 15. In den Judicial Statistics 1884 Seite XLII werden auf Grund mehrfacher Umrechnungen, welche sich aber einer Kontrolle entziehen, die Durchschnittskosten pro Kopf des Gefangenen folgendermaßen angegeben:

1872/3:	27 £	8 s	4 d
1873/4:	27 „	7 „	4 „
1874/5:	25 „	19 „	1 „
1875/6:	25 „	16 „	6 „
1876/7:	24 „	12 „	5 „
1878/9:	21 „	12 „	7 „
1879/80:	21 „	14 „	6 „
1880/1:	23 „	17 „	10 „
1881/2:	21 „	2 „	— „
1882/3:	21 „	14 „	6 „
1883/4:	22 „	1 „	3 „
1884/5:	22 „	8 „	2 „

[2]) Du Cane II Seite 100 ff.

Auch heute sind es die kleinen Anstalten, welche den Etat besonders belasten. Wenn man von den Erträgnissen der Gefangenenarbeit ganz absieht [1]) und bloß die Ausgaben betrachtet, so stellen sich dieselben im Jahre 1885/6 pro Kopf des Gefangenen durchschnittlich auf 22 £ 7 s 1 d, aber in der kleinsten Anstalt (Brecon mit einer Durchschnittsbelegung von 17 Gefangenen) auf 56 £ 8 s 5 d, anderseits in der großen Anstalt von Wandsworth (mit einer Durchschnittsbelegung von 1014 Gefangenen) auf 17 £ 14 s 4 d. Dem Durchschnittssatze am nächsten kommen die Anstalten mit 100 bis 200 Gefangenen, z. B. Northampton (131 Gefangene): 21 £ 5 s 11 d; Lincoln (199 Gefangene): 21 £ 18 s 6 d; Norwich (181 Gefangene): 22 £ 8 s 3 d.

Die baren Ausgaben für Neubauten und Umbauten in den local prisons beliefen sich im Jahre 1885/6 auf 40 122 £. In der Zeit seit Übernahme der local prisons in die Staatsverwaltung war der Mindestbetrag für diesen Posten 30 891 £ im Jahre 1881/2; der Höchstbetrag: 41 995 £ im Jahre 1879/80. Zu diesen baren Ausgaben würde der Wert der von den Gefangenen bei den Bauten geleisteten Arbeiten (cf. oben Seite 171, 172) hinzuzurechnen sein, welcher auf jährlich rund 50 000 £ angeschlagen wird. [2])

Über die durchschnittlichen Kosten von Gefängnisbauten liegen Angaben nicht vor. Bei Übernahme der local prisons in die Staatsverwaltung wurde, wie oben Seite 87 ausgeführt worden ist, der Satz von 120 £ pro Einzelzelle zu Grunde gelegt. Dieser Satz bleibt zweifellos hinter den wirklichen Herstellungskosten zurück. Bei dem vor Übernahme in die Staatsverwaltung zuletzt vollendeten local prison zu Maidstone betrugen die Baukosten pro Einzelzelle 185 £. [3])

---

[1]) Diese stehen natürlicherweise in den ganz kleinen Anstalten, wo eine regelmäßige einträgliche Beschäftigung der Gefangenen schwer durchzuführen ist, weit hinter denjenigen der großen Anstalten zurück.

[2]) Du Cane II Seite 98.

[3]) Du Cane II Seite 79.

Kapitel III.

Die Behandlung der Gefangenen.

Titel I.

Darstellung der Behandlung nach allgemeinen Gesichtspunkten in ihren Hauptmomenten.

§ 10. Klassifizierung der Gefangenen.

Unter den Gefangenen werden zwei Klassen scharf geschieden: Strafgefangene („criminal prisoners") und solche Gefangene, welche gesetzlich nicht als Verbrecher anzusehen sind („who shall not be deemed to be a criminal prisoner"). Die Strafgefangenen zerfallen in diejenigen, welche zur penal servitude verurteilt worden sind („convict prisoners", Sträflinge), und in die zu einer Gefängnisstrafe Verurteilten („prisoners", Gefangene schlechthin). Die andre Kategorie von Gefangenen zerfällt in die Untersuchungsgefangenen („prisoners awaiting trial"), die Schuldgefangenen („debtors") und die „Übelthäter"[1]) („first-class misdemeanants").

Diese drei zuletzt genannten Klassen befinden sich sämtlich in den local prisons. In diese Anstalten werden ferner — abgesehen von den zu einer Gefängnisstrafe Verurteilten — convict prisoners

---

[1]) „Übelthäter" dürfte wohl der entsprechendste deutsche Ausdruck für „first-class misdemeanants" sein.

auf kurze Zeit aufgenommen, bis der Transport derselben nach einem convict prison erfolgen kann.[1])

Für die Behandlung der zur penal servitude Verurteilten einerseits und der zu einer Gefängnisstrafe Verurteilten anderseits bestehen getrennte Vorschriften. Dieselben enthalten jedoch eine Reihe allgemeiner, für beide Klassen in fast gleicher Weise gültiger Gesichtspunkte, welche hier im Titel I zusammen behandelt werden. Auf die Spezialbestimmungen für den Vollzug der beiden Strafarten soll dann im Titel II und III noch des nähern eingegangen werden.

Über die Behandlung der not criminal prisoners enthalten die Prison Acts von 1865 und 1877 einige allgemeine Grundsätze, welche im wesentlichen darauf hinauslaufen, daß die Freiheitsentziehung bei diesen Klassen von Gefangenen nur mit solchen Einschränkungen bezüglich der Beschäftigung, der Beköstigung, der Bekleidung u. s. w. verbunden sein soll, wie sie die Notwendigkeit der Aufrechterhaltung der Ordnung in den Anstalten unbedingt erfordert. Im übrigen wird der Staatssekretär des Innern ermächtigt, die erforderlichen Regulative über die Behandlung der Untersuchungsgefangenen, der Schuldgefangenen und der first-class misdemeanants zu erlassen. Die Regulative sind dann unter dem 19. Februar 1878 für jede dieser drei Klassen gesondert ergangen.[2]) Die Bestimmungen sind aber im wesentlichen dieselben und sollen im folgenden Paragraphen zur Darstellung gebracht werden.

Unter „Untersuchungsgefangenen" werden nach englischem Rechte sowohl diejenigen verstanden, welche von dem Friedensrichter nach Abschluß des Vorverfahrens zur Aburteilung an die Quarter Sessions oder an die Assizes überwiesen sind („committed for trial"), wie diejenigen, bei denen das Vorverfahren noch schwebt und die Fortdauer der Haft angeordnet ist („remanded"). Die von der Polizei Festgenommenen verbleiben bis zum ersten richterlichen Verhör im Polizeigewahrsam in sogenannten

---

[1]) Es handelt sich hier lediglich um Ersparung der Kosten für Eskortierung und Transportierung. Auch bei den zu einer Gefängnisstrafe Verurteilten findet bezüglich derjenigen, welche aus irgendwelchen Gründen in Spezialanstalten geschafft werden sollen, ein derartiges Ansammeln behufs gemeinsamen Transportes statt. Du Cane II Seite 103.

[2]) Die Regulative sind abgedruckt in den Local Prison Rules Seite 9—33.

lock-up cells. Der Richter hat, wenn er nicht ohne weiteres die Sache abweist („discharged"), darüber zu befinden, ob die Sache zur Aburteilung reif ist — in welchem Falle er das Urteil sofort selbst spricht resp., wenn die Sache außerhalb seiner Kompetenz liegt, den Angeschuldigten den kompetenten Gerichten überweist — oder ob noch weitere Vernehmungen erforderlich sind. Soweit die Sache nicht im ersten Termine zu Ende geführt wird, entscheidet der Richter weiter, ob der Angeschuldigte gegen oder ohne Sicherheitsleistung aus der Haft zu entlassen ist, oder ob die Haft fortzudauern hat. In dem letzteren Falle erfolgt die Überführung des Angeschuldigten in ein local prison.[1]) Bis zu seiner Verurteilung aber wird der Angeschuldigte gesetzlich für „Nichtschuldig" erachtet („in law presumably innocent"), und daraus ergibt sich, daß die Haft, welche nur den Zweck hat, sich der Person des Angeschuldigten zu versichern, so wenig drückend als möglich zu gestalten ist („to make the confinement as little as possible oppressive").[2])

Was die „Schuldgefangenen" (debtors) anbetrifft, so ist nach der jetzigen Lage der Gesetzgebung Schuldhaft nur noch in Ausnahmefällen zulässig.[3]) Ohne auf Einzelheiten einzugehen, läßt sich kurz sagen, daß Schuldhaft auf Antrag des Gläubigers angeordnet werden kann, wenn der Schuldner einem richterlichen Befehl zur Zahlung einer bestimmten Summe Geldes an den Gläubiger nicht nachkommt und der Gläubiger den Beweis dafür liefert, daß der Schuldner zur Zeit des Erlasses des Befehls thatsächlich die

---

[1]) Wenn bei der Verhandlung vor den Assizes oder den Quarter Sessions eine Vertagung eintritt, so wird der Untersuchungsgefangene nicht in das local prison zurückgebracht, sondern wird in einem der sog. court houses in gerichtlichem Gewahrsam gehalten. Die Einrichtung der court houses, welche Sache der Lokalbehörden ist, wird als in vielen Fällen wenig befriedigend geschildert, und man ist z. B. bestrebt, hier Besserungen anzubringen. Cf. Report of Commissioners 1885 Seite 15, 1886 Seite 12. Der Staatssekretär des Innern hat zur näheren Untersuchung eine besondere Kommission eingesetzt, welche im Februar 1887 ihren Bericht erstattete. Nach Zeitungsnachrichten sind dabei recht böse Zustände aufgedeckt worden, und es wird ein energisches Einschreiten gegenüber den Lokalbehörden, welche ihre Pflicht vielfach auf das Gröblichste versäumt haben, verlangt.

[2]) Prison Act 1877 s. 39.

[3]) 32 a. 33 Vict. c. 62; 33 a. 34 Vict. c. 76.

Mittel zur Zahlung gehabt hat. Die Sache wird hier gleichsam als eine ungehörige Mißachtung eines richterlichen Befehls („act of contempt of court") aufgefaßt und die Schuldhaft als eine Art Ordnungsstrafe behandelt. Die Strafe soll im wesentlichen nur in Freiheitsentziehung bestehen, und es gelten dabei fast die gleichen Bestimmungen wie für die Untersuchungsgefangenen.[1])

In gleicher Weise wie die Schuldgefangenen wurden früher diejenigen behandelt, welche eine ihnen auferlegte Friedensbürgschaft nicht leisteten[2]) und infolgedessen verhaftet wurden („imprisoned in default of finding sureties"). Bezüglich dieser Surety Prisoners ist aber durch eine Verordnung des Staatssekretärs des Innern vom 1. Juli 1881 angeordnet worden, daß dieselben fortab gleich den andern zu einer Gefängnisstrafe Verurteilten behandelt und ihnen nur einige Vergünstigungen bezüglich des Empfanges von Besuchen und des Schreibens von Briefen zugestanden werden sollen.[3])

Unter first-class misdemeanants endlich werden diejenigen zu einer Gefängnisstrafe ohne harte Arbeit Verurteilten verstanden, bei denen der Richter eine bessere Behandlung, als sie den gewöhnlichen Gefangenen zu teil wird, für angemessen erachtet. Es hängt im allgemeinen lediglich von dem Ermessen des Richters ab, ob er einen Verurteilten zu einem first-class misdemeanant erklären will oder nicht. Nur bei dem Delikte der Verbreitung aufrührerischer Schriften (seditious libel) ist gesetzlich vorgeschrieben, daß der Verurteilte stets als first-class misdemeanant zu behandeln sei.[4]) Es ist ferner gesetzlich angeordnet, daß diejenigen, welche wegen Mißachtung des Gerichts (contempt of court) zu einer Ordnungsstrafe verurteilt worden sind, als first-class misdemeanants erachtet werden sollen.[5]) Der first-class misdemeanant erleidet im wesentlichen nur Freiheitsentziehung. Es ist auf diese Weise ein Strafmittel geschaffen, welches etwa der deutschen Strafe der Haft entspricht.

---

[1]) Prison Act 1865 s. 17; Prison Act 1877 s. 38.

[2]) Cf. oben Seite 100.

[3]) Die Verordnung ist abgedruckt in den Local Prison Rules Seite 34.

[4]) Prison Act 1865 s. 67, Prison Act 1877 s. 40.

[5]) Prison Act 1877 s. 41.

§ 11. Die Spezialvorschriften bezüglich der Behandlung der not criminal prisoners.

Die not criminal prisoners sind streng getrennt von den criminal prisoners zu halten. Die Untersuchungsgefangenen sind stets in Einzelzellen unterzubringen, und es ist jeder Verkehr mit andern Gefangenen verboten.[1]) Von den Schuldgefangenen erhält zwar jeder einzelne eine abgesonderte Schlafzelle; zum Aufenthalte während des Tages ist jedoch ein gemeinsamer Raum bestimmt, und es ist den Schuldgefangenen sowohl hier, wie auf dem Spazierhofe freier Verkehr untereinander gestattet.[2]) Für die first-class misdemeanants sind besondere Zellen reserviert, welche meist etwas größer sind als die andren Zellen.

Den drei Klassen der not criminal prisoners kann es auf ihren Antrag von dem Visiting Committee gestattet werden, die ihnen zugewiesene Zelle auf ihre eignen Kosten besser auszustatten oder, falls besser ausgestattete Zellen in der Anstalt vorhanden sind,[3]) dieselben gegen Zahlung festnormierter Summen zu beziehen. In gleicher Weise können diese Gefangenen gegen Zahlung bestimmter Beträge von der Reinigung ihrer Zellen und ähnlichen Arbeiten entbunden werden. Der Governor bestimmt alsdann eine geeignete Person zu ihrer Aufwartung.

Die not criminal prisoners sind berechtigt, sich selbst zu beköstigen. Diese Selbstbeköstigung unterliegt jedoch einer gewissen Aufsicht, um Schwelgereien zu verhindern („to prevent luxury or waste"), insbesondere ist das erlaubte Quantum von Getränken genau vorgeschrieben. Es ist diesen Gefangenen ferner gestattet, sich auf ihre Kosten Lektüre zu verschaffen, vorausgesetzt daß dieselbe nicht nach Ansicht des Governor resp. des Visiting Committee von anstößigem Charakter ist.

---

[1]) Ausnahmsweise kann einem Untersuchungsgefangenen von dem Visiting Committee gestattet werden, gemeinsam mit einem andern Untersuchungsgefangenen auf dem Spazierhofe zu promenieren.

[2]) Die Schuldgefangenen kommen fast sämtlich in besondere, für diese Klasse bestimmte Anstalten; die meisten befinden sich in Holloway.

[3]) Das letztere ist z. B. in Holloway der Fall, wo eine besondere Abteilung für first-class misdemeanants besteht.

Die Gefangenen der benannten Kategorieen tragen ihre eigne Kleidung, soweit dieselbe nicht an sich eine ungenügende ist.[1]) In gleicher Weise kommen die mit Rücksicht auf die Reinlichkeit getroffenen Anordnungen — wie das Nehmen von Bädern, das Schneiden des Haares und des Bartes — bei diesen Kategorieen nur ausnahmsweise, wenn es erforderlich erscheint, zur Anwendung.

Was weiter die Art und Weise der Beschäftigung anbetrifft, so ist dem Gefangenen die freie Wahl gelassen. Soweit thunlich, soll ihm die Möglichkeit gegeben werden, seine bisherige Beschäftigung fortzusetzen. Von dem Erträgnisse der Arbeit behält der Governor einen von der Zentralbehörde festgesetzten Betrag als Entschädigung für die der Verwaltung entstandenen Kosten der Verpflegung u. s. w. zurück,[2]) das übrige erhält der Gefangene.

Das Visiting Committee kann diesen Gefangenen außer den allgemeinen, für die Anstalt bestimmten Besuchszeiten noch den Empfang von Besuchen zu andern Zeiten oder auf längere Dauer als gewöhnlich gestatten[3]) und dieselben von den beschränkenden Vorschriften bezüglich der Korrespondenz entbinden. Der Untersuchungsgefangene ist berechtigt, täglich seinen Anwalt zu sehen.[4]) Es soll möglichste Vorsorge getroffen werden, daß die Besucher in keinerlei Berührung mit den Besuchern der criminal prisoners kommen.

Dies dürften die wichtigsten der für die not criminal prisoners getroffenen Spezialbestimmungen sein;[5]) im übrigen unterliegen auch diese Personen den allgemeinen Anstaltsvorschriften.

---

[1]) Wenn dies letztere der Fall ist und infolgedessen Kleidungsstücke von der Anstalt geliefert werden müssen, so sollen die Kleidungsstücke für die Untersuchungsgefangenen von einer andern Farbe sein, als diejenigen der Strafgefangenen.

[2]) Wird der Untersuchungsgefangene freigesprochen, so erhält er auch diesen Betrag ausgezahlt.

[3]) Auch kann gestattet werden, daß mehrere Besucher zu gleicher Zeit zu dem Gefangenen gelassen werden.

[4]) Es kann aber stets angeordnet werden, daß bei dem Besuche ein Gefängniswärter zugegen sei.

[5]) Es könnte noch die Befugnis der not criminal prisoners, sich von ihrem eignen Arzte behandeln zu lassen, hervorgehoben werden.

§ 12. Allgemeine Gesichtspunkte bei der Behandlung der criminal prisoners.

Ehe auf die Einzelheiten bei der Behandlung der Gefangenen eingegangen wird, erscheint es zweckmäßig, die dabei zur Geltung gebrachten allgemeinen Gesichtspunkte in möglichster Kürze vorweg hervorzuheben. Die einzelnen Momente in der Behandlung der Gefangenen erscheinen auf diese Weise als Glieder eines Systems, und es ist die Möglichkeit gegeben, bei jedem einzelnen Momente zu beurteilen, ob die bezüglichen Bestimmungen nicht nur an sich rationell sind, sondern auch weiter, ob dieselben in den Rahmen des Systems hineinpassen. [1])

Bei der Behandlung der criminal prisoners erscheinen folgende drei Hauptgesichtspunkte als maßgebend:

1) die Freiheitsentziehung muß so gestaltet sein, daß sie eine abschreckende Wirkung ausübt; [2])

2) es muß Vorsorge dafür getroffen werden, daß weder der körperliche noch der sittliche Zustand des Gefangenen am Straforte Schaden leidet;

---

[1]) Ich muß dabei allerdings zugeben, daß die Aufstellung eines derartigen Systems englischen Anschauungen nicht entspricht. In keinem einzigen englischen Buche ist eine Zusammenstellung der allgemeinen Gesichtspunkte, wie sie im Texte zu geben versucht wird, auch nur andeutungsweise enthalten; es kommt lediglich bei der Erörterung spezieller praktischer Fragen vor, daß englische Schriftsteller auch einmal auf allgemeine Gesichtspunkte eingehen. Der Engländer ist eben weder ein Freund der Systematik, noch interessiert er sich allzusehr für theoretische Betrachtungen. In Deutschland ist dies bekanntlich durchaus anders, und aus Rücksicht auf den deutschen Leser glaubte ich, den Paragraphen geben zu sollen. Ich habe mich dabei bemüht, so objektiv zu sein, wie es in einem Falle möglich ist, wo man lediglich aus der persönlichen Beobachtung von Einrichtungen und Verhältnissen die dabei maßgebenden allgemeinen Gesichtspunkte ermitteln kann.

[2]) Dies gilt in England als das Hauptprinzip der Strafe. Cf. Du Cane I Seite 18: „the object of reforming the offender must not be effected in such a manner as to interfere with the principle of deterring from the commission of crime, because punishment is primarily to prevent crime by the warning held up to those who might, but for such influences, fall into it“. Cf. ferner Sir Joshua Jebb, Report of Directors 1860 Seite XVIII: „the deterring features of a sentence should predominate“.

3) es muß Rücksicht darauf genommen werden, daß der Gefangene nach Verbüßung der Strafe in erfolgreicher Weise in die bürgerliche Gesellschaft zurücktreten kann.[1])

Für alle drei Gesichtspunkte sind die Interessen der Gesellschaft nicht weniger maßgebend, als diejenigen des betreffenden Individuum. Ja, man kann wohl sagen, daß, insoweit eine Kollision dieser Interessen stattfinden sollte, die Interessen der Gesellschaft vorgehen: die Strafe soll abschreckend wirken nicht nur, um den Bestraften selbst von der Begehung neuer Rechtswidrigkeiten abzuhalten, sondern vor allem, um — soweit es durch Furcht vor der Strafe möglich ist — zu verhindern, daß andre Glieder der bürgerlichen Gesellschaft sich zu Rechtswidrigkeiten verleiten lassen;[2]) Gesundheit, Leben und Moralität des Gefangenen darf nicht geschädigt werden, nicht bloß aus humanen Rücksichten für seine Person, sondern in erster Linie aus Rücksicht für die bürgerliche Gesellschaft, welcher andernfalls der Gefangene nach seiner Entlassung zur Last fallen würde; die Vorbereitung des Gefangenen zum erfolgreichen Wiedereintritt in die bürgerliche Gesellschaft hat nicht nur im Auge, daß dem Gefangenen selbst damit etwas Gutes geschieht, sondern vor allem, daß das Gemeinwesen durch die erfolgte Besserung des Gefangenen vor neuen Rechtsverletzungen geschützt werde und in dem Entlassenen ein nützliches Glied der bürgerlichen Gesellschaft erhalte.

Die aus den angegebenen drei Hauptgesichtspunkten gezogenen praktischen Konsequenzen sind die folgenden.

a) Der Ernst und die Strenge der Strafe ist un-

---

[1]) Du Cane I Seite 21: „it has always been held that we must bear in mind that the prisoner should not only be punished and taught what is right, but should be returned to society fitted both morally and physically to fulfil his proper duties in the battle of life“.

[2]) Cf. Du Cane II Seite 2: „if by punishing those who have an incurable tendency to crime we can deter fresh recruits from joining the ranks of the criminal class, the object of punishment is effected“. Cf. auch Du Cane in der Law Quarterly Review 1886 Seite 230: „the punishment of an offender has of course the object of preventing him from offending again, but a much more important object is to deter others“, „by checking the first we exhaust the supply of the second“.

bedingt aufrecht zu erhalten.[1]) Die Strafe muß so gestaltet sein, daß sie sich sowohl für das einzelne Individuum wie nach der allgemeinen Auffassung als ein Übel charakterisiert. Strenge Zucht, harte Arbeit, Gewährung nur desjenigen, was zur Erhaltung des Lebens und der Gesundheit unbedingt erforderlich ist, sind die maßgebenden Prinzipien. Es ist alles zu vermeiden, was die Lage des Gefangenen zu einer bessern machen könnte, als diejenige des schlechtest gestellten freien Arbeiters ist.[2])

b) Zur Vermeidung moralischer Ansteckung während der Haft ist ständige Beaufsichtigung des Gefangenen notwendig. Dieselbe läßt sich während der Nachtzeit nicht anders durchführen, als dadurch, daß jedem Gefangenen eine besondere Schlafzelle angewiesen wird.

c) Für die Gestaltung der Haft während der Tageszeit kommen verschiedene und zum Teil widerstreitende Momente in Betracht.

Wenn man zunächst den Standpunkt in's Auge faßt, daß der Gefangene von allen bösen Anregungen, welche von Mitgefangenen ausgehen können, ferngehalten werden soll, so empfiehlt sich die vollständige Isolierung des Gefangenen als das sicherste Mittel. Die Isolierhaft hat ferner den Vorzug, daß der Gefangene in der Isolierung am leichtesten zur Selbsterkenntnis, zur Reue über die Vergangenheit und zu guten Vorsätzen für die Zukunft umgestimmt werden kann und den Zusprüchen der Religion am besten zugänglich ist.

Allein anderseits entspricht die Isolierhaft dem dritten, oben aufgestellten Gesichtspunkte der Vorbereitung des Gefangenen zum

---

[1]) Cf. Paper for the National Conference on Charities Prisons etc. at St. Louis, prepared by the Howard Association 1885: „there has been maintained, in this country, a vigilant care that the penal element, which is so essential a part of wise criminal treatment, shall not be lost sight of".

[2]) Du Cane I Seite 54: „Public opinion demanded that prisoners throughout their sentence should have only the barest necessaries in the way of food, and just sufficient money on discharge to enable them to maintain themselves while seeking employment", „this sound principle was adopted".

Wiedereintritte in die bürgerliche Gesellschaft nur wenig.[1]) Der Gefangene, welcher zu der verbrecherischen That, wegen derer er bestraft wurde, gekommen ist, weil er gar keine Herrschaft über seinen Willen besitzt, oder weil er aus Unlust zur Arbeit den Kampf um das Dasein in redlicher Weise zu führen nicht vermag, bedarf, um ein nützliches Glied der bürgerlichen Gesellschaft zu werden, vor allem einerseits einer Stärkung seiner Willenskraft, seiner Selbstbeherrschung, seiner Widerstandsfähigkeit gegen Versuchungen, und anderseits einer Erziehung zu ehrlicher, ausdauernder und anstrengender Arbeit. Beides kann in der Isolierzelle nicht erreicht werden: der Gefangene wird hier von jeder Versuchung, an welcher sich seine Willenskraft erproben könnte, vollständig ferngehalten, er ist von der Außenwelt, von denjenigen Gesellschaftskreisen, in welche er nach seiner Entlassung zurücktritt, vollständig abgeschlossen, er verläßt hier vielfach die Anstalt nicht wirklich gekräftigt für den Kampf um das Dasein, sondern unselbständig und verzogen wie ein Kind. In gleicher Weise ist eine geeignete Beschäftigung des Gefangenen, welche ihn zu vernünftiger voller Ausnutzung seiner Arbeitskraft, zu geordneter, angestrengter Arbeit, wie er sie in der Freiheit fortsetzen kann, erzieht, in der Einzelzelle schwer möglich. Hieraus ergibt sich, daß die Isolierhaft den Anforderungen des oben sub 3 aufgestellten Gesichtspunktes nicht genügt.

d) Wo es daher einer wirklichen Erziehung des Gefangenen zum vernünftigen Gebrauche der Freiheit bedarf, erweist sich die Isolierhaft als unzureichend. Die Isolierhaft bleibt aber das geeignete Strafmittel bei leichteren Straffällen, wo eine derartige Erziehung überflüssig erscheint, wie überhaupt bei allen kurzzeitigen Strafen, bei welchen es zur Vollendung einer derartigen Erziehung an Zeit gebrechen würde.[2]) Es genügt hier, daß

---

[1]) Cf. insbesondere Du Cane I Seite 21: „Perpetual seclusion in a cell for years with no communications with his fellows is an artificial state of existence so entirely unlike that which he is to be prepared to follow on his discharge from prison that it cannot be expected to fulfil the required object".

[2]) Du Cane I Seite 58: „for those prisoners in local prisons, whose sentences are very short, there is obviously not time to do much in the way of reforming by the influence of labour and the exclusively penal stage will occupy the whole of their time in prison".

der Gefangene durch die strenge Gefängnisdisziplin zu Gehorsam und Ordnung angehalten und dadurch auch an eine Unterordnung unter die bestehende allgemeine Rechtsordnung gewöhnt wird. Es ist ferner bei dieser Klasse von Gelegenheitsverbrechern vor allem darauf Rücksicht zu nehmen, daß der Gefangene unter keinen Umständen dem demoralisierenden Einflusse schlechterer und mehr verderbter Mitgefangener ausgesetzt ist. Für die Erreichung dieses letzteren Zieles bietet nur die Isolierhaft eine sichere Garantie.

Wo es dagegen einer wirklichen Erziehung des Gefangenen bedarf, um ihn darauf vorzubereiten, als ein nützliches Glied in die bürgerliche Gesellschaft einzutreten, da darf der Gefangene nicht als Treibhauspflanze behandelt werden, sondern seine Erziehung ist dem wirklichen Leben entsprechend zu gestalten: er ist während der Strafzeit durch verschiedene Läuterungsstadien hindurch zu dem Endziele, dem vernünftigen Gebrauche der Freiheit, hinzuleiten. Auf dem Wege zu diesem Ziele sind all die Einflüsse, welche auf die menschliche Natur zu wirken geeignet erscheinen, mitheranzuziehen: [1]) Furcht vor Strafe, geistlicher Zuspruch, Erwecken der Hoffnung auf Besserung der eignen Lage, Gewährung von Vorteilen für pflichtmäßige Arbeit und pflichtmäßiges Verhalten und endlich Wiederbelebung des Vertrauens in die eigne Kraft.

Die Erziehung beginnt in der Isolierzelle, wo die Sprache des Gewissens durch nichts unterbrochen wird und das Gemüt für die Ermahnungen des Geistlichen empfänglich ist. Hier soll sich der Gefangene das Maß seiner Verschuldung klar machen und den Vorsatz zur Besserung fassen. [2]) Die Art und Weise seiner Beschäftigung darf ihn nicht von diesen Gedanken abziehen; sie muß trocken und

[1]) Cf. Wines Seite 603 ff.: „Now, what are the essential bases of a reformatory prison discipline? Such a system must work with nature, not against it“ „Hope is the master-spring of human action, without it the bad cannot possibly regain their virtue. It must be implanted in the breast of the prisoner the first hour of his incarceration and kept there as an ever-present and living force“. „Hope is just as truly, just as vitally, just as essentially at the root of all right prison discipline as it is of all vigorous and successful effort in free life“.

[2]) Du Cane I Seite 20: „he is likely to feel sorrow for the past and to welcome the words of those who shew him how to avoid evil for the future“.

langweilig sein. Hat der Gefangene dieses erste Stadium, welches aus physischen und psychischen Rücksichten nicht zu lange dauern darf,[1]) hinter sich, so wird dem von ihm empfundenen Freiheitsbedürfnisse Rechnung getragen: die strenge Isolierung wird aufgehoben, er darf hin und wieder einen Besuch empfangen, einen Brief schreiben, seine Beschäftigung wird eine anregendere, er erhält für seine Arbeit eine kleine Vergütung, und dabei wird ihm in Aussicht gestellt, daß, wenn er sich gut führt und fleißig arbeitet, all diese Vergünstigungen weiter ausgedehnt werden. Wie ihn diese Hoffnung auf eine bessere Zukunft belebt, so hält ihn die Erinnerung an die überstandene Isolierung, in welche er bei schlechtem Verhalten zurückversetzt zu werden fürchten muß, davon ab, den an ihn herantretenden Verführungen und Verlockungen nachzugeben.

So wird seine Willenskraft gestärkt, er lernt sich selbst beherrschen. Diese Gymnastik des Willens dauert während der ganzen Strafzeit fort: der Strafzwang wird immer mehr und mehr gemildert, die Vergünstigungen nehmen zu. Der Gefangene wird in fortwährender Spannung erhalten:[2]) die Anerkennung, welche sein pflichtmäßiges Verhalten und seine fleißige Arbeit finden, erfüllt ihn mit Befriedigung. Er gewinnt das Zutrauen zu sich selbst, daß er im stande sein wird, die von ihm während der Isolierung gefaßten guten Vorsätze in der vollen Freiheit auszuführen, da er gelernt und bereits bewiesen hat, daß er den Versuchungen Widerstand entgegensetzen kann, und da ihm die angestrengte Arbeit zur zweiten Natur geworden ist.

Das sind die Gedanken, welche den englischen Strafvollzug beherrschen: Erziehen des Gefangenen zum Gebrauche der Freiheit durch allmähliches Nachlassen des Strafzwanges. Nur die ganz kurzzeitigen Strafen können in gleichförmiger Weise vollstreckt werden. Bei Strafen von längerer Dauer würde sich der Gefangene im Falle einförmiger Strafvollstreckung

---

[1]) Die Zeit der völligen Isolierung muß ferner in einem rationellen Verhältnisse zu der Gesamtdauer der Strafe stehen. Du Cane I Seite 22: „the treatment of prisoners should be adapted to the length of their sentences“.

[2]) Cf. Wines Seite 614: „Thus would the bracing stimulus of hope be kept ever active, and the prisoner would be encouraged and quickened in a course of vigorous self-restraint, self-conquest and self-culture“.

an das Strafleiden gewöhnen, die Strafe würde ihre innere sittliche Reaktion auf den Charakter des Sträflings verlieren, da sich derselbe allmählich vollständig indifferent dem Zustande unterwerfen würde, an dem er doch nichts ändern könnte.[1]) Deshalb ist Abstufung der Strafe in mehrere, in sich verschiedene Perioden erforderlich, damit der Gefangene in der Hoffnung erhalten wird, durch sein eignes Verhalten seine Lage bessern zu können, und damit er dadurch veranlaßt wird, fortdauernd an sich selbst zu arbeiten.[2]) Der Gefangene soll nicht bloß dulden, sondern ein vernünftiges Selbstwollen entwickeln.

Die einzelnen Stufen des Strafvollzugs stehen in einem inneren Zusammenhang: durch das erste Stadium soll der Wille des Gefangenen zur Unterwerfung gebracht werden; er soll einen höheren Willen über sich anerkennen lernen; in den folgenden Stadien soll der Gefangene lernen, einen eignen Willen zu haben, welcher sich den bestehenden Einrichtungen unterordnet.[3])

Bei den zu einer Gefängnisstrafe Verurteilten wird davon ausgegangen, daß nicht alt-eingerottete verbrecherische Neigungen, nicht ein wirklicher Hang zum Bösen vorliegt[4]) und daß es deshalb genügt, wenn dem Betreffenden während der Strafverbüßung sein Unrecht zum Bewußtsein gebracht, sein Pflichtgefühl neu belebt, seine Selbstbeherrschung gekräftigt, sein Sinn für Ordnung und Arbeitsamkeit von neuem geweckt wird. Man kann alsdann die Hoffnung hegen, daß der Betreffende, von der Notwendigkeit und Nützlichkeit einer Rechtsordnung überzeugt, in seiner bürgerlich guten Gesinnung gehoben, als nützliches Glied in die

---

[1]) Cf. v. Holtzendorff II Seite 34.

[2]) Cf. Du Cane II Seite 77: „The principle on which this system (viz. the system of progressive stages) is founded is that of setting before prisoners the advantages of good conduct and industry by enabling them to gain certain privileges or modifications of the penal character of the sentence by the exertion of these qualities".

[3]) Dies ist es, was in England unter reforming the prisoner verstanden wird: der Gefangene soll zu bürgerlich guter Gesinnung erzogen werden. Dies ist nicht zu verwechseln mit moralischer Besserung, von der es dahingestellt bleiben muß, inwieweit man hierin bei erwachsenen Menschen auf einen Erfolg rechnen kann.

[4]) Momentane Willensschwäche, Genußsucht, Unlust zur Arbeit sind hier als leitende Motive anzusehen. Durch die Strafe soll diesen Motiven in energischer Weise Halt geboten werden.

Gesellschaft zurücktreten werde. Es ist deshalb nicht nötig, daß der Gefangene in der Anstalt selbst den in der Gemeinschaftshaft liegenden Versuchungen probeweise ausgesetzt wird. Einzelhaft mit fortschreitender Abnahme des Strafzwanges und der Isolierung erscheint hier das richtige Strafmittel.

Anders bei den zur penal servitude Verurteilten. Hier handelt es sich um Personen, welche sich als für das Gemeinwesen gefährliche Subjekte erwiesen haben und bei denen eine vollständige Umbildung des Charakters vorgenommen werden muß, um aus ihnen nützliche Glieder der bürgerlichen Gesellschaft zu machen. Im Interesse der Gesellschaft müssen diese Personen während der Strafverbüßung selbst auf die Probe gestellt werden, ob sie Versuchungen, wie dieselben zweifellos bei ihrer Entlassung an sie herantreten werden, zu widerstehen vermögen.[1] Die Gemeinschaftshaft ist hier Notwendigkeit.

Der Verbrecher muß erst praktisch beweisen, daß er die guten Vorsätze, welche in der Isolierhaft in ihm wachgerufen sind, auszuführen die Kraft besitzt, daß sein verbrecherischer Wille wirklich gebrochen ist und daß er nicht nur gezwungen, sondern aus eignem Entschlusse der bestehenden Ordnung sich unterzuordnen gewillt ist. Erst wenn er diese Prüfung bestanden hat, kann man ihm vertrauen, daß er von der wiedererlangten Freiheit einen vernünftigen Gebrauch machen wird. Und diese Prüfung muß, wenn man der Gefahr einer Täuschung durch Heuchelei entgegenwirken will, eine langjährige sein, bei welcher mit immer mehr nachlassendem Strafzwange immer stärkere Versuchungen zu überwinden sind. Hier gilt der Satz „festina lente". In der Gemeinschaftshaft ist die Straferziehung zu vollenden, für welche die Isolierhaft nur vorbereiten soll.[2]

---

[1]) An diesem Punkte tritt die Rücksicht auf die Allgemeinheit deutlich und scharf in den Vordergrund vor derjenigen für das einzelne Individuum.

[2]) Cf. Wines Seite 614: „to prepare men for society in society appears to be just as necessary as to prepare them to be seamen on the sea or engineers in the woods and fields. Moral lessons, like navigation and engineering, require a field of progressive experimental application. Books, counsels, exhortations are not enough. There must be friction, the contact with temptation, and the toning up and hardening of the character, which result from the habitual and successful resistance of temptation".

Hiermit ist gleichzeitig die **Lösung** gegeben, **welche die Frage nach der Anwendung der Einzelhaft in England gefunden** hat: Isolierung während der Nacht für alle Gefangenen; während der Tageszeit Trennung mit allmählich nachlassendem Strafzwange und allmählicher Erleichterung in der Isolierung bei denjenigen Gefangenen, welche von vornherein als besserungsfähig erscheinen, dagegen bei den andern Gefangenen prinzipiell Gemeinschaftshaft mit ebenfalls allmählich nachlassendem Strafzwange und nur als Vorbereitung für das Hauptstadium der Strafe Einzelhaft, welche fortfällt, sobald der Betreffende für die Gemeinschaftshaft reif zu erachten ist.

Zu welcher von diesen beiden Klassen der Verurteilte gehört, bestimmt der Richter, indem er zwischen den beiden Strafmitteln des Gefängnisses und der penal servitude die Wahl trifft. Die Dauer der in Einzelhaft zu verbüßenden Gefängnisstrafe darf — von unbedeutenden Ausnahmen abgesehen — die Dauer von zwei Jahren nicht überschreiten, die Dauer der eine vollständige Erziehung zum vernünftigen Gebrauche der Freiheit enthaltenden Strafe der penal servitude darf nicht unter fünf Jahren zurückbleiben.

Für beide Arten der Strafe aber gilt — was hier nochmals hervorgehoben werden soll — der Grundsatz, daß ihre Vollstreckung nicht während der ganzen Strafdauer eine gleichförmige sein darf,[1]) sondern daß vielmehr der Strafzwang je nach dem Verhalten des Gefangenen nach bestimmter Zeit nachzulassen hat. Dieses Verhalten, welchem hiernach für den Inhalt der Strafe eine bestimmende Bedeutung beigelegt ist, wird nach festgeregelten Grundsätzen unter Anwendung des nun näher zu besprechenden Markensystems überall gleichmäßig konstatiert.

§ 13. Das Markensystem.

Das Markensystem ist zuerst von **Kapitän Maconochie bei den zur Transportationsstrafe Verurteilten** zur Anwendung gebracht worden. **Sir Walter Crofton** hat es so-

[1]) Auch bei Einzelfragen in der Behandlung der Gefangenen, wie Beschäftigung, Beköstigung u. s. w., tritt durchweg das Prinzip hervor, die körperliche und geistige Spannkraft wach zu erhalten und alle Einförmigkeit zu vermeiden, durch welche diese Kraft geschwächt werden könnte.

dann in Irland bei den zur penal servitude Verurteilten eingeführt, und von Irland ist es nach England übernommen worden. Das System hat zunächst in England auf Grund der Empfehlungen der Royal Commission von 1863 eine etwas veränderte Gestalt angenommen. Dieses neue System, welches in Gemäßheit der Vorschläge der Royal Commissioners von 1879 noch einige Veränderungen erhalten hat, ist jetzt auch in Irland eingeführt worden.[1]) Alsbald nach Übernahme der local prisons im Jahre 1878 in die Staatsverwaltung ist das Markensystem sodann weiter auch bei den zur Gefängnisstrafe Verurteilten zur Anwendung gebracht worden.[2]) Es bildet somit jetzt eine charakteristische Erscheinung des englischen Gefängniswesens überhaupt.

Die wesentlichsten Grundsätze des Systems sind die folgenden:

Jeder Gefangene erhält eine Karte, auf welcher ihm täglich in Gestalt der Eintragung einer bestimmten Markenzahl ein Attest über sein Verhalten ausgestellt wird.

Dieses Attest bezieht sich zunächst lediglich auf den bei der zugeteilten Arbeit bewiesenen Fleiß. Der Gefangene erhält 8 Marken täglich bei fortgesetzter angestrengter Arbeit (steady hard labour) und Vollendung des ganzen ihm zugeteilten Arbeitspensums; 7 Marken bei einem geringeren Grade von Emsigkeit; 6 Marken, wenn er nur ein mäßiges Tageswerk (a fair but moderate day's work) vollbringt. Die Zuteilung der Marken liegt unter fortgesetzter Kontrolle durch den Gefängnis-Oberaufseher und den Governor, sowie unter gelegentlicher Abschätzung der Tagespensa durch spezielle Sachverständige demjenigen Wärter ob, welcher die Beschäftigung des betreffenden Gefangenen zu beaufsichtigen hat.

Wenn ein Gefangener infolge seines körperlichen Zustandes zu harter, angestrengter Arbeit nicht angehalten werden kann und aus diesem Grunde nur zu leichter Arbeit (light labour) verwendet wird, so konnte er früher den Maximalbetrag von 8 Marken täglich überhaupt nicht erhalten, sondern es wurden ihm nur, je nach seinem Fleiße, 6 oder 7 Marken zugeteilt. Diese Bestimmung, welche von

[1]) Report of the Royal Commission on Prisons in Ireland 1884 Seite 27.

[2]) Report of Commissioners 1878 Seite 39 ff.

Sir Edm. Du Cane aus dem Grunde verteidigt wurde, daß sie den Verhältnissen des gemeinen Lebens entspräche, wo auch ein schwächlicher Mensch nicht soviel Lohn verdienen könnte, als ein kräftiger, wurde in dem Berichte der Royal Commissioners von 1879 für irrationell erklärt: bei Abmessung der Gefangenenarbeit sei nicht sowohl auf die Arbeitsfähigkeit, als auf die Arbeitswilligkeit Rücksicht zu nehmen.[1]) Seitdem ist diese Bestimmung dahin geändert, daß Gefangene, welche mit light labour beschäftigt sind, zunächst zwar nur 6 resp. 7 Marken täglich zugeteilt erhalten können, daß jedoch der Governor die Befugnis hat, die zugeteilte Markenzahl nachträglich zu erhöhen, wenn er überzeugt ist, daß der Betreffende in der That schwere Arbeit nicht zu verrichten vermag.[2])

Wenn ein Gefangener krank ist, so erhält er während der Zeit, wo er nicht arbeiten kann, regelmäßig nur 6 Marken täglich. Nur ausnahmsweise kann ihm von dem Governor eine höhere Markenzahl bewilligt werden, wenn die Krankheit die Folge eines Unfalls bei der Arbeit ist.

Wenn somit in erster Linie die Arbeitsleistung für die Zuteilung der Marken maßgebend ist, so wird jedoch weiter auch dem Betragen des Gefangenen ein gewisser Einfluß beigelegt.[3])

Zunächst richtet sich die Zahl der für den Sonntag, wo nicht gearbeitet wird, gegebenen Marken lediglich nach dem Betragen des Gefangenen mit der Maßgabe, daß der Gefangene in keinem

---

[1]) Report of the Royal Commissioners 1879 Seite XLV: „the test being not the capacity but the willingness to work".

[2]) Die Royal Commissioners erkennen an, daß es nötig ist, etwaigen Simulationen vorzubeugen „no doubt it is necessary to take the strictest precautions to prevent convicts from evading hard labour on the pretence of infirmity and sickness, but it cannot be impossible to combine such precautions with the allotment of full marks to all who satisfy the prison authorities that they are doing their best".

[3]) Die Royal Commission von 1863 (Report § 39) betonte mit Entschiedenheit, daß gutes Betragen allein nicht ausschlaggebend für die Zuteilung der Marken sein dürfte „credit should not be given for general good conduct as well as for industry, good conduct in prison apart from industry being merely abstinence from misconduct".

Falle für den Sonntag mehr Marken erhalten kann, als er im Durchschnitte der vorhergehenden Wochentage verdient hat. [1])

Sodann ist der Governor berechtigt, wegen jedes Verstoßes gegen die Gefängnisdisziplin auf Entziehung bereits verdienter Marken zu erkennen; und es bildet dies für leichtere Disziplinarstraffälle die zumeist angewandte Art der Strafe. Der Governor kann je nach Lage des Falles auf Verwirkung von 6—84 Marken erkennen. [2])

Man wird hiernach sagen können, daß in der Zahl der verdienten Marken dem Gefangenen ein Attest gegeben wird über den bei der Arbeit bewiesenen Fleiß unter Berücksichtigung seines guten Verhaltens. [3])

Die praktische Bedeutung der verdienten Marken liegt nun zunächst darin, daß die Zeitdauer der einzelnen Stufen der Strafvollstreckung abgemessen wird nach bestimmten Beträgen von Marken. Ein Beispiel wird dies klar machen. Bei den zu einer Gefängnisstrafe Verurteilten besteht die Bestimmung, daß dieselben auf der ersten Stufe der Strafvollstreckung — der strengen Isolierhaft, wo der Gefangene keine andre Person als die Gefängnis-Beamten zu sehen bekommt — verbleiben müssen, bis sie 224 Marken verdient haben. Da der Gefangene in maximo täglich 8 Marken verdienen kann, so beträgt die Zeitdauer dieser ersten Strafstufe bei vorzüglichem Fleiße des Gefangenen und unter der Voraussetzung, daß er sich keinerlei Übertretung der Gefängnisordnung zu Schulden kommen

---

[1]) In der Praxis ist die Zahl der für den Sonntag gegebenen Marken regelmäßig dieselbe, wie diejenige an den vorhergehenden Wochentagen.

[2]) Die Verwirkung von Marken kann sowohl als alleinige Strafe in Anwendung gebracht werden, wie als Nebenstrafe neben anderweitigen Disziplinarstrafen.

[3]) Das früher in Irland zur Geltung gebrachte Prinzip war ein anderes. Hier erfolgte die Zuteilung der Marken monatlich, und der Gefangene konnte monatlich in maximo 9 Marken verdienen, und zwar 3 Marken für seinen Arbeitsfleiß, 3 Marken für sein gutes Betragen und 3 Marken für den beim Unterricht bewiesenen Fleiß. Das Betragen hatte somit einen positiven Einfluß auf die Zuteilung von Marken, während es jetzt im wesentlichen nur negativ: in der Entziehung von verdienten Marken bei schlechtem Verhalten, zum Ausdruck gelangt. Dem Fleiß beim Unterricht wird jetzt überhaupt bei der Zuteilung der Marken eine Berücksichtigung nicht mehr geschenkt.

läßt, 28 Tage. [1]) Wenn der Gefangene weniger fleißig ist und z. B. nur 7 Marken durchschnittlich per Tag verdient, so verlängert sich diese Zeitdauer auf 32 Tage; wenn der Betreffende dann noch etwa wegen eines Verstoßes gegen die Gefängnisordnung 35 Marken verwirkt hat, so muß er 37 Tage in strenger Isolierhaft bleiben. Dem Gefangenen wird auf diese Weise durch seine Markenkarte stets vor Augen gehalten, daß er durch sein eignes Verhalten die Zeitdauer der einzelnen Strafstufen bestimmt.

Bei den zur penal servitude Verurteilten, bei welchen eine Strafkürzung zulässig ist, wird durch die Zahl der verdienten Marken nicht nur die Dauer der einzelnen Strafstufen, sondern weiter auch die Gesamtdauer der Strafzeit innerhalb gewisser Grenzen bestimmt. [2])

Eine weitere Bedeutung erlangen die verdienten Marken bei der Bemessung der Arbeitsbelohnung (gratuity). Die Höhe dieser Arbeitsbelohnung wird, wie unten im § 22 noch näher auszuführen ist, berechnet durch Multiplikation der Ziffer der Marken mit bestimmt festgesetzten Wertbeträgen.

Dies sind die wesentlichsten Grundsätze des Markensystems, welches auf den ersten Anblick vielleicht etwas kompliziert erscheint, sich aber in der Praxis äußerst leicht handhabt. Es sei dabei noch besonders hervorgehoben, daß die praktische Handhabung eine äußerst gewissenhafte ist; von den Gefängnis-Oberbeamten wird strenge Kontrolle darüber geführt, daß volle Gerechtigkeit bei der Zuteilung der Marken herrscht. [3])

---

[1]) Personen, welche zu weniger als 28 Tagen Gefängnis verurteilt worden sind, müssen hiernach die ganze Strafzeit in strenger Isolierhaft verbringen. Diese Gefangenen bilden diejenige Klasse, auf welche — wie im vorhergehenden Paragraphen näher ausgeführt wurde — nur strafend, nicht auch erziehend eingewirkt werden soll.

[2]) Das Nähere hierüber sowie über andre Spezialbestimmungen bezüglich der Anwendung des Markensystems auf die convict prisoners wird unten im Titel III behandelt werden.

[3]) Cf. Report of the Royal Commissioners 1879 Seite XXVI f.: „the allotment of marks is no mere form“, „the convicts themselves jealously watch the record“. Es mag dabei bemerkt werden, daß ein Gefangener, welcher glaubt, bei Zuteilung der Marken ungerecht behandelt zu sein, ein Beschwerderecht hat. Cf. auch Du Cane I Seite 133 f., wo ein Zirkularbrief der Zentralbehörde abgedruckt ist, in welchem die Gefängnisbeamten auf die Bedeutung des Markensystems hingewiesen werden und denselben möglichste Sorgfalt bei Ausführung des Systems besonders an das Herz gelegt wird.

Die Vorzüge des Markensystems lassen sich kurz dahin zusammenfassen:

1) Das Verhalten der Gefangenen wird fortlaufend auf gerechte und zuverlässige Weise konstatiert; es wird dadurch eine sichere Unterlage für ein Gesamturteil über den einzelnen Gefangenen geschaffen.[1])

2) Der Gefangene wird in steter Spannung gehalten, er hat ständig vor Augen, wie durch sein eignes Verhalten seine Lage bestimmt wird: daß er, ohne sich selbst zu schaden, keinen Tag in seiner Arbeit träge sein darf. Es ist ihm in jedem Momente eine anschauliche Feststellung der Folgen seines Verhaltens gegeben: er sieht, welche Fortschritte er in der Strafverbüßung macht, welches seine Aussichten für die Zeit der Entlassung sind. Dadurch, daß ihm diese Ziele ständig in das Gedächtnis gerufen werden, wird Hoffnung und eignes Streben in ihm wachgerufen: er geht aus eignem Interesse, nicht bloß gezwungen, an die Arbeit. Das Leben in der Anstalt wird auf diese Weise dem wirklichen Leben ähnlicher.[2])

3) Soweit die Gefangenen in Gemeinschaftshaft gehalten werden, hat das System noch den besonderen Vorteil, daß dem einzelnen Gefangenen seine Sonderinteressen klar vor Augen gehalten werden und daß damit der Tendenz auf Genossenschaftlichkeit unter den Gefangenen in wirksamer Weise entgegengewirkt ist.

§ 14. Die Aufnahme der Gefangenen.

Der Gefangene wird bei seinem Eintritt in die Anstalt zunächst in einen besonderen Aufnahmeflügel (reception ward) gebracht. Hier werden seine Papiere nachgesehen und ein

---

[1]) Es ist dies besonders bei langjährigen Strafen von der größten Wichtigkeit. Ohne eine solche sichere Unterlage wird regelmäßig — besonders bei inzwischen stattgehabtem Wechsel des Beamtenpersonals — nur ein oberflächliches Gesamturteil möglich sein.

[2]) Cf. Report of the Royal Commissioners 1879 Seite XXVII: „we need scarcely point out the value of this system both as an aid to the

vollständiges Nationale (nominal record) von ihm aufgenommen. Er wird sodann von dem Gefängnisarzt sobald als möglich untersucht. Der Arzt hat den allgemeinen Gesundheitszustand des Gefangenen in dem Nationale zu vermerken. Wenn der Gefangene krank befunden wird, so ist event. seine Aufnahme in die infirmary zu verfügen oder es sind sonst geeignete Anordnungen zu treffen. [1])

Der Gefangene wird sodann in die Badezelle gebracht, hat sich hier vollständig zu entkleiden und wird nackt von einem Gefängnisbeamten untersucht, [2]) ob er etwa Verbotenes bei sich führt. Nachdem er ein Bad genommen hat, hat er die Gefangenenkleidung anzuziehen.

Die eignen Kleidungsstücke des Gefangenen werden, falls es erforderlich erscheint, zur Reinigung und Desinfektion geschickt und sodann aufbewahrt; der Gefangene erhält darüber einen Empfangsschein. Wenn der Gefangene Geld oder sonst Sachen von Wert bei sich hat, so soll er dieselben an einen besondern Beamten abliefern, welcher darüber eine Eintragung in das Prisoner's Property Book zu machen hat.

Tabak, Spirituosen und sonst verbotene Artikel werden dem Gefangenen fortgenommen und zerstört.

Convict Prisoners werden sofort bei ihrer Aufnahme in die

---

maintenance of order and discipline and as a means of inducing the prisoners to acquire habits of steady industry and obedience".

[1]) Wenn der Gefangene an einer ansteckenden Krankheit leidet oder sonst sein Gesundheitszustand ein Verbleiben in der Anstalt unthunlich erscheinen läßt, so hat der Arzt darüber sofort dem Governor Bericht zu erstatten, damit der letztere bezüglich der Fortschaffung des Gefangenen das weitere verfügen kann.

[2]) Bei weiblichen Gefangenen hat dies durch eine Gefängniswärterin zu geschehen. — Bei der Entkleidung und der Untersuchung darf niemals ein andrer Gefangener zugegen sein. Die Untersuchungen (naked searches) werden während des Aufenthaltes in der Anstalt periodisch — in Pentonville wöchentlich einmal, in Portland sechsmal im Jahre — wiederholt und finden ferner auf besondere Anordnung des Governor oder eines Gefängnis-Oberaufsehers dann statt, wenn Verdacht vorliegt, daß der Gefangene verbotene Artikel versteckt hält. Der Report of the Royal Commissioners 1879 Seite XLVII spricht sich gegen die allzuhäufige Wiederholung der naked searches aus, welche weder im Interesse der Sicherheit noch der Disziplin für notwendig erachtet werden könnten. In den Public Works Prisons werden die Gefangenen, sobald sie von der Arbeit zurückkommen, durch einfaches Befühlen des Körpers darauf untersucht, ob sie irgend etwas zu sich gesteckt haben.

Anstalt photographiert. [1]) Dieses kann auch bei andern Gefangenen geschehen, soweit es von dem Governor angeordnet wird.

Jeder Gefangene wird ferner bei seiner Aufnahme in die Anstalt gewogen und sein Gewicht in dem Nationale vermerkt. Den Convict Prisoners wird das Kopfhaar kurz geschnitten [2]) und der Bart abrasiert.

Der Gefangene erhält sodann seine Zelle angewiesen. [3]) Für jede Klasse von Gefangenen (cf. oben § 1) besteht eine besondere Abteilung (ward oder division). Wenn jedoch die Zellen der einen Abteilung besetzt sind, während in einer andern noch Zellen leer stehen, so kann der Governor dem Gefangenen für kurze Zeit eine Zelle in der Abteilung, in welche derselbe eigentlich nicht hineingehört, anweisen. Es sind jedoch alsdann besondere Vorkehrungen zu treffen, damit der Gefangene niemals — insbesondere nicht bei den Spaziergängen — Gefangene der andern Klasse zu Gesicht bekommt. Nur aus besonderen, in jedem Einzelfalle speziell zu vermerkenden Rücksichten kann ein Gefangener mit einem andern Gefangenen in dieselbe Zelle gebracht werden. Der Fall, wo dies am häufigsten vorkommt, ist bei epileptischen Personen, welche nach Ansicht des Arztes nicht allein gelassen werden dürfen. Soweit ein Gefangener hiernach mit andern dieselbe Zelle zu teilen hat, sucht man es aus naheliegenden Gründen einzurichten, daß drei Personen zusammen in eine Zelle kommen.

Eine weibliche Gefangene, welche ihr Kind selbst nährt, kann dasselbe mit in die Anstalt bringen. Das Kind ist so lange bei der Mutter zu lassen, als die letztere dasselbe an der Brust behält. Sobald das Kind neun Monate alt ist, hat der Gefängnisarzt speziell darüber zu berichten, ob es mit Rücksicht auf den Zustand des Kindes notwendig ist, dasselbe noch länger bei der Mutter zu belassen; doch soll das Kind in keinem Falle über das Alter von einem Jahre hinaus in der Anstalt behalten werden. [4])

---

[1]) Cf. oben Seite 16, Anm. 2.

[2]) Den weiblichen Gefangenen darf nur aus besonderen Rücksichten der Gesundheit oder Reinlichkeit das Haar abgeschnitten werden.

[3]) Die besonderen, für die not criminal prisoners geltenden Bestimmungen sind bereits im § 11 angeführt worden und sollen im folgenden nicht nochmals wiederholt werden. Das hier Gegebene gilt somit uneingeschränkt nur für die criminal prisoners.

[4]) Solange sich das Kind in der Gefängnisanstalt befindet, ist es auf Kosten der Gefängnisverwaltung mit den erforderlichen Kleidungsstücken u. s. w.

Der Gefängniswärter, unter dessen spezielle Aufsicht der einzelne Gefangene gestellt ist, hat denselben sofort nach der Aufnahme über die ihm obliegenden Pflichten zu unterrichten. Der Gefangene ist dabei mit den wesentlichsten Bestimmungen der Gefängnisordnung bekannt zu machen. In jeder Zelle ist ein Auszug dieser Bestimmungen angeschlagen und derselbe soll dem des Lesens unkundigen Gefangenen vorgelesen werden.

§ 15. Die Pflichten der Gefangenen und die Tagesordnung.

Der Gefangene hat allen Anordnungen der Gefängnisbeamten unbedingten Gehorsam zu leisten. Im Falle des Ungehorsams macht er sich straffällig.

Anderseits ist aber dafür umfassende Sorge getragen, daß der Gefangene sich über jede ungerechtfertigte Anordnung beschweren kann. Er kann die Beschwerde zunächst bei dem Anstaltsvorsteher, dem Governor oder Deputy Governor, anbringen, welcher täglich wenigstens einmal die ganze Anstalt zu durchwandern hat. Der Gefangene kann sodann ferner seine Beschwerde dem Visiting Committee — bei den convict prisons den Visitors — vortragen. Wie schon ausgeführt wurde, liegen für diese außerhalb der Gefängnisverwaltung stehenden Aufsichtsorgane in der Anstalt besondere Bücher — Minute Books — aus, in welche sofort eine Eintragung zu machen ist, sobald ein Gefangener den Wunsch ausspricht, ein Mitglied dieser Organe zu sprechen. Endlich hat jeder Gefangene das Recht, sich unter Umgehung der unteren Instanzen an die Zentralgefängnisbehörde — ja sogar direkt an den Staatssekretär — zu wenden. Wenn auch von dieser letzteren Befugnis wegen der dabei notwendigerweise eintretenden Verzögerung der Entscheidung thatsächlich wenig Gebrauch gemacht werden wird, so enthält dieselbe doch eine wichtige Schutzmaßregel gegen eine etwaige ungerechte Behandlung eines Gefangenen in der Anstalt. Der Governor ist durch die Gefängnisordnung ausdrücklich verpflichtet, derartige Eingaben sofort an die betreffende Adresse gelangen zu lassen.

---

zu versehen. Wenn das Kind von der Mutter fortgenommen wird und Verwandte, zu denen es gebracht werden könnte, nicht vorhanden sind, so ist es in das nächstgelegene workhouse zu schaffen.

Bei diesem weitgehenden Beschwerderecht muß natürlicherweise für frivol angebrachte, unbegründete Beschwerden eine Bestrafung eintreten.

Betreffs des allgemeinen Verhaltens der Gefangenen seien sodann noch folgende Vorschriften hervorgehoben.

Der Gefangene hat sich jeder Beschädigung oder Beschmutzung der Gefängnisräume und der darin enthaltenen Utensilien zu enthalten. Er ist verpflichtet, täglich seine Zelle zu säubern und sein Bett ordentlich zu machen. Seine Kleidungsstücke, sein Werkzeug und die ihm übergebenen Bibliotheksbücher hat er sorgfältig zu behandeln; mutwillige Beschädigungen werden auf das strengste bestraft.

Dem Gefangenen wird für seine eigne Person die größte Reinlichkeit zur Pflicht gemacht. Soweit nicht ärztliche Anordnungen entgegenstehen, muß jeder Gefangene in bestimmten Zwischenräumen — zumeist alle 14 Tage — ein Bad nehmen. [1])

Es wird streng darauf gehalten, daß in der Anstalt die größte Ruhe herrscht. Singen, Pfeifen und sonstiges Lärmen wird streng bestraft. Der Gefangene darf weder in der Zelle noch auf dem Spaziergange mit andern Gefangenen durch Sprechen, Zeichen, Klopfen u. dgl. in Verkehr treten. [2]) *Für Gefangene, welche in gemeinsamen Arbeitssälen beschäftigt werden, besteht kein absolutes Schweiggebot.* Die Gefangenen müssen jedoch, wenn sie sprechen, so laut reden, daß der Gefängniswärter die Worte verstehen kann. Der letztere darf ferner nicht dulden, daß die Gefangenen eine eigentliche Unterhaltung über etwas führen, was nicht zur Arbeit selbst gehört. Der Aufforderung zum Stillschweigen ist sofort Folge zu geben. [3])

---

[1]) Gewöhnlich erhält der Gefangene abwechselnd in der einen Woche ein Fußbad, in der andern ein Vollbad.

[2]) Es ist schon hervorgehoben worden, daß Übertreibungen der Isolierung — wie Masken, stalls in Kirchen oder Schule — in den englischen Anstalten nicht bestehen. Auch Isolierspazierhöfe kommen nur noch ausnahmsweise zur Verwendung.

[3]) Englische Gefängnisbeamte haben mir gegenüber wiederholentlich ihre Ansicht dahin ausgesprochen, daß es unmöglich sei, in der Gemeinschaftshaft ein absolutes Schweiggebot durchzuführen. Ein derartiges *unnatürliches* Gebot provoziere zu Übertretungen und führe beständige Disziplinarstrafen mit

Das tägliche Leben der Gefangenen verläuft in der größten Regelmäßigkeit. Die Tagesordnung ist unter Berücksichtigung der Jahreszeiten für das ganze Jahr genau bestimmt.[1])

In den local prisons wird um 6 resp. $6^1/_2$ Uhr durch das Läuten der Glocken das Signal zum Aufstehen der Gefangenen gegeben; in den Public Works Prisons geschieht dies schon um 5 resp. $5^1/_2$ Uhr. Den Gefangenen werden 40 Minuten zur Reinigung des Körpers und zum Ankleiden, sowie zur Säuberung ihrer Zelle gewährt. Für das Frühstück[2]) haben die Gefangenen sodann $^3/_4$ Stunden Zeit.

In den Public Works Prisons findet, bevor die Gefangenen zur Arbeit gehen, noch ein kurzer Gottesdienst statt. Die Arbeit beginnt in den Public Works Prisons um 7 resp. $7^1/_2$ Uhr. Um $11^1/_2$ Uhr kehren die Gefangenen zum Mittagessen zurück; die Pause für das Mittagessen beträgt $1^1/_4$ Stunde. Die Nachmittagsarbeit dauert sodann bis 4 resp. $5^1/_4$ Uhr. Die eigentliche Arbeitszeit abzüglich der Zeit für den Weg zur Arbeitsstelle beträgt während des größten Teils des Jahres — vom 16. Februar bis 31. Oktober — $8^1/_2$ Stunden täglich. Für das Spazierengehen ist hier, wo die Gefangenen im Freien arbeiten, eine besondere Zeit nicht gegeben.

In den local prisons ist die Tageseinteilung je nach der Strafstufe, in welcher sich der Gefangene befindet, eine verschiedene. Die Arbeitszeit beträgt hier zwischen 9 und 10 Stunden. Einen Teil dieser Zeit verbringt der Gefangene im Freien. Der Gefangene soll täglich wenigstens zwei Stunden, sei es arbeitend sei es spazieren gehend, sich im Freien aufhalten.

Für den Unterricht sind regelmäßig nur die Abendstunden be-

---

sich. Was man erreichen könne und erstreben müsse, sei, zu verhindern, daß verfängliche Themata besprochen und daß längere Unterhaltungen, welche von der Arbeit abziehen, geführt würden.

[1]) Nach Maßgabe der Tagesordnung vollzieht sich auch der Dienst der Beamten mit großer Präzision: innerhalb 25 Minuten, nachdem das Signal zum Aufstehen der Gefangenen gegeben ist, versammeln sich die Gefängnisbeamten zum Appell. Der Dienst dauert mit genügenden Unterbrechungen für die Mahlzeiten bis 7 Uhr abends. Alsdann treten besondere, für den Nachtdienst bestimmte Aufseher (evening [reserve] guards, night patrols) an, deren Dienst bis zum Appell am andern Morgen dauert.

[2]) Das Frühstück, wie die andern Mahlzeiten, wird in großen Speisekesseln auf Handwagen von Zelle zu Zelle gefahren und von Gefängnisaufsehern an die Zelleninsassen verteilt.

stimmt. Das Abendessen wird in den Public Works Prisons um $4^1/_2$ resp. $5^1/_2$ Uhr, in den andern Anstalten um $5^3/_4$ Uhr gereicht. In den Public Works Prisons wird das Licht um 8, in den andern Anstalten um $8^1/_2$ Uhr ausgedreht. Die Schlafenszeit beträgt somit in den Public Works Prisons 9 resp. $9^1/_2$, in den andern Anstalten $9^1/_2$ resp. 10 Stunden.

Die hier beschriebene Tagesordnung erleidet nur am Sonnabend eine Ausnahme. An diesem Tage bleibt der Nachmittag von jeder Arbeit frei und wird zur Reinigung des Anstaltsgebäudes, zur Ablieferung der schmutzigen und Ausgabe der reinen Wäsche,[1]) zum Baden, Haarschneiden, zur ärztlichen Untersuchung der Gefangenen, zur Verteilung der Bibliotheksbücher u. s. w. benutzt.

An Sonn- und Feiertagen ruht jede Arbeit. Das Signal zum Aufstehen wird erst um 7 Uhr morgens gegeben und schon um 7 Uhr abends werden die Lichter ausgelöscht. Am Vormittage und am Nachmittage ist Gottesdienst und es finden vormittags wie nachmittags längere Spaziergänge[2]) statt.

§ 16. Verkehr der Gefangenen nach außen.

In der Erlaubnis zum Verkehr mit der Außenwelt durch Korrespondenz und Besuche besteht eine der Hauptvergünstigungen, welche sich ein Gefangener durch sein gutes Verhalten beim Aufsteigen in die höheren Strafstufen erwerben kann.[3])

I. Die zur penal servitude Verurteilten dürfen zunächst bei ihrer Aufnahme in die Anstalt einen Brief schreiben, in welchem sie ihre Angehörigen von ihrem Aufenthaltsorte unterrichten. Diese Befugnis wiederholt sich, sobald der Gefangene von einer Anstalt in eine andere geschafft wird. Im übrigen dürfen die convict prisoners auf der ersten Strafstufe (der probation class) Briefe

---

[1]) Die Leibwäsche wird wöchentlich, die Bettwäsche monatlich einmal gewechselt.

[2]) Den local prisoners werden während der ersten Strafstufe — der strengen Isolierhaft — Spaziergänge am Sonntag nicht gestattet.

[3]) Wegen der dem Visiting Committee zustehenden Befugnis auf Zulassung eines Verkehrs mit der Außenwelt bei den not criminal prisoners cf. oben Seite 186.

weder schreiben noch empfangen. Auf der zweiten Strafstufe dürfen sie, vorausgesetzt daß ihr Verhalten während der letzten zwei Monate ein gutes war, alle sechs Monate, auf der dritten Strafstufe alle vier Monate, auf der vierten Strafstufe alle drei Monate je einen Brief schreiben und empfangen.

Die zu einer Gefängnisstrafe Verurteilten dürfen erst auf der vierten Strafstufe — d. i. frühestens nach Verbüßung von drei Monaten — einen Brief schreiben und empfangen. Von da ab dürfen sie, vorausgesetzt daß ihr Verhalten ein gutes bleibt, alle drei Monate einen Brief schreiben und empfangen.

Der Governor hat das Recht, im Disziplinarwege die dem Gefangenen hiernach zustehenden Befugnisse zur Korrespondenz einzuschränken, anderseits kann der Governor unter besonderen Verhältnissen — z. B. bei gefährlicher Erkrankung eines Familienangehörigen — ausnahmsweise eine weitergehende Befugnis zur Korrespondenz gewähren. In beiden Fällen sind die Gründe in dem von dem Governor zu führenden Journal zu vermerken.

Wenn ein Gefangener nicht schreiben kann, so übernimmt es im Anfange der Strafzeit der Lehrer, den Brief für den Gefangenen aufzusetzen. Diese Hilfe fällt jedoch — von Ausnahmefällen abgesehen — nach einiger Zeit fort. Es soll dadurch dem Gefangenen ein in der That wirksamer Anlaß gegeben werden, das Schreiben zu erlernen.

Über den Inhalt der Korrespondenz führt der Governor eine strenge Kontrolle. Jeder ankommende oder abgehende Brief wird zunächst von ihm durchgesehen, und die Weiterbeförderung erfolgt erst dann, wenn der Governor zum Zeichen dafür, daß der Inhalt keinem Bedenken unterliegt, den Brief mit seinen Initialen versehen hat.[1]) Im Falle der Beanstandung des Inhalts werden die Briefe zurückbehalten, und die Gründe hierfür sind in dem Journal besonders anzugeben.

Außer dem Governor erhält auch der Gefängnisgeistliche Kenntnis von der Korrespondenz: die von dem Gefangenen geschriebenen Briefe werden zunächst dem Geistlichen vorgelegt, welcher zum Zeichen dafür, daß er Bedenken gegen

[1]) Eine Ausnahme besteht für die an die Zentralgefängnisbehörde resp. an den Staatssekretär gerichteten Eingaben, welche ohne weiteres an ihre Adresse zu befördern sind.

den Inhalt des Briefes nicht hat, seine Initialen beisetzt. Die an einen Gefangenen adressierten Briefe werden dem Geistlichen erst vorgelegt, nachdem der Governor Entscheidung dahin getroffen hat, daß dieselben nicht beanstandet werden. Der Geistliche erhält auf diese Weise, ohne daß ihm selbst eine Kontrollbefugnis beigelegt ist, einen Einblick in die Korrespondenz, wie ein solcher für eine richtige Beurteilung des Gefangenen durch den Geistlichen wünschenswert erscheint.

II. Was den Empfang von Besuchen anbetrifft,[1]) so darf der zur penal servitude Verurteilte auf der ersten Strafstufe gar keinen Besuch empfangen, auf der zweiten Strafstufe ist alle sechs Monate ein Besuch von 20 Minuten, auf der dritten Strafstufe alle vier Monate ein Besuch von 20 Minuten, auf der vierten alle drei Monate ein Besuch von 30 Minuten gestattet. Die zu einer Gefängnisstrafe Verurteilten dürfen erst auf der vierten Strafstufe — also frühestens nach Verbüßung von drei Monaten — einen Besuch von 20 Minuten empfangen; von da ab steht ihnen dieses Recht alle drei Monate einmal zu.

Wie bei der Korrespondenz, so hat auch hier der Governor das Recht, Ausnahmen zu gestatten resp. das Privileg zu entziehen.

Für jeden einzelnen Besuch hat der Governor eine schriftliche Erlaubnisordre auszustellen. Diese Erlaubnis wird in der Regel nur an Angehörige oder an nahe und gut beleumundete Freunde des Gefangenen gegeben. Der Besucher hat seinen Namen, seine Adresse und seine Beziehungen zu dem Gefangenen vor Erteilung der Ordre in ein besonderes Buch einzutragen. Wenn der Governor es für angezeigt hält, den Besucher abzuweisen, so hat er die Gründe hierfür in seinem Journal zu vermerken. Der Governor kann anordnen, daß der Besucher vor Zulassung daraufhin untersucht werde, ob er nicht verbotene Artikel bei sich führe.

Am Sonntag dürfen nur in Ausnahmefällen Besucher zugelassen werden. Im übrigen sind in den Hausordnungen der einzelnen Anstalten bestimmte Tage und Stunden für die Besuche festgestellt.

Bei den Besuchen muß stets ein Gefängnisbeamter — bei Besuchen weiblicher Gefangener eine Aufseherin — zugegen sein. Die

[1]) Über die regelmäßigen Besuche der Gefangenen durch die Gefängnisbeamten und die Mitglieder des Visiting Committee ist schon oben das Nähere angegeben worden.

Besuche finden regelmäßig nicht in der Zelle, sondern in besonderen, für den Empfang von Besuchen bestimmten Räumen statt.

In den local prisons hat das Visiting Committee die Befugnis, in Einzelfällen Ausnahmen von den vorstehenden Bestimmungen anzuordnen. Jedes Mitglied des Visiting Committee kann ferner selbständig eine Ordre auf Zulassung eines Besuchers ausstellen.

Zur Ausstellung einer Zulassungsordre ist ferner die Zentral-Gefängnis-Behörde befugt. Ohne eine derartige Ordre ist die Besichtigung einer Anstalt nicht zulässig.[1]) Es ist oben Seite 161 schon erwähnt worden, daß die Zentralbehörde in sehr liberaler Weise die Erlaubnis zum Besuche der Anstalten an die Vorstandsmitglieder der Vereine für entlassene Sträflinge, sowie an Mitglieder einzelner geistlicher Orden gewährt. Eine generelle Befugnis zum Besuche der Anstalten haben die Polizeibeamten des Criminal Investigation Department — einer besonderen Polizei-Abteilung in London — welche sich durch derartige Besuche die Möglichkeit zur späteren Identifizierung der Verbrecher verschaffen.

§ 17. Die Sorge für die Gesundheit der Gefangenen.

Die englische Zentral-Gefängnis-Behörde hat von jeher ein Hauptaugenmerk darauf gerichtet, daß in den unter ihrer Aufsicht stehenden Anstalten alle erforderlichen Vorkehrungen getroffen werden, um den mit der Gefangenschaft stets verbundenen nachteiligen Einwirkungen auf die Gesundheit und das Leben der Gefangenen nach Möglichkeit vorzubeugen.

Es ist eine nicht nur in England, sondern überall konstatierte Thatsache, daß die Morbiditäts- und Mortalitäts-Ziffer unter den

---

[1]) Sir Edmund Du Cane hatte die Güte, mir eine Ordre zur Besichtigung aller englischen Anstalten auszustellen, und ich schulde den Governors der von mir besuchten Anstalten, welche mich mit großer Liebenswürdigkeit durch ihre Anstalt begleiteten und mir jede wünschenswerte Auskunft erteilten, verbindlichsten Dank; besonders dem Governor von Wormwood Scrubs, Kapitän W. T. Harvey, und dem Governor von Borstal, Kapitän A. G. Dawson, möchte ich meinen Dank hiermit noch einmal öffentlich aussprechen.

Insassen der Gefängnisanstalten eine größere ist, als in der Gesamtbevölkerung. [1]) Es hängt dies zunächst damit zusammen, daß die Gefängnisbevölkerung durchweg, was ihre körperliche, geistige und sittliche Gesundheit anbetrifft, infolge ererbter Fehler, vernachlässigter Erziehung und bisheriger Lebensweise hinter der Gesamtbevölkerung zurücksteht. Dazu treten die Einwirkungen der Gefangenschaft selbst, welche durch die Beschränkung des persönlichen Willens, durch die mit dem Erleiden der Strafe verbundene Gemütserregung, durch die Sorge um die zurückgebliebene Familie, durch die einförmige Lebensweise in der Anstalt einen je nach der Individualität des einzelnen Gefangenen mehr oder minder nachteiligen Einfluß auf seinen Gesundheitszustand ausüben müssen. Der Staat hat somit nicht nur im Interesse der Humanität, sondern auch mit Rücksicht auf die freie Bevölkerung, in welche der Gefangene bei seiner Entlassung als arbeitsfähiges Mitglied zurückkehren soll, alle Veranlassung, für den Gesundheitszustand der Gefangenen, soweit es das Wesen der Strafe irgend zuläßt, [2]) Sorge zu tragen.

Dieser Verpflichtung kommt England im vollen Maße nach.

Es wird zunächst bei der Anlage und Einrichtung der Anstalten allen Vorschriften der Gefängnis-Hygiene mit Bezug auf die Lage der Anstalt, die Heizung, Beleuchtung, Ventilation, Wasserversorgung, Beseitigung der Fäkalstoffe u. s. w. in vollem Maße Rechnung getragen. Die genaue Prüfung aller Baupläne durch den sachverständigen Rat in der Zentralbehörde bietet hierfür volle Garantie. Durch die fortwährenden, zu vorher nicht bestimmten Zeiten stattfindenden Inspektionen der einzelnen Anstalt wird sodann genügende Aufsicht darüber geführt, daß alle diese Einrichtungen in gutem Zustande erhalten bleiben und daß Reinlichkeit und Ordnung in der Anstalt herrscht. [3]) Eine Überfüllung der Anstalt kann nicht vor-

---

[1]) Ausführliche Angaben und Untersuchungen hierüber sind in dem ausgezeichneten Buche von Baer „Die Gefängnisse, Strafanstalten und Strafsysteme, ihre Einrichtung und Wirkung in hygienischer Beziehung" Berlin 1871 mitgeteilt.

[2]) Diese Schranke ist allerdings strenge innezuhalten, damit nicht die Lage des Insassen eines Gefängnisses sich zu einer besseren gestalte, als diejenige eines freien, wenn auch noch so armen Arbeiters ist. Andernfalls würde die Strafe ihre abschreckende Wirkung verlieren. Cf. Du Cane I Seite 89.

[3]) In den englischen Gefängnissen, welche ich besucht habe, war nirgends etwas von dem eigentümlichen, sich anderswo in derartigen Anstalten so häufig vorfindenden „Gefängnisgeruch" zu bemerken.

kommen, da ja für jeden Gefangenen unter allen Umständen eine Einzelschlafzelle vorhanden sein muß, die Maximalbelegung der Anstalt somit ein für allemal gegeben ist.

Was sodann die Sorge für den einzelnen Gefangenen betrifft, so wird dessen Gesundheitszustand sofort bei seiner Aufnahme genau ermittelt. Leidet derselbe bei seiner Aufnahme oder im Verlaufe der Gefangenschaft an einer ansteckenden Krankheit, oder erscheint ein Verbleiben in der Anstalt bei seinem Gesundheitszustande sonst mit Gefahr verbunden, so erfolgt seine alsbaldige Fortschaffung. [1]) Es ist schon erwähnt worden, daß dafür gesorgt ist, daß der Gefangene seinen Körper rein halte und daß er sich täglich eine genügende Zeitlang in freier Luft bewege. In § 19 wird ferner des nähern ausgeführt werden, in welcher Weise die Beköstigung mit Rücksicht auf den Gesundheitszustand des Gefangenen eingerichtet ist.

All die erwähnten, aus sanitären Rücksichten getroffenen Einrichtungen stehen — was hier nochmals [2]) besonders hervorgehoben werden mag — unter der ständigen, strengen Kontrolle des Gefängnisarztes. Er ist für den Gesundheitszustand in der Anstalt in erster Linie verantwortlich. Jedes Vierteljahr hat er wenigstens einmal einen ausführlichen Bericht (Sanitary Report) an die Zentralbehörde darüber zu erstatten, ob die Vorschriften für Reinlichkeit, für genügende Temperatur und Ventilation, für Wasserversorgung u. s. w. eingehalten werden, ob die den Gefangenen gereichte Kost von guter Qualität und Zubereitung gewesen ist und welche Vorschläge er mit Rücksicht auf den allgemeinen Zustand der Gefangenen zu machen hat. In jeder Woche hat er wenigstens einmal alle Gefangenen zu besuchen. Über den dabei vorgefundenen Gesundheitszustand hat er allmonatlich einen ausführlichen Bericht an den Governor zu erstatten. Er läßt periodisch den Gefangenen wiegen und überzeugt sich durch Vergleichung mit dem bei Aufnahme des Gefangenen festgestellten Gewichte, ob eine bemerkenswerte Körperabnahme stattgefunden hat. Wenn der Gefängnisarzt glaubt, daß der Gesundheitszustand eines Gefangenen eine ander-

---

[1]) Cf. die oben Seite 150 Anm. 2 erwähnte Verfügung des Staatssekretärs. Bei der Aufnahme eines Gefangenen soll der Arzt jedesmal dessen „fitness for the ordinary discipline of a prison" attestieren. Cf. Report of Commissioners 1885 Seite 12.

[2]) Cf. oben Seite 150 f.

weitige Beschäftigung oder eine bessere Beköstigung oder längere Bewegung im Freien erfordere, [1]) so hat er sofort eine diesbezügliche Eintragung in dem Medical Book, welches dem Governor täglich vorzulegen ist, zu machen. Der Governor vermerkt daneben die von ihm getroffene Entscheidung. In dem — übrigens sehr seltenen — Falle, daß der Governor den Vorschlägen des Arztes nicht Folge gibt, hat derselbe ausführlich die Gründe hierfür schriftlich auszuführen.

Wenn so der Gefängnisarzt in der Lage ist, in wirksamer Weise für die Erhaltung der Gesundheit der Gefangenen zu sorgen, so wird er es doch bei seiner Stellung im Gefängnisdienste [2]) stets vor Augen behalten, daß es sich um Leute handelt, welche eine Strafe verbüßen und denen die Entbehrungen und Leiden, welche sie von Rechts wegen zu ertragen haben, durch ärztliche Anordnungen nicht ohne zwingende Notwendigkeit erleichtert werden sollen.

Die Befugnisse des Gefängnisarztes erweitern sich naturgemäß in dem Augenblicke, wo ein Gefangener wirklich ernst erkrankt. Es hängt hier zunächst von seinem Urteile ab, ob der Gefangene überhaupt in der Anstalt belassen werden kann, oder fortgeschafft werden muß. Die Fortschaffung tritt regelmäßig ein, wenn ein Gefangener geisteskrank wird (cf. den folg. §); sie tritt ferner bei ansteckenden Krankheiten und bei Krankheiten, welche schwierige Operationen erfordern, [3]) in den kleineren Anstalten ein, welche für eine Behandlung derartiger Kranken keine genügenden Einrichtungen

[1]) Wenn ein Gefangener den Arzt zu sehen wünscht, so hat der Gefängniswärter dies sofort in das Medical Book einzutragen, damit ein alsbaldiger Besuch stattfindet, über dessen Ergebnis von dem Arzte dann neben der Eintragung ein Vermerk gesetzt wird. Wenn ferner der Gefängnisgeistliche irgendwelche auffälligen Erscheinungen bei einem Gefangenen wahrnimmt, so hat er dies in dem von ihm zu führenden und täglich dem Governor vorzulegenden Buche zu vermerken, worauf dann der Governor den Besuch des Arztes anordnet.

[2]) Mit Recht weist Baer l. c. S. 206 darauf hin, wie wichtig es ist, daß der Gefängnisarzt neben seiner ärztlichen Wissenschaft auch genügende Kenntnis von den Prinzipien des Gefängniswesens besitzt. Und eine derartige Kenntnis findet sich durchweg bei dem englischen Gefängnisarzte, dessen Thätigkeit nicht — wie anderswo häufig — bloß auf die Behandlung der erkrankten Gefangenen beschränkt ist, sondern welcher eine feste Stellung in der gesamten laufenden Gefängnisverwaltung einnimmt.

[3]) Ebenso bei weiblichen Gefangenen im Falle bevorstehender Entbindung.

haben. Wohin der Gefangene zu schaffen ist — ob in eine andre Gefängnisanstalt [1]) oder in ein Hospital —, oder ob der Gefangene, bei welchem eine unmittelbare Lebensgefahr besteht, zu seiner Familie zu entlassen ist, darüber entscheidet — soweit nicht generelle Verfügungen Platz greifen — die Zentralbehörde.

Wenn die Behandlung des erkrankten Gefangenen in der Anstalt selbst erfolgt, so wird derselbe in besondere, von der übrigen Anstalt getrennte Räume gebracht. Früher waren dies fast durchweg gemeinsame Krankensäle; in neuerer Zeit hat man daneben in vielen Anstalten auch noch einzelne Zellen zur Aufnahme von Kranken eingerichtet. [2]) Es hängt alsdann von dem Ermessen des Arztes ab, wohin der einzelne Gefangene gebracht werden soll; maßgebend ist dabei, wo die erforderliche Krankenpflege am besten stattfinden kann, und ob nach Lage der Krankheit es angemessen erscheint, dem kranken Gefangenen das Gefühl der Einsamkeit zu ersparen.

Für den Krankenpflegedienst sind regelmäßig besondere Wärter (Hospital Warders) bestimmt, welche den Anordnungen des Arztes Folge zu leisten haben. Der Arzt ist es überhaupt, welcher über die Behandlung des erkrankten Gefangenen im wesentlichen allein zu bestimmen hat; dem Governor steht nur eine gewisse Kontrolle zu, wie sie erforderlich ist, um Simulationen nach Möglichkeit vorzubeugen.

Die Gesamtzahl der Fälle eines natürlichen

---

[1]) Es ist schon hervorgehoben worden, daß das convict prison zu Parkhurst ausschließlich zur Aufnahme von Epileptischen, Altersschwachen, Blödsinnigen und sonstigen Personen, welche einer dauernden ärztlichen Beaufsichtigung bedürfen, bestimmt ist. Eine derartige vollständige Absonderung derjenigen Gefangenen, welche der gewöhnlichen Gefängnisdisziplin nicht unterworfen werden können, wird in dem Report der Royal Commissioners 1879 Seite XLII ff. auf das lebhafteste gebilligt. Auch in den local prisons ist mit Rücksicht auf die Gefängnisdiziplin sowie auf eine geeignete ärztliche Behandlung von epileptischen, blödsinnigen sowie solchen Gefangenen, welche einen Hang zum Selbstmorde zeigen, das Bedürfnis einer Absonderung hervorgetreten, und es ist jetzt die Bestimmung getroffen worden, daß derartige Gefangene, wenn sie eine Gefängnisstrafe von mindestens zwei Monaten zu verbüßen haben, in besondere, eigens für dieselben eingerichtete Gefängnisse geschafft werden sollen. Report of Commissioners 1885 Seite 14.

[2]) Diese Krankenzellen sind geräumiger und luftiger, als die gewöhnlichen Zellen.

Todes in den local prisons[1]) betrug im Jahre 1885/6: 116, d. i. 0,75% der durchschnittlichen Tagesbelegung. Die beiden Krankheiten, welche die meisten Todesfälle herbeiführten, waren Pneumonia resp. Pleuro Pneumonia (22 Todesfälle) und Phtisis (18 Todesfälle). Aus ärztlichen Gründen wurden aus den local prisons im Laufe des Jahres entlassen: 182 Personen, von denen schätzungsweise angenommen wird,[2]) daß etwa die Hälfte noch im Laufe des Jahres gestorben sein mag, so daß sich eine Mortalitätsziffer von 1,34 % ergeben würde. In dem vorhergehenden Jahre 1884/5 betrug bei einer um ca. 1300 höheren Durchschnittsbelegung die Zahl der Todesfälle 130, die Zahl der Entlassungen aus ärztlichen Gründen 203.

Bei allen Todesfällen in den local prisons wird in dem jährlich erscheinenden Report genaue Mitteilung darüber gemacht, wie lange Zeit der Betreffende in der Anstalt gewesen ist, in welchem Gesundheitszustande er sich bei Aufnahme in die Anstalt befunden hat und ob insbesondere das Leiden, an welchem er gestorben ist, als schon vorher vorhanden konstatiert werden kann. Von den 130 im Jahre 1884/5 eines natürlichen Todes Gestorbenen war bei 27 der Gesundheitszustand bei Aufnahme in die Anstalt schon als „schlecht" oder gar „sehr schlecht" angegeben.[3])

Die Zahl der Selbstmorde in den local prisons betrug im Jahre 1884/5 bei einer durchschnittlichen Tagesbelegung von 16619: 14.[4])

---

[1]) Wenn ein Gefangener in der Anstalt stirbt, so hat eine post mortem examination stattzufinden, bei welcher neben dem Gefängnisarzte ein andrer Arzt zuzuziehen ist.

[2]) Report of Commissioners 1886 Seite 50.

[3]) Läßt man die Zahl der aus ärztlichen Gründen Entlassenen, bei denen eben nicht genau festzustellen ist, inwieweit sie dem Leben erhalten geblieben sind, außer Betracht, so haben die Todesfälle in den local prisons seit der Übernahme derselben in die Staatsverwaltung ständig abgenommen. Die Zahl der Todesfälle auf die tägliche Durchschnittsbelegung berechnet ergibt, daß seit dem Jahre 1878 der höchste Prozentsatz 0,89 % (in den Jahren 1880 und 1881) gewesen ist, während vor der Übernahme der Anstalten in die Staatsverwaltung der Prozentsatz nur in zwei Jahren (1874: 0,97 %; 1877: 0,91 %) unter 1 % geblieben und mehrfach über 1,3 % betragen hat (1875: 1,47 %; 1866: 1,38 %: 1867: 1,37 %; 1865: 1,32 %). Cf. Report of Commissioners 1886 Seite 69.

[4]) Report of Commissioners 1885 Seite 52. Auch die Ziffer der Selbst-

Die Gesamtzahl der Krankheitsfälle im Jahre 1884/85 wird auf 175 233 angegeben, wovon jedoch 170 370 als leichte Unwohlsein (slight indispositions) bezeichnet werden, so daß nur 4863 Fälle übrig bleiben, in welchen eine Behandlung in der Krankenabteilung erforderlich war (infirmary cases).[1])

Bezüglich der convict prisons fehlt es an einer Zusamenstellung der Todes- und Krankheitsfälle. Es werden hier bei jeder Anstalt gesondert von dem Gefängnisarzte Mitteilungen über den Gesundheitszustand gemacht. Das aus diesen Mitteilungen zu entnehmende Gesamtbild ist ein durchweg günstiges.[2])

§ 18. Die Behandlung geisteskranker Gefangener insbesondere.

Die geisteskranken Gefangenen (Criminal Lunatics) zerfallen in 5 Hauptklassen:

a) Personen, die von den Friedensrichtern überwiesen werden als ergriffen unter Umständen, welche Geisteskrankheit und eine Neigung zur Begehung von Verbrechen anzeigen (dangerous lunatics committed by Justices).[3]) Die Zahl derselben ist eine sehr kleine; im Jahre 1883/4 fielen insgesamt nur 4 Personen unter diese Kategorie.[4])

---

morde ist seit Übernahme der local prisons in die Staatsverwaltung geringer geworden. Du Cane II Seite 105 gibt die Durchschnittsziffer der Selbstmorde in den $5^1/_2$ Jahren vor Übernahme der Anstalten auf jährlich 17,6 an. In einigen Anstalten, welche ich besucht habe, schien mir nur geringe Sorgfalt darauf gelegt zu werden, die Möglichkeit zum Selbstmorde zu nehmen; die Gasflammen waren zum Teil ungenügend geschützt und es fanden sich eiserne Haken u. s. w. in den Zellen. Cf. auch Report of the Royal Commission on Prisons in Ireland 1884 S. 26 f.

[1]) Judicial Statistics 1884 Seite XI.

[2]) Es verdient erwähnt zu werden, daß sich die Leiter des größten englischen Vereins für entlassene Sträflinge, der Royal Society for the Assistance of Discharged Prisoners, Mr. Ranken und Major Tillbrook, vor den Royal Commissioners 1879 (Report Seite XXXVIII) auf das allergünstigste über den Gesundheitszustand aussprachen, in welchem sich die entlassenen Sträflinge durchweg befänden.

[3]) 1 a. 2 Vict. c. 14.

[4]) Die im folgenden gegebenen Ziffern sind entnommen den Judicial Statistics 1884 Seite LII—LIV.

b) Personen, welche wegen eines Verbrechens bereits zum Hauptverfahren überwiesen sind und bei denen der Staatssekretär des Innern auf Grund ärztlicher, eine Geisteskrankheit bescheinigender Atteste im administrativen Wege die Überführung in eine Irrenanstalt anordnet (removed by warrant of the Secretary of State as insane after committal for trial).[1]) Die Zahl dieser Personen belief sich im Jahre 1883/4 auf 253.

c) Angeklagte, bei denen eine zu diesem Zwecke besonders gebildete Jury ein Verdikt dahin ausspricht, daß dieselben zur Zeit geisteskrank sind (found insane by jury).[2]) Im Jahre 1883/4 wurde ein solches Verdikt gegen 184 Personen ausgesprochen.

d) Angeklagte, bei denen die Jury im gewöhnlichen Gange des Verfahrens durch Verdikt feststellt, daß dieselben bereits zur Zeit der Begehung des Verbrechens geisteskrank waren, und welche infolgedessen freigesprochen, aber einer Irrenanstalt überwiesen werden (acquitted insane). Die Zahl der so Freigesprochenen belief sich 1883/4 auf 320.

e) Verurteilte, welche während der Strafverbüßung geisteskrank befunden und durch eine Ordre des Staatssekretärs des Innern einer Irrenanstalt überwiesen werden (convicts becoming insane, removed by warrant of the Secretary of State as insane after trial).[3]) Die Zahl dieser Personen wird für 1883/4 auf 222 angegeben.

Die Gesamtzahl der hiernach im Jahre 1883/4 als criminal lunatics neu überwiesenen Personen belief sich auf 983.[4])

Die letzte oben erwähnte Klasse ist von den vier andern dadurch wesentlich verschieden, daß es sich bei ihr um wirkliche Verbrecher handelt, um Verurteilte, welche sich im Stadium der Strafverbüßung befinden, während den Personen der andern Klassen der Charakter als Verbrecher keineswegs beiwohnt. Die letzteren sind mit Rücksicht auf die allgemeine Sicherheit für die Dauer der Geisteskrankheit in den Anstalten untergebracht und werden in Freiheit gesetzt, sobald

[1]) 3 a. 4 Vict. c. 54.

[2]) 39 a. 40 George III c. 94.

[3]) 27 a. 28 Vict. c. 29, 47 a. 48 Vict. c. 64.

[4]) Die Zahl ist im Zunehmen begriffen; im Jahre 1882/3 betrug sie nur 914. Es mag nebenbei bemerkt werden, daß die Zahl der Geisteskranken überhaupt in England eine ständig steigende ist. Cf. Aschrott, Englisches Armenwesen Seite 358.

sie von ihrer Krankheit genesen sind; die Personen der fünften Klasse dagegen befinden sich in der Strafverbüßung und sind mit Rücksicht auf ihre Geisteskrankheit aus einer Strafanstalt in eine andre Anstalt gebracht, aus welcher sie im Falle ihrer Genesung wieder in die Strafanstalt zurückkehren.[1]) Während es hiernach im Interesse der Personen der ersten Kategorie liegt, möglichst bald als genesen aus der Anstalt entlassen zu werden, haben die Personen der andern Kategorie ein Interesse, möglichst lange in der Anstalt zu verbleiben, um so der Rückversetzung in die Strafanstalt und der in der letzteren bestehenden strengen Zucht zu entgehen. Es sind hier ganz besondere Maßregeln notwendig, um Simulationen vorzubeugen.

Was nun die **Behandlung der criminal lunatics** anbetrifft, so wurden dieselben bis zum Jahre 1863 in besonders eingerichteten Abteilungen einzelner öffentlicher Irrenanstalten untergebracht, und es waren hier die erforderlichen Sicherungsmaßregeln gegen Fluchtversuche getroffen. **Im Jahre 1863 wurde eine besondere Anstalt für criminal lunatics zu Broadmoor eröffnet.**[2]) In diese Anstalt wurden fortab alle Personen der fünften Klasse und, soweit Raum vorhanden war, auch die Personen der vier andern Klassen geschafft. Diejenigen criminal lunatics, welche in Broadmoor nicht untergebracht werden konnten, kamen wie früher in die öffentlichen Irrenanstalten. Die criminal lunatics der fünften Klasse bildeten etwa $^1/_3$ der Gesamtzahl der Insassen von Broadmoor.

Nachdem der Direktor dieser Anstalt in seinen Jahresberichten

---

[1]) Wenn bei Ablauf der Strafzeit eine Genesung noch nicht eingetreten ist, und wenn der Direktor der Anstalt der Ansicht ist, daß der Betreffende nicht in Freiheit gesetzt werden dürfe („that such person is insane and unfit to be at large“), so hat der Direktor ein Verfahren vor dem Friedensrichter einzuleiten. Wenn der Friedensrichter findet, „that such person is insane and a proper person to be detained under care and treatment“, so erläßt er eine Ordre, auf Grund deren der Betreffende fortab als pauper lunatic in einer Anstalt detiniert wird. 47 a. 48 Vict. c. 64 s. 7.

[2]) Eine Beschreibung dieser Anstalt findet sich in der Allg. Zeitschr. für Psychiatrie Band 27 Heft 1 u. 2. Die Anstalt wurde auf Grund eines im Jahre 1857 gefaßten Parlamentsbeschlusses errichtet; sie besteht aus einer Männer- und einer Frauenabteilung, deren jede in getrennte Unterabteilungen für ruhige und für unruhige Geisteskranke zerfällt.

wiederholt auf die Schwierigkeiten hingewiesen hatte,[1]) welche dadurch entständen, daß dieses eine Drittel der Insassen in andrer Weise als die übrigen behandelt und bewacht werden müßte, entschloß sich der Staatssekretär des Innern im Jahre 1873 zu einer Änderung, indem er bestimmte, daß vom Jahre 1874 ab die schweren Verbrecher männlichen Geschlechts, welche während der Strafverbüßung geisteskrank werden, nicht mehr nach Broadmoor, sondern nach einer besonders eingerichteten Abteilung (lunatic division) des convict prison von Woking geschafft werden sollten.

Es bestehen also nunmehr die drei, bei der Behandlung geisteskranker Gefangener überhaupt in Frage kommenden Systeme in England nebeneinander: die geisteskranken Männer, welche schwere Verbrecher sind — die Zahl derselben belief sich von 1874 bis 1886 insgesamt auf 422 — kommen in eine besondere Abteilung einer Strafanstalt; die Mehrzahl der übrigen criminal lunatics — im Jahre 1883/4: 595 Personen — kommen in die Spezialanstalt von Broadmoor; der Rest — im Jahre 1883/4: 388 Personen — in besondere Abteilungen der öffentlichen Irrenanstalten.[2])

Die Behandlung der Insassen von Broadmoor bietet kein weiteres Interesse;[3]) sie ist im wesentlichen dieselbe wie in allen

---

[1]) In besonders scharfer Weise geschieht dies in dem Report von 1872, wo noch hervorgehoben wird, daß die Mehrzahl der Fluchtversuche und der thätlichen Angriffe gegen die Wärter von den convicts becoming insane ausgingen. Es wird ferner auf den nachteiligen Einfluß in moralischer Hinsicht aufmerksam gemacht, welchen diese Klasse auf die übrigen Insassen ausübt.

[2]) Auf die Frage selbst, welche Art der Unterbringung als die geeignetste erscheint oder ob sich ein kombiniertes System empfiehlt, soll hier nicht eingegangen werden. Es kommt hierbei neben dem Kostenpunkte auf die Höhe der Gesamtzahl der geisteskranken Gefangenen an.

[3]) Es mag erwähnt werden, daß für die ruhigen Geisteskranken gemeinsame Schlafsäle bestehen. — Die Anstalt besitzt ausgedehntes Ackerland, wo viele Geisteskranke mit landwirtschaftlichen Arbeiten beschäftigt werden. In der Anstalt werden ferner Schuhmacher-, Schneider- und Tischlerarbeiten ausgeführt; auch in der Bäckerei finden viele Geisteskranke Beschäftigung. — Die Durchschnittskosten betrugen 1883/4 in Broadmoor pro Kopf: 45 £ 18 s 4 d. Dies ist erheblich viel mehr, als die Kosten in irgend einer der öffentlichen Irrenanstalten, in welchen criminal lunatics untergebracht werden. In diesen Anstalten variierten die Kosten im Jahre 1883/4 zwischen 21 £ 4 s 8 d (Norfolk County Asylum) und

andern Irrenanstalten, nur sind die Sicherungsmaßregeln gegen Fluchtversuche verstärkt. [1])

Dagegen ist das Verfahren in der lunatic division von Woking in sofern bemerkenswert, als hier auch bei den geisteskranken Verbrechern die allgemeinen Prinzipien des englischen Gefängniswesens zur Anwendung gebracht werden: die Insassen sind in Klassen abgeteilt, das Vorrücken in die höheren Klassen wird durch das Markensystem geregelt, die Marken werden erworben durch fleißige Arbeit bei gutem Betragen; die Vorteile des Vorrückens in die höheren Klassen bestehen in besonderer Kleidung und in Diätzulagen.

Die in Woking erreichten Resultate werden als vorzügliche bezeichnet: während eines Zeitraums von 11 Jahren ist keine Entweichung eines Insassen vorgekommen, die Gesamtzahl der Selbstmorde belief sich auf nur 2, diejenigen der thätlichen Angriffe auf nur 4. Von den insgesamt aufgenommenen 422 Geisteskranken konnten 105 als genesen in ein Public Work Prison zurückgeschickt werden. Es bleibt daher zu bedauern, daß — offenbar nur aus Rücksichten auf den Kostenpunkt [2]) — vom 21. Juni 1886 ab die Sendung von convicts becoming insane nach Woking eingestellt worden ist; dieselben sollen fortab wiederum nach Broadmoor gebracht werden.

Eine besondere, noch erwähnenswerte Einrichtung ist die Überführung aller Gefangenen, bei denen ein Zweifel über die Gesundheit ihrer Geisteskräfte besteht, nach Millbank behufs Beobachtung des Geisteszustandes durch einen hierfür in dieser Anstalt eigens angestellten hervorragenden Irrenarzt. [3])

39 £ 8 s 8 d (Cornwall Asylum) pro Kopf. Cf. Judicial Statistics 1884 Seite LIV.

[1]) Daß derartige Sicherungsmaßregeln erforderlich sind, ergibt sich schon daraus, daß unter den im Jahre 1883/4 neu aufgenommenen criminal lunatics bei den Frauen 42,5 %, bei den Männern 23,1 % unter der Anklage des Mordes gestanden hatten.

[2]) Der Generalbericht, welchen der Medical Inspector der Zentralbehörde, Dr. Gover, über die günstigen Resultate der lunatic division zu Woking erstattet hat — derselbe ist abgedruckt im Report of Directors 1886 Seite XLI ff. —, läßt es auffällig erscheinen, daß die Zentralbehörde ihren Beschluß, diese Abteilung eingehen zu lassen, ohne eine eingehendere Begründung in dem Report of Directors 1886 Seite X bekannt gibt.

[3]) In dem Report of the Royal Commissioners 1879 Seite XLII findet dies Verfahren lebhafte Billigung.

§ 19. Die Beköstigung.

Die Frage nach einer richtigen Beköstigung der Gefangenen hat in England die eingehendste Behandlung erfahren. Zur Prüfung derselben wurden in den Jahren 1864 und 1867 besondere Parlamentskommissionen niedergesetzt. [1]) Als sodann die local prisons in die Staatsverwaltung übernommen wurden, war es eines der ersten Ziele der Zentralgefängnisbehörde, eine einheitliche Speiseordnung in diesen Anstalten einzuführen. Zu diesem Zwecke legte der Staatssekretär des Innern einer hierfür besonders berufenen Kommission von Ärzten einen Fragebogen vor, in welchem wohl alle für die Materie der Beköstigung wichtigen Punkte enthalten waren. [2]) Den Vorschlägen dieser Kommission entsprechend ist denn auch unter Genehmigung des Parlamentes die jetzt in allen local prisons geltende Speiseordnung eingeführt worden. Die dabei maßgebenden Grundsätze, welche auch in den convict prisons zur Geltung gelangt sind, sollen hier unter der Rubrik einzelner Fragen, wie sie der Kommission vorgelegen haben, dargestellt werden.

1) Ist eine gleichartige Diät für alle Distrikte einzuführen, oder ist der verschiedenartigen Beköstigungsweise der arbeitenden Klasse in den einzelnen Distrikten Rechnung zu tragen?

Die Kommission entschied sich für eine einheitliche Diät. Es wird dabei ausgeführt, daß die Verschiedenheiten in der Beköstigung der arbeitenden Klasse in den einzelnen Distrikten in Wirklichkeit gar keine so großen seien, falls man nur überall Personen derselben oder ähnlicher Beschäftigungsweise in Betracht zöge. Man dürfe

---

[1]) Schon im Jahre 1843 hatte der Staatssekretär des Innern eine Speiseordnung aufgestellt, welche er zur Einführung in den local prisons empfahl. Es waren jedoch nur wenige Anstalten, welche dieser Empfehlung Folge leisteten. In der That bildeten die Speiseordnungen in den local prisons bis zur Übernahme in die Staatsverwaltung eine sehr bunte Musterkarte.

[2]) Cf. Report of Commissioners 1878 Seite 41 ff., Du Cane II Seite 84 ff. Die Kommission war dabei ausdrücklich darauf hingewiesen worden, daß die in den Gefängnissen zu gewährende Beköstigung genügend — aber auch nicht mehr als genügend — sein solle, um Gesundheit und Arbeitskraft zu erhalten „that the diet shall be sufficient and not more than sufficient to maintain health and strength".

nicht die Beköstigung eines Distriktes, in welchem die arbeitende Klasse vorzugsweise in Eisenwerken beschäftigt sei, mit derjenigen eines ländlichen Distriktes vergleichen. Der Arbeiter in einem Eisenwerke brauche naturgemäß, um bei voller Arbeitskraft zu bleiben, andre und reichlichere Nahrungsmittel als der landwirtschaftliche Arbeiter. Die Quantität der Nährstoffe, welche die freien Arbeiter der verschiedenen Branchen zu sich nähmen, stände regelmäßig in Proportion zu dem durch die Beschäftigungsart herbeigeführten Kräfteverbrauch. Und dieser Maßstab — ohne Rücksicht auf die Beköstigungsart des Gefangenen in der Freiheit — sei auch in der Strafanstalt einzuhalten.

Es sei zwar zuzugeben, daß provinzielle Verschiedenheiten in der Art und Weise, wie die Nährstoffe genommen und wie die Speisen zubereitet würden, vorhanden seien; allein dies könne in der Anstalt nicht berücksichtigt werden. Für die Beköstigung in der Anstalt sei das allein maßgebende Prinzip, daß die Nährstoffe in einer Form und in einer Zubereitung gereicht würden, welche unter Berücksichtigung der ganzen Lebensweise der Gefangenen[1]) als die geeignetste erscheine, um die Nahrungsbestandteile durch die Verdauung möglichst auszunutzen und für den Organismus zu verwerten.

2) Soll bei der Beköstigung die Individualität des einzelnen Gefangenen, sein Alter, seine Körpergröße, sein Gewicht, sein Gesundheitszustand u. s. w. berücksichtigt werden?[2])

Die Kommission verneinte diese Frage. Es sei bei der Beköstigung der Gefangenen stets im Auge zu behalten, daß dieselbe aus öffentlichen Mitteln erfolge und daß deshalb größte Sparsamkeit verlangt werden müsse. Wo immer es sich um die Verpflegung großer Massen handele, wie z. B. auch in der Armee und Marine,

[1]) Es ist eine allen Gefängnis-Praktikern bekannte Thatsache, daß die sitzende Lebensweise, die geringe Bewegung im Freien, die vielfach deprimierte Gemütsstimmung der Gefangenen und andre mit der Gefangenschaft verbundene Momente die Verdauungsthätigkeit herabstimmen, so daß es notwendig erscheint, durch die Art und Weise der Zubereitung der Speisen, durch Gewürze 2c. Lust zum Essen zu erwecken und die Verdauungsorgane zu regerer Thätigkeit anzuregen. Baer in dem schon citierten Buche Seite 123 ff. hat diese Frage in eingehender wissenschaftlicher Weise behandelt. Cf. auch Blätter für Gefängniskunde Band 21 Seite 233 ff.

[2]) In Deutschland ist es besonders Professor Voit, welcher eine Bildung von Kostgruppen nach derartigen individuellen Momenten verlangt.

da könne nicht auf jeden Einzelnen Rücksicht genommen werden, sondern der Einzelne müsse sich dem nach Durchschnittsberechnungen für genügend erachteten Maßstabe unterwerfen. Wenn dabei der Einzelne in Ausnahmefällen wirklichen Schaden leide, so sei es Sache des Arztes, die erforderlichen Maßregeln anzuordnen.

3) Wie sind die Befugnisse des Arztes bezüglich der Anordnung einer besonderen Beköstigung zu regeln?

Die Kommission erklärte sich dafür, dem Gefängnisarzte, welcher ja in England für den Gesundheitszustand der Gefangenen in erster Linie verantwortlich ist, möglichst umfassende Freiheit zu gewähren.

Der Arzt kann, wenn es ihm erforderlich erscheint, neben der gewöhnlichen Gefängniskost die Gewährung von besonderen Speisen und Getränken, sog. Extras, anordnen. Er hat jede derartige Anordnung unter detaillierter Angabe des Grundes dafür in das Medical Book einzutragen, so daß eine Prüfung der von ihm getroffenen Anordnungen stets möglich ist. Es wird dabei ausdrücklich bemerkt, daß die Abnahme des Gewichtes bei einem Gefangenen als solche keinen Beweis dafür liefere, daß die gewöhnliche Kost unzureichend sei, da eine Gewichtsabnahme aus sehr verschiedenen Gründen resultieren könne.[1]) Die Anordnung von Zusatznahrungsmitteln ist ferner höchstens für die Dauer eines Monats zulässig. Wenn der Arzt es für nötig hält, dieselbe länger fortdauern zu lassen, so hat er eine neue Anordnung mit ausführlichen Gründen zu erlassen.[2])

Für die Gefangenen in den Krankenabteilungen bestehen vier besondere Speiseordnungen (Ordinary-, Extra-Fish-, Low-, Hospital-Diet), unter welchen der Gefängnisarzt die im Einzelfalle geeignetste zu wählen hat.

4) Ist die Speiseordnung je nach der Länge der zu verbüßenden Strafe verschieden zu gestalten?

Die Kommission hielt dies für erforderlich, um der Strafe die nötige Strenge und damit die abschreckende

---

[1]) „to lose sight of the fact that fluctuations in weight are due not to one but to many causes is unreasonable and must lead to unnecessary interference with discipline".

[2]) Auf diese Weise ist eine fortgesetzte Prüfung der ärztlichen Anordnung möglich, wofür event. das Gutachten des bei der Zentralbehörde angestellten Arztes eingeholt wird.

Wirkung zu bewahren. Wenn es einerseits bei längeren Strafen zur Erhaltung der Gesundheit und der Arbeitskraft notwendig sei, dem Gefangenen eine Kost zu reichen, welche möglicherweise eine reichlichere und bessere sei, als sie der Betreffende in der Freiheit gehabt hätte, so sei anderseits bei kurzzeitigen Strafen die zu reichende Kost auf ein Minimum zu beschränken. Es geschehe der Gesundheit kein Nachteil, wenn dem Gefangenen für die Dauer von 7 Tagen nur das Allernotdürftigste gereicht würde, und eine derartige Beschränkung sei notwendig, um bei kurzzeitigen Strafen dem Gefangenen die Strafe als ein Übel erscheinen zu lassen. Würde man hier dem Gefangenen an Speisen so viel oder gar noch mehr reichen, als er in der Freiheit erhalten hat, so würde bei vielen Personen die kurze Strafe jede abschreckende Wirkung verlieren, [1]) ja unter Umständen würden neue Rechtsverletzungen bloß zu dem Zwecke begangen werden, um auf bequeme Weise Verpflegung auf öffentliche Kosten zu erhalten.

5) Ist den Gefangenen mit längerer Strafzeit sofort die höhere Kost zu gewähren, oder sollen dieselben mit der niederen Kost beginnend allmählich zu der höheren aufsteigen?

Diese Frage hatte bereits den in den Jahren 1864 und 1867 eingesetzten Kommissionen vorgelegen und eine verschiedenartige Beantwortung gefunden.

Die Kommission von 1864 hatte sich für eine allmählich reichlicher werdende Speiseordnung (progressive dietary) erklärt, indem sie ausführte, es würde als eine Ungerechtigkeit erachtet werden, wenn der schwere Verbrecher eine bessere Kost erhalte, als der gleichzeitig eingelieferte leichte Delinquent. Alle Gefangenen sollten deshalb mit der niedrigsten Kost beginnen; erst wenn der Zeitraum, für welchen die niedrigste Kost angeordnet sei, abgelaufen wäre, sollte für die Gefangenen mit längerer Strafzeit die bessere Kost eintreten.

Die Kommission von 1867 dagegen hatte sich dafür erklärt, dem Gefangenen mit langer Strafzeit sofort die bessere Kost zu gewähren. Die ersten Wochen der Strafverbüßung seien für den

[1]) „it would constitute an encouragement to the commission of petty crimes; and, by thus paving the way to indulgence in the more serious class of offences would assist in the manufacture of the habitual criminal".

Gefangenen stets die härtesten: die Scham über die Verurteilung, die Reue und die Gewissensbisse träten hier am stärksten auf; die Beschränkung der persönlichen Freiheit, die Gewöhnung an die strenge Hausordnung und Disziplin wären hier am meisten fühlbar. Wenn hierzu noch eine schmale Kost hinzukomme, so würde der Gefangene in vielen Fällen die Kraft zu der von ihm mit Recht im Gefängnisse geforderten harten Arbeit verlieren.

Man wird zugeben müssen, daß in beiden Argumentationen etwas Wahres liegt. Dies erkannte denn auch die Kommission von 1878 an und schlug deshalb einen Mittelweg vor, welcher im Prinzip an der progressive dietary festhielt, dabei aber auf die Gesamtdauer der Strafe Rücksicht nahm.

Es geschah dies in folgender Weise. Es werden vier verschiedene, sich von Klasse zu Klasse verbessernde Speiseordnungen aufgestellt, welchen die Gefangenen gemäß folgender Tabelle unterworfen werden:

Zeitdauer der Strafe	Speiseordnung der Klasse I,	der Klasse II	der Klasse III	der Klasse IV
a) bis zu 7 Tagen	für die ganze Zeit	—	—	—
b) zwischen 7 Tagen und 1 Monat	für 7 Tage	für den Rest der Strafzeit	—	—
c) zwischen 1 und 4 Monaten	—	für 1 Monat	für den Rest der Strafzeit	—
d) mehr als 4 Monate	—	—	für 4 Monate	für den Rest der Strafzeit

Man sieht: je kürzer die Strafzeit ist, je geringer ist die gereichte Kost, und bei allen Strafen — abgesehen von den Strafen bis zu 7 Tagen — nimmt die Kost während der Dauer der Strafzeit allmählich zu.

In der Klasse I besteht die Kost aus folgendem:

Frühstück: 8 ounces[1]) Brot.

Mittagessen: $1\frac{1}{2}$ pints[2]) stirabout, d. i. eine aus Hafermehl und Mais bereitete Speise.

Abendessen: 8 ounces Brot.

---

[1]) 16 ounces sind = 1 ℔.

[2]) 1 pint etwa = $\frac{4}{7}$ Liter.

In der Klasse II besteht neben der gewöhnlichen Kost a) eine etwas bessere Kost b) für männliche Gefangene, welche in harter Arbeit beschäftigt sind. [1])

Frühstück: 1 pint Haferschleim (gruel) und a) 5 ounces, b) 6 ounces Brot.

Mittagessen: 3mal die Woche: $^1/_2$ pint Suppe,
2mal die Woche: 8 ounces Kartoffeln,
2mal die Woche: Nierenfett-Pudding (suet pudding), und zwar a) 6 ounces, b) 8 ounces.
Dazu an jedem Tage a) 5 ounces, b) 6 ounces Brot.

Abendessen: 1 pint Haferschleim und a) 5 ounces, b) 6 ounces Brot.

In der Klasse III bestehen ebenfalls zwei Speiseordnungen: a) die gewöhnliche; b) diejenige für männliche Gefangene in harter Arbeit. [2])

Frühstück: 1 pint Haferschleim, a) 6 ounces, b) 8 ounces Brot.

Mittagessen: 3mal die Woche: $^3/_4$ pint Suppe und a) 6 ounces, b) 8 ounces Kartoffeln und 6 resp. 8 ounces Brot,
2mal die Woche: Nierenfett-Pudding, und zwar a) 6 ounces, b) 8 ounces, dazu a) 6 ounces, b) 8 ounces Kartoffeln und 4 ounces Brot,
2mal die Woche: Rindfleisch, und zwar a) 6 ounces, b) 8 ounces, dazu 8 ounces Kartoffeln und a) 6 ounces, b) 8 ounces Brot.

Abendessen: 1 pint Haferschleim, 6 ounces Brot.

---

[1]) Es entspricht dies dem oben dargelegten Prinzip, daß die Kost bestehen solle in dem Minimum des je nach der Beschäftigung zur Erhaltung der Gesundheit und der Arbeitskraft Erforderlichen, „to apportion the quantity of the food to the amount of the labour required“.

[2]) Die not criminal prisoners, welche von dem Rechte der Selbstbeköstigung keinen Gebrauch machen, erhalten die Speiseordnung der Klasse IIIa, haben aber dabei das Recht, als Frühstück und Abendessen anstatt 1 pint Haferschleim $^1/_2$ pint Kakao zu verlangen.

In der **Klasse IV** sind die Speiseordnungen die folgenden:

Frühstück: a) 1 pint Haferschleim und 6 ounces Brot,
b) 1 pint dicke Mehlsuppe (porridge) und 8 ounces Brot.

Mittagessen: 3mal die Woche: 1 pint Suppe, 8 ounces Kartoffeln und a) 6 ounces, b) 8 ounces Brot,
2mal die Woche: Nierenfett-Pudding, und zwar a) 10 ounces, b) 12 ounces, dazu 8 ounces Kartoffeln und a) 4 ounces, b) 6 ounces Brot,
2mal die Woche: Rindfleisch, und zwar a) 3 ounces, b) 4 ounces, dazu 8 ounces Kartoffeln und a) 6 ounces, b) 8 ounces Brot.

Abendessen: a) 1 pint Haferschleim und 6 ounces Brot,
b) 1 pint dicke Mehlsuppe und 8 ounces Brot.

In den **convict prisons**, wo es sich um Gefangene mit einer Strafzeit von wenigstens fünf Jahren handelt und wo mit Rücksicht auf diese lange Dauer der Freiheitsentziehung von vornherein eine ausreichendere Kost gewährt werden muß, bestehen vier **Speiseordnungen**:

a) für die Zeit der Isolierhaft,
b) bei harter Arbeit
c) bei industrieller Arbeit
d) bei leichter Arbeit
} in den Public Works Prisons.

Dazu tritt noch eine besondere Speiseordnung für die Frauen.

Das Prinzip ist hier, wie bei den Speiseordnungen in den local prisons, nur das zur Erhaltung der Gesundheit und der Arbeitskraft unbedingt Erforderliche zu gewähren. Es mag genügen, die in Chatham — wo alle Sträflinge schwere Arbeiten zu verrichten haben — bestehende Speiseordnung [1]) anzuführen.

Frühstück: 1 pint Haferschleim.

Mittagessen: 2mal die Woche 5 ounces Rindfleisch,
1mal die Woche 5 ounces Hammelfleisch,
1mal die Woche 1 ℔ Nierenfett-Pudding,

---

[1]) Du Cane I Seite 120.

2mal die Woche 1 pint Suppe und 2 ounces
Käse,
am Sonntage 1 pint Suppe.
An jedem Wochentage [1]) erhält der Sträfling ferner 1 *ℓℓ*. Kartoffeln.

Abendessen: 1 pint Kakao.

An Brot erhält der Sträfling täglich 24 ounces und an Salz $\frac{1}{2}$ ounce.

Sowohl für die local prisons wie für die convict prisons ist sodann durch besondere punishment dietaries festgesetzt, inwieweit und auf wie lange Zeit die gewöhnliche Kost einem Gefangenen im Disziplinarstrafwege entzogen werden kann. [2]) So ist beispielsweise angeordnet, daß ein Gefangener niemals länger als drei Tage hintereinander bei Wasser und Brot gehalten werden darf; soll diese Strafe eine längere Zeit hindurch dauern, so hat jedesmal, nachdem der Gefangene drei Tage bei Wasser und Brot gehalten worden ist, wiederum für einen Tag die gewöhnliche Kost einzutreten. Während der Zeit, wo der Gefangene nur Wasser und Brot erhält, ist er täglich vom Arzte zu besuchen. [3]) Wenn der Arzt die Fortsetzung der Strafe als der Gesundheit schädlich erklärt, hat dieselbe zu unterbleiben. Es ist auf diese Weise genügende Vorsorge getroffen, daß durch die strafweise erfolgende Entziehung der Kost die Gesundheit des Gefangenen unter keinen Umständen geschädigt werde. [4])

Die Frage, ob einem Gefangenen bei gutem Verhalten eine Kostverbesserung gewährt werden soll und ob es dem Gefangenen gestattet sein mag, sich selbst aus seiner Arbeitsbelohnung Zusatzartikel zu der gewöhnlichen Kost anzuschaffen, wurde

[1]) In der geringeren Kost am Sonntage kommt deutlich das Prinzip zum Ausdrucke, daß durch die gereichte Kost nur die verbrauchten Nährstoffe ersetzt werden sollen.

[2]) In dem Report of Commissioners 1878 Seite 10 wird dieser Einrichtung „of putting in the hands of the authorities a wholesome means of promoting good conduct and industry without running any risk of acting injuriously on the health of those subject to them" große Bedeutung beigelegt.

[3]) Cf. Report of the Royal Commission on Prisons in Ireland Seite 25, wo diese Einrichtung auch für Irland empfohlen wird.

[4]) Dies ist das Urteil, zu welchem die Royal Commissioners (Report 1879 Seite XXXIX) nach eingehender Untersuchung gelangten.

bereits von der im Jahre 1864 niedergesetzten Kommission verneint. Dieselbe war der Ansicht, daß es unrichtig wäre, in dieser Weise als Motiv für gutes Verhalten an niedere, rein materielle Neigungen zu appellieren,[1]) und daß man ferner durch derartige Verbesserungen der Kost unliebsame Vergleichungen mit der Beköstigung, welche sich der freie, redliche aber arme Arbeiter zu verschaffen in der Lage wäre, hervorrufen würde.[2]) Demgemäß besteht in England die Regel, daß Zusätze zu der gewöhnlichen Kost nur auf Grund ärztlicher Anordnungen gereicht werden dürfen.[3])

In letzterer Hinsicht ist noch eine, erst in neuerer Zeit eingeführte Bestimmung hervorzuheben. Es waren Klagen darüber laut geworden, daß Gefangene, welche langzeitige Freiheitsstrafen verbüßt hätten, sich zuweilen bei ihrer Entlassung nicht in einem derartigen Kräftezustande befänden, um als ebenbürtige Bewerber auf dem Arbeitsmarkte auftreten zu können. Obwohl konstatiert wurde, daß derartige Fälle sehr selten vorkämen, so ordnete die Zentral-Gefängnisbehörde doch an, daß fortab alle Gefangenen mit längeren Strafzeiten 21 Tage vor ihrer Entlassung von dem Gefängnisarzte daraufhin untersucht werden sollten, ob sie im Vollbesitze ihrer Arbeitskraft wären; wenn dies nicht der Fall ist, so soll der Arzt befugt sein, für die letzten drei Wochen der Strafzeit eine bessere Beköstigung vorzuschreiben.[4])

Wenn man die Gesamtheit der hier angeführten Bestimmungen zusammenhält, so wird man sagen können, daß in England die Beköstigung so eingerichtet ist, daß der Gefangene in seiner Gesundheit und Erwerbsfähigkeit erhalten bleibt.[5]) Anderseits aber wird umfassende Rücksicht darauf genommen, daß die Strafe nicht durch

[1]) „to appeal to the baser feelings, such as a good moral education should endeavour to suppress."

[2]) Du Cane I Seite 28 f. spricht seine volle Übereinstimmung mit dieser Argumentation aus.

[3]) Selbstbeköstigung ist — was nochmals hervorgehoben werden mag — nur bei den not criminal prisoners gestattet.

[4]) Cf. Report of the Royal Commission on Prisons in Ireland Seite 24, wo die Einführung einer ähnlichen Bestimmung für die irischen Anstalten in Vorschlag gebracht wird.

[5]) Sehr eingehende Untersuchungen sind hierüber von den Royal Commissioners (Report 1879 Seite XXXVII ff.) angestellt worden, und es sind dabei eine Anzahl früherer Sträflinge über diese Frage vernommen worden.

die Art und Weise der Beköstigung den Charakter eines Übels verliert.

Der Zubereitung der Speisen, welche überall in den Gefängnis-Anstalten selbst erfolgt,[1]) wird die größte Aufmerksamkeit geschenkt. Für jede einzelne Speise hat die Zentralbehörde ein genaues Kochrezept erlassen. Als die neue Speiseordnung in den local prisons eingeführt wurde, ernannte die Zentralbehörde einen besonders qualifizierten Arzt, welcher in allen Anstalten systematischen Unterricht in der Art der Zubereitung der Speisen zu geben hatte.[2])

Wie schon hervorgehoben wurde, ist es die Pflicht des Gefängnisarztes, fortlaufend die Gefängniskost zu revidieren und darüber besonderen Bericht zu erstatten. Jeder Gefangene hat ferner das Recht, wenn er glaubt, daß die ihm gereichte Kost nicht vorschriftsmäßig sei, zu verlangen, daß die Kost in seiner Gegenwart untersucht und gewogen werde.

§ 20. Seelsorge, Unterricht, Lektüre.

Während die Sorge für die Gesundheit und eine geeignete Beköstigung der Gefangenen in erster Linie dem Gefängnisarzte obliegt, ist die Sorge für das geistige Wohl der Gefangenen vorzugsweise in die Hände des Gefängnisgeistlichen gelegt.

Es wurde bereits oben Seite 148 angeführt, daß in jeder Gefängnisanstalt ein der englischen Hochkirche angehöriger Geistlicher angestellt ist. Es ist aber auch dafür Sorge getragen, daß Gefangene andrer Glaubensbekenntnisse des geistlichen Zuspruchs nicht entbehren.[3]) In einigen der größeren Anstalten findet sich neben dem der Hochkirche angehörigen Chaplain ein katholischer oder jüdischer Geistlicher (Roman Catholic Priest

---

[1]) Alle größeren Anstalten haben ein eignes Backhaus. Für die kleineren Anstalten wird das Brot in bestimmten großen Anstalten mitgebacken.

[2]) Report of Commissioners 1878 Seite 10.

[3]) Zur Regelung der diesbezüglichen Anordnungen wurde im Jahre 1870 eine besondere Kommission eingesetzt, cf. Report of the Select Committee 1870 appointed to inquire into the Prison Acts and Prison Ministry Acts as far as respects the religious instructions provided for prisoners others than those belonging to the established church.

resp. Jewish Minister) in der Stelle eines Assistant Chaplain. [1]) In den andern Anstalten sind besondere Abkommen mit Geistlichen der katholischen resp. jüdischen Religion getroffen, wodurch den diesen Bekenntnissen angehörigen Gefangenen ein geistlicher Zuspruch gesichert ist. Bei Strafen von längerer Dauer werden diese Gefangenen jedoch regelmäßig in eine der Anstalten geschafft, wo ein Geistlicher ihres Bekenntnisses angestellt ist. [2])

Der Chaplain hat an jedem Sonn- und Festtage vormittags und nachmittags in der Anstalt Gottesdienst abzuhalten, welchem alle der Hochkirche angehörigen Gefangenen beiwohnen müssen, soweit sie nicht durch Krankheit [3]) oder aus sonstigen Gründen von der Teilnahme besonders entbunden sind. In den Public Works Prisons findet ferner an jedem Vormittage in der Woche ein kurzer Gottesdienst statt, ehe die Sträflinge zur Arbeit ausrücken.

Im übrigen soll der Chaplain durch periodische Besuche [4]) des einzelnen Gefangenen in seiner Zelle eine moralische und religiöse Einwirkung auf Sinn und Gemüt desselben auszuüben suchen: dem Gefangenen soll seine Verschuldung vor Augen gehalten, das Gefühl der Reue und das Verlangen nach Vergebung geweckt, Trost und Hoffnung gespendet werden. Damit der Geistliche hierbei dem Gefangenen lediglich als teilnehmender Seelsorger erscheine, ist dem Geistlichen keinerlei Einwirkung auf die Behandlung des Gefangenen eingeräumt: der Geistliche hat auf die Gewährung von Vergünstigungen, insbesondere auf die Verteilung der Marken, keinerlei Einfluß. Es ist hierdurch

---

[1]) In dem großen Public Works Prison von Portsmouth ist ein Chaplain, ein Roman Catholic Priest und ein Jewish Minister angestellt.

[2]) Diese Einrichtung kann allerdings auch ihre Schattenseiten haben. So erzählte mir der Governor von Borstal Prison, daß die in dieser Anstalt mehrfach vorgekommenen Übertritte von Gefangenen zur katholischen Religion nach seiner Überzeugung wohl durchweg darauf zurückzuführen wären, daß die Betreffenden aus dem einen oder dem anderen Grunde wünschten, aus der Anstalt von Borstal, in welcher ein katholischer Geistlicher nicht angestellt ist, fortzukommen.

[3]) Wenn ein Gefangener in erheblicher Weise erkrankt, so ist der Chaplain resp. der Geistliche des betr. Glaubensbekenntnisses sofort davon zu benachrichtigen und hat den Gefangenen in der Krankenabteilung zu besuchen.

[4]) Hierfür wird besonders der Sonntag, an welchem nicht gearbeitet wird, benutzt. Aber auch sonst stattet der Geistliche, insbesondere wenn der Gefangene einen dahin gehenden Wunsch ausspricht, in der arbeitsfreien Zeit Besuche ab.

in wirksamer Weise der Heuchelei vorgebeugt: der Gefangene hat keinen Anlaß, sich dem Geistlichen gegenüber anders zu geben, als er wirklich ist.

Neben dieser seelsorgerischen Thätigkeit führt der Chaplain nun ferner die Aufsicht über den in der Anstalt erteilten Unterricht.

In dem Berichte, welchen die kurz nach Übernahme der local prisons in die Staatsverwaltung eingesetzte Kommission über die Neugestaltung des Unterrichts in diesen Anstalten erstattete, sind die für den Unterricht in den Strafanstalten in England maßgebenden Grundsätze niedergelegt.[1]) Dieselben sollen hier kurz zur Darstellung gebracht werden.

Die Grundidee, aus welcher dann alle weiteren praktischen Schlußfolgerungen hergeleitet werden, ist die folgende: die Strafanstalt ist nicht der geeignete Ort für eine Bildungsstätte „not a place of literary education".[2]) Soweit überhaupt den Gefangenen Unterricht erteilt wird, darf derselbe unter keinerlei Umständen den Charakter der Strafe als eines Übels irgendwie abschwächen.[3]) Der Zweck des Unterrichts besteht lediglich darin, ein ehrliches Fortkommen des Gefangenen nach seiner Entlassung zu erleichtern und zu sichern.[4])

Die aus diesen Sätzen gezogenen praktischen Konsequenzen sind die folgenden:

1) Während des ersten Monats der Strafzeit wird Unterricht nicht gegeben, damit nicht die abschreckende

---

[1]) Cf. über den Bericht der Kommission Reports of Commissioners 1885 Seite 7 ff. 1886 Seite 6 ff., sowie Du Cane II Seite 79 ff. Das von der Kommission vorgeschlagene Unterrichtssystem wurde zunächst im Jahre 1880 probeweise in sieben großen Gefängnis-Anstalten eingeführt. Nachdem es sich hier bewährt hatte, wurde es auch für die übrigen Anstalten angenommen. Heute gilt dies Unterrichtssystem für die gesamte Gefängnis-Verwaltung in England.

[2]) Report of Commissioners 1885 Seite 10.

[3]) Du Cane II Seite 79 „it would be bad policy to diminish the deterrent influence of penal discipline in favour of those who are ignorant".

[4]) Du Cane I Seite 24 „As a knowledge of reading and writing affords so much opportunity for mental and moral improvement and may have so important an effect on a prisoner's well-being in after-life".

Wirkung, welche gerade von dem ersten Stadium der Strafe erwartet wird, irgendwie abgeschwächt werde.[1])

2) Unterricht wird ferner nicht erteilt an Gefangene, deren Strafzeit unter 4 Monaten beträgt. Die für den Unterricht nach Abzug des ersten Monats hier verbleibende Zeit würde zu kurz sein, um einen dauernden Nutzen des Unterrichts für das weitere Fortkommen des Betreffenden erwarten zu können.

3) Gefangene, welche über 40 Jahre alt sind, erhalten nur in Ausnahmefällen Unterricht.[2]) Auch hierfür ist die Auffassung maßgebend, daß der Unterricht schwerlich für das weitere Fortkommen dieser Gefangenen von Bedeutung sein kann.

4) Soweit hiernach überhaupt Unterricht erteilt wird, darf durch denselben niemals die vorgeschriebene Arbeitszeit verkürzt oder in irgend einer Weise störend in die Arbeit, welche neben der Freiheitsentziehung den wesentlichsten Faktor der Strafe bildet, eingegriffen werden.[3]) Der Unterricht wird daher in den local prisons während der Mittagspause und in den Abendstunden, in den Public Works Prisons lediglich in den Abendstunden erteilt. Während der Sommermonate findet der Unterricht in den Public Works Prisons mit Rücksicht auf die lange Arbeitszeit überhaupt nur in beschränktem Umfange statt.[4])

---

[1]) Du Cane II Seite 81 „prisoners are not allowed the relaxation which the class or the visits of the schoolmasters afford".

[2]) Vielfach wird schon diese Altersgrenze für zu hoch gehalten, cf. Report of Commissioners 1886 Seite 8 „A general impression prevails that it is useless to attempt school instruction with persons past 30."

[3]) Report of Commissioners 1885 Seite 8: „Assuming that the prevention and repression of crime are the first objects of imprisonment, it would not be wise, then, as a rule, to conduct the education in such manner as to sacrifice the fundamental and well established principles of prison management by which these results are attained, under any idea that the good which might result from any probable increased amount of learning would more than counterbalance the certain harm which must result, as repeated experience has shown, from setting these principles aside".

[4]) Cf. Report of the Royal Commissioners 1879 Seite XV.

5) Der Unterricht beschränkt sich auf Lesen, Schreiben und Rechnen.[1]) Ein Unterricht in andern Gegenständen findet nicht statt. Gefangene, welche diese elementaren Kenntnisse bereits besitzen, erhalten überhaupt keinen Unterricht.

Das Vorstehende bezieht sich nur auf erwachsene Gefangene; für jugendliche, d. h. unter 16 Jahre alte Gefangene finden die Bestimmungen keine Anwendung. Bei den jugendlichen Gefangenen ist der Unterricht — was gar keiner weiteren Erläuterung bedarf — von einer viel größeren Bedeutung für das weitere Fortkommen; dieselben befinden sich ferner in einem Alter, wo sie für den Unterricht besonders aufnahmefähig sind, und endlich tritt bei ihnen überhaupt das strafende Element gegenüber dem erziehenden in den Hintergrund. Die jugendlichen Gefangenen erhalten daher stets — auch bei kurzen Strafen und während des ersten Monats der Strafverbüßung — täglich während der Tagesstunden eine Stunde Unterricht. Dieser Unterricht wird in Klassen erteilt, in welche die Jugendlichen gemäß ihren Fähigkeiten und Kenntnissen eingeteilt werden.

Bei den Erwachsenen ist man von jeher für Erteilung des Unterrichts in der Zelle gewesen.[2]) Für diejenigen Gefangenen, welche des Lesens und Schreibens noch ganz unkundig sind und welche daher einer ständigen Anleitung bedürfen, besteht jedoch mit Rücksicht darauf, daß für eine Erteilung des Unterrichts an jeden Einzelnen die vorhandenen Lehrkräfte nicht ausreichen würden, Klassenunterricht. Doch werden die Klassen so klein wie möglich gebildet; es sind selten mehr als sechs Gefangene in einer Klasse, so daß der Lehrer eine genügende Aufsicht darüber führen kann, daß die Gefangenen dem Unterrichte ihre ganze Aufmerksamkeit schenken und daß keinerlei Verkehr zwischen den Ge-

---

[1]) Die Gefangenen sollen bis zum Standard III der englischen Volksschule — es gibt 6 Standards — gebracht werden, d. h. fließend lesen schreiben und in den 4 Spezies rechnen können.

[2]) Es wird hierfür vor allem angeführt, daß die erwachsenen Gefangenen bei einem Unterrichte in Klassen leicht nervös würden und in beständiger Besorgnis schwebten, von den andern Gefangenen wegen ihrer Antworten verspottet oder ausgelacht zu werden. Cf. die in den Reports of Commissioners 1885 Seite 9 ff. und 1886 Seite 6 ff. wiedergegebenen Urteile einer größeren Anzahl von Gefängnisgeistlichen.

fangenen stattfindet. Sobald diese Gefangenen über die Elemente des Lesens und Schreibens hinaus sind, erhalten sie den ferneren Unterricht in der Zelle.

Der Unterricht in der Zelle wird dem Gefangenen zweimal wöchentlich, jedesmal auf die Dauer von $^1/_4$ Stunde erteilt. Der Lehrer gibt dabei dem Gefangenen eine Anzahl Aufgaben, welche derselbe bis zur nächsten Unterrichtsstunde zu vollenden hat.[1] Der Klassenunterricht findet ebenfalls zweimal wöchentlich statt und dauert jedesmal wenigstens $^1/_2$ Stunde.

Die Resultate des Unterrichts innerhalb der von vornherein gesteckten Grenzen werden als durchaus befriedigende dargestellt.[2] Es ist die Pflicht des Chaplain, sich durch halbjährliche Prüfungen jedes einzelnen Gefangenen über die von demselben gemachten Fortschritte zu vergewissern. Ein Hauptmittel, den Gefangenen zu Fortschritten im Lesen und Schreiben anzuregen, besteht in den Vorschriften über die Korrespondenz,[3] wonach der Lehrer — von Ausnahmefällen bei älteren oder des Lernens gänzlich unfähigen Gefangenen abgesehen — nur zweimal im Anfange der Strafzeit einen Brief für den Gefangenen schreiben resp. einen, von dem Gefangenen empfangenen Brief demselben vorlesen soll. Wenn der Gefangene also des Privilegs der Korrespondenz teilhaftig sein will, so muß er sich Mühe geben, selbst lesen und schreiben zu lernen.

Eine sehr wesentliche Ergänzung erhält der Unterricht durch die Lektüre. Während man in England den Unterricht auf die bloßen Elementarkenntnisse beschränkt, bemüht man sich auf das eifrigste, die Kenntnisse des Gefangenen durch Ge-

---

[1]) Wenn Gefangene die ihnen aufgegebenen Arbeiten nicht machen oder während des Unterrichts unaufmerksam sind, so kann der Governor auf Anzeige des Lehrers hin als Strafe eine Entziehung bereits verdienter Marken anordnen.

[2]) Du Cane I Seite 25; Report of Commissioners 1886 Seite 6. Es sei dabei bemerkt, daß seit Übernahme der local prisons in die Staatsverwaltung die Ausgaben für die Besoldung der Lehrer in diesen Anstalten um 2230 £ per Jahr erhöht worden sind.

[3]) Weiter ist zu erwähnen, daß convict prisoners, welche nicht lesen und schreiben können, von der Beförderung in die special class und den damit verbundenen besonderen Vergünstigungen ausgeschlossen bleiben.

währung guter lehrreicher Bücher zu erweitern.[1]) Auch hier tritt uns das allgemeine Prinzip des englischen Gefängniswesens entgegen: das Fortschreiten des Gefangenen auf dem Wege der Besserung soll durch eigne Thätigkeit bewirkt werden, und nicht dadurch, daß der erwachsene Gefangene wie ein Schulkind oder wie eine Treibhauspflanze behandelt wird. Der Gefangene erhält in England keinen Unterricht in Geographie oder Geschichte, wohl aber werden ihm gut und interessant geschriebene Geschichtswerke, Reisebeschreibungen, Lebensbilder hervorragender Männer u. s. w. geliefert,[2]) aus denen er sich als erwachsener Mensch, wenn er nur den festen Willen dazu hat, allein fortbilden kann. Die Zahl der dem Gefangenen gelieferten Bücher,[3]) welche wöchentlich einmal gewechselt werden, hängt davon ab, in welcher Strafklasse sich der Betreffende befindet. Die Vergünstigung, beim Fortschreiten in die höheren Klassen mehr Bücher zu erhalten, wird von vielen Gefangenen als eine ganz besonders erstrebenswerte geschätzt. Die Bibliotheken, welche unter der Leitung des Chaplain stehen, sind selbst in den kleinsten Anstalten reichlich ausgestattet und entsprechen allen berechtigterweise zu stellenden Anforderungen.

Zum Schlusse dieses Paragraphen sei noch ausdrücklich darauf hingewiesen, daß in England ein scharfer — und zwar auch sprachlich scharfer — Unterschied gemacht wird zwischen Unterricht (instruction) und Erziehung (education).[4])

---

[1]) Dies wird wesentlich dadurch erleichtert, daß es in England populär und leicht verständlich geschriebene Bücher von den ersten Autoritäten auf allen Gebieten des Wissens gibt. In England gilt es bekanntlich nicht als unwissenschaftlich, populär zu schreiben.

[2]) Ich habe ferner vielfach in den Zellen technische Unterrichtsbücher gefunden, aus denen sich der Gefangene in seinem Gewerbe weiter zu bilden suchte. — In den convict prisons werden mit besonderer Vorliebe Bücher, welche eine Schilderung des Lebens und der Verhältnisse in den Kolonieen enthalten, verteilt und, wie mir versichert wurde, auch gelesen. In dem Sträfling soll auf diese Weise der Wunsch, nach Verbüßung der Strafe nach den Kolonieen auszuwandern, rege gemacht werden.

[3]) In jeder Zelle liegt eine Bibel und zumeist noch ein anderes Andachtsbuch aus.

[4]) Es mag hier gestattet sein, der vortrefflichen Ausführungen über Unterricht und Erziehung in den Gefängnis-Anstalten bei Ad. Prins, Criminalité et répression. Bruxelles 1886 (besonders Kap. V) Erwähnung zu thun.

Es ist oben im § 12 des ausführlichen dargelegt worden, wie für die gesamte Behandlung der Gefangenen in England die Erziehung derselben zum vernünftigen Gebrauche der Freiheit ein leitendes Prinzip bildet. Zur Erreichung dieses Zieles wird nur bei den jugendlichen Gefangenen der Unterricht als ein wesentliches Moment angesehen, bei erwachsenen Gefangenen dagegen sind die Mittel der Erziehung: Stärkung der Willenskraft und der Selbstbeherrschung, sowie Anhalten zu ausdauernder und angestrengter Arbeit. Der Staat unterläßt es aber dabei nicht, denjenigen Gefangenen, welche die elementarsten Kenntnisse des Lesens, Schreibens und Rechnens nicht besitzen und in noch bildungsfähigem Alter sind, während der Zeit, welche zur Arbeit nicht verwendet werden kann,[1]) diese Kenntnisse soweit als möglich beizubringen, davon ausgehend, daß durch eine derartige Erweiterung der Kenntnisse die Möglichkeit eines redlichen Lebenserwerbs erhöht wird.[2])

Als ein eigentliches Erziehungs- und Besserungsmittel dagegen wird der Unterricht hier nicht behandelt, wie man denn auch in England weit davon entfernt ist, die mangelnden Kenntnisse als eine Ursache der Verbrechen zu betrachten. In der auch in England bestehenden Thatsache, daß sich unter den Verbrechern eine große Zahl völlig Ungebildeter befindet,[3]) wird nicht ein ursächliches Moment für die Verbrechen gefunden, sondern diese Thatsache wird ebenso wie das Verbrechen

---

[1]) Diese Einschränkung ist wesentlich, wenn anders nicht der Verbrecher besser gestellt sein soll, als der freie Arbeiter, welchem es doch nicht möglich ist, seinen Arbeitstag abzukürzen, um seine Kenntnisse zu erweitern.

[2]) Diese Auffassung des Unterrichts als einer dem Gefangenen gewährten Vergünstigung tritt deutlich in der Bestimmung hervor, daß der Unterricht erst beginnt, nachdem der Gefangene infolge seines Arbeitsfleißes und seines guten Verhaltens in die zweite Strafstufe befördert worden ist.

[3]) Die im Jahre 1884/5 Verurteilten verteilen sich nach ihrem Bildungsgrade folgendermaßen:

a) es konnten weder lesen noch schreiben:	30	%
b) es konnten nur unvollkommen lesen und schreiben:	65,7	„
c) es konnten gut lesen und schreiben:	3,7	„
d) eine höhere Bildung besaßen:	0,1	„
e) der Bildungsgrad wurde nicht ermittelt bei:	0,5	„

Judicial statistics 1884 Seite XXXVII.

selbst als eine Folgeerscheinung auf den Pauperismus, vernachlässigte Erziehung und andere soziale Momente zurückgeführt. [1])

§ 21. Die Beschäftigung der Gefangenen.

Für die Beschäftigung der Gefangenen in den englischen Strafanstalten sind drei Hauptgesichtspunkte maßgebend: [2])

1) Der Charakter der Strafe als eines Übels soll durch den ausgeübten Zwang zu angestrengter Arbeit verschärft werden.

2) Durch die Gewöhnung des Gefangenen an eine volle Ausnutzung seiner Arbeitskräfte in streng geregelter Thätigkeit soll ein wesentlicher Faktor für die in der Strafanstalt vorzunehmende Erziehung des Gefangenen zu einem erfolgreichen Wiedereintritte in die bürgerliche Gesellschaft gegeben werden.

3) Durch die Arbeitsleistungen der Gefangenen soll dem Staate soweit als möglich ein Ersatz für die Kosten der Unterhaltung der Strafanstalten gewährt werden.

Die Erreichung der hier angegebenen drei Ziele der Beschäftigung der Gefangenen bildet ein überaus schwieriges Problem, an dessen Lösung in England fortgesetzt gearbeitet wird. Die Schwierigkeit

---

[1]) Cf. Du Cane Law Quarterly Review 1886 Seite 234 f.: „The circumstance, that the prisons contain a great number of illiterate people does not arise from any connexion of cause and effect between them, but rather that both are due to the same cause, in fact that the criminal classes are recruited from among those who are least favourably placed for obtaining instruction, people who by their own fault or the fault of their parents have degenerated and have had no moral training". Cf. ferner Herbert Spencer (citiert im Report of Commissioners 1886 Seite 10): „The partisans of instruction triumph when they prove by statistics that the number of uninstructed criminals is the most considerable. — They never dream of asking themselves whether other statistics, established on the same system, would not prove, in a manner quite as conclusive, that crime is caused by the absence of linen; the uncleanness of the skin; the dwelling in narrow alleys, &c. — He who would teach geometry by giving lessons in Latin—or who expected to learn the piano by drawing—would be judged fit to be placed in a mad-house. He would not be more unreasonable though than those who pretend to improve the moral sense by instrcution in grammar, arithmetic, &c."

[2]) Cf. Du Cane I Seite 50.

liegt einmal in der Natur eines Arbeitsbetriebes in den Strafanstalten, wo sich Personen jeglichen Alters mit verschiedenen Fähigkeiten und Vorkenntnissen bald für längere bald nur für kurze Zeit befinden, anderseits in dem Umstande, daß die drei bei der Beschäftigung maßgebenden Gesichtspunkte vielfach miteinander in Kollision geraten. Für diese letztere Thatsache sei hier ein Beispiel angeführt.

Vom finanziellen Standpunkte aus würde es entschieden am vorteilhaftesten sein, den Gefangenen ausschließlich in demjenigen Arbeitszweige zu beschäftigen, in welchem derselbe schon vorher thätig war, für welchen er also die nötigen Vorkenntnisse mitbringt. Aber abgesehen davon, daß dies überhaupt nur in beschränktem Maße möglich ist, weil die Zahl der in den Anstalten durchführbaren Beschäftigungen naturgemäß eine kleine ist, würde die Strafe vielfach den Charakter eines Übels verlieren, wenn der Gefangene, befreit von all den Sorgen des täglichen Lebens, seine frühere Beschäftigung in der Anstalt einfach fortsetzen würde.[1]) Der Gefangene würde die Arbeit vielfach lediglich als eine Wohlthat betrachten, welche ihm die Qual der Langeweile erspart und ihn seine Lage vergessen läßt.

Ganz aus demselben Grunde erscheint es unthunlich, bei der Beschäftigung ausschließlich auf die Wünsche des Gefangenen Rücksicht zu nehmen und denselben lediglich zu der Arbeit anzuhalten, zu welcher er schon an sich Neigung hat, obgleich es auch hier keinem Zweifel unterliegen kann, daß die Resultate der Arbeit, an welche der Gefangene von vornherein mit Lust und Liebe geht, höhere sein würden, als diejenigen der erzwungenen Arbeit.

Wie hier der finanzielle Gesichtspunkt mit der Aufrechterhaltung des Charakters der Strafe als eines Übels in Widerspruch tritt, so findet in andern Fällen eine Kollision zwischen diesem letzteren Momente und der Benutzung der Gefangenenarbeit als eines Erziehungsmittels statt. Wenn der Gefangene zwangsweise zu einer Beschäftigung angehalten wird, für welche er keine Neigung oder gegen welche er geradezu Widerwillen empfindet, so wird er diese Arbeit schwerlich nach seiner Entlassung aus der Anstalt fortsetzen.

[1]) Cf. Report of the Royal Commissioners 1879 Seite XLVI „the mere saving of expense or convenience of employing skilled men in their own trade are not sufficient to justify the exemption af a particular class of artisans from the severe penal labour which is undergone by other less fortunate prisoners“.

Die erhoffte Gewöhnung an die Arbeit wird nicht eintreten, da die geleistete Arbeit in dem Gefangenen ein Gefühl der Freude nicht hervorgerufen hat: er hat die Arbeit nicht liebgewonnen, sondern sie ist ihm geradezu zur Pein geworden. Wenn die Beschäftigung in der Strafanstalt ein wirksames Erziehungsmittel zur Arbeit bilden soll, so muß auch der Gefangene dabei diejenige Befriedigung empfinden, welche jede pflichtmäßig geleistete Arbeit gewährt. Nur in diesem Falle wird es dem Gefangenen zum Bewußtsein gebracht werden, daß der Mensch nicht zum Müßiggange da ist, sondern daß ein jeder die Pflicht hat, sich nach dem Maße seiner Kräfte nützlich zu machen.

Man ist in England zu der Überzeugung gelangt, daß der hiernach bestehende Widerspruch zwischen dem oben sub 1) und dem sub 2) angeführten Ziele sich nur in der Weise heben läßt, daß für jedes dieser Ziele ein besonderes Strafstadium ausschließlich bestimmt wird.[1]) Der Gefangene soll zunächst in einer Weise beschäftigt werden, daß er die von ihm verlangte Arbeit als eine wirkliche Verschärfung der Strafe empfindet. Erst nachher wird der Gefangene unter gleichzeitiger Androhung, im Falle schlechten Verhaltens in das frühere Strafstadium zurückversetzt zu werden, zu einer Arbeit angehalten, von welcher anzunehmen ist, daß der Gefangene sie auch in der Freiheit fortsetzen werde, und durch welche er sich alsdann in redlicher Weise ernähren kann.

Die Beschäftigung während des ersten Strafstadiums hat in einförmiger, schwerer Arbeit zu bestehen; das finanzielle Erträgnis der Arbeit kommt hier kaum in Betracht. Während des folgenden Strafstadiums besteht das Bestreben, die Arbeit in der Anstalt der freien Arbeit möglichst ähnlich zu gestalten. Um hierbei dem Gefangenen einen äußeren Sporn zur Thätigkeit zu geben, wie ihn der freie Arbeiter durch die für seine Arbeitsleistung erhaltene Vergütung besitzt, wird dem Gefangenen eine nach dem Werte seiner Arbeit bemessene Arbeitsbelohnung gewährt (cf. den folgenden §). Der finanzielle Gesichtspunkt wird während dieses zweiten Straf-

[1]) Du Cane I Seite 56 f. „the most practical way of achieving the three objects to be attained by the industrial employment of prisoners is to divide the period of punishment into different stages, during one of which the penal or deterrent object should be principally considered, during the other, the reformatory and pecuniary may prevail in various degrees“.

stadiums stets im Auge behalten, er tritt aber zurück, soweit es der erzieherische Zweck der Beschäftigung erfordert.

In dieser Weise werden die drei an die Spitze dieses Paragraphen gestellten Gesichtspunkte im englischen Strafvollzuge berücksichtigt. Es ist dabei noch an die oben im § 13 hervorgehobene Unterscheidung zwischen drei Klassen von Delinquenten zu erinnern.

Bei den zu kurzer Gefängnisstrafe — bis zu einem Monat — Verurteilten kommt es auch bei der Beschäftigung lediglich darauf an, die abschreckende Wirkung der Strafe aufrecht zu erhalten. Die Gefangenen dieser Klasse werden ausschließlich in anstrengender, einförmiger Arbeit beschäftigt.[1]) Als eigentliche hard labour wird hier vorzugsweise die Tretmühle (tread wheel), sowie das Drehen einer Kurbel (labour on cranks) verwendet. Diejenigen Gefangenen, welche nicht zu hard labour verurteilt worden sind, sowie diejenigen Gefangenen, deren Gesundheitszustand nach dem Urteile des Gefängnisarztes die Anwendung dieser sogenannten first-class hard labour nicht thunlich erscheinen läßt,[2]) werden mit Wergzupfen (oakum picking), Steineklopfen oder sonstiger second-class hard labour beschäftigt.[3]) Ein wirklicher finanzieller Ertrag der Arbeit wird hier nicht erzielt.

Was sodann die zu längerer Gefängnisstrafe Verurteilten betrifft, welche nicht nur gestraft, sondern auch zu Ordnung und Arbeitsamkeit erzogen werden sollen, so folgt bei ihnen auf eine Periode der hard labour von wenigstens einem Monate[4])

---

[1]) „hard, dull, useless, uninteresting, monotonous labour; it is necessary to resort to this for its penal effect". Du Cane I Seite 57.

[2]) Jeder Gefangene ist, bevor er zu first-class hard labour gesetzt wird, von dem Arzte zu untersuchen. Es soll unter allen Umständen vermieden werden, daß der Gesundheitszustand des Gefangenen durch die Art und Weise seiner Beschäftigung geschädigt werde.

[3]) Beschäftigung in der Anstalts-Verwaltung soll nur ausnahmsweise als second-class hard labour gelten; eine derartige Beschäftigung soll lediglich als Belohnung für Fleiß und gutes Verhalten gestattet werden.

[4]) Nach dem Prison Act 1865 sollte die Dauer der first-class hard labour wenigstens 3 Monate betragen. Man fand jedoch allgemein, daß dies zu lange sei, und setzte die Periode der first-class hard labour in dem Prison Act 1877 auf 1 Monat herab, jedoch mit dem Zusatze, daß der Gefangene im Falle von Unfleiß oder schlechtem Verhalten während der Dauer der Haft stets wieder in das Stadium der first-class hard labour zurückversetzt werden könnte.

Beschäftigung in industrieller Arbeit. Die Zweige dieser Beschäftigung sind sehr mannigfaltige. Die frühere Thätigkeit des Gefangenen, seine Kenntnisse und Fertigkeiten werden dabei berücksichtigt, und es wird im Auge behalten, ob die Zeit der Inhaftierung lang genug ist, um den Gefangenen einen neuen Erwerbszweig mit Erfolg erlernen zu lassen. Wo dies nicht der Fall ist, wird eine Beschäftigung gewählt, welche angestrengte Thätigkeit erfordert und zu gleicher Zeit ein finanzielles Erträgnis liefert. Als ein Beispiel hierfür sei die Verwendung von Gefangenen zum Mattenflechten sowie zum Ausklopfen von Teppichen u. s. w. erwähnt; die letztere Arbeit, welche in manchen Bezirken ausschließlich in den Gefängnissen geschieht, gewährt befriedigende Einnahmen.

Das Strafübel liegt während des zweiten Stadiums der Gefängnisstrafe weniger in der Art und Weise der Beschäftigung, als in dem strengen Zwange zu anstrengender und anhaltender Thätigkeit in der anbefohlenen Arbeit. Und in dieser Gewöhnung an Ordnung und voller Ausnutzung der Arbeitskraft liegt zu gleicher Zeit ein erzieherisches Moment, zu welchem in vielen Fällen die Erlernung eines Handwerks während der Haftzeit hinzutritt. Die Erzielung möglichst hoher Erträgnisse der Arbeit wird dabei durchweg im Auge behalten.

Bei den zur penal servitude Verurteilten endlich, bei welchen eine vollständige Umbildung (reformation) für erforderlich erachtet wird, um aus ihnen nützliche Glieder der menschlichen Gesellschaft zu machen, bildet die Beschäftigung den Hauptfaktor in diesem Erziehungsprozesse. Nachdem hier während des ersten Stadiums der Strafverbüßung, während der neunmonatlichen Isolierhaft, das strafende, abschreckende Element auch in der Art und Weise der Beschäftigung genügend zum Ausdruck gebracht ist, [1]) tritt während der gesamten übrigen Strafzeit der erziehende Zweck der Beschäftigung in den Vordergrund.

Das durch den Aufenthalt des Sträflings in dem Public Works Prison erstrebte Ziel besteht im wesentlichen darin, den Sträfling, welcher sich als ein schäd-

[1]) Während dieses Stadiums werden die Sträflinge in ähnlicher Weise beschäftigt wie die zu einer Gefängnisstrafe Verurteilten: zunächst während 4—6 Wochen in wirklicher hard labour, sodann in industrieller Arbeit, soweit dieselbe in der Isolierzelle vorgenommen werden kann, so insbesondere in Schneiderei, Schuhmacherei und Anfertigung von Postsäcken.

liches Glied der bürgerlichen Gesellschaft erwiesen hat, an Arbeitsamkeit zu gewöhnen und in die Lage zu bringen, nach Verbüßung der Strafe sich durch ehrliche Arbeit zu ernähren und damit ein brauchbares Glied der Gesellschaft zu werden. [1]) Die Beschäftigung muß daher eine solche sein, von welcher anzunehmen ist, daß der entlassene Sträfling sie fortsetzen kann, und daß er in ihr seinen Lebenserwerb finden wird. Dabei wird vor allem die Lage, in welcher sich der Sträfling bei seiner Entlassung befindet, im Auge behalten: derjenige, welcher vor seiner Bestrafung einen Vertrauensposten innegehabt hat, darf, nachdem er zu der schweren Strafe der penal servitude verurteilt gewesen ist, nicht darauf rechnen, nach seiner Entlassung aus der Strafverbüßung eine ähnliche Stellung wiederzuerlangen. Er kann — jedenfalls für die erste Zeit — nur daran denken, durch volle Ausnutzung seiner körperlichen Kräfte, durch seiner Hände Arbeit sich seinen Lebensunterhalt zu schaffen. *Die Arbeit, an welche der Sträfling zu gewöhnen ist, soll daher eine derartige sein, wie sie — ohne Rücksicht auf die Antezedenzien des Arbeiters — auf dem allgemeinen Arbeitsmarkte jederzeit leicht einen Abnehmer findet.*

Dabei wird ferner auf allgemeine soziale Gesichtspunkte die gebührende Rücksicht genommen. Man hat insbesondere die Nachteile scharf im Auge, welche für die Allgemeinheit dadurch entstehen würden, wenn der Sträfling in der Strafanstalt ein Handwerk lernen würde, für welches er nur in den größeren Städten mit Sicherheit auf einen Abnehmer rechnen kann, so daß naturgemäß ein Zuzug nach den großen Städten gerade von denjenigen Elementen erfolgen würde, welche hier mehr als irgendwo anders infolge der geringeren Möglichkeit einer strengen Beaufsichtigung gefährlich sind.

Nachdem hiermit in großen Zügen zur Darstellung gebracht ist, nach welchen Grundsätzen man in England bei den verschiedenen Klassen von Gefangenen die an die Spitze dieses Paragraphen gestellten Ziele der Beschäftigung zu erreichen sucht, ist noch darauf einzugehen, in welcher Weise dabei den natürlichen Schwierigkeiten eines Arbeitsbetriebes in den Strafanstalten begegnet wird. Diese Schwierigkeiten lassen sich unter zwei Haupt-

---

[1]) Es sei an den schönen Ausspruch John Howards erinnert „make men diligent and they will be honest“.

gesichtspunkte bringen: Schwierigkeiten, welche durch das zur Verfügung stehende Arbeitspersonal gegeben sind, und Schwierigkeiten infolge der mit Recht geforderten Rücksichtnahme auf den allgemeinen Arbeitsmarkt.

Die zu einer Freiheitsstrafe Verurteilten bestehen aus Personen von sehr verschiedenen körperlichen und geistigen Fähigkeiten. Eine große Anzahl derselben hat überhaupt ein Handwerk, wie es in einer Strafanstalt betrieben werden kann, nicht, oder wenigstens nicht ordnungsmäßig erlernt. Wie ist es möglich, hier einen Arbeitsbetrieb derartig durchzuführen, daß sich nennenswerte finanzielle Erträgnisse ergeben? Die Antwort, welche auf diese Frage in England gegeben wird, lautet: durch Klassifikation der Strafanstalten nach der darin hauptsächlich betriebenen Arbeit und durch Zuteilung des einzelnen Gefangenen je nach seiner Individualität an die entsprechende Strafanstalt.

Betrachten wir zunächst, wie dieses Prinzip in den local prisons durchgeführt wird.

Die Gefangenen mit ganz kurzen Strafzeiten kommen dabei nicht in Betracht: Gefangene bis zu einem Monate werden lediglich zu Strafarbeiten angehalten; sie verbüßen ihre Strafe in der nächstgelegenen Strafanstalt. Dasselbe gilt auch von denjenigen Gefangenen, deren Strafzeit den Umfang eines Monats zu wenig überschreitet, um das Erlernen eines Handwerks möglich erscheinen zu lassen; sie werden nach Ablauf des ersten Monats mit Wergzupfen, Mattenflechten, Teppichausklopfen und andern Arbeiten beschäftigt, wie solche in jeder Anstalt ausgeführt werden können. Eine Anzahl Gefangener findet außerdem in Beschäftigungen für die Anstalt selbst Verwendung; die Reinigung und Instandhaltung der Anstaltsräume, der Wäsche und Bekleidungsstücke, die Zubereitung der Gefangenenkost u. s. w. erfordern überall Arbeitskräfte.

Bei den übrig bleibenden Gefangenen mit relativ längeren Strafzeiten erfolgt unter Berücksichtigung der bisherigen Beschäftigungsart und etwaiger schon vorhandener gewerblicher Kenntnisse Bestimmung darüber, in welcher Weise die Leute nach Abbüßung des ersten Monats der Strafzeit zu beschäftigen sind. Soweit für die im Einzelfalle geeignet erscheinende Beschäftigung resp. für die Erlernung derselben in der betreffenden Anstalt keine Gelegenheit ist, wird der Gefangene in eine andere hierfür geeignete

Anstalt gebracht. Es ist bereits oben Seite 165 hervorgehoben worden, daß es sich die Zentralbehörde angelegen sein läßt, einzelne größere Gefängnisse ausschließlich für bestimmte Gewerbszweige zu reservieren, und zwar immer für dasjenige Gewerbe, welches in dem betreffenden Distrikte vorzugsweise betrieben wird. In diese Anstalten werden dann auch aus andern Distrikten geeignete Gefangene geschafft. Die Gefängniswärter in diesen Spezialanstalten haben selbst genügende Kenntnisse in dem Gewerbszweige, um die Anleitung der darin noch unkundigen Gefangenen selbst zu übernehmen. Die Kosten für die Transportierung der Gefangenen nach der Spezialanstalt und von dort zurück [1]) werden reichlich durch die erhöhten Einnahmen gedeckt, wie sie durch diese rationelle Einrichtung des Arbeitsbetriebes erzielt werden. Die Spezialanstalt ist in der Lage, größere Aufträge für längere Zeit hinaus anzunehmen, da sie stets eines genügenden Stammes qualifizierter Arbeiter sicher ist.

In den convict prisons ist die Durchführung des Prinzips eine noch einfachere. Alle zur penal servitude Verurteilten verbüßen die ersten neun Monate ihrer Strafe in London — früher in den Penitentiaries von Pentonville und Millbank, jetzt in der neuen Anstalt von Wormwood Scrubs. Hier erfolgt dann unter Berücksichtigung des Alters, des Gesundheitszustandes, der körperlichen Kräfte u. s. w. die Bestimmung desjenigen Public Works Prison, welches nach den darin vorgenommenen Arbeiten für den einzelnen Sträfling am geeignetsten erscheint. Sträflinge im Vollbesitze der Arbeitskraft kommen nach Chatham, Portland oder Portsmouth, weniger kräftige nach Dartmoor, kranke oder gebrechliche nach Parkhurst oder Woking u. s. w. Indem so jeder einzelne Sträfling an die für ihn geeignete Stelle geschafft wird, werden seine Arbeitskräfte voll ausgenutzt.

Die zweite oben erwähnte Schwierigkeit des Arbeitsbetriebs in den Strafanstalten liegt darin, daß der freien Arbeit durch

---

[1]) Es fehlen genaue Angaben über die Höhe dieser Transportkosten. Um an diesen Kosten möglichst zu sparen, besteht die Anordnung, daß die zur Überführung in eine andre Anstalt bestimmten Gefangenen angesammelt und erst, nachdem ihre Zahl eine größere geworden ist, transportiert werden sollen. Es läßt sich dies während des ersten Monats der Strafverbüßung leicht durchführen, wo die Gefangenen mit Strafarbeiten, wie sie ja überall ausgeführt werden können, beschäftigt werden.

die Gefangenen-Arbeit keine ungebührliche Konkurrenz gemacht werden darf.[1]) Von diesem Gesichtspunkte aus ist Vorsorge zu treffen, daß weder die Arbeitskräfte der Gefangenen noch die von denselben hergestellten Arbeitsprodukte zu Preisen veräußert werden, welche außer Verhältnis zu denjenigen des freien Arbeitsmarkts stehen und dadurch mittelbar den Lohn des freien Arbeiters herabdrücken würden.

Diese Vorsorge ist jetzt in England in so umfassendem Maße getroffen, daß Klagen über Konkurrenz der Gefangenen-Arbeit kaum mehr vorkommen. Es ist dies in folgender Weise erreicht:

1) Eine Vermietung der Arbeitskräfte der Gefangenen an Privatunternehmer findet überhaupt nicht statt. Das anderswo so beliebte Entreprisesystem besteht in England nirgends. Der Arbeitsbetrieb geschieht ausschließlich auf Staatsrechnung.

2) Die Arbeitsprodukte sind fast ausschließlich für Staatsinstitute bestimmt,[2]) insbesondere für den Gefängnisdienst selbst, für Armee und Marine, für Post und Polizei u. s. w. Nur ein ganz geringfügiger Prozentsatz der Arbeitsprodukte wird an Private abgesetzt, und der Preis dafür wird durch öffentliche Lizitation festgestellt. Es handelt sich durchweg um Arbeiten, wie Mattenflechten, Wergzupfen u. s. w., bei denen von einer wirklichen

---

[1]) Eine gewisse Konkurrenz wird der freien Arbeit durch jede Beschäftigung der Gefangenen gemacht werden, indem Leute, welche bisher ein träges Leben geführt und keinen ehrlichen Lebensberuf gehabt haben, zur Ausnutzung ihrer Arbeitskräfte angehalten und dabei zu gleicher Zeit in einem Handwerke unterrichtet werden. Allein soweit wird wohl kein Verständiger gehen, zu verlangen, daß die in den Gefangenen vorhandenen Arbeitskräfte unbenutzt gelassen werden sollen. Es sind immer nur einzelne Gewerbe, welche sich darüber beklagen, daß ihrem Gewerbszweige durch die Gefangenen-Arbeit Konkurrenz gemacht werde. Eine Beachtung verdienen derartige Klagen — abgesehen von dem im Texte Auszuführenden — nach der Richtung hin, daß in den Anstalten Vorsorge für eine möglichst mannigfaltige Beschäftigung getroffen werden muß, damit nicht durch Beschäftigung einer zu großen Zahl von Gefangenen in einem einzelnen Gewerbszweige eine Überfüllung des letzteren mit Arbeitskräften hervorgerufen wird.

[2]) In dem Report of the Royal Commissioners 1879 Seite LVI wird dies Prinzip als das allein richtige warm verteidigt „by manufacturing for the Government, instead of for sale, many difficult questions as to competitain with private enterprise and industry and as to the sale to the best odvantage of the articles made will be avoided“.

Privat-Konkurrenz nicht die Rede ist. Unter Benutzung von Maschinenkräften hergestellte Gegenstände werden niemals auf den freien Arbeitsmarkt gebracht. [1])

Die hiernach in den englischen Strafanstalten vorgenommenen Arbeiten lassen sich unter fünf Rubriken bringen: a) Arbeiten für den Anstaltsbedarf, b) industrielle Arbeiten, c) Gefängnisbauten, d) öffentliche Arbeiten (public works), e) landwirtschaftliche Arbeiten.

Die erste Rubrik umfaßt alles, was zur Instandhaltung der Anstalt und für den laufenden Betrieb erforderlich ist, also alle Reinigungs- und Reparaturarbeiten, sowie die Arbeiten in der Küche, Bäckerei und Waschanstalt. Es liegt in der Natur der Sache, daß alle diese Arbeiten von Insassen der Anstalt verrichtet werden. Dabei besteht übrigens eine gewisse Zentralisation, indem in einigen großen Anstalten das Brot für die in der Nähe liegenden kleinen Gefängnisse mitgebacken wird; andre Anstalten — insbesondere die großen Weibergefängnisse — besorgen die Wascharbeiten für eine größere Anzahl von Gefängnissen.

Unter den Arbeiten der zweiten Kategorie nimmt Schuhmacherei und Schneiderei die Hauptstelle ein. Alles Schuhwerk und alle Kleidungsstücke, welche in den Strafanstalten erforderlich sind, sowie die Uniformen u. s. w. der Beamten, werden in den Anstalten selbst verfertigt. Ferner beziehen Armee, Marine und Polizei Schuhwerk und vielfach auch Uniformstücke aus den Strafanstalten. Einen andern Hauptartikel bildet die Anfertigung von Postsäcken („mail bags"), welche ausschließlich in den Strafanstalten hergestellt werden. Auch Druck- und Buchbinderarbeiten für die verschiedenen Regierungsdepartements finden besonders in den in London gelegenen Strafanstalten vielfach statt. Mattenflechten und Körbemachen wird jetzt zwar weit weniger wie früher betrieben, spielt aber immerhin bei der großen Zahl kurzzeitiger Gefängnisstrafen eine Rolle. Wergzupfen, womit eine relativ sehr große Zahl von Gefangenen beschäftigt wird, kann kaum unter die gewinnabwerfenden Arbeiten gerechnet werden, das Reinerträgnis dafür wird pro Woche auf $2^1/_2 - 4^1/_2$ pence angegeben.

---

[1]) Es würde hier, wie Du Cane II Seite 184 mit Recht bemerkt, „not merely competition against prison labour, but against Government capital" vorliegen.

Die zur penal servitude Verurteilten werden mit den bisher angeführten Arbeiten während des zweiten Strafstadiums nur in soweit beschäftigt, als die Körperkräfte derselben Arbeiten der folgenden Rubriken unthunlich erscheinen lassen.[1])

Über die Verwendung der Gefangenen, und zwar sowohl der zu Gefängnis wie der zur penal servitude Verurteilten, zur Herstellung von Gefängnisbauten ist bereits oben Seite 171 ausführlich gesprochen worden. Es sei gestattet, hier lediglich auf das dort Angeführte Bezug zu nehmen.

Landwirtschaftliche Arbeiten finden in den local prisons nur in geringem Umfange statt, da nur mit einzelnen Anstalten etwas Acker- oder Gartenland verbunden ist. In den Public Works Prisons von Parkhurst und Woking, wo ältere, kränkliche oder schwächliche Sträflinge untergebracht sind, wird neben industrieller Beschäftigung Ackerbau betrieben. Ein landwirtschaftlicher Betrieb von großem Umfange findet nur in dem Public Works Prison von Dartmoor statt; es werden hier durchschnittlich 230 Sträflinge ausschließlich zur Urbarmachung des umliegenden Moorlandes verwendet.

Was endlich die öffentlichen Arbeiten betrifft, so werden zu denselben nur die zur Strafe der penal servitude Verurteilten angehalten. Für diese Klasse der Gefangenen aber werden die öffentlichen Arbeiten als die geeignetste Beschäftigung gehalten, und es werden daher alle convict prisoners, deren Körperkräfte dies irgend zulassen, während des zweiten Strafstadiums, welches direkt die Bezeichnung Public Works Stage führt, in dieser Weise beschäftigt.

Die ersten öffentlichen Arbeiten wurden in Portland ausgeführt, wo ein fast 2 Meilen langer und 50—60 Fuß tiefer Steindamm errichtet wurde. Alle hierfür erforderlichen Arbeiten von dem Brechen der Steine in den Steinbrüchen bei Portland bis zur Vollendung des großen Werkes geschahen durch Sträflinge. Als der Steindamm fertig war, wurde die Beseitigung der alten Festungswerke bei Port-

---

[1]) Es trifft dies insbesondere bei den weiblichen Sträflingen zu. Für dieselben besteht neben anderweitigen Arbeiten ein besonderer Beschäftigungszweig in der Herstellung von Mosaikarbeiten, welche besonders in der Anstalt von Woking in ausgedehntem Maßstabe erfolgt. Der Mosaikboden um die Gräber von Nelson und Wellington in der Krypta von St. Paul ist in Woking angefertigt worden.

land und die Errichtung neuer Festungswerke in Angriff genommen. Von ähnlicher Art sind die Arbeiten, welche in den andern Public Works Prisons von Chatham, Portsmouth und Borstal vorgenommen werden: Neuerrichtung von Festungswerken, Anlage oder Erweiterung von Docks und sonstige umfassende Hafenbauten. Das bei der Auswahl derartiger Arbeiten maßgebende Prinzip ist, daß die Unternehmungen groß genug sind, um eine langdauernde Beschäftigung einer größeren Zahl von Leuten zu sichern, und daß die Unternehmungen selbst einen wirklichen öffentlichen Nutzen versprechen. [1])

Die Ausführung derartiger Arbeiten [2]) wird dadurch erleichtert,

---

[1]) Die in Portland, Portsmouth und Chatham unternommenen öffentlichen Arbeiten nähern sich jetzt der Vollendung. Es wurde aus dieser Veranlassung im Jahre 1881 eine besondere Kommission niedergesetzt, um Vorschläge über die in Zukunft vorzunehmenden öffentlichen Arbeiten zu machen. Der Bericht der Kommission, in welchem ausführlich auf die Vorteile, welche die Beschäftigung der Sträflinge mit öffentlichen Arbeiten mit sich bringt, und auf die bei der Auswahl der Arbeiten maßgebenden Prinzipien eingegangen wird, ist abgedruckt im Report of Directors 1885 Seite XXXIV ff. Die Kommission empfahl die Anlage eines Hafens in Dover und in Filey (Yorkshire); mit den Arbeiten in Dover würden 900 Sträflinge auf 12 Jahre und mit den Arbeiten in Filey 720 Sträflinge ebenfalls auf 12 Jahre hinaus Beschäftigung finden. — Es werden auf diese Weise für die Allgemeinheit nützliche Arbeiten ausgeführt „which the public might not be always willing otherwise to undertake" (Du Cane II Seite 178).

[2]) Es würde zu weit führen, auf die Art und Weise der Ausführungen der Arbeiten hier im einzelnen einzugehen. Es sei nur hervorgehoben, in welcher Weise dabei Vorkehrungen gegen Fluchtversuche und Aufstände der Sträflinge getroffen werden. Ich nehme dabei als konkretes Beispiel die Einrichtungen in Borstal, wo die Sträflinge mit Festungsbauten beschäftigt sind. Dieselben werden hier in Einzelabteilungen von 18—24 Mann mit je 2 Wärtern eingeteilt. In jeder Abteilung befinden sich eine Anzahl Sträflinge, deren Strafzeit binnen kurzem abläuft und von denen daher anzunehmen ist, daß sie sich an Fluchtversuchen oder Aufständen schon aus dem Grunde nicht beteiligen werden, weil sie sonst der bereits angesammelten Arbeitsbelohnungen verlustig gehen würden. Es kommen ferner in jede Abteilung einige Sträflinge aus der special class, auf welche man sich unbedingt verlassen kann und welche überhaupt einen günstigen Einfluß auf die übrigen Sträflinge ausüben werden. Auf diese Weise enthält jede Abteilung eine Zahl von Leuten, welche nicht nur einer besonderen Beaufsichtigung nicht bedürfen, sondern auch im Notfalle bei Aufständen und Fluchtversuchen die Wärter unterstützen würden. Diese Maßregeln haben sich als durchaus genügende erwiesen: es sind in den letzten 25 Jahren Gefangenen-Aufstände in den Public Works

daß, wie schon früher hervorgehoben wurde, die Governors all dieser Public Works Prisons dem Ingenieurkorps entnommen sind und somit die nötigen technischen Kenntnisse zur Leitung der Unternehmungen besitzen. Auch von den Wärtern, welche ebenfalls zum großen Teile aus der Armee hervorgegangen sind, werden genügende technische Kenntnisse verlangt, um die Sträflinge in den Arbeiten unterweisen und bei den Arbeiten beaufsichtigen zu können.[1])

Man ist in England darüber einig, daß öffentliche Arbeit sich in jeglicher Richtung als die geeignetste Form der Beschäftigung der Sträflinge empfiehlt.[2]) Die Arbeitskräfte werden voll ausgenutzt, die im Freien und in Gemeinschaft vorgenommene Arbeit ist die für Körper und Geist zuträglichste, es ist eine Beschäftigung der Sträflinge in den mannigfaltigsten Arbeitszweigen möglich, und die Arbeitszweige sind solche, in welchen der Sträfling auch nach seiner Entlassung Beschäftigung zu finden hoffen kann.

Prisons überhaupt nicht vorgekommen, und die Zahl der Fluchtversuche ist eine sehr geringe. Im Jahre 1885 fand in der Anstalt Borstal mit einer Belegung von 482 Sträflingen ein Fluchtversuch statt (der Betreffende wurde noch an demselben Tage ergriffen); im Jahre 1886 kam bei einer Belegung mit 505 Sträflingen kein Fluchtversuch vor. Interessant ist es, die Beförderung der Sträflinge von der Anstalt in Borstal nach dem Arbeitsplatze zu sehen. Diese Beförderung erfolgt in Borstal, wo die einzelnen Arbeitsplätze ziemlich entfernt von der Anstalt liegen, mittelst tramway; wenn die Arbeit an dem einen Platze vollendet ist, so werden die Schienen nach einem andern Platze gelegt. Bei dem Ausrücken zur Arbeit stellen sich die Sträflinge abteilungsweise auf dem Hofe der Anstalt auf; jede Abteilung wird einzeln nach einem Waggon geführt und der Waggon dann verschlossen. In gleicher Weise erfolgt am Arbeitsorte abteilungsweise die Entleerung der Waggons.

[1]) Es liegt auf der Hand, daß es nicht nur vom finanziellen Standpunkte aus, sondern auch im Interesse der Disziplin von großem Werte ist, wenn die Wärter zu gleicher Zeit die Funktionen von Werkmeistern ausfüllen.

[2]) Cf. Du Cane II Seite 177 „outdoor employment is healthiest both for body and mind, and it is the more appropriate to the circumstances, as generally it requires less skill", „severe labour on public works is most beneficial in teaching criminals habits of industry and training them to such employment as digging, road-making, quarrying, stone-dressing, building and brickmaking". „It is found that employment of this nature is most easily obtained by convicts on their release, since men are taken on for rough work without the strict inquiries as to previous character which are made in other cases".

Es sei nun noch kurz auf die finanziellen Erträgnisse der Gefangenenarbeit eingegangen. Hier ist vor allen Dingen im Auge zu behalten, daß nur ein kleiner Teil der Arbeitsprodukte wirklich verkauft und auf diese Weise der Wert mit Bestimmtheit ermittelt wird. [1]) Die weit überwiegende Mehrheit der Arbeit wird, wie schon ausgeführt wurde, für andre Regierungsdepartements geleistet und die Festsetzung des Wertes dieser Arbeit ist an sich lediglich eine Rechnungsoperation zwischen den beteiligten Departements. Eine derartige Festsetzung kann niemals als eine genaue Wertsermittelung angesehen werden, sondern bleibt immer nur eine Schätzung.

In den letzten Jahren hat die Zentralgefängnisbehörde alle möglichen Anstrengungen gemacht, um diese Schätzung der Wahrheit möglichst nahe zu bringen: [2]) man hat einzelne Arbeiten von freien Arbeitern anfertigen lassen, um so die Höhe des für die Fertigstellung der Arbeit zu zahlenden Lohnes genau zu ermitteln; man hat dann dieselben Arbeiten von Sträflingen mit mittlerer Arbeitstüchtigkeit machen lassen und durch Verrechnung der ermittelten Lohnsumme auf die von dem Sträfling gebrauchten Arbeitstage den Wert der Tagesarbeit eines Sträflings in der betreffenden Branche festgestellt. [3])

Dieser Wert beträgt, [4]) um einige Beispiele anzuführen,

für Schneiderarbeiten	1 s	9	d	per Tag,
für Schuhmacherarbeiten	1 s	9	d	=
für Mattenflechten		$9\frac{1}{2}$	d	=

---

[1]) Von dem Erlöse dieser Arbeitsprodukte wird ein bestimmter Prozentsatz für Abnutzung der Werkzeuge und sonstige mit der Herstellung verbundene Kosten in Abzug gebracht.

[2]) Das Verfahren der Abschätzung der Arbeit in den local prisons ist seit Übernahme derselben in die Staatsverwaltung zu gleicher Zeit mit dem damals eingeführten verbesserten Arbeitsbetriebe neu geregelt worden; cf. den im Report of Commissioners 1878 Seite 49 abgedruckten Bericht des Committee appointed to inquire into the system of accounts in the local prisons.

[3]) In dem Report of the Royal Commissioners 1879 Seite LVI ff. wird das Verfahren bei Abschätzung der Gefangenenarbeit eingehend besprochen und der hohe Wert einer möglichst genauen Abschätzung scharf hervorgehoben. Infolgedessen sind mannigfache Verbesserungen bei der Abschätzung vorgenommen worden.

[4]) Die nachstehenden Angaben sind einem Zirkularbriefe der Zentralgefängnisbehörde vom 17. März 1886 entnommen.

für Maurerarbeiten	2 s 6 d	per Tag,
für Arbeiten in der Anstaltsküche	1 s 11 d	„
für Arbeiten im Backhause	1 s 9 d	„
für Arbeiten in der Waschanstalt	1 s 3 d	„

Überall ist hierbei ein zehnstündiger Arbeitstag zu Grunde gelegt. Soweit im Einzelfalle der Gefangene eine geringere Anzahl Stunden am Tage arbeitet, wird auch ein entsprechend geringerer Wertbetrag eingesetzt.

Unter Zugrundelegung der erwähnten Bestimmungen ist für das Jahr 1884 der Gesamtwert der Gefangenenarbeit in den local prisons

auf 126 093 £

ermittelt worden.[1]) Da die Durchschnittsbelegung der local prisons in diesem Jahre 16 619 betrug, so ergibt sich auf den Kopf des Gefangenen ein jährliches Arbeitserträgnis von 7 £ 11 s 8 d.

Erheblich viel größer ist das Arbeitserträgnis in den convict prisons. Im Jahre 1884 wird der Wert der Gefangenenarbeit, und zwar hier unter Nichteinrechnung der in der Verwaltung der Anstalt selbst (in the service of the prison) geleisteten Arbeit

auf 191 501 £

angegeben.[2]) Dies ergibt hier auf den Kopf des Gefangenen ein jährliches Arbeitserträgnis von 20 £ 14 s 2 d.

Ein noch viel höheres Erträgnis stellt sich für diejenigen Public Works Prisons heraus, in denen sich ausschließlich voll arbeitsfähige männliche Sträflinge befinden; so betrug das tägliche Arbeitserträgnis

in Chatham	2 s $4^1/_4$ d,
in Portsmouth	2 s $3^1/_4$ d,
in Portland	2 s 1 d,

---

[1]) Judicial Statistics 1884 Seite 70; die obige Summe verteilt sich auf

a) manufactures buildings etc.:	66 235 £.
b) labour in the ordinary service of the prison:	57 335 „
c) incidental receipts, sale of old stores etc.:	2 522 „

[2]) Report of Directors 1885 Seite XXVI. Im Jahre 1886 betrug infolge der Abnahme der Zahl der Sträflinge der Wert der Arbeit nur 165 413 £. Report of Directors 1886 Seite XXVIII.

dagegen

in Parkhurst 1 s $^1/_2$ d,
in Woking — 8$^3/_4$ d.

In den Anstalten von Chatham und Portsmouth decken die Arbeitserträgnisse fast vollständig die Unterhaltskosten der Anstalt: die Nettokosten — d. h. abzüglich der Arbeitserträgnisse — pro Kopf des Sträflings betragen in Chatham jährlich 2 £ 14 s 9 d; in Portsmouth 3 £ 9 s 2 d.[1])

Wenn man in Berücksichtigung zieht, daß sich unter den Insassen einer Strafanstalt eine erhebliche Anzahl nur halb arbeitsfähiger Personen befindet,[2]) und daß es bei der Gefängnisarbeit überhaupt, speziell aber bei der Arbeit in der Einzelzelle, an dem Wetteifer fehlt, wie er bei der freien Arbeit die Leistungen fördernd sich entwickelt,[3]) so wird man die in den englischen Anstalten erreichten finanziellen Erträgnisse als befriedigende bezeichnen müssen.

Es verdient dabei noch hervorgehoben zu werden, daß nach der in England herrschenden Ansicht die finanzielle Seite bei der Beschäftigung der Gefangenen gegenüber den beiden andern an der Spitze des Paragraphen erwähnten Gesichtspunkten überhaupt zurückzutreten hat.[4])

---

[1]) Die vorstehenden Ziffern sind auf Grund der Angaben im Report of Directors 1885 Seite XXIX und XXVI berechnet.

[2]) Interessant ist in dieser Richtung ein im Jahre 1881 in den convict prisons aufgenommener Zensus. Es ergab sich dabei, daß 3,6 % der Insassen für jede Arbeit unfähig waren und daß 21,5 % nur zu leichter Arbeit verwendet werden konnten.

[3]) „Prison labour must always be carried on under the disadvantage of being without that stimulus to industry which is afforded by the prospect of immediate resulting benefit.“ Du Cane I Seite 53. Auf Seite 150 ff. ist daselbst ein Gutachten von Kapitän Percy Smith über den Wert der Gefangenenarbeit abgedruckt; cf. auch die in dem Report of the Royal Commissioners 1879 Seite LVIII f. angeführten Gutachten der Ingenieure Bernays und Henry Wood. Unter Berücksichtigung der zumeist kürzeren Arbeitszeit in den Strafanstalten wird das Verhältnis der Arbeitsleistung eines Sträflings zu derjenigen eines freien Arbeiters auf 3 : 2, teilweise sogar nur auf 2 : 1 angegeben.

[4]) Lord Carnavon bezeichnete auf dem Gefängnis-Kongresse von 1872 die Ansicht, daß die Strafanstalten self-supporting gemacht werden müßten, als „as dangerous as it is attractive“. Das Committee of the House of Lords vom Jahre 1863 sprach sich dahin aus „a profitable return from the indu-

§ 22. Die Arbeitsbelohnung.

Unter denjenigen Vorteilen, welche sich der Gefangene durch seine Arbeitsamkeit bei gutem Verhalten — wie beides durch das Markensystem in der im § 13 dargestellten Weise konstatiert wird — erwerben kann, nimmt die Arbeitsbelohnung (gratuity)[1]) die erste Stelle ein.

Indem dem Gefangenen eine materielle Belohnung für fleißige Arbeit gewährt wird, erhält derselbe eine Aufmunterung zum Fleiße. Die Gefangenenarbeit wird der freien Arbeit ähnlicher, wenn dem Gefangenen ein materieller Vorteil derselben vor Augen gehalten wird. Der Gefangene vermag auf diese Weise die Arbeit, obwohl sie eine erzwungene ist, liebzugewinnen. Und nicht nur der erzieherische Einfluß der Beschäftigung wird hierdurch gefördert werden, sondern auch das finanzielle Interesse des Staates, welcher bei einer derartigen Aufmunterung des Fleißes bessere Arbeitsleistungen erhalten wird.

Aber noch nach einer andern Richtung hin empfiehlt sich die Gewährung einer Arbeitsbelohnung. Wenn der Gefangene die Anstalt verläßt, so wird er, was für Vorkehrungen auch immer getroffen sein mögen, für die erste Zeit ganz besonders schwierigen Verhältnissen entgegengehen. Abgesehen davon, daß der Entlassene nicht immer sofort Beschäftigung finden kann, bedarf derselbe auch zu einem gedeihlichen Fortkommen eines gewissen Anlagekapitals, um sich die nötigen Handwerkszeuge anzuschaffen und sonst notwendige Ausgaben zu bestreiten. Ein erfolgreicher Übertritt aus der Anstalt in die Freiheit, woran der Staat das allergrößte Interesse

---

strial employment ought not to be made the test of prison efficiency". In dem Report of the Royal Commissioners 1879 Seite XLVI heißt es: „we are fully alive to the importance of obtaining as large a return as possible from the labour of convicts; but the first object ought to be the infliction as nearly as may be of equal punishment".

[1]) Die Bezeichnung „Arbeitsbelohnung" erscheint die sachgemäßeste. In dem Worte „Arbeitsgeschenk" kommt es zu wenig zum Ausdrucke, daß der Gefangene ein wirkliches Recht auf eine gratuity nach festbestimmten Regeln hat. Der Ausdruck „Arbeitsverdienst" oder „Arbeitsverdienstanteil" anderseits erscheint mit Rücksicht auf die vorhandene Befugnis der Entziehung einer bereits zugesprochenen gratuity nicht treffend.

hat, wird daher erheblich erleichtert, wenn dem Gefangenen die Möglichkeit gegeben ist, versehen mit einem kleinen Kapitale, das er sich aus den Arbeitsbelohnungen angesammelt hat, in die Freiheit zurückzukehren.

Den vorstehenden Erwägungen gemäß wird in England, sowohl den zur penal servitude wie den zu einer Gefängnisstrafe Verurteilten, eine Arbeitsbelohnung gewährt. Dabei hat man von vornherein Vorkehrungen getroffen, damit aus einer derartigen Belohnung nicht Nachteile nach anderer Richtung hin entstehen können.

Es ist zunächst die Gefahr zu beachten, daß die Lage des Gefangenen sich nicht durch hohe Arbeitsbelohnungen zu einer besseren gestalte, als diejenige des freien Arbeiters ist. All der für den Lebensunterhalt notwendigen Ausgaben ist der Gefangene enthoben; wenn er Familie besitzt, so wird dieselbe während seiner Gefangenschaft im Falle der Bedürftigkeit aus öffentlichen Mitteln unterstützt. Es würde eine Ungerechtigkeit sein, dem Gefangenen bei seiner Entlassung ein Kapital von einer Höhe auszuhändigen, wie er es als freier Arbeiter, wenn er die erwähnten Ausgaben aus seinem Arbeitslohne hätte bestreiten müssen, nicht erspart haben könnte. Es ergibt sich daraus die Notwendigkeit, von vornherein die Maximalhöhe der Arbeitsbelohnung niedrig zu bemessen. [1])

Sodann aber erscheint es auch erforderlich, eine gewisse Aufsicht über die Verwendung der Arbeitsbelohnung auszuüben und bezüglich der Verwendung Einschränkungen aufzuerlegen, damit der in das Auge gefaßte Zweck der Belohnung erreicht und damit auch hier der Gefahr vorgebeugt wird, daß die gewährte Summe nicht zu Genüssen und sonstigen Ausgaben verwendet werde, welche sich der freie, aber unvermögende Arbeiter versagen muß.

Die thatsächliche Gestaltung, welche das Institut der Arbeits-

---

[1]) Diese Notwendigkeit wird besonders in dem Report of the Royal Commission 1863 scharf betont. Die bis dahin bestehende Praxis bezüglich der Gewährung der gratuities wird verworfen „as having the effect of rendering penal servitude in some rare cases an object of desire rather than of apprehension, and as having the further serious fault of enabling those whose sentences were the longest and whose crimes were therefore presumably the gravest to earn the largest sum of money“.

belohnung unter Berücksichtigung all dieser Momente in England gefunden hat, ist die folgende.[1])

Die zu einer Gefängnisstrafe bis zu einem Monate Verurteilten erhalten keinerlei Arbeitsbelohnung, da sie ja auch keine gewinnbringende Arbeit verrichten, sondern zu reinen Strafarbeiten angehalten werden. Bei den zu längeren Gefängnisstrafen Verurteilten richtet sich die Höhe der Arbeitsbelohnung nach der Summe der verdienten Marken unter Berücksichtigung der Strafklasse, in welcher sich der Betreffende befindet. In der untersten Strafklasse kann der Gefangene eine Arbeitsbelohnung nicht erwerben; in der zweiten Klasse erhält er für je 20 Marken, in der dritten Klasse für je 12 Marken, in der obersten Klasse für je 10 Marken: 1 Penny Arbeitsbelohnung. Die Maximalhöhe der Arbeitsbelohnung, welche hiernach der Gefangene in der zweiten Strafklasse erwerben kann, beträgt: 1 Schilling, in der dritten Klasse: 1 Schilling 6 Pence, in der vierten Klasse: 10 Schilling, zusammen also 12 Schilling 6 Pence. Ein Gefangener, welcher in der Anstalt besonders wertvolle Arbeiten ausführt, kann dafür zusätzlich zu den obigen Beträgen eine Extravergütung erhalten; es darf jedoch der sich danach ergebende Gesamtbetrag in keinem Falle die Summe von 2 £ übersteigen.

Die zur penal servitude Verurteilten erhalten während der neunmonatlichen Einzelhaft keinerlei Arbeitsbelohnung. Während des Public Works Stage erhalten sie, vorausgesetzt daß sie täglich acht Marken verdienen, in der untersten Strafklasse monatlich 1 Schilling, in der zweiten Klasse monatlich 1 Schilling 6 Pence, in der obersten Klasse monatlich 2 Schilling 6 Pence. Die Maximalhöhe der Arbeitsbelohnung beträgt hiernach 3 £. Wenn ein Sträfling aus der obersten Strafklasse in die special class befördert worden ist, so kann ihm zusätzlich zu den obigen Beträgen noch eine Extravergütung von 3 £ gewährt werden.

Die Extravergütungen werden stets behufs zweckentsprechender Verwendung für den Entlassenen an denjenigen Verein zur Fürsorge für entlassene Strafgefangene ausgezahlt, unter dessen Obhut

[1]) Cf. Report of Commissioners 1878 Seite 39 f., Report of the Royal Commission 1879 Seite XVII.

sich der Betreffende begibt. Die Unterwerfung unter eine derartige Schutzaufsicht ist Bedingung für die Gewährung der Extravergütung.

Was die regelmäßigen Arbeitsbelohnungen betrifft, so hängt es von dem Ermessen des Governor ab, ob der Gesamtbetrag bei der Entlassung ausgezahlt wird oder nur ein Teil unter Übersendung des Restes an einen Schutzverein behufs zweckentsprechender Verwendung, resp. für den Fall, daß sich der Entlassene der Aufsicht eines Schutzvereins nicht unterwirft, an die Polizeibehörde des zukünftigen Aufenthalts behufs ratenweiser Auszahlung. Bei den zur penal servitude Verurteilten, deren Arbeitsbelohnung mehr als 10 Schilling beträgt, bildet die Auszahlung nur eines Teiles der Summe die Regel.

Durch diese Bestimmungen wird *nicht nur eine rationelle Verwendung der Arbeitsbelohnung sichergestellt, sondern zu gleicher Zeit auch die Thätigkeit der Vereine zur Fürsorge für entlassene Strafgefangene in wirksamer Weise gefördert.*

Eine Auszahlung der verdienten Arbeitsbelohnung während der Strafzeit findet nur in Ausnahmefällen statt. Derartige Ausnahmen werden insbesondere dann gemacht, wenn die Summe zur Unterstützung zurückgebliebener Angehöriger verwendet werden soll. *Eine Verwendung der Arbeitsbelohnung zum Ankaufe von Lebensmitteln oder zu sonstiger Verbesserung der Gefangenenkost ist unter allen Umständen unzulässig.*[1])

Die Entziehung der bereits verdienten Arbeitsbelohnung oder eines Teiles derselben kann im Wege der Disziplinarstrafe erfolgen. In den Public Works Prisons wird von diesem Strafmittel vielfach Gebrauch gemacht, doch besteht die ausdrückliche Bestimmung, daß dem zur penal servitude Verurteilten bei seiner Entlassung stets ein Betrag von 10 Schillingen als Arbeitsbelohnung verbleiben soll.

---

[1]) In dem Report of the Royal Commission 1863 wird hervorgehoben, daß die Gefängniskost an sich genügend sein müsse, um den Gefangenen in Gesundheit und Arbeitskraft zu erhalten. Wenn es dem Gefangenen alsdann noch gestattet werde, sich aus der Arbeitsbelohnung Extragenüsse zu verschaffen, so müsse dies zu nachteiligen Vergleichen zwischen der Lage des freien Arbeiters und derjenigen des Gefangenen veranlassen. („led to a comparison favourable to a dishonest instead of an honest career“).

Der Gesamtbetrag der im Jahre 1885/6 zur Auszahlung gelangten Arbeitsbelohnungen belief sich in den convict prisons auf 7123 £, in den local prisons auf 6914 £.[1])

§ 23. Die Disziplinarstrafen.

Zur Aufrechterhaltung der Hausordnung in einer Anstalt, deren Insassen in der Freiheit den gesetzlichen Vorschriften und den Forderungen des Lebens in der staatlichen Gemeinschaft die Unterwerfung verweigert haben, bedarf es naturgemäß strenger Disziplinarstrafen. Die Hausordnung enthält die eigentümlichen Pflichten, welche der Strafzustand mit sich führt; die Disziplinarstrafen sollen die unbedingte Unterwerfung des Gefangenen unter die durch die Hausordnung geregelte Strafzucht sicher stellen.

Wie die eigentliche Strafe, so muß auch die Disziplinarstrafe zwei Hauptbedingungen genügen; sie muß einerseits ein wirkliches Übel enthalten, welches als solches empfunden wird und welches geeignet ist, sowohl die speziell betroffene Person wie andre von weiteren Übertretungen abzuhalten; sie darf anderseits aber auf den davon Betroffenen nicht einen nachteiligen Einfluß, insbesondere nach gesundheitlicher Richtung ausüben.[2])

Diesem letzteren Momente wird in England die größte Beachtung geschenkt. Bei Feststellung der zulässigen Disziplinarstrafmittel sind ärztliche Sachverständige darüber gehört worden, ob und inwieweit das eine oder andre Strafmittel ohne Schädigung der Gesundheit zur Anwendung gebracht werden kann. Insofern bei einem Strafmittel eine Schädigung der Gesundheit möglicher Weise zu befürchten ist, soll der Gefängnisarzt in jedem Einzelfalle vor Vollstreckung der Strafe konstatieren, daß bei der betreffenden Person die Gefahr einer Schädigung der Gesundheit

---

[1]) Report of Directors 1886 Seite XXVIII. Report of Commissioners 1886 Seite 39.

[2]) In dem Report of the Royal Commissioners 1879 Seite XLVII ff. wird bezüglich der einzelnen Disziplinarstrafmittel sehr eingehend erörtert, inwieweit dieselben den im Texte angeführten zwei Hauptbedingungen gerecht werden. — Eine ebenfalls ausführliche Darstellung der zulässigen Disziplinarstrafen findet sich in dem Report of the Royal Commission on Prisons in Ireland 1884 Seite 25 ff.

durch Vollstreckung der erkannten Strafe ausgeschlossen erscheint („that such prisoner is in a fit condition of health to undergo such punishment"). Der Gefängnisarzt hat in diesen Fällen ferner die Vollstreckung der Strafe fortlaufend zu überwachen und ohne weiteres die Aussetzung der weiteren Vollstreckung anzuordnen, sobald er davon eine Schädigung der Gesundheit befürchtet.

Die in England zulässigen Disziplinarstrafmittel sind folgende:

1) Entziehung derjenigen Vorteile, welche der Gefangene durch seine Beförderung in eine höhere Strafklasse erlangt hat (loss of privilege),

2) Zurückversetzung des Gefangenen in eine niedere Strafklasse (loss of stage),

3) Schmälerung der gewöhnlichen Kost oder völlige Entziehung derselben unter bloßer Gewährung von Wasser und Brot (dietary punishment),

4) Einsperrung in eine Strafzelle (punishment cell),

5) körperliche Züchtigung (corporal punishment).

Jedes dieser Strafmittel kann sowohl einzeln wie in Verbindung mit einem der andern Strafmittel verhängt werden.[1]) In der Praxis wird insbesondere häufig neben einer der sub 3—5 erwähnten Strafen auf Verlust von Privilegien erkannt, und es wird ferner mit der Einschließung in eine Strafzelle meistenteils eine Kostentziehung verbunden. Bei ganz leichten Übertretungen der Hausordnung kann von einer eigentlichen Disziplinarstrafe Abstand genommen werden, indem dem Schuldigen lediglich ein Verweis erteilt wird.

Die Anwendung der Zwangsjacke oder die Fesselung eines Gefangenen (mechanical restraints,

---

[1]) Für das Jahr 1885/6 seien folgende Ziffern über die Anwendung der einzelnen Disziplinarstrafmittel angeführt. Es wurde erkannt:

	in den convict prisons	in den local prisons
a) auf corporal punishment	69 mal	138 mal
b) auf punishment cell	71 „	784 „
c) auf dietary punishment	9040 „	18041 „
d) auf loss of stage or privilege	8083 „	16393 „

Report of Directors 1886 Seite XXXII, Report of Commissioners 1886 Seite 47.

irons, handcuffs) findet nicht als Strafmittel statt, sondern nur als eine Sicherungsmaßregel, falls zu befürchten ist, daß der betreffende Gefangene Hand an sich legen oder andre Personen verletzen oder einen Fluchtversuch machen werde.[1]) In all diesen Fällen bedarf es einer schriftlichen Ordre des Governor, in welcher genau anzugeben ist, welche Zwangsmittel und für welche Zeit dieselben angewandt werden sollen.[2]) Eine derartige Ordre darf erst nach eingehender Untersuchung des Falles und unter genauer Angabe der Umstände, welche die Anwendung des Zwangsmittels erforderlich erscheinen lassen, ausgestellt werden. Während der Zeit, wo das Zwangsmittel angewendet wird, ist der Betreffende täglich sowohl von dem Gefängnisarzte wie von dem Gefängnisgeistlichen zu besuchen, und es sind ferner besondere Vorkehrungen für eine unausgesetzte Beaufsichtigung des Gefangenen zu treffen. Im Jahre 1885/6 sind in den convict prisons Zwangsmittel 229mal (186mal bei Männern, 43mal bei Frauen), in den local prisons 159mal (54mal bei Männern, 105mal bei Frauen[3]) angewendet worden.[4])

Das Recht der Verhängung von Disziplinarstrafen steht in erster Linie dem Governor zu, welcher jeden Einzelfall in ein dem revidierenden Director oder Inspector vorzulegendes punishment book in detaillierter Weise einzutragen hat.[5]) Die Disziplinar-

[1]) Report of the Royal Commissioners 1879 Seite XLIX „as a means of restraint either for the protection of the convict or the protection of others or in cases of attempted escape".

[2]) Die Handeisen und sonstigen Zwangsapparate werden den einzelnen Anstalten von der Zentralgefängnisbehörde geliefert. Eine Zirkularverfügung vom 21. Februar 1870 (abgedruckt bei Du Cane I Seite 126 f.) gibt genaue Vorschriften darüber, in welcher Weise bei Anwendung der einzelnen Zwangsapparate zu verfahren ist.

[3]) Diese letztere Ziffer ist ganz ungewöhnlich hoch und resultiert lediglich dadurch, daß in dem Gefängnisse von Westminster gegen einige weibliche Gefangene 58mal von Zwangsmitteln Gebrauch gemacht werden mußte.

[4]) Report of Directors 1886 Seite XXXII, Report of Commissioners 1886 Seite 46, 47. Im Jahre 1884/5 betrugen die Ziffern 250 resp. 96.

[5]) Dem eines Disziplinarvergehens beschuldigten Gefangenen ist Kenntnis von der Beschuldigung und Gelegenheit zu seiner Verteidigung zu geben. In dem Report of the Royal Commissioners 1879 Seite LI ff. wird die Frage eingehend erörtert, inwieweit es hierbei gestattet sein soll, Mitgefangene als Zeugen zu hören. Die Ansichten über diese Frage gingen auseinander. Ein hervorragender Gefängnisbeamter erklärte, er würde niemals das Zeugnis eines Gefangenen gegen dasjenige eines Wärters zulassen, einem Strafgefangenen dürfe

strafgewalt des Governor ist jedoch nur eine beschränkte: er kann einzelne Strafmittel nur für kurze Dauer und körperliche Züchtigung überhaupt nicht verhängen. Die Befugnis, auf diese, der Kompetenz des Governor entzogenen Strafen zu erkennen, steht in den local prisons dem Visiting Committee, in den convict prisons den Directors zu, welchen der Governor den Fall zu unterbreiten hat, falls er eine über seine Kompetenz hinausgehende Strafe für erforderlich erachtet.

Gehen wir noch kurz auf die einzelnen Strafmittel ein.

Von dem Rechte der Entziehung der durch bisheriges gutes Verhalten und durch Arbeitsamkeit erworbenen Privilegien wird jetzt in der Praxis ein sehr großer Gebrauch gemacht. Dieses Strafmittel bietet den erheblichen Vorteil, in seiner Anwendung eine große Mannigfaltigkeit zu gestatten: es kann die Vergünstigung der Korrespondenz, des Empfangs von Besuchen, der Gewährung von Lektüre, des Spaziergangs am Sonntage genommen, oder es kann auf Verlust einer angemessenen Zahl bereits verdienter Marken und damit gleichzeitig der entsprechenden Arbeitsbelohnung erkannt werden. Je nach dem Einzelfalle wird so eine Strafe gewählt werden, welche dem davon Betroffenen als ein wirkliches Übel erscheint. Dem Gefangenen wird damit in deutlicher Weise vor Augen geführt, daß seine ganze Lage in der Anstalt von seinem eignen Verhalten abhängig ist; er wird an eine Unterwerfung seines Willens unter eine bestehende Ordnung gewöhnt. Die Gefahr einer Schädigung der Gesundheit kann bei diesem Strafmittel nie vorliegen.

Eine härtere Strafe besteht in der Zurückversetzung in eine niedere Strafklasse, weil damit eo ipso der Verlust aller durch das bisherige Verhalten erworbener Vorteile eintritt. Eine Zurückversetzung aus dem Public Works Stage in die Isolierhaft

---

niemals dieselbe Glaubwürdigkeit beigemessen werden, wie einem freien Manne und besonders einem Beamten. Die Royal Commissioners erkannten die Bedenken gegen eine formelle Zulassung von Mitgefangenen als Zeugen an, sprachen sich jedoch dafür aus, daß es unter gewissen Umständen ratsam sei, von Mitgefangenen auf indirektem Wege — nicht durch formelle Vernehmung als Zeugen — Näheres über die der Beschuldigung zu Grunde liegenden Thatsachen zu ermitteln. Inwieweit dann auf derartige Ermittelungen für die Beurteilung der Sachlage Wert gelegt werden könnte, müßte dem Ermessen des erkennenden Beamten im Einzelfalle überlassen bleiben.

findet, wie schon oben Seite 66 angeführt wurde, statt, wenn ein Sträfling durch sein Verhalten fortgesetzt Anlaß zu Tadel gibt und in dieser Weise einen nachteiligen Einfluß auf die übrigen Sträflinge ausübt. [1]) Das Recht, auf eine derartige Zurückversetzung zu erkennen, steht nicht den Governors, sondern nur den Directors zu.

Entziehung oder Schmälerung der Kost bildet ein wichtiges und viel angewandtes Strafmittel, bei welchem jedoch mit Recht umfassende Vorkehrungen getroffen worden sind, damit keine nachteilige Einwirkungen auf die Gesundheit und Arbeitskraft des Bestraften eintreten. Es gibt drei punishment dietaries: die erste besteht lediglich aus Wasser und Brot, bei der zweiten wird dazu mittags 1 pint stirabout, [2]) bei der dritten $1^1/_2$ pints stirabout gereicht. Die bread and water diet darf in den local prisons höchstens auf 15, in den convict prisons auf 18 Tage verhängt werden, und es hat nach je drei Tagen für einen Tag die gewöhnliche Kost einzutreten. Auf die zweite punishment dietary, die stirabout diet, darf nur auf die Dauer von 21 Tagen, auf die dritte punishment dietary, die full stirabout diet, nur auf die Dauer von 42 Tagen erkannt werden. [3]) Während der Zeit, wo der Gefangene nur Wasser und Brot erhält, ist er täglich vom Arzte zu besuchen, und es darf von ihm keinerlei Arbeit verlangt werden. Der Governor kann die bread and water diet nur auf 24 Stunden verhängen; [4]) die Anordnung für eine längere Zeitdauer bleibt den Visiting Justices resp. den Directors vorbehalten. Wenn ein Gefangener länger als 24 Stunden bei Wasser und Brot gehalten werden soll, so hat der Gefängnisarzt vorher ein schriftliches Attest dahin abzugeben, daß der Gesundheitszustand des Betreffenden eine derartige längere Dauer der Strafe gestattet.

---

[1]) In dem Report of the Royal Commissioners 1879 Seite XLVIII wird ausdrücklich hervorgehoben, daß eine Zurückversetzung in die Isolierhaft nicht von allen Sträflingen als eine Strafe betrachtet werden würde, da sicherlich manche die Isolierung der harten Arbeit in den Public Works Prisons vorzögen.

[2]) Eine aus Hafermehl und Mais bereitete Speise, cf. oben Seite 224.

[3]) Cf. Report of Commissioners 1878 Seite 45 und Du Cane I Seite 123 f.

[4]) Bis vor wenigen Jahren konnte der Governor diese Strafe für die Dauer von drei Tagen verhängen. Cf. Report of the Royal Commission on Prisons in Ireland 1884 Seite 26.

Von dem Strafmittel der Einsperrung in eine Strafzelle wird jetzt nur noch wenig Gebrauch gemacht.[1]) Über die Einrichtung der Strafzellen sind besondere Vorschriften erlassen worden, in denen bestimmt ist, wie groß die Zelle sein muß, in welcher Weise und in welchem Maße Licht zugeführt wird u. s. w. Als Strafzellen dürfen nur solche Zellen benutzt werden, bei welchen durch Attest des Surveyor festgestellt ist, daß sie diesen Vorschriften entsprechen. Die Maximaldauer der Einsperrung in eine Strafzelle beträgt 14 Tage. Während der Einsperrung ist der Gefangene täglich sowohl von dem Gefängnisarzte wie von dem Gefängnisgeistlichen zu besuchen.

Körperliche Züchtigung ist nur bei männlichen Gefangenen gestattet und kann nur von den Visiting Justices resp. den Directors für bestimmte, in der Hausordnung angegebene Vergehen verhängt werden. Die Maximalzahl der Schläge beträgt bei Gefangenen unter 18 Jahren: 18, bei älteren Gefangenen: 36. Bei den jugendlichen Gefangenen werden die Schläge stets mittels einer Rute (birch rod) beigebracht, bei den andern Gefangenen kann statt der Rute eine Peitsche gebraucht werden.[2]) Es dürfen nur solche Ruten und Peitschen verwendet werden, welche der Zentralgefängnisbehörde vorgelegen haben und von dieser als zweckentsprechend erachtet worden sind. Bei der Ausführung der Züchtigung muß der Governor und der Gefängnisarzt zugegen sein. Wenn der Gefängnisarzt besondere Anordnungen zur Vermeidung nachteiliger Einwirkung der Bestrafung auf den Gesundheitszustand des betreffenden Gefangenen für nötig erachtet, so ist denselben unbedingte Folge zu geben.

In den jährlichen Reports der Directors und der Commissioners wird stets die Zahl der im Laufe des Jahres verhängten

---

[1]) Dieses Strafmittel wird relativ häufiger bei weiblichen als bei männlichen Gefangenen angewendet (im Jahre 1885/6 wurde in den local prisons auf Einsperrung in eine Strafzelle gegen 234 Frauen und gegen 550 Männer erkannt). In dem Report of the Royal Commissioners 1879 Seite XLVIII wird hervorgehoben, daß die Einsperrung in eine mehr oder minder dunkle Zelle bei weiblichen Gefangenen in besonders hohem Maße eine beruhigende Wirkung ausübe.

[2]) In letzter Zeit wird auch bei den älteren Gefangenen die körperliche Züchtigung fast stets mit der Rute beigebracht. In dem Report of the Royal Commissioners Seite XLVIII wird dies als das wirkungsvollere bezeichnet „convicts feeling it to be a humiliation to be punished like children.“

Disziplinarstrafen unter Hervorhebung der Gründe der Bestrafung angegeben. Diese Gründe werden unter vier Rubriken zusammengefaßt: a) Gewaltthätigkeiten (violence), b) Fluchtversuch, c) Faulheit bei der Arbeit, d) andre Übertretungen der Hausordnung. Die letztere Rubrik umfaßt naturgemäß die größten Ziffern. [1])

Es ist dabei bemerkenswert, daß sich die Menge der Übertretungen auf eine relativ kleine Zahl von Personen verteilt, welche sich mehrfacher Übertretungen schuldig machen. Die 13 232 prison offences in den convict prisons im Jahre 1885/6 wurden begangen von 4575 Sträflingen, während 9197 Sträflinge während des Jahres keinerlei Bestrafung erhielten. Die 43 771 prison offences in den local prisons im Jahre 1885/6 wurden begangen von 23 085 Gefangenen, während 174 789 Gefangene während des Jahres unbestraft blieben.

Als ein günstiges Zeichen muß es hervorgehoben werden, daß die Zahl der Disziplinarbestrafungen von Jahr zu Jahr abgenommen hat. Im Jahre 1871 wurden in den convict prisons 7036 Sträflinge bestraft, während die Zahl derjenigen, welche keine Strafe während des Jahres erhielten, 8210 betrug. [2]) Das Verhältnis der Zahl der Bestraften zu der Gesamtzahl der Insassen hat sich also seitdem erheblich gebessert. Eine noch größere Besserung ist in den local prisons bemerkbar. Im Jahre 1878, vor Übernahme dieser Anstalten in die Staatsverwaltung, kamen auf je 1000 Insassen durchschnittlich täglich 8,2 Disziplinarstrafen; im Jahre 1885/6 betrug diese Ziffer nur 6,2. [3]) Und dies

[1]) Im Jahre 1885/6 verteilten sich die prison offences folgendermaßen:

	in den convict prisons	in den local prisons
a) Gewaltthätigkeiten	1309	863
b) Fluchtversuch	11	10
c) Faulheit	2189	17725
d) andre Übertretungen	9723	25173
Totalsumme	13232	43771

Report of Directors 1886 Seite XXXII. Report of Commissioners 1886 Seite 47.

[2]) Cf. Du Cane II Seite 165.

[3]) Report of Commissioners 1878 Seite 11. Besonders bemerkenswert ist die erhebliche Abnahme der beiden Strafmittel punishment cell und dietary punishment. Im Jahre 1877 wurde: 39 159mal auf dietary punishment

günstige Resultat ist erreicht worden, obwohl die Disziplin und die Aufrechterhaltung der Ordnung heutzutage sicherlich nicht weniger stramm gehandhabt wird, als in früheren Jahren. [1])

§ 24. Die Entlassung der Gefangenen.

Es ist wiederholentlich bei der Darstellung der Behandlung der Gefangenen in England hervorgehoben worden, wie hier stets das Ziel vor Augen behalten wird, daß der Gefangene als nützliches Glied in die bürgerliche Gesellschaft zurücktrete. Je mehr sich die Strafhaft ihrem Ende naht, eine um so eingehendere Rücksicht ist auf dies Endziel zu nehmen.

Alle Gefangenen mit längeren Strafzeiten werden, wie bereits auf Seite 228 ausgeführt wurde, 21 Tage vor ihrer Entlassung von dem Gefängnisarzte genau auf ihren Gesundheitszustand untersucht. Ergibt sich dabei, daß der Gefangene sich nicht im Vollbesitze seiner Arbeitskraft befindet, und ist danach zu befürchten, daß derselbe außer stande sein wird, sich in der Freiheit auf ehrliche Weise ein Fortkommen zu verschaffen, so ist der Gefängnisarzt befugt, die Gewährung von Zusatznahrungsmitteln anzuordnen. Wenn der Gefangene zur Zeit der Entlassung krank ist, so wird seine Familie davon benachrichtigt und er selbst

---

erkannt, im Jahre 188/56: 18 041mal; im Jahre 1877 wurde die Strafe punishment cell: 18 263mal, im Jahre 1885/6 nur 784mal verhängt. An Stelle dieser beiden Strafmittel, bei deren Anwendung so leicht eine nachteilige Einwirkung auf die Gesundheit eintreten kann, wird jetzt in umfassendem Maße von den Strafmitteln ,des loss of stage or privilege Gebrauch gemacht, Strafmittel, welche in den local prisons vor Übernahme in die Staatsverwaltung gar nicht bestanden.

[1]) Du Cane I Seite 34 bemerkt in dieser Richtung treffend: „The result is not due to an easy and slack system, under which offences are passed over without report and without punishment; on the contrary, it will be apparent, even to a casual visitor and is well known to those who are more intimately acquainted with the interior of prisons that order is strictly maintained in them, and that the discipline is exact without being severe. The result shews, in fact, that in this respect, at all events our system produces the result it is intended to do, but more especially it shews that the organisation of the department is effective and that the staff of officers perform their duties with resolution and with judgment."

entweder in der Krankenabteilung des Gefängnisses noch einige Tage weiter behandelt oder sofort nach einem Hospitale gesandt.

Es wird ferner bei der Entlassung dafür Sorge getragen, daß der Betreffende ordentliche Kleidungsstücke besitzt, um eine für ihn geeignete Beschäftigung aufsuchen zu können. Die bei der Aufnahme in das Gefängnis abgenommenen Kleidungsstücke werden zurückgegeben.[1]) Erscheinen dieselben mangelhaft oder ungenügend, so werden dem Gefangenen aus seiner Arbeitsbelohnung neue Kleidungsstücke angeschafft. Wenn sich der Gefangene unter den Schutz eines Vereins zur Fürsorge für Strafentlassene stellt, so wird die Anschaffung geeigneter Kleidungsstücke regelmäßig dem Vereine überlassen. Im andern Falle werden in der Anstalt selbst die notwendigsten Kleidungsstücke angefertigt oder fertig gekaufte Sachen für den Gefangenen passend gemacht.

Der Gefangene erhält sodann, falls die Zeit seiner Inhaftierung nicht bloß eine ganz kurze war oder er aus sonstigen Gründen von einer Arbeitsbelohnung (gratuity) ausgeschlossen geblieben ist, bei seiner Entlassung stets so viel bares Geld, um davon wenigstens einige Tage leben zu können, bis er Arbeit gefunden hat. Es wird mit dem Gefangenen vor seiner Entlassung eine Abrechnung über die ihm zustehende Arbeitsbelohnung aufgemacht. Soweit ihm die letztere nicht sofort bar ausgezahlt wird, sind die näheren Bestimmungen über die weiteren Zahlungen zu treffen (cf. oben Seite 256).

Die Entlassung erfolgt[2]) regelmäßig in der Anstalt selbst. Ist die Anstalt von dem Orte, wo die Verurteilung erfolgte, entfernt gelegen, so hat der Gefangene einen Anspruch auf freie Rückbeförderung nach diesem Orte.[3]) Er wird zur nächsten Bahnstation gebracht und dort mit einem Fahrbillet versehen. In neuerer Zeit hat man es dem zur Entlassung Kommenden gestattet, den Ort, wohin er befördert sein will, selbst zu wählen; von seiten der Ge-

---

[1]) Ebenso erfolgt Rückgabe all der andern Sachen, welche etwa dem Gefangenen bei der Aufnahme abgenommen worden sind.

[2]) Wenn die Strafzeit am Sonntage abläuft, so wird der Gefangene bereits am Sonnabend abend entlassen. 28 a. 29 Vict. c. 126 s. 41.

[3]) Cf. Wilkinson Seite 43 ff. Die Zentralgefängnisbehörde ist berechtigt, einen Gefangenen, welcher wünscht, nach seinem Heimatsorte zurückbefördert zu werden, auch die hierfür erforderlichen Reisekosten zu gewähren. 28 a. 29 Vict. c. 126 s. 43.

fängnisanstalt wird jedoch in diesem Falle das Fahrbillet nur soweit gezahlt, als der gewählte Ort von der Anstalt nicht weiter entfernt liegt als der Ort der Verurteilung. Wenn die Entfernung eine größere ist, so wird der Mehrbetrag für das Fahrbillet aus der Arbeitsbelohnung bestritten.

Besondere Bestimmungen gelten bezüglich derjenigen convict prisoners, welche in London verurteilt worden sind. Dieselben werden einige Tage, bevor ihre Strafzeit abgelaufen ist, aus dem Public Works Prison nach London zurücktransportiert und bis zu ihrer Entlassung in ein dortiges Gefängnis untergebracht. Hier erfolgt an bestimmten Tagen der Woche eine Art Kontrollversammlung über all diese zur Entlassung Kommenden durch Beamte des Metropolitan Convict Supervision Office (cf. oben Seite 122) sowie durch Detektives. Diese Kontrolle hat den Zweck, die Polizeibeamten mit der Persönlichkeit des wieder freigelassenen convict prisoner bekannt zu machen. Die Beamten vergleichen dabei die von den Sträflingen aufgenommenen Photographieen, prüfen die Richtigkeit der in dem Zentralregister und dem Distinctive Marks Register enthaltenen Eintragungen (cf. oben Seite 15, 16) und machen sich etwa erforderlich erscheinende Notizen.[1])

---

[1]) Dies Verfahren dient zweifellos dazu, der Polizeiaufsicht über die Entlassenen eine möglichst sichere Unterlage zu geben, und liegt im Interesse der öffentlichen Sicherheit, deren Aufrechterhaltung in London eben besondere Maßregeln erfordert. Allein es soll nicht verschwiegen werden, daß gegen dies Verfahren in der schon früher einmal citierten Schrift „Five years' penal servitude“ Bedenken erhoben worden sind, denen eine gewisse Berechtigung nicht abzusprechen ist. Es wird hier Seite 173 ff. ausgeführt: es sei wenig human und dem Fortkommen des Entlassenen direkt hinderlich, wenn derselbe wenige Tage vor seiner Entlassung durch die unter Aufsicht eines Gefängnisbeamten zusammen mit andern Sträflingen erfolgende Beförderung nach London den neugierigen Blicken Fremder ausgesetzt werde. (Die Sträflinge wurden bei diesem Transporte bisher sogar gefesselt; eine Bestimmung, welche jetzt mit Recht aufgehoben worden ist.) Es läßt sich in der That nicht verkennen, daß eine große Härte darin liegt, in einem Sträflinge, welcher seine Strafzeit bis auf wenige Tage abgebüßt hat, durch die im Texte angegebene Behandlung noch einmal das Gefühl der Erniedrigung wachzurufen. Und der Sträfling wird das Erniedrigende der ihm zu teil werdenden Behandlung um so mehr empfinden, als er äußerlich keine Spuren seiner Bestrafung mehr an sich trägt: er hat die Gefängniskleider mit seiner gewöhnlichen Kleidung vertauscht; er hat sich Haar und Bart wieder wachsen lassen.

Soviel über die bezüglich der Entlassung der Gefangenen geltenden Bestimmungen.

In England wird es nun aber weiter für eine staatliche Pflicht erachtet, auch nach der Entlassung besondere Vorkehrungen zu treffen, um einen erfolgreichen Rücktritt in die bürgerliche Gesellschaft nach Möglichkeit sicherzustellen. Es geschieht dies zunächst in negativer Hinsicht dadurch, daß ein Teil der Gefangenen noch für die erste Zeit nach der Entlassung einer besonderen Aufsicht unterstellt wird, um die Betreffenden dadurch abzuhalten, in die frühere verbrecherische Laufbahn zurückzufallen oder sich mit früheren schlechten Genossen wieder zu verbinden. Das Nähere über die in dieser Weise zur Verhütung des Rückfalls in das Verbrechen angeordnete Polizeiaufsicht ist im Abschnitte I § 10 gegeben worden.

Aber auch nach der positiven Seite hin wird für ein gedeihliches Fortkommen des Entlassenen Sorge getragen, indem die Vereine zur Fürsorge für Strafentlassene vom Staate gefördert und unterstützt werden. Die in dieser Richtung getroffenen Bestimmungen bezüglich der Überweisung von Arbeitsbelohnungen und Extravergütungen an die Vereine wurden bereits auf Seite 256 angeführt. Der Staat gewährt sodann weiter den Vereinen einen jährlichen Zuschuß von 4000 £. Durch diese Unterstützung erkennt der Staat die Thätigkeit der Vereine als im öffentlichen Interesse liegend an.

Mit Rücksicht auf den inneren Zusammenhang, in welchem die Wirksamkeit der Vereine zur Fürsorge für Strafentlassene mit dem von uns behandelten Thema steht, und in Anbetracht der Wichtigkeit des Gegenstandes erscheint es angemessen, hier kurz noch auf diese Vereine einzugehen.

Anhang.

Die Vereine zur Fürsorge für Strafentlassene.

Mit jeder Gefängnisanstalt in England ist jetzt ein Verein zur Fürsorge für Strafentlassene (Discharged Prisoners' Aid Society) verbunden. Selbst derjenige Gefangene, welcher seine Strafe in einem ganz kleinen Gefängnisse verbüßt, befindet sich jetzt bei seinem Austritte aus der Anstalt nicht mehr hilf- und schutzlos, sondern kann sich an den mit der betreffenden Anstalt verbundenen Fürsorgeverein wenden, welcher ihn in dem Bestreben, als nützliches Glied in die bürgerliche Gesellschaft zurückzutreten, mit Rat und That unterstützt.

Diese erfreuliche Entwickelung des Fürsorgewesens in England ist erst in allerneuester Zeit erreicht worden. Die erste derartige Gesellschaft wurde im Jahre 1802 gegründet. Im Jahre 1878 existierten in England 29 Fürsorgevereine; [1]) heute beträgt deren Zahl 60. [2])

All diese Vereine verdanken ihre Entstehung zunächst privater Initiative. [3]) Die Entwickelung würde aber wohl schwerlich eine so schnelle gewesen sein, wenn die Bewegung nicht von seiten des Staates auf das wirkungsvollste gefördert worden wäre.

Es erging zunächst im Jahre 1862 ein Gesetz — 25 a. 26 Vict. c. 44 — durch welches die Lokalgefängnisbehörden (die Visiting Justices) ermächtigt wurden, den Fürsorgevereinen eine Beihilfe bis zur Höhe von 2 £ pro Kopf für diejenigen Strafentlassenen zu gewähren, welche sich unter den Schutz des Vereins stellten.

Seit dem Jahre 1878 hat sodann das Parlament alljährlich eine Summe von 4000 £ zur Unterstützung der Fürsorgevereine bewilligt. Die Verteilung unter die Vereine erfolgt in der Art, daß die Höhe der dem einzelnen Vereine gewährten Beihilfe nach der Zahl der von dem Vereine im vorhergehenden Jahre unterstützten Strafentlassenen bemessen wird, jedoch mit der Einschränkung, daß der Gesamtbetrag der Staatsbeihilfe in jedem einzelnen Jahre nicht mehr betragen darf, als diejenige Summe, welche von dem betreffenden Vereine durch wohlthätige Gaben und durch Beihilfen der Lokalbehörden aufgebracht wurde. Auf diese Weise erhält die Privatwohlthätigkeit eine besondere Anregung zum Eingreifen und

---

[1]) Du Cane II Seite 196 ff.

[2]) Third Report of the Central Committee of Discharged Prisoners' Aid Societies Seite 8 ff.

[3]) Besondere Verdienste um die Entstehung und rationelle Wirksamkeit der Discharged Prisoners' Aid Societies haben sich Mrs. Fry und in neuerer Zeit Mr. Barwick Baker erworben.

auch das Interesse der Lokalbehörden an den Vereinen wird wachgehalten.

Eine fernere Einnahmequelle ist den Fürsorgevereinen durch den Prison Charities Act vom Jahre 1882 erschlossen worden. Es bestanden nämlich in England eine erhebliche Anzahl von Stiftungen, welche für mannigfache Zwecke auf dem Gebiete des Gefängniswesens bestimmt waren, Zwecke, welche heutzutage teils völlig obsolet geworden sind, teils schon auf andre Weise eine genügende Erfüllung finden.[1]) Durch das Gesetz von 1882 wurde nun die zur Neuregelung des Stiftungswesens überhaupt eingesetzte Kommission, die Charity Commissioners, ermächtigt, auf Antrag des Staatssekretärs des Innern Bestimmungen über eine anderweitige Verwendung dieser Stiftungseinkünfte, insbesondere zur Unterstützung der Fürsorgevereine, zu treffen. Es handelt sich dabei um recht beträchtliche Summen; doch schreitet die Aufstellung der neuen Verwendungsbestimmungen (new schemes of appropriation) infolge der mancherlei Rücksichten, welche auf den Stiftungswillen und andre Verhältnisse zu nehmen sind, bis jetzt nur langsam vorwärts.[2])

Es ist endlich noch ein weiterer Faktor für die erfreuliche Entwickelung der Fürsorgevereine zu erwähnen: nämlich die im Jahre 1877 in das Leben gerufene Zentralisation aller in England bestehenden Fürsorgevereine. Diese höchst segensreich wirkende Maßregel wurde von der Reformatory and Refuge Union,[3]) deren bereits oben Seite 113 Anm. 2 gedacht worden ist, angeregt.

---

[1]) Im Report of Commissioners 1885 Seite 87 ff. ist eine im Jahre 1878 aufgestellte Liste dieser obsoleten Gefängnisstiftungen abgedruckt. Unter den obsoleten Stiftungszwecken seien folgende häufig vorkommende hervorgehoben: Verwendung der Einkünfte zur Besoldung eines Geistlichen, welcher Gottesdienst in der Anstalt halten soll, zum Ankaufe von Brot oder sonstigen Lebensmitteln für die Gefangenen, zur Bezahlung gewisser Schulden von Schuldgefangenen u. s. w.

[2]) Cf. Report of Commissioners 1885 Seite 6. Die Gesamthöhe der jährlichen Einkünfte der obsoleten Gefängnisstiftungen wird auf 3348 £ angegeben. Die bis jetzt erlassenen new schemes behandeln Einkünfte im Gesamtbetrage von jährlich 260 £.

[3]) Der Präsident dieser unter dem Protektorate des Prinzen von Wales stehenden Gesellschaft war bis zu seinem im Jahre 1885 erfolgten Tode der Earl of Shaftesbury. Der jetzige Präsident ist Lord Aberdare. Das Büreau der Gesellschaft, welches zu gleicher Zeit auch als Büreau des Central Committee of Discharged Prisoners' Aid Socities dient, befindet sich 32 Charing Cross London S. W. Ich benutze diese Gelegenheit, für die mir von dort übersandten interessanten Schriften hier meinen Dank abzustatten.

Auf Grund der Beschlüsse, welche die von der Union im April 1877 nach London berufene Konferenz faßte, ist ein Central Committee of Discharged Prisoners' Aid Societies in London niedergesetzt. Dieser Ausschuß soll die Zentralstelle für alle Bestrebungen auf dem Gebiete des Fürsorgewesens bilden und alle wünschenswerten Reformen sollen hier beraten werden. Durch Schriften und Versammlungen soll das Interesse weiterer Kreise für die Sache angeregt und da, wo es nötig ist, auf Vornahme von Verbesserungen eingewirkt werden. Der Ausschuß gibt einen Jahresbericht heraus, in welchem ausführliche Auskunft über die eigne Thätigkeit wie über diejenige der einzelnen Fürsorgevereine gegeben wird.

Auf die energische Thätigkeit dieses Zentralausschusses ist die erfreuliche Thatsache in erster Linie zurückzuführen, daß sich heutzutage bei jeder Gefängnisanstalt ein Fürsorgeverein befindet. Infolge der Anregung des Ausschusses ist ferner zwischen der Mehrzahl der Fürsorgevereine eine Art Kartellverhältnis zustande gekommen, auf Grund dessen der eine Verein für den andern als Agent thätig ist: er zieht Auskünfte über Personen und Verhältnisse ein, übernimmt die Überwachung derjenigen Entlassenen, welche in seinen Bezirk aus dem andern übersiedeln, zahlt im Auftrage des andern Vereins Gelder aus u. s. w. Endlich hat es der Ausschuß zuwege gebracht, daß fast überall in den großen Städten Zufluchtsstätten für Strafentlassene (Homes for Discharged Prisoners) entstanden sind, wo der Entlassene für kurze Zeit gegen eine Arbeitsleistung Wohnung und Kost erhält, bis er eine Beschäftigung gefunden hat.[1]) Die Homes sind teils von größeren Fürsorgevereinen,

[1]) Wohl die größte derartige Anstalt ist die Royal Society's Memorial Refuge in London. Es wurden hier im Jahre 1885: 93 Strafentlassene aufgenommen, von denen 63 die Anstalt vor Ablauf des Jahres wieder verließen. Die Einnahmen aus den Arbeitsleistungen der Insassen betrugen 1570 £; die Gesamtausgaben 2914 £. — Eine besonders berühmte Anstalt ist ferner mit dem Gefängnisse zu Wakefield verbunden. Die Insassen haben hier für Wohnung und Beköstigung wöchentlich 7 s 2 d zu zahlen, während sie für ihre Arbeitsleistungen durchschnittlich 11—12 s erhalten, so daß ihnen ein wöchentlicher Überschuß von 4—5 s verbleibt. Die für die Arbeitsleistungen gezahlten Preise sind unter den auf dem freien Arbeitsmarkte bestehenden angesetzt, damit der Strafentlassene einen Anlaß hat, möglichst schnell aus der Anstalt heraus und in freien Arbeitsdienst zu kommen. Cf. Wines Seite 225 ff.

teils von besonders für diesen Zweck gebildeten Gesellschaften errichtet worden.[1])

Gehen wir nunmehr kurz auf die Art und Weise ein, wie sich die Fürsorgevereine der Strafentlassenen annehmen.

In jeder Gefängniszelle ist eine Notiz angeschlagen, wodurch dem Gefangenen Kenntnis gegeben wird, welcher Verein die Fürsorge für ihn bei seiner Entlassung zu übernehmen bereit ist,[2]) in welcher Weise er sich an den Verein zu wenden hat u. s. w. Ein Mitglied des Vereins pflegt periodisch in kurzen Zwischenräumen die Gefängnisanstalt zu besuchen und dort mit denjenigen Gefangenen nähere Rücksprache zu nehmen, welche demnächst zur Entlassung kommen werden und den Wunsch ausgesprochen haben, sich unter den Schutz des Vereins zu stellen. Der betreffende Gefangene wird dabei veranlaßt, sich frei darüber auszusprechen, was er selbst über die Gestaltung seiner Zukunft denkt, wie er sich auf redlichem Wege seinen Lebensunterhalt zu verschaffen hofft, ob und von welcher Seite er hierin eine Unterstützung erwarten kann. Der Verein zieht dann etwa erforderliche Erkundigungen ein, setzt sich mit der Familie und Freunden des Gefangenen in Verbindung, und sucht es, soweit irgend möglich, durchzusetzen, daß der Gefangene alsbald nach seiner Entlassung eine zweckmäßige Beschäftigung findet.

Wenn der Gefangene aus der Gefängnisanstalt entlassen ist, so hat er sich auf dem Büreau des Vereins zu melden. Ist schon eine Arbeitsgelegenheit für ihn ermittelt worden, so wird nunmehr erörtert, in welcher Weise die dem Vereine überwiesene Arbeitsbelohnung verwendet werden soll. Es werden dem Betreffenden aus der Arbeitsbelohnung Kleidungsstücke und das notwendige Handwerkszeug angeschafft. Soweit es erforderlich ist und angemessen erscheint, leistet der Verein zu diesem Zwecke Zuschüsse. Wenn eine Arbeitsgelegenheit noch nicht gefunden ist, so wird für die vorläufige Unterbringung des Entlassenen Sorge getragen, event. durch

---

[1]) Es bestehen mehr Homes für weibliche Strafentlassene als für männliche. Bei bestraften Weibern dürfte auch in der That mit Rücksicht auf die erhöhte Schwierigkeit eines Rücktritts in geordnete Verhältnisse ein größeres Bedürfnis für derartige Anstalten vorhanden sein.

[2]) Einige Vereine sind ausschließlich für convict prisoners, andre ausschließlich für local prisoners bestimmt; zu den ersteren zählt in London die Royal Society for the Assistance of Discharged Prisoners, zu den letzteren die Metropolitan Discharged Prisoners' Aid Society.

Aufnahme in ein Home for discharged prisoners. Dem Entlassenen werden Ratschläge erteilt, in welcher Weise er sich am zweckmäßigsten um Arbeit verwenden kann, und er wird hierbei von einem Agenten des Vereins unterstützt, dessen spezielle Aufgabe es ist, eine fortlaufende Verbindung mit Arbeitgebern aufrecht zu erhalten, welche bereit sind, Strafentlassene zu beschäftigen.

Ist der Strafentlassene alsdann in Arbeit gebracht, so sucht der Verein noch eine Zeitlang eine Kontrolle über ihn auszuüben, damit er nicht wieder auf Abwege gerät. Eine wirksame Handhabe hierfür ist dadurch gegeben, daß, wenn möglich, ein Teil der Arbeitsbelohnung von dem Verein zurückgehalten und erst nach Verlauf einiger Zeit ausgezahlt wird. Diese zurückgehaltene Summe dient gewissermaßen als eine Kaution, welche der Schützling für sein gutes Verhalten dem Vereine stellt.

Bei denjenigen Entlassenen, welche unter Polizeiaufsicht stehen, übernimmt der Verein eine Art Vermittlerrolle zwischen dem Betreffenden und der Polizeibehörde: die vorgeschriebenen Meldungen können anstatt an die Polizei an den Verein erstattet werden, Kontrollbesuche durch einen Beauftragten des Vereins ersetzen die sonst durch die Polizei ausgeübte Kontrolle (cf. oben Seite 122). Damit werden die naheliegenden Gefahren, welche aus einer Einmischung der Polizei in das Leben und Treiben des Entlassenen entstehen können, beseitigt, ohne daß die für notwendig erachtete Kontrolle selbst fortfällt.

Der Verein geht bei seiner gesamten Thätigkeit mit voller Offenheit, ohne die Antezedenzien seines Schützlings zu verheimlichen, vor. Dem Vereine werden von dem Gefängnisvorsteher die Personalakten des Gefangenen jederzeit zur Verfügung gestellt, damit sich der Verein ein richtiges Urteil darüber bilden kann, in welcher Weise die Fürsorge für den Betreffenden am besten zu gestalten ist. Wenn alsdann der Strafentlassene bei einem Arbeitgeber untergebracht wird, so gibt der Verein dem letzteren jede erwünschte Auskunft über das Vorleben seines Schützlings. Man hält dies Verfahren für zweckmäßig, um die Arbeitgeber zu werkthätiger Mitarbeit an der großen Aufgabe, aus dem Bestraften ein nützliches Glied der menschlichen Gesellschaft zu machen, heranzuziehen.

Bei der Unterbringung der Strafentlassenen wird das Prinzip verfolgt, den Bestraften soweit als möglich von seiner früheren Umgebung fern zu halten. Einerseits wird dadurch dem Entlassenen die Versuchung genommen, sich mit früheren schlechten Genossen wieder zu verbinden und damit in das frühere Leben zurückzufallen, anderseits wird die Ausführung des guten Vorsatzes, ein neues Leben zu beginnen, erheblich erleichtert, wenn der Entlassene nicht stets zu befürchten hat, daß ihm sein früherer Fehltritt vorgeworfen und daß er dadurch aus der Stellung, in welche er sich durch redliches Streben hineingearbeitet hat, wieder ausgestoßen werde. Die Durchführung des Prinzipes wird dadurch ermöglicht, daß, wie schon hervorgehoben wurde, die Mehrzahl der Vereine miteinander in Kartellverbindung stehen: der eine Verein übernimmt die Schützlinge des andern, falls deren Unterbringung in dem betreffenden Bezirke wünschenswert erscheint.[1])

Von dem angegebenen Gesichtspunkte aus empfiehlt sich aber vor allem die Unterstützung der Auswanderung Strafentlassener: in der fernen Kolonie wird der Bestrafte ohne weiteres frei von all den Fesseln, welche seine Vergangenheit ihm im alten Vaterlande anlegt, und er findet dort Raum zur freien Entfaltung seiner Kräfte. Einer derartigen Unterstützung freiwilliger Auswanderung kann auch der Vorwurf nicht gemacht werden, daß sie, wie die Transportationsstrafe, ein Unrecht gegen die Kolonie enthalte: nicht ein Schuldiger wird hinübergesandt, dessen sich die alte Welt auf bequeme Weise entledigen will, sondern der Betreffende ist für seine Schuld bereits in der alten Welt bestraft und dabei nach Möglichkeit zu einem nützlichen Gliede der bürgerlichen Gesellschaft erzogen. Es erscheint hiernach durchaus gerechtfertigt, wenn die Fürsorgevereine, soweit als die Mittel dafür reichen, die Auswanderung derjenigen Strafentlassenen unterstützen, von welchen nach ihrem Alter, ihrer körperlichen Gesundheit und ihren sonstigen Eigenschaften zu erwarten ist, daß sie sich eine neue Existenz in den Kolonieen zu gründen vermögen.[2]) Nach dem letzten Jahres-

---

[1]) Von der Royal Society for the Assistance of Discharged Prisoners wurden im Jahre 1884 nur 94 Strafentlassene in dem Bezirke von London in Beschäftigung gethan, während 156 zu diesem Zwecke nach andern Bezirken gesandt wurden. Report of the Royal Society etc. 1885 Seite 3.

[2]) Eine in gleicher Weise empfehlenswerte Maßregel ist die Unterbringung von Bestraften in jüngerem Lebensalter auf Seeschiffen. In dem Report der Metropolitan Discharged Prisoners' Aid Society 1885 Seite 8 heißt es darüber „this means of disposing offers peculiar advantage, since by the long absence and apprenticeship old influences or associations are broken through“. Die Zahl der von diesen Vereinen zur See gesandten Strafentlassenen betrug im Jahre 1884: 70 gegenüber einer Gesamtzahl von 391 Unterstützten.

berichte des Central Committee of Discharged Prisoners Aid Societies sind im Jahre 1885: 73 Strafentlassene mit Unterstützung der Vereine ausgewandert.

Die Zahl der im Jahre 1885 von den Fürsorgevereinen unterstützten local prisoners betrug 14764 gegenüber einer Gesamtzahl von 171036 in diesem Jahre Entlassenen; die Zahl der unterstützten convict prisoners betrug 1332 gegenüber einer Gesamtzahl von 2055 Entlassenen.[1]) Man sieht aus diesen Ziffern, in welchem umfassenden Maße in England durch die Thätigkeit der Fürsorgevereine der Strafvollzug ergänzt und „pour le lendemain de la peine“ Sorge getragen wird. Es ist ein stolzes, aber den thatsächlichen Verhältnissen entsprechendes Wort, welches sich in dem letzten Jahresberichte der Royal Society for the Assistance of Discharged Prisoners (Seite 5) findet: „The Committee are able confidently to state, that no discharged convict, who is physically capable and willing to work, has any excuse for relapsing into crime.“ [2])

Titel II.

§ 25. Die Bestimmungen über die Behandlung der zu einer Gefängnisstrafe Verurteilten im speziellen.

Die Gefängnisstrafe kann in einer verschärften Form als imprisonment with hard labour und in einer gemilderten Form als imprisonment as a misdemeanant of the first division erkannt werden. Die Behandlung der zu einer Gefängnisstrafe mit harter Arbeit Verurteilten ist, abgesehen

[1]) Report of Commissioners 1886 Seite 81 und 84. Report of Directors 1886 Seite XXXIX ff.

[2]) Es mag nicht überflüssig sein, besonders darauf hinzuweisen, daß die großen Erfolge auf dem Gebiete des Fürsorgewesens für Strafentlassene wie auf dem nahe verwandten Gebiete der Besserung und Erziehung jugendlicher Übelthäter (cf. Abschnitt I § 9) erreicht worden sind durch private Wohlthätigkeit unter wesentlicher Förderung und materieller Unterstützung seitens des Staates.

von der Beschäftigung, worüber das Nähere oben Seite 240 angegeben ist,[1]) genau dieselbe, wie diejenige der zu einfacher Gefängnisstrafe (imprisonment ohne weiteren Zusatz) Verurteilten. Dagegen ist die Behandlung der als first-class misdemeanants im Urteile Bezeichneten eine durchaus abweichende.

Die näheren Bestimmungen über die Behandlung der first-class misdemeanants ist oben im § 11 gegeben worden; man kann dieselben kurz dahin zusammenfassen: die Strafe besteht hier in einfacher Freiheitsentziehung, welche nur mit solchen Einschränkungen verbunden ist, wie sie durch die Rücksicht auf die in der Anstalt aufrecht zu erhaltende Disziplin und Ordnung geboten erscheinen.[2])

Was die übrigen zu einer Gefängnisstrafe Verurteilten anbetrifft, so haben dieselben, soweit die Strafzeit dazu ausreicht, vier Strafstufen durchzumachen. Das jedesmalige Vorrücken in eine höhere Strafstufe erfolgt auf Grund des Markensystems, nachdem die für die betreffende Strafstufe festgesetzte Markenzahl verdient worden ist.[3])

Auf der ersten Strafstufe wird der Gefangene täglich 10 Stunden lang mit first- resp. second-class hard labour beschäftigt. Er wird in strenger Isolierung gehalten; er erhält keinen Schulunterricht und außer der Bibel keine Bibliotheksbücher. Am Sonntage bleibt er in seiner Zelle eingeschlossen; Bewegung im Freien ist ihm an diesem Tage nicht gestattet. Er schläft auf harter Lagerstätte ohne Matratze (plank bed without a mattress).[4])

---

[1]) Nur die zu einer Gefängnisstrafe mit harter Arbeit Verurteilten werden zu eigentlicher Strafarbeit (first-class hard labour, wie tread wheel, labour on cranks) angehalten.

[2]) Es könnte noch hinzugefügt werden, daß auch das finanzielle Interesse des Staates in sofern gewahrt wird, als der first-class misdemeanant diejenigen Ausgaben, welche durch die ihm gewährten Privilegien — wie bessere Kost, Aufwartung in der Zelle u. s. w. — entstehen, aus eignen Mitteln zu bestreiten hat. Wenn er unvermögend ist, so fallen diese Privilegien fort.

[3]) Cf. für das Folgende Report of Commissioners 1878 Seite 39 ff.

[4]) Der Gebrauch harter Lagerstätten für die erste Periode der Strafzeit war zuerst in dem Report of the Committee of the House of Lords on gaols vom Jahre 1863 empfohlen worden, und zwar unter besonderem Hinweise auf die guten Erfahrungen, welche hiermit in den Militärgefängnissen gemacht worden seien. In dem Prison Act von 1865 wurden dann die plank beds eingeführt (28 a. 29 Vict. c. 126 s 26), jedoch nicht in obligatorischer Weise, sondern die Anwendung derselben wurde in das Ermessen der Gefängnisbehörden gestellt. Die darauf basierende Praxis war eine sehr un-

Der Gefangene verbleibt in der ersten Strafstufe, bis er 224 Marken verdient hat, also wenigstens vier Wochen. Gefangene mit einer Strafzeit bis zu vier Wochen verbleiben daher während der ganzen Zeit in der ersten Strafstufe. Solange sich der Gefangene in der ersten Strafstufe befindet, erhält er keine Arbeitsbelohnung (gratuity).

Auf der zweiten Strafstufe wird der Gefangene mit second-class hard labour resp. mit industrial labour beschäftigt. Er erhält Schulunterricht und Schulbücher und darf sich am Sonntage im Freien bewegen. Für fünf Nächte in der Woche wird ihm eine Matratze gegeben, während er an zwei Nächten noch auf harter Lagerstätte zu schlafen hat. Der Gefangene verbleibt in der zweiten Strafstufe, bis er hier 224 Marken verdient hat, also wiederum wenigstens vier Wochen. Er erhält für je 20 Marken eine Arbeitsbelohnung von 1 d.

Auf der dritten Strafstufe braucht der Gefangene nur noch eine Nacht in der Woche auf harter Lagerstätte zu schlafen. Er erhält neben den Schulbüchern noch andre Bücher aus der Bibliothek, und es wird ihm für je 12 Marken eine Arbeitsbelohnung von 1 d gegeben. Auch die Zeitdauer der dritten Strafstufe beträgt wenigstens vier Wochen; es müssen auch hier 224 Marken verdient werden.

Den Rest der Strafzeit[1]) verbringt der Gefangene in der vierten Strafstufe. Der Gebrauch von plank beds fällt hier ganz fort. Der Gefangene kann zu Diensten in der Anstaltsverwaltung herangezogen werden. Er darf ferner alle drei Monate einen Brief schreiben und erhalten, und es ist ihm gestattet, alle drei Monate einen Besuch für die Zeit von 20 – 30 Minuten zu empfangen.

---

gleiche: in einigen Anstalten wurden die plank beds bei allen Insassen angewendet, in andern nur bei den erwachsenen männlichen Gefangenen, wiederum in andern Anstalten wurde zwar auch bei den weiblichen Gefangenen von den plank beds Gebrauch gemacht, es wurde den letzteren jedoch noch ein Strohsack (small mattress) und ein Kopfkissen gegeben. Seit 1878 besteht überall die gleichmäßige Praxis, daß bei allen Gefangenen während der ersten Periode der Strafzeit plank beds zur Anwendung kommen, Kinder unter 13 Jahren, männliche Gefangene über 60 Jahre und alle weiblichen Gefangenen erhalten jedoch dabei noch einen Strohsack und ein Kopfkissen. Cf. Report of Commissioners 1878 Seite 9.

[1]) Cf. die oben Seite 93 angeführten statistischen Notizen über die Höhe der Strafmaße, welche im Jahre 1885 bei Verurteilungen zu Gefängnisstrafen zur Anwendung gebracht worden sind.

Dies sind die Spezialbestimmungen über die Anwendung des in dem progressiven Strafensysteme liegenden pädagogischen Gedankens bei den zu einer Gefängnisstrafe Verurteilten.[1]) Es ist noch hinzuzufügen, daß im Disziplinarwege, insbesondere wegen Trägheit bei der Arbeit, Unaufmerksamkeit beim Unterrichte oder ungebührlichen Verhaltens die Zurückversetzung in eine niedere Strafstufe oder die Entziehung einzelner Privilegien der höhern Strafstufe, insbesondere der Arbeitsbelohnung, oder ein längeres Verbleiben in der niederen Strafstufe angeordnet werden kann.

Titel III.

Die Bestimmungen über die Behandlung der zur Strafe der penal servitude Verurteilten im speziellen.

§ 26. Charakteristik der Behandlung.

Es ist oben im § 12 versucht worden, die Klasse der zur Strafe der penal servitude Verurteilten im Gegensatze zu den zu einer Gefängnisstrafe Verurteilten im allgemeinen zu charakterisieren. Die zur penal servitude Verurteilten wurden dabei als Personen bezeichnet, welche zu der verbrecherischen That gelangt sind, weil sie aus Unlust zu ausdauernder und anstrengender Arbeit nicht vermocht haben, den Kampf um das Dasein in redlicher Weise zu führen, oder weil sie aus Mangel an jeder Herrschaft über sich selbst, über den eignen Willen, den an sie herantretenden Versuchungen einen Widerstand nicht entgegensetzen konnten. Es wurde weiter ausgeführt, daß hier die Notwendigkeit einer vollständigen Erziehung zum vernünftigen Gebrauche der Freiheit und einer gänzlichen Umbildung des Charakters bestände, und daß nach der in England herrschenden Ansicht gemeinsame Zwangsarbeit für wesentlich zur Erreichung dieser

---

[1]) Auch die Gefängniskost wird allmählich mit der Dauer der Strafzeit eine bessere. Cf. über das System der progressive dietary oben Seite 224 ff.

Ziele erachtet würde, weil nur hier die Möglichkeit einer zweckentsprechenden Beschäftigung und einer wirklichen Bildung des Charakters vorhanden wäre, während es an den Erfordernissen hierfür in der Einzelzelle fehlte.

Die gemeinschaftliche Zwangsarbeit, deren Handhabung in der Praxis im § 21 ausführlich dargestellt worden ist, bildet somit den Kernpunkt des Strafmittels der penal servitude. Um jedoch die Erreichung der in das Auge gefaßten Ziele sicherzustellen, bedarf es einerseits einer Vorbereitung für die gemeinschaftliche Zwangsarbeit, anderseits einer Prüfung, ob die versuchte Umbildung des Charakters gelungen ist, so daß dem Betreffenden ohne Gefahr für die Allgemeinheit die volle Freiheit und Selbständigkeit wiedergegeben werden kann.

Das Vorbereitungsstadium besteht in neunmonatlicher strenger Isolierhaft, welche der gemeinschaftlichen Zwangsarbeit vorausgeht, das Prüfungsstadium besteht in der vorläufigen Entlassung des Sträflings aus der Strafanstalt vor Ablauf der Strafzeit unter Ausübung einer strengen Kontrolle darüber, daß der Betreffende einen vernünftigen Gebrauch von der Freiheit macht, und unter der Möglichkeit, den Entlassenen sofort ohne alle Weitläufigkeiten in die Anstalt zurückzuversetzen, sobald zu befürchten ist, daß er wieder auf Abwege gerät.

Der Vollzug der Strafe der penal servitude beginnt also mit der Isolierhaft. Zweck derselben ist zunächst, den Sträfling durch vollständige Absonderung von allem, was seinen Geist zerstreuen könnte, zur Einkehr in sich selbst zu bringen. Dabei soll ihm durch strenge Behandlung die volle Wucht der Strafe zum Bewußtsein gebracht werden, er wird so den Wert der Freiheit und Selbständigkeit zu würdigen beginnen, welche er dadurch verscherzt hat, daß er einen vernünftigen Gebrauch davon zu machen nicht verstand. Je mehr ihm dabei das Maß seiner Verschuldung klar wird, um so aufmerksamer wird er für die Ermahnungen des Geistlichen sein, um so eher wird er zu guten Vorsätzen für die Zukunft gelangen. Erst wenn man annehmen kann, daß es der Sträfling mit diesen guten Vorsätzen wirklich ernst meint, wird man ihn den Versuchungen aussetzen können, wie sie bei der gemeinschaftlichen Zwangsarbeit durch das Zusammensein mit andern Sträflingen an ihn herantreten. Sobald die Isolierung als Vorbereitung ür die gemeinschaftliche Zwangsarbeit nicht mehr notwendig erscheint, hat sie fortzufallen, um demjenigen Strafstadium Platz zu machen, welches den Kernpunkt der Straferziehung bildet.

Nun ist es ja natürlich, daß der hiernach für die Isolierhaft notwendige Zeitraum je nach der Individualität des Sträflings ein sehr verschiedener sein wird. Es dürfte aber kaum weniger selbstverständlich erscheinen, daß eine geordnete Gefängnisverwaltung sich gar nicht durchführen läßt, wenn der Strafvollzug für jeden einzelnen Gefangenen verschieden gestaltet wird, und es ist ferner auch darauf aufmerksam zu machen, daß es einem schlauen Sträflinge vielfach gelingen kann, die Gefängnisbeamten über seine wahre Gesinnung und über den Ernst seiner guten Vorsätze zu täuschen. Hieraus ergibt sich, daß Durchschnittserfahrungen bei der Festsetzung der Zeitdauer der Isolierhaft maßgebend sein müssen. In England gilt es nun als fester — vielleicht zu starr festgehaltener — Erfahrungssatz, daß neun Monate Isolierhaft für die in das Auge gefaßten Ziele notwendig, aber auch in der Regel ausreichend sind. Die Zeitdauer der Isolierhaft ist daher zunächst für alle Sträflinge gleichmäßig auf neun Monate festgesetzt.[1]) Bei denjenigen Sträflingen, bei welchen sich dieser Zeitraum der Isolierhaft später in dem Stadium der gemeinschaftlichen Zwangsarbeit als nicht ausreichend herausstellt, tritt die Isolierhaft von neuem ein.[2])

Soviel über das Vorbereitungsstadium! Wir gehen nunmehr auf den Schluß der Strafzeit, auf das Stadium der vorläufigen Entlassung ein. Über die Art und Weise, in welcher während dieses Stadiums die Aufsicht über den aus der Gefängnisanstalt Entlassenen geführt wird, ist auf die im Abschnitte I § 10 gegebene Darstellung zu verweisen.

---

[1]) Die Royal Commissioners von 1878/9 lehnten einen Antrag, die Periode der Isolierhaft zu verlängern, mit dem Bemerken ab „the limitation of the period to nine months is the result of much experience“. Dieselben gaben aber dabei ausdrücklich zu (Report Seite XXVII), daß der früher für die Beschränkung der Isolierhaft auf neun Monate häufig angeführte Grund, wonach eine längere Isolierhaft für die körperliche und geistige Gesundheit schädlich wirken würde, nicht als stichhaltig anerkannt werden könnte. Es wurde anerkannt, daß vom sanitären Gesichtspunkte aus eine Verlängerung der Isolierhaft auf zwölf Monate keinerlei Bedenken unterliegen würde.

[2]) Cf. oben Seite 66.

Die für die vorläufige Entlassung maßgebenden Gesichtspunkte sind im wesentlichen folgende.

Es soll für den Bestraften, welcher während des Stadiums der gemeinschaftlichen Zwangsarbeit zum vernünftigen Gebrauche der Freiheit erzogen worden ist und von welchem nach seinem gesamten Verhalten während dieser Zeit angenommen werden kann, daß das Erziehungswerk ein erfolgreiches gewesen ist, ein vermittelnder Übergang aus dem Zustande der vollständigen Freiheitsentziehung in denjenigen der vollen Selbständigkeit und Freiheit geschaffen werden. Dies geschieht in erster Linie im Interesse der Gesellschaft. Es ist eine in England wie in andern Ländern festgestellte Thatsache, daß die Zahl der Rückfälle in der ersten Zeit nach der Entlassung aus der Strafanstalt am größten ist. Die Gesellschaft ist also während dieses Zeitraums einer besonders starken Gefahr ausgesetzt. Eine scharfe Kontrolle über die wieder in die Freiheit Gesetzten während der ersten Zeit erscheint daher geeignet und notwendig, um das allgemeine Sicherheitsbewußtsein zu erhöhen. Durch eine derartige Aufsicht wird nicht nur direkt die Begehung von Verbrechen erschwert, sondern der Strafentlassene wird auch in seinen guten Vorsätzen und in seinem Bestreben, sich auf redliche Weise durch das Leben durchzuschlagen, gestärkt und gekräftigt, wenn er sich bewußt bleiben muß, daß er bei dem geringsten Schritte abseits vom richtigen Wege, ja schon bei bloßem Verdachte hierfür,[1]) in die Strafanstalt zurückzuwandern hat.

Wenn hiernach eine strenge Beaufsichtigung der Strafentlassenen an sich wünschenswert erscheinen muß, so liegt es anderseits auf der Hand, daß der Betreffende damit erheblichen Beschränkungen in seiner Freiheit unterworfen wird. Die ihm geschenkte Freiheit ist nicht nur eine bedingte, sondern auch eine bloß teilweise: von der Strafzucht, wie sie in der Strafanstalt ausgeübt worden ist, bleibt noch ein gut Teil bestehen. Demgemäß ist es durchaus gerechtfertigt, die Zeit, während welcher die Aufsicht ausgeübt wird, in die Strafzeit einzurechnen. Der Zeitraum, für welchen der Bestrafte vorläufig entlassen wird, charakterisiert sich dann als *versuchsweise bewilligte Milderung des Strafzwanges*. Würde man den Bestraften nach voller Abbüßung der Strafe noch einer

---

[1]) z. B. wenn er Verkehr mit Personen von notorisch schlechtem Charakter unterhält, cf. oben Seite 122.

strengen Beaufsichtigung unterwerfen, so müßte dies als eine neue Strafe angesehen werden.

Auch darauf ist noch aufmerksam zu machen, daß die Lage des Strafentlassenen schon durch das abstoßende Verhalten, welches er fast regelmäßig von der bürgerlichen Gesellschaft erfährt, in erheblicher Weise erschwert wird. Der Strafentlassene, welcher seine verbrecherische That durch Abbüßung der richterlich erkannten Strafe gesühnt hat, wird faktisch von der Gesellschaft einer nochmaligen Bestrafung unterworfen. So sehr man auch dahin streben mag, in dieser Beziehung auf die öffentliche Meinung aufklärend einzuwirken, und so sehr man auch den Entlassenen durch die Fürsorgevereine zu helfen sucht, es wird immer die Thatsache bestehen bleiben, daß die Lage eines Bestraften unmittelbar nach seiner Entlassung aus der Strafanstalt eine erheblich schlechtere und schwierigere ist, als diejenige der übrigen Menschen.

Es ist somit nicht nur im Interesse der Gesellschaft wünschenswert, sondern auch gegenüber dem Bestraften gerecht, wenn dem letzteren gestattet wird, das letzte Stadium seiner Strafzeit außerhalb der Strafanstalt, aber unter strenger Beaufsichtigung seines Verhaltens zuzubringen.

Eine derartige Einrichtung ist nun aber weiterhin von dem größten Werte für das bei der Strafe der penal servitude in das Auge gefaßte Ziel der Erziehung des Bestraften zum vernünftigen Gebrauche seiner Freiheit. Wenn diese Zwangserziehung von nachhaltigem Erfolge begleitet sein soll, so muß dafür Sorge getragen werden, daß das reine Zwangselement während der Strafzeit immer mehr und mehr in den Hintergrund tritt, und daß der Sträfling selbst sich aus eignem Antriebe an dem Erziehungswerke beteiligt. Der Sträfling soll sich nicht lediglich unter dem ausgeübten Zwange fügen, nicht ausschließlich dulden, sondern er soll auch handeln lernen. Dazu ist notwendig, daß bei dem Sträflinge, soweit möglich, all die Hebel in Bewegung gesetzt werden, welche bei der Thätigkeit des freien Menschen in Wirksamkeit treten. Kein stimulus ist aber hierbei so mächtig als die Aussicht, durch die Thätigkeit die eigne Lage zu verbessern.

Es ist schon mehrfach hervorgehoben worden, daß bei der Gestaltung des gesamten Strafvollzugs in England diesem Gesichtspunkte der Aufrechterhaltung einer Hoffnung auf Besserung stete

Berücksichtigung zu teil wird. [1]) Die Vergünstigungen, welche sich der zu einer Gefängnisstrafe Verurteilte durch sein gutes Verhalten erwerben kann, wie Erhöhung der Arbeitsbelohnung, größere Freiheit im Verkehr mit der Außenwelt u. s. w., finden nun zwar auch bei den zur penal servitude Verurteilten Anwendung. Allein diese Vergünstigungen erscheinen hier, wo es sich um lange Strafzeiten handelt, nicht ausreichend. Die Aussicht auf frühere Entlassung aus der Strafanstalt muß hier als stimulus eintreten; sie allein ist geeignet, auf den zu langer Freiheitsstrafe Verurteilten tief einzuwirken und ihn zu gutem Verhalten und zu Anstrengungen anzuspornen. [2])

Auch der zu langer Freiheitsstrafe Verurteilte wird so vor stumpfem Hinbrüten und vor geistiger Erschlaffung bewahrt, sein Geist wird auf die Zukunft hingelenkt und das Freiheitsbedürfnis in ihm rege gehalten. Die Aussicht, durch Arbeitsamkeit und gutes Verhalten die Zeit des Aufenthaltes in der Strafanstalt abkürzen zu können, wirkt einerseits der für Geist und Körper leicht gefährlichen Monotonie der Isolierhaft entgegen und gibt anderseits dem Sträflinge einen kräftigen Impuls, den bei der gemeinschaftlichen Zwangsarbeit an ihn herantretenden Versuchungen nachhaltigen Widerstand entgegenzusetzen. Dadurch, daß es dem Sträflinge ständig vor Augen bleibt, wie sein eignes Verhalten bestimmend für die Dauer seiner Inhaftierung ist, wird das Selbstinteresse in ihm wachgerufen und er wird für den Kampf gegen die Gefahren der Gemeinschaft gestählt.

Nachdem die Institution der vorläufigen Entlassung nach den verschiedensten Seiten hin hier gerechtfertigt worden ist, soll noch die Frage berührt werden, wie sich dieselbe mit streng juristischen Grundsätzen verträgt, ob durch dieselbe nicht eine faktische Änderung des richterlichen Urteils herbeigeführt wird, indem die in dem Urteile festgestellte Strafzeit thatsächlich eine Verkürzung erfährt. Für die Beantwortung der Frage ist maßgebend einerseits die Art und Weise der Handhabung, anderseits das grundsätzliche Wesen der vorläufigen Entlassung.

---

[1]) Dem Gefangenen sollen die Stadien des Strafvollzugs als Stufen der Hoffnung erscheinen.

[2]) „He is made perfectly to see and feel that his fate is in his own hands and that he has something more to work and to hope for than the mere avoidance of punishment." Du Cane II Seite 169.

Die vorläufige Entlassung tritt in England ein als unmittelbare Folge des unter Anwendung des Markensystems ermittelten Verhaltens des Sträflings. Nach der Länge der zu verbüßenden Strafe wird festgestellt, welche Zahl von Marken der betreffende Sträfling zu verdienen hat; je nach seinem Verhalten kann er sich diese Markenzahl in längerer oder kürzerer Zeit erwerben. Sobald er sich die festgesetzte Markenzahl verdient hat, muß seine vorläufige Entlassung eintreten. Dieselbe hängt also weder von dem Belieben irgend einer Behörde ab, noch von dem Umstande, ob der Betreffende ein ordentliches Unterkommen zu finden vermag.[1]) Wenn der Sträfling stets nur die Mindestzahl von 6 Marken pro Tag erwirbt, so muß er während der ganzen Strafzeit in Haft bleiben. Wenn sich der Sträfling stets die Maximalzahl von 8 Marken pro Tag verdient und ihm niemals im Disziplinarwege Marken entzogen werden, so verkürzt sich die Zeit, welche er in dem Public Works Prison zu verbringen hat, um ein Viertel. Die Minimal- wie die Maximalzeit der Inhaftierung steht also von vornherein fest; innerhalb dieser Grenzen entscheidet das Verhalten des Sträflings über die Zeitdauer der Inhaftierung.[2])

Was sodann das grundsätzliche Wesen der Institution betrifft, so kann es nach den jetzt in England geltenden Bestimmungen kaum einem Zweifel unterliegen, daß die vorläufige Entlassung als ein Stadium des Strafvollzugs, nicht aber als ein Akt der Begnadigung aufzufassen ist.

Gegen die Auffassung als Begnadigung spricht zunächst der Umstand, daß jeder Sträfling ein Recht auf vorläufige Entlassung

---

[1]) In Irland war früher die vorläufige Entlassung von dem Nachweise abhängig, daß ein Dienstverhältnis für den Entlassenen in sicherer Aussicht stände. Mit Recht machte Colonel Sir Joshua Jebb hiergegen geltend, daß diese Bestimmung einen Widerspruch mit dem ganzen Systeme enthielte, da die Freilassung nicht bloß von dem guten Verhalten des Sträflings, sondern auch von dem Wohlwollen dritter Personen bedingt erschiene und die in dem Sträflinge geweckte und mit Absicht unterhaltene Hoffnung auf Freilassung in vielen Fällen unerfüllt bleiben müßte. (Cf. Holtzendorff II Seite 106) Der Nachweis eines Dienstverhältnisses wird denn auch jetzt in Irland nicht mehr gefordert.

[2]) Die Strafzeit zerfällt somit in einen festen und einen beweglichen Teil; aber auch die Ausfüllung des beweglichen Teiles erfolgt nach ein für allemal festbestimmten Grundsätzen und ist nicht der Willkür der Gefängnisverwaltung anheimgegeben.

nach bestimmter Strafzeit besitzt, falls er sich durch sein Verhalten die festgesetzte Markenzahl erworben hat. Der Natur eines Begnadigungsrechtes würde es entsprechen, wenn ausnahmsweise in besonders gearteten Fällen Sträflingen eine vorläufige Entlassung gewährt würde, nicht aber, wenn diese Institution, wie es thatsächlich geschieht, regelmäßig nach feststehenden Grundsätzen zur Anwendung gelangt.

Sodann erscheint die Bestimmung, daß die vorläufige Entlassung sowohl von dem Staatssekretär wie von dem Richter widerrufen werden kann, und daß der Betreffende im Falle des Widerrufs für die gesamte bei der vorläufigen Entlassung noch restierende Strafzeit in die Strafanstalt zurückversetzt wird, mit der Auffassung der Institution als eines Gnadenaktes unvereinbar. Der Widerruf könnte hier nur die Zurückversetzung für die zur Zeit des Widerrufs noch nicht abgelaufene Strafzeit zur Folge haben, nicht aber, wie es hier der Fall ist, mit rückwirkender Kraft ausgestattet sein.

Die vorläufige Entlassung ist somit als ein Stadium der Strafvollstreckung anzusehen.[1]) Der staatliche Strafzwang bleibt auch gegenüber dem vorläufig Entlassenen bestehen, wenn auch nur bedingungsweise und in veränderter Form. Dem Sträflinge wird für die letzte Zeit seiner Strafe versuchsweise ein Urlaub, aber unter erheblichen Einschränkungen und mit der Maßgabe bewilligt, daß die regelrechte Art der Strafverbüßung wieder in Kraft tritt, sobald der Versuch als mißlungen anzusehen ist. *Diese Urlaubszeit erscheint als innerhalb der Strafzeit liegend.*

Von einer Verkürzung der richterlich erkannten Strafzeit kann somit nicht die Rede sein; es handelt sich vielmehr bei der vorläufigen Entlassung um eine nach festbestimmten Regeln erfolgende und dem Richter bei Abmessung der Strafe von vornherein bekannte Modifikation des Strafvollzuges. Es er-

[1]) Es ist hierbei allerdings zuzugestehen, daß die Form der order of license, durch welche die vorläufige Entlassung vom Staatssekretär bewilligt wird, eher für die Auffassung der Institution als eines Gnadenaktes spricht. Allein nicht diese Form, sondern das Wesen der Sache muß für die Auffassung entscheidend bleiben, und dies um so mehr, als die Form der license aus einer Zeit herrührt, in welcher die Entlassung sich thatsächlich als Akt der Begnadigung charakterisierte. Cf. oben Seite 53.

scheint daher auch gerechtfertigt, wenn in dem Report of the Royal Commissioners 1879 Seite XXXII f. der Wunsch ausgesprochen wird, daß der Richter im Urteilsspruche selbst denjenigen Betrag der Strafzeit ausdrücklich hervorhebe, um welchen der regelrechte Strafvollzug im Falle guten Verhaltens gekürzt wird.[1])

Nicht die Zeit der Strafe, sondern nur die Art und Weise der Strafvollstreckung bleibt innerhalb gewisser Grenzen bei der Urteilsfällung unbestimmt und regelt sich je nach dem Verhalten des Sträflings.[2]) Dies widerspricht in keinerlei Weise den Prinzipien der Gerechtigkeit: der Richter berücksichtigt bei Abmessung der Strafe nicht nur die objektive That, sondern auch die Persönlichkeit des Thäters, wie sie insbesondere in dem Verhalten bei und nach der That hervortritt, und so erscheint auch eine Ergänzung dieser subjektiven Beurteilung in Gestalt einer Modifikation des Strafvollzugs je nach dem Verhalten des Bestraften während der Strafverbüßung durchaus gerechtfertigt.

Nachdem die drei Stadien der Strafe der penal servitude in der ihnen beigelegten strafpolitischen Bedeutung und ihrem Wesen nach charakterisiert worden sind, soll nunmehr die Art und Weise, wie die Strafe selbst zur Ausführung gelangt, des näheren dargelegt werden.

---

[1]) Es heißt in dem Report: „We think it desirable that the judges in passing sentence should declare in open court the amount of remission which the convict may earn by industry and good conduct. Such a statement would tend to remove much of the misapprehension and suspicion which now exists in the public mind as to the release of convicts before their sentences have expired; and the judges having thus clearly before them on each occasion the exact effect of the sentence would be better able to apportion the penalty to the sentence".

[2]) Auf diese Weise wird dasjenige im wesentlichen erreicht, was man bei der häufig aufgestellten, aber vom juristischen Standpunkte aus unerfüllbaren Forderung, daß das Strafmaß vom Richter unbestimmt gelassen und erst nachher von der Gefängnisverwaltung bestimmt werde, vor Augen hat: das bei der Kürze der Gerichtsverhandlung der Natur der Sache nach nur oberflächliche Urteil des Richters über die Persönlichkeit des Verbrechers wird durch die während der Haftzeit mögliche nähere Ermittelung des Charakters in zweckentsprechender und mit den Prinzipien der Gerechtigkeit übereinstimmender Weise ergänzt. Die Strafe ist „indefinite in effect but not in form".

§ 27. Die Art und Weise der praktischen Durchführung.

Sämtliche männlichen Sträflinge werden für das Stadium der neunmonatlichen Isolierhaft in der neuerrichteten, in unmittelbarer Nähe von London gelegenen Anstalt von Wormwood Scrubs untergebracht. [1])

Die Isolierung ist hier eine strenge: die Sträflinge sehen einander nur während des Gottesdienstes und beim Spaziergange. Bei beiden Gelegenheiten wird scharf auf Beobachtung des absoluten Schweiggebots gehalten. Ein Gottesdienst in der Kapelle der Anstalt findet täglich statt. Jeder Sträfling hat sich täglich eine Stunde im Freien auf dem Spazierhofe zu bewegen. In der Zelle werden die Sträflinge zunächst für kurze Zeit mit Wergzupfen, dann in industrieller Arbeit beschäftigt. Hauptbeschäftigungen bilden die Anfertigung von Kohlensäcken und Hängematten für die Admiralität, sowie von Postsäcken für das Generalpostamt; eine größere Anzahl von Sträflingen wird auch mit Buchbinderei für das Ministerium des Innern beschäftigt. Diejenigen Sträflinge, welche noch des Unterrichts im Lesen, Schreiben oder Rechnen bedürfen, erhalten zweimal die Woche Elementar-Unterricht. Es ist dafür gesorgt, daß der Sträfling häufig von dem Geistlichen in der Zelle besucht wird; jeder diesbezügliche Wunsch des Sträflings ist dem Geistlichen durch den Gefängniswärter alsbald mitzuteilen und, wenn irgend möglich, noch an demselben Tage zu erfüllen.

Nach Ablauf der neun Monate wird der Sträfling in dasjenige Public Works Prison geschafft, für welches er qualifiziert erscheint. Über die in dieser Richtung bestehende Spezialisierung der Anstalten ist oben im § 7 das Nähere angeführt worden.

Der Sträfling verbleibt, vorausgesetzt daß er nicht wegen schlechten Verhaltens in die Isolierhaft zurückversetzt wird, in dem Public Works Prison bis zu seiner vorläufigen Entlassung. [2]) Dieser Zeitraum wird ausgedrückt durch eine be-

---

[1]) Pentonville, welches früher diesem Zwecke diente, wird seit 1886 als local prison benutzt.

[2]) Bei den zu lebenslänglicher Strafe Verurteilten wird nach Ablauf von 20 Jahren dem Staatssekretär ein Spezialbericht eingereicht, in welchem Vorschläge darüber gemacht werden, ob der Betreffende zur Begna-

stimmte Anzahl von Marken, welche sich der Sträfling verdienen muß. Die Zahl wird in der Weise berechnet, daß von der Gesamtstrafzeit zunächst neun Monate als die Zeit der Isolierhaft in Abzug gebracht und die alsdann übrig bleibende Zahl von Tagen mit 6 multipliziert wird. Während der Isolierhaft erhält der Sträfling keine Marken, während der übrigen Zeit kann er sich nach den oben im § 13 angeführten Grundsätzen pro Tag 6—8 Marken verdienen. Ein Sträfling, welcher stets pro Tag nur 6 Marken erhält, hat somit die volle Strafzeit in der Anstalt zu verbringen; ein Sträfling, welcher sich stets pro Tag 8 Marken verdient und niemals im Disziplinarwege Marken entzogen erhält (cf. oben Seite 260), kürzt die Zeit, welche er in dem Public Works Prison zuzubringen hat, um ein Viertel ab.

Nehmen wir den Fall, daß der Sträfling zu dem niedrigsten Strafmaße der penal servitude von fünf Jahren verurteilt worden ist, so beträgt der Zeitraum des Public Works Stage vier Jahre und drei Monate. Dieser Zeitraum kann höchstens abgekürzt werden um ein Viertel, d. i. ein Jahr und 23 Tage, so daß der zu fünfjähriger Strafe Verurteilte schon etwas vor Ablauf von vier Jahren zur vorläufigen Entlassung kommen kann. [1])

---

digung zu empfehlen sei oder ob event. eine Milderung in der Behandlung angebracht erscheine. Die Behandlung der zu lebenslänglicher Strafe Verurteilten muß bei dem geltenden Systeme besondere Schwierigkeiten bereiten: eine Straferziehung hat hier keinen rechten Sinn, und auf der andern Seite bilden diese Personen eine für die gemeinschaftliche Arbeit sehr leicht gefährliche Klasse. Sir Edmund Du Cane und Sir Walter Crofton haben deshalb vorgeschlagen, daß die zu lebenslänglicher Strafe Verurteilten in einer besonderen Anstalt untergebracht werden sollten. Report of the Royal Commissioners 1879 Seite XXXIII. Die Gesamtzahl der im Jahre 1886 in Strafanstalten befindlichen Sträflinge, welche zu lebenslänglicher Strafe verurteilt waren, betrug 309. Report of Directors 1886 Seite XXXIII.

[1]) Zur Beurteilung der ganzen Institution dürften folgende Ziffern von Interesse sein. Es betrug die Zahl der männlichen convict prisoners, welche entlassen wurden,

im Jahre	überhaupt	davon mit einer order of license	und zwar mit der gesetzlich zulässigen Minimalfrist
1881	1558	1347	438
1882	1510	1315	453
1883	1922	1719	542
1884	1988	1753	669
1885	1698	1456	430

Die Sträflinge in den Public Works Prisons zerfallen in fünf streng von einander gesonderte Klassen.[1])

Bei ihrer Überführung in das Public Works Prison kommen sie zunächst in die probation class; sie verbleiben in derselben, bis sie sich 720 Marken verdient haben, also wenigstens drei Monate. Während dieser Zeit erhalten sie keinerlei Arbeitsbelohnung, dürfen keine Besuche oder Briefe empfangen und nur bei ihrer Ankunft in dem Public Works Prison einen Brief schreiben. Am Sonntage ist ihnen während einer Stunde Bewegung im Freien gestattet.[2]) Wenn der Arbeitsfleiß oder das sonstige Verhalten des Sträflings ein schlechtes ist, so kann die Periode der probation class verlängert werden, indem die Beförderung in die folgende Klasse, die third class, erst verfügt wird, nachdem eine bestimmte Anzahl von Zusatz-Marken in der probation class verdient worden sind.

In der third class haben die Sträflinge 2920 Marken zu verdienen; die Zeitdauer ihres Verbleibens in dieser Klasse ist also wenigstens ein Jahr. Während dieser Zeit erhalten sie eine Arbeitsbelohnung von 1 d für je 20 verdiente Marken, der Gesamtbetrag der in dieser Klasse erlangbaren Arbeitsbelohnung beträgt 12 s. Die Sträflinge dieser Klasse dürfen alle sechs Monate einen Besuch von 20 Minuten empfangen und je einen Brief schreiben und erhalten. Als Unterscheidungsmerkmale sind bei ihnen an der Gefängniskleidung weiße Litzen angebracht.

Report of Directors 1886 Seite XXXV. Durchschnittlich etwa 18 % erhielten somit keine license, sondern wurden erst am Ende ihrer Strafzeit aus dem Public Works Prison entlassen; 25—30 % erwarben sich die Maximalstrafkürzung von $1/4$; der Rest erhielt zwar eine vorläufige Entlassung, aber nicht zu der frühesten gesetzlich zulässigen Zeit.

[1]) In die Arbeitskolonnen (gangs) werden jedoch, wie schon oben Seite 248 Anm. 2 angeführt wurde, Sträflinge der verschiedensten Klassen eingeteilt.

[2]) Über die Klassifikation der Sträflinge in den Public Works Prisons ist unter dem 22. Juli 1864 eine Verfügung ergangen, welche abgesehen von kleinen Abänderungen noch heute in Kraft ist. Cf. Du Cane I Seite 116 ff. Für diejenigen Sträflinge, welche aus Gesundheitsrücksichten nicht zu öffentlichen Arbeiten verwendet werden können und deshalb in industrieller Weise beschäftigt werden, ist auch in der Woche täglich eine Stunde zur Bewegung im Freien bestimmt.

Nach der Beförderung in die second class erhalten die Sträflinge für je 20 verdiente Marken eine Arbeitsbelohnung von $1\frac{1}{2}$ d und dürfen alle vier Monate einen Besuch von 20 Minuten empfangen und je einen Brief schreiben und erhalten. Sie haben außerdem das Recht, anstatt dem zum Frühstücke gereichten Haferschleime eine Portion Thee zu verlangen. [1]) Am Sonntage ist ihnen ein zweimaliger Spaziergang von je einer Stunde gestattet. Die Sträflinge, welche in dieser Klasse gelbe Litzen an der Gefängniskleidung tragen, verbleiben hier wenigstens ein Jahr, bis sie 2920 Marken verdient haben. Die Beförderung in die first class wird jedoch verschoben, wenn der Sträfling bis dahin Lesen und Schreiben noch nicht genügend gelernt hat. [2])

In der first class erhalten die Sträflinge, welche hier blaue Litzen an der Gefängniskleidung tragen, eine Arbeitsbelohnung von $2\frac{1}{2}$ d für je 20 verdiente Marken und dürfen alle drei Monate einen Besuch von 30 Minuten empfangen und je einen Brief schreiben und erhalten. Auf ihren Wunsch wird ihnen zum Frühstücke Thee an Stelle von Haferschleim und einmal in der Woche gedämpftes Fleisch an Stelle von gekochtem Fleische gereicht. Am Sonntage ist ihnen ein dreimaliger Spaziergang gestattet.

Eine Beförderung aus der first class in die special class findet nur bei exemplarisch gutem Verhalten und innerhalb zwölf Monaten vor der Entlassung statt. Die Sträflinge der special class legen die gewöhnliche Gefängniskleidung ab und erhalten einen blauen Anzug; sie werden zu Vertrauensposten, insbesondere in der Anstaltsverwaltung selbst, verwendet. Es kann ihnen ferner zu dem für die übrigen Sträflinge bestehenden Maximalbetrage der Arbeitsbelohnung von 3 £ eine zusätzliche Belohnung bis zur Höhe von 3 £ zugebilligt werden.

Soviel über die Behandlung der männlichen Sträflinge. Bei den weiblichen Sträflingen sind einige Modifikationen in der Ausführung des Systems, an dessen prinzipiellen Punkten man festgehalten hat, für nötig erachtet worden. [3])

---

[1]) Cf. im übrigen bezüglich der gereichten Gefängniskost oben Seite 226 f. Eine eigentliche Kostverbesserung ist jetzt mit der Beförderung in eine höhere Klasse nicht mehr verbunden; es wird nur eine gewisse Abwechselung in der Kost gestattet.

[2]) Schulunterricht findet in den Public Works Prisons nur in beschränktem Umfange und nur während der Abendstunden statt.

[3]) Cf. Report of the Royal Commissioners 1879 Seite XX ff.

Es sind dafür mehrere Gesichtspunkte maßgebend gewesen. Man glaubte zunächst, daß die Durchführung strenger Isolierung und des Schweiggebotes hier bei dem leichter erregbaren und von Natur mitteilsameren Charakter des Weibes eine sehr schwierige sein würde und leicht eine ernstliche Schädigung der Gesundheit herbeiführen könnte. Es wurde ferner nicht verkannt, daß eine an sich schwere und zur Arbeit erziehende Beschäftigung für das weibliche Geschlecht nur schwer zu finden ist. Und endlich war man sich bewußt, daß ein erfolgreicher Rücktritt in die Freiheit und in geordnete Verhältnisse bei einer mit der schweren und langdauernden Strafe der penal servitude belegten Frauensperson erheblich größeren Schwierigkeiten unterliegt, als bei einem männlichen Sträflinge, und man hielt es daher für nötig, das Übergangsstadium aus der Gefangenschaft in die Freiheit anders zu gestalten.

Die Behandlung der weiblichen Sträflinge während der neunmonatlichen Isolierhaft [1]) weicht nur in sofern von derjenigen der männlichen Sträflinge ab, als der Schulunterricht hier stets klassenweise gegeben wird und die Zellenthür während der Arbeit geöffnet bleibt. Die Beschäftigung besteht neben Wergzupfen vorzugsweise in Näh- und Flickarbeiten.

Für das Stadium der gemeinschaftlichen Zwangsarbeit sind die beiden Anstalten von Woking und Fulham bestimmt. Die für die weiblichen Sträflinge bestehenden Strafklassen sind: probation-, third-, second-, first- und refuge-class. Die letztere, hier noch speziell zu behandelnde Klasse tritt an die Stelle der special class bei den männlichen Sträflingen. Die Minimalzeitdauer für die third- und second-class ist von einem Jahre auf neun Monate verkürzt.

In der second class tritt zu den oben angeführten Vergünstigungen noch die hinzu, daß den weiblichen Sträflingen dieser Klasse zweimal in der Woche ein Spaziergang in Gruppen von dreien oder vieren gestattet wird. In der first class finden derartige gruppenweise Spaziergänge täglich statt.

Die weiblichen Sträflinge werden während des Public Works

---

[1]) Bisher verbüßten die weiblichen Sträflinge die Isolierhaft in Millbank. Infolge des Umbaues dieser Anstalt wurden die weiblichen Sträflinge gegen Ende 1885 von dort nach andern Anstalten fortgeschafft (cf. oben Seite 164). Der jetzige Zustand ist nur ein Provisorium; voraussichtlich werden in Zukunft auch die weiblichen Sträflinge zur Verbüßung der Isolierhaft nach Wormwood Scrubs gebracht werden.

Stage vorzugsweise mit Schneiderarbeiten, in der Waschanstalt, mit Arbeiten in der Anstaltsverwaltung (Kochen, Backen, Reinigen) beschäftigt. Ein besonderer Industriezweig besteht in der Anstalt von Woking in der Herstellung von Mosaikarbeiten.

Das Markensystem und die darauf basierende Kürzung der Haftzeit durch vorläufige Entlassung erleiden zweierlei Modifikationen: weibliche Sträflinge erhalten auch während der Isolierhaft Marken zugeteilt und der Maximalbetrag der zulässigen Haftkürzung ist durch anderweitige Verrechnung der Marken von ein Viertel auf ein Drittel der Strafzeit erhöht. Ein weiblicher Sträfling, welcher zu fünf Jahren penal servitude verurteilt worden ist, kann also schon nach Ablauf von drei Jahren und vier Monaten vorläufig entlassen werden. Die Maximalhöhe der zulässigen Arbeitsbelohnung endlich ist um 1 £ höher; sie beträgt 4 £.

Die Beförderung aus der first- in die refuge-class kann frühestens neun Monate vor Ablauf der Haftzeit erfolgen. Von dieser Beförderung sind diejenigen ausgeschlossen, welche schon einmal mit penal servitude bestraft und in einer refuge gewesen sind. Die Sträflinge der refuge class kommen, sobald in einer der bestehenden refuges eine Stelle frei ist, aus der Strafanstalt heraus in die refuge.

Die refuges entsprechen den in Irland vorhandenen und im nächsten Abschnitte noch zu besprechenden Zwischenanstalten. Wie diese verfolgen sie den Zweck, den Sträfling am Ende der Strafzeit dem Leben in der Freiheit näher zu bringen. Eine Reihe von Beschränkungen, wie sie in der Strafanstalt bestehen, fallen fort: die Insassen tragen keine Gefängniskleidung mehr, sind völlig frei in ihrem Verkehr miteinander, können mit der Außenwelt korrespondieren und in der arbeitsfreien Zeit Besuche empfangen; die Disziplin wird ohne eigentliche Strafmittel dadurch aufrecht erhalten, daß im Falle schlechten Betragens die Rücksendung in die Strafanstalt erfolgt. Man sucht sodann eine Verbindung der Anstalt mit dem freien Arbeitsmarkte herzustellen, auf Grund derer sich die Insassen schon vor ihrer Entlassung eine Beschäftigung für die Zukunft sichern können. Unter Rücksichtnahme hierauf werden auch die Arbeitsleistungen der einzelnen Insassen eingerichtet: einige von ihnen werden zu häuslichen Diensten in der Anstalt, wie Kochen, Waschen, Reinigen, verwendet, andre beschäftigen sich mit Schneiderei oder weiblichen Handarbeiten.

Die erste Anregung zur Gründung von refuges ist im Jahre 1853 von Lord Palmerston ausgegangen. Derselbe ließ im November 1853 durch seinen Sekretär, Mr. Waddington, ein Schreiben veröffentlichen, in welchem es als wünschenswert bezeichnet wurde, daß Wohlthätigkeitsinstitute die Errichtung von Anstalten in die Hand nähmen, in welchen weibliche Sträflinge in der angegebenen Weise auf den Rücktritt in die Freiheit vorbereitet würden.[1]) Da jedoch diese Aufforderung zunächst ohne Erfolg blieb, so errichtete die Regierung selbst eine derartige Anstalt zu Fulham. Als dann später zweckentsprechende Privatinstitute entstanden, gab die Regierung die refuge zu Fulham auf (das Gebäude wurde als Gefängnis benutzt) und beschränkte sich darauf, den Privatinstituten gegen die Verpflichtung zur Aufnahme einer bestimmten Anzahl von weiblichen Sträflingen Unterstützungen zu gewähren.[2])

Es bestehen zur Zeit drei refuges, nach welchen weibliche Sträflinge gesandt werden: Carlisle Memorial Refuge zu Winchester, Westminster Memorial Refuge zu Streatham und East End Refuge zu Finchley; die letztere Anstalt nimmt nur katholische Sträflinge auf. Die Regierung zahlt für jeden in dieser Anstalt befindlichen Sträfling pro Woche 10 s.[3]) Im übrigen bestreiten die refuges ihre Ausgaben durch Verwertung der Arbeitsleistungen der Insassen, sowie aus freiwilligen Spenden. Es findet somit hier in ähnlicher Weise wie in den im Abschnitte I § 9 besprochenen Reformatory und Industrial Schools ein harmonisches Zusammenwirken des Staates mit der Privatwohlthätigkeit statt.

---

[1]) Cf. Report of Directors 1861 Seite LVII ff.

[2]) Cf. die Verfügung vom 1. Februar 1873 (abgedruckt Du Cane I Seite 110); es werden hier nähere Bestimmungen darüber gegeben, unter welchen Voraussetzungen weibliche Sträflinge in die refuges gesandt werden sollen. Außer dem in dieser Richtung schon im Texte Angeführten verdient noch hervorgehoben zu werden, daß von der Aufnahme in eine refuge ausgeschlossen bleiben Personen, welche wegen Mord oder gefährlicher Körperverletzung verurteilt worden sind oder welche sich während der letzten sechs Monate der Haftzeit eine Gewaltthätigkeit haben zu Schulden kommen lassen, sowie endlich Personen, deren Gesundheitszustand ein schlechter ist, so daß sie schon hierdurch gehindert sind, sich selbständig ihren Lebensunterhalt zu erwerben.

[3]) Die Zahl der im Jahre 1885 nach Streatham gesandten weiblichen Sträflinge betrug 43, die Zahl der nach Finchley gesandten 50; eine Sendung von Sträflingen nach Winchester fand in diesem Jahre nicht statt. Report of Directors 1886 Seite XL.

Abschnitt III.*)

Das irische Gefängnissystem.

*) Quellen: Report of the Royal Commission on Prisons in Ireland 1884. — Die jährlichen Berichte des General Prisons Board Ireland. — Rules for the treatment of prisoners confined in convict prisons in Ireland 1885. — Die Schriften von Crofton I und II. — Die Schriften von v. Holtzendorff II–IV.

Die hervorragende Stellung, welche Irland in der Gefängniswissenschaft einnimmt, verdankt dasselbe ausschließlich den ausgezeichneten Leistungen eines einzelnen Mannes, Sir Walter Croftons. Es ist daher auch durchaus der Sache entsprechend, wenn in der Litteratur das irische System vielfach direkt als Croftonsches System bezeichnet wird.[1])

Wenn nun hier anhangsweise noch ganz kurz auf das Gefängniswesen in Irland eingegangen wird, so soll sich die Darstellung im wesentlichen auf eine Klarlegung desjenigen beschränken, was Crofton leistete, unter Hervorhebung der Entwickelung, welche die Sache bis auf den heutigen Tag genommen hat. Eine eigentliche Behandlung des Gefängniswesens in Irland wird hier um so weniger beabsichtigt, als dieselbe bis auf einen hier näher zu erörternden Punkt auf eine bloße Wiederholung des im Abschnitte II bezüglich Englands Ausgeführten hinauslaufen würde.[2])

---

[1]) Die nachfolgende Darstellung wird ergeben, ob es überhaupt gerechtfertigt ist, von einem irischen Systeme zu sprechen. Ich wende diese Bezeichnung hier lediglich deshalb an, weil dieselbe in Deutschland gebräuchlich ist. Dies ist aber in England durchaus nicht der Fall; ja man kann sagen, die Bezeichnung „irisches System“ wird heutzutage in England kaum verstanden. Als ich im Beginne meiner Studien in England mit Fachleuten von dem irischen Systeme sprach, erhielt ich mehrfach die Antwort: „Sie meinen das intermediate system, welches von Baron Holtzendorff durch seine Schrift „Das irische Gefängnissystem“ in Deutschland bekannt gemacht ist.“

[2]) Aus diesem Grunde wird die Vollstreckung der Gefängnisstrafe in Irland hier gar nicht behandelt, da hierfür dieselben Bestimmungen gelten wie in England. Zu erwähnen wäre nur, daß in Irland eine gemeinschaftliche Zentralbehörde für die Vollstreckung der Gefängnisstrafe und der Strafe der penal servitude, das General Prisons Board, besteht, während in England die Vereinigung bisher lediglich durch Personalunion hergestellt worden ist

Als Sir Walter Crofton — damals Kapitän Crofton — im Jahre 1854 die Leitung des irischen Gefängniswesens übernahm, war dasselbe in einem traurigen Zustande: die Anstalten genügten in keinerlei Weise den zu stellenden Anforderungen und waren mehr als überfüllt; das Beamtenpersonal der Anstalten befand sich auf einer sehr niederen Bildungsstufe und war für die Erfüllung der ihm obliegenden Pflichten durchaus unqualifiziert; die Arbeitsleistungen der Gefangenen waren höchst ungenügende; für einen geeigneten Unterricht war fast gar nicht gesorgt; die Zahl der zur Anwendung gebrachten Disziplinarstrafen war eine sehr hohe; der Gesundheitszustand mußte als durchaus unbefriedigend bezeichnet werden. All dieses war von der Parlamentskommission, welche im Jahre 1853 zur Untersuchung der irischen Gefängniszustände niedergesetzt worden war und welcher auch Kapitän Crofton angehört hatte, als das Resultat eingehender Prüfung festgestellt worden.

Die wesentlichste Maßregel, welche zur Abhilfe dieser Zustände von der Parlamentskommission empfohlen und durch das Gesetz 17 a. 18 Vict. c. 76 eingeführt wurde, war die Einsetzung einer Zentralgefängnisbehörde für Irland. An die Spitze dieser Behörde wurde Kapitän Crofton gestellt. In verhältnismäßig kurzer Zeit gelang es diesem ausgezeichneten Manne, die Gefängniszustände Irlands vollständig umzugestalten und die von der Parlamentskommission vorgefundenen Mängel im wesentlichen zu beseitigen.[1]) Soweit es sich dabei um Verwaltungsmaßregeln handelt, deren günstiger Erfolg lediglich der Energie und persönlichen Tüchtigkeit zuzuschreiben ist, mit welcher dieselben von Crofton zur Ausführung

---

Das General Prisons Board ist zusammengesetzt aus einem Vorsitzenden (Gehalt 1200 £), zwei bezahlten Mitgliedern (Gehalt 1000 und 1100 £), einem unbezahlten Mitgliede mit beratender Stimme, und drei Inspektoren (Gehalt 500—600 £) als Hilfsarbeitern.

[1]) Über die Größe der von Crofton erzielten Resultate ist seinerzeit ein sehr erbitterter litterarischer Streit geführt worden (cf. v. Holtzendorff IV). Es mag dabei ohne weiteres anerkannt werden, daß die von Crofton und seinen Anhängern angeführten Ziffern nicht durchweg beweiskräftig sind. Es hat keinen Zweck, auf diesen Streit hier näher einzugehen; es genügt, die Thatsache zu konstatieren, daß es Crofton gelungen war, das Gefängniswesen in Irland, welches sich in einem so traurigen Zustande befunden hatte, derartig zu verbessern, daß dasselbe in dem Berichte der Royal Commission von 1863 als in vielen Punkten nachahmungswertes Beispiel für England hingestellt werden konnte.

gebracht wurden, kann hier darüber hinweggegangen werden. Von dauerndem Interesse ist nur die Art und Weise, in welcher Crofton das damals in Irland wie in England neu eingeführte Strafmittel der penal servitude praktisch gestaltete, da eben diese Gestaltung die Veranlassung für die Bezeichnung „irisches System" gewesen ist.

Die zur Strafe der penal servitude Verurteilten kamen in Irland zunächst auf neun Monate nach der Strafanstalt von Mountjoy bei Dublin in strenge Isolierhaft.[1]) Nach Ablauf der neun Monate kamen die arbeitsfähigen männlichen Gefangenen nach Spike Island, einer befestigten Insel im Hafen von Cork, wo dieselben mit Erd- und Wasserarbeiten, Festungsbauten u. s. w. in einzelnen Arbeitsgruppen auf angestrengte Weise beschäftigt wurden.[2]) Die Sträflinge waren hier in fünf Strafklassen geteilt; sie rückten nach gewisser Zeit und unter gewissen Bedingungen von der niederen Klasse zu einer höheren auf;[3]) mit jedem Vorrücken in eine höhere Klasse war eine Milderung des Strafzwangs durch Gewährung von besonderen Vergünstigungen verbunden. Aus der höchsten Strafklasse (advanced class) wurden Sträflinge unter gewissen Voraussetzungen in eine der beiden Zwischenanstalten (intermediate prisons) von Lusk und Smithfield gebracht, wo den Insassen ein viel größeres Maß von Freiheit und Selbständigkeit gelassen war; Lusk war dabei für landwirtschaftliche, Smithfield für industrielle Arbeiten bestimmt. In wesentlich gleicher Weise wie in England erfolgte die vorläufige Entlassung des Sträflings aus der Haft, nachdem er sich eine nach der Länge der Strafzeit bemessene Anzahl Marken erworben hatte.

---

[1]) An Stelle dieses festen Zeitraums für das Stadium der Isolierhaft wurde später die Bestimmung getroffen, daß je nach dem Verhalten des Sträflings die Isolierhaft acht Monate bis ein Jahr dauern sollte. Report of the Royal Commissioners 1879 Seite XXIII.

[2]) Diejenigen Sträflinge, welche sich nach ihrem Gesundheitszustande zu den Arbeiten in Spike Island nicht qualifizierten, verblieben in Mountjoy und wurden dort zu gemeinschaftlichen Arbeiten industrieller Art angehalten.

[3]) Der in Irland für das Verbleiben in den unteren Strafklassen bestimmte Minimalzeitraum war kürzer bemessen als in England. Auch hinsichtlich des Markensystems finden sich einige abweichende Bestimmungen. Da hier im Texte lediglich die Prinzipien des irischen Strafensystems besprochen werden sollen, so sei bezüglich aller Detailfragen ausdrücklich auf die Schriften von v. Holtzendorff II—IV verwiesen.

Über den vorläufig Entlassenen wurde bis zum vollen Ablaufe der Strafzeit eine strenge Polizeiaufsicht geführt.

Man sieht, abgesehen von dem zwischen das Stadium der gemeinschaftlichen Zwangsarbeit und der vorläufigen Entlassung eingeschobenen Stadium der Zwischenanstalt, war die Gestaltung der Strafe der penal servitude genau dieselbe, wie sie auch damals schon in England bestand[1]) und oben im Abschnitte I § 4 dargestellt wurde. Lediglich die Institution der Zwischenanstalt bedarf daher hier einer näheren Darstellung; denn auch die praktische Ausführung der Strafe während der übrigen Stadien wich nur in wenigen Punkten von derjenigen in England ab. Wie in England, so wurde auch in Irland während des Stadiums der Isolierhaft dem Unterrichte und der geistlichen Ermahnung die größte Aufmerksamkeit geschenkt, dagegen bildete körperliche Anstrengung und Gewöhnung an ausdauernde Arbeit den Kernpunkt des Stadiums der gemeinschaftlichen Zwangsarbeit. Die

---

[1]) Als Beleg hierfür mag noch der nachfolgende Absatz aus dem im Jahre 1855 erstatteten Berichte der schon erwähnten Parlamentskommission, deren Mitglied Crofton war, hier angeführt werden:

„It appears to us beyond all question, that by such measures as have been in operation for some years in England, and which are now being introduced into the Convict Service in this country, the following results are clearly to be attained, viz., the application of the labour of able-bodied convicts to the production of works of permanent utility and profit in the country; a considerable return for the outlay and expense incurred in the maintenance of convicts, derived from the value of the work actually performed by them; the establishment of habits of steady industry, and in most cases, a determination to lead an honest life, and a desire to obtain a respectable position in society.

We believe these results to have been fully produced of late years in England, and we do not see that any greater difficulties are presented to their attainment in this country; on the contrary, the character of the Irish convict is in very many cases less seriously depraved, their crime having been produced in some measure by extreme distress, and the want of industrial employment: there is, therefore, greater ground to hope for a speedy and complete reformation. These objects being, as we hope, obtained by the reformatory system adopted towards the convict during his detention, it remains to offer him facilities for securing a respectable social position, by affording him the opportunity to exercise the habits of industry which he has acquired, and confirm the reformation effected in his character.“

über die vorläufig Entlassenen ausgeübte Polizeiaufsicht war in Irland von vornherein eine bessere und wirkungsvollere als in England: zunächst stand in Irland eine viel stärkere und weit besser organisierte Polizeimacht zu Gebote, als sie England in der damaligen Zeit besaß, sodann war in der Hauptstadt Irlands, in Dublin, einem besonders tüchtigen Beamten, Mr. Organ, die Beaufsichtigung der vorläufig Entlassenen übertragen worden, welcher neben der nötigen Strenge es sich auch angelegen sein ließ, den Leuten soweit als möglich den Weg zu einem redlichen Lebensunterhalte zu ebnen, und welcher daher von denselben als wirklicher Berater angesehen wurde. Es unterliegt keinem Zweifel, daß bezüglich der richtigen Handhabung der Polizeiaufsicht Irland für die in dieser Richtung später in England eingeführten Verbesserungen das Beispiel abgegeben hat.

Gehen wir nunmehr auf die Einrichtung der Zwischenanstalten des nähern ein. Der Zweck dieser Institution wird von Crofton selbst in einer im November 1855 an die Regierung gerichteten Denkschrift[1]) dahin angegeben: es soll eine Annäherung zwischen der bürgerlichen Gesellschaft und dem Sträflinge vor seiner Entlassung geschaffen werden. Eine solche Annäherung ist notwendig, um einen erfolgreichen Rücktritt des Sträflings in die Freiheit sicherzustellen. Die große Schwierigkeit, mit welcher die entlassenen Sträflinge zumeist zu kämpfen haben, besteht in dem Mangel an Arbeit. Infolge der Vorurteile, welche die Gesellschaft gegen eine Beschäftigung entlassener Sträflinge hegt, wird es den letzteren vielfach unmöglich gemacht, die guten Vorsätze, mit welchen sie in die Freiheit zurückgetreten sind, zur Ausführung zu bringen; notgedrungen fallen sie wieder dem Verbrechen anheim. Es kommt also darauf an, diese Vorurteile zu beseitigen. Dazu ist zunächst notwendig, daß der Sträfling während des letzten Teils seiner Strafe durch zweckentsprechenden Unterricht und geeignete Beschäftigung auf das Leben in der Freiheit und die ihm dabei bevorstehenden Kämpfe besonders vorbereitet werde. Zu gleicher Zeit muß dem Sträflinge ein größeres Maß von Freiheit gegeben werden, damit er zu zeigen vermag, daß er nicht nur unter dem Zwange einer strengen Hausordnung, wie sie in den eigentlichen Strafanstalten besteht, sich gut zu führen gelernt habe, sondern daß er

[1]) Cf. v. Holtzendorff II Seite 76 ff.

auch sein gutes Verhalten fortsetzen werde, wenn dieser äußere Zwang von ihm genommen ist. Wenn auf diese Weise der bürgerlichen Gesellschaft der Beweis geliefert wird, daß derjenige Teil der Sträflinge, welchen die Gefängnisverwaltung als vollkommen vertrauenswürdig aus der Strafanstalt in die Zwischenanstalt versetzt, hier das ihm geschenkte Vertrauen rechtfertigt, so wird auch die Gesellschaft ihrerseits das bisher dem entlassenen Sträfling entgegengebrachte Mißtrauen aufgeben und ihm bereitwillig Beschäftigung gewähren.

Die sich hiernach für die Gestaltung der Zwischenanstalten ergebenden Hauptmomente bestehen in besonderen Einrichtungen für Unterricht und Arbeit und in einer Hausordnung, welche von derjenigen der Strafanstalten verschieden ist und dem Sträflinge ein möglichst großes Maß von Freiheit gewährt.

Die Art und Weise, in welcher von Crofton diese Idee praktisch zur Ausführung gebracht wurde, war die folgende.

Unter den Sträflingen der advanced class werden diejenigen ausgewählt, von welchen anzunehmen ist, daß ein Übergangsstadium zur Freiheit für ihr weiteres Fortkommen nützlich sein werde, und welchen die Gefängnisverwaltung zu gleicher Zeit das Vertrauen schenkt, daß sie den ihnen gewährten Nachlaß im Strafzwange nicht mißbrauchen werden. Von diesen Gesichtspunkten aus sind nicht wählbar: a) Sträflinge, deren frühere Lebensstellung weit über diejenige eines gewöhnlichen Arbeiters hinausgegangen ist, und von denen daher anzunehmen ist, daß sie auch nach ihrer Entlassung in andrer Weise ihr Brot zu erwerben in der Lage sein werden. b) Sträflinge, welche sich nicht im Vollbesitze ihrer Arbeitskraft befinden, und von denen daher befürchtet werden muß, daß sie so wie so außer stande sein werden, sich selbständig in der Freiheit durchzubringen. c) Sträflinge, welche nach den von ihnen begangenen Verbrechen (z. B. Mord, widernatürliche Unzucht) als besonders gefährliche anzusehen sind. d) Sträflinge, welche bereits früher einmal in einer Zwischenanstalt gewesen sind, und bei denen sich somit durch ihre spätere Verurteilung gezeigt hat, daß die ihnen zu teil gewordene Behandlung den beabsichtigten Zweck nicht erreicht hat. e) Sträflinge, welche für Lebenszeit verurteilt sind.

Die nach Maßgabe dieser Grundsätze ausgewählten Sträflinge kommen für die letzte Zeit vor ihrer vorläufigen Entlassung in eine der beiden Zwischenanstalten, und zwar beträgt der hier zu verbringende Zeitraum für die zu fünf Jahren Verurteilten sechs Monate und erhöht sich für jedes weitere Strafjahr um einen Monat. Nach Lusk werden diejenigen Sträflinge geschafft, welche bereits früher in der Landwirtschaft thätig gewesen sind oder welche sich aus sonstigen Gründen für landwirtschaftliche Arbeiten besonders qualifizieren. Lusk selbst ist eine große Farm, auf welcher die dorthin Gebrachten in der mannigfaltigsten Weise und in gleicher Art beschäftigt werden wie gewöhnliche Landarbeiter. Es besteht keine weitergehende Aufsicht während der Arbeit, als sie auch bei freien Arbeitern stattfindet, und es fehlen alle Vorkehrungsmaßregeln gegen Fluchtversuche. Die nach Smithfield Gebrachten werden dort in denjenigen Handwerken beschäftigt, für welche sie sich besonders qualifizieren, und werden ebenfalls wie freie Arbeiter behandelt. In beiden Anstalten tragen die Insassen keine Gefängniskleidung.[1] Sie werden von dem Verkehr mit der Außenwelt nicht abgeschlossen, sondern werden vielfach absichtlich mit der freien Bevölkerung zusammengebracht, indem ihnen Botengänge oder sonstige Besorgungen in der Nachbarschaft aufgetragen werden. Für ihre Arbeitsleistungen erhalten sie wöchentlich durchschnittlich 1 s, hiervon werden ihnen 6 d zur vollständig freien Verfügung ausgezahlt; den Rest sowie die in der Strafanstalt erworbene Arbeitsbelohnung erhalten die Sträflinge erst bei ihrer Entlassung.

Gegenüber der großen, den Insassen der Zwischenanstalten hiernach gewährten Freiheit besteht als einziges Strafmittel für Zuwiderhandlungen gegen die Hausordnung oder sonstige Vergehen die Zurückversetzung des Betreffenden in die Strafanstalt. Eigentliche Disziplinarstrafen werden in der Zwischenanstalt nicht verhängt.

Was den Unterricht in den Zwischenanstalten betrifft, so wird davon ausgegangen, daß ein das Leben in der Freiheit vorbereitender praktischer Unterricht hier wichtiger sei, als die Erlangung von Elementarkenntnissen. Demgemäß tritt der Elementarunterricht hinter Vorlesungen über Geographie und „praktische Moral" zurück. Bei dem Unterrichte in der Geographie werden besonders die englischen Kolonieen behandelt und die Aussichten dar-

---

[1]) Es bestehen ferner auch für die Nachtzeit keine Einzelzellen, sondern größere Schlafsäle.

gestellt, welche sich dort den Einwanderern eröffnen, unter Hervorhebung der Voraussetzungen, unter welchen das einzelne Individuum auf einen günstigen Erfolg rechnen kann. Der Unterricht wird in dieser Weise benutzt, um den Sträflingen die freiwillige Auswanderung zu empfehlen und dabei gleichzeitig Rat und die nötige Information zu geben, wohin der Einzelne am besten seine Schritte lenkt. In den Vorlesungen über „praktische Moral" werden den Sträflingen in der Form allgemeiner Erörterungen die sozialen Pflichten, wie Unterordnung unter die bestehende Gewalt, Arbeitsamkeit, Sparsamkeit, Nüchternheit u. s. w. vor Augen gehalten; irgend einen konfessionellen Charakter sollen diese Vorträge nicht tragen. All diese Vorlesungen werden in den Abendstunden gehalten. An jedem Sonnabend abend findet unter Leitung des Lehrers eine Art Examinatorium der Sträflinge unter sich (competitive examination) über die im Laufe der Woche gehörten Vorträge statt, wobei sich der Lehrer überzeugt, inwieweit das von ihm Vorgetragene verstanden worden ist, um event. noch erforderliche Ergänzungen oder Erläuterungen zu geben.

In dieser Weise suchte Crofton seine Idee zu verwirklichen. Mit der größten Liebe zur Sache überwachte er selbst die Ausführung, verwendete fortgesetzt seine volle Aufmerksamkeit auf eine richtige Auswahl der in die Zwischenanstalten gebrachten Sträflinge und — last not least — suchte und fand die geeigneten Kräfte für die Leitung dieser Anstalten und den Unterricht in denselben. In erster Linie ist dabei des Mr. Organ zu gedenken, welcher in der sehr schwierigen Organisation des Unterrichts in den Zwischenanstalten Vortreffliches leistete.

Auf Grund der in Irland gemachten günstigen Erfahrungen machte Lord Naas — später Earl Mayo — in der im Jahre 1863 zur Untersuchung der Wirksamkeit der Penal Servitude Acts eingesetzten Königlichen Kommission den Vorschlag, die Institution der Zwischenanstalten auch in England einzuführen. Der Vorschlag wurde in erster Lesung angenommen, in der zweiten mit einer Majorität von bloß einer Stimme abgelehnt.[1]) So blieben die Zwischen-

---

[1]) Es ist weiter anzuführen, daß es in dem Report of the Royal Commissioners 1879 Seite LXIII fernerer Überlegung anheimgegeben wird, ob nicht bei der von den Commissioners vorgeschlagenen Bildung einer star class

anstalten als eine Eigentümlichkeit des irischen Gefängniswesens bestehen.

Allein in Irland selbst nahm die Bedeutung und Wertschätzung der Institution bald ab, nachdem im Jahre 1864 Sir Walter Crofton von der Leitung des Gefängniswesens zurückgetreten war [1]) und nachdem später auch der Tod der segensreichen Thätigkeit des Mr. Organ ein Ende gesetzt hatte. [2]) Im Jahre 1869 wurde

von Neulingen im Verbrechen (cf. oben Seite 60 f.) für die letzte Periode der Strafzeit eine besondere Anstalt nach dem Muster von Lusk einzurichten sei.

[1]) Im Jahre 1878 trat Sir Walter Crofton vorübergehend — für die Dauer von etwa einem Jahre — wieder an die Spitze des irischen Gefängniswesens.

[2]) Die Geschichte des irischen Gefängniswesens, auf welche — soweit nicht die Institution der Zwischenanstalt in Frage kommt — hier nicht näher eingegangen werden soll, liefert einen treffenden Beleg dafür, wie sehr es auf dem Gebiete des Gefängniswesens auf die persönliche Tüchtigkeit der Leiter desselben ankommt. Das Gefängniswesen in Irland ist seit dem Rücktritte von Sir Walter Crofton von Jahr zu Jahr zurückgegangen. Während man in England fortgesetzt bestrebt war, Verbesserungen im Strafvollzuge vorzunehmen, blieb man in Irland ruhig beim alten: man unterließ es, aus den Erfahrungen, welche man über die Gefängniskost gesammelt hatte, irgendwelchen Nutzen zu ziehen; man unterließ es, das Markensystem, dessen Reformbedürftigkeit offenkundig war und für dessen richtige Gestaltung England ein Beispiel gab, abzuändern; man verblieb bei der bisherigen Art der Gewährung von Arbeitsbelohnungen und machte sich die hierin in England eingeführten Verbesserungen nicht zu nutze. Ja noch weiter, man verschob die auf das Allerdringendste notwendigen Maßregeln: während Crofton schon im Jahre 1862 darauf aufmerksam gemacht hatte, daß man auf die Auffindung anderweitiger öffentlicher Arbeiten für die Sträflinge Bedacht nehmen müsse, da eine rationelle Beschäftigung derselben in Spike Island auf die Dauer nicht mehr möglich sein würde, schickte man noch fast 18 Jahre lang ruhig die Sträflinge nach Spike Island und beschäftigte sie dort so gut wie es eben ging. — Es ist daher begreiflich, daß das in dem Report of the Royal Commissioners 1879 über das irische Gefängniswesen ausgesprochene Urteil ein recht ungünstiges war, und daß umfassende Reformen auf das dringendste gefordert wurden.

Das Public Works Prison von Spike Island ist denn auch geschlossen worden; allein man ist noch nicht dazu gekommen, ein neues Public Works Prison zu eröffnen, sondern hat sich vorläufig notdürftig damit beholfen, die Mehrzahl der Sträflinge auch für das Public Works Stage in Mountjoy zu lassen und dort teils mit industriellen Arbeiten, vorzugsweise aber mit den in dieser Anstalt recht notwendigen Umbauten zu beschäftigen. Um in Mountjoy Platz zu schaffen, hat man für die bisher dort untergebrachten weiblichen Sträflinge die Anstalt von Grangegorman Prison bestimmt, so daß sich jetzt in Mountjoy nur männliche Sträflinge befinden. Auch bei dem ebenfalls not-die Anstalt von Smithfield gänzlich aufgehoben, so daß nur noch die Zwischenanstalt von Lusk bestand, in welcher lediglich landwirtschaftliche Arbeiten betrieben werden. Allein auch für diese Anstalt fehlte es immer mehr und mehr an geeignetem Materiale von Sträflingen: die tägliche Durchschnittsbelegung sank bis unter 21 herab. Naturgemäß wurden dadurch die auf den Kopf des Sträflings entfallenden Ausgaben immer höhere. Das führte zu der Erörterung der Frage, ob es nicht geraten sei, auch die Zwischenanstalt von Lusk aufzuheben. [1]) In dem Berichte der zur Beratung über das Gefängniswesen in Irland eingesetzten Königlichen Kommission vom Jahre 1884 wird hervorgehoben, [2]) daß die über diese Frage gehörten Sachverständigen geteilter Meinung gewesen seien. Die Kommission wolle sich deshalb jeglichen Urteils über die Institution der Zwischenanstalten an sich enthalten, sei aber der Ansicht, daß sich die in der Anstalt von Lusk bestehenden hohen Ausgaben pro Kopf des Sträflings nicht weiter rechtfertigen ließen, und sehe anderseits kein Mittel, wie diese Ausgaben vermindert werden könnten; insbesondere erscheine eine Vermehrung der aus den Strafanstalten nach Lusk zu sendenden Sträflinge ausgeschlossen,

---

wendigen Umbau vieler local prisons finden Sträflinge Verwendung. Sodann ist eine Royal Commission eingesetzt worden, um Vorschläge zu machen, ob und wie der Strafvollzug in Irland in größere Gleichmäßigkeit mit demjenigen in England gebracht werden könnte. Die Royal Commission on Prisons in Ireland kommt nach eingehenden Untersuchungen in ihrem Report vom Jahre 1884 zu dem Schlusse, daß es ratsam sei, die in England getroffenen Einrichtungen auch in Irland einzuführen und lediglich bezüglich der Beköstigung einige Abweichungen zu machen, wie dieselben durch die verschiedene Lebensweise der Bevölkerung in Irland gerechtfertigt erscheinen. Demgemäß gelten jetzt nach den Rules for the treatment of prisoners confined in convict prisons in Ireland 1885 auch für Irland diejenigen Bestimmungen über die Behandlung der Gefangenen, welche im Abschnitte II Kapitel 3 für England angegeben worden sind.

Wer sich jedoch ein richtiges Urteil über das Progressivsystem verschaffen will, dem kann nach meinen persönlichen Erfahrungen nur der Rat gegeben werden, dasselbe nicht in Irland zu studieren, sondern in England, wo das System unter der energischen und tüchtigen Leitung von Sir Edmund Du Cane zu seiner vollen Entwickelung gebracht worden ist.

[1]) Vorträge, wie sie zur Zeit Croftons in Lusk gehalten wurden, finden jetzt nicht mehr statt.

[2]) Report of the Royal Commission on Prisons in Ireland 1884 Seite 29 f.

da ja das ganze Prinzip der Zwischenanstalt auf einer sorgfältigen Auswahl der zuzulassenden Sträflinge beruhe.

Hiernach muß auch die Aufhebung der Zwischenanstalt in Lusk lediglich als eine Frage der Zeit angesehen werden.[1]) Wenn somit die ganze, von Crofton in das Leben gerufene Institution eine praktische Bedeutung kaum mehr besitzt, so verdient dieselbe doch hier noch nach zweifacher Richtung einer Erörterung: es ist zunächst festzustellen, in welchem Verhältnisse dieselbe zu dem in England bestehenden und in dem vorhergehenden Abschnitte ausführlich behandelten Strafensysteme steht, und sodann noch kurz auf den Wert und die Zweckmäßigkeit von Zwischenanstalten überhaupt einzugehen.

Die Zwischenanstalt erscheint in dem oben angegebenen Croftonschen Ideengange als wünschenswerter Abschluß des auf Umbildung des Charakters und Vorbereitung für eine erfolgreiche Rückkehr in die bürgerliche Gesellschaft gerichteten progressiven Strafvollzugs. Wie in England, so soll auch in Irland der Bestrafte durch mehrmonatliche Isolierhaft und sich daran anschließende gemeinschaftliche Zwangsarbeit, während welcher ihm je nach seinem guten Verhalten fortschreitend immer größere Vergünstigungen gewährt werden, zur Unterwerfung unter eine bestehende Ordnung, zu voller Ausnutzung seiner Körperkräfte, zu andauernder angestrengter Arbeit erzogen werden. Wie in England, so wird in Irland bei gutem Verhalten des Sträflings der Strafzwang am Schlusse der Strafzeit soweit

[1]) Aus dem Report des General Prisons Board Ireland 1885 Seite 151 f., 186 f. seien folgende Ziffern für das Jahr 1884/5 angeführt. — Am 1. April 1884 befanden sich in der Anstalt von Lusk 23 Sträflinge. Im Laufe des Jahres kamen hinzu 46 Sträflinge. Von diesem Bestande von 69 gelangten 41 zur vorläufigen Entlassung, 5 wurden nach Mountjoy zurückgesandt, 4 mußten in die Krankenabteilung zu Mountjoy gebracht werden, so daß ein Bestand von 19 am 31. März 1885 verblieb. Die tägliche Durchschnittsbelegung hatte während des Jahres 20,84 betragen. Von den während des Jahres Neueingelieferten war bei 28, also mehr als der Hälfte, die Bestrafung wegen Diebstahls erfolgt; 26 von den Neueingelieferten waren im Alter von 20—30 Jahren.

Die Kosten beliefen sich pro Kopf des Sträflings auf 90 £ 13 s (in Mountjoy: 55 £ 2 s 7 d), dabei betrugen die Ausgaben für das Beamtenpersonal pro Kopf des Sträflings 47 £ 17 s 3 d (in Mountjoy 21 £ 9 s 8 d). Die Arbeitsleistungen ergaben ein Erträgnis von 562 £ 14 s, also pro Kopf etwas über 27 £.

nachgelassen, daß der Sträfling unter Auferlegung bestimmter Einschränkungen versuchsweise in Freiheit gesetzt wird. Crofton war nur der Ansicht, daß es für das Gelingen dieses Versuches wünschenswert sei, wenn schon am Schlusse der Haftzeit ein erhebliches Nachlassen des Strafzwanges erfolge als Prüfung für die Fähigkeit des Sträflings, einen vernünftigen Gebrauch von der Freiheit zu machen.

Für dieses Prüfungsstadium sollte die besondere Institution der Zwischenanstalt dienen, da es mit der Ordnung und Disziplin in einer gewöhnlichen Strafanstalt unvereinbar sein würde, einem Teile der Insassen so große Freiheit im Verkehr miteinander und mit der Außenwelt zu gewähren, wie es für das Prüfungsstadium erforderlich wäre. Gleichzeitig sollte in der Zwischenanstalt Unterricht und Beschäftigungsart mit Rücksicht auf den demnächstigen Übertritt des Sträflings in die Freiheit besonders gestaltet werden; doch erscheint dies in sofern als nebensächlich, als derartiges ohne alle Schwierigkeit auch in einer gewöhnlichen Strafanstalt durch eine einfache Verwaltungsmaßregel eingerichtet werden könnte.

Die Rechtfertigung für die Einrichtung besonderer Zwischenanstalten liegt also ausschließlich in der von Crofton angenommenen Notwendigkeit, dem Sträflinge vor seiner Entlassung ein größeres Maß von Freiheit zu geben und ihn absichtlich Versuchungen auszusetzen, um so einen erfolgreichen Rücktritt in die Freiheit noch sicherer zu stellen.

Dies aber charakterisiert sich als dem Wesen nach mit den Prinzipien des englischen Strafensystems, wie dieselben im Abschnitte II dargelegt worden sind, völlig übereinstimmend. Es wurde daselbst insbesondere im § 12 ausgeführt, daß man es in England für nötig erachtet, die zu der harten Strafe der penal servitude Verurteilten während der Haftzeit selbst auf die Probe zu stellen, und daß die gemeinschaftliche Zwangsarbeit gerade mit Rücksicht auf die Versuchungen, welche dabei an den Einzelnen herantreten und an denen derselbe die während der Isolierhaft gefaßten guten Vorsätze zu erproben hat, als wichtiger Erziehungsfaktor für die Charakterumbildung angesehen würde. Es ist auch weiter darauf aufmerksam gemacht worden, daß mit dem Vorrücken in die höheren Strafklassen und den dabei gewährten Vergünstigungen der Strafzwang allmählich gemildert wird, so daß die in die special class

Beförderten selbst zu Vertrauensstellungen in der Anstaltsverwaltung herangezogen werden.

Die irischen Zwischenanstalten enthalten somit nichts dem Wesen nach von dem englischen Systeme Verschiedenes, sondern nur quantitativ etwas mehr: einem Teile der Sträflinge wird noch mehr Vertrauen geschenkt, der Strafzwang wird ihnen gegenüber noch mehr gemildert, und gleichzeitig damit werden die Versuchungen, deren erfolgreiches Bestehen die Probe für die erreichte Charakterumbildung abgeben soll, erhöht.

Aber die Institution der Zwischenanstalt charakterisiert sich nicht nur als eine einfache Fortbildung des in dem englischen Progressivsysteme liegenden Ideenganges, sondern sie war auch schon vor ihrer Einführung in Irland für eine Klasse von Sträflingen in England angeregt worden.

In dem auf Veranlassung von Lord Palmerston im Jahre 1853 veröffentlichten Schreiben war die Errichtung von refuges für weibliche Sträflinge als ein Übergangsstadium zwischen Strafanstalt und vorläufiger Entlassung für eine wünschenswerte Maßregel erklärt worden (cf. oben Seite 292). Es sollte damit genau derselbe Zweck erstrebt werden, welchen Crofton bei der Errichtung von Zwischenanstalten vor Augen hatte: es sollten die thatsächlich für die Strafentlassenen bestehenden Schwierigkeiten, eine geeignete Beschäftigung zu finden, nach Möglichkeit durch die Errichtung von Anstalten, welche eine Zwischenstellung zwischen dem Gefängnisse und der Freiheit einnehmen, beseitigt werden.

In England wurde dies nur für die weiblichen Sträflinge für notwendig erachtet, während Crofton ein gleiches Bedürfnis für die männlichen Sträflinge anerkannte. Die ganze Eigentümlichkeit des Crofton'schen oder irischen Systems läuft somit auf eine abweichende Beantwortung einer einfachen Bedürfnisfrage hinaus. Und auch dieser abweichenden Ansicht liegen nicht Verschiedenheiten in prinzipiellen Punkten, sondern faktische Momente zu Grunde.

Während es nämlich in England Fürsorgevereine für Strafentlassene gab, welche in erfolgreicher Weise den Strafentlassenen Beschäftigungen zu verschaffen suchten, existierte zu Crofton's Zeiten trotz der hohen Zahl von Sträflingen in Irland kein einziger derartiger Verein. Heutigentags, nachdem die Zahl der Sträflinge in

Irland sich erheblich vermindert hat und nachdem wenigstens zwei Fürsorgevereine (in Dublin und Belfast) entstanden sind,[1]) wird auch für Irland von sehr kompetenter Seite das Bedürfnis für eine Zwischenanstalt verneint.[2])

Betrachten wir nun zum Schlusse noch ganz kurz den Wert und die Zweckmäßigkeit der Institution von Zwischenanstalten im allgemeinen ohne Rücksicht auf die besonderen Verhältnisse in England oder Irland.

Daß gegen die Institution vom Standpunkte des Progressivsystems aus prinzipielle Einwendungen nicht erhoben werden können, ist bereits dargelegt worden. Wer ein Gegner des Progressivsystems ist, wer es für unzulässig erachtet, daß der Sträfling absichtlich Versuchungen ausgesetzt werde, wer der Ansicht ist, daß der Strafzwang während der ganzen Dauer der Strafe ein gleichmäßiger sein müsse und nicht von der Gefängnisbehörde je nach dem Verhalten des Sträflings allmählich gemildert werden dürfe, der muß ein entschiedener Gegner der Zwischenanstalten sein, in welchen der Strafzwang auf ein Minimum reduziert und die an den Sträfling herantretenden Versuchungen auf das höchste gesteigert sind. Lassen wir diese prinzipiellen Gegner hier außer Betracht, so werden für die Frage: „ob Zwischenanstalten oder nicht" drei Momente zu berücksichtigen sein.

Die erste zu beantwortende Frage muß die sein: Besteht ein thatsächliches Bedürfnis, für die Sträflinge oder einen Teil derselben eine besondere Anstalt nach dem Muster der irischen Zwischenanstalt für die letzte Periode der Strafzeit zu errichten?

Niemand wird leugnen, daß es dem Bestraften schwer fällt, sofort bei seiner Entlassung eine geeignete Beschäftigung, ja häufig überhaupt irgend eine Beschäftigung zu finden, von welcher er sich in redlicher Weise ernähren kann. Es liegt auf der Hand, daß durch die auf diese Weise dem redlichen Fortkommen des Strafentlassenen

---

[1]) Allerdings steht auch heute noch dasjenige, was in Irland durch diese Vereine für die Strafentlassenen geschieht, weit zurück hinter den Leistungen der englischen Fürsorgevereine.

[2]) Sowohl Sir Edmund Du Cane wie der jetzige Leiter des irischen Gefängniswesens Charles F. Bourke haben dies mir gegenüber ausgesprochen.

bereiteten Schwierigkeiten die guten Vorsätze, welche demselben während seiner Strafzeit anerzogen worden sind, auf eine schwere Probe gestellt werden, und daß der Strafentlassene vielfach in dem schweren Kampfe, welchen er zu bestehen hat, unterliegen wird, ohne daß ein vernünftiger Mensch, welcher von seiner begünstigten Lebensstellung aus nicht mit pharisäischem Hochmute auf diese armen Subjekte herabblickt, dem Betreffenden einen allzuschweren Vorwurf aus diesem Unterliegen machen kann. Die Gesellschaft begeht eben ein schweres Unrecht, indem sie den Bestraften, nachdem er durch Abbüßung der über ihn erkannten Strafe seine Strafthat gesühnt hat, von sich abstößt und auf diese Weise ohne alle Berechtigung eine neue Strafe verhängt. Und man muß sich leider sagen, daß trotz alles Bestrebens, die Gesellschaft in dieser Richtung aufzuklären, die traurigen Vorurteile für absehbare Zeit noch fortbestehen werden.

Demgegenüber erscheint es als Pflicht der Gesellschaft und des Staates, dem Sträflinge durch alle nur möglichen Maßregeln den Übergang in geordnete Verhältnisse zu erleichtern. Und dies geschieht, wenn der Sträfling vor seiner Haftentlassung in der Art und Weise, wie es unter Crofton's und Organ's Leitung geschah, nicht nur in intensiver Weise auf das Leben in der Freiheit vorbereitet, sondern auch mit der Außenwelt in Verbindung gebracht wird, damit dieselbe das gehegte Mißtrauen aufgibt. *Die Zwischenanstalt wird also einem thatsächlich bestehenden Bedürfnisse gerecht.*

Aber nun entsteht die zweite Frage: *kann dieses Bedürfnis nicht in andrer Weise besser und mit geringeren Kosten befriedigt werden?*

Es ist ohne weiteres klar, daß eine Anstalt, in welche ihrer ganzen Anlage nach nur eine kleine Zahl besonders ausgewählter Sträflinge gebracht werden kann, und für welche anderseits ein ganz hervorragend tüchtiges Beamtenpersonal nötig ist, sehr große Ausgaben verursachen muß. Es läßt sich aber auch ferner nicht bestreiten, daß in vielen Fällen der in das Auge gefaßte Zweck nicht erreicht werden wird. Zunächst werden auch bei der sorgfältigsten Auswahl einzelne Sträflinge in die Zwischenanstalt hineinkommen, welche das ihnen entgegengebrachte Vertrauen nicht verdienen und durch die in Strafanstalten häufig vorkommende und oft schwer entdeckbare Heuchelei die Anstaltsbeamten getäuscht haben. Es besteht die Gefahr, daß durch solche Subjekte die Sicherheit in der Anstalt und in der Umgebung derselben gefährdet werde, und vor allem, daß bei dem nahen und fast gar nicht beaufsichtigten Verkehr der Sträflinge untereinander die schlechten Elemente die guten verderben. Sodann aber ist es doch noch recht zweifelhaft, ob die Bevölkerung durchweg den aus der Zwischenanstalt Entlassenen ein volles Vertrauen entgegenbringen wird, ob sie nicht die in der Zwischenanstalt abgelegte Probe für die Vertrauenswürdigkeit deshalb für ungenügend erachtet, weil die Insassen der Zwischenanstalten sich stets unter dem Damoklesschwerte der Rückversetzung in die Strafanstalt und des Verlustes der für ihre Verhältnisse erheblichen Arbeitsbelohnung befinden.

Billiger und sicherer dürfte es sein, dem Strafentlassenen durch Fürsorgevereine den Rücktritt in die Freiheit zu erleichtern. England bietet ein klassisches Beispiel dafür, was sich durch derartige Vereine erreichen läßt, wenn dieselben überall im Lande festen Fuß fassen und beim Staate sowohl in ihrer Thätigkeit wie in materieller Hinsicht die erforderliche Unterstützung finden. Die von diesen Vereinen in das Werk gesetzte freiwillige Auswanderung Bestrafter nach den englischen Kolonieen, wo die Vergangenheit der Betreffenden nicht bekannt ist und wo sich ein weites Gebiet für ihre Thätigkeit eröffnet, führt zweifellos sicherer zu dem erstrebten Ziele als alle Zwischenanstalten.

Wenn hiernach die Institution der Zwischenanstalten zwar an sich gerechtfertigt ist, es aber aus mehrfachen Gründen wünschenswert erscheint, sie durch andre Maßregeln zu ersetzen, so darf endlich noch ein drittes Moment nicht außer acht gelassen werden. Der Erfolg der ganzen Institution hängt im wesentlichen ab von der Tüchtigkeit derjenigen Beamten, welche zu ihrer praktischen Durchführung berufen sind. Dieses persönliche Moment, welches überall im Strafvollzuge von der höchsten Bedeutung erscheint,[1]) ist nirgends so wichtig wie hier. Man denke sich, was aus einer Zwischenanstalt werden würde, wenn die Auswahl der dorthin Geschafften eine leichtfertige wäre, man

[1]) Es sei an den Ausspruch erinnert, welchen der Dezernent für das Gefängniswesen im kgl. preuß. Ministerium des Innern, Geh. Rat Illing, in das Souvenir du troisième Congrès Pénitentiaire internationale, Rome 1885 eingetragen hat:

„Les meilleurs systèmes pénitentiaires restent sans effet, si l'on n'est pas en état de trouver pour leur application des personnes assez intelligentes et devouées.“

stelle sich vor, was aus dem Unterrichte über „praktische Moral" durch einen nicht auf der Höhe seiner Aufgabe stehenden Lehrer gemacht werden könnte u. s. w.! Wo immer man die Errichtung einer Zwischenanstalt in praktische Überlegung ziehen will, muß man zunächst völlig sicher sein, daß man ein Beamtenpersonal besitzt und stets haben wird, welches den ihm hier obliegenden, besonders schwierigen Pflichten vollauf gewachsen ist.

Darin eben besteht auch das große Verdienst Crofton's, daß er nicht nur persönlich an der praktischen Gestaltung der von ihm in das Leben gerufenen Institution regen Anteil nahm, sondern daß er es auch verstand, für die Durchführung derselben im Einzelnen die richtigen Kräfte zu wählen.

Und damit ist derjenige Punkt berührt, auf welchem Crofton's hervorragende Bedeutung beruht. Wie schon hervorgehoben wurde, ist das Progressivsystem zwar nicht von Crofton erfunden,[1]) aber erst durch ihn zur vollen Würdigung gebracht worden: erst durch die Ausführung, welche dies System infolge der eignen praktischen Tüchtigkeit Crofton's und seiner Fähigkeit, die geeigneten Leute zu wählen und sie mit dem richtigen Verständnisse und mit Liebe zur Sache zu erfüllen, gefunden hat, ist dasselbe in seinem vollen Werte erkannt worden. Die hohen Verdienste Crofton's werden deshalb auch kaum geschmälert werden, wenn man der Überzeugung Ausdruck gibt, daß es nicht ein eigentlich neues System ist, welches seinen Namen trägt. Der Name Crofton's, als eines der größten praktischen Gefängnisreformatoren, wird für immer in den Annalen der Gefängniswissenschaft eingezeichnet verbleiben.

[1]) Folgender Passus in einer Besprechung, welche Sir Edmund Du Cane in der Law Quarterly Review April 1886 Seite 227 ff. dem Werke von Adolphe Prins, Criminalité et Répression, Bruxelles 1886, widmet, verdient hier hervorgehoben zu werden:

„The author falls into a singular mistake when describing, with approval, the progressive system in force in the United Kingdom, under which the cellular stage of discipline is followed by work in association under supervision, the cell being occupied except when at work, and finally by conditional release. He says that it was first applied in Ireland, under Sir Walter Crofton; a claim which Sir Walter would assuredly be the first to repudiate, because it would deprive some most distinguished prison reformers of the credit due to them; and he certainly has taken among prison reformers a sufficiently high place to be able to dispense with any which is not due to his own merits."

Otto Hauthal (G. Pätz'sche Buchdr.) Naumburg a/S.

Kolibri 360 GmbH
Ferdinand-Jühlke-Straße 7
99095 Erfurt, Deutschland
produktsicherheit@kolibri360.de